Behaviour of Steel Structures in Seismic Areas

OTHER BOOKS ON SEISMIC ENGINEERING

Approximate Methods in Structural Seismic Design
A.S. Scarlat

Behaviour of Steel Structures in Seismic Areas
Edited by F.M. Mazzolani and V. Gioncu

Concrete Shear in Earthquake
Edited by T.C.C. Hsu and S.T. Mau

Concrete under Severe Conditions: Environment and Loading
Edited by K. Sakai, N. Banthia and O.E. Gjorv

Disaster Planning, Structural Assessment, Demolition and Recycling
Edited by C. De Pauw and E.K. Lauritzen

Earthquake, Blast and Impact: Measure and Effects of Vibration
Edited by SECED

Earthquake Engineering
Y.X. Hu, S.C. Liu and W. Dong

Earthquake Resistant Concrete Structures
G.G. Penelis and A.J. Kappos

Handbook of Seismic Design of Buildings
F. Naeim

International Handbook of Earthquake Engineering: Codes, Programs and Examples
Edited by M.H. Paz

Making Buildings Safer for People during Hurricanes, Earthquakes and Fires
A.S. Nowak and T.V. Galambos

Natural Risk and Civil Protection
Edited by T. Horlick-Jones, A. Amendola and R. Casale

Nonlinear Dynamic Analysis of Structures
Edited by K.S. Virdi

Nonlinear Seismic Analysis and Design of Reinforced Concrete Buildings
Edited by P. Fajfar and H. Krawinkler

Soil Dynamics and Earthquake Engineering V
Edited by I B F

Structural Design for Hazardous Loads
Edited by J.L. Clarke, F.K. Garas and G.S.T. Armer

Structures Subjected to Dynamic Loading: Stability and Strength
Edited by R. Narayanan and T.M. Roberts

Theory and Design of Seismic Resistant Steel Frames
F.M. Mazzolani and V. Piluso

For more information, contact E & FN Spon, 2-6 Boundary Row, London SE1 8HN, UK
Tel: Intl +44 171-865 0066, Fax: Intl +44 171 522 9623

Behaviour of Steel Structures in Seismic Areas

STESSA '94

Proceedings of the International Workshop, organised by the European Convention for Constructional Steelwork

Timisoara, Romania
26 June – 1 July 1994

Edited by

FEDERICO M. MAZZOLANI
Professor of Structural Engineering,
University 'Federico II' of Naples, Italy

and

VICTOR GIONCU
Professor for Design of Structures,
Department of Architecture,
Technical University of Timisoara, Romania

E & FN SPON
An Imprint of Chapman & Hall

London · Glasgow · Weinheim · New York · Tokyo · Melbourne · Madras

**Published by E & FN Spon, an imprint of Chapman & Hall,
2–6 Boundary Row, London SE1 8HN, UK**

Chapman & Hall, 2–6 Boundary Row, London SE1 8HN, UK

Blackie Academic & Professional, Wester Cleddens Road, Bishopbriggs, Glasgow G64 2NZ, UK

Chapman & Hall GmbH, Pappelallee 3, 69469 Weinheim, Germany

Chapman & Hall USA, 115 Fifth Avenue, New York, NY 10003, USA

Chapman & Hall Japan, ITP-Japan, Kyowa Building, 3F,
2-2-1 Hirakawacho, Chiyoda-ku, Tokyo 102, Japan

Chapman & Hall Australia, Thomas Nelson Australia, 102 Dodds Street, South Melbourne, Victoria 3205, Australia

Chapman & Hall India, R. Seshadri, 32 Second Main Road, CIT East, Madras 600 035, India

First edition 1995

© 1995 E & FN Spon

Printed in Great Britain at the University Press, Cambridge

ISBN 0 419 19890 3

Apart from any fair dealing for the purposes of research or private study, or criticism or review, as permitted under the UK Copyright Designs and Patents Act, 1988, this publication may not be reproduced, stored, or transmitted, in any form or by any means, without the prior permission in writing of the publishers, or in the case of reprographic reproduction only in accordance with the terms of the licences issued by the Copyright Licensing Agency in the UK, or in accordance with the terms of licences issued by the appropriate Reproduction Rights Organization outside the UK. Enquiries concerning reproduction outside the terms stated here should be sent to the publishers at the London address printed on this page.

The publisher makes no representation, express or implied, with regard to the accuracy of the information contained in this book and cannot accept any legal responsibility or liability for any errors or omissions that may be made.

A catalogue record for this book is available from the British Library

Printed on acid-free text paper, manufactured in accordance with ANSI/NISO Z39.48-1992 (Permanence of Paper).

Contents

Preface		xi
Conference organisation		xv

KEYNOTE PAPERS   1

1   Development and design of seismic-resistant steel structures in Romania   3
V. Gioncu

2   Development and design of seismic-resistant steel structures in Japan   28
B. Kato

PART ONE BASIC PROBLEMS   43

3   Basic problems: general report   45
L.-W. Lu

4   Recent seismic research on steel bridges   53
A. Astaneh-Asl

5   Damage assessment in steel members under seismic loading   63
G. Ballio and C. A. Castiglioni

6   Toward the definition of a consistent seismic-resistant design method based on damage criteria   77
E. Cosenza and G. Manfredi

7   Influence of distributed inertial mass and stress state on the seismic response   89
I. Dimoiu and M. Ivan

8   Research on seismic behavior of steel and composite structures at Lehigh University   97
L.-W. Lu, J. M. Ricles and K. Kasai

9   Quality control of material properties for seismic purposes   111
B. Calderoni, F. M. Mazzolani and V. Piluso

| 10 | Dynamic behavior of border columns of a cantilever braced wall
M. Shibata | 121 |

PART TWO LOCAL DUCTILITY 131

11	Local ductility in steel structures subjected to earthquake loading A. S. Elnashai	133
12	On the ductility of two welded steel beam–column connections I. Dimoiu and S. Dan	149
13	Cyclic behaviour of steel-to-concrete end-plate connections L. Dunai, Y. Ohtani and Y. Fukumoto	159
14	Contributions to the study of plastic rotational capacity of I-steel sections V. Gioncu, G. Mateescu and A. Iuhas	169
15	Alternative methods for assessing local ductility V. Gioncu and F. M. Mazzolani	182
16	Influence of cyclic actions on the local ductility of steel members M. A. Aiello and L. Ombres	191
17	Ductility of steel–concrete mixed section beams V. Pacurar	201
18	Increase of buckling resistance and ductility of H-sections by encased concrete A. Plumier, A. Abed and B. Tiliouine	211
19	Structural performance aspects on cyclic behavior of the composite beam–column joints A. M. Pradhan and J. G. Bouwkamp	221
20	Local cyclic behaviour of steel members I. Vayas and I. Psycharis	231
21	Cyclic behaviour of thin-walled welded knee joints H. Pasternak and I. Vayas	242

PART THREE GLOBAL DUCTILITY 251

| 22 | An overview of global ductility
H. Akiyama | 253 |

23	**Statistical evaluation of the behaviour factor for steel frames** B. Calderoni, D. Rauso and A. Ghersi	278
24	**Contribution to the study of the ductility of unbraced frames** V. Gioncu, G. Mateescu and A. Iuhas	289
25	**A new method to design steel frames failing in global mode including P–Δ effects** F. M. Mazzolani and V. Piluso	300
26	**Remarks on behaviour of concentrically and eccentrically braced steel frames** F. M. Mazzolani, D. Georgescu and A. Astaneh-Asl	310
27	**Resistance deterioration during earthquakes compiled in earthquake response test database on steel frames** K. Ohi and K. Takanashi	323
28	**On the seismic resistance of light gauge steel frames** B. Calderoni, A. de Martino, R. Landolfo and A. Ghersi	333
29	**A method for the evaluation of the behaviour factor for steel regular and irregular buildings** I. Vayas, C. Syrmakezis and A. Sophocleous	344

PART FOUR EXPERIMENTAL METHODOLOGY 355

30	**Experimental methodologies: general report** D. Diaconu and A.-C. Diaconu	357
31	**Error effects in predictor–corrector algorithms for pseudodynamic tests** O. S. Bursi and P. B. Shing	371
32	**Cyclic behaviour of steel beam-to-column connections – an experimental research** L. Calado and J. Ferreira	381
33	**Experimental testing for the analysis of the cyclic behaviour of steel elements: an overview of the existing procedures** E. Cosenza, G. Manfredi and A. de Martino	390
34	**PSD-testing of a full-scale three-storey steel frame** A. Kakaliagos, G. Verzeletti, G. Magonette and A. V. Pinto	401

viii *Contents*

35	**Experimental dynamic measurement of tall buildings with steel structures** A. La Tegola and W. Mera	411
36	**Shear behaviour of steel corrugated panels** R. Landolfo and F. M. Mazzolani	421
37	**The tests of the column–base components** F. Wald, I. Šimek and J. Seifert	431

PART FIVE CODIFICATION   439

38	**Codification: general report** A. Plumier	441
39	**Seismic design of composite structures in the United States** A. Astaneh-Asl	448
40	**Proposals for improving the Romanian seismic code. Provisions concerning steel structures** C. Dalban, P. Ioan and St. Spanu	458
41	**Technical norms on steel plate girders damaged by earthquakes** M. Dinculescu and D. Georgescu	472
42	**Romanian seismic provisions for one-storey industrial buildings** D. Georgescu and M. Bugheanu	478
43	**On the evaluation of the global ductility of steel structures: proposals of codification** V. Gioncu and F. M. Mazzolani	485
44	**Safety levels in seismic design** F. M. Mazzolani, D. Georgescu and A. Astaneh-Asl	495
45	**Some considerations on the needs of future development of earthquake resistant design practice and codes** H. Sandi	507

PART SIX SEMI-RIGIDITY   517

46	**Semi-rigidity: general report** D. Georgescu	519

47	Behaviour of unbraced semi-rigid composite frames under seismic actions C. Amadio, F. Benussi and S. Noe	535
48	Seismic behavior and design of steel semi-rigid structures A. Astaneh-Asl	547
49	Moment–rotation behaviour of top-and-seat angle connections C. Bernuzzi, M. de Stefano, E. d'Amore, A. de Luca and R. Zandonini	557
50	Semirigid top-and-seat cleated connections: a comparison between Eurocode 3 approach and other formulations M. de. Stefano, C. Bernuzzi, E. d'Amore, A. de Luca and R. Zandonini	568
51	Numerical simulation concerning the response of steel frames with semi-rigid joints under static and seismic loads D. Dubina, D. Grecea and R. Zaharia	580
52	Connection influence on the seismic behaviour of steel frames C. Faella, V. Piluso and G. Rizzano	590
53	Connection influence on the elastic and inelastic behaviour of steel frames C. Faella, V. Piluso and G. Rizzano	600
54	Behaviour of semi-rigid connections under alternate static loads D. Grecea and G. F. Mateescu	610
55	COST-C1 – seismic working group activities A. V. Pinto	621

PART SEVEN BASE ISOLATION AND ENERGY DISSIPATION   629

56	State of the art report on base isolation and energy dissipation E. Mele and A. de Luca	631
57	Serviceability and ultimate performance of base-isolated steel frames A. de Luca, G. Faella and E. Mele	659
58	Design of base isolation devices for steel structures M. Imbimbo and G. Serino	669

x *Contents*

59	Development of vibration control system using U-shaped water tank K. Shimizu and A. Teramura	681
60	Seismic base isolation for framed structures P. Lenza, M. Pagano and P. P. Rossi	691
61	Design methodologies for energy dissipation devices to improve seismic performance of steel buildings G. Serino	703

PART EIGHT DESIGN AND APPLICATIONS 715

62	Strengthening and repairing of lattice structures N. Balut and A. Moldovan	717
63	Use of steel structures for consolidating a multi-storey reinforced concrete structure damaged by seismic actions C. Dalban, S. Dima, R. Angelescu, V. Mustata and A. Leonte	727
64	On the seismical behaviour of two building models with suspended storeys I. Dimoiu and A. Botici	736
65	Seismic upgrading of churches by means of dissipative devices F. M. Mazzolani and A. Mandara	747
66	Models, simulations and condensations in the design of a steel–concrete composite structure placed in seismic zone V. Stoian and I. Olariu	759

Author index	769
Subject index	771

Preface

It is generally accepted in design practice that steel is an excellent material for seismic-resistant structures, thanks to its performance in terms of strength and ductility. Due to this reputation, both research and codification for seismic design were rather limited for many years.

But in the last decade, specialists have recognized that the so-called good ductility of steel structures can be exploited only if certain conditions are accomplished.

The need to assess particular design provisions was confirmed after some major accidents occurred in steel structures, the most famous being the Pino Suarez highrise building which collapsed during the Mexico City Earthquake. Additional confirmation came from the analysis of the behaviour of steel structures during the Californian earthquakes at Loma Prieta (17 October 1989) and, more recently, Northridge (17 January 1994).

During the 1980s the ECCS Committee TC 13 'Seismic Design' carried out extended research activity on the behaviour of steel structures and components under seismic actions. With the publication of the European Recommendations for Steel Structures in Seismic Zones in 1988 the already existing 'steel culture' reached a significant meeting point with the 'earthquake culture'. In a very short period of time the gap in knowledge with traditional reinforced concrete construction was completely filled.

The latest issue of ECCS Committee TC 13, the Manual of Design of Steel Structures in Seismic Zones, was published in 1994, and provided designers and practitioners with the basic principles upon which the latest generation of modern seismic codes are based. The contents of this Manual show that in the last few years, due to the major developments in the field of seismic-resistant steel structures on an international level, the scenario is completely changed: new areas must now be explored and many problems require adequate solutions.

For these reasons we decided to promote a summit meeting of qualified specialists in this field in order to point out the main directions in research and codification.

The Conference, "STESSA '94", included a Workshop and Seminar, and was devoted to the specific subject 'Behaviour of Steel Structures in Seismic Areas'. The great success of this specialty Conference demonstrated the advantages of such a focused formula, in contrast with the more general Seismic Conferences dealing with all constructional materials and all phenomenological aspects, such as the 10th World Conference in Earthquake Engineering (Madrid, July 1992) and the 10th European Conference on Earthquake Engineering (Vienna, September 1994), where a very small part was devoted to steel structures.

xii *Preface*

The organization of the Conference, in the range of activity of ECCS Committee TC 13, was supported by the following bodies:

University of Naples
Imperial College, London
University of Liège
Romanian Academy
Technical University of Timisoara
Building Research Institute (INCERC) of Timisoara

The city of Timisoara, Romania, was selected for the Conference venue, as a fair acknowledgement of the great contribution of the Romanian scientists in this field.

The main outstanding experts from all over the world were invited by the Scientific Committee, covering 17 countries. The contributions prepared by more than one hundred specialists are collected in this Proceedings volume, following the same subdivision in eight working sessions as they were presented during the Workshop, dealt with the following topics:

1	Basic problems	Chairman: F. M. Mazzolani
		General Reporter: Le-Wu Lu
2	Local ductility	Chairman: J. M. Aribert
		General Reporter: A. S. Elnashai
3	Global ductility	Chairman: E. Cosenza
		General Reporter: H. Akiyama
4	Experimental methodology	Chairman: I. Vayas
		General Reporter: I. Diaconu
5	Codification	Chairman: F. Wald
		General Reporter: A. Plumier
6	Semi-rigidity	Chairman: L. Calado
		General Reporter: D. Georgescu
7	Base isolation and energy dissipation	Chairman: K. Takanashi
		General Reporter: A. De Luca
8	Design and applications	Chairman: A. Ghersi

During the Workshop, three invited lectures were presented:

A. Astaneh: Development and design of seismic resistant steel structures in the USA: Lessons learned from Northridge Earthquake (oral presentation only)
B. Kato: Development and design of seismic resistant steel structures in Japan (the lecture was presented by Professor Mazzolani due to the absence of the Author)
V. Gioncu: Development and design of seismic resistant steel structures in Romania

The Honorary Chairman of the Conference was the Academician Dan Mateescu.

The contents of this volume demonstrate that STESSA '94 was a very successful event in this field. This encourages us to continue in this direction by considering STESSA '94 as the first of a series of regular specialty Conferences to be held every three years. For the future we already have the invitation of the Universities of Berkeley (California) and of Tokyo (Japan). So, we are at the beginning of very important new activity in the field of seismic-resistant steel structures.

December 1994
Federico M. Mazzolani and Victor Gioncu
Chairmen of the International Scientific Committee

Conference organisation

HONORARY CHAIRMAN OF THE CONFERENCE

Professor **Dan Mateescu**, Member of the Romanian Academy

CONFERENCE DIRECTORS

Professor **Antonella De Luca**, University of Reggio Calabria, Italy
Professor **Liviu Gadeanu**, Technical University of Timisoara, Romania

Professor **Dan Dubina**, *Deputy Director*, Technical University of Timisoara, Romania

Res. C. Eng. Daniel Grecea, *General Secretary*, INCERC-Timisoara, Romania

INTERNATIONAL SCIENTIFIC COMMITTEE

Professor **Federico M. Mazzolani**, *Chairman*, University of Naples, Italy
Professor **Victor Gioncu**, *Co-Chairman*, INCERC-Timisoara, Romania

Professor **H. Akiyama**, University of Tokyo, Japan
Professor **J. M. Aribert**, I.N.S.A., Rennes, France
Professor **A. Astaneh**, University of California at Berkeley, USA
Professor **G. Ballio**, University of Milan, Italy
Professor **J. G. Bouwkamp**, Technical University Darmstadt, Germany
Professor **L. Calado**, Technical University of Lisbon, Portugal
Professor **E. Cosenza**, University of Naples, Italy
Res. C. Eng. D. Diaconu, INCERC-Iasi, Romania
Professor **A. S. Elnashai**, Imperial College London, United Kingdom
Professor **D. Georgescu**, Civil Engineering Institute of Bucharest, Romania
Professor **A. Plumier**, University of Liège, Belgium
Professor **A. Reinhorn**, University of New York, Buffalo, USA
Professor **G. Sedlacek**, University of Aachen, Germany
Professor **I. Vayas**, National Technical University of Athens, Greece

INTERNATIONAL ADVISORY COMMITTEE

Dr. Eng. H. Sandi, *Chairman*, INCERC-Bucharest, Romania
Professor A. Ghersi, *Co-Chairman*, University of Catania, Italy
Professor J. Azevedo, Technical University of Lisbon, Portugal
Professor B. Csak, Technical University Budapest, Hungary
Professor C. Dalban, Civil Engineering Institute of Bucharest
Professor C. Faella, University of Salerno, Italy
Professor M. Ivan, Technical University of Timisoara, Romania
Professor M. Ivanyi, Technical University of Budapest, Hungary
Professor Le-Wu Lu, Lehigh University, Bethlehem, USA
Professor G. Penelis, Aristotle University of Thessaloniki, Greece
Professor R. Ramasco, University of Naples, Italy
Professor K. Takanashi, University of Tokyo, Japan

Professor I. Dimoiu, *Scientific Secretary*, Technical University of Timisoara, Romania

CONFERENCE SECRETARIAT

Res. C. Eng. B. Calderoni, Italy
Professor A. De Martino, Italy
Res. C. Eng. V. Piluso, Italy
C. Eng. D. Bulzesc, Romania
Assist. Prof. M. Georgescu, Romania
Res. C. Eng. G. Mateescu, Romania

KEYNOTE PAPERS

1 DEVELOPMENT AND DESIGN OF SEISMIC-RESISTANT STEEL STRUCTURES IN ROMANIA

V. GIONCU
Building Research Institute, INERC, Timisoara, Romania

Abstract
The frequent occurrence of strong earthquakes in Romania poses some direct impact on engineering activities. The report is intended to present the experience of Romanian structural engineers in the field of seismic-resistant steel structures. The last very strong earthquake from 1977 and the produced damages in steel structures are related. The evolution of the approaches on design codes and the presentation of theoretical and experimental research works performed last time in Romania shows that this activity is one of the most important in the field of structural engineering.
Keywords: Accelerogram record, Design code, Earthquake, General ductility, Local ductility, Seismic behaviour, Territory seismicity.

1 Seismicity of Romanian territory

Exist many very important differences between other well-known and well-studied earthquakes and the ones produced in Romania. These differences involve some special anti-seismic design approaches, especially for the steel structures. This is the reason for a short presentation of the Romanian earthquakes characteristics.

The experience of seismic events shows the frequent occurrence of earthquakes of relative high magnitude over extensive zones of Romanian territory. These earthquakes affect also large areas of neighbouring countries.

Historically the informations concerning seismic events are starting with the year 1091. During the period 1091 - 1900, 78 seismic events with the intensities $I > VI$ are related in ancient documents. For 1900-1981, 91 earthquakes with magnitudes $M > 5$ are produced on the Romanian territory (Balan et al,1982). About the events for past centuries one may conclude that the informations have rather qualitative sense. It is reasonable to assess a high credibility only to magnitude estimates for events of this century (Sandi, 1992). The main characteristic of the seismicity of

Behaviour of Steel Structures in Seismic Areas. Edited by F. M. Mazzolani and V. Gioncu.
Published in 1995 by E & FN Spon, 2-6 Boundary Row, London SE1 8HN. ISBN: 0 419 19890 3.

Romanian territory is the existence of multiple sources (Fig.1), with very different behaviours. From these different notified epicentres, the Vrancea and Banat ones seem to be of the first importance.

The <u>Vrancea earthquakes</u> are intermediare depth (h=80-140 km) and generate intensities up to VIII over extensive areas. The source of this zone delivers about 97% of the whole seismic energy from Romania. The strongest earthquake seems to be event of 1802, with a presumed magnitude M=7.5. In our century it were generated: 3 events of M⟩ 7.0; 6 events of M⟩ 6.5; 11 events of M⟩ 6.0; 23 events of M⟩ 5.5 (Sandi, 1992). The time history is given in Fig.2a, while Fig.2b shows a picture of the recurrence law. One can noticed that in the last fifty years Romania was affected by five destructive earthquakes: 10 November 1940 (M=7.4), 4 March 1977 (M=7.2), 30 August 1986 (M=7.0), 30 May 1990 (M=6.7) and 31 May 1990 (M=6.2). The first two earthquakes were very destructive, leading to very heavy losses. The latter ones were somewhat surprising if compared with the expectancy of major events (Radu et al.,1990), showing the increase of source activity. The damages were not so important as the former ones. Their importance consists in the

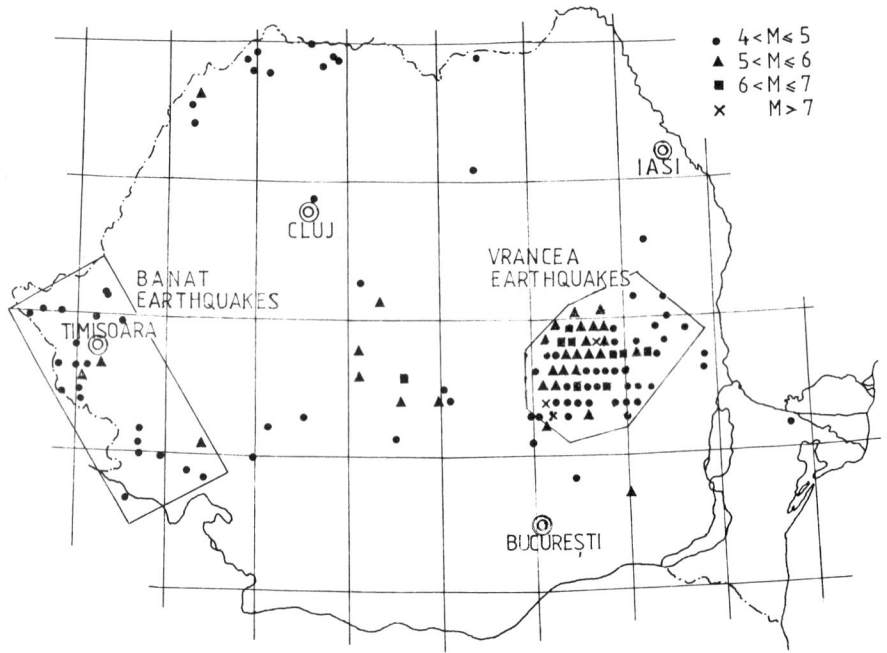

Fig.1. Seismicity of Romanian territory.

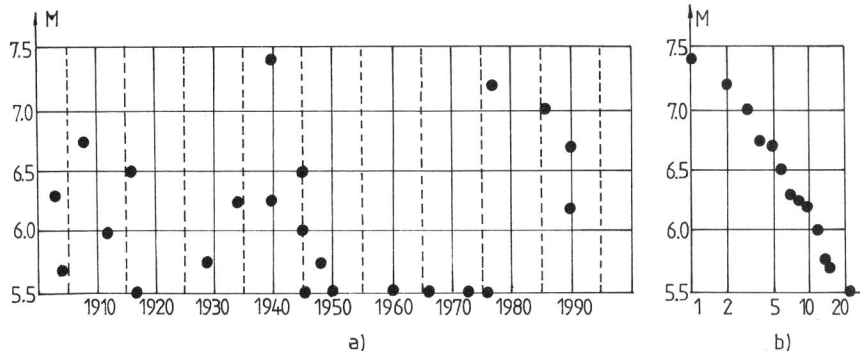

Fig.2. Time history of Vrancea earthquakes.

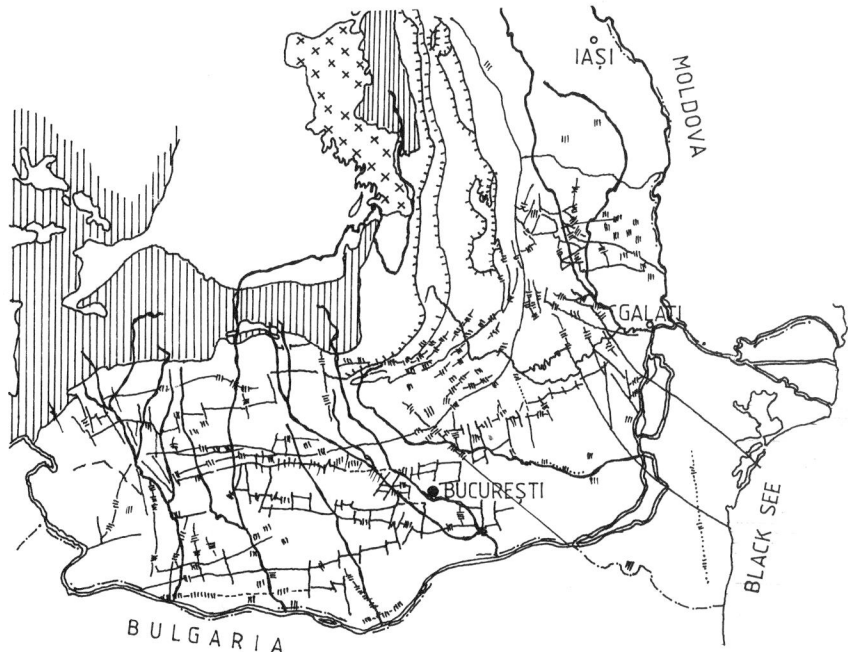

Fig.3. Faults and cracks of East and South zones of Romania.

fact that after 1977 a very dense instrumental network was achieved and very important amount of informations were obtained. If in 1977 only two accelerogram records were performed, in 1986 the number exeeded 60 and almost 120 for the two events of 1990 were recorded.

The seismic events with the Vrancea epicentre have a very large area of influence on the East and South zones of

Romania due to the existence of a lot of important faults and cracks in the earth crust (Fig.3), which can be actived by the Vrancea earthquake or can be the source of a new movement. The accelerogram records of 1977 earthquake is very uncommon. In Fig.4 a are plotted the accelerogram records N-S and E-W of Vrancea - 77 earthquake (INCERC records) and Fig.4 b shows the linear spectra. One can see very important amplifications in the interval 1...1,6 sec. for N-S direction and 0.7...1.2 sec. interval for E-W direction.

The instrumental informations after the events of 1977 are very rich and, what is the most important, one dispose of sequences of three successive records on the same point and several other points, which make it possible to have an insight into the feature of the phenomena, fact which is very rarely available in other regions (Sandi, 1992). When one compare different records at the same point, on may remark some similarities but important differences also. The

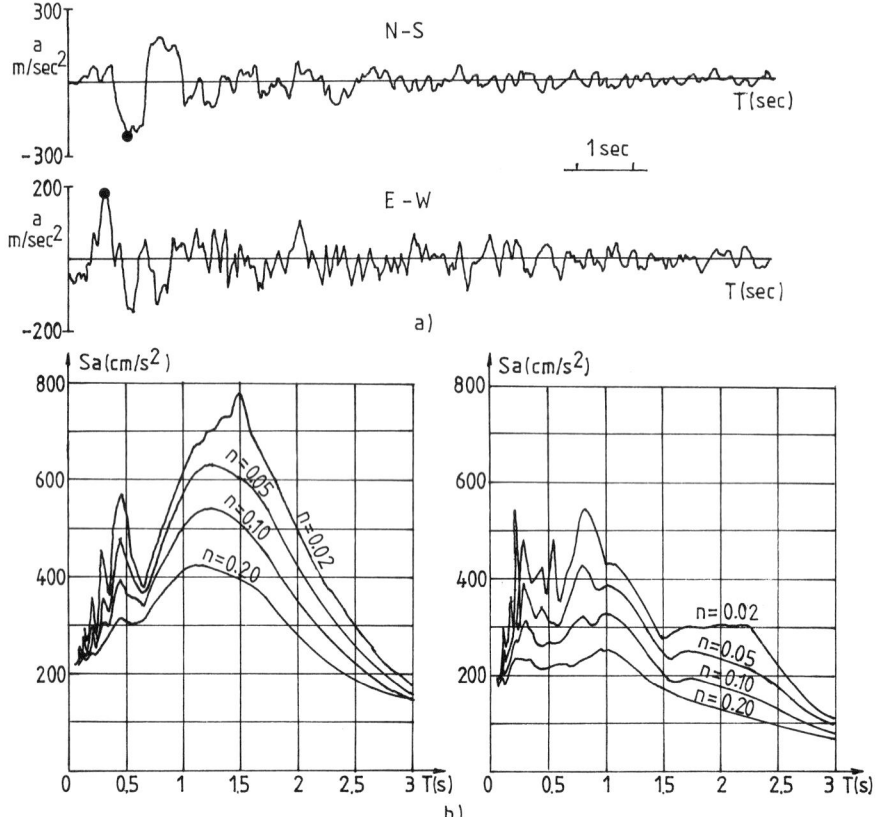

Fig.4. Accelerogram records and spectra for 1977 Vrancea earthquake.

very marked peak in N-S spectrum for 1977 event for long
period of 1.5 sec is reduced for 1986 earthquake and disappears totally in 1990 (Fig.5). On the other hand, the
shapes and response spectra for E-W direction are quite
similar for all three events. For the same event, the
differences between records for closer points are important,
due to strong influence of local conditions (Radu et al.
1990).

The Banat earthquakes are the most active crustal events.
The crust of Banat region is characterized by the presence
of very important faults and fissures (Fig.6) which produce
seismic event with low depth (10-20 km) and reduced area.
The informations concerning seismic events start from the
1739 earthquake and till 1900 there are noticed 217 events:
3 events with I⩾VIII, 13 events with I>VII, 35 events with
I>VI. In the period 1900...1991, 102 events were produced,
2 events with I>VIII, 5 events with I>VII, 15 events with
I>VI. The more recently were those of 12 July 1991
(M=5.7), 18 July 1991 (M=5.6) and 2 December 1991 (M=5.5)
(Sandi, 1992). These events produced heavy damages in the
villages close to the epicentres. The instrumental date
present unexpectedly high differences of spectral contents.
While for the event of 12 July 1991 the predominant periods
tend to be short (0.2 to 0.3 sec) characteristic for crustal
events for the event of 2 December 1991 the initial spectral
accelerations were high up to 0.8 sec, followed by numerous
short periods.

The examination of all these aspects concerning the
seismicity of Romanian territory put in to evidence a strong
diversity of possible seismic events. It is possible to
note that exist a tendency to higher importance of longer
periods in the South of Romania and of shorter periods in

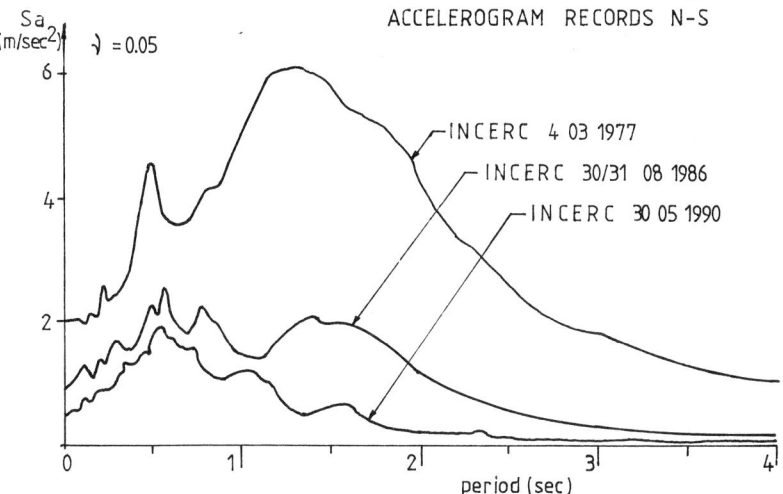

Fig.5. Comparison of 77, 86 and 90 earthquakes.

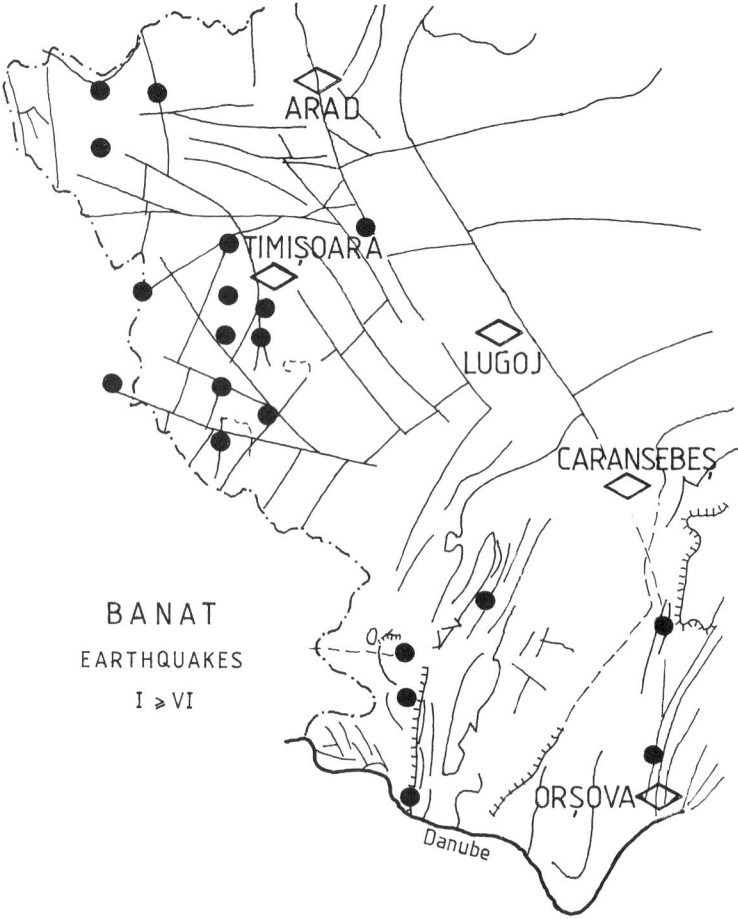

Fig.6. Faults and cracks of Banat region.

the East and West. These differences underline the danger to use in the structural design in a "time-history" method of an only one accelerogram record, even if this record was performed in the place of the building.

Other very important characteristic of the Romanian earthquakes are the short periods between two major events. So each structure can be affect by two or more strong earthquakes and the cumulative damages represents a first importance problem in antiseismic structural design. One must mention that 28 buildings from the 31 ones collapsed during 77 earthquake were old, passing through two seismic events.

2 Behaviour of steel structures during 4 March 1977 earthquake

Romania suffered greately from the earthquake of March 4, 1977, in which some 1,6oo persons were killed, over 11,000 were injured, 33,000 buildings collapsed or were severely damaged, industrial facilities were seriously damaged, and the total damage was over $2 billion. The great majority of these injuries were in the masonry and reinforced concrete buildings. Generally, the steel structures reacted very well during this earthquake. However some damages were noted, showing that if some inadequate structural conceptions are used, even the steel structures can be affected. One must mention that the number of steel structures was not so high, because long time in Romania this structural type was prohibited due to a false political reason. Only for mill-buildings with heavy travelling crane bridge steel structures were accepted. For long spans, steel roofs were provided, but the columns were designed in reinforced concrete. So the conclusions refering to the behaviour of steel structures during the Romanian earthquakes have limited values.

2.1 One story buildings

Over 80% of steel structures for industrial buildings were built after 1950 and designed on the base of the first Romanian code, which ensure an earthquake resistance. So the steel industrial buildings have behaved satisfactorily during the earthquake. The produced damages are due to (Dalban, Dragomirescu, 1985):
 - the use of a heavy roof deck (reinforced concrete panels) introduce lateral seismic forces significantly higher than the values resulting from the codes. Due to this effect, the upper part of some stepped columns had lateral plastic displacements of 8-85 cm (Dalban, Dragomirescu, 1985);
 - the vertical bracings in mill-buildings with traveling crane bridge sized for forces resulting from crane, had a good behaviour. However some bracings with very slender cross diagonals failed through the buckling of the compressed diagonal, and to yielding of the connections;
 - the connections between steel members and concrete members sometimes suffered damages resulted from over stressing during the earthquake. In the same time some pulling out of steel plates from concrete columns were produced, due to the damage of the concrete around the plate anchor bolts;
 - some steel roofs were collapsed due to unadequate support solutions: collapse of reinforced concrete columns, collapse of masonery walls, collapse of some roller expansion bearings etc.

2.2 Multi-story buildings

Due to a tradition in Romania, the reinforced concrete was used for the structure of multi-story buildings. So, there were a little number of steel structures involved in the 77 earthquake, and for all of these, this event was the second one, because the earthquake from 1940 already produced some damages.

Victoria Telephone Exchange Office was erected between 1931-33, conforming to an American project. It is composed by an initial building with 12 levels (for that time the highest building of Bucharest), which was enlarged latter (1939) with two adjacent buildings with 6 levels each (Ferbinteanu et al.,1985, Mazilu et al, 1985). The steel structure is built of columns and girders jointed with rivets, and the floors of reinforced concrete. The external walls from bricks were covered with stone pieces, resulting a very heavy façades. After the 77 earthquake some damages were observed. The remanence lateral displacements extended up to the third floor, in the front wall facing were about 15 cm, showing that the frames undergo very important plastic deformations. The measures of strengthening consist in introducing double diagonal steel bracings in the façade.

The Adriatica building was erected before the Second World War. The joints between columns and girders are rivered and the floors are reinforced concrete. The ground floor was free of walls, while the six upper floors were infilled with heavy brick walls. During the earthquake, very important lateral displacements were produced and due to plastic deformations some remaining displacements about 18-20 cm were measured.

The Transport Ministry building (12 stories), erected in 1940, was designed in composite steel-concrete frames. Due to the great deformability of structure, the brick panels were damaged by very important crecks and fissures.

3 Codification activities

The activity in the field of codification of seismic protection falls in two directions, the first concerning the protection of buildings and the second to the zoning of Romanian territory.

3.1. Codification for buildings protection

The first specifications for the protection of public works were published in 1942 (Fig.7) based on the experience of the great earthquake from 1940. It was the base of the design activity untill 1953 when the Russian code was introduced used till 1963. On the base on knowledges of the community of design engineers aquired in the early sixties, the first Romanian code was elaborated P13-63

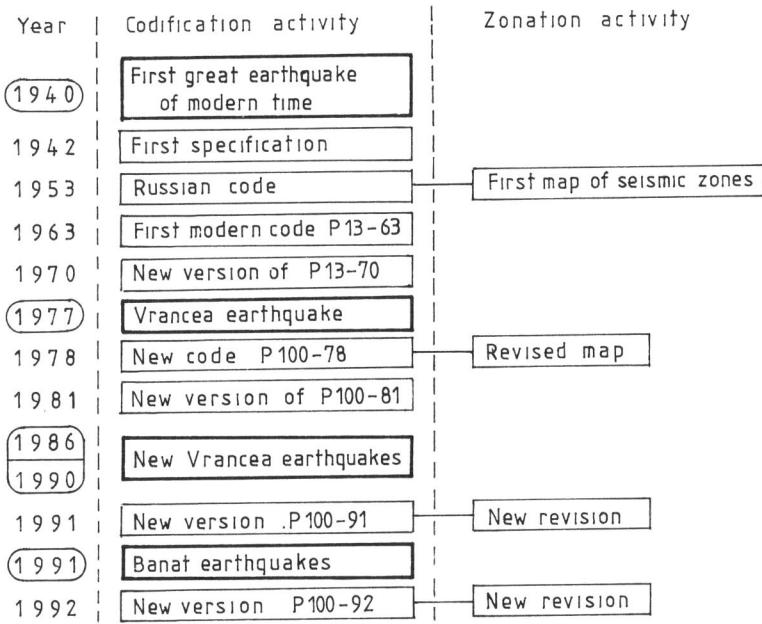

Fig.7. Codification activities.
(INCERC Bucuresti, I.C.Bucuresti)

"Design of civil and industrial buildings in seismic regions". It was the first modern Romanian code containing world level provisions. A new version is performed in 1970. The experience of 1977 great earthquake in design spectra and the requirement to provide ductility for reinforced concrete was used in the elaboration of a new Romanian code P100-78 "Code for Aseismic Design of Residential, Public, Agrozootechnical and Industrial Buildings". This code was improved in a new version in 1981, P100-81.

But all these versions of the Romanian code contained only few provisions for steel structures, due to the poor level of the knowledges, at that time, in the field of behaviour of these structures in seismic zones.

The impact of more recent strong eartquakes from 1986 and 1990 for the reinforced structures, and the contacts, with the works of ECCS TC13 Working Groupe which delivered the "European Recommendations for Steel Structures in Seismic Zones", lead to a new version P100-91, in which the new knowledges in this field were introduced. A new version of this code is the P100-92. So, the Romanian code is now at the level of EUROCODE 8 and other modern codes.

At present, based on the new theoretical and experimental results, a new version is prepared for the next year.

In the activity of codification the main contributions

were performed by INCERC Bucuresti and Civil Engineering Institute Bucharest.

3.2. Codification of seismic zones

Concerning the codification of seismic zones the first standard was elaborated in 1952, with new version after each earthquake, in 1963, 1977, 1991, 1992, containing, a new improve obtained by the new experimental data. For instance, the last Banat earthquakes from 1991 imposed the last modification of seismic zones.

4. Development of research activity

4.1. Main research centres and the existing equipments

In Fig.1. are also marked the four Romanian research centres involved in the studies on seismic behaviour of steel structures. Each of these centres are characterized by the existence of an important Technical University with some academic staff concerned in these problems. Closed to universities were developed a network of branches of Romanian Building Research Institute - INCERC, which belong to the Ministry of Public Works. The Table 1 presents the equipments of these centres and their main characteristics.

Due to the attribute of staffs and experimental possibilities, each centre is characterized by a distinct feature.
Bucharest centre is involved on general studies concerned engineering seismology, vulnerability and risk analysis, and elaboration of codes.

The research works of Iasi centre, due to the very important existent and in erection equipments, are devoted to the experimental studies of dynamic-seismic behaviour of elements and structures.

The activity of Timisoara centre is characterized by theoretical studies, using the computer possibilities to modelling the seismic behaviour of structures and experimental verifications of these results.

The research works of Cluj centre are also devoted to experimental studies, using the pseudo-dynamic method for structure tests.

The main theoretical and experimental results are presented in the following.

4.2 Basic problems

Studies on engineering seismology have been developed by INCERC Bucharest, particularly concerning for an intensity scale based on accelerographic data and response data (Sandi 1990). The work on seismic zonation was based on instrumental information (about 150 strong motion accelerograms). The development of a full probabilistic approach made it possible to derive by analytical means accurrence law for intensities (Sandi, 1992). To illustrate this,

Table 1 Equipments of Romanian research centres

Institution	Testing system	Loading system	Dimens LxIxh	Accel. horiz. vert. m/sec^2
INCERC Bucharest	ST-IB1	DS	6x6	27.3 / 27.3
	ST-IB2	DS	3x3	38.3 / 38.3
	RW	S,MA,D	24x24x12	-
INCERC Iasi	ST I-140	DS	10x10	10.0 / -
	ST I-15	DS	3.2x3.2	15.0 / -
	ST I-1000*	DS	14.4x14.4	5.3 / -
	ST I-200*	DS	12x12	7.4 / 4.6
	ST I-30*	DS	6x7.5	12.0 / 6.0
	ST I-5*	DS	3.5x3.5	19.5 / 13.0
	TS	S,MA,D,DS	6x6x2.7	-
	TS	S,MA,D,DS	14x14x5.5	-
INCERC Timisoara	TS	S,MA	12x3x5	-
	TS*	S,MA,D	18x8x5	-
INCERC Cluj	RW1	S,MA,D	8x5x3.5	
	RW2	S,MA,D	7.8x6.5x6.5	-
	RW3	S,MA,D	7x6.5x3.5	

Testing systems
ST-shaking table
RW-reaction wall
TS-testing stand
* - in erection

Loading systems
s - static
MA- monotonic alternate
D- dynamic
DS - dynamic-seismic

the recurrence law of magnitudes assumed for the Vrancea zone is given in Fig.8a. The recurrence law of intensities in Bucharest derived for various assumptions statistical analysis aimed to calibrate the attenuation parameters is presented in Fig.8b. The dramatic influence of high scatter of about one intensity unit is visible there.

Analytical work on vulnerability characteristics of structures was carried out also at INCERC in connection with the analytical research on intensities and with the

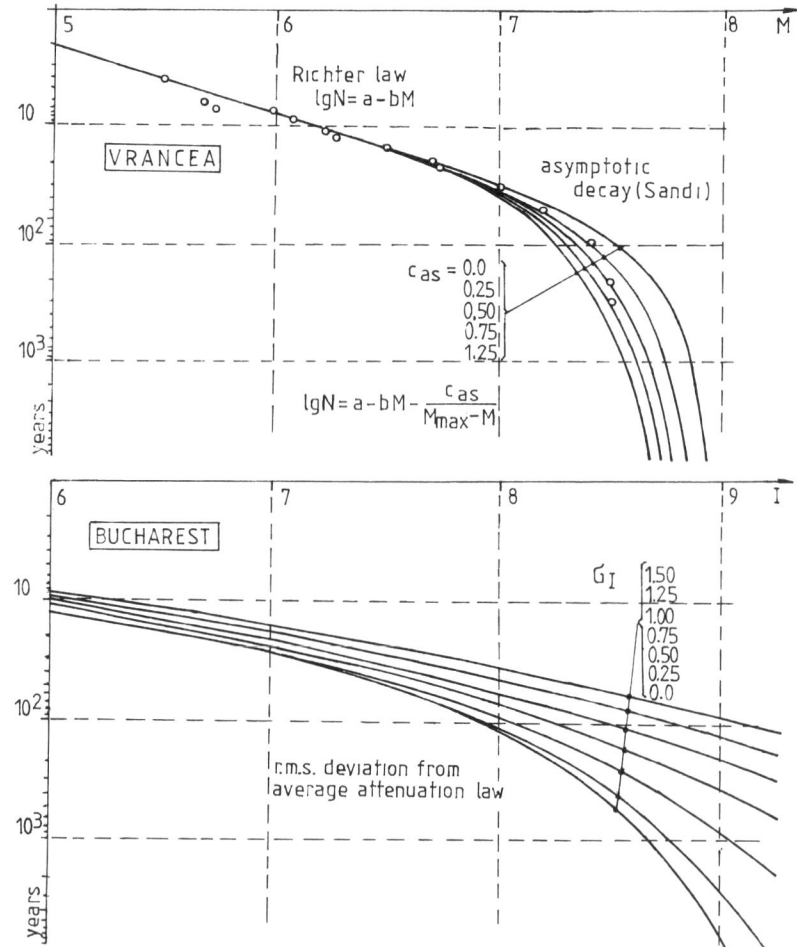

Fig.8. Reccurence laws.

development of appropiate regression techniques (Sandi,1986). Vulnerability studies were organized on the basis of post-earthquake surrey of damages and of parametric Monte-Carlo engineering analyses of typical structures.

Risk analysis, aimed to quantify the risk factory for various categories of structures, were performed. They put for evidence (Sandi, Floricel, 1993) the probability of exceeding various damage degrees, as functions of hazard and vulnerability characteristics and of exposure time. They were aimed to be used for calibration and differentiation of design parameters, with special attention for the problems of unsufficiently protected existing building stock.

4.3. Joints

A cooperation between INCERC Timisoara and INCERC Iasi was performed in the aim to study the behaviour of different joint types.

Column bases were studied at INCERC Iasi (1989a) (Fig.9). The connections are characterized by a stiffened base plate, so that only the anchor bolts have relevant influence on joint behaviour. Eight steel column-fundation subassemblages were tested under cyclic loads. The variable parameters were the anchored system (by hooks, hooks with transversal bars, end plates), length of anchor bolts, and exterior bolt length. Relevant data have been obtained concerning the influence of these parameters on joint behaviour, shoving that even this one with stiffened base plate has a feature corresponding to semi-rigid or partially restrained joint.

Marginally beam to column connections were tested at INCERC Timisoara (1991a) (Grecea, Mateescu, 1994) (Fig.10). The five bolted semi-rigid joints were compared with the reference welded joint, a rigid one. The characteristics of these semi-rigid joints were the beam-to-column connection systems, using end plates, partial end plates, cover plates, flange and web angles. High strength bolts were used. For each joint type, one monotonic test were completed with two cyclic tests in the aim to compare the two different behaviour type. The determination of moment - rotation characteristics of joints were one of the most important purpose of the research works.

Centrally beam-to-column connections were studied at INCERC Iasi (1987, 1989a) (Acatrinei et al, 1988)(Fig11). Four welded rigid joints, with different configurations of column panel (without and with stiffeners) were tested. Three bolted semi-rigid joints, using normal or high strength bolts were also investigated. Static and dynamic loads were applied. The experimental results show the great influence of the connection type on the joint failure, particularly on the presence of panel stiffeners.

4.4 Local ductility

Intensive theoretical and experimental research works were performed at INCERC Timisoara (1987, 1993a) (Gioncu et al, 199, 1994a) (Fig.12). Based on experience evidence, the plastic rotation capacity was studied using a collapse mechanism formed by yielding lines and plastic zones. A plastic moment - plastic rotation relationship was obtained containing the main factors influencing the phenomenon. For beams, a computer program POSTEL, and experimental studies were performed, having the purpose to verify the ductility classes proposed in EC3 and EC8. Very important differences were obtained, showing the scarsity of codes provisions. The studies were completed with the case of beam-column members, for which the computer program DUCTROT 93

No	TYPE OF ANCHOR	CHARACTERISTICS	TEST
F1		○ bars with hooked anchorages; ○ bolt length 35 φ	cyclic alternant
F2		○ bars with hocked anchorages and transversal bars φ 25; ○ bolt length 20 φ	idem
F3 F4 F5 F6 F7 F8		○ bars with end plates with different elements to prevent the slip; ○ bolt lengths φ 17, φ 13, φ 11, φ 9	idem

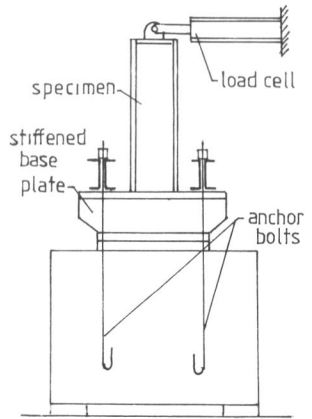

Fig.9. Column bases (INCERC IAsi).

No	TYPE OF CONNECTION	CHARACTERISTICS	TESTS
A		Rigid connection Welded I – plates connected by welding	monotonic cyclic alternant
B		Semi-rigid connection Partial end plates connected with HSB 24	idem
C		Semi-rigid connection End plate connection with 12 HSB 24	idem
D		Semi-rigid connection Flange angles 160×160×12 with HSB 24	idem
E		Semi-rigid connection Welded cover plate	idem
F		Semi-rigid connection Flange (150×150×14) and web angles (120×120×12) with HSB 24	idem

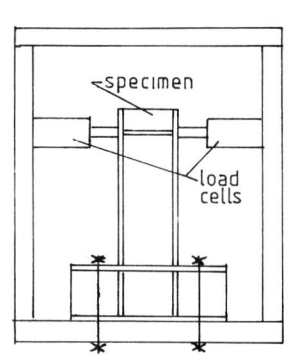

Fig.10. Beam-to-column marginally connections (INCERC-Timisoara).

No	TYPE OF CONNECTION	CHARACTERISTICS	TESTS
SM1	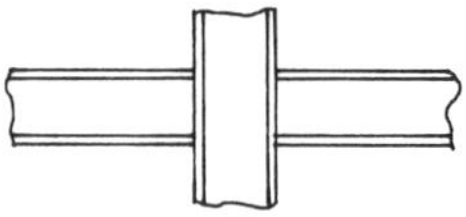	Welded, unstfifened joint	static and dynamic
SM2	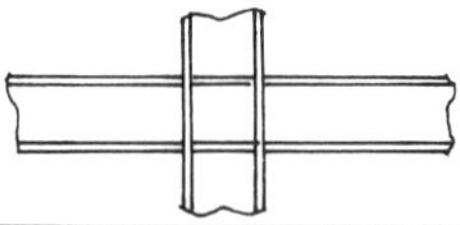	Welded, with horizontal diaphragms	idem
SM3	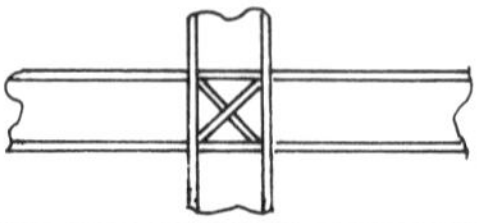	Welded, with horizontal and diagonal diaphragms	idem
SM4	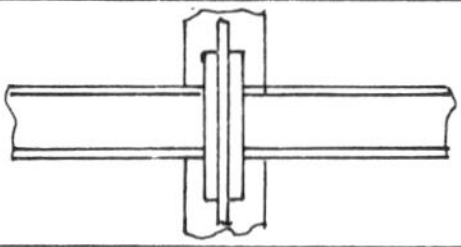	Welded, beams to column web connection	idem
SM5	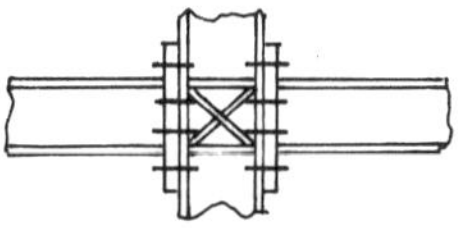	Bolted, flange connections with normal bolts	idem
SM6	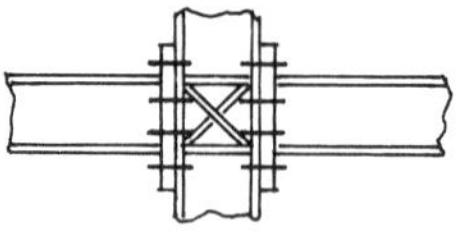	Bolted with HSB; flange connections	idem
SM7 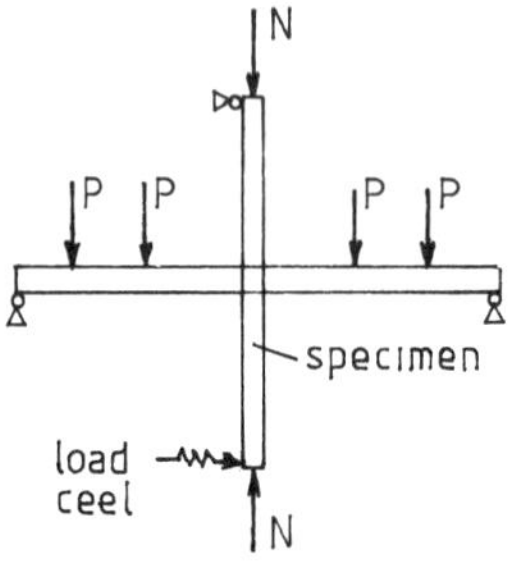	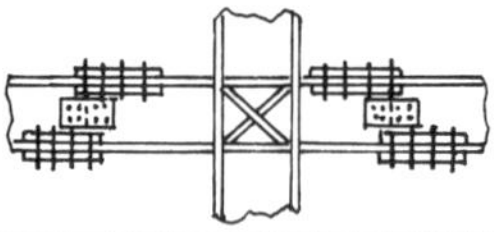	Bolted, eccentrically connections 	idem

Fig.11. Beam-to-column centrally connections (INCERC Iasi).

a) THEORETICAL RESEARCH WORKS

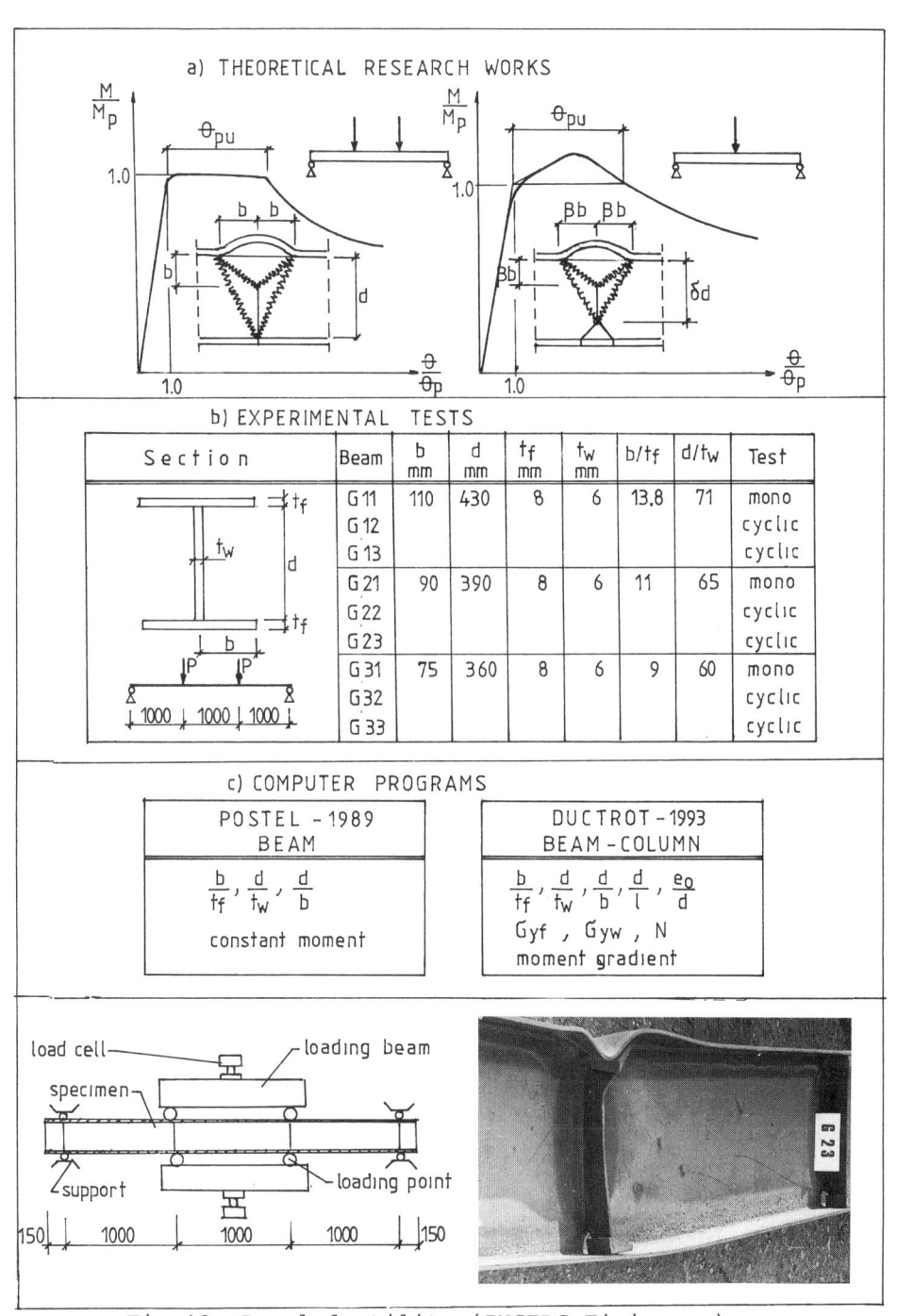

b) EXPERIMENTAL TESTS

Section	Beam	b mm	d mm	t_f mm	t_w mm	b/t_f	d/t_w	Test
	G 11	110	430	8	6	13.8	71	mono
	G 12							cyclic
	G 13							cyclic
	G 21	90	390	8	6	11	65	mono
	G 22							cyclic
	G 23							cyclic
	G 31	75	360	8	6	9	60	mono
	G 32							cyclic
	G 33							cyclic

c) COMPUTER PROGRAMS

POSTEL - 1989 BEAM	DUCTROT - 1993 BEAM - COLUMN
$\frac{b}{t_f}, \frac{d}{t_w}, \frac{d}{b}$ constant moment	$\frac{b}{t_f}, \frac{d}{t_w}, \frac{d}{b}, \frac{d}{l}, \frac{e_0}{d}$ G_{yf}, G_{yw}, N moment gradient

Fig.12. Local ductility (INCERC Timisoara).

was elaborated. The influence of main factors was studied and tables containing the values of ultime rotation capacity for I cross - section were obtained.

4.5 Moment – resisting frames

<u>Theoretical research</u> works were performed at INCERC Timisoara (1992a-c, 1993 b,c) (Gioncu et al, 1994b) (Fig.13) with the purpose to calibrate the structural behaviour factor q. Various structure types were analysed: a) one story frames with one or two spans and simple or stepped columns: b) multi - story frames with one or two spans and various level number. Two methods were used, one based on Ballio - Setti procedure which takes into account the lateral displacements, the second which consider the plastic rotation capacities of cross-sections as the main behaviour factor. Two computer programs were used: ANALISE and DRAIN-2D, both using the "time-history" analysis. The second method, considering as the collapse criterium is the ultimate plastic rotation of frame members, seems to be preferable, but this procedure requires a correct determination of ultimate plastic rotation, which is not one easy task. The developed method allows the study of influence of main factors: accelerogram record types, column rigidity natural period, story number, degradation, semi-rigid joints (Dubina et al, 1994), etc.

<u>Experimental tests</u> on moment-resisting frames were performed at INCERC Iasi (1988, 1989b, 1990) (Acatrinei, Rotaru, 1991) (Fig.14). One - story spatial frames with two spans with hinged or, rigid beam-to-column joints were tested on shaking table. Various natural or artificial accelerogram records were used. A three levels spatial frame was also tested at INCERC Iasi (Diaconu, 1970) on the shaking table. Historically speaking this was the first steel structure tested in Romania for seismic actions. Two -story frame was tested at INCERC Timisoara (1991b) (Gioncu et al, 1991) under cyclic loads. The conclusions of these experimental tests confirm very well the theoretical results, but the high costs of experiments leads to the necessity to develope the numerical methods for calibration of the q - factor.

4.6 Braced frames

Intensive theoretical and experimental works were performed at INCERC Cluj (Georgescu et al, 1987, 1992)(Fig.15). After the studies on the behaviour of a bracing member under cyclic loading, different ten bracing types were tested. The first five refer to a one story braced frames, with concentrically, eccentrically and especially bracing systems. The last five were composed on two story braced frames, with steel or reinforced concrete columns. An other parameters of study is the influence of runway girder on bracing system behaviour. The tests have shown that the

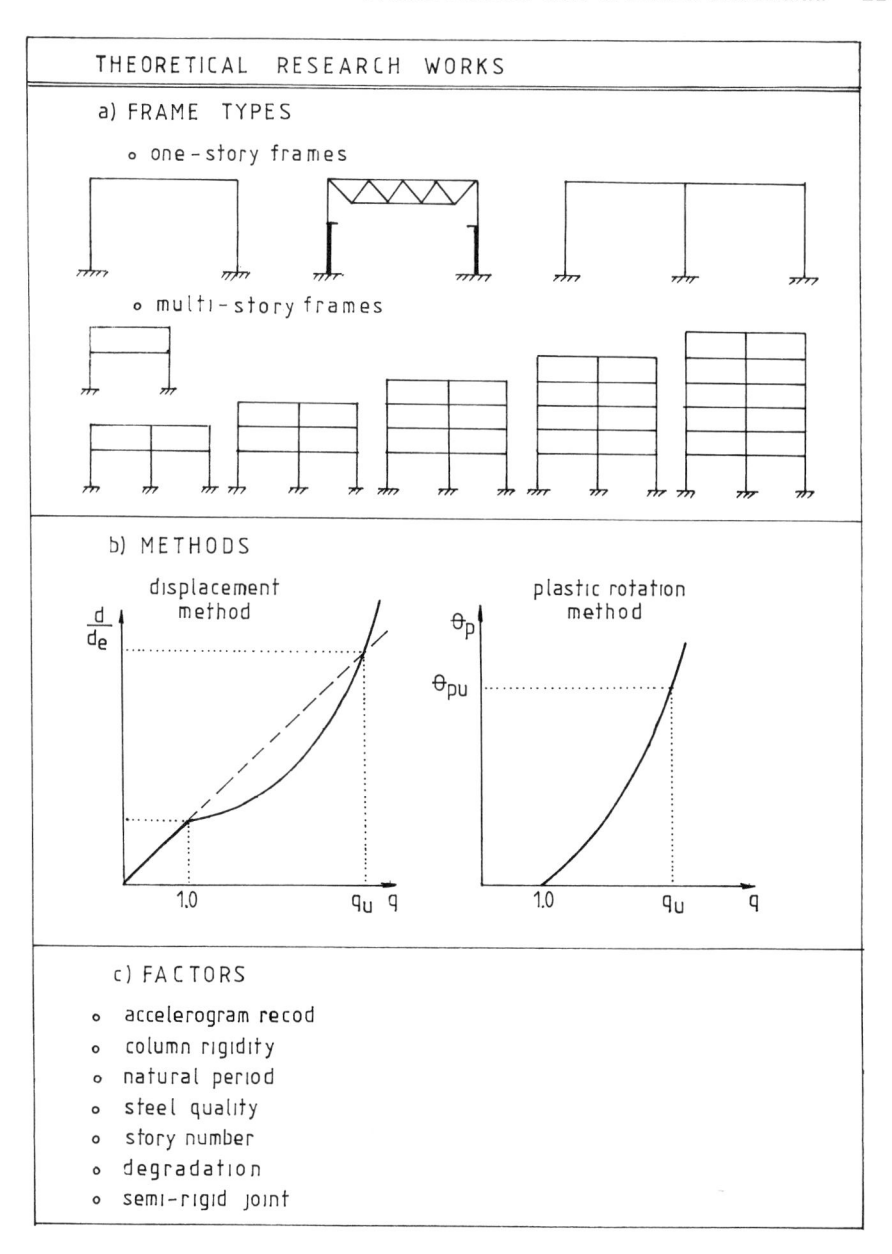

Fig.13. Theoretical studies on global ductility (INCERC Timisoara).

N	STRUCURE TYPE	CHARACTERISTICS	TEST

1	(dimensions 5×1800, 4500, 4500, 2920)	○ dimensions 9 × 9 × 2.95 m
		○ beam-column joint
		  -variant 1: hinged
		  -variant 2: rigid
		INCERC Iași
			dynamic-seismic using the shaking table

2	(1.5F, 40kN, 40kN, 40kN, 40kN, F, 4000, 1500, 2000)	○ two story frame ○ rigid joints ○ cross-section 2[16 INCERC Timișoara	cyclic alternant
3	(3075, 3075, 4700, 1600, 1600)	○ three story frames ○ rigid joint ○ beams I 14 ○ columns I 16 INCERC Iași	dynamic-seismic using the shaking table

Fig.14. Experimental tests on moment-resisting frames (INCERC Iasi, INCERC Timisoara).

	TYPE OF BRACING	CHARACTERISTICS	TEST
P1 P2 P3	⟵3400⟶ 2900	Concentrically V braced frames with different cross-sections	cyclic alternant
P4	⊢1400⊣	Eccentrically V braced frame	idem
P5	⊢1000⊣	Especially bracing system	idem
CL1		Concentrically two story V braced frame; r.c. columns	idem
CL2		Concentrically two story X braced frame r.c. columns	idem
CL3+ CL5	800	Eccentrically two story V braced frame; r.c. and steel columns	idem
CL4		Eccentrically two story V braced frame with double beams (runway girder)	idem

Fig.15. Braced frames (INCERC Cluj).

braced steel frames possesses a certain capacity to dissipate energy under seismic loading. A good concordance exists between theoretical and experimental results.

4.7 Controled energy dissipation

In the aim to control the energy dissipation, at INCERC - Timisoara (1993d) a solution to reduce the cross-section of beam near the marginally columns was studied. In this case it is sure that the plastic hinge is formed in beams, that the story collapse mechanism is prevent. Three moment--resisting frames were studied using time history analysis. The proceeding to reduce the cross-section of beams was proved to be efficiently only for the case of two spans frames.

In the same purpose at INCERC Iasi (Palamar et al,1994) experimental tests were performed on story dissipative systems with dampers with regressive rigidity, made by parallel steel plates. The dissipation of seismic energy was very important, showing the efficiently of this system.

5. Conclusions

Due to the seismicity of Romanian territory, the activity in the field of research elaboration of codes and design is very intensely. The report is intended only to inventory the main aspects take into account, without to be too insistent on the results, due to the restricted disposal space and the fact that many of these theoretical, experimental or design aspects are presented in our conference.

5 References

Acatrinei, L., Mihai, C., Precupanu, D., Ibanescu, M.(1988) Static and dynamic behaviour of three welded metal joints in **Proceedings of V Conference on Metal Structure,** Timisoara, vol.3, 38-43.

Acatrinei, L., Rotaru, M., (1991), Global seismic responce and collapse of an one-story frame with different bracing types, in **Proceedings of VI Metal structure Conference,** Timisoara, vol.3, 83-90.

Balan, St., Cristescu, St., Cornea, I., (1982), **The Romania earthquake of 4 March 1977** (in Romanian) Editura Academiei, Bucharest.

Crainic, L., Dalban, C., Postelnicu, T., Sandi, H., Teretean, T., (1992) A new approach of the Romanian seismic code for building **10 WCEE,** Madrid.

Dalban, C., Dragomirescu, (1985); Some conclusions on the behaviour of the steel and composite structures subjected to seismic action,in **Proceedings of Joint US-Romanian seminar on earthquakes and energy,**vol.2 Bucharest 77-84.

	TYPE OF BRACING	CHARACTERISTICS	TEST
V1	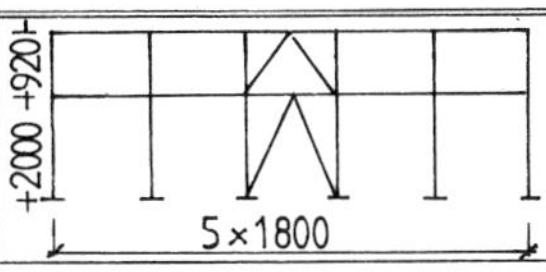 5×1800 (+2000, +920)	Concentrically two story V braced frame	dynamic-seismic on shaking table
V2	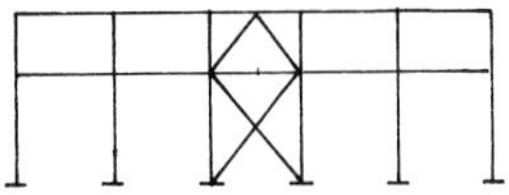	Concentrically two story V and X braced frame	idem
V3	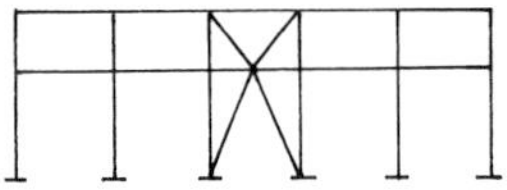	Concentrically two story V and Λ braced frame	idem
V4	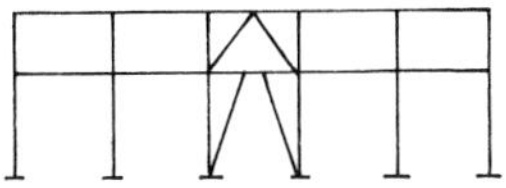	Eccentrically two story V braced frame	idem

Fig.16. Longitudinal braced frames (INCERC Iasi).

	TYPE OF STRUCTURES	CHARACTERISTICS	OBS.
C 1-3	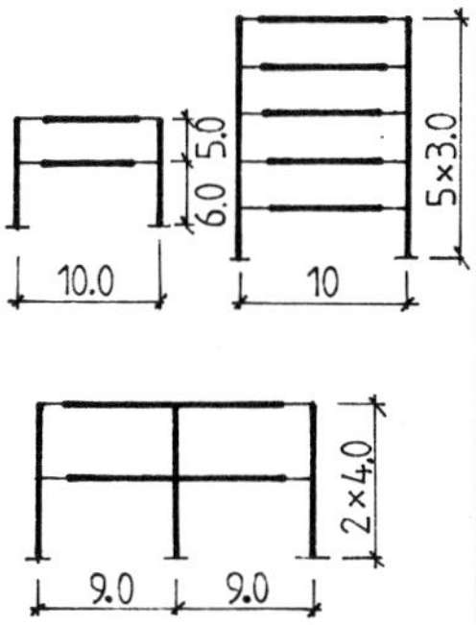 10.0 / 10 (6.0, 5.0, 5×3.0); 9.0 / 9.0 (2×4.0)	Moment resisting frames with reduced section near marginally columns INCERC Timișoara	Theoretical studies
MOD 1-5	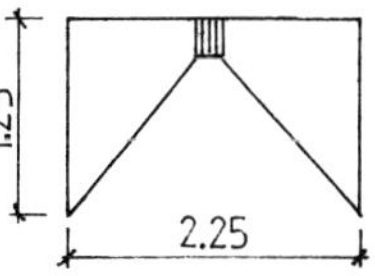 2.25 (1.25)	Story dissipative systems INCERC Iași	Experimental works

Fig.17. Energy dissipation systems
(INCERC Timisoara, INCERC Iasi).

Diaconu et al, (1970), Reponse d'une ossature metallique a pleusieurs etajes, essayee sur la plate-forme sismique de grande capacite, in **Proceedings of Earthquake Analysis of Structures**, Iasi.

Dubina, D., Grecea, D., Zaharia, R., (1994), Numerical simulation concerning the response of steel frames with semi-rigid joints under static and seismic loads, in **Proceedings of STESSA 94,** Timisoara.

Ferbinteanu, V., Balcu, M., Petrovici, D., Dima, M. (1985), Choice of strengthening solution by the evaluation of post-shaking strength capacity of a multistory building having a steel skeleton structure and brick masonry infilling, in **Proceedings of Joint US-Romanian seminar on earthquakes and energy,** vol.2, Bucharest, 272-284.

Georgescu, D., Toma,C., Gosa,O.(1992): Post-critical behaviour of K braced frames. **J.Construct. Steel Research** 21 (1992), 115-133.

Georgescu, D., Fulea, D., Gosa, O., Manoiu, O., Toma, C., (1987), Seismic response of braced steel frames used to stiffen one-floor industrial halls. **Atti del XV Convergo Nazionale** AIPADS, Pisa, 1987.

Gioncu, V., Mateescu, G., Orasteanu, S., (1989), Theoretical and experimental research regarding the ductility of welded I sections subjected to bending, in **Proceedings of SSRC Int. Symp. on Stability, Beijing.**

Gioncu, V., Mateescu, G., Biro, A., Grecea, D.(1991) Behaviour of two-story unbraced frame to cyclic loads, in **Proceedings of VI Metal Structure** Conference, vol.3, 76-82.

Gioncu, V., Mateescu, G., Iuhas, A., (1994a): Contributions to the study of plastic rotational capacity of I steel sections, in **Proceedings of STESSA 94,** Timisoara.

Gioncu, V., Mateescu, G., Iuhas, A., (1994b): Contribution to the study of the ductility of unbraced frames, in **Proceedings of STESSA 94,** Timisoara.

Grecea, D., Mateescu, G. (1994), Behaviour of semi-rigid connections under alternate static loads, in **Proceedings STESSA 94,** Timisoara.

Mazilu, P., Ieremia, M., Neicu, M., Rosca, L.,(1985): Seismic behaviour of some industrial and telecomunication structures: solutions for seismic risk reduction, in **Proceedings of Joint US-Romanian seminar on earthquakes and energy** vol.2, Bucharest, 67-76.

Palamar, G., Mihai, C., Rotaru,. (1994) Story dissipative systems for steel structures, in **Proceedings of STESSA 94,** Timisoara.

Radu, C., Radulescu, D., Sandi, H., (1990) Some data and considerations on recent strong earthquakes of Romania, **Cahier Technique** no.3, AFPS.

Sandi, H., (1986): Vulnerability and risk analysis for individual structures and for systems. **Report to 8-th ECEE,** Lisbon. Idem for 9th conference, 1990, Moskow.

Sandi, H. (1992), Use of instrumental data for evoluation of ground motion and for specification of some conditions. Some data on recent Romanian experience. **Proc. International Symp. on Earthquake Disaster Prevention**, CENAPRED, Mexico City.

Sandi, H., Floricel, I., (1993), Analysis of seismic risk affectory the building stock (in Romanian). **Gazeta AICR**, no.2-3.

INCERC reports (in Romanian).

INCERC Bucuresti (1991): Studies for improve the Banat seismic zonation. Report.

INCERC-Iasi (1987): Static and dynamic behaviour of three welded joint types. Report.

INCERC-Iasi (1988): Seismic behaviour of new steel frame types for industrial buildings. Report.

INCERC-Iasi (1989a), Experimental research works on beam-to-beam joints and column-fundation joints. Report.

INCERC-Iasi (1989b): Experiments on one-story frames. Report.

INCERC-Iasi (1990), Experimental tests on one-story frame atlongitudinal static and dynamic actions. Report.

INCERC-Timisoara (1987), Ductility of bent elements. Theoretical and experimental research works. Report.

INCERC-Timisoara (1991a): Experimental research works on joint systems. Report.

INCERC-Timisoara (1991b), Experiments and conclusions concerning the behaviour of unbraced frames subjected to cyclic loads. Report.

INCERC-Timisoara,(1992a), Steel structure safety for dynamic-seismic actions. Report.

INCERC-Timisoara, (1992b), Parametric studies concerning q-factor for braced and unbraced frames. Report.

INCERC-Timisoara, (1992c), Calibration of q-factor for steel structures. Conclusions on theoretical studies and numerical tests. Report.

INCERC- Timisoara (1993a), Theoretical studies and methods for energy dissipation evaluation for one-story buildings. Report.

INCERC-Timisoara (1993b), Theoretical studies concerning dynamic-seismic instability of slenderness members. Report.

INCERC- Timisoara, (1993c), Numerical tests on influence of local buckling in seismic behaviour of steel structures. Report.

INCERC-Timisoara (1993d), Theoretical studies and numerical tests on moment-resisting frames with controled energy dissipation. Report.

2 DEVELOPMENT AND DESIGN OF SEISMIC-RESISTANT STEEL STRUCTURES IN JAPAN

B. KATO
Professor Emeritus, University of Tokyo, Tokyo, Japan

Abstract
Theoretical and experimental background of Japanese seismic code were reviewed laying the emphasis on 1) design criteria, 2) design earthquake load, 3) structure type-coefficient, and 4) rotation capacity of members. Some comparisons were made with EC codes.
Keywords: Limit state design, Response spectra, Energy dissipation, Width-to-thickness-ratio.

1 Introduction

The Japanese Building Standard Law for seismic design was revised in 1981, and The Architectural Institute of Japan (AIJ) had specified the detailed design procedure on the basic of this standard law making use probabilistic approach. In Europe, ECCS-TC13 "European Recommendations for Steel Structures in Seismic Zones" was published in 1988, and Eurocode 8 "Structures in Seismic Regions" was published also in 1988. Since these codes or specifications are the condensation of efforts devoted by a large number of scientists and engineers, the deliberate study of them will bring a very valuable lesson.

From this point of view, the theoretical and experimental background of the regulations are investigated laying the emphasis on the magnitude of design earthquake, structure type coefficient (behaviour factor) and the rotation capacity of members which are still debatable. And compare the basic concepts of the regulations in both regions in hopes of eventual harmonization.

2 Design criteria (Safety verifications)

The safety format of Japan (AIJ [1]) is based on the first order second moment method (level 2 format) and that of EC(ECCS-TC13[2], Eurocode 8 [3]) is based on the partial safety factor method (level 1 format), and therefore the direct comparison among these are difficult. But it will be of interest to observe what difference will be resulted in in the design practice when these rules are applied.

2.1 Ultimate limit state
A major design earthquake which is used for the ultimate limit state design is unlikely to occur within the life of a structure, but is used to examine the ultimate structural safety.

2.1.1 EC

$$\frac{1}{\gamma_{rd}} R_d \geq \gamma_I E_e + G + \sum_i \varphi \psi_{2i} Q_{ik} \qquad (1)$$

in which,
 R_d : design resistance of structure
 E_e : Load effect of earthquake according to EC
 Q_{ik} : effect of characteristic variable load i
 γ_{rd} : partial safety factor for the resistance of members to be bucking, and equal to 1.1
 γ_I : importance factor
 φ : intensity factor of Q_{ik}, and equal to 0.5 except for top story
 ψ_{2i} : combination factor, and equal to 0.3

2.1.2 Japan(AIJ)

$$\Phi R_n \geq 2.0 E_j + 1.1D + 0.4L \qquad (2)$$

in which,
 R_n : nominal resistance of structure
 E_j : load effect of earthquake according to AIJ
 D : load effect of dead load
 L : load effect of live loads
 Φ : resistance factor, and equal to 0.85~0.9 for structural members

If the symbols used in EC are to be unified to those of Japan, and taking the typical office buildings for example ($\gamma_I=1$), eq.(1) and eq.(2) are rewritten as

$$0.91 R_n \geq E_e + D + 0.15L \qquad (1)'$$
$$(0.85 \sim 0.9) R_n \geq 2.0 E_j + 1.1D + 0.4L . \qquad (2)'$$

Since the characteristic values in EC and nominal values in Japan might differ in their definition, the direct comparisons of design resistance and of design actions of dead load and live loads may be less interesting. Rather, the principal objective of this paper should be the comparison of design earthquakes E_e and $2.0 E_j$.

2.2 Serviceability limit state

Moderate earthquakes are expected with a reasonably high probability during the life of a structure. The structure should be proportioned to resist the moderate earthquake without significant damage to the skeleton structure and to the non-structural elements of it.

2.2.1 EC (Eurocode 8)

General remarks are stated that, the structure as a whole, including structural and non-structural elements shall be planned, designed and constructed such that it is protected against the occurrence of damages and limitations of use as a consequence of an earthquake having a larger probability of occurrence than the one to be considered in the ultimate limit state design.

2.2.2 Japan

$$\Phi R_e \geq 0.4 E_j + 1.0D + 0.4L \qquad (3)$$

in which,
 R_e : nominal yield (elastic limit) strength of structure
 Φ : resistance factor, and equal to 0.9

Comparing with eq.(2), it can be seen that the design earthquake intensity is 1/5 of that for the ultimate limit state design (about 80 cm/sec² in peak ground acceleration).
 The structures have to resist elastically against this intensity of earthquake, in addition, the story deflection angle must not exceed $1/120 \sim 1/200$ depending on the flexibility of non-structural elements.

3 Design earthquake load

The design horizontal shear forces at the base of buildings are specified as followings;

3.1 EC

$$V = \frac{1}{q} A R_e M \tag{4}$$

in which,
 V : nominal design horizontal shear force at the base
 A : the design value of the peak ground acceleration, and is given as (g is the acceleration of gravity)

seismicity zones		
high	medium	low
0.35g	0.25g	0.15g

M : total mass of a building
q : behaviour factor
R_e : normalized design response spectrum, and is specified as,

$$\left. \begin{array}{ll} R_e = R_o & \text{for } T \leq T_o \\ R_e = \dfrac{R_o}{(T/T_o)^k} & \text{for } T > T_o \end{array} \right\} \tag{5}$$

R_o is the maximum response spectral value normalized to the peak ground acceleration, T is the fundamental period of structure.
R_o, T_o and k are given in Table 1 depending on the soil profiles.

Table 1. R_o, T_o and k values

soil profile	T_o (sec)	R_o	k
a	0.4	3.0	2/3
b	0.6	3.0	2/3
c	0.8	2.5	2/3

Soil profiles are defined as, a: firm (rock), b: medium (stable deposit), c: soft (soft to medium stiff deposit).

3.2 Japan

$$V = 2.5 \, A_e \, Z \, R_j \, D_s \, M \qquad (6)$$

in which,
- A_e : expectation of ground acceleration in the return period of 50 years, and is estimated to be 0.2g
- Z : Zone coefficient assigned according to the seismic activity (1~0.7)
- D_s : structure type coefficient related to the ductility of building, and is given in Table 2

Table 2. D_s-values

type of structure	D_s
moment resisting frames with excellent ductility	0.25
moment resisting frames with moderate ductility with or without diagonal bracings	0.35
moment resisting frames with poor ductility with or without diagonal bracings	0.45

R_j : design response spectrum normalized to the maximum response spectral value, and is specified as

$$\left. \begin{array}{ll} R_j = 1 & \text{for} \quad T < T_c \\ R_j = 1 - 0.2 \left(\dfrac{T}{T_c} - 1 \right)^2 & \text{for} \quad T_c \leq T < 2T_c \\ R_j = 1.6 \left(\dfrac{T_c}{T} \right) & \text{for} \quad 2T_c \leq T \end{array} \right\} \qquad (7)$$

T_c is given as depending on the soil profiles

soil profile	1 (firm)	2 (medium)	3 (soft)
T_c (sec)	0.4	0.6	0.8

3.3 Comparison of design elastic response

Both q in eq.(4) and D_s in eq.(6) are reduction factors for elastic shear response taking the dissipation of earthquake input energy by the plastic work of the structure into consideration. The assessment of q and D_s is a debatable problem, and will be discussed in the next chapter.

For the present, the design seismic forces in terms of response spectra are compared for EC and Japanese rules removing q and D_s factors respectively.
Assuming high seismicity zone (A=0.35g in eq.(4) and Z=1 in eq(6)), eqs.(4) and (6) are written as

$$V = 0.35 \, R_e \, Mg \qquad (4)'$$
$$V = 2.5 \times 0.2 \, R_j \, Mg = 0.5 \, R_j \, Mg . \qquad (6)'$$

Noting that R_e is normalized to the peak acceleration, and that R_o in eq.(5) is the amplification factor, $AR_o=0.35g \times 3.0=1.05g$ is the peak acceleration response in EC rule. And nothing that R_j is normalized to the maximum response spectral value, and that 2.5 in eq.(6) is the amplification factor (similar to R_o), $2.5 \times 0.2g=0.5g$ is the nominal peak acceleration response in Japan. Since the load factor in EC is unity, and that in Japan is 2.0 as seen in eqs.(1)' and (2)' respectively, the design peak acceleration for each is $1 \times 1.05g=1.05g$ for EC and $2.0 \times 0.5g=1.0g$ for Japan. Thus it can be concluded that the design peak accelerations for high seismicity zone are almost the same for both regions except for soft soil profile condition, though the spectral shapes related to the fundamental period, T, are somewhat different as shown in Fig.1.

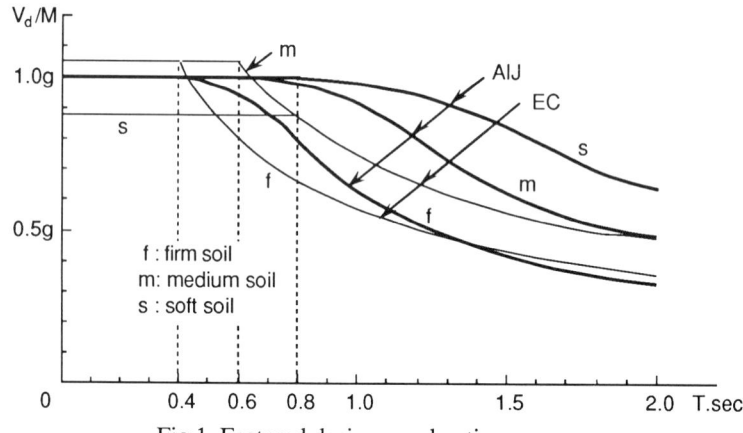

Fig.1 Factored design acceleration response

4 Structure type coefficient (behaviour factor)

The structure type coefficient, D_s, in Japan and the behaviour factor, q, in EC are the response reduction factors taking the energy dissipation by the plastic work of structures into consideration.

In the following, the structure type coefficient, D_s, will be related to the structure's ductility using a simple rigid-plastic model.

4.1 Balance of energy

The earthquake input energy into a structure, E_i, must balance to the kinetic elastic energy, E_e, and dissipating plastic energy, E_p, of a structure as

$$E_i = E_e + E_p. \qquad (8)$$

E_i is given as

$$E_i = \frac{1}{2} M S_v^2 \qquad (9)$$

where, M is the total mass of structure, and S_v is the spectral pseudo-velocity response for a damped system (including the damping effect of sub-soil).

E_e is given as

$$E_e = \frac{1}{2} M S_{ev}^2. \qquad (10)$$

Since the design acceleration response is $R_j \cdot g$ as was discussed in the preceding section, S_v can be written as

$$S_v = \frac{R_j \cdot g}{\omega} = \frac{T}{2\pi} R_j \cdot g. \qquad (11)$$

Similarly, the elastic pseudo-velocity response, S_{ev}, is given as

$$S_{ev} = \frac{T}{2\pi} \alpha_y \cdot g \qquad (12)$$

where, $\alpha_y = V_o/M \cdot g$ is the yield base shear coefficient and V_o is the horizontal yield shear capacity at the base of a structure.

From eqs.(8) through (12), E_p is written as

$$E_p = \frac{MT^2 g^2}{8\pi^2} (R_j^2 - \alpha_y^2). \qquad (13)$$

4.2 Capacity of plastic work, W_p

Referring to the ultimate state of rigid-plastic deformation model as shown in Fig.2, the cumulative plastic work in the two directions done by the inertial horizontal forces, W_p, is

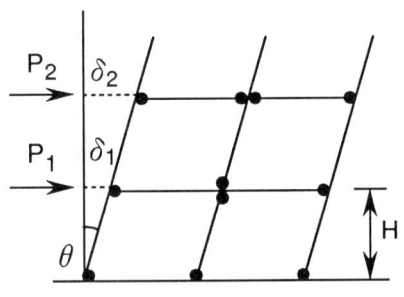

Fig.2 Plastic work

$$W_p = 2\sum_i P_i \delta_i \qquad (14)$$

where, P_i is the inertial force acting at i-th story, and δ_i is the cumulative deformation at i-th story.
If P_i and δ_i are normalized as $p_i = P_i/V_o$ and $d_i = \delta_i/\delta_{min}$ respectively, eq.(14) is reduced to

$$W_p = 2V_o \delta_{min} \sum_i p_i d_i \qquad (15)$$

in which,
$V_o = \sum_i P_i$: the yield shear capacity at the base of a structure

$\delta_{min} = H\theta$: minimum value of δ_i except for zero
H : height of the story beneath the floor which yields δ_{min}
θ : rotation of the story which yields δ_{min}.

Then eq.(15) is rewritten as

$$W_p = 2V_o H\theta \sum_i p_i d_i . \qquad (16)$$

If the moment-rotation relationship of a member subjected to double curvature bending under repeated and reversed loading is reduced to dynamically equivalent monotonic elastic-perfectly plastic relationship as shown in Fig.3, the ultimate cumulative rotation in one direction, θ_u, can be written as

$$\theta_u = \theta_p \eta \qquad (17)$$

where, θ_p is the elastic slope when the end moment reaches the full plastic moment, M_p. Equating θ_u from eq.(17) to θ from eq.(16), W_p can be related to the rotation capacity of the representative member, η, as

$$W_p = 2V_o H\theta_p \eta \sum p_i d_i . \qquad (18)$$

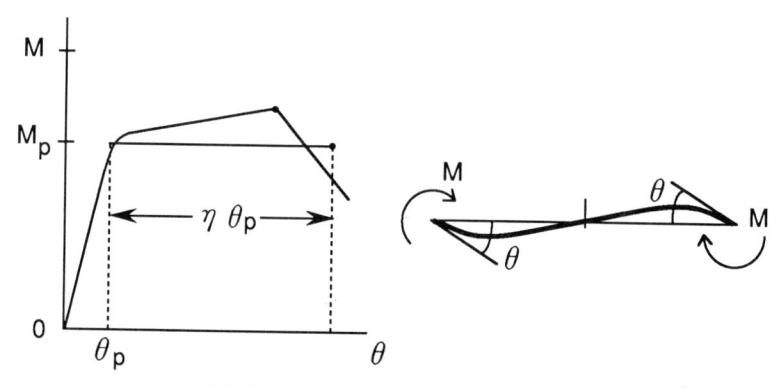

Fig.3 Cumulative rotation capacity

4.3 D_s-value related to η

The capacity of plastic work of a structure, W_p, must not be less than the dissipating plastic energy corresponding to the earthquake input energy, E_p, that is

$$W_p \geq E_p. \qquad (19)$$

Using eq.(13) and eq.(18), eq.(19) is written as

$$2V_o H\theta_p \eta \sum_i p_i d_i \geq \frac{MT^2g^2}{8\pi^2} (R_j^2 - \alpha_y^2). \qquad (20)$$

From eq.(6), the factored design shear force at the base is

$$\gamma V = 2.5 \, \gamma \, A_e \, Z \, R_j \, D_s \, M. \qquad (21)$$

Considering that, $Z = 1$, $\gamma = 2$, and $A_e = 0.2g$, eq.(21) is reduced to

$$\gamma V = D_s \, R_j \, Mg. \qquad (22)$$

In the design, the yield shear capacity, V_o, should at least be equal to the factored design shear force, γV, and eq.(22) is reduced to

$$V_o = \alpha_y \, Mg = D_s \, R_j \, Mg.$$

And it follows that

$$D_s = \frac{\alpha_y}{R_j}. \qquad (23)$$

Namely, another definition of D_s is the ratio of the yield base shear coefficient to the normalized design response.
Introducing the relation of eq.(23) into eq.(20), D_s is related to the member ductility, η, as

$$\frac{1}{D_s} - D_s \leq \frac{16\pi^2 H \theta_p \eta}{T^2 R_j g} \sum_i p_i d_i . \qquad (24)$$

The alternative expression of eq.(24) is given as

$$\frac{1}{D_s} - D_s \leq \left(\frac{\Delta_s}{\Delta_r}\right) \kappa \eta \qquad (25)$$

in which,

$\Delta_s = H \theta_p$: the smallest elastic limit story deflection in a building

$\Delta_r = \dfrac{T^2 R_j g}{16\pi^2}$: a kind of deflection response, and can be calculated using eq.(7)

$\kappa = \sum_i p_i d_i$: the regularity factor of a structure.

The Δ_r - T relations are shown in Fig.4 for each soil profile condition. It can be observed that Δ_r value increases with the increase of fundamental period, T. This means that when a building becomes taller, the large ductility, η, is needed for the assumed D_s coefficient.

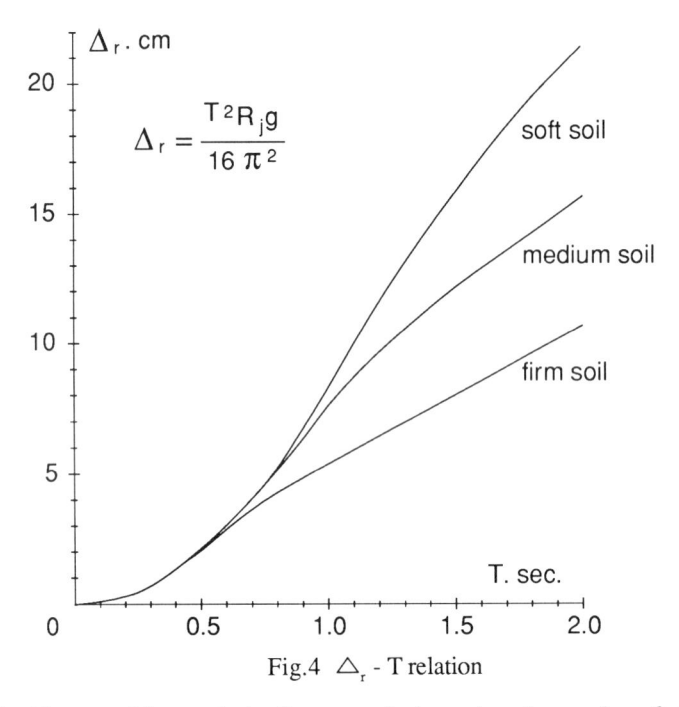

Fig.4 Δ_r - T relation

The significance of the regularity factor, κ, is that, when the number of plastified story increases, which means the increase of plastic energy dissipation, this factor becomes larger, and thus D_s-value becomes smaller. κ is calculated for 10 and 5 story buildings assuming that floor mass is the same throughout all stories and that the shear force coefficient distribution along the height of building is in accordance with AIJ rule, and is depicted in Fig.5. In this figure, the number of collapsed stories from the base is taken in horizontal axis.

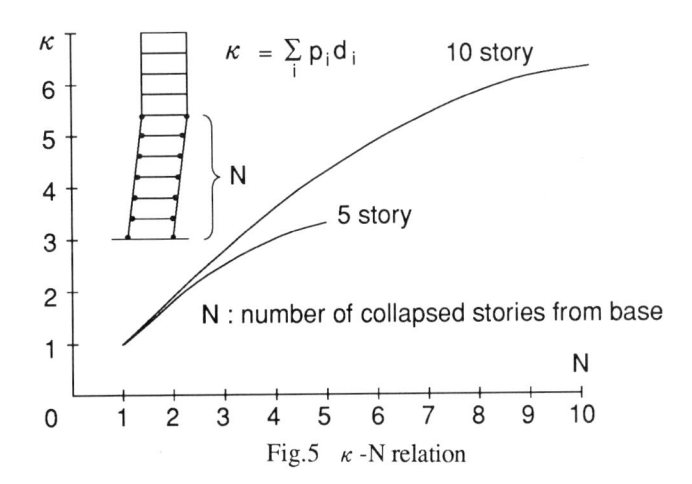

Fig.5 κ-N relation

In order to obtain the sound regularity of building, AIJ specification recommends to observe the following items;
1. the optimum design shear force distribution along the height of building
2. the optimum stiffness distribution along the height of building
3. minimizing the eccentricities in plan and elevation

In practical design, however, it is difficult to realize the ideal regularity, and therefore some inevitable imperfection must be taken into account.

According to eq.(25), D_s-η relations are evaluated for a 10-story sample building in which T = 1sec, H = 400cm, and θ_p = 1/150 (according to the serviceability limit rule of AIJ) are assumed. Two levels of regularity are considered, one is the case where 6 stories from the base are plastified (formed the collapse mechanism) and the other is the case where 4 stories from the base are plastified. The results are shown in Fig.6. The regularity factor, κ, is very influential to the D_s coefficient.

In AIJ standard, D_s-values are specified as shown in Table 3 depending on the rotation capacity of members, and are depicted by bold dashed lines in the figure.

In Table 3, q-values specified by EC are also shown. These q-values are not explicitly related to the rotation capacities of member in the EC rule.

Table 3. Classification of D_s and q values

AIJ			EC	
η	D_s	$1/D_s$	ductility class	q
4<	0.25	4.00	1	<6
2<	0.35	2.86	2	<4
0<	0.45	2.22	3	<2

Fig.6 D_s-η relation for a 10-story building

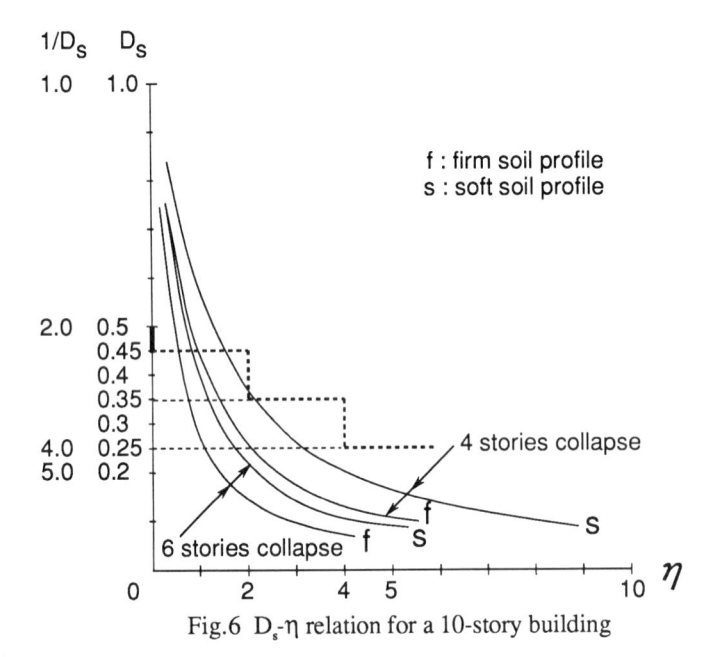

5 Rotation capacity

In order to secure the frame ductility, various provisions are made in the specifications.

5.1 ECCS, TC13
The classifications of q-factor are made from the structure type's view point and from member ductility's view point. Since the member ductility is mostly controlled by the occurrence of local buckling, ductility classes are specified by the width-to-thickness ratios of cross-section. For I-shaped section, the width-to-thickness ratio limitations for each ductility class are as following [2];

Table 4. Width-to-thickness ratio limitation (EC)

ductility class	b/t_f	h/t_w	
		beam	beam-column *
1	10 ε	66 ε	33 ε
2	11 ε	78 ε	39 ε
3	13 ε	82 ε	57 ε

$\varepsilon = \sqrt{235/f_y}$,   * $\rho = P/P_y = 0.3$ is assumed

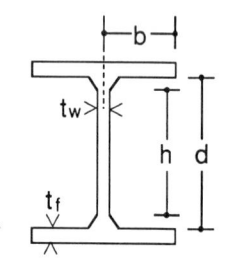

Fig.7. Dimension of I-Section

Symbols for dimensions of section are shown in Fig.7.

Seismic-resistant steel structures in Japan 39

As for the deformability of each class of section, the following definitions are given, but the quantitative information on the rotation capacity is not given.
Class 1. cross sections are those which can form a plastic hinge with the rotation capacity required for plastic analysis.
Class 2. cross sections are those which can develop their plastic moment resistance, but have limited rotation capacity.
Class 3. cross sections are those in which the calculated stress in the extreme compression fibre of the steel member can reach the yield strength, but local buckling is liable to prevent developments of the plastic moment resistance.

Though these classified cross sections correspond to q-factors as seen in Table 3, its theoretical or experimental backgrounds are not given.
q-factors determined by structure types also fall within the range of 2~6.

5.2 AIJ

Firstly, slenderness limitations on beams and beam-columns are specified to avoid the premature lateral-torsionel buckling. For members which meet these requirements, width-to-thickness ratio limitations of cross-sections are given as the function of rotation capacity, η.

The followings are the general construction of the width-to-thickness ratio limitation formulae. Their theoretical and experimental backgrounds are given in reference 4.

For I-shaped section, the interaction formulae for the width-to-thickness ratio limitation are expressed as;

Beam-column

$$\frac{(b/t_f)^2}{\left(\sqrt{\frac{D-C}{A}}\sqrt{\frac{E}{f_{yf}}}\right)^2} + \frac{(d_e/t_w)^2}{\left(\sqrt{\frac{D-C}{B}}\sqrt{\frac{E}{f_{yw}}}\right)^2} = 1 \qquad (26)$$

in which,

$$D = \frac{\mu - \sqrt{\mu^2 - e\xi}}{\xi}, \quad \mu = e - \varepsilon + (1-\rho)\eta_\theta, \quad \xi = e - 2\varepsilon + 2\rho(1-\rho)\eta_\theta,$$

$$e = E/E_{st}, \quad \varepsilon = \varepsilon_p/\varepsilon_y$$

f_{yf} : yield stress of flange, f_{yw} : yield stress of web, $\rho=P/P_y$: axial stress ratio, η_θ: rotation capacity of the member in terms of slope, E_{st} : strain-hardening modules of steel material, ε_p : strain of plateau part of $\sigma - \varepsilon$ curve, ε_y : strain at yield stress of steel material

$$\left.\begin{array}{ll} d_e = [2\rho A_f \gamma + (1+\rho)A_w]/2t_w & \text{for } \rho < \dfrac{A_w}{2A_f\gamma + A_w} \\[2mm] d_e = d & \text{for } \dfrac{A_w}{2A_f\gamma + A_w} \le \rho \le 1 \end{array}\right\}$$

where, $A_f = 2bt_f$ = sectional area of one flange, $A_w = dt_w$ = sectional area of web, and $\gamma = f_{yf}/f_{yw}$.

Symbols for dimensions of section are shown in Fig.7.

Beam

$$\frac{(b/t_f)^2}{\left(\sqrt{\frac{D'-C}{A}}\sqrt{\frac{E}{f_{yf}}}\right)^2} + \frac{(d/t_w)^2}{\left(2\sqrt{\frac{D'-C}{B}}\sqrt{\frac{E}{f_{yw}}}\right)^2} = 1 \qquad (27)$$

in which,

$$D' = \frac{\zeta - \sqrt{\zeta^2 - e(e-2\varepsilon)}}{e - 2\varepsilon}, \quad \zeta = e - \varepsilon + \frac{1}{2}\eta_\theta$$

Specific values of A, B, C are given in Table 5 depending on the grades of steel.

Table 5. A, B, C values

steel grade	f_y (MP$_a$)	A	B	C
SS400	235	0.4896	0.0460	0.7606
SM490	325	0.2868	0.0588	0.7730
SM570	441	0.1999	0.0748	0.7672

For a particular case of SM490 steel with $\rho=0.3$ and $A_f/A_w=1$, the interaction formulae for width-to-thickness ratio limitation are given as following depending on the rotation capacity demand, η_θ;

ductility calss	η_θ	beam column	beam
P-I-1	4	$\left(\frac{b/t_f}{14.4}\right)^2 + \left(\frac{d/t_w}{33.5}\right)^2 \leq 1$	$\left(\frac{b/t_f}{14.6}\right)^2 + \left(\frac{d/t_w}{65.0}\right)^2 \leq 1$
P-I-2	2	$\left(\frac{b/t_f}{18.4}\right)^2 + \left(\frac{d/t_w}{42.6}\right)^2 \leq 1$	$\left(\frac{b/t_f}{18.4}\right)^2 + \left(\frac{d/t_w}{81.3}\right)^2 \leq 1$
P-II	0	$\left(\frac{b/t_f}{22.4}\right)^2 + \left(\frac{d/t_w}{52.2}\right)^2 \leq 1$	$\left(\frac{b/t_f}{22.4}\right)^2 + \left(\frac{d/t_w}{99.2}\right)^2 \leq 1$
		and $d/t_w \leq 39.1$	and $d/t_w \leq 55.1$

Thus, it can be seen that Ds-coefficients are related to the cross-section classifications of members through the medium of rotation capacity, η_θ.

Although the definitions of classification for ductility are somewhat different between the two specifications, a comparison is made for I-shaped section made of grade SM490 steel in Fig.8 for beam-columns and in Fig.9 for beams.

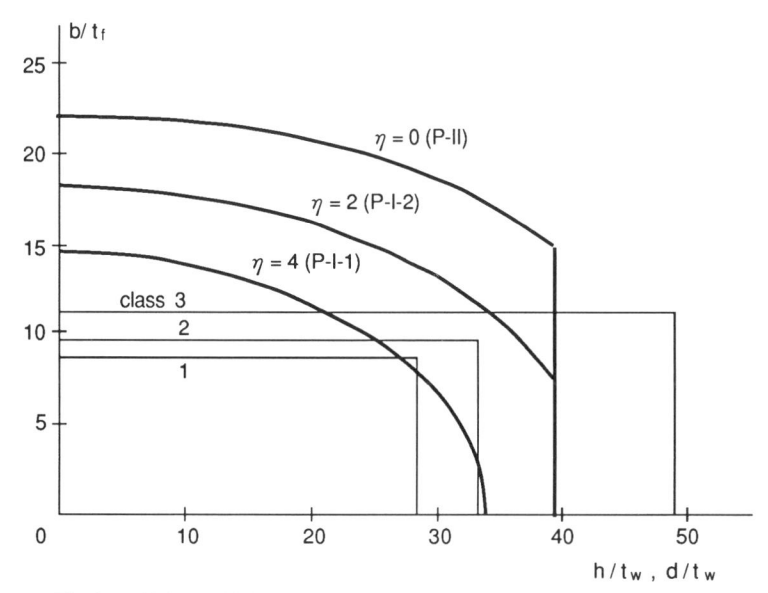

Fig.8 Width-to-thickness ratio limitations (beam-column) : SM490

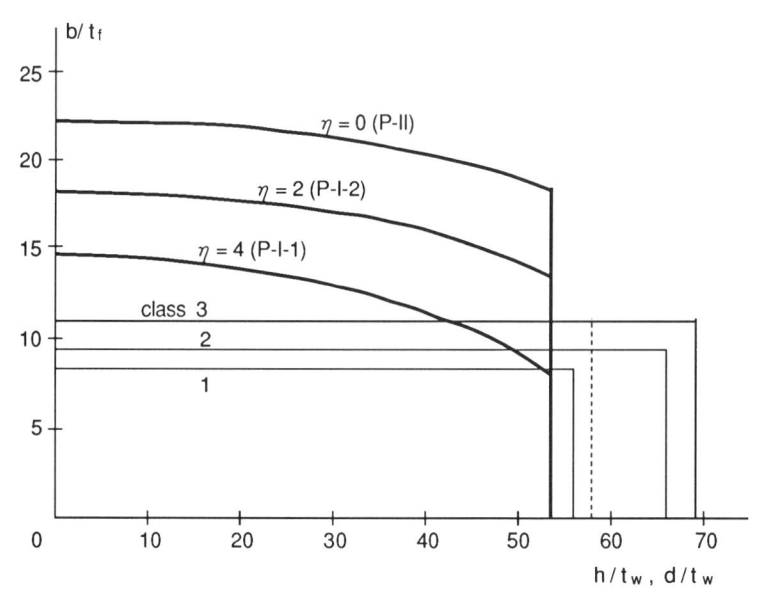

Fig.9 Width-to-thickness ratio limitations (beam) : SM490

The European width-to-thickness ratio limitation for web, h/t_w, given in Table 4 seems to be the results of investigation which are mainly taking the buckling due to normal stresses (bending and / or axial compression) into consideration, and there are no remarks on the buckling by shear stress. But considering that the effect of shear stress on the web buckling could not be ignored for the seismic loading case, the precaution against it must be necessary. The dashed line in Fig.9 was referred from the provision of Shear (section 5,4,6) of Eurocode 3 in which the following limitation is stated ;

$$h/t_w \leq 69\varepsilon, \quad \varepsilon = \sqrt{\frac{235}{f_y}}$$

6 Conclusions

Theoretical and experimental background of Japanese seismic code were investigated laying the emphasis on 1) design criteria, 2) design earthquake load, 3) structure type coefficient, D_s, or behaviour factor, q, and 4) rotation capacity of members. Comparison with EC code was made for each item.

As the result of study, D_s-value was clearly related to the rotation capacity of member, η, and η-value was evaluated as the function of width-to-thickness ratios of cross-section elements. And thus the D_s coefficients were related to the cross section classifications through the medium of rotation capacity of members.

It was also pointed out that the regularity of structure is quite influential to the D_s-value.

7 References

1. Standard for Limit State Design of Steel Structures (draft), Architectural Institute of Japan (AIJ), 1990 (in Japanese).
2. ECCS-TC13, European Recommendations for Steel Structures in Seismic Zones, ECCS, 1988.
3. Eurocode 8, Structures in Seismic Regions, Commission of the European Committees, 1988.
4. Ben Kato and M. Nakao, Strength and Deformation Capacity of H-shaped Steel Members Governed by Local Buckling, Journal of Structural and Construction Engineering, Trans. AIJ, No.458, 1994 (in Japanese).

PART ONE
BASIC PROBLEMS

3 BASIC PROBLEMS: GENERAL REPORT

L.-W. LU
Lehigh University, Bethlehem, Pennsylvania, USA

Abstract
This paper presents a review of the advances made on such issues as basic understanding of seismic behavior of steel structures, establishing analytical models for behavior prediction and damage evaluation, and developing improved techniques for dynamic response analysis. Comments are given on all the written contributions of the working session related to these issues.
Keywords: Seismic behavior, Structural System, Ductility, Dynamic Analysis, Damage Criteria, Quality Control, Seismic-Resistant Design.

1 Introduction

There are, basically, three major concerns in the design of a seismic-resistant structure: strength, stiffness, ductility and detailing. Much research has been conducted on these problems and adequate criteria are now available for designing structures to achieve an expected level of performance. In design as well as in developing code requirements, engineers should always recognize the interrelation between the ground motion input and level of structural performance or damage. This interrelation is shown in Figure 1. A structure may experience no damage, minor damage, major damage or collapse during an earthquake, depending on the dynamic effect of the earthquake and the strength, stiffness and ductility of the structure. In the current practice, the required strength is determined by applying a response reduction factor, also known as ductility factor or behavior factor, to the estimated elastic response (base shear). This factor is related to the overall performance of a structured system and the implied ductility that is available when it is loaded into the inelastic range. To achieve the implied ductility, careful attention must be paid to problems which are known to impair the inelastic deformability of a structure. Most of these problems have been studied in detail in the recent years.

Damage models and failure criteria for materials and structural components and connections subjected to cyclic loading need to be established in order to rationally assess the available ductility and energy absorption capacity. Much progress has been made in this area and quantitative evaluations of local damage or ultimate failure for simple structural elements are now possible.

Behaviour of Steel Structures in Seismic Areas. Edited by F. M. Mazzolani and V. Gioncu.
Published in 1995 by E & FN Spon, 2–6 Boundary Row, London SE1 8HN. ISBN: 0 419 19890 3.

Dynamic Effect of Earthquake	Structural Response and Performance
Characteristics of Earthquake	No Damage
Soil Condition	Minor Damage
Soil-Structure Interaction	Major Damage
	Collapse

Figure 1 Earthquake Effect and Structural Performance

Significant advances have also been made in developing methods for performing non-linear seismic response analysis of structures. Gross idealizations with regard to mass distribution and damping are no longer necessary. Analytical models capable of representing real behavior of various types of structural components and connections have been developed and incorporated in powerful computer programs.

The recent years also saw major advances being made in use of dampers, energy absorbing devices, base isolators and passive and active systems to reduce or control seismic response of structures. Some of the concepts have already been implemented in actual structures and show considerable promise.

This report first presents brief summaries of selected issues and concepts important in the development of improved seismic-resistant steel structures. Some of these issues are dealt with in the contributions of this working session (as well as those of Sessions 2, 3, 6 and 7). Comments on the individual contributions of this session are then made in light of the results presented and/or conclusions drawn.

2 Structural systems and their seismic performance

The structural systems which are generally used in construction of seismic-resistant buildings are:

  Moment-resisting frame
  Concentrically braced frame
  Eccentrically braced frame
  Frame with infilled shear panels
  Composite frame
  Dual or combination system

All these systems, except the concentrically braced frame (CBF), have shown, by laboratory testing (supplemented by theoretical analyses), good overall ductility if all the design details are properly taken into account. The relatively poor performance of the CBF is due primarily to buckling, and subsequent fracture, of the braces. Recently, the concept of buckling inhibited brace (BIB) or ductile brace has been developed (Chen and Lu, 1990) to improve the seismic performance of CBF.

In all the framing systems semi-rigid connections may be used in place of the traditional fully-rigid moment connections (Astaneh-Asl, et al. 1991).

Use of infilled shear panels can significantly increase the lateral stiffness of a moment-resisting frame and reduce its drift (due to earthquake or wind). This also allows the use of more semi-rigid connections or even simple connections (Xue and Lu, 1994).

3 Basic issues affecting ductility

The global or overall ductility of a structure is related to the local ductility or deformability of the critical sections, members and connections. The following are some of the important issues or factors that affect ductility: material properties, cross sectional behavior, member behavior (including instability effects), connection behavior, and quality control (fabrication and erection). Extensive research conducted during the last 40 years has examined these issues and rational design provisions have been developed to insure ductile behavior of structural members and connections. Early local buckling of cross sectional elements may be prevented by using sections with small flange and web slenderness ratios. Member instability effects may be controlled by limiting the amount of axial compression applied and the length of the member (or its slenderness ratio).

Although the behavior of beam-to-column connections has been the subject of intensive research, fail-safe design methodologies, which can be applied to all types of connections, are still not available. Numerous full-moment connections with welded flanges and bolted web developed cracks during the January, 1994 Northridge earthquake in southern California. The cracked connections are generally much larger than the connections tested in the laboratories, which did not show sign of premature fracture. Most of the cracks appeared to have initiated at the welds connecting the lower beam flange to the column. These welds are generally more difficult to make than the welds connecting the upper flange. These observations lead to two important concerns: (1) reliability of extrapolating test results from medium-size laboratory specimens to large-size connections existing in the field, and (2) quality control of field welding. An intense research effort is currently underway to determine the causes of fracture and to develop effective repair schemes for the cracked connections. Also being examined are ways to enhance the ductility of the connections without significantly increasing their cost, especially that associated with field welding.

4 Damage models and failure criteria

To assess the strength, deformability and energy absorption capacity of materials, members and connections under cyclic loading, empirical or semi-empirical damage models and failure criteria are often used. Three types of damage models have been proposed. The first type is based on the concept of low cycle fatigue damage under constant or variable stress (or strain) amplitudes. The second considers damage as a function of cumulative deformation (displacement or rotation) and assumes that failure occurs when the accumulated cyclic deformation has researched a critical value. In the third type damage and failure are evaluated based on an usable life criterion which represents an accumulation of both deformation and hysteretic energy (Kim and Lu, 1992). Since they were established or formulated empirically using data from experiments on specific types of materials, members or connections, considerable additional research will be necessary before their applicability can be extended or generalized.

5 Comments on contributions of working session 1

A total of 10 contributions were assigned to this session, three from Italy, two each from Romania and USA, and one each from Brazil, Japan and Moldova. The subjects dealt with include: seismic behavior and analysis of concentrically braced frames, eccentrically braced frames, moment-resisting frames with infilled plate panels, composite moment-resisting frames, member and system ductility, seismic damage and retrofit of bridges, quality control, damage models and failure criteria, methods of dynamic analysis, and seismic response reduction and control. Because some of the contributions deal with more than one subject, in the following the contributions are commented on, not completely on an individual basis, but under some common subject headings.

5.1 Seismic behavior of structural members

Experimental investigations of the strength and compressive ductility of selected chord members of the main stiffening truss and the legs of the towers of approach structures of the Golden Gate Bridge are reported by ASTANEH-ASL. The riveted test specimens failed by either local buckling or overall column buckling and exhibited no significant post-buckling ductility. Care must, therefore, be exercised in proportioning members which are subjected to primarily axial compression. He recommends that critical members in bridges in seismic regions be designed with slenderness ratio and width-to-thickness ratios limited to those currently used in building structures.

The contribution of LU, RICLES and KASAI presents results of tests on concrete encased composite columns subjected to combined axial compression and cyclic lateral loading. Among the several parameters included in the test program, the spacing of transverse reinforcement was found to have a strong influence on the strength and ductility of composite columns. The same observation has also been made from testing reinforced concrete columns under cyclic loading. Design provisions with regard to transverse reinforcement need to be developed for composite columns.

5.2 Analytical modeling of member behavior

Analytical models for three types of structural elements are described by LU, RICLES, and KASAI. The models are for predicting cyclic behavior of (1) shear links in eccentrically braced frames, (2) composite beam-columns, and (3) infilled shear panels. With the capability of analytically modeling the full-range behavior of shear links, it is possible to achieve more efficient EBF designs with improved performance. For composite beam-columns a fiber model, which takes into the account strength and ductility enhancements due to confinement, is available and good agreement between analytical predictions and test results has been reported. The analytical model for infilled plate panels considers the effects of large post-buckling deformation and yielding and the constraint offered by the boundary frame. The newly developed analytical models have already been implemented in some frame analysis programs and can be used in design of such structural systems as EBF's, composite frames, and moment-resisting frames with infilled plate panels.

5.3 Seismic behavior of structural systems

ACATRINEI, MIHAI and ROTARU describes shaking table tests of a three-dimensional, single-story factory building model with several variations of framing systems. In the transverse direction, the test model had two open bays and the structural system consisted of parallel moment-resisting frames. The model was tested in two ways, both with earthquake excitation applied in the transverse direction. In the first series of tests, all the interior girder-to-column connections were hinged, and in the second series of tests, the connection were fully continuous. The maximum ground acceleration attained before failure in the first series of tests was 8.62 m/s^2 and that in the second series of tests was 6.89 m/s^2. Apparently the more flexible system existed in the initial tests exhibited better performance. Tests were also conducted with earthquake excitation applied in the longitudinal direction of the model. The structural system in the longitudinal direction had concentric X braces in the two exterior frames and concentric V braces in the interior frame in one series of tests. In another series, the concentric X and V braces were replaced by eccentric V braces; thus it was possible to obtain a direct comparison of the performance of the two bracing systems. The frames having the eccentric braces behaved very ductilely and resisted a maximum peak acceleration of 7.08 m/s^2 which is more than twice of that resisted by the frames having the concentric braces. The shear links between the eccentric braces in the longitudinal girders did yield extensively.

The paper by SHIBATA deals with boundary columns in concentrically braced frames subjected to seismic loading. The boundary columns on the compression side of the frame may be subjected to larger than expected compressive forces due to the excess resistance of the tension side columns. This may cause local failure in the frame. This problem, which has not been discussed extensively in the literature, becomes more critical in the lower-story columns. The author proposes a method to control the maximum compression present in the boundary columns by reducing the strength of the braces in the lower stories. The concept is illustrated by two examples, an eight-story and a sixteen-story frame.

5.4 Quality control of material properties

The paper by CALDERONI, MAZZOLANI and PILUSO examines the requirements for quality control and the acceptance of structural steels with attention on the statistical tests regarding the conformity between material properties determined by the manufacturer's internal tests (or mill tests) and those by the official tests. The Italian code has a very restrictive conformity requirement between the results of the two types of tests. The effects of variation of material properties on such structural parameters as ultimate strength, displacement at 90% of the calculated ultimate strength, and global ductility of multi-story, multi-bay frames were evaluated in detail. Also studied was the statistical variation of the seismic response reduction factor q. A hybrid Monte Carlo simulation technique was used in the investigation. One of the major findings is that strict conformity to the code requirements does not necessarily lead to an improvement of structural reliability (in terms of the parameters examined). The authors recommend that the conformity requirements in the Italian code be removed, because they are not significant.

5.5 Damage models and failure criteria

When a severe earthquake ground shaking occurs, the critical elements of a structure would experience inelastic strain or deformation cycles. Since information on structural properties, such as strength and deformability, are often obtained from monotonic tests, it is useful to develop a procedure that can predict the maximum achievable inelastic deformation (displacement) under cyclic loading based on the maximum deformation determined from a monotonic test. Such a procedure is presented in the contribution by COSENZA and MANFREDI. It makes use of the concept of an "equivalent ductility" factor which can be developed by considering low-cycle fatigue damage. The maximum achievable displacement under cyclic loading is then given as the product of the maximum monotonic displacement and the equivalent ductility factor. In developing the coefficients contained in the expression of the ductility factor, the following input parameters were considered: fundamental period of structure and its yield level, effective earthquake duration and a structural damage (low cycle fatigue) parameter. Numerical values of the coefficients are determined from statistical analyses of the results of numerous seismic simulation studies.

BALLIO and CASTIGLIONI proposed an unified approach to assess damage of structural members due to hihg- and low-cycle fatigue. The latter is usully characterized by failure after application of a small number of plastic strain cycles of relatively large amplitude and is an important problem in designing structures subjected to earthquake loading. In this approach the usual S-N curve concept (S = stress range and N = number of cycles to failure) for high-cycle fatigue is extended to predict low-cycle fatigue life by using a hypothetical equivalent stress range defined by $\Delta\delta^* = E\Delta\epsilon$, where E is the Young's modulus and $\Delta\epsilon$ the strain range. The same S-N curves with a slope of -3 (on a log-log plot) can be used. The approach was verified by extensive tests on cantilever beams and beam-columns of different cross sections (HE220A, HE220B, IPE300 and HE120A). The sections had different flange and web slenderness ratios and would buckle locally at different strain levels. The study also shows that local buckling may be regarded as a notch effect and its influence can be considered by selecting or specifying an appropriate S-N curve. In the case of variable amplitude cycles the Miner's rule can be adopted in conjunction with a cycle counting method (such as the rainflow method). The findings reported are very significant in low-cycle fatigue research and may lead to a better way of evaluating ductility and energy absorption capacity of seismic structures.

5.6 Dynamic response analysis

CLARET and VENANCIO-FILHO describe an improved procedure for frequency domain analyis of single-degree-of-freedom systems. The procedure makes use of a new matrix, called IFT matrix, whose derivation and physical interpretation are given in the paper. The procedure has several advantages over the classical approach using the FFT algorithm.

DIMOIU and IVAN examine the influence of distributed intertia mass and stress state on seismic response of structures. The study was based on the classical time-domain analysis procedure. The major findings are: (1) the response of a structure is generally stiffer from a distributed mass analysis than a lumped mass analysis, and (2) when the stress state is included in an dynamic analysis, a more flexible response is observed.

Seismic analysis of liquid storage tanks is the subject dealt with in the contribution by PUKHOVSKI, MARYAMIS and RUSNAC. The authors made some important behavioral observations from dynamic tank tests. These observations were then incorporated in the development of a refined discrete model for analysis. It would be of interest to compare seismic response predictions provided by the refined model with those using the currently available models.

5.7 Seismic response reduction

In his paper on bridge research, ASTANEH-ASL advocates that for seismic retrofit of existing bridges serious considerations should be given to use of semi-rigid connections (in lieu of the traditional rigid connectors) in order to reduce the local as well as overall earthquake induced forces. This concept can also be applied to other types of structures (new and existing), such as buildings and towers if lateral drift requirement can be met. Another effective approach to reduce seismic response is to install energy absorbing dampers. The paper by LU, RICLES and KASAI describes the use of viscoelastic dampers to achieve damage free structures even under severe earthquake shaking. New design methodologies for damped structures are currently under development at Lehigh University.

6 Summary

A brief state-of-art summary of the current status of research on issues related to seismic behavior, response prediction, design methodology, and means of reducing or controlling seismic forces has been presented. The issues and subjects selected reflect the contents of the contributions of the working session.

Although steel structures generally performed well during the past earthquakes, severe damages were observed in the steel frames of the Morisada building in Sendai, Japan after the 1978 Miyagi-Oki earthquake (Wang and Lu, 1984). Total collapse of the 21 story Pino Suarez tower occurred during the 1985 Mexico City earthquake (Ger, Cheng and Lu, 1993). This building used a combination of concentrically braced frames and moment-resisting frames. The causes of these damages are now well understood. With continuing research on the various issues discussed, it is certain that improved seismic analysis and design methodologies for steel structures will evolve. These methodologies will help insure that severe damage or collapse will not occur in the future.

7 References

Astaneh-Asl, A., Nader, M. and Harriott, J.D. (1991) Behavior and design considerations in semi-rigid frames. **Proceedings, AISC National Steel Construction Conference,** June 1991, Washington, D.C.

Chen, C.C. and Lu, L.W. (1990) Development and experimental investigation of a ductile CBF system. **Proceedings, 4th U.S. National Conference on Earthquake Engineering,** May, 1990, Palm Springs, CA. Vol. 2, 575-584.

Ger, J. F., Cheng, F. Y. and Lu, L.W., (1993) Collapse behavior of Pino Suarez building during 1985 Mexico City earthquake. **Journal of Structural Engineering, ASCE,** Vol. 119, No. 3, pp. 852-870.

Kim, W. and Lu, L.W. (1992) An usable life criterion for seismic failure analysis of structures. **Proceedings, NSF EHM/BCS/ENG Research Grantee Workshop,** Phoenix, Arizona, pp. 53-54.

Wang, S. J. and Lu, L. W. (1984) Seismic damage analysis of the Morisada building. **Proceedings, Annual Technical Session, Structural Stability Research Council,** San Francisco, CA, pp. 281-290.

Xue, M. and Lu, L.W. (1994) Interaction of steel plate shear panels with surrounding framing members. **Proceedings, Annual Technical Session, Structural Stability Research Council,** June, 1994, Bethlehem, PA, pp. 339-354.

4 RECENT SEISMIC RESEARCH ON STEEL BRIDGES

A. ASTANEH-ASL
Department of Civil Engineering, University of California at Berkeley, USA

Abstract
After the October 17, 1989 Loma Prieta earthquake, a comprehensive study of seismic behavior of the eastern steel truss part of the San Francisco-Oakland Bay Bridge was conducted. The project resulted in establishing seismic vulnerabilities of the bridge as well as development of new seismic retrofit concepts and strategies including the use of semi-rigidity and rocking of the main towers to control and reduce the seismic response. During the January 17, 1994 Northridge earthquake, a number of short span steel bridges sustained damage. The damage consisted of failure of the connection of superstructure to substructure and buckling of diaphragm members. Some of the results of the ongoing research on these bridges and lessons learned are presented.
Keywords: Steel Bridges, Seismic, Loma Prieta Earthquake, Northridge Earthquake, San Francisco Bay Bridge, Golden Gate Bridge.

1 Seismic studies of long span steel bridges

During the October 1989 earthquake, seven long span, major steel bridges in Northern California were shaken and some sustained damage. The most serious damage to long span steel bridges was the collapse of a 17-m-long deck portion of the San Francisco-Oakland Bay Bridge. The collapse resulted in closure of the bridge, which has daily traffic of about 250,000 cars. Since the earthquake, the California Department of Transportation and other agencies have funded several projects to study the seismic behavior of these existing major bridges in order to understand seismic behavior of the bridges and to develop rational and economical retrofit concepts. Summaries follow for two of these projects, one on the San Francisco-Oakland Bay Bridge and one on the Golden Gate Bridge.

2 The San Francisco-Oakland Bay Bridge

The San Francisco-Oakland Bay Bridge, a 13.4-km-long structure connecting the cities of San Francisco and Oakland, was constructed from 1933-1937. The study summarized here focused on the East Bay

Crossing portion of the bridge (Figure 1). There are eight expansion joints in the East Bay Crossing that divide the steel truss superstructure into ten segments. The expansion joints are identified as "E.J." in Figure 1. In the East Bay Crossing, Piers YB1, YB2, YB3, YB4, E1, and E2 are supported on the bedrock and all other piers are supported on the reinforced concrete hollow foundations and several hundred Douglas fir timber piles. A typical cross section of the bridge is provided in Figure 2.

During the Loma Prieta earthquake, two parts of the bridge to the east and west of Pier E9 moved in opposite directions and two, 17-m segments of the upper and lower decks dropped off. The damage resulted in one fatality and 13 injuries and closure of the bridge for a month. Following the Loma Prieta earthquake, a research project was granted to the University of California at Berkeley by the California Department of Transportation to document and study the damage (Astaneh-Asl, Mac Cracken and Somantari 1994). This project was followed by a project to conduct a comprehensive study of the seismic behavior of the bridge and to develop seismic retrofit concepts and recommendations. The research resulted in establishing seismic vulnerabilities and developing efficient seismic retrofit strategies (Astaneh-Asl et al. 1993).

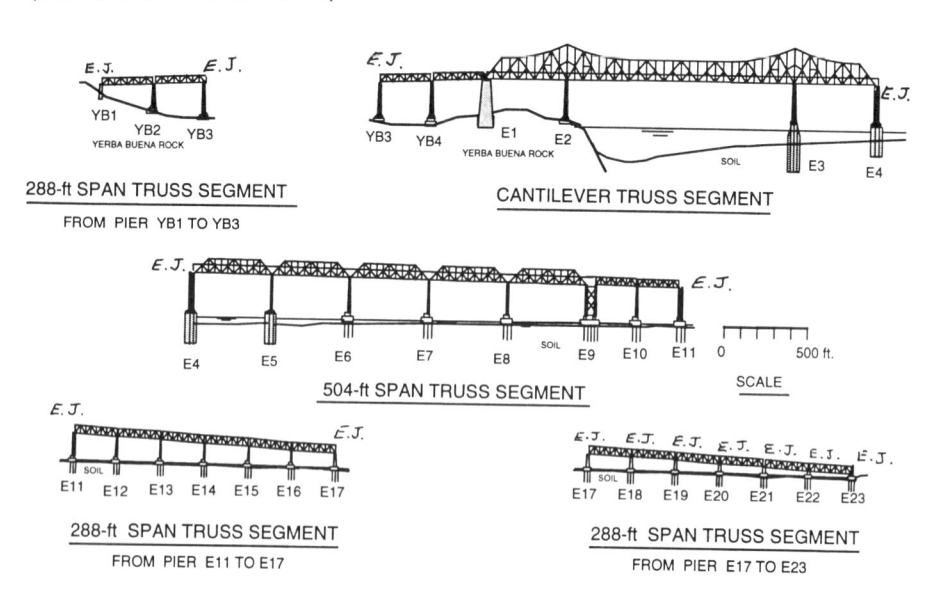

Fig. 1. A view of the East Bay Crossing of the San Francisco-Oakland Bay Bridge.

2.1 Research project to study the Bay Bridge

The research methodology can be summarized as follows. First, by studying the character of the nearby San Andreas and Hayward faults, series of ground motions are established at the likely hypocenters. Then, by using information available on the seismological and geological characteristics of the ground between the fault and the Bay Bridge, bedrock acceleration time histories at a number of points beneath the bridge are established. By using the bedrock motions and conducting dynamic analyses of the response of the soil to the bedrock motions, a series of free field ground motions are established at the site. Then by conducting soil-structure interaction studies, the free field motion is modified to include the effects of the presence of the piles, foundations and the structure. In the last stage, a realistic computer model of the superstructure is built, the model is subjected to base excitations developed in previous stages, and a series of time history dynamic analyses is conducted. The process is iterated as many times as necessary until a refined and realistic model of the behavior of the bridge emerges.

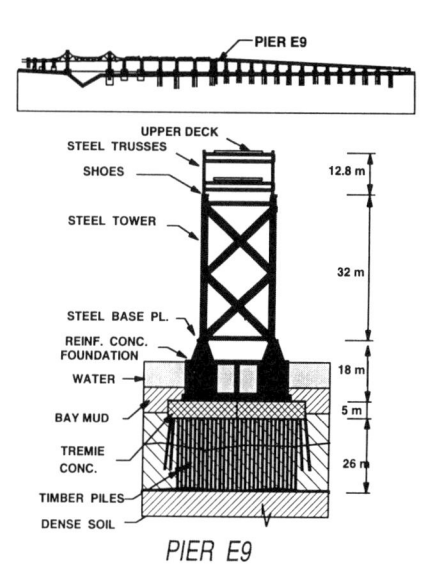

Fig. 2. Typical cross section of the East Bay Crossing.

In order to evaluate the seismic condition of the Bay Bridge, the following equation was developed and applied to each limit state of components for the Bay bridge.

$$\text{Demand/Supply} \leq (R)(I) \qquad (1)$$

The demand term in the equation resulted from dynamic analyses. The supply term was established for each limit state by studying the seismic behavior of the corresponding component and/or by conducting experimental research (Astaneh-Asl and Shen 1993a; Cho and Astaneh-Asl 1993). The performance and limit states of each component of the bridge are given in Astaneh-Asl and Shen (1993b).

The term R in the above equation is the Response Modification

Factor and is introduced to represent the modification necessary to convert elastic response to a more realistic inelastic response. If inelastic response analyses are conducted, the value of R is equal to 1.0. However, if elastic response analyses are used to establish the demand, such demand values should be modified by using appropriate R values for each component of the bridge. Note that unlike steel structures for buildings, where R (Reduction Factor) is applied to global base shear and not to each component, in the case of bridges, the reduction factor R is specific to each component of the bridge and is applied to component forces at the local level.

The term I in the above equation is the Importance Factor; it represents the importance of each limit state and the consequences of reaching that limit state. In current design codes for the seismic design of new structures, all limit states related to member performance are assigned the same importance. Even though a margin of safety is built into all codes, this margin of safety is due to uncertainties in the load, material properties, geometry, workmanship, modeling and computational techniques. However, in evaluating seismic performance of an existing steel structure, the author felt an importance factor should be assigned to each limit state based on the past performance of the component and the consequences of the occurrence of the limit state of failure.

2.2 Seismic vulnerabilities and retrofit strategies for the Bay Bridge

The studies of the East Bay Crossing of the Bay Bridge indicated that the existing structure of the bridge has the following seismic vulnerabilities:

(a) The R/C tower at Pier E1 (Figure 1) can rock during a major quake due to longitudinal forces.
(b) The eyebar bottom chord member of trusses from Pier E4 to Pier E9 can buckle under compression.
(c) The timber pile foundations of all piers from E6 to E23 need strengthening.
(d) The steel towers at Pier E9 and the R/C pedestal at Pier E17 need strengthening.
(e) The transverse movement of the upper deck relative to the lower deck for the 300-ft (87-m) span trusses is on the order of one meter and is unacceptable.
(f) Expansion joints need to be improved in order to tolerate large longitudinal displacements.

To mitigate the above vulnerabilities, several traditional and innovative seismic retrofit strategies were studied or developed. The most promising retrofit concept for each segment was selected and recommended. The studies clearly indicated that strengthening alone is not the answer. The selected retrofit strategy is based on strengthening some elements and improving and reducing the seismic response by the use of semi-rigid devices at strategic locations throughout the bridge. Figure 3 shows the most promising retrofit strategy for the East Bay Crossing of the Bay Bridge (Astaneh-Asl and Shen 1993a).

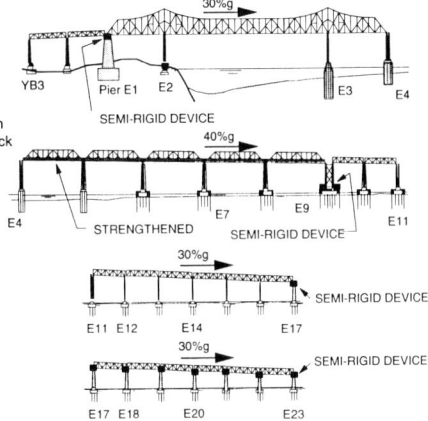

Fig. 3. Seismic retrofit strategies for the East Bay Crossing of the Bay Bridge.

3 Compressive ductility of Golden Gate Bridge members

After the 1989 Loma Primate earthquake, a project began at T.Y. Lin International of San Francisco to evaluate the seismic vulnerabilities and to develop seismic retrofit strategies for the Golden Gate Bridge. Part of the effort has been establishing the compressive ductility of two of the critical members of the bridge, the chord members of the main stiffening truss and the legs of the towers of the approach structures. A research project was conducted (Cho and Astaneh-Asl 1993) to establish the compressive behavior of these riveted members and their local buckling and post-buckling ductility.

The project consisted of constructing realistic and nearly full-size riveted specimens of a typical main truss chord member and the leg member of the approach towers shown in Figure 4. The materials used for the specimens were as similar as possible to the original materials used in the construction of the bridge in the 1930s. The riveted specimens were instrumented and subjected to axial compressive force. The results of the tests (Figure 5) are in the form of axial load-axial displacement. The truss chord developed a relatively sudden local buckling while the failure mode of the approach tower leg was overall buckling. The post-buckling ductility of the truss chord was not significant. The studies have continued and have been expanded to include the establishment of post-buckling ductility of laced and riveted members used in many existing steel bridges.

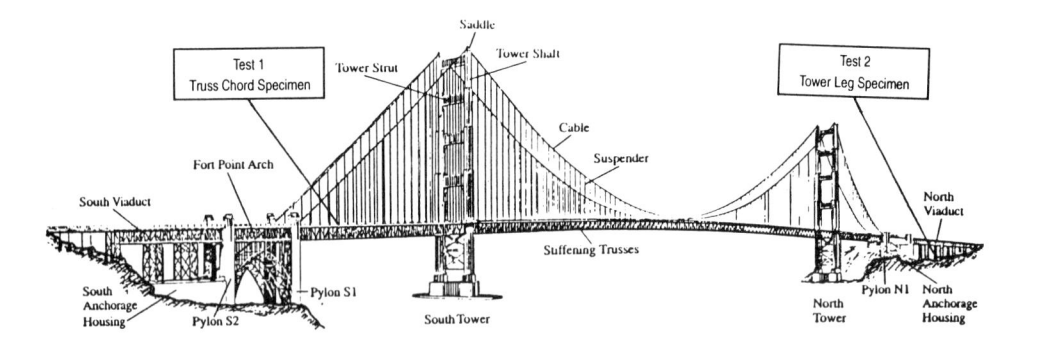

Fig. 4. Location of test specimens in the Golden Gate Bridge.

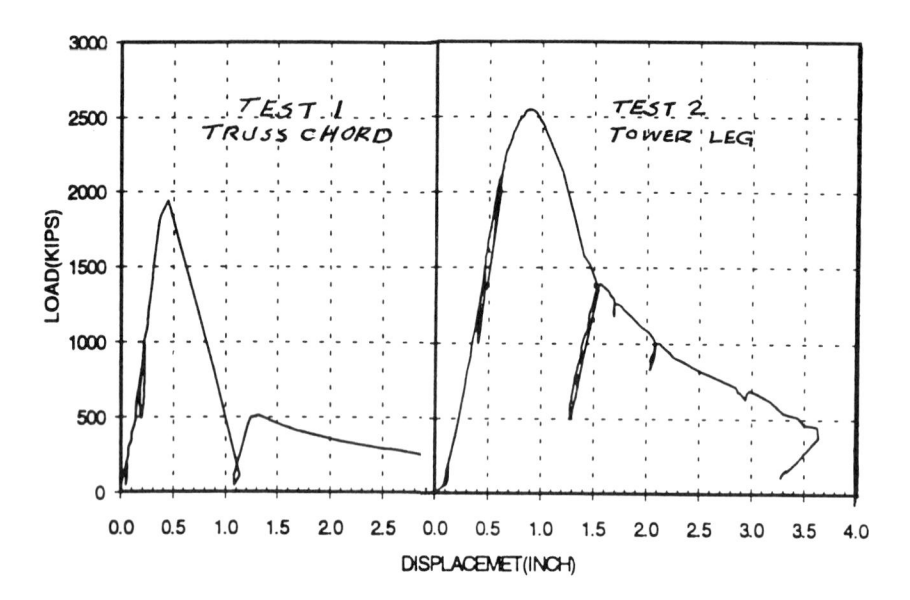

Fig. 5. Axial compressive behavior of critical members of the Golden Gate Bridge.

4 Seismic performance of steel bridges during the 1994 Northridge earthquake

On January 17, 1994, a magnitude Mw=6.7 earthquake shook the northwestern part of greater Los Angeles. The quake caused five reinforced concrete overpasses to collapse and several others to be damaged. Following the earthquake, a team of researchers from the Department of Civil Engineering of the University of California at Berkeley undertook an intense documentation, study and analysis of the seismic performance of steel bridges in the area (Astaneh-Asl, Bolt et al. 1994). The results are summarized in this section.

Four short span steel bridges or overpasses were damaged; all were located approximately 15 km northwest of the epicenter. This was the area where apparently the thrust fault rupture would have surfaced if the fault had reached the ground surface. A strong motion instrument in the area recorded maximum peak accelerations of about 0.60 g in all three directions.

The main type of damage to steel structures was failure of the reinforced concrete support of the steel superstructure (Figure 6).

Fig. 6. Damage to the support of steel girder during 1994 Northridge earthquake.

In some cases, particularly in skew bridges, a combination of longitudinal, vertical and transverse movements of the superstructure had caused shear or tension failure of the anchor bolts or crushing of the lightly reinforced concrete support of the girder, as shown in Figure 7. In addition, in some cases the diagonal bracing members of the end diaphragms had buckled or the gusset plates connecting the bracings had fractured. The dynamic analyses of the bridges indicated that the bulk of seismic forces are transferred by the end diaphragms to the pier. In one bridge, the reinforced concrete piles had failed and the bridge had settled about 200 mm. In general there was no serious damage to steel gravity load carrying systems. The worst damage was in skewed bridges.

Fig. 7. Damage to top of the pier in a skew bridge.

The main lessons learned from this earthquake regarding seismic behavior of steel short span bridges are:

(a) In skewed bridges, the superstructure moves in a direction perpendicular to the pier and abutment and not along the longitudinal axis of the bridge.
(b) The end diaphragms transfer almost the entire seismic load and should be designed accordingly.
(c) The connection of steel superstructure to concrete substructure needs serious attention.
(d) The use of "upset" anchor bolts where the shank of the bolt has less diameter than the threaded part can make the connection very ductile.

5 Summary

This paper summarizes some of the studies on the seismic behavior and design of steel bridges conducted at the Department of Civil Engineering of the University of California at Berkeley since the 1989 Loma Prieta earthquake. The research projects included analytical and experimental studies of seismic behavior and design of steel long span bridges as well as short span overcrossings. The main findings of these studies can be summarized as:

(a) The reinforced concrete substructure of most bridges designed prior to the 1970s is brittle and may not be able to withstand severe earthquakes. In new designs, the substructures should be designed to remain essentially elastic because, in the aftermath of a major seismic event, it is very difficult or impossible to inspect the seismic condition and damage status of the substructure. In addition, the repair of damaged substructures is very costly and may require closure of the bridge. For existing brittle substructures, adding reinforced concrete or composite jackets and additional composite piles can add to the strength and stiffness of the substructure and prevent it from sustaining damage during a major earthquake.

(b) The connection of substructure to superstructure in steel bridges is a potentially vulnerable area. The anchor bolts, the support concrete, bearing supports, and other elements of connection should be made ductile to tolerate seismic displacements without damage to brittle elements. The use of "upset" anchor bolts where the shank is smaller than the threaded area is strongly encouraged to prevent brittle failure of threaded areas of anchor bolts.

(c) In all steel bridges, special attention should be paid to seismic local buckling of steel elements. The slenderness ratio, b/t, of main non-redundant elements should be limited to those currently used in building structures.

(d) In short span bridges and overcrossings, one of the important considerations is the skewness of the bridge. Research (Astaneh-Asl, Bolt et al. 1994) has indicated that skewness tends to stiffen the bridge. In addition, the primary mode of vibration of a skew bridge is movement of the superstructure in a direction perpendicular to the piers and not in the longitudinal or transverse direction of the bridge. This mode of vibration should be considered in the design of bearing supports as well as for the entire bridge.

(e) During the recent earthquakes in California, damage has been observed in the end diaphragms of steel bridges. Also, studies by Astaneh-Asl, Bolt et al. (1994) have indicated that the bulk of inertia forces in the transverse direction of steel bridges is transferred to the piers by the end diaphragms. Therefore, the end diaphragms should be designed not only for wind bracing but also to transfer seismic forces.

(f) The use of base isolation in short span overpasses resulted in elongation of period, reduction of seismic force, and increase of seismic displacements, as expected. However, since relatively large displacements on the order of 25 cm were recoverable, the use of base isolation, if done properly, could potentially be useful in reducing seismic response.

(g) More research on the actual cyclic behavior of steel bridge components is needed. Unlike building components, that have been studied extensively in the past 20 years, the data on seismic behavior of steel bridges is very limited and needs to be expanded.

8 Acknowledgments

The information provided in this paper was based on several projects conducted at the Department of Civil Engineering of the University of California at Berkeley. The main sponsor was the California Department of Transportation. The opinions expressed in this paper are those of the author and do not necessarily reflect the views of the sponsor or individuals whose names appear in this paper.

Many individuals have contributed to these studies including those whose names appear in the References. Their contributions are acknowledged and sincerely appreciated.

9 References

Astaneh-Asl, A., Bolt, B., Fenves, G.L., Lysmer, and Powell, G. (1993) Summary of seismic studies of the East Bay Crossing of the Bay Bridge, Vol. 12 in Seismic condition assessment of the East Bay Crossing of the San Francisco-Oakland Bay Bridge. **Rep. No. UCB/CE-Steel-93/14,** Dept. of Civil Engrg., Univ. of California, Berkeley, CA.

Astaneh-Asl, A., Bolt, B., McMullin K., Donikian, R., Modjtahedi, D., and Cho, S. (1994) Seismic performance of steel bridges during the 1994 Northridge earthquake. **Rep. No. UCB/CE-Steel-94/01,** Dept. of Civil Engrg., Univ. of California, Berkeley, CA.

Astaneh-Asl, A., Mac Cracken, W., and Somantari, D. (1994) Damage to the San Francisco-Oakland Bay Bridge during the October 17, 1989 earthquake. **Rep. No. UCB/CE-Steel-94/04,** Dept. of Civil Engrg., Univ. of California, Berkeley, CA.

Astaneh-Asl, A., and Roberts, J., eds. (1993) **Proceedings, First U.S. Seminar on Seismic Evaluation And Retrofit Of Steel Bridges,** October 18, 1993, San Francisco, CA. Dept. of Civil Engrg., Univ. of California, Berkeley, CA.

Astaneh-Asl, A., and Shen, J.-H. (1993a) Seismic evaluation and retrofit concepts. Vol. 10 in Seismic condition assessment of the East Bay Crossing of the San Francisco-Oakland Bay Bridge. **Rep. No. UCB/CE-Steel-93/12,** Dept. of Civil Engrg., Univ. of California, Berkeley, CA.

Astaneh-Asl, A., and Shen, J.-H. (1993b) Seismic performance and design considerations in steel bridges, in **Proceedings, First U.S. Seminar on Seismic Evaluation And Retrofit Of Steel Bridges,** October 18, 1993, San Francisco, CA. Dept. of Civil Engrg., Univ. of California, Berkeley, CA.

Cho, S.W., and Astaneh-Asl, A. (1993) Experimental studies of compression behavior of Golden Gate Bridge members. **Rep. No. UCB/CE-Steel 93/01,** Dept. of Civil Engrg., Univ. of California, Berkeley, CA.

5 DAMAGE ASSESSMENT IN STEEL MEMBERS UNDER SEISMIC LOADING

G. BALLIO and C. A. CASTIGLIONI
Department of Structural Engineering, Milan Polytechnic, Milan, Italy

Abstract

A method is presented trying to unify both design and damage assessment methods for high and low cycle fatigue, based on the results of an extensive experimental research programme. Interpreting the stress range $\Delta\sigma$ as associated to the real strain range $\Delta\varepsilon$ in an ideal perfectly elastic material, high and low cycle fatigue test data can be fitted by the same Wöhler (S-N) lines usually given in Recommendation for (high cycle) fatigue design of steel structures. Local buckling can be regarded as a notch effect, intrinsic to the various shapes, and related to their geometrical properties (c/t_f and d/t_w slenderness ratios of the flanges and the web). In the case of variable amplitude loading histories, a linear damage accumulation rule together with the previously defined S-N curves can lead to a reliable collapse criterion also for members under seismic loading.

Keywords: Damage assessment, damage accumulation models, local buckling, S-N curves.

1. Introduction

Eurocode-3 (CEN 1993), defines fatigue as "damage in a structural part, through gradual crack propagation caused by repeated stress fluctuations".
Depending on a number of factors, these load excursions may be introduced either under stress or strain controlled conditions. Depending on the number of cycles sustainable to failure, and on their amplitude, we can distinguish failure for high or low cycle fatigue.
Failure by high cycle fatigue is characterised by a large number of withstandable cycles with a nominal stress range $\Delta\sigma$ in the elastic range (i.e. with $\Delta\sigma < 2f_y$, f_y being the yield stress of the material). This is a well known effect, and was studied since 19th century for mechanical engineering applications (Wöhler, 1860). Although only a limited number of typologies of connections and of structural details can be considered at present thoroughly investigated, the general aspects of this problem, and in particular the basic methodologies for assessment and design, can be considered well established.
Low cycle fatigue is characterised by a small number of cycles to failure, with large plastic deformations (i.e. with strain range $\Delta\varepsilon > 2\varepsilon_y = 2f_y/E$, E being the Young modulus of the material). In general, low cycle fatigue problems in civil engineering structures arise either under seismic loading or in pressure vessels or under severe thermal cycling. Cycles with large amplitudes in the plastic range are usually connected with local

Behaviour of Steel Structures in Seismic Areas. Edited by F. M. Mazzolani and V. Gioncu.
Published in 1995 by E & FN Spon, 2-6 Boundary Row, London SE1 8HN. ISBN: 0 419 19890 3.

buckling in structural members. At present, knowledge of low cycle fatigue behavior of civil engineering connections is not jet as well established as the high cycle fatigue one. In particular, there is no generally recognised design or damage assessment method for low cycle fatigue, and a clear definition of a collapse criterion is lacking.

In this paper a procedure is described, trying to unify design and damage assessment methods for structural details under high and low cycle fatigue. After discussing the proposed approach, its experimental validation based on constant and variable amplitude cyclic test results will be presented. By transforming the nominal strain range $\Delta\varepsilon$ in an equivalent stress range ($\Delta\sigma^* = \Delta\varepsilon\,E$) computed by considering the material as indefinitely linear elastic, the experimental test data obtained under cycles with a constant amplitude in the plastic range can be interpolated by the same Stress range-Number of cycles to failure (S-N) lines usually given in recommendations for the (fatigue) design of steel structures (e.g. Eurocode-3). Furthermore, a linear damage accumulation model (Miner's rule), together with the rainflow cycle counting method, can be adopted for the damage assessment under variable amplitude loading.

2. The proposed approach

The proposed approach to unify the design and damage assessment procedures for steel structures under low and/or high cycle fatigue, originally proposed by Ballio & Castiglioni (1994a) is based on the two following assumptions:

1. To know, for a given structural detail (cycled under strain controlled conditions), the relationships between the number of cycles to failure N_f and the cycle amplitude Δs, expressed in terms of generalised displacement components (i.e. of displacements Δv or of rotation $\Delta\phi$ or of deformation $\Delta\varepsilon$). These relationships have the same meaning in high and in low cycle fatigue with the following difference:
   - in high cycle fatigue the component is subjected to cycles in the elastic range, with a cycle amplitude $\Delta s < 2\,s_y$ where s_y is the value of the displacement component corresponding at first yield in the material;
   - in low cycle fatigue the component is subjected to cycles in the plastic range, i.e. with an amplitude $\Delta s > 2\,s_y$.

2. Damage accumulation in a structural detail is a linear function of the number of cycles sustained by the component itself. This means to assume that Miner's rule, usually applied in high cycle fatigue damage assessment, can be applied also in low cycle fatigue.

Immediate consequence of the second assumption is the definition of a unified failure criterion for both high and low cycle fatigue: a structural component fails when Miner's damage index reaches unity. Of course, the problem of the definition of the failure conditions is transposed to that of the identification of appropriate S-N curves corresponding to the desired safety level.

Consequence of the first assumption is the possibility to interpret low cycle fatigue with the same laws commonly accepted for high cycle fatigue.

In fact, in high cycle fatigue (under stress controlled conditions):
- a structural component is subjected to load cycles having a constant amplitude ΔF;
- the maximum value of the load excursion ΔF must be lower than the value F_y associated with the attainment of the yield stress in the material. F_y may be theoretically computed or experimentally evaluated;
- the nominal stress induced by the external load F may computed either theoretically or with conventional methods, leading to a relationship of the type $\sigma=\sigma(F)$;
- the stress range $\Delta\sigma=\Delta\sigma(\Delta F)$ is finally correlated to the number of cycles to failure N_f, independently from the yield strength of the material.

In order to generalise this approach, under the assumption of indefinitely linear elastic material, it can be written:

$$\Delta\sigma = \frac{\Delta F}{F_y}\sigma(F_y) \qquad (1)$$

In low cycle fatigue (under strain controlled conditions):
- a structural component is subjected to displacement cycles having a constant amplitude Δs;
- the maximum value of the excursion Δs of the generalised displacement component s is greater than the value s_y associated with the attainment of the yield stress in the material. s_y may be theoretically computed or experimentally evaluated;
- if the material can be regarded as an elastic perfectly plastic one (as in the case of steel), and the hypothesis of concentrated plastic hinge can be considered realistic (as shown by Ballio & Castiglioni (1994b) for steel members under seismic loading), it can conventionally be assumed that strains are proportional to the generalised displacement component s, and it can be stated that:

$$\frac{\Delta\varepsilon}{\varepsilon_y} = \frac{\Delta s}{s_y} \qquad (2)$$

This equation defines the nominal strain range in a particular way, as discussed in detail by Ballio & Castiglioni (1994a), taking into account the local reduction of stiffness at plastic hinge location by an equivalent uniform reduction of stiffness along the total beam length.
- For an ideally linear elastic material, the relationship between strains and the load causing the displacement s can be written as:
$$E\varepsilon = \sigma(F) \qquad (3)$$
it follows that, at yield:
$$E\varepsilon_y = \sigma(F_y) \qquad (4)$$
- The value $E\Delta\varepsilon$, which can be interpreted as the stress range associated in an ideally linear elastic material to the strain range $\Delta\varepsilon$, can be finally correlated to the number of cycles to failure N_f.

- Because of (2) and (4), it can be noticed that:

$$\Delta \sigma^* = E \Delta \varepsilon = E \frac{\Delta s}{s_y} \varepsilon_y = \frac{\Delta s}{s_y} \sigma(F_y) \qquad (5)$$

- Equation (5) is similar to equation (1), that is valid for high cycle fatigue. The difference consists in having considered cyclic displacements instead of cyclic forces.

3. Experimental validation

In order to experimentally validate the proposed approach, tests were performed by Ballio & Castiglioni (1994b) at the Structural Engineering Department of Politecnico di Milano, on full scale cantilever members 1.6 m long (fig.1), of the commercial shapes HE220A, HE220B and IPE300, using an equipment designed by Ballio and Zandonini (1985), capable of applying horizontal cyclic actions in a quasi-static way. Presently, the testing programme is continuing, in order to enlarge the data-base; tests are carried out also considering the presence of an axial load, on both HE220A and HE120A specimens, characterised by different slenderness ratios of both the flanges and the web with respect to HE220A (tab.1).

Shape	Area (cm²)	Flange			Web			
		b=2c (mm)	t_f	c/t_f	h	d (mm)	t_w	d/t_w
HE 220A	64.3	220	11.0	10.0	210	152.0	7.0	21.7
HE 220B	91.0	220	16.0	6.9	220	152.0	9.5	16.0
IPE 300	53.8	150	10.7	7.0	300	248.6	7.1	35.1
HE 120A	25.3	120	8.0	7.5	114	74.0	5.0	14.8

Table 1- Geometrical properties of specimen shapes

3.1 Constant amplitude tests

3.1.1 Description of the test results

To date, 37 tests were performed (11 on HE220A shapes, 12 on HE220B and 11 on IPE300, 3 on HE120A) imposing to the specimens displacement cycles with a constant amplitude. Furthermore, 8 tests were performed on HE220A specimens, subjected to an axial load; three values of the axial load P were considered: P/P_y = 0.05, 0.10 and 0.125, $P_y = f_y A$ being the plastic strength of the cross section.
In addition to the usual hysteresis loops in terms of force applied on the top vs. top displacements, for each specimen the experimental measurements obtained by means of a set of displacement transducers positioned as shown in fig. 1, were processed following the ECCS (1986) Recommended procedures in order to obtain informations regarding

Damage assessment in steel members 67

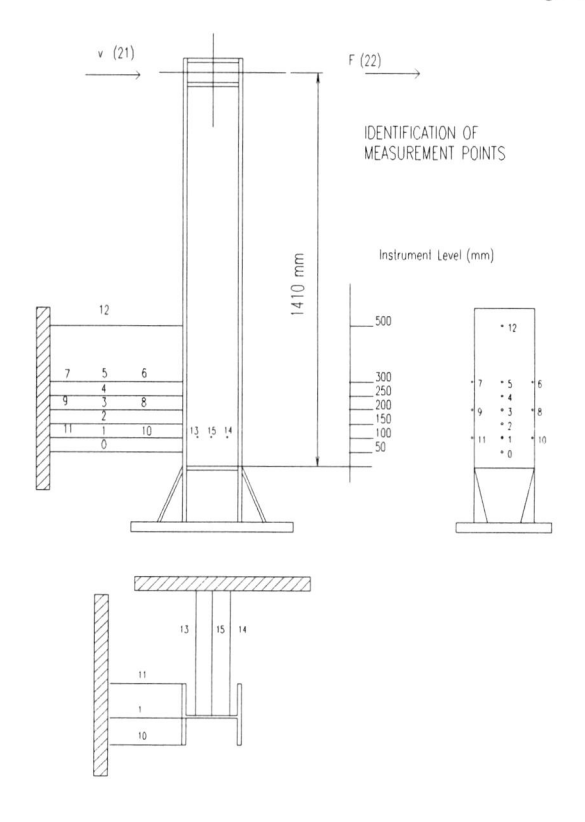

fig. 1 Specimen setup

resistance ratio, rigidity ratio, cumulative energy ratio and buckle size which were plotted vs. the number of applied cycles. Before any comment, it is important to remember that, according to the definitions given in EC-3, HE220A profile has a c/t_f ratio of the flanges (10.0) larger than that of HE220B (6.9), of IPE300 (7.0) and of HE120A (7.5), while its width to thickness ratio for the web (d/t_w= 21.7) is intermediate between those of HE120A (14.8), which is similar to that of HE220B (d/t_w =16.0), and of IPE300 (d/t_w =35.1) (tab. 1).

Based on the test results by Ballio Castiglioni (1994a) the following considerations can be drawn, having a general validity, for the shapes examined in this study:

1. Deterioration effects, causing a reduction in load carrying capacity, stiffness and hysteresis loops area begin earlier in HEA specimens (having larger c/t_f ratio) than in both IPE and HEB ones. IPE beams, characterised by larger d/t_w ratios, although initially following an intermediate behavior between HEA and HEB, experience much faster degradation;
2. Local buckling starts a few cycles earlier in IPE beams than in HEA ones, whilst HEB specimens can sustain a larger amount of cycles without buckling. A major role in governing local buckling effects is played by the web slenderness ratio d/t_w. Buckling develops completely within a few cycles, whose number seems to depend on the c/t_f ratio, then stabilisation of the buckles size occurs.
3. For all types of profile, once local buckling takes place and buckle size stabilisation occurs, also the hysteresis loops stabilise, and the rate of reduction in load carrying capacity decreases with increasing the number of cycles imposed to the specimen, until a final stage is reached, when the deterioration rate suddenly increases again and the specimen collapses after a few cycles. In particular, both IPE and HEA beams clearly evidence this type of behavior, while HEB specimen show a smoother transition from the (longer) phase of cycle stabilisation to that leading to collapse;

4. These differences in behavior at the final stage, between HEB specimens and HEA and IPE ones is associated with a difference in their failure modes. HEA and IPE beams generally collapse by steady crack propagation due to low-cycle fatigue effects; the HEB specimens, on the contrary, evidence some kind of brittle fracture of both the flange and the web, either at the specimen-to-base welds or at the plastic hinge where, due to large localised distortions, surface cracks usually develop a few cycles after local buckling of the flange plates.

3.1.2 Re-processing the test data

When test data are re-processed according to equation (5), the following parameters must be defined: the number of cycles to failure N_f, the values of the generalised displacement component (s_y) and of the stress level (σ (F_y)) corresponding at first yield. Of course, various operative choices are available for the definition of these parameters; in order to clearly identify the consequences of these operative choices, the following considerations should be taken into account.

- To define the number of cycles to failure N_f, a collapse criterion must be adopted, either assumed conventionally a-priori, or identified test by test corresponding to failure. In the first case, for example, failure might be associated with the reduction of the load carrying capacity to a given percentage γ of the yield strength (e.g. γ =50%, or γ=70%). In the second case, for example, as a consequence of the last two considerations of the previous paragraph, the number of cycles to failure N_f can be assumed either as the one corresponding to complete separation of one flange (generally this is the case for HEB shapes) or as the one corresponding to the sudden (final) increase in the deterioration trend after cycle stabilisation (generally applicable to HEA and IPE shapes).
- The generalised displacement component can be assumed either as a displacement v or as a rotation ϕ. Accordingly, the value corresponding to first yield can be defined as the yield displacement v_y (or the yield rotation ϕ_y), and can be theoretically computed or conventionally defined reprocessing test data, for example adopting the ECCS (1986) Recommended procedures.
- The nominal stress level (σ (F_y)) associated to an external load (F_y) causing first yield in the material can be determined experimentally, by means of tensile tests, or theoretically, by applying conventional methods of structural mechanics. In the first case, σ $(F_y) = f_y$. In order to apply the second procedure, both the yield strength (F_y) should be determined (e.g. according to ECCS Recommended procedures) and, for flexural members, the dimension of the plastic hinge.

Once determined the number of cycles to failure N_f, test data can be re-processed to plot in a log-log scale N_f vs. $\Delta\sigma^*$ given by eq. (5). The domain log $(\Delta\sigma^* = E\Delta\varepsilon)$ vs. log N is the usual domain for the Wöhler (1960) (S-N) curves adopted by various International Codes and Standards for (high cycle) fatigue design of steel structures. In fact, the strain range $\Delta\varepsilon$ (having the same physical meaning in both high and low cycle fatigue) has been correlated to the number of cycles to failure N_f and is then multiplied by the Young modulus E in order to deal with the same parameters commonly used by designers dealing with high cycle fatigue. For high cycle fatigue design, the most common structural details have been grouped into a number of categories (a same category for different details having similar fatigue strength), and to each category has been associated an *S-N* curve. It has recently been recognised the possibility to unify the slope of such curves which, in the most recent Recommendations (e.g. EC-3, ECCS 1985) are defined by an equation of the type:

$$N\Delta\sigma^3 = A \qquad (6)$$

A being a constant assuming different values for each S-N curve.

In order to verify the first assumption introduced in the previous point 2 of this paper (equivalence of $E\Delta\varepsilon$ - N_f curves for high and low cycle fatigue), it is tried to interpret the experimental test data of the low cycle fatigue tests performed during the present study, by means of the *S-N* curves proposed by Eurocode-3, whose validity is extrapolated in the low cycle fatigue range (i.e. for number of cycles *N* ranging from 10 to 500, and corresponding stress ranges $E\Delta\varepsilon = \Delta\sigma^* > 2 f_y$) by means of eq. (5).

In high cycle fatigue, strength categories implicitly account for different notch effects, i.e. for different local stress concentrations due to geometry of the detail and/or defects caused by fabrication procedures. It is supposed that the same consideration holds also in the case of low cycle fatigue: local buckling can in fact be regarded as a notch effect, because it induces local stress concentrations in the buckled area (at plastic hinge location). In fact, as already discussed, and in good agreement with previous results by Yamada (Yamada, 1969; Yamada & Shirakawa, 1971; Yamada et al. 1988), the different geometries of the cross sections make the specimens more or less vulnerable by local buckling effects. This means that each shape, as a function of its geometrical properties, can be considered as belonging to a definite fatigue strength category, because intrinsically affected by a more or less pronounced notch effect.

If this assumption is true, it must be expected that the different shapes considered in this

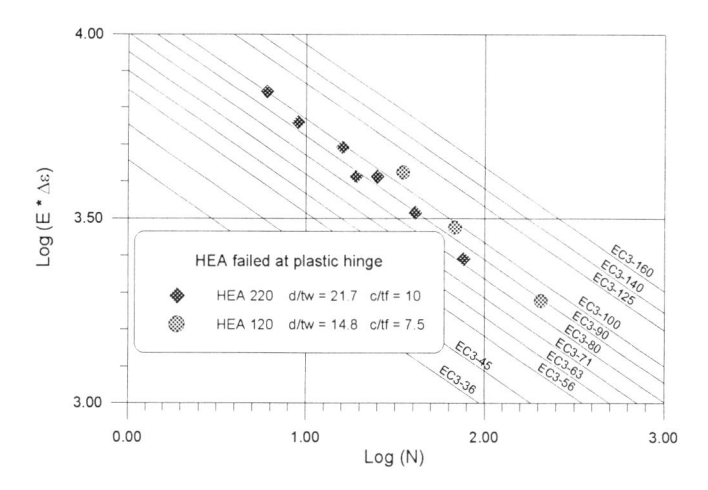

fig. 2 HEA test data fitted by EC-3 S-N curves

study belong to different fatigue strength categories: HE220B to a higher one, IPE300 to a lower one and HEA to intermediate ones, with HE120A specimens showing a higher

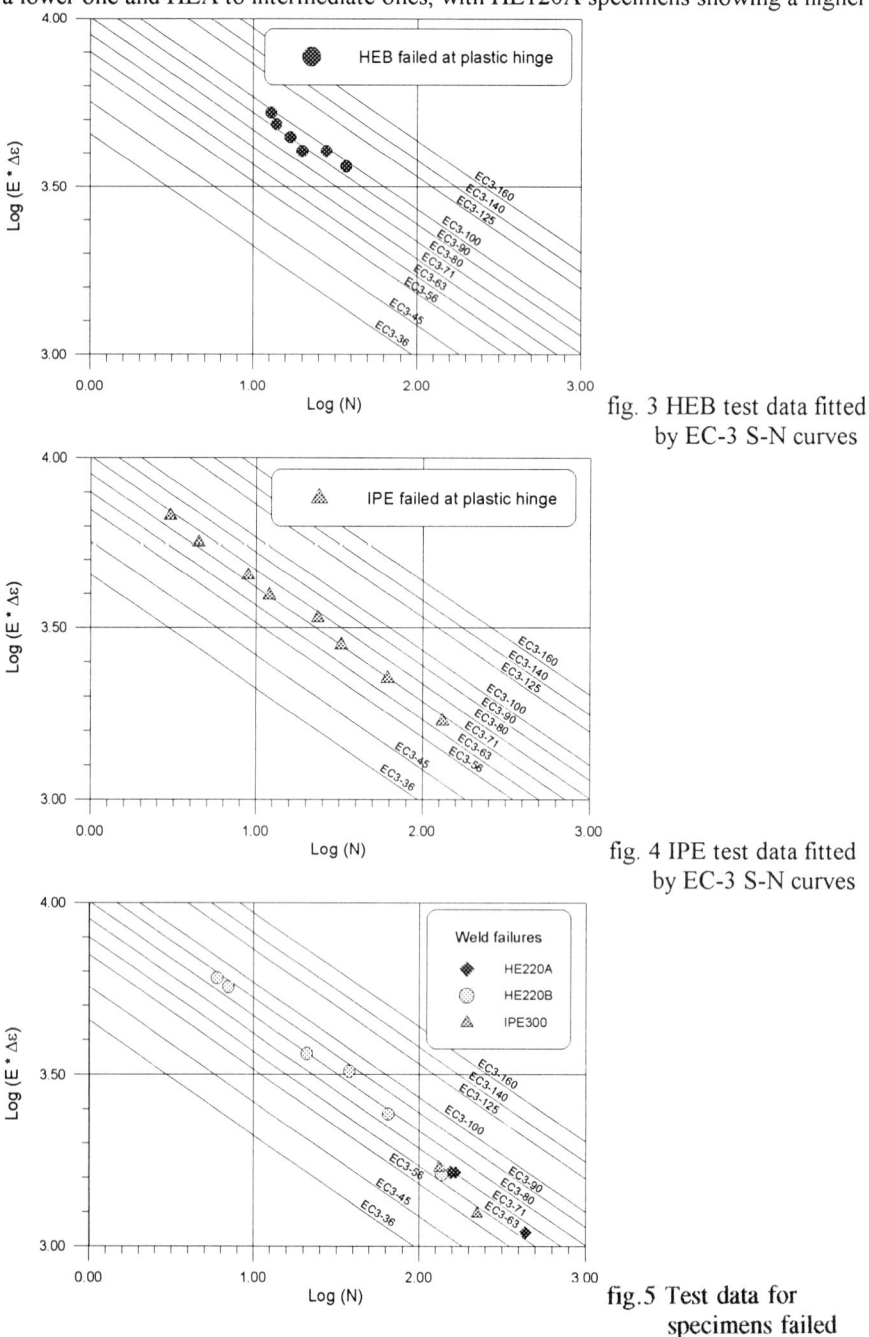

fig. 3 HEB test data fitted by EC-3 S-N curves

fig. 4 IPE test data fitted by EC-3 S-N curves

fig.5 Test data for specimens failed at weldings

specimens failed at weldings fatigue strength than HE220A, because of the lower c/t_f and d/t_w ratios. Furthermore, the tested specimens evidenced two different failure modes: by cracking in the base material at the plastic hinge locations, or by cracking at the welding of the reinforcement plates to the specimens. It must than be expected that different fatigue strength curves apply to the different failure modes.

The following figures 2-4 respectively show the test data for HE220A, HE220B and IPE300 specimens failed for cracking in the base material at plastic hinge locations, fitted by the S-N lines of Eurocode-3.

As expected, test data for HE220B can be fitted by EC-3 line for category 80, and those for IPE300 by that for category 63; those for HE220A specimens can be fitted by the line for category 71, intermediate between the two previous ones, while HE120A specimens can be fitted by line for category 80.

Figure 5 refers to test data for specimens failed at weldings; HE220A and HE220B specimens can be fitted by category 63 line, while IPE300 specimens, despite the same welded detail was adopted for all shapes, are fitted by line 56 and show a lower fatigue strength. This is probably caused by the formation of the plastic hinge nearer to the base (i.e. nearer to the weldment) in IPE specimens rather than in HE ones. This fact seems again to confirm the hypothesis that local buckling can be considered as a notch effect reducing the fatigue strength of the profile.

However, independently on the category of fatigue resistance pertinent to each shape, it is important to notice that the slope of the line fitting (in a log-log plot) the low cycle fatigue test data, reprocessed according to eq. (5), is nearly -3. This is in good agreement with the results of research on high cycle fatigue. It follows that both high and low cycle fatigue test data can be fitted by S-N lines having the same slope -3, the fatigue category depending on the notch effect associated with the various structural details.

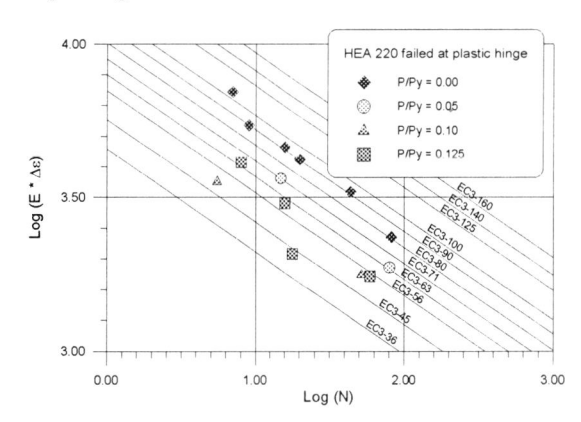

fig. 6 Effect of axial load on HE220A beams

Fig. 6 shows the effect of an axial load applied on top of HE220A specimens. Increasing the applied axial load results in a reduction of the fatigue strength of the member, but also in a scattering of the results, although the average slope of the fitting line remains nearly -3. It can however be noticed that the line fitting test data for P/P_y=0.10 and 0.125 nearly coincide. The test data available at present are too few to allow definitive conclusions to be drawn; it seems however that, beyond a certain level, the detrimental effect of the axial load on the fatigue strength is reduced, probably because it reduces the tensile strains due to bending, hence "retarding" fatigue crack opening and propagation.

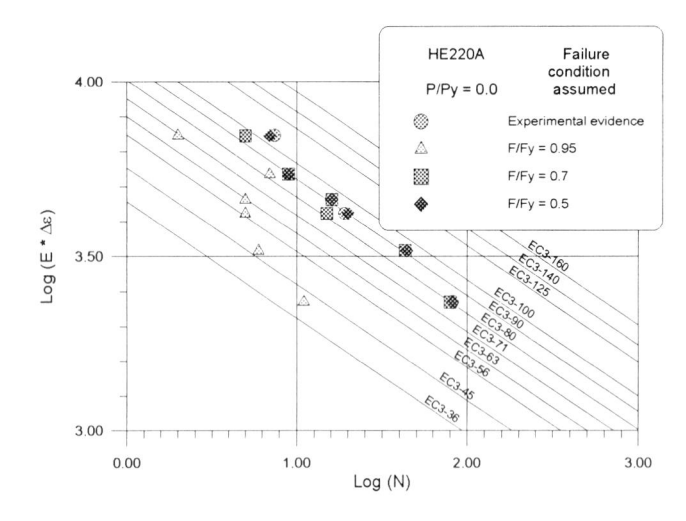

fig. 7 Effect of assumed failure condition on damage assessment

For realising the effects of the operative choices in the definition of the various parameters when re-processing test data, previously discussed, the following figs. 7 and 8 are presented. Fig. 7 shows, for HE220A specimens, the effect of the assumed failure condition, comparing results obtained defining N_f:
- based on experimental evidence, as the one corresponding to the final sudden increase in the deterioration rate (leading to failure)
- based on a condition defined a-priori, associated to a reduction of the load carrying capacity to a given percentage of the yield strength F_y.

$F/F_y=0.95$ leads to conservative assessments of the fatigue strength, as well as to scattering of results, while assuming as failure condition a reduction of strength to $0.50F_y$ leads to assessments similar to those based on test evidence.

Fig. 8 shows the effects of the definition of the stress range $\Delta\sigma^*$ (i.e. of the definition of the yield stress f_y and of the assumed generalised displacement component s) on the assessment of the fatigue strength of HE220A specimens, in the case of a failure condition assumed a-priori corresponding to a reduction of the specimen strength F to $0.50F_y$. It can be noticed that scattering in the results is connected to the definition of the yield strength, while smaller is the influence of the assumed displacement component; in fact, if top displacements v or rotations at plastic hinge location ϕ are assumed as generalised displacement component s in eq. 5, a small scattering can be noticed in the results. On the contrary, if the stress level corresponding at yield ($\sigma(F_y)$) is defined through a tensile test (i.e. coincident with f_y) or by means of structural mechanics (i.e. as $F_y L'/W$, where W is the section modulus and $L'= L-(\xi h/2)$ the cantilever member length minus half the dimension of the plastic hinge), due to the uncertainties connected with the latter definitions of both L' and F_y, a larger scatter in the results can be noticed. Fig. 8.1, 8.2 and 8.3 respectively refer to different values of parameter ξ (respectively assumed equal 1.0, 1.5 and 2.0). It can be noticed how scattering of the results is strongly influenced by this parameter.

Damage assessment in steel members 73

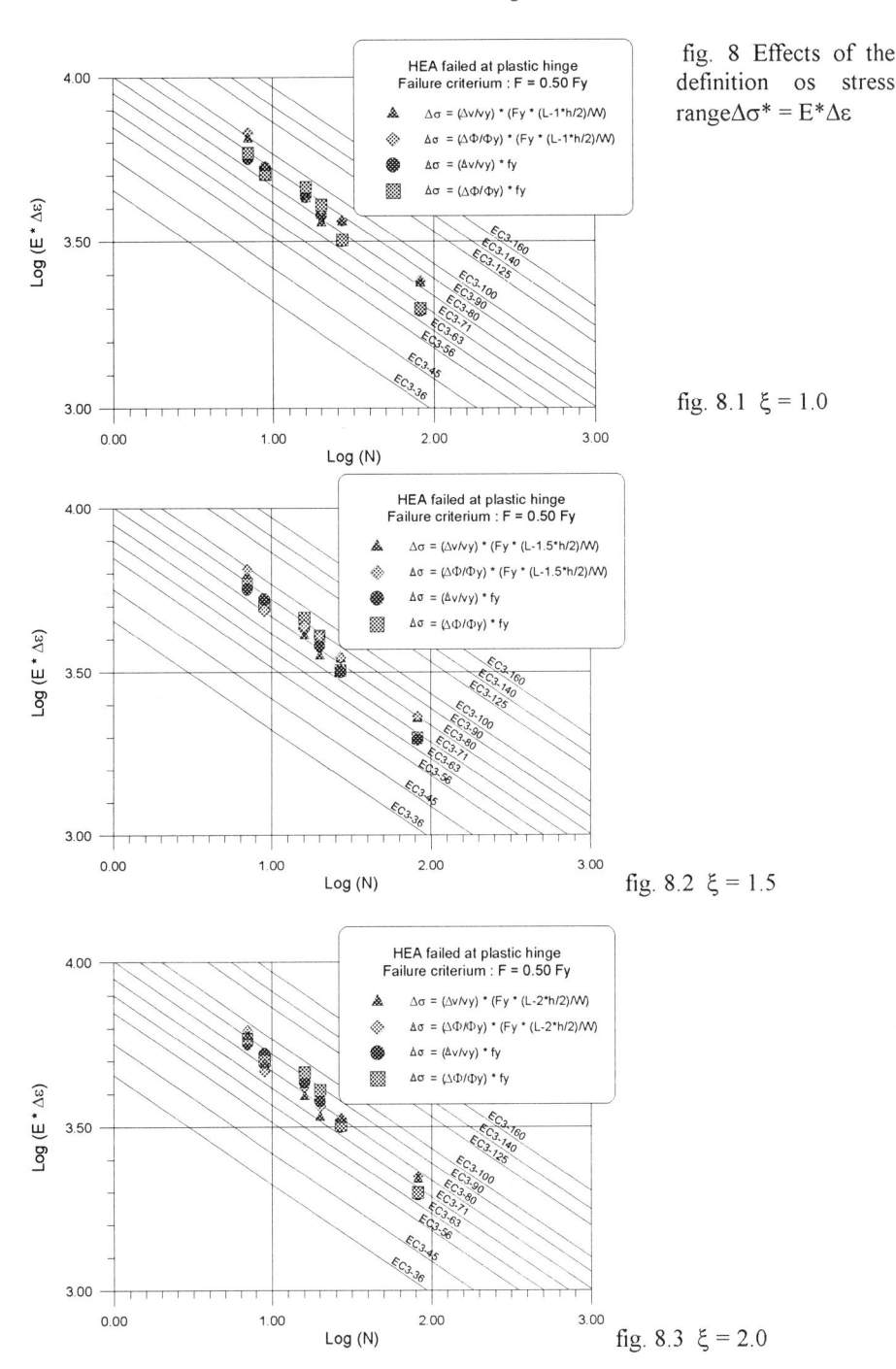

fig. 8 Effects of the definition os stress range $\Delta\sigma^* = E^*\Delta\varepsilon$

fig. 8.1 $\xi = 1.0$

fig. 8.2 $\xi = 1.5$

fig. 8.3 $\xi = 2.0$

Hence, in order to avoid biased results from re-processing test data, extreme accuracy and consistency must be adopted when defining the various parameters. In particular, when possible, the yield strength should be determined by tensile tests.

3.2 Variable amplitude tests

The second assumption, previously introduced in chapter 2, deals with the possibility of adopting Miner's rule for defining an acceptable failure criterion in low cycle fatigue.

Some random displacement histories were numerically obtained by means of a dynamic numerical simulation code by Ballio, Castiglioni and Perotti (1988), adopting artificial accelerograms obtained on the basis of Eurocode-8 (1988) recommended spectra. Various oscillograms were numerically obtained under increasing values of the peak ground acceleration, and Miner's damage index was computed using the Rainflow method for cycle counting. When a time history giving a Miner's damage index greater than 1 (i.e. indicating failure) was obtained as output of the numerical simulation, such a displacement history was imposed in a quasi-static way to the specimens under testing.

At present, a total of 11 random tests have been performed (4 on HE220A shapes, 4 on IPE300 and 3 on HE220B). The experimental results were re-processed by means of the rainflow cycle counting method and Miner's damage index associated with collapse of each specimen was computed, based on the transformation given by equation (5) and on the EC-3 fatigue strength lines previously identified for the various profiles. The obtained results are summarised in table 2 where the failure mode (S= at plastic hinge, W= at welding) and the damage index corresponding to the EC-3 lines are given.

TEST	EC-3 FATIGUE CURVE				Failure
	90	80	71	63	
HEA1	0.542	0.772	1.104	1.580	S
HEA9	0.854	1.202	1.740	2.489	S
HEA10	0.599	0.853	1.220	1.746	S
HEA11	0.773	1.040	1.490	2.135	S
HEB1	1.420	2.030	2.900	4.152	S
HEB12	0.717	1.020	1.460	2.090	W
HEB13	0.523	0.744	1.060	1.524	W
IPE9	0.461	0.657	0.939	1.340	S
IPE10	0.385	0.547	0.783	1.120	S
IPE11	0.460	0.660	0.940	1.346	S
IPE15	0.478	0.680	0.973	1.394	S

Table 2 - Damage indexes corresponding to specimen collapse in variable amplitude tests

The following considerations can be drawn:
- Miner's rule gives damage index values with scatters similar to those commonly accepted in random high cycle fatigue.
- For HEA specimens Miner's rule correctly allows prediction of failure in association with EC-3 line for fatigue strength category 71.

- For HEB specimens, depending on the failure mode, Miner's rule leads to a correct prediction of failure respectively in association with EC-3 lines for fatigue strength categories 80 and 63. In particular, the adopted fatigue strength lines lead to damage assessments largely on the safe side; increasing the fatigue strength of HE220B specimens by one category results in damage index values nearer to unity.
- For IPE specimens, Miner's rule correctly allows prediction of failure in association with EC-3 line for fatigue strength category 63.

4. Conclusions

The obtained results show the validity of the two assumptions on which the proposed approach is based:
1. the same S-N curves are valid in high and low cycle fatigue, if an equivalent stress range $\Delta \sigma^* = E \Delta \varepsilon$ is considered, associated with an ideal indefinitely elastic behavior of the material; in particular, the slope of these S-N curves in a (log-log) plot is -3.
2. Miner's rule can be adopted, together with the previously defined S-N curves and with a cycle counting method (e.g. Rainflow) to define a unified collapse criterion, valid for both high and low cycle fatigue.

The application of these results and of the proposed method for damage assessment, to steel structures under seismic loading, may lead to an overcoming of seismic design methods based on the behavior factor, as shown by Ballio & Castiglioni (1994c).

5. References

Ballio G., Castiglioni C.A. (1994a) A unified approach for the design of steel structures under low and/or high cycle fatigue, to appear on **Jour. of Constr. Steel Research**.

Ballio G., Castiglioni C.A. (1994b) Seismic behavior of steel sections", **Journal of Constructional Steel Research**, 1994, n.29

Ballio G., Castiglioni C.A. (1994c) An approach to the seismic design of steel structures based on cumulative damage criteria, to appear on **Earthquake Engineering & Structural Dynamics**.

Ballio G., Castiglioni C.A., Perotti F. (1988) On the assessment of structural design factors for steel structures, **IX W.C.E.E., Tokyo**, 5, 1167-1171.

Ballio G., Zandonini R. (1985) An experimental equipment to test steel structural members and subassemblages subject to cyclic loads, **Ingegneria sismica**, 2, 25-44.

CEN (1992), Eurocode-3: Design of Steel Structures - Part 1-1: General rules and rules for buildings, **ENV 1993-1-1:1992 E**.

ECCS (1985), Recommendations for the fatigue design of structures, 1st Ed.

ECCS, T.C.1, T.W.G. 1.3, (1986) Recommended testing procedure for assessing the behavior of structural steel elements under cyclic loads, **Rept. n. 45**.

Eurocode-8 (1988), Design of structures in seismic regions, Part 1: General rules and rules for buildings.

Wöhler A. (1860), **Zeitschrift für Bauwesen,** vol.10.

Yamada M.,(1969), Low cycle fatigue fracture limits of various kinds of structural members subjected to alternately repeated plastic bending under axial compression as an evaluation basis or design criteria for aseismic capacity, **IV W.C.E.E., Santiago,** Chile, Vol.1, B-2, Jan. 69, 137-151.

Yamada M., Kawamura H., Tani A., et al. (1988), Fracture ductility of structural elements and of structures, **IX W.C.E.E., Tokyo,** IV, IV219-IV224.

Yamada M. Shirakawa H. (1971), Elasto-plastic bending deformation of wide-flange steel beam-columns subjected to axial load, Part II, **Stahlbau**, 40,H.3 65-74, H.5 143-151

6 TOWARD THE DEFINITION OF A CONSISTENT SEISMIC-RESISTANT DESIGN METHOD BASED ON DAMAGE CRITERIA

E. COSENZA and G. MANFREDI
Istituto di Tecnica delle Costruzioni, University of Naples, Naples, Italy

Abstract

In this paper a consistent method that considers the low cycle damage to evaluate the part of the monotonic ductility that the structure can develop under seismic loading is proposed. It is based on the estimation of the number of inelastic cycles and on a statistical characterization of the distribution of plastic excursions; the parameters chosen in the analysis are the fundamental period of the structure, the yielding level, the effective duration of earthquake and a damage structural parameter.

Keywords: Seismic-resistant design, Inelastic cycles, Damage

1 Introduction

The first and fundamental step in seismic-resistant design of structures is the reliable evaluation of design forces in relation with the possible ground motions that can occurs in the site of the structure. Currently, it is accepted that the structures can tolerate a certain degree of damage and therefore the design is conducted using a Smoothed Inelastic Design Response Spectra obtained thought the use of a reduction factor R depending on the displacement ductility μ. These inelastic spectra give only the value of the maximum global ductility demand and don't are able to give a reliable definition of the damage potential of the earthquake due to low cycle damage. Some researchers, in fact, have underlined that the real inelastic response of structures is depending on the entire inelastic displacement history (related to the structural damage) and a powerful analytical tool to define the effective required strength can be represented by the damage functionals [Bertero and Uang (1992), Krawinkler and Nassar (1992), Cosenza and Manfredi (1992b)].

These functionals are based on the simultaneous knowledge of ductlity μ and hysteretic energy E_H or on the definition of an accumulative ductility [Cosenza and Manfredi (1993), Park and Ang (1985), Krawinkler and Zohrei (1983)]. This last concept, related with the low cycle fatigue criteria, seems the most promising and powerful, as detailed in the following. Therefore it is necessary to develop simple design methods based on the damage functionals [Fajfar (1993), Fajfar and Vidic (1994)]. In this paper a consistent methods to evaluate the cumulative ductility from the inelastic ductility is proposed trough the introduction of an equivalent factor p. It is based on the estimation of the number of inelastic cycles and on the statistical characterization of the distribution of these. After a comprehensive statistical analysis the number of cycles is related with the effective duration of earthquake t_d with a simple formulation: therefore the global damage potential of ground motion is strictly related with this parameters so as observed from others researchers [Uang and Bertero (1990), Hadidi and Krawinkler(1988), Fajfar et al. (1988), Yayong and Minxian (1990), Jeong and Iwan (1988)]

2 Low cycle fatigue and the concept of "equivalent ductility"

In the following the concept of "damage functionals" will be widely used [Cosenza and Manfredi 1992b]. In particular we will consider the damage functional related to the concept of low cycle fatigue D_F:

$$D_F = A \sum_{i=1}^{n} (\Delta x_i)^b \quad \text{with} \quad b > 1 \qquad (1)$$

where A and b are parameters necessary to define the model, n is the total number of plastic cycles and Δx_i is the plastic amplitude of a generic cycle. The case $D_F=0$ characterizes the absence of plastic damage and the case $D_F=1$ the collapse under cyclic actions.

A different form of eqn. (1) can be introduced by using the result of a test where the specimen collapse with only one cycle of amplitude $\Delta x_{1,u}$; eqn (1) becomes:

$$1 = A(\Delta x_{1,u})^b \Rightarrow A = \frac{1}{(\Delta x_{1,u})^b} \Rightarrow D_F = \sum_{i=1}^{n} \left(\frac{\Delta x_i}{\Delta x_{1,u}}\right)^b \qquad (2)$$

The absence of collapse under cyclic loads follows from the achievement of the relation:

$$\sum_{i=1}^{n} \left(\frac{\Delta x_i}{\Delta x_{1,u}}\right)^b = \left(\frac{\Delta x_{max}}{\Delta x_{1,u}}\right)^b \left(1 + \sum_{i=1}^{n-1} x_i^b\right) \leq 1 \qquad (3)$$

having defined the maximum required ductility Δx_{max}, and having introduced the non-dimensional ratio $x_i=\Delta x_i/\Delta x_{max}$. The achievment of the collapse provides:

$$D_F = 1 \Leftrightarrow \frac{\Delta x_{max}}{\Delta x_{1,u}} = \frac{1}{\left(1 + \sum_{i=1}^{n-1} x_i^b\right)^{1/b}} \tag{4}$$

From eqn (4) follows that, considering the low cycle fatigue, it is possible to reach only the fraction $p<1$ of the value $\Delta x_{1,u}$:

$$\Delta x_{max} = p \Delta x_{1,u} \quad \text{with} \quad p = \frac{1}{\left(1 + \sum_{i=1}^{n-1} x_i^b\right)^{1/b}} \tag{5}$$

The value of p, which defines the "equivalent ductility" that the structure can develop under a cyclic displacement history, depends on the damage parameter b, on the number of plastic cycles n, and more generally on the statistical distribution of these cycles. The problem of the statistical characterization of the plastic cycles will be extensively analyzed in the following.

3 The optimum choice of the probability law

In the following the statistical characterization of the distribution of plastic excursions of the elasto-plastic SDOF will be provided. More in detail, the random variable $x = \Delta x / \Delta x_{max}$, Δx being the generic plastic deformation and Δx_{max} the maximum plastic excursion (x belongs to the range $(0,1)$) will be analyzed from a probabilistic point of view.

The analysis of the response of the SDOF model provides to a single seismic input, beside Δx_{max} and the number n of the plastic cycles, a random sample with dimension $n-1$ defined by the value of the random variable x:

$$(x_1, x_2, ..., x_i, ..., x_{n-1}) \quad \text{with} \quad x_i = \frac{\Delta x_i}{\Delta x_{max}} \tag{6}$$

In [Cosenza and Manfredi (1992a)] a wide comparison between different probabilistic models is provided; the probabilistic models: truncated normal, lognormal, beta, gamma, exponential and pareto were extensively analyzed. The main results are that the normal and lognormal are not satisfactorily in analyzing the problem; the beta and gamma are very accurate, but these models require two parameters (mean and standard deviation) to define the problem; the exponential and pareto probability distributions are defined by only one parameter (i.e. in the case of the exponential model by the variable ν that is equal to the inverse of the mean number of plastic cycles) and are sufficiently accurate for technical aims. In the following we will consider the exponential model, that is defined by the following density function $f_X(x)$ ($f_X(x)$ is equal to the probability that the random variable assumes values in the range x, x+dx) and by the following distribution function $F_X(x)$ ($F_X(x)$ is equal to the probability that the random variable assumes values less or equal to x):

$$f_X(x) = \nu e^{-\nu x} \quad ; \quad F_X(x) = 1 - e^{-\nu x} \quad (\nu, x \geq 0) \tag{7}$$

The use of this probability model and of the damage functional of low cycle fatigue allow to obtain useful results, as we will show in the following.

It is well known that, if the random variables x_i are distributed with an exponential law, x_i^b is characterized by a Weibull probabilistic law, with the following mean value:

$$E[X^b] = m^b \Gamma(b+1) \tag{8}$$

where Γ is the well known "gamma function" (if x is a integer number: $\Gamma(x+1) = x!$).

Assuming that n is deterministically known, the eqns. (5) and (8) provide the following result for the equivalent factor p:

$$p = \frac{1}{[1+(n-1)m^b \Gamma(b+1)]^{1/b}} \tag{9}$$

As a consequence we can say that the "equivalent ductility" depends on the two parameters n (number of plastic cycles) and m (mean value of the plastic excursions, excluding the maximum), beside the damage parameter b. In the following the problem to evaluate the values of n and m is analyzed.

4 The evaluation of the number and of the mean amplitude of the plastic cycles: the choice of the reference parameters

The elasto-plastic SDOF model considered in the analysis is characterized by mass m, period T, viscous damping ν, yielding force F_y.

The first problem is to define the parameters that govern the problem. In a general framework we can individuate model parameters: elastic response parameters and inelastic response parameters; and input parameters: earthquake parameters.

For what concerns the "model parameters", the elastic response of the model, for a given accelerometric record, is completely defined by the period T; the inelastic response of the model is fully described by defining the "yielding level" η:

$$\eta = \frac{F_y}{m S_{a,el}(T,\nu)} \quad . \tag{10}$$

where $S_{a,el}(T,\nu)$ is the elastic acceleration spectra; the percentage of viscous damping will be considered as a variable in the statistical analysis described in the following.

More complicated is the problem of defining the seismic input. More in detail the problem is the definition of a single appropriate seismic parameter to define the values of n and m; the following seismic parameters will be analyzed:

- total duration t [s]
- peak ground acceleration a_g [m/s^2]
- peak ground velocity v_g [m/s]
- peak ground displacement d_g [m]
- effective duration t_D after Trifunac e Brady [s]
- intensity after Arias I_A [m,s]
- intensity after Fajfar et al. I_F [m,s]

The optimum choice of the seismic parameters is made on the basis of an exponential correlation of the following type:

$$n = A \cdot p_s^\alpha \cdot \eta^\beta \cdot T^\gamma \qquad (11)$$

$$m = \overline{A} \cdot p_s^{\overline{\alpha}} \cdot \eta^{\overline{\beta}} \cdot T^{\overline{\gamma}} \qquad (12)$$

where p_s is the generic parameters previously defined. It is assumed that the best parameter is the one that provides the maximum correlation coefficient.

In the present analysis all the Italian records, starting from 1972, with maximum ground acceleration greater than 0.1 g., were analyzed. All the records were subdivided in 3 type: S1, S2 and S3 (rock, alluvium with depth less than 20 m. and alluvium with depth more than 20 m.). The analyzed period ranges between 0 and 3 s., considering low period range, medium period range and long period range [Newmark and Hall (1982)] with the same mean number of periods. The yielding level varies between 0.05 and 0.5; the percentages of viscous damping considered in the analysis are 2%, 5%, 10% and 50%.

The results of the statistical analysis are given in tables 1 and 2 analyzing the influences of damping, period range and soil type; in the tables the correlation factor for each seismic parameter is provided. The best seismic parameters is evidenced with the shading and it is the effective duration t_D, even if in some cases other parameters provide better results for m, but always with low correlations.

It is worth to underline that this analysis is made to provide the optimum correlation for evaluating n and m and not to evaluate the best parameter to define the damage potential of the earthquake.

Table 1. Correlation coefficient between the number of plastic cycles n and different seismic parameters.

	t_d	a_g	I_F	a_g/v_g	I_A	v_g	d_g
2%	0.551	0.227	0.103	0.178	0.110	0.061	0.011
5%	0.549	0.247	0.076	0.158	0.104	0.090	0.009
10%	0.535	0.259	0.058	0.146	0.106	0.105	0.019
50%	0.573	0.243	0.156	0.272	0.185	0.005	0.115

	t_d	a_g	I_F	a_g/v_g	I_A	v_g	d_g
low	0.633	0.197	0.157	0.281	0.170	0.017	0.089
medium	0.545	0.368	0.081	0.095	0.012	0.294	0.224
long	0.525	0.215	0.007	0.058	0.082	0.184	0.182

	t_d	a_g	I_F	a_g/v_g	I_A	v_g	d_g
Soil 1	0.432	0.010	0.190	0.137	0.253	0.090	0.038
Soil 2	0.507	0.284	0.055	0.112	0.062	0.138	0.028
Soil 3	0.744	0.257	0.258	0.306	0.216	0.032	0.159

Table 2. Correlation coefficient between the mean amplitude of plastic cycles n and different seismic parameters.

	t_d	a_g	I_F	a_g/v_g	I_A	v_g	d_g
2%	0.178	0.023	0.213	0.237	0.022	0.180	0.212
5%	0.206	0.013	0.203	0.253	0.002	0.161	0.204
10%	0.229	0.022	0.206	0.255	0.004	0.156	0.203
50%	0.328	0.087	0.231	0.331	0.023	0.154	0.234

	t_d	a_g	I_F	a_g/v_g	I_A	v_g	d_g
low	0.402	0.075	0.253	0.423	0.010	0.158	0.245
medium	0.012	0.064	0.067	0.005	0.078	0.064	0.083
long	0.039	0.014	0.028	0.048	0.098	0.015	0.006

	t_d	a_g	I_F	a_g/v_g	I_A	v_g	d_g
Soil 1	0.261	0.350	0.384	0.312	0.313	0.383	0.381
Soil 2	0.139	0.074	0.043	0.097	0.015	0.008	0.033
Soil 3	0.184	0.117	0.105	0.192	0.048	0.054	0.069

5 The evaluation of the number and of the mean amplitude of the plastic cycles: the choice of the reference parameters

Assuming the seismic parameters that provide the best correlation with the number of cycles and with the mean value of plastic deformation is the effective duration t_D, the form of equations to predict n and m is the following:

$$n = A \cdot t_D^\alpha \cdot \eta^\beta \cdot T^\gamma \tag{13}$$

$$m = \overline{A} \cdot t_D^{\overline{\alpha}} \cdot \eta^{\overline{\beta}} \cdot T^{\overline{\gamma}} \tag{14}$$

Typical values of A, α, β, γ and $\overline{A}, \overline{\alpha}, \overline{\beta}, \overline{\gamma}$, provided by experimental correlations, are given in tables 3; in these tables the whole parametric range is examined; the influence of the viscous damping is shown.

Table 3. Parameter of eqns. (13), (14); influence of viscous damping.

	A	α	β	γ	$\overline{A}$	$\overline{\alpha}$	$\overline{\beta}$	$\overline{\gamma}$
2%	1.702	-0.398	-0.818	0.314	.458	.154	.320	-.087
5%	1.506	-0.406	-0.860	0.322	.489	.167	.311	-.094
10%	1.416	-0.402	-0.883	0.321	.520	.170	.307	-.097
50%	0.657	-0.488	-1.019	0.352	.722	.196	.301	-.120

It should be noted that:
- the values of the parameters α, β, and γ are practically constant;
- the value of β is close to 1;
- the value of α and γ are almost equal in value and opposite in sign; this fact means that a good parameter is, in a simpler way, the ratio t_D/T.

To test statistically this last sentence, in the table 4 it is provided the percentage difference in terms of correlation coefficient and mean error considering independently t_D and T and considering just the parameter t_D/T. The approximation is practically the same, even if t_D/T is

Table 4. Percentage of variation of mean error and of the global correlation coefficient using the parameter t_D/T instead t_D and T.

	2%	5%	10%	50%		2%	5%	10%	50%
%	3.0	3.3	3.2	7.4	%	-0.4	-0.4	-0.4	-1.0

a meaningful parameter [Mason and Iwan (1983), Jeong and Iwan (1988)]. In the following we will consider a simplest formula to evaluate n and m that considers only the influence of t_D/T and η.

6 The evaluation of the number and of the mean amplitude of the plastic cycles: a simplified formulation

The analysis briefly described in the previous paragraph shows that it is possible to use the simpler formulas:

$$n = A \cdot \left(\frac{t_D}{T}\right)^\alpha \cdot \eta^\beta \qquad (15)$$

$$m = \overline{A} \cdot \left(\frac{t_D}{T}\right)^{\overline{\alpha}} \cdot \eta^{\overline{\beta}} \qquad (16)$$

Optimal values of the parameters A, α, β and $\overline{A}, \overline{\alpha}, \overline{\beta}$ are given in tab. 5 (viscous damping influence; all soils and all period ranges), tab. 6 (period range influence; 5% of viscous damping and all the soils) and tab. 7 (influence of the type of soil; 5% of viscous damping and all period ranges).

Table 5. Influence of viscous damping.

	A	α	β		$\overline{A}$	$\overline{\alpha}$	$\overline{\beta}$
2%	1.640	0.345	-0.817	2%	0.472	-0.113	0.319
5%	1.453	0.353	-0.858	5%	0.504	-0.120	0.309
10%	1.370	0.351	-0.881	10%	0.535	-0.124	0.305
50%	0.639	0.402	-1.004	50%	0.734	-0.148	0.293

Table 6. Influence of the range of periods.

	A	α	β		$\overline{A}$	$\overline{\alpha}$	$\overline{\beta}$
low	0.968	0.423	-0.935	low	0.711	-0.193	0.422
medium	1.457	0.349	-0.829	medium	0.382	-0.001	0.334
long	1.633	0.344	-0.826	long	0.364	-0.024	0.219

We can affirm that, for what concerns n:
- the power of α and β is practically constant; mean values are 0.35 and -0.86;
- the parameter A is influenced by the amount of viscous damping, range of period, type of soil.

Table 7. Influence of the type of soil.

	A	α	β		$\overline{A}$	$\overline{\alpha}$	$\overline{\beta}$
Soil 1	1.401	0.305	-0.808	Soil 1	0.554	-0.147	0.389
Soil 2	1.749	0.275	-0.896	Soil 2	0.496	-0.077	0.323
Soil 3	1.409	0.409	-0.874	Soil 3	0.469	-0.108	0.260

For what concerns m:
- the value of $\overline{\alpha}$ is very low and practically equal to 0;
Considering the significant case of medium periods and viscous damping of 5%, we can conclude that approximate previsions of n and m are:

$$n \cong 1.45 \left(\frac{t_D}{T}\right)^{0.35} \cdot \eta^{-0.86} \quad ; \quad m \cong 0.38 \cdot \eta^{0.33} \tag{17}$$

7 An application of equivalent ductility concept to steel structures

The equivalent parameter p expressed by eqn.(9) correctly provides that the available ductility depends on the number of yielding reversal n, the average value of their statistical distribution m and, also, on the structural degradation properties trough the damage parameter b. It is worth to notice that in the cases characterized by $n \to 1$, $m \to 0$ or also $b \gg 1$ results that p is equal to 1: in other words the ductility and the low cycle fatigue criteria provide the same results.

Analyzing the eqn.(2), it is possible to notice that the structural elements, characterized by an hysteretic energy depending collapse, have a low value of b, while the elements, characterized by a ductility depending collapse, have an high value of b [Cosenza and Manfredi 1993]. Reliable values of b for different elements and materials can be obtained from the Yamada experimental results [Yamada et al. (1988)]. Typical values of b for steel elements are summarized in table 8.

Table 8. Typical values of b parameter fo steel elements.

TYPES	b
Wide flange steel	5
Steel box	2÷2.4
Steel, b/t=10	1.9
Steel, b/t=20	1.6
Steel, b/t=30	1.3

With reference to double T beams, it is clearly possible to observe that sections with wide flanges have a value of b equal to 5 and the collapse is practically depending from ductility, therefore these steel elements have good resources of energy dissipation.

On the other hand, sections characterized by an high flange slenderness (high ratio b/t) have a value of b equal to 1.3 with a significative degradation due to local buckling and, therefore, the yielding reversal cycles reduce significantly the available ductility.

The influence of low fatigue phenomena on structural response is summarized in table 9 for some representative ground motions, often used in seismic analysis; particularly the values of parameter p (from eqns. 9 and 17) are reported, assuming a yielding level $\eta=0.25$ and an elastic period $T=0.5$ s, with reference to steel elements with high and low damage features.

Table 9. Values of parameter p for earthquakes with different effective duration (yielding level = 0.25; T=0.5 s.).

Earthquake	a_g [1/g]	t_d [s]	p [b=1.3]	p [b=5.0]
Rocca	.603	2.55	0.510	0.890
Tolmezzo	.313	4.92	0.451	0.864
Petrovac	.438	9.30	0.398	0.841
El Centro	.348	24.4	0.325	0.804
Hachinoke	.229	27.9	0.315	0.799
Mexico City	.171	38.9	0.293	0.785
Calitri	.166	47.5	0.280	0.777

It is worth to notice that the response of a model representative of steel structures without local buckling (i.e $b=5.0$) is low depending from cyclic inelastic accumulation and, therefore, a ductility based design method is suitable without substantial errors ($p>75\%$). Conversely, the response of a model representative of steel structures with large local buckling problems (i.e. $b=1.3$) is very depending on low cycle fatigue phenomena and a damage criteria design method is required because the ductility methods is strongly not conservative: in fact the available ductility is a low fraction (varying from the 28% to 51%) of monotonic ductility.

These remarks are confirmed by fig.1, where the global variation of p parameter versus t_d/T and η is drawn for $b=1.3$ and $b=5.0$. The available ductility is very close to monotonic ductility (i.e $p \cong 1$) for steel elements with low cyclic damage phenomena and the maximum possible reduction is lower than 25% for very high values of t_d/T and for very low yielding level. On the contrary the steel structures with possible large buckling phenomena are very sensitive to duration of ground motion and to the required inelastic performance: in fact, for values of t_d/T greater than 50 and yielding level lower than 0.30, the reduction of monotonic ductility is always greater than 50%.

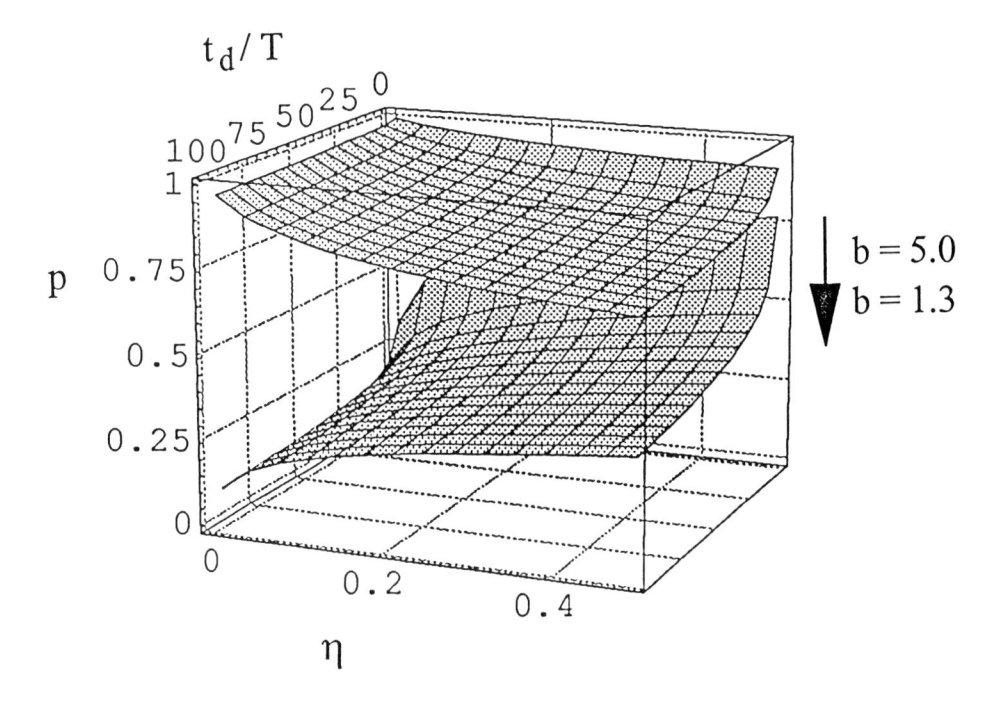

Fig. 1. p function surfaces for b equal to 5.0 and 1.3.

8 Concluding remarks

This paper includes a part of a study on the development of a SDOF seismic-resistant design method based on damage criteria. This methods is based on the definition of a reduction factor p that allows to obtain the effective available ductility under seismic actions. Using this reduced ductility, it is possible to obtain damage controlled spectra in the same simple way used for the inelastic displacement spectra. The p parameter is depending on elastic and inelastic characteristics of SDOF system (through elastic period and yielding level), damage sensitivity of structure (trough the low-cycle fatigue parameter b), and from the seismic input factor (trough the effective duration t_d). Therefore, as shown in an application to steel structures, the global damage effect of earthquake is a function both of intrinsic damage potential of ground motion and of the damage characteristics of structures. In fact for structures with low sensitivity to damage (high b) it is possible to use, also for earthquake with high damage potential (high t_d), ductility based design methods without large errors; on the contrary, for structures with high sensitivity to damage (low b), it is necessary the introduction of a damage based design method.

9 References

Bertero, V.V. and Uang, C.M. (1992) Issues and Future Directions in the Use of Energy Approach for Seismic-Resistant Design of Structures, in **Nonlinear Seismic Analysis of RC Buildings** (eds: P. Fajfar and H. Krawinkler), Elsevier, pp.3-22.

Cosenza, E. and Manfredi G. (1992a) Low Cycle Fatigue: Characterization of the Plastic Cycles due to Earthquake Ground Motion, in **Testing of Metals for Structures** (eds F.M.Mazzolani), E&FN Spon - Chapman & Hall, pp. 116-131.

Cosenza, E. and Manfredi G. (1992b) Seismic Analysis of Degrading Models by means of Damage Functional Concept, in **Nonlinear Seismic Analysis of RC Buildings** (eds: P. Fajfar and H. Krawinkler), Elsevier, pp.77-93 .

Cosenza, E. Manfredi, G. and Ramasco, R. (1993) The Use of Damage Functionals in Earthquake-Resistant Design: a Comparison Among Different Procedures, **Structural Dynamics and Earthquake Engineering**, 22, 855-868.

Fajfar, P. (1993) Equivalent Ductility Factors Taking into Account Low-Cycle Fatigue, **Earthquake Engineering and Structural Dynamics**, 21, 837-848.

Fajfar, P. and Vidic, T. (1993) Consistent Inelastic Design Spectra: Hysteretic and Input Energy, submitted to publication on **Earthquake Engineering and Structural Dynamics**.

Fajfar, P. Vidic, T. and Fischinger M. (1989) Seismic Demand in Medium- and Long-Period Structures, **Earthquake Engineering and Structural Dynamics**, 18, 1133-1144.

Hadidi-Tamjed, H. Lee, D.G. and Krawinkler, H. (1988) Statistical Study on Seismic Response Parameters for Damage Evaluation, in **IX WCEE Proceedings**, Tokyo-Kyoto, paper 7-3-1, Vol. V.

Jeong, G.D. and Iwan, W.D. (1988) The Effect of Earthquake Duration on the Damage of Structures, **Earthquake Engineering and Structural Dynamics**, 16, 1201-1211.

Krawinkler, H. and Nassar, A.A. (1992) Seismic Design Based on Ductility and Cumulative Damage Demands and Capacities, in **Nonlinear Seismic Analysis of RC Buildings** (eds: P. Fajfar and H. Krawinkler), Elsevier, pp.23-40 .

Krawinkler, H. and Zohrei, M. (1983) Cumulative Damage in Steel Structures Subjected to Earthquake Ground Motion, **Computers & Structures**, 16, 531-541.

Mason, A.B. and Iwan, W.D. (1983) An Approach to the First Passage Problem in Random Vibration, **Journal of Applied Mechanics ASME**, 50, 641-646.

Newmark, N.M. and Hall, W.J. (1982) **Earthquake Spectra and Design**, EERI, Berkeley, CA.

Park, Y.J. and Ang, A.H-S.(1985) Mechanistic Seismic Damage Model for Reinforced Concrete, **Journal of the Structural Engineering, ASCE**, 111, 722-739.

Uang, C.M. and Bertero, V.V. (1990) Evaluation of Seismic Energy in Structures, **Earthquake Engineering and Structural Dynamics**, 19, 77-90.

Yamada, M. et al. (1988) Fracture Ductility of Structural Elements and of Structures, in **IX WCEE Proceedings**, Tokyo-Kyoto, paper 6-3-3, Vol. IV.

Yayong, W. and Minxian, C. (1990) Dependence of Structural Damage on the Parameters of Earthquake Strong-Motion, **European Earthquake Engineering**, 4, 13-23.

7 INFLUENCE OF DISTRIBUTED INERTIAL MASS AND STRESS STATE ON THE SEISMIC RESPONSE

I. DIMOIU and M. IVAN
Civil Engineering Faculty, Timisoara, Romania

Abstract
There are two different ways of taking into account the inertial mass in the seismic response of a structure: the equivalent mass matrix and the lumped inertia mass matrix The former mains a more appropriate situation to the real state of the structure. Inertia mass is often lumped at nodal points. This leads to a diagonal mass matrix and to an excessive elasticity of the structure.
 The influence of fixed end forces on dynamic parameters of the structure is presented in the paper. Assigning the real stress state, the seismic response is like for a soften structure.
 Using the equivalent inertia mass matrix, considering very term of stiffness matrix and the stress state, the seismic response is more realistic.
Keyword: Inertia Mass Matrix, Stiffness Matrix, Fixed End Forces.

1 Introduction

The inertia masses are often considered like the lumped ones. Only the translation degrees of freedom are also assigned. The coefficients of stiffness matrix corresponding to the bending moment are only used. These assumptions allow to obtain a seismic response which is different from the real one. A more correct seismic reaction could be attained if a real distribution of the inertia mass, and very component of stiffness matrix is taken into account. On the same way, the very component of the resultant stress must be introduced in the computing lines.

2 Influence of translational distributed inertial mass

The inertia mass of structure elements can be introduced in dynamic analysis using two possibilities. The first is in the lumped mass matrix form. The inertia mass is concentrated in denoted points. The mass matrix shape becomes a diagonal one if only the translational freedom degrees are considered. The inertia mass matrix element is stated by equalities:

$$m_{ii} = m_i \; ; \; m_{ij} = m_{ji} = 0 \tag{1}$$

There is no need to specify the reference system, the lumped mass matrix keeps distinct numerical values only on the main diagonal.

The equivalent mass matrix is another alternative to consider inertia mass of a structure. It is computed on the base of the assumptions: the inertia mass of an element is uniformly distributed and the displacement of a current cross section is given by the translational displacement of the member ends. The influence coefficient of the mass located at the i-th node on the j-th freedom degree is determined by the calculus formula.

$$m_{ij} = \int m(x) N_i(x) N_j(x) dx \tag{2}$$

where $m(x)$ is uniformly distributed mass over the entire length of the element; $N_i(x)$, $N_j(x)$ are interpolating functions refered to the displacement of the current cross section of the beam on the i-th degree of freedom and the j-th degree of freedom, respectively. In a local coordinate reference system, the equivalent mass matrix is of 12 rows and 12 columns for a space beam having 6 degree at freedom at very end. The equivalent mass matrix is of 6 rows and 6 columns for a plane beam having 3 freedom degree at very end. Replacing the equivalent mass matrix in the global reference system needs the directory cosines matrix to be done.

Modal analysis has been performed on a plane cantilever and a space one, respectively. The beam is made of IPE 400 steel profile. It is discretized in 5 finite elements. Fig.1. Six natural vibrating modes are studied. The equivalent inertia mass matrix and the lumped inertia mass matrix have been considered. The period and modal inertia mass have been noted in Table 1 and Table 2 respectively.

3 Influence of stiffness matrix content of dynamic response

The dynamic response of a structure is find out by integrating the diferential equation system:

$$M\ddot{X} + C\dot{X} + KX = P(t) \tag{3}$$

where M is mass matrix, C is damping matrix, K is stiffness matrix, $P(t)$ is disturbing force vector. X, $\dot{X}$ and $\ddot{X}$ represent displacement, velocity and acceleration vector, respectively.

Dynamic stiffness of a structure can be expressed in the expended form as:

$$K = K_o + \omega^4 K_4 + \omega^8 K_8 + \dots \tag{4}$$

where K_o is static stiffness matrix, the other terms being the dynamic corrections. The first term of formula (4), K_o is always used in dynamic analysis of structures.

Performing the nodal analysis on the previous examples but vanishing each time one of the cross section characteristic, the natural period and the modal mass have been obtained for the first 6 oscillation modes.

Table 1 ilusstrates the modal analysis results for the plane steel cantilever at which the equivalent inertia mass matrix has been used. There is a great difference between the period of the real cantilever (having cross area A, shear area A_s) noted in the (7) th column and the periode at the suppositional cantilevers, written in the column. number: (1), (3) and (5) when the cross area, shear area or both areas are canceled.

Concerning the modal mass M^*, it should be emphasized a different modal mass distribution among the modes considered at the real cantilever. That is presented by the figures of the column number (8). New and changed modal mass distributions are noted for the suppositional cantilevers in the column number: (2), (4) and (6).

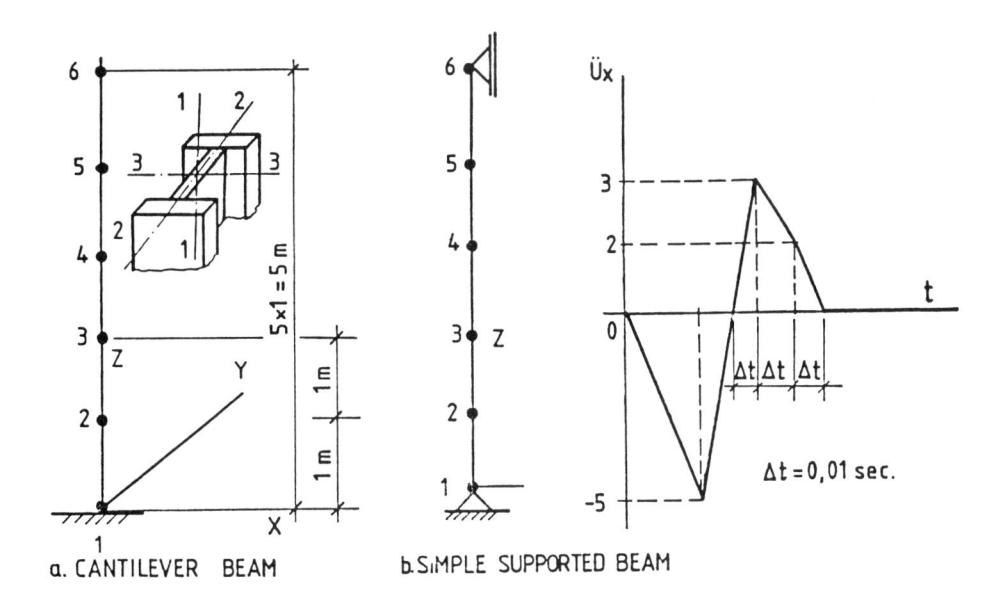

a. CANTILEVER BEAM   b. SIMPLE SUPPORTED BEAM

FIG.1 UNIDIMENSIONAL ELEMENT   FIG. 2 ACCELERATION DIAGRAM

The modal analysis results for the plane steel cantilever but considering lumped inertia mass matrix are presented in the Table 2. The above mentioned conclusions are valid for the numerical values of Table 2 included.

From the periode point of view there is insignificant difference between the numerical values obtained on the two ways: considering distributed inertia mass and lumped inertia mass. In the latter case the structure is slight more elastic than in the former one.

Table 1. Natural period and modal mass.

Mode number	Vanished characteristic						Not vanished characteristics	
	$A=A_s=0$		$A=0$		$A_s=0$			
	T(s)	$M^*(\%)$	T(s)	$M^*(\%)$	T(s)	$M^*(\%)$	T(s)	$M^*(\%)$
(0)	(1)	(2)	(3)	(4)	(5)	(6)	(7)	(8)
1	.89	87x	.29	67x	.39	78x	.29	67x
2	.82	88z	.17	88z	.12	18x	.05	21x
3	.31	10x	.06	8z	.06	3x	.02	7x
4	.28	8z	.05	21x	.05	1x	.01	3x
5	.19	2z	.04	2z	.04	.1x	.009	88z
6	.18	2z	.03	1z	.01	.9z	.008	.9x

Table 2 Natural period and modal mass

Mod number	Vanished characteristic						Not vanished characteristic	
	$A=A_s=0$		$A=0$		$A_s=0$			
	T(s)	$M^*(\%)$	T(s)	$M^*(\%)$	T(s)	$M^*(\%)$	T(s)	$M^*(\%)$
(0)	(1)	(2)	(3)	(4)	(5)	(6)	(7)	(8)
1	.94	87x	.26	67x	.35	78x	.257	67x
2	.86	88z	.15	89z	.11	18x	.044	21x
3	.32	10x	.05	9z	.06	3x	.017	7x
4	.28	8z	.04	21x	.04	.8x	.010	3x
5	.20	2x	.03	2z	.04	.2x	.008	88z
6	.19	2z	.03	.6z	.01	.01	.007	1x

If a space steel cantilever beam is considered the previous conclusions are stand out. Considering a space cantilever with distributed mass the values of period are noted in Table 3. The values of period for the same space cantilever but having lumped mass, are shown by Table 4.

Table 3   Natural period

Mod number	Characteristic vanished					Not vanished characteristic
	I_{1-1}; A_{2-2}; A_{3-3}	I_{1-1}	A_{2-2}	A_{3-3}	A	
(0)	(1)	(2)	(3)	(4)	(5)	(6)
1	1.615	.663	.663	.714	.663	.663
2	.898	.287	.391	.288	.287	.281
3	.824	.110	.118	.159	.170	.111
4	.552	.049	.110	.077	.110	.048
5	.340	.041	.065	.055	.058	.041
6	.309	.022	.050	.049	.049	.022

Table 4   Natural period

Mod number	Characteristic vanished					Not vanished characteristic
	I_{1-1}; A_{2-2}; A_{3-3}	I_{1-1}	A_{2-2}	A_{3-3}	A	
(0)	(1)	(2)	(3)	(4)	(5)	(6)
1	1.684	.898	.898	.966	.898	.898
2	.936	.389	.530	.389	.389	.389
3	.859	.249	.160	.2?5	.230	.149
4	.575	.149	.149	.105	.149	.067
5	.354	.086	.088	.074	.079	.056
6	.321	.067	.068	.066	.067	.030

4   Influence of stress state on dynamic response

The real structures undergoes static loads. Consequently, there is on initial stress state in very element when the dynamic loads, especially the seismic forces overload the bearing system.

The previous steel IPE 400 has been used again but as the plane simple supported beam shown in Fig.1b. The beam is acted on the ground acceleration shown in Fig.2. The external forces are considered sequentialy: axial forces, transversal force and applied bending moment. The forces are acting on very node. The horizontal displacements of node 3 is the primary aim of computing stage. They are written in Table 5.

Table 5. Deflection of node number 3.

Time	Applied load at node 2 up to 5					
	X=-10	X=10	Y=1000	Y=-1000	M=1000	M=-100
(0)	(1)	(2)	(3)	(4)	(5)	(6)
.02	21	23	44	44	94	7
.06	44	97	67	65	118	17
.10	43	87	66	61	116	15
.14	48	91	70	59	121	20
.18	46	94	69	49	120	19
.36	42	39	64	- 3	115	14
.40	42	87	63	-14	115	14
.44	42	87	63	-20	115	14
.48	40	86	61	-37	113	12
.66	35	81	55	-64	108	7
.70	34	79	53	-65	107	6
.74	32	78	52	-63	106	5
.96	24	70	41	-21	97	- 4

The results shown in Table 5 allow to emphasize the influence of different sort of loads to the flexural deformation of a plane beam. Only a great axial load modifies the magnitude of the flexural deformation - column (1) and (2). The horizontal load (perpendicular to the beam axes) increases greatly the deflection if the sens of the load coincides to the excitation sens, column (4). If sens of the force is against to the excitation, a decrease of the deflection is obtained - column (5). A similar conclusions is valid in the case of applied bending moment - column (7) and (8). The beam considered, looks like "a soften element" when it is of an acute stress state.

Fig.3. presents a plane frame of a real industrial plant, which belongs to a heating station. Two different situation are considered: 1) a time history process of the frame having only lumped masses; 2) a time history process of the real frame with lumped masses and loods. The same acceleration has been considered in the two mention cases. The relative displacement of nodes 1 and 42 located on the left stanchion at different instant are noticed in Table 6.

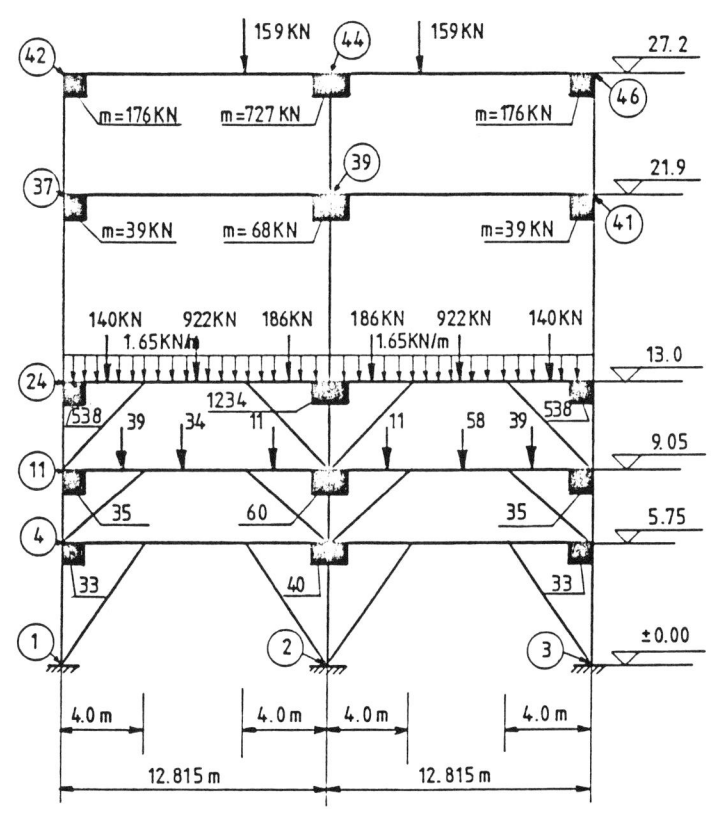

FIG.3 PLANE FRAME OF AN INDUSTRIAL PLANT

Table 6   Horizontal relative displacements of node 42

Time	Frame undergoes inertia mass only	Frame undergoes mass and loads
.025	-.0011	.0089
.050	-.0084	.0016
.075	-.00261	-.0161
.100	-.0500	-.0401
.125	-.0739	-.0639
.150	-.0912	-.0813
.175	-.0975	-.0876
.200	-.0965	-.0865

Table 7 presents the absolut displacement and rotation of the nodes located on the left stanchon at time .05 sec. The results out line the greater values at the real frame than the hypothetic one.

Table 7 Absolute displacements and rotation

Node	Frame undergoes inertia mass only			Frame undergoes mass and loads		
	x	y		x	y	
4	-.006	.000	.00057	-.007	-.001	-.0065
11	-.008	.000	.00030	-.009	-.001	-.0013
24	-.008	.000	.00009	-.008	-.001	-.0010
37	-.008	.000	-.0072	-.009	-.001	-.0077
42	-.008	.000	.0000	-.010	-.001	.0000

5 Conclusions

Using the inertia distributed mass matrix shows that a beam is slight more stiffner than the same beam if the lumped mass matrix is used. The higher degree of interpolating polinoms $N_i(x)$, $N_j(x)$ are used in equivalent mass matrix, the more approach value to the exact natural period is obtained.

If the cross section area is considered, an insignificant increase of the period value is attained at the first mode, the oscillating modes of higher orders are affected. A similar conclusion is lined out with respect to the use at shear area of cross section. Taking into account the both shear area and cross section a great decrease of period value has been resulted. This conclusion is valuable for very mode of oscillation. The both plane cantilever and space cantilever show the validity of the conclusion. Thus the real element is a stiffner one.

Concerning the stress state of on element or a structure is should be mentioned that an acute stress state makes the element or the structure to look softer than in the case of neglicting the stress state.

The seismic computing of structures provides a great value of acceleration spectrum to a stiffen structures while the least acceleration spectrum is destinated to the elastic ones.

Taking into account the chaotic and hazardous nature of the seismic motion, the stress state of the structure and the equivalent mass matrix have to be considered.

6 References

1. Cheng F.Y., Botkin M.E., Tseng W.H.(1970) - Matrix Calculations of Structural Dynamic Characteristics and Response. Proc. Conf. Earth. and Struc. Iassy.
2. Przemienieki J.S. (1968) - Theory of Matrix Structural Analysis Mc.Graw Hill.

8 RESEARCH ON SEISMIC BEHAVIOR OF STEEL AND COMPOSITE STRUCTURES AT LEHIGH UNIVERSITY

L.-W. LU, J. M. RICLES and K. KASAI
Lehigh University, Bethlehem, Pennsylvania, USA

Abstract
Many issues remain related to the response, design, and analysis of structural steel and composite steel-concrete structures under seismic loading. Several analytical and experimental studies have been pursued by the authors at Lehigh University, addressing major issues on the basic seismic behavior of these structures. These include the seismic performance of steel and composite frames, as well as the use of supplemental damping devices in steel frames to mitigate seismic damage.
<u>Keywords:</u> Analysis, Composite Construction, Experiment, Structural Steel, Supplemental Damping.

1 Introduction

This paper presents a description of several studies conducted at Lehigh University on the seismic performance of steel and composite structural frames. Significant results and their implications on current and future design are discussed.

2 Modeling of EBF shear links for seismic loading

Research has been conducted involving the development and verification of inelastic modeling of shear links in seismically resistant eccentrically braced frames (EBFs) (Ricles and Popov 1987b, 1994). In properly designed EBFs the links play a key role during seismic attack, yielding and dissipating energy while allowing the other structural elements to remain essentially elastic. In order to reliably predict the inelastic seismic response of an EBF, an accurate and efficient link element is therefore required. Short links yield predominantly in shear, with some flexural yielding also occurring, and hence are called shear links. Shear links are typically used in EBFs because they result in a smaller weight structure (Sec. 3).

A reliable link model for general random inelastic cyclic loading, such as that imposed to a link during an earthquake, was developed using a lumped plasticity stress-resultant formulation. The element consists of an elastic component in series with two rigid-plastic hinges at each end (see Fig. 1). Each hinge has several subhinges, which individually possess a concentric yield surface in moment-shear force space (see Fig. 2). The rigid-plastic action-deformation relationships for a series of subhinges combine to produce a multi-linear function with strain hardening for each hinge, and therefore the element. Mroz theory for yielding of metals is used to establish relationships between each hinge's action (forces) and deformation. These relationships

Behaviour of Steel Structures in Seismic Areas. Edited by F. M. Mazzolani and V. Gioncu.
Published in 1995 by E & FN Spon, 2-6 Boundary Row, London SE1 8HN. ISBN: 0 419 19890 3.

are utilized to establish the element's tangent flexibility matrix in element deformation coordinates, which is subsequently inverted and expanded to form the element's complete stiffness matrix. It has been observed in experiments (Ricles and Popov 1987a and 1989) that shear links develop a combined isotropic and kinematic strain hardening. Hence, this effect in addition to flexural kinematic strain hardening was included in the element's formulation. Kinematic hardening in each subhinge was based on Mroz strain hardening for metals, considering the relative alignment and positions of the yield surfaces of each consecutive subhinge. Isotropic shear hardening was formulated as being a function of the length of the plastic shear strain trajectory. More complete details of the element's formulation can be found in Ricles and Popov (1987b).

The element has been shown to accurately predict the response of both short (e.g. shear) and longer links. Shown in Fig. 3 is a comparison of the element's and experimental response of a shear link tested in the laboratory under cyclic loading by Kasai and Popov (1986). The shear-end displacement (V-d) relationship of the experimental test is shown to possess both kinematic and isotropic shear strain hardening, and the specimen's behavior accurately predicted by the element. Both the element and specimen developed shear and flexural inelastic response.

The link element has been used to conduct several parametric studies of EBFs with shear links subjected to seismic loading (Ricles and Popov 1987b; and Popov et al. 1992). Fig. 4(a) illustrates the extent of link shear force developed among floors in a 6-story EBF subjected to the 1940 El Centro accelerogram scaled by a 1.5 factor to produce a peak ground acceleration of 0.5g. Three analysis cases are shown, namely: 1) elastic-perfectly plastic modeling of links; 2) link modeling with strain hardening and EBF non-proportional viscous damping; and 3) link modeling with strain hardening and EBF Rayleigh proportional viscous damping. Envelopes of brace forces in the EBF model corresponding to the three analysis cases are shown in Fig. 4(b). These results illustrate the significance of proper link modeling, particularly the effects of including link strain hardening in EBF seismic analysis.

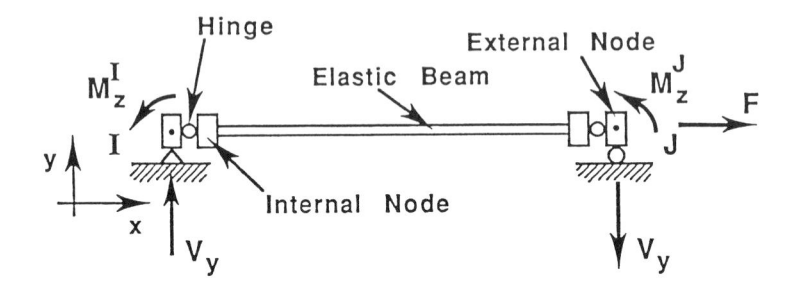

Fig. 1. Inelastic single-component link element

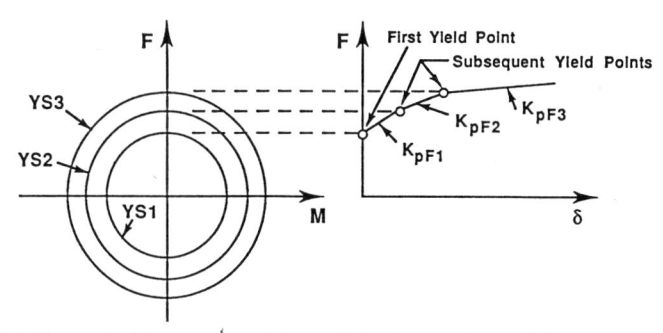

Fig. 2. Subhinge yield surfaces with resulting hinge force-deformation response

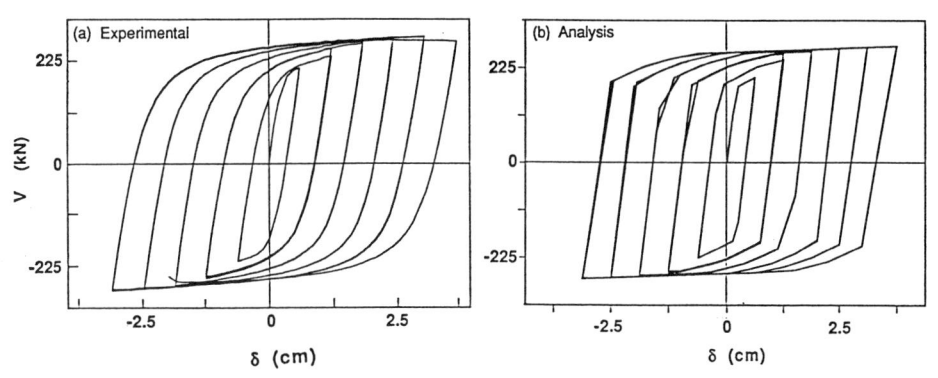

Fig. 3. Comparison of (a) experimental; and (b) analytical shear-deformation hysteretic response of link

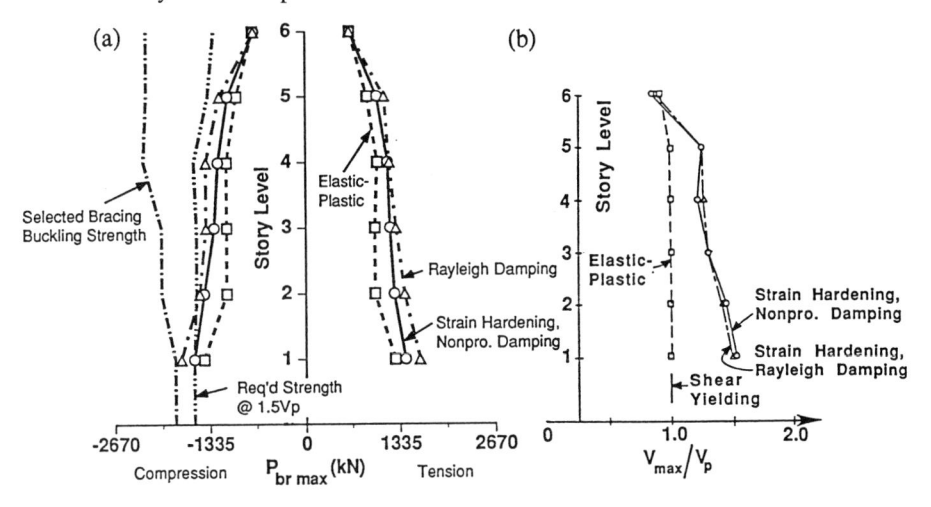

Fig. 4. (a) Axial brace force envelopes; and (b) normalized maximum link shears of a six-story EBF frame

3 Refined design methods for conventional and new eccentrically braced frames

Simplified step-by-step design method for a variety of EBFs is completed (Kasai and Goyal 1993). The method clarifies the unique EBF design philosophy, produces efficient EBF design, and reduces misunderstanding of EBF concept. It is applied to redesign the 6-story EBF real-size model tested in mid 1980's as Phase II of U.S.-Japan Cooperative Research Program. The redesigned EBF satisfying LRFD seismic requirements (1992) appears to have smaller steel sections. Using Ricles' shear link element (Sec. 2), the frames are dynamically analyzed. The newly designed EBF (with a concentric brace frame CBF at top) performs much better (Fig. 5) than the original U.S.-Japan EBF by distributing the link inelastic activities uniformly throughout the EBF height and avoiding concentration of excessive damage as well as drift at particular story levels that was observed during the U.S.-Japan experiment (Kasai and Wei 1994).

Also, more than 20 EBFs having different heights (Fig. 6(a)) and link lengths were designed using the simplified method. Each frame was elasto-plastically analyzed using static forces as well as several different earthquakes. The results indicate that the best economy (Fig. 6(b)) and seismic performance can be achieved by using the links having length of about 1/9 to 1/7 times the span length (Kasai and Goyal 1993).

The conventional EBFs discussed above use welded fully restrained (FR-) link-column connections. A new EBF with bolted partially restrained (PR-) connections (Fig. 7(a)) is being developed (Kasai and Popov 1991), and has significant advantages as follows: its yield capacity as well as stiffness, using relatively short links, become the same (Fig. 7(b)) as those of the FR-EBF; it is much more economical than FR-EBF; the concept can be applied to retrofit weak moment resisting frame (MRF) or gravity frame using PR-connections, and; the mode of failure of the PR-connection, if it occurs, is more gradual than that of FR-connection (many brittle weld failures occurred in FR-connections of MRFs during 1994 Northridge earthquake, Los Angeles). Full-scale tests of the PR-EBF and PR-MRF will be conducted at Lehigh (Sec. 8).

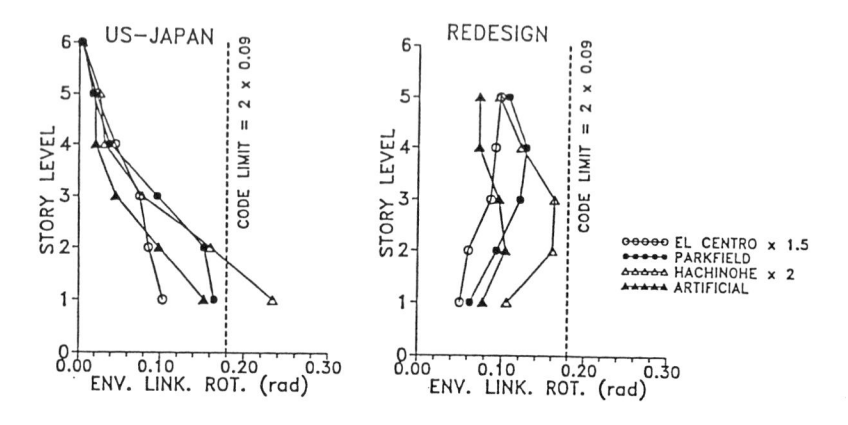

Fig. 5. Comparison of link deformation for U.S.-Japan frame and redesigned frame

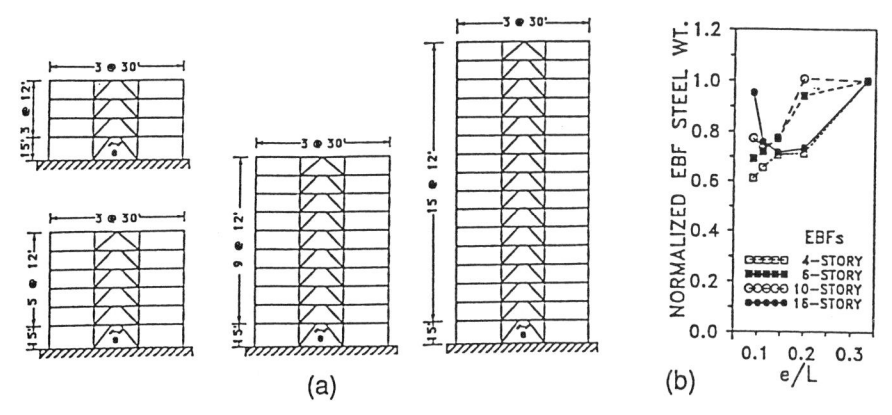

Fig. 6. (a) Elevation of designed EBFs, and (b) comparison of their steel weights

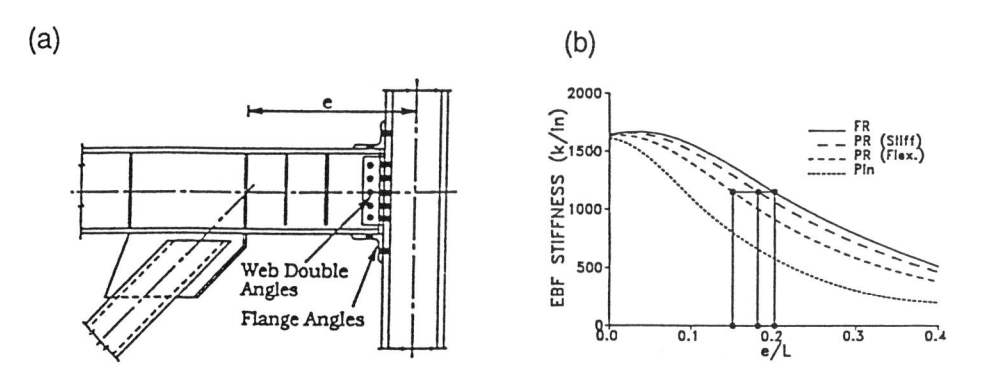

Fig. 7. (a) PR-connection detail, and (b) EBF lateral stiffness with various connections

4 Steel frames with infilled plate panels

Steel frame structures incorporating infilled plate panels have shown definite economic and performance advantages over the conventional rigid frame structures, especially when the structures are required to resist large lateral loads due to earthquake motions. Multi-story buildings with steel shear panels have been constructed in such high seismic regions as Japan and the West Coast of the U.S. The Olive View Medical Center, rebuilt after the 1971 San Fernando Valley earthquake, is an example. It experienced horizontal accelerations as high as 2.31g (g = acceleration of gravity) during the January 17, 1994 Northrige earthquake, but showed no sign of structural damage. It is, however, recognized that current approaches of designing plate panels, which allows no shear buckling, are excessively conservative. Furthermore, without complete knowledge of the behavior of panels in the entire range of loading, it is impossible to evaluate the true seismic performance of a panel-strengthened structure. For this reason, a research project was initiated at Lehigh University to develop the needed knowledge and to formulate a rational design approach. The overall research consists of the following five interrelated studies: (1) frame-panel interaction and selection of the "most suitable" panel connection arrangement, (2) monotonic behavior of individual panels and development of formulas for predicting their shear strength and stiffness, (3) hysteretic behavior of panels and development of empirical hysteretic models, (4) implementation of the hysteretic model in seismic time-history analyses, and (5) evaluation and assessment of frame-panel systems and formulation of design provisions. Selected results obtained in the first three studies are presented here and more complete information can be found in papers by Xue and Lu (1994a and 1994b).

A detailed study of the effect of panel connection arrangement on the strength and stiffness of frame-panel systems was carried out. Three different arrangements connecting the panels to framing members were considered: connecting to both girders and columns, only to girders, and only to columns. In addition, the connection (simple vs. full moment) between columns and girders of the infilled bay was also included as a variable. When the different arrangements are judged on the basis of structural performance and construction efficiency, the arrangement shown in Fig. 8 has been found to be the most desirable. Note that the frame-panel system shown constitutes only part of a multi-story, multi-bay structure, in which full-moment connections are used to connect girders to columns in the bays without panels.

The strength and load-deformation characteristics of the frame-panel system were studied in details, including large post-buckling deformation and yielding. The principal parameters were the width-to-thickness ratio and the panel aspect ratio. Figure 9 shows some typical results and the key structural properties defining the behavior of a frame-panel system. Simple formulas have been developed to provide predictions of the post-buckling stiffness, significant yield point and post-yield stiffness. A special finite element analysis program was prepared for analyzing panels subjected to cyclic shear. Based on the results of the monotonic and cyclic analyses, a hysteretic model has been proposed for the frame-panel systems. A comparison of predictions by the hysteretic model and finite element analysis for a sample panel is shown in Fig. 10. Further work on cyclic load analysis and hysteretic modeling is underway.

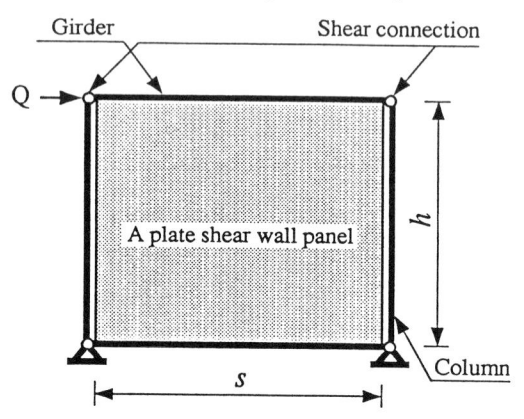

Fig. 8. Connection arrangement

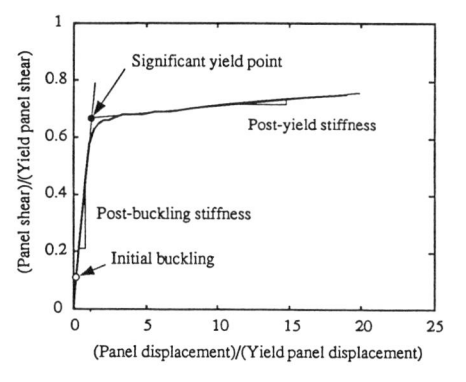

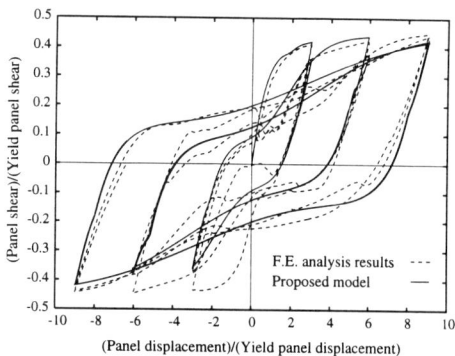

Fig. 9. Characteristics of shear panel   Fig. 10. Hysteretic behavior of shear panel

5 Seismic performance of composite beam-columns

Several large-scale experimental studies have been conducted at Lehigh involving composite steel sections encased in concrete beam-columns (Ricles and Paboojian 1992, 1994). The cross-section of a typical specimen is shown in Fig. 11, consisting of a W-shape encased in a square reinforced concrete section. These columns would exist in the perimeter of a steel MRF, in which the columns have been encased in reinforced concrete to increase the building's lateral stiffness and strength. The columns at the base of these structures are exposed to the largest forces, where the bottom end of the ground floor columns are susceptible to inelastic deformations. The objective of these studies has been to assess the effects of transverse reinforcement, concrete strength (31 MPa to 69 MPa compressive strength), flange shear studs, and the longitudinal reinforcement on column behavior under combined axial and lateral cyclic loading. In addition, the experimental results provided the opportunity to assess the reliability of several analysis methods for predicting composite column capacity. Each test specimen involved an individual column which was subjected to combined axial and lateral loading.

The hysteretic behavior of two specimens (Specimens 1 and 3) is shown in Fig. 12, which illustrates their lateral load-displacement response (H-D). Specimens 1 and 3 both consisted of Detail A, shown in Figure 11, however the transverse reinforcement that surrounded the column in Specimen 1 was spaced at 127 mm along the column, as opposed to 95 mm in Specimen 3. In addition, Specimen 1 was 2489 mm in height whereas Specimen 3 was 1930 mm. The larger transverse reinforcement spacing of Specimen 1 caused it to develop buckling of its longitudinal reinforcement near the base of the column, following the spalling of the outer core of concrete. This buckling led to a loss of the inner concrete core and local buckling of the flanges of the W-shape, which in turn resulted in a loss of member capacity and ductility μ (μ is defined in Fig. 12, where D_{max} and D_y are horizontal and yield displacement, respectively). The closer transverse reinforcement spacing in Specimen 3 enabled it to achieve a greater cyclic strength and ductility due to the inner concrete core remaining intact.

The result of several tests has indicated that the current ACI provisions (1992) for reinforced concrete provide good seismic resistant details for composite beam-columns, as long as the spacing of the transverse reinforcement is limited to 6 longitudinal reinforcement bars diameters. This was also found to be consistent with composite columns having high strength concrete. The experimental program also showed that shear studs placed on the flanges of the W-shape were found to be ineffective in enhancing the stiffness and strength of the composite column. Other results obtained from tests indicated that prior to developing concrete cracking ($\mu < 0.7$) the concrete (V_c) resisted a majority of the applied shear V. However, with greater ductility demand as expected during an earthquake ($\mu > 1.0$), the steel W-shape (V_s) resisted almost all of the applied shear. This phenomenon is illustrated in Fig. 13 for Specimen 1.

The capacities of the specimens subjected to combined axial and lateral loading were found to be accurately predicted by fiber models based on full strain compatibility and which accounted for concrete strength and ductility enhancement due to confinement by the transverse steel reinforcement. Strength prediction based on superimposing the capacities of the steel W-shape and reinforced concrete portions of the column was also found to be reasonably accurate (Ricles and Paboojian, 1992). The ACI method (1992) and AISC LRFD (1986) provisions were found to be conservative, particularly LRFD which had an experimental-to-predicted capacity ratio average of 1.48; as opposed to 1.19, 1.08, and 1.02 for ACI, the superimposed strength method, and fiber analysis results, respectively.

Research is currently continuing, involving testing and analysis of column concrete filled tubes (CFTs) and column CFT-to-steel beam connections under seismic loading. The objective of this study is to assess the affects of various parameters on the cyclic behavior of these elements and to develop seismic design provisions.

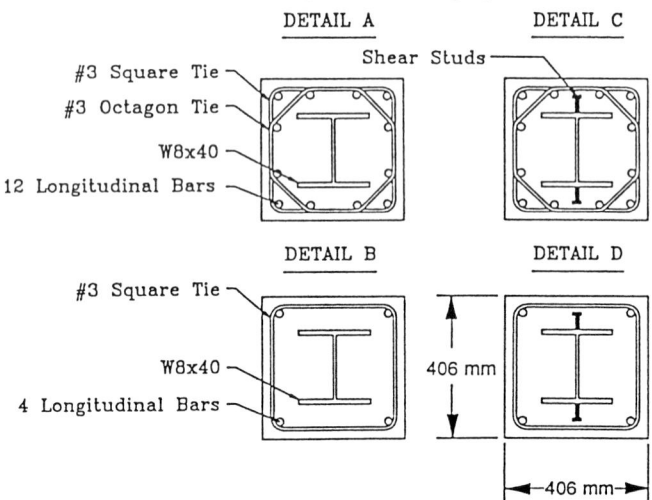

Fig. 11. Composite column cross section details

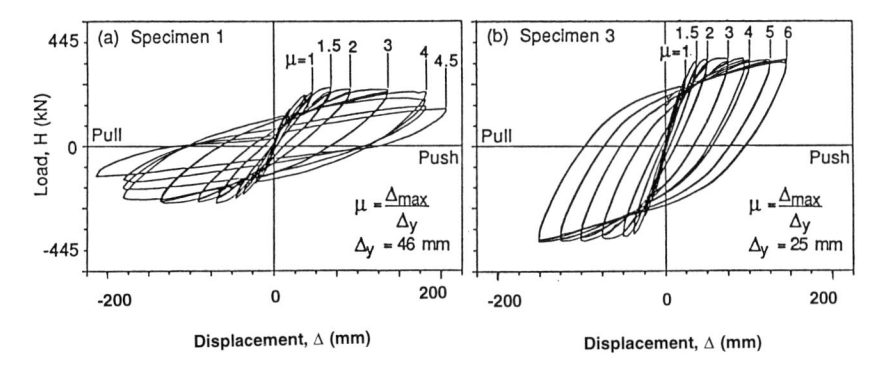

Fig. 12. Hysteretic response of composite columns subjected to lateral cyclic loading

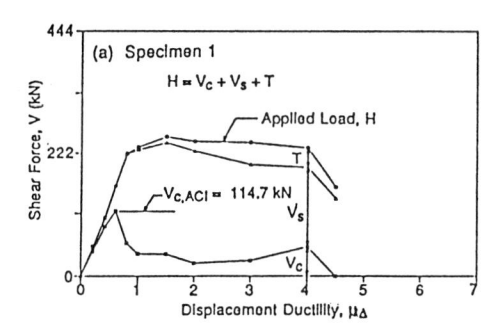

Fig. 13. Transverse shear force distribution in column at first peak of each displacement ductility

6 Seismic response of composite frames

A type of composite framing system consisting of steel beams (acting compositely with a metal-deck reinforced slab) and encased composite columns (Fig. 14) has been recognized as a viable alternative to the conventional steel or reinforced concrete system for high-rise construction. The composite system has the desired characteristics of the conventional systems, such as high strength, stiffness, ductility and fire resistance, and has been found to be cost effective for a range of building heights and under certain design conditions. In Japan, the system has been used as a seismic-resistant system in buildings.

Only limited studies have been conducted to investigate the performance of such a system under seismic ground excitation. Before the overall response of the frame can be accurately analyzed, it is first necessary to develop a full understanding of the cyclic behavior of its individual components and to be able to represent the behavior by analytical models. Three types of components have been studied in detail in a recent investigation: encased beam-column, deck-supported steel beam, and beam-to-column joints. A nonlinear cyclic composite column analysis program, based on the fiber concept, has been developed and validated by comparing the predicted responses of 38 test columns with the experimental results (Kim and Lu, 1994). Hysteretic models have also been proposed for seismic response analysis. Similarly, hysteretic models for deck-supported composite beams have been developed to include the nonlinear effects of stiffness degradation, pinching, and strength deterioration (Kim, 1991). A major research effort was on the hysteretic modeling of the shear vs. angular deformation relationships of composite joints. This work resulted in simple representations of joint behavior for both monotonic and cyclic loading (Kim and Lu, 1992). The component models developed have been implemented in a dynamic frame analysis program (Kim and Lu, 1993), and Fig. 15 shows sample results of analyses performed on a six story composite frame, using the first 10 second of 1952 Taft earthquake record (S21W component, scaled up to a peak acceleration of 0.6g). The effect of frame instability was included in the analyses. The research also examined, in detail, the general issues involved in developing the dynamic response factor for composite structures within the framework of the current U.S. codes.

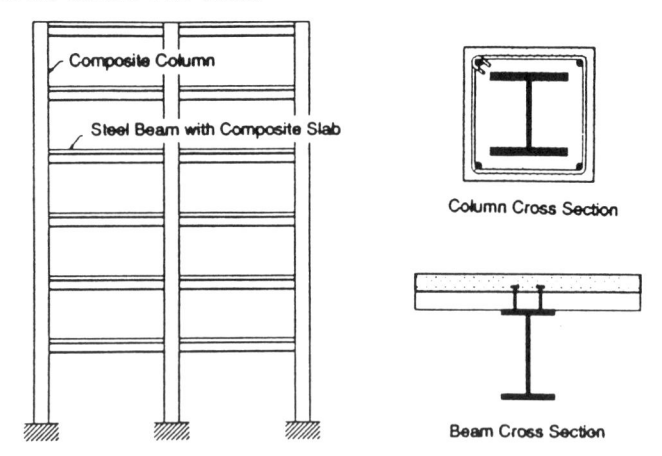

Fig. 14. Type of composite building frame studied

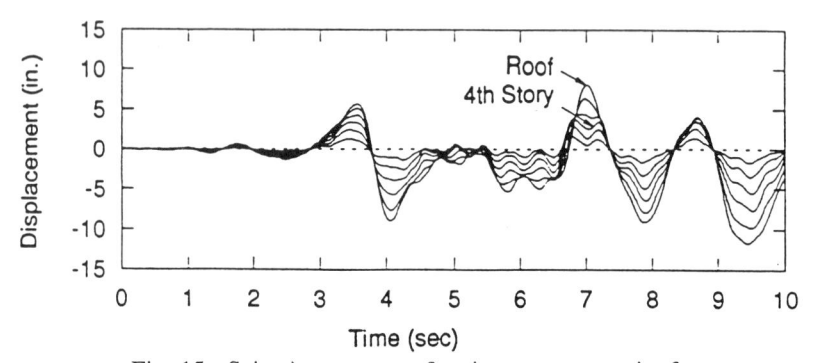

Fig. 15. Seismic response of a six-story composite frame

7 Damage-free steel frame with viscoelastic dampers

Analytical element to simulate the nonlinear cyclic behavior of viscoelastic (VE-) damper is developed (Kasai et al. 1993). Effects of excitation frequency as well as temperature on stiffness and damping of VE-damper are accurately modeled. Temperature rise of the damper due to cyclic energy dissipation causes non-linear response of the VE-material, and is predicted using the thermo-mechanics principle. Analytical result correlate well with the test result (Fig. 16) (Kasai et al. 1993, 1994a). Using the analytical element, systematic study on seismic application of VE-damper has been conducted at Lehigh. The topics include: accuracy evaluation of a conventional elastic modal analysis procedure for an analysis of VE-frame (Munshi and Kasai 1994); behavior of VE-frame with yielding members (Kasai and Munshi 1994b); retrofit of weak steel frame using VE-dampers (Maison and Kasai 1994); simplified seismic response prediction of VE-frame (Sause, Hemmingway, and Kasai 1994); special VE-frame design method to minimize the temperature effect (Kasai and Lai 1994), comparative study on the effects of metal inelastic damper and VE-damper; use of VE-damper to mitigate collision damage of adjacent buildings (Kasai and Munshi 1993), and; a design guideline for VE-frames.

Substantial reduction of drifts and inelastic deformation of an existing weak steel frame is possible when retrofitted by VE-dampers (Fig. 17). Likewise, new steel VE-frame properly designed by minimizing temperature sensitivity can behave elastically, developing small drifts even against a catastrophic earthquake. Such a VE-frame has substantial advantages over the conventional steel frames that are typically designed to behave inelastically during the overload. With relatively smaller amount of steel, the VE-frame protects structural as well as non-structural components even under the major seismic event, well reserving the buildings' socioeconomical values that have increased significantly lately. See Sec. 8 for experimental study of the VE-frame.

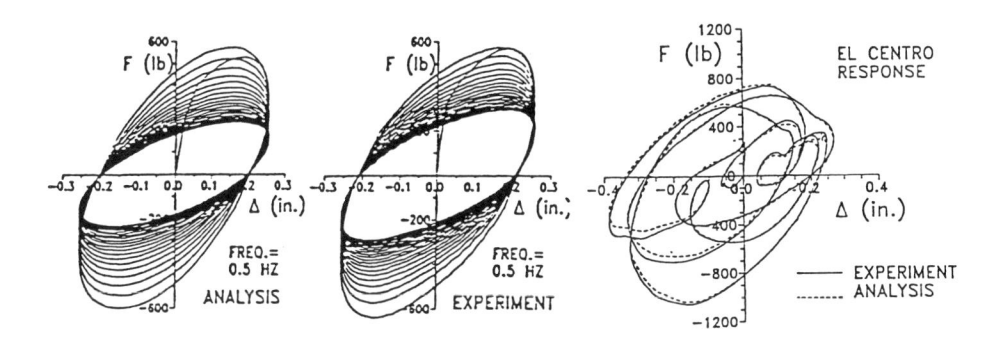

Fig. 16. Comparison between analysis and test for VE-damper hysteresis

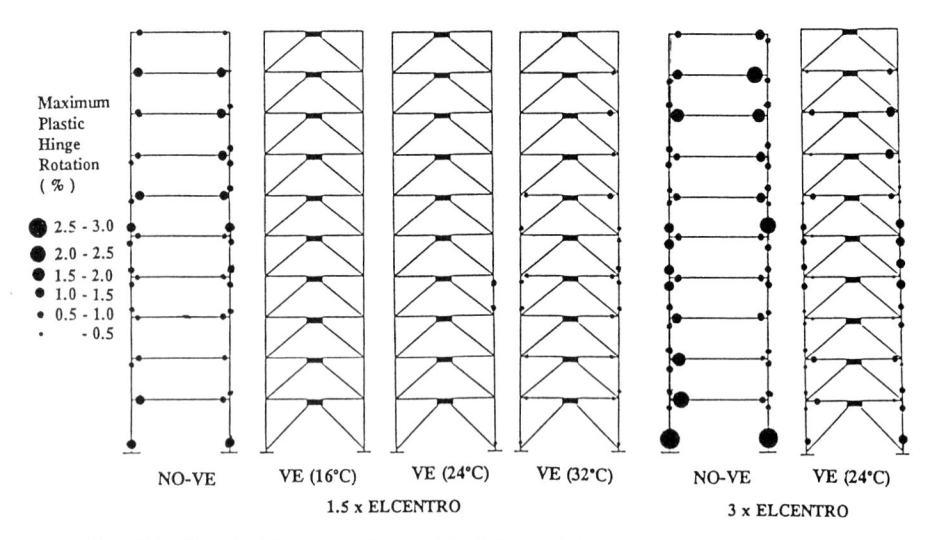

Fig. 17. Plastic hinge rotations of building with and without VE-dampers

8 Full-scale tests for attractive steel frames

Extensive experiments for a variety of real-size steel frames have begun at the Center for Advanced Technology for Large Structural Systems (ATLSS), Lehigh University. Each test frame represents the bottom 3-story of a 10-story building frame (Fig. 18).

The frames will be subjected to the simulated dead, live, earthquake, and wind loads having extremely large magnitudes. The frame is loaded using the strong reaction wall that is 50 feet high and the largest in U.S. Up to 80% of the wall height will be used to load and support the test frames (Fig. 18).

The frames considered are: (1) VE-frame (Sec. 7); (2) PR-MRF; (3) PR-EBF (Sec. 3); (4) MRF with ATLSS connections developed for automated construction; (5) CBF with non-buckling braces; (6) CBF with conventional braces, and; (7) other frames receiving a substantial interest. Frame-dependent unresolved key design issues are addressed, by providing unique realistic test and analysis data for the performance of members, connections, and overall systems.

9 Summary

This paper presented an overview of several studies conducted at Lehigh University aimed at acquiring knowledge of the basic behavior of structural steel and composite components and systems under seismic attack, through investigations involving experimental and analytical effort.

Several of these studies have included parametric studies that have resulted in the formulation and recommendation of design criteria. A number of these recommendations have been adapted and incorporated into current seismic building codes.

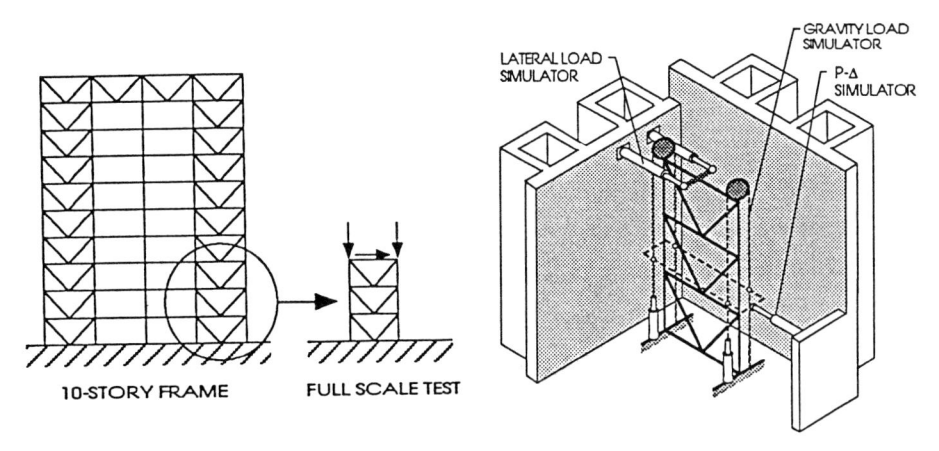

Fig. 18. Bottom 3-story segment of 10-story frame and test set-up

10 Acknowledgments

The authors of this paper would like to acknowledge the sponsors of the studies from which the results of this paper were presented. The continued interest and sponsorship of the National Science Foundation, Dr. S.C. Liu and Dr. K. Chong cognizant program officials, and the Engineering Foundation, Mr. C.V. Freiman Director, is gratefully appreciated. The opinions expressed herein are those of the authors and do not necessarily reflect the views of the sponsors.

11 References

Building Code Requirements for Reinforced Concrete (ACI 318), (1992) ACI, Committee 318, American Concrete Institute, Detroit, Michigan.

Kasai, K. and Popov, E.P. (1986) General Behavior of WF Steel Shear Link Beams, **J. of Str. Eng.**, ASCE, Vol. 112, No. 2.

Kasai, K. et al. (1993) Viscoelastic damper hysteresis model: Theory, experiment, and application, **ATC 17-1 Semin.: Seis. Isol., Pass. Energ. Dissip., and Act. Control**, App. Tech. Council (ATC), Vol.2, pp.521-532, March 11-12, San Francisco, CA.

Kasai, K., Munshi, J., and Maison, B.F. (1993) Viscoelastic dampers for seismic pounding mitigation, **Proc. '93 Structural Congress**, ASCE, Vol.1, pp.730-735, April 19-21, Irvine, CA.

Kasai, K., and Goyal, A (1993) Link length design and EBF seismic performance, **Proc, '93 Structural Congress**, ASCE, Vol.1, pp.397-402, Apr. 19-21, Irvine, CA.

Kasai, K., and Lai, M.L. (1994) Finding of temperature insensitive viscoelastic frames, **Proc. 1st World Conf. on Str. Contrl**, August 3-5, Los Angeles, CA.

Kasai K., and Munshi, J. (1994a) Application of viscoelastic dampers to high-rise buildings", discussion to appear in **J. of Str. Eng.**, ASCE.

Kasai, K., and Munshi, J. (1994b) Seismic response of viscoelastic frame with yielding members", **Proc. 5th Nat. Conf. on Earthq. Eng.**, EERI, July, Chicago, IL.

Kasai, K. and Popov, E.P. (1991) EBFs with PR flexible link-column connection, **Proc. '91 Structural Congress**, ASCE, Apr. 29 - May 1, Indianapolis, Indiana.

Kasai K., and Wei, B. (1994) Redesign and analysis of US-Japan EBF, **Proc. 50th Anniversary Conf.**, Struct. Stab. Res. Council, Lehigh Univ., June, Bethlehem, PA.

Kim, W. (1991) Seismic response analysis and design of composite building structures, **Ph.D. dissertation**, Lehigh University, Bethlehem, Pennsylvania.

Kim, W. and Lu, L.W. (1992) Hysteretic analysis of composite beam-to-column joints, **Proc., 10th World Conf. on Earthq. Eng.**, Madrid, Spain, pp.4541-4546.

Kim, W. and Lu, L.W. (1993) Cyclic lateral load analysis of composite frames, in **Composite Construction in Steel and Concrete II**, edited by W.S. Easterling and W.M.K. Roddis, American Society of Civil Engineering, pp. 366-381

Kim. W. and Lu, L.W. (1994) Hysteretic analysis and modeling of composite beam-columns, to appear in **Proc., 5th National Conf. on Earthq. Eng.**, Chicago, Illinois.

Load and Resistance Factor Design, (1986, 1992) American Institute of Steel Construction, Manual of Steel Construction, Chicago, Illinois.

Maison, B.F., and Kasai, K. (1994) Study of building retrofit using viscoelastic dampers", **Proc. 5th Nat. Conf. on Earthq. Eng.**, EERI, July, Chicago, IL.

Munshi, J., and Kasai, K. (1994), Modal Analysis Procedures for Viscoelastic Frames, **Proc. 5th Nat. Conf. on Earthq. Eng.** EERI, July.

Popov, E.P., Ricles, J.M., and Kasai, K. (1992) EBF Design Methodology for Optimized Inelastic Link Behavior, **Proc. of the 10th World Conf. on Earthq. Eng.**, Vol. 7, Madrid, Spain.

Ricles, J.M. and Popov, E.P. (1987a) Experiments on Eccentrically Braced Frames with Composite Floors,**Earthq. Eng. Res. Ctr. Rpt. No. 87/06**, U. of Calif., Berkeley,CA.

Ricles, J.M. and Popov, E.P. (1987b) Dynamic Analysis of Seismically Resistant Eccentrically Braced Frames, **Earthquake Engineering Research Center Report No. 87/06**, Univ. of California, Berkeley, CA.

Ricles, J.M. and Popov, E.P. (1989) Composite Action in Eccentrically Braced Steel Frames, **J. of Str. Eng.**, ASCE, Vol. 115, No. 8.

Ricles, J.M. and Paboojian, S.D. (1992) Seismic Performance of Composite Beam-Columns, **ATLSS Report No. 93-1**, Lehigh University, Bethlehem, PA.

Ricles, J.M. and Paboojian, S.D. (1994) Seismic Performance of Steel Encased Composite Columns, **J. of Str. Eng.**, ASCE, Vol. 120, No. 8.

Ricles, J.M. and Popov, E.P. (1994) Inelastic Link Element for EBF Seismic Analysis, **J. of Str. Eng.**, ASCE, Vol. 120, No. 2.

Sause, R., Hemmingway, G., and Kasai, K. (1994), Simplified Seismic Response Analysis of Viscoelastic-Damped Framed Structures, **Proc. 5th Nat. Conf. on Earthq. Eng.**, EERI, July.

Xue, M. and Lu, L.W. (1994a) Monotonic and cyclic behavior of infilled steel shear panels, to appear in **Proceedings, Seventeenth Czech and Slovak International Conference on Steel Structures and Bridges**, Bratislava, Slovakia, September.

Xue, M. and Lu, L.W. (1994b) Interaction of steel plate shear panels with surrounding frame members, to appear in **Proc., 50th Anniversary Conference of Structural Stability Research Council**, Lehigh University, Bethlehem, PA, U.S.A., June.

9 QUALITY CONTROL OF MATERIAL PROPERTIES FOR SEISMIC PURPOSES

B. CALDERONI and F. M. MAZZOLANI
Istituto di Tecnica delle Costruzioni, University of Naples, Naples, Italy
V. PILUSO
Department of Civil Engineering, University of Salerno, Italy

Abstract
In this paper, the requirements for the quality control and the acceptance of structural steels are examined. The attention is focused on the statistical tests regarding the conformity between the manufacturer internal tests and the official tests. The consequences of the conformity requirement are discussed through the numerical simulation of the inelastic static response of some frames, including the effects of the random material variability. The analysis method is the Montecarlo simulation.
<u>Keywords:</u> Quality Control, Probabilistic Methods, Inelastic Behaviour

1 Introduction

Seismic-resistant steel structures are usually designed relying on their ability to sustain plastic deformations. Therefore, the earthquake input energy is dissipated through the hysterethic behaviour of the material. However, the plastic deformations have to be limited to values compatible with the structural local and global ductility in order to prevent collapse and to safeguard the human lives.

As the available global ductility is strictly related to the collapse mechanism, assuming the maximum value in case of global type mechanism, the modern seismic codes, such as ECCS Recommendations (1988) and EC8 (1993) provide simplified design criteria in order to control the failure mode.

Under this point of view, the randomness of yield strength of members plays a very important role, affecting the plastic hinge formation process and the ultimate behaviour of structures. In particular, it always leads to an energy dissipation capacity and to an available ductility different from the predicted ones.

As soon as the coefficient of variation of the material yield strength is concerned, it can be recognized that the occurrence probability of local failure modes, which reduce both ductility and energy dissipation capacity, increases as far as the coefficient of variation increases. Therefore, it is clear that the random variability of the material properties has to be limited in order to improve the reliability of the inelastic behaviour, as it is usually predicted by means of deterministic analyses.

The study of the inelastic behaviour of steel frames with random material variability has been already faced by different authors (Kuwamura and Kato, 1989; Kuwamura and Sasaki, 1990; Elnashai and Chryssanthopoulos, 1990). In (Kuwamura and Kato, 1989; Kuwamura and Sasaki, 1990) the statistical simulation of the load bearing capacity of a 6 storey-3 bay frame is presented and the influence of the random material variability has been pointed out for different values of the coefficient of variation of the material yield stress. On the base of the obtained results, the advantages of a more severe quality control procedure, which produces a reduction of the variability range of the material mechanical properties, are demonstrated.

Similar conclusions are obtained in (Elnashai and Chryssanthopoulos, 1990), even though they are based upon the analysis of a simple portal frame.

The «weak beam-strong column» design philosophy represents the case in which the plastic hinges formation process has the most decisive influence on the ultimate behaviour, so that the random variability of material properties can lead to significant effects on the design criteria, which state the column-to-beam strength ratio.

For this reason, the influence of the random material variability on the failure mode of frames has been investigated in (Mazzolani, Mele and Piluso, 1991a and 1991b) by considering a frame designed in order to obtain, on the base of the nominal yield strength of the material, a global mechanism. The Authors have evidenced that, for such frames, the random material variability leads to a reduction of the available ductility and to an increase of the ultimate resistance under horizontal forces. As the second effect is predominant with respect to the first one, the Authors have found an improvement on the structural ability to withstand severe earthquakes.

In this paper a different aspect, related to the random material variability, is examined. The attention is focused on the statistical requirements which are currently used in the material certification activity. In particular, the consequences of the restrictions to the differences between the results of official tests and manufacturer internal tests are investigated. Particular reference is made to the italian recommendations for material acceptance, which are particularly severe, independently of the kind of application (structures in seismic as well as in non seismic areas).

2 Acceptance requirements of structural steels in Italy

The present Italian technical code (D.M. 14/2/92) regarding the quality control and acceptance requirements of structural steels appears very restrictive. In Italy, steels of different grades can be used only if they are «qualified», i.e. the entire production process, and not only each separated supply, is to be submitted to a continuous quality control.

This assures the maintenance of mechanical properties meeting the requirements established by law. Yield stress, ultimate tensile strength, coefficient of yield stress variation, toughness, ultimate elongation and chemical composition are the material features to be checked by quality control.

In order to obtain the qualification for his products, a manufacturer (italian as well as foreigner) has to:
- demonstrate the reliability of the production process;
- perform continuous «internal tests» on the usual production to verify, with statistical tools, that it fulfils the acceptance requirements;
- submit the production site to periodical visits (two times a year) carried out by controllers of an Official Laboratory; during this visit, for each kind of product to be qualified, at least thirty additional specimens are tested (official tests).

The fulfilment of all acceptance requirements both for internal and official tests is necessary, but not sufficient for acquiring and keeping qualification. In fact, with reference to yield stress and ultimate strength, the results of internal tests must be also compared with the ones of official tests in order to prove their statistical conformity, i.e. the two sets of specimens have to strictly belong to the same population.

In order to carry out the conformity control, the Student's and the Snedecor's tests have to be performed. The Student's test compares the two series of data by means of the significance of the difference between the averages of the two series, while the

Snedecor's test is based on the significance of the difference between the variances of the two series.

By accepting a probability level of 99% (the least restrictive allowed by the Code) the production fulfils the requirements as:

$$u_{st} = (\bar{x}_1 - \bar{x}_2) \left(\frac{s_1^2}{n_1} + \frac{s_2^2}{n_2} \right)^{-1/2} < 2.58 \qquad (Student's\ test) \qquad (1)$$

$$u_{sn} = (s_1^2 - s_2^2) \left(\frac{2\, s_1^4}{n_1} + \frac{2\, s_2^4}{n_2} \right)^{-1/2} < 2.33 \qquad (Snedecor's\ test) \qquad (2)$$

where $\bar{x}$ is the average and s^2 the variance of the sample population. Subscripts 1 and 2 indicate internal and official tests (or viceversa).

3 Remarks on the statistical controls required by Italian Code

On the basis of the expertise gathered in the last decades by the «Istituto di Tecnica delle Costruzioni» of the University of Naples during its testing and certification activity regarding the structural steels produced in various countries of the European Community, remarkable difficulties have been found in the fulfilment of the statistical conformity requirement between internal and official tests, as requested by the Italian Code.

Even small discordances between the averages and/or the variances of the two data sets may lead to u_{st} and u_{sn} values exceeding the limits established by the code. It may happen that, even though the samples of a given material have passed both internal and official tests, the material cannot be qualified, because the average or the variance of the two tests are slightly different. Thus, a misleading feeling of steel bad quality may result. It is worth noticing that this may happen even if the material is satisfactory from a structural point of wiew, because all the mechanical properties of the single samples meet the Code requirements.

In Figure 1, some typical gaussian curves corresponding to results of internal and official tests are shown. They refer to the yield stress measured during the tests performed on some steel products. The statistical conformity tests (Student and Snedecor) applied to these numerical results are not satisfactory according to the italian code, because they do not fulfil its requirements given by equations (1) and (2). It is due to very small discordances in the values of sample means and/or of sample variance.

We believe that tests suggested by the code are general purpose statistical tests for quality control, but they set too strict limits when they are applied for the control of the «steel production reliability», because they do not take into account some aspects.

In fact, differences between official and internal tests do not necessarily represent a lack of uniformity in the production process. It has to be taken into account:
- the great difference between the number of specimens of internal tests (up to thousand or more) and of official tests (usually thirty specimens);
- the width of the range of the thickness categories ($t \leq 16$ mm; 16 mm $< t \leq 40$ mm; $t > 40$ mm);
- the association of different kind of steels (grades A/B/C) in an unique class, that is usually preferred by the manufacturer in order to reduce qualification burden;
- the existence, in a group of a given steel resistance class, of specimens belonging to a

Fe510 B 16 mm < t ≤ 40 mm

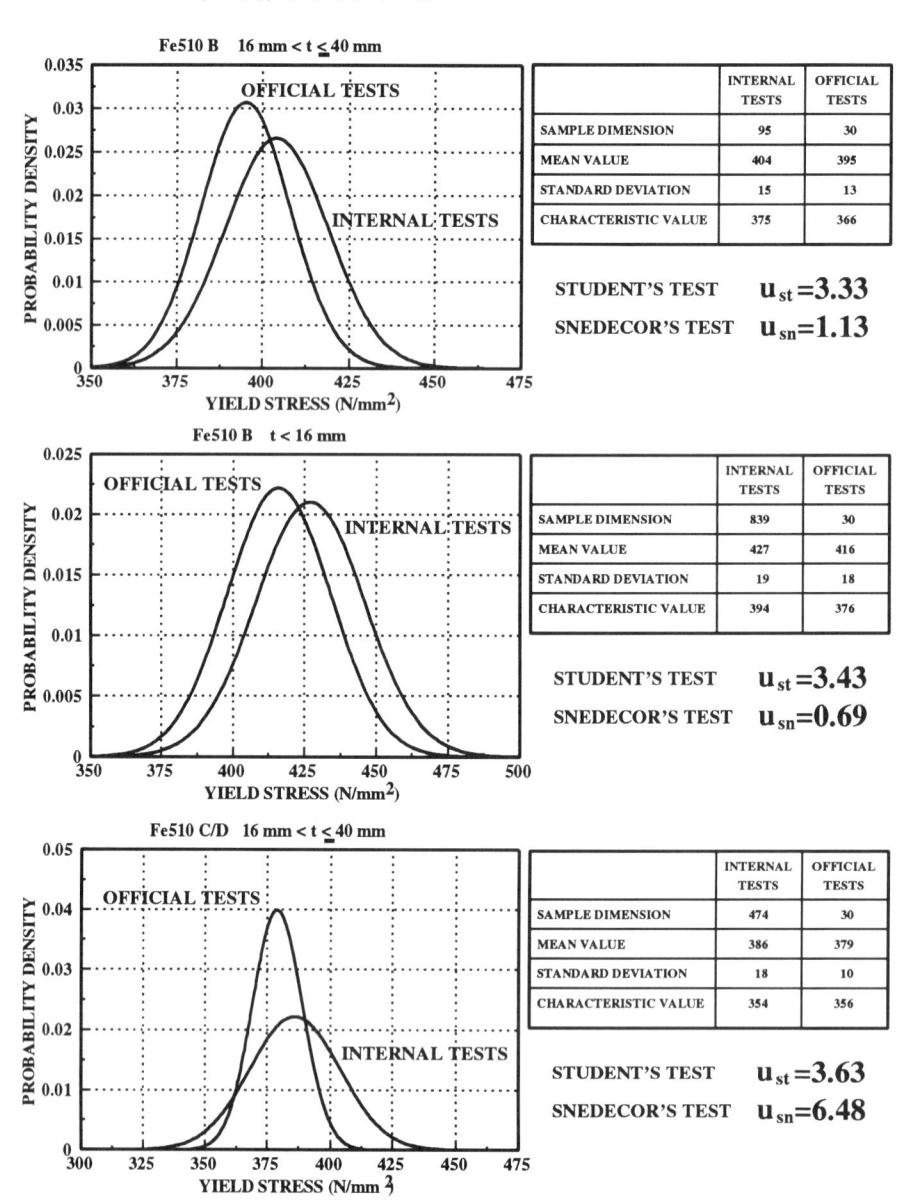

	INTERNAL TESTS	OFFICIAL TESTS
SAMPLE DIMENSION	95	30
MEAN VALUE	404	395
STANDARD DEVIATION	15	13
CHARACTERISTIC VALUE	375	366

STUDENT'S TEST $u_{st} = 3.33$
SNEDECOR'S TEST $u_{sn} = 1.13$

Fe510 B t < 16 mm

	INTERNAL TESTS	OFFICIAL TESTS
SAMPLE DIMENSION	839	30
MEAN VALUE	427	416
STANDARD DEVIATION	19	18
CHARACTERISTIC VALUE	394	376

STUDENT'S TEST $u_{st} = 3.43$
SNEDECOR'S TEST $u_{sn} = 0.69$

Fe510 C/D 16 mm < t ≤ 40 mm

	INTERNAL TESTS	OFFICIAL TESTS
SAMPLE DIMENSION	474	30
MEAN VALUE	386	379
STANDARD DEVIATION	18	10
CHARACTERISTIC VALUE	354	356

STUDENT'S TEST $u_{st} = 3.63$
SNEDECOR'S TEST $u_{sn} = 6.48$

Fig.1
Examples of theoretical gaussian curves corresponding to the results of internal and official tests

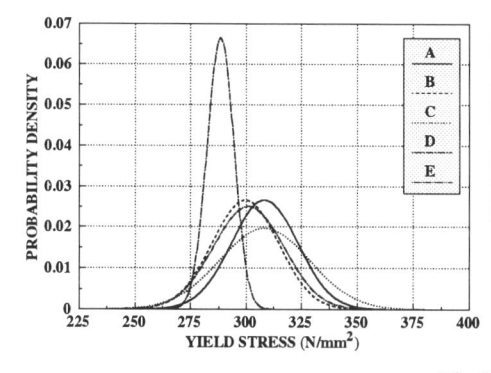

	A	B	C	D	E
N	30	1500	1500	30	1500
$\bar{x}$	308.30	299.6	308.3	288.32	301.08
s	15	15	20.3	6.0	15.9
k	2.22	1.64	1.64	2.22	1.64
f_{yk}	275	275	275	275	275
COV	0.0487	0.0501	0.0658	0.0208	0.0528

	B-A	C-A	E-A	B-D	C-D	E-D
u_{st}	3.15	0	2.60	9.71	16.45	10.9
u_{sn}	0	3.12	0.47	15.24	21.26	16.55

Fig.2
Probability density functions of samples leading to the same value of f_{yk}

higher quality material, which do not meet the requirements of the upper class, so that they have been degraded in a lower class.

The statistical methods, required by the code, cannot take into account these problems; they probably are made for more numerous samples. The results of the internal tests, involving a great number of samples, are not affected by the above-mentioned variabilities; but the official tests, checking only small size samples, can give different results according to the choose of the specimens. Thus, the fulfilment of the rule on statistical conformity between internal and official tests appears nearly a random event rather than an indicator of good quality.

Furthemore such a severe and punitive quality control is requested just for structural steel. For instance, in the case of concrete, only the testing of a small number of specimens and the fulfilment of the characteristic compression strength is just requested. Additional informations, such as concrete composition, gravel provenance, type of cement and so on, are not explicitly referred to.

In order to verify the effectiveness of such statistical method, the most effective case has been considered: the one of the «structural safety» of steel constructions in seismic zones. With this purpose, an extensive numerical analysis on some frames including the effects of the random material variability has been performed. Each frame has been analysed by assuming five different probability density functions for the material yield stress, characterized by a proper average and variance, but leading to the same characteristic resistance ($f_{yk} = \bar{x} - k\,s = 275\ N/mm^2$ for Fe 430 steel, where k is a numerical coefficient depending on the sample size N). The values of these parameters have been selected in such a way that the mechanical requirements prescribed by law are satisfied for each population, but the Student's and/or Snedecor's tests are not verified, when each possible couple of populations is considered.

The gaussian curves of the chosen statistical populations are shown in Fig.2, in which the sample size N, the mean value $\bar{x}$, the standard deviation s, the tensile characteristic yield stress f_{yk} and its coefficient of variation COV are indicated. In the same figure, the values of u_{st} and u_{sn}, obtained by the comparison of the different populations, are also reported.

It has be noted that the averages and the variances of yield stress have been selected in a realistic way, so that they could be the actual values resulting from tests performed on real good quality steel. In particular, A and D samples are representative of official test results, whereas B, C and E samples exemplify internal test results.

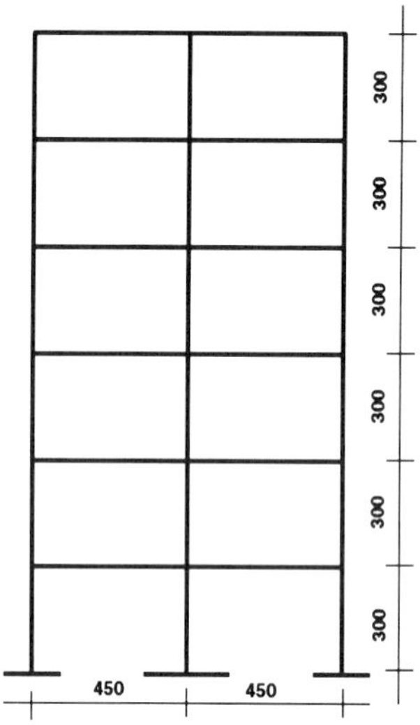

Fig.3
Analysed frames

DM86 FRAME		
STOREY	BEAMS	COLUMNS
1	IPE330	HE220B
2	IPE330	HE220B
3	IPE330	HE200B
4	IPE300	HE200B
5	IPE300	HE180B
6	IPE300	HE180B

EC8 FRAME		
STOREY	BEAMS	COLUMNS
1	IPE300	HE240B
2	IPE300	HE240B
3	IPE300	HE220B
4	IPE270	HE220B
5	IPE270	HE200B
6	IPE270	HE200B

4 The adopted methodology

4.1 Analysed parameters

The inelastic response of the frames, shown in Fig.3, subjected to horizontal forces has been investigated. The following parameters characterizing the behavioural curve relating the multiplier of horizontal forces to the top sway displacement have been considered:
- α_u is the ultimate multiplier of horizontal forces (maximum value);
- δ_u is the ultimate displacement at 90% of α_u (10% strength degradation);
- μ is the ductility ratio computed by assuming δ_{u1} as ultimate displacement and by characterizing the yield state by means of the top displacement δ_y^* which the indefinitely elastic structure exhibits under the horizontal forces corresponding to the ultimate multiplier α_u.

Figure 4 illustrates the definition of the various parameters on a typical curve relating the multiplier of horizontal forces to the top displacement.

In addition, as the q-factor represents the most synthetic parameter for evaluating the inelastic performance of seismic-resistant structures, the following parameters have been also analysed:
- q is the q-factor evaluated with the energy method proposed in (Como and Lanni, 1983). Its practical evaluation can be performed by considering the work done by a system of equivalent horizontal forces, statically applied and distributed according to a combination of a selected number of vibration modes:

$$q = \left(\frac{W_u}{W_y} \right)^{1/2} \tag{3}$$

being W_y the elastic strain energy stored by the system when first yielding occurs and

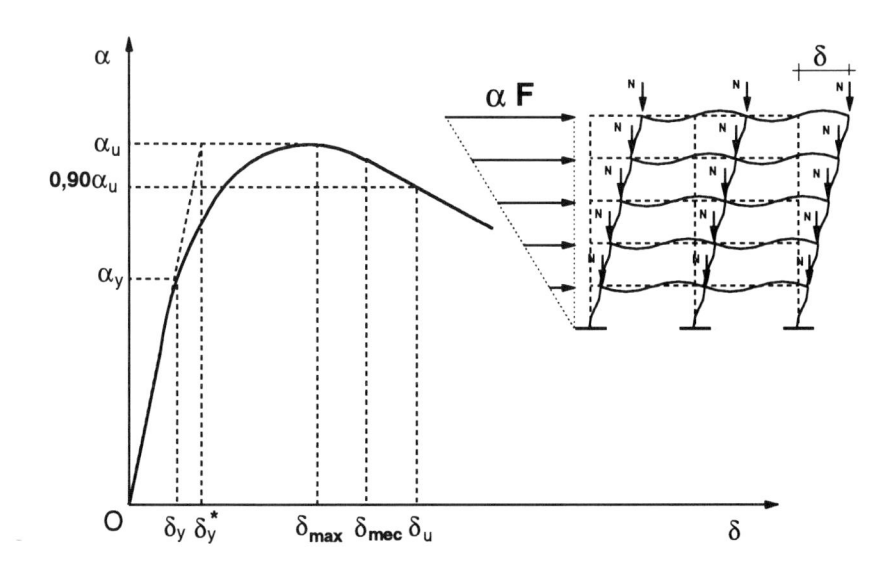

Fig.4
Description of the analysed parameters

W_u the total energy stored and dissipated up to failure, assuming δ_u as ultimate condition.

As the member sections have to be chosen among the standard shapes, the first yielding is attained under horizontal forces greater than the design ones, i.e $\alpha_y \geq 1$. Therefore, the structure is able to withstand the severest design earthquake provided that $\alpha_y q$ is greater than q_d (being q_d the design value of the q-factor). For this reason, the «modified» q-factor:

$$q^* = \alpha_y \ q = \left(\frac{W_u}{W_1}\right)^{1/2} \quad (4)$$

has to be added to the previous list of parameters characterizing the curve relating the multiplier of horizontal forces to the top displacement (being W_1 the elastic strain energy stored by the system under the design horizontal forces, i.e. for $\alpha = 1$).

4.2 Simulation method

The adopted numerical procedure is the Monte Carlo Simulation method in which the generation of the random yield strength of frame members and the inelastic structural analysis are carried out in sequence. The first step is represented by the generation of numbers uniformly distributed between 0 and 1. At each random number corresponds a random value of the yield strength, for a given probability distribution law of the structural steels. The trasformation of the random numbers in random values of the yield strength satisfying a given probability distribution law has been performed by means of the Box and Müller method.

In order to simplify the analysis, the geografical distribution of yield strength along the cross-section has been neglected and just the mean value for a given thickness has been assumed. In addition, considering that the flexural behaviour of double T sections is mainly governed by the stress state of flanges, the yield strength of flanges can be rationally assumed as the yield strength of the member.

Two frames have been analysed. Both are two bay-six storey frames, but the first one (EC8) has been designed according to Eurocode 8 (Commission of the European Communities, 1993), while the second one (DM86) has been designed according to the italian seismic code (D.M. 24/1/1986).

For each probability density function representing the result of official or manufacturer internal controls, being the analysed frames composed by 30 members, 100 series of 30 random yield stresses have been generated: what means 100 different frames with the same geometrical configuration and loading condition. Therefore, the complete numerical investigation has requested 1000 inelastic analyses.

Random yield stress generation and inelastic structural analysis are carried out in sequence. As final result, the relationship between the multiplier of horizontal forces and the top displacement is available for each frame. The statistical characterization of the inelastic response of the structures requires also a further statistical analysis on the data previously obtained for the investigated parameters, as defined in Section 4.1.

This kind of methodology has to be considered a so-called «Hybrid Monte Carlo Simulation» (Augusti, Baratta and Casciati, 1984), in which the numerical simulation is used in order to obtain a statistical sample of the structural response, whose dimension is specified in order to obtain sufficient accuracy in the estimation of the «central values» of the variables which constitute the object of the investigation.

Table 1
Main results of the Montecarlo simulation

		DM 86 FRAME					EC8 FRAME				
		A	B	C	D	E	A	B	C	D	E
α_u	min.	1.9300	1.8600	1.8700	1.8702	1.8999	1.9301	1.8702	1.9204	1.8605	1.8798
	max.	2.1700	2.0900	2.1400	1.9304	2.1098	2.0905	2.0508	2.1309	1.9101	2.0504
	$\bar{x}$	2.0590	1.9840	2.0388	1.8977	1.9963	2.0245	1.9608	2.0179	1.8821	1.9666
	s	0.0406	0.0378	0.0513	0.0146	0.0437	0.0336	0.0353	0.0423	0.0139	0.0332
	fractile	1.9798	1.9103	1.9386	1.8693	1.9110	1.9588	1.8918	1.9353	1.8549	1.9017
μ	min.	2.4319	2.4245	2.3428	2.4933	2.3618	3.4744	3.6071	3.3676	3.6136	3.5431
	max.	3.0745	3.0191	3.0193	2.7285	2.9829	4.1175	4.9327	4.3378	3.8838	4.2208
	$\bar{x}$	2.6208	2.6236	2.6437	2.5915	2.6270	3.7491	4.1878	3.7613	3.7580	3.7793
	s	0.1104	0.1124	0.1362	0.0443	0.1224	0.1215	0.2262	0.1770	0.0521	0.1323
	fractile	2.4051	2.4042	2.3776	2.5051	2.3880	3.5118	3.7462	3.4157	3.6563	3.5209
q	min.	2.4396	2.4477	2.4651	2.5000	2.4561	3.7483	3.6071	3.6434	3.9355	3.7324
	max.	3.0745	3.0596	3.2571	2.7947	3.2695	4.8070	4.9327	5.7447	4.4561	4.9904
	$\bar{x}$	2.6979	2.7151	2.7909	2.6084	2.7332	4.1736	4.1878	4.3188	4.1101	4.2704
	s	0.1220	0.1355	0.1746	0.0563	0.1581	0.2139	0.2262	0.3419	0.1229	0.2599
	fractile	2.4596	2.4506	2.4501	2.4984	2.4245	3.7560	3.7462	3.6512	3.8702	3.7630
q^*	min.	4.3200	4.1200	4.2100	4.0701	4.0999	5.1101	4.9201	5.0509	4.8801	4.9809
	max.	4.9500	4.7500	4.9200	4.2803	4.9198	5.7702	5.5699	5.8599	5.1488	5.5491
	$\bar{x}$	4.5559	4.3951	4.5354	4.1779	4.4268	5.3704	5.2008	5.3657	5.0001	5.2408
	s	0.1271	0.1318	0.1590	0.0502	0.1501	0.1203	0.1242	0.1524	0.0485	0.1128
	fractile	4.3077	4.1377	4.2249	4.0799	4.1338	5.1355	4.9584	5.0681	4.9054	5.0205

Table 2
Percentage scatters of the fractiles for the analysed couples of distributions

	DM86 FRAME						EC8 FRAME					
	B-A	C-A	E-A	B-D	C-D	E-D	B-A	C-A	E-A	B-D	C-D	E-D
u_{st}	3.15	0	2.60	9.71	16.45	10.90	3.15	0	2.60	9.71	16.45	10.90
u_{sn}	0	3.12	0.47	15.24	21.26	16.55	0	3.12	0.47	15.24	21.26	16.55
Δ_{cov}	2.88	35.11	8.42	140.87	216.35	153.85	2.88	35.11	8.42	140.87	216.35	153.85
α_u	-3.51	-2.08	-3.48	2.19	3.70	2.23	-3.42	-1.20	-2.91	1.99	4.33	2.52
μ	-0.04	-1.14	-0.71	-4.03	-5.09	-4.67	6.67	-2.74	0.26	2.46	-6.58	-3.70
q	-0.37	-0.39	-1.43	-1.91	-1.93	-2.96	-0.26	-2.79	0.19	-3.20	-5.66	-2.77
q^*	-3.95	-1.92	-4.04	1.42	3.55	1.32	-3.45	-1.31	-2.24	1.08	3.32	2.35

5 Discussion of results

For each frame (EC8 and DM86), a statistical sample of the parameters characterizing the inelastic behaviour has been obtained by assuming as input for the material yield strength, from the statistical point of view, the five different distributions described by the probability density functions denoted, in Section 3, with A, B, C, D and E.

The statistical analysis of the random values of the inelastic parameters governing the frame behaviour has provided, for each inelastic parameter, the sample maximum value, the sample minimum value, the mean value, the standard deviation and the 5% fractile. These results are given in Table 1 for the two analysed frames. As reference values, the 5% fractiles are considered the most significant data in order to analyse the consequences, on the structural reliability, due to the fact that the conformity tests are not satisfied. It can be immediately recognized that the differences arising from the various distributions are not significant from the structural point of view.

In table 2, for each parameter, the percentage scatters evaluated for all possible couples of the five distributions are given. The corresponding values of u_{st} and u_{sn} are recalled in the upper rows of the same table. In addition, for each couple of distributions, the corresponding percentage difference between their coefficients of variation is given.

It can be observed that the values of the percentage scatters between the fractiles of a given parameter are related not only to the values of u_{st} and u_{sn}, but also to the percentage difference Δ_{cov} between the coefficients of variation of the distributions constituting each couple. As the highest scatters correspond to the distributions having the highest difference for the coefficient of variation, it is clear that the code provisions for the quality control and acceptance of structural steels should provide a limit to this difference between internal and official tests, rather than thei statistical conformity tests of Student and Snedecor.

In fact, we can observe that the variations of α_u are always less than 5%, which corresponds to the degree of approximation usually accepted in structural design. The variations of the q-factor are always below 6%, which is negligible considering the uncertainties in both definition and evaluation of such parameter. Also the ductility parameter μ has a variation limited to 7%.

It is important to note that these variations, even if negligible, from the structural point of view, are not at all related to the values of the Student and Snedecor parameters (u_{st} and u_{sn}).

It is also interesting to remark that the couple E-A, which is the only one almost in line with the code requirements, does not exhibit the best results in term of variations, as it should be expected as a rational consequence of the fulfilment of the regulations.

It means that the strict respect of this code requirement, based on the statistical conformity, does not necessarily lead to an improvement of the structural reliability. As a consequence, our proposal is to remove these conformity tests, because they are not significant, being at the same time too severe.

As an alternative for checking the congruity between internal and official tests, a limit for the difference of the two values of the *COV* coefficient should be recommended.

6 References

Augusti, G., Baratta, A. and Casciati, F. (1984) **Probabilistic Methods in Structural Engineering**, Chapman and Hall, London, New York.
Commission of the European Communities (1993) **Eurocode 8: European Code for Seismic Regions**, October (Draft).
Como, M. and Lanni, G. (1983) Aseismic Toughness of Structures, **Meccanica**, N.18, pp.107-114.
Decreto Ministeriale 14/2/92 (1992), **Gazzetta Ufficiale della Repubblica Italiana**, N.65, 18/3/92.
Decreto Ministeriale 24/1/86 (1986), Norme tecniche relative alle costruzioni sismiche **Gazzetta Ufficiale della Repubblica Italiana**, N.108, 12/5/86.
ECCS (European Convention for Constructional Steelwork) (1988) **European Recommendations for Steel Structures in Seismic Zones**.
Elnashai, A.S. and Chryssanthopoulos, M. (1990) Effect of Random Material Variability on Seismic Design Parameters of Steel Frames, **Earthquake Engineering and Structural Dynamics**.
Kuwamura, H. and Kato, B. (1989) Effect of Randomness in Structural Members' Yield Strength on the Structural Systems' Ductility, **Journal of Constructional Steel Research**, N.13.
Kuwamura, H. and Sasaki, M. (1990) Control of Random Yield-Strength for Mechanism-Based Seismic Design, **Journal of Structural Engineering**, ASCE, Vol.116, pp.98-110.
Mazzolani, F.M., Mele, E. and Piluso, V. (1991) Analisi Statistica del Comportamento Inelastico di Telai in Acciaio con Resistenza Casuale, **V Convegno Nazionale, L'Ingegneria Sismica in Italia**, Palermo, 29 Settembre-2 Ottobre.
Mazzolani, F.M., Mele, E. and Piluso, V. (1991) On the Effect of Randomness of Yield Strength in Steel Framed Structures Under Seismic Loads, **ECCS Document TC13.01.91**.

10 DYNAMIC BEHAVIOR OF BORDER COLUMNS OF A CANTILEVER BRACED WALL

M. SHIBATA
Setsunan University, Neyagawa, Japan

Abstract
Buildings designed as their seismic resistant property depend upon their energy dissipation capacity, should not cause its local failure at its maximum strength. In the multi-story braced wall, border columns are designed to hold axial forces at the maximum strength of the system. As the compression of the tension side column is not so large as the other side, the horizontal resistance of the tension side column introduced by its axial surplus may cause the unexpected increase of the maximum strength of the system. The more is the maximum strength, the larger is the column axial force. This paper presents how to control the maximum axial forces of border columns in the seismic excitation.
Keywords: Cantilever Braced Wall, Border Columns, Response Analysis, Maximum Compression,

1 Introduction

No local failure should occur at the maximum strength of a structure, whose seismic resistant property depends upon its energy dissipation capacity. Some structural elements are often designed so as to resist the stresses governed by the maximum strength of the system, that the poor estimation of the structural strength may introduce serious damages on such elements.

In an isolated cantilever type braced wall, border columns are subjected to large axial forces due to the overturning moment in seismic excitation, so border columns are designed to resist the axial forces at the maximum strength of the system. What is the maximum strength and how to estimate it?

Not a few investigators discussed about the design load for braces, but there are few on the interaction between braces and border columns in cantilever type braced wall (Shibata et al. (1985), Shibata (1988)). This paper presents how to control the maximum compression of border columns in cantilever type braced wall at the seismic excitation.

Behaviour of Steel Structures in Seismic Areas. Edited by F. M. Mazzolani and V. Gioncu.
Published in 1995 by E & FN Spon, 2-6 Boundary Row, London SE1 8HN. ISBN: 0 419 19890 3.

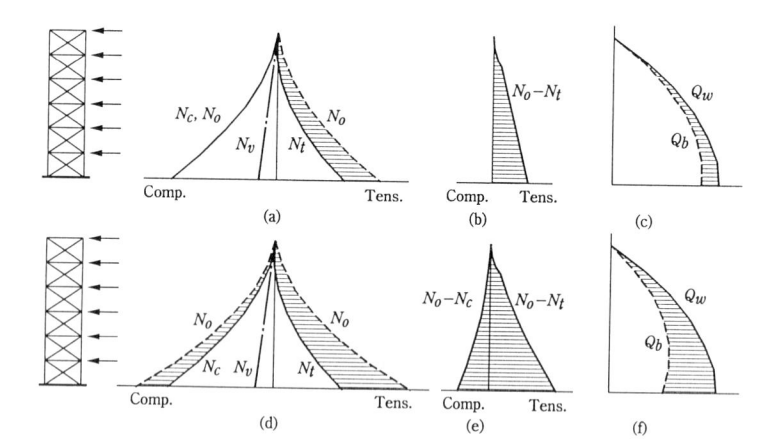

Fig.1 Axial forces of border columns

2. Axial Forces of Border Columns

In Fig.1(a), the dash and dot line N_v shows the axial force of the border column due to the dead weight of a cantilever braced wall, and solid lines N_c, N_t denote axial forces of the compression and tension side at the design earthquake load, respectively. Although, the compression side column is subjected to large axial force, the axial force on the tension side is not so large as the other side, because some of the tension force due to the overturning moment is canceled out by the compression due to the dead load. If the axial strength of the border column N_o is designed to resist the axial force of compression side N_c, the reserved strength generated in the tension side column, hatched in Fig.1(a) and (b), will introduce the unexpected increase of the horizontal resistance of the system, which may cause local failures of border columns. As border columns are usually designed stronger than expected, to leave a reasonable margin to the design load [See Fig.1(d)], the risk of local failure increases du to the increase of reserved strength.

Braces in a braced wall are often designed to carry the full design load of the braced wall. In this case, above mentioned axial surplus of border columns may introduce their local failures. In order to prevent such local failures, design strength of braces Q_b should be cautiously reduced, so as to cancel out the increase of horizontal resistance of border columns [dashed lines in Fig.1(c), (f)], and to carry the difference between the design load Q_w of the braced wall [solid lines in Figs.1(c),(f)] and the horizontal strength of border columns, after the exact estimation of column strength.

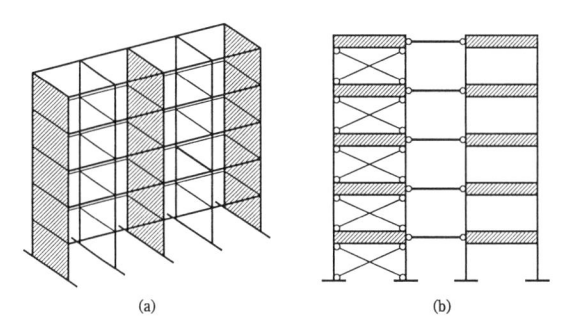

Fig.2 Analytical model

3. Response Analysis

3.1 Assumptions
1) The objective frame is shown in Fig.2. It is composed of an isolated braced wall and an open frame. They are connected to realize the same story drift in each other.
2) The braced wall is composed of rigid and strong girders, fixed-ended border columns and pin-ended X-type braces.
3) The axial force of a brace is estimated by the piece wise linear function of its elongation [See Fig.3], which is based on the hysteresis function proposed by Wakabayashi et al. (1977).
4) The border column has an ideal I-section, with linear isotropic hardening material whose initial yield condition is shown in Fig.4, and is short enough to ignore buckling. N_o and M_o in Fig.4 denote the limit axial force and the full plastic moment in pure bending, respectively.
5) The elastic elongation of the middle portion of border columns and the plastic elongation at the yield hinge are taken into account to estimate the total bending deformation of the wall.
6) The open frame portion is assumed to be a multi-mass shear system with the hysteretic characteristics shown in Fig.5, where x_o and Q_o denote the elastic limit story drift and strength, respectively.

3.2 Equations of Motion
The equation of motion for undamped multi-mass system is given by
$$\{\ddot{x}\}+[\mu]\{Q\}=-\{a\} \tag{1}$$

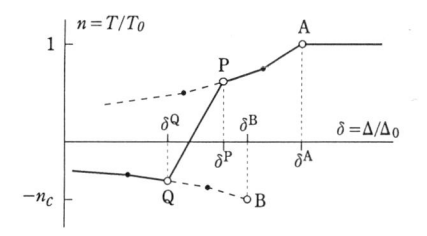

Fig.3 Hysteretic rules of braces

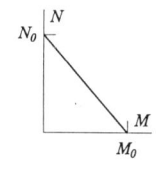

Fig.4 yield locus of border columns

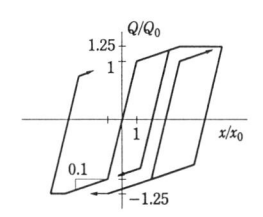

Fig.5 Hysteresis model of open frame

where,

$$[\mu]=\begin{bmatrix} \dfrac{1}{m_1} & \dfrac{-1}{m_1} & 0 & 0 & \cdots & & 0 \\ \dfrac{-1}{m_1} & \dfrac{1}{m_1}+\dfrac{1}{m_2} & \dfrac{-1}{m_2} & 0 & \cdots & & 0 \\ 0 & \dfrac{-1}{m_2} & \dfrac{1}{m_2}+\dfrac{1}{m_3} & \dfrac{-1}{m_3} & \cdots & & 0 \\ 0 & 0 & \ddots & \ddots & \ddots & & \vdots \\ \vdots & \vdots & \ddots & \ddots & \ddots & & \dfrac{-1}{m_{N-1}} \\ 0 & 0 & 0 & \cdots & & \dfrac{-1}{m_{N-1}} & \dfrac{1}{m_{N-1}}+\dfrac{1}{m_N} \end{bmatrix}$$

$\{a\}=\{a(t)\ 0\ 0\ \cdots\ 0\}^T$
$\{x\}$: story drift $\{Q\}$: restoring force
m_k : mass of k-th story $a(t)$: ground acceleration.

Substituting Eq.(1) into the basic equation of Newmark's β-method,
$$\{x\}=\{x^*\}+\{\dot{x}^*\}\tau+(1/2-\beta)\{\ddot{x}^*\}\tau^2+\beta\{\ddot{x}\}\tau^2$$
we get following equation,
$$\{F\}=\{x\}+\beta\tau^2[\mu]\{Q\}-\{x^*\}-\{\dot{x}^*\}\tau-\bigl((1/2-\beta)\{\ddot{x}^*\}-\beta\{a\}\bigr)\tau^2=\{0\} \tag{2}$$

where, $\{x^*\}$: story drift at the previous step, τ : time step of integration, $\beta=0.25$: constant. Numerical integration of Eq.(1) is conducted by solving nonlinear simultaneous equations (2) at each step. The essential properties of this method is the same as the Newmark's β-method, and its adequacy is confirmed by comparing the results with those by Runge-Kutta method for multi-mass shear system.

3.3 Restoring Force of Braced Wall
1) Equilibrium
In Fig.6, following relationships are given,

$$Q_k^W = Q_k^B + Q_k^L + Q_k^R \tag{3}$$
$$Q_k^B = \bigl(T_k^A - T_k^B\bigr)\cos(\alpha_k) \tag{4}$$
$$N_k^L + N_k^R = -N_k^B - W_k \tag{5}$$
$$N_k^B = \bigl(T_k^A + T_k^B\bigr)\sin(\alpha_k) \tag{6}$$
$$M_k^W = (N_k^L - N_k^R)\cdot L/2 \tag{7}$$

where, Q_k^W : restoring force of braced wall at the k-th story
Q_k^B : horizontal force shared by braces
Q_k^L, Q_k^R : horizontal forces shared by border columns
T_k^A, T_k^B : axial forces of braces
N_k^L, N_k^R : axial forces of left and right border columns
N_k^B : sum of vertical components of brace axial forces
W_k : wall dead weight carried by the k-th story columns
M_k^W : the overturning moment of the k-th story.

2) Geometry
In Fig.7, following geometrical relationships are given,
$$x_k = x_k^S + x_k^B \tag{8}$$

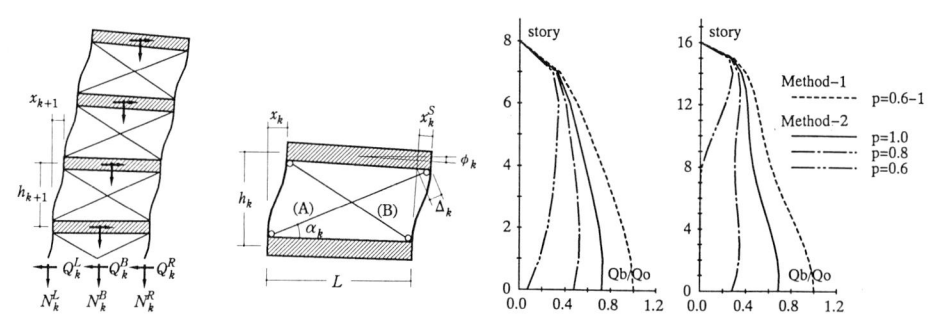

Fig.6 Braced wall   Fig.7 Geometry   Fig.8 Horizontal force shared by braces

$$x_k^B = h_k \cdot \left(\sum_{i=0}^{k-1} \phi_i + \phi_k / 2 \right) \quad (9) \qquad \phi_k = M_k^W / (EI)_k^W \quad (10)$$

where,
 x_k : story drift of k-th story
 x_k^S : shear deformation between stories
 x_k^B : bending deformation $\phi_0 = M_0^W / K_0$: base rotation
 K_0 : rotational rigidity of basement
 ϕ_k : relative rotation between stories

and the bending rigidity $(EI)_k^W$ of the wall is estimated from the elastic axial rigidity of border columns.

3.4 Solution of Nonlinear Simultaneous Equation
Provided that the approximate solution of Eq.(2) is given as $\{x^S\}^\#$, the correction vector $\{\Delta x^S\}$ for $\{x^S\}$ is given by

$$\{F(\{x^S\}^\#)\} = [\partial F_i / \partial x_j^S] \{\Delta x^S\} \quad (11)$$

where $[\partial F_i / \partial x_j^S]$ is the Jacobian matrix of Eq.(2).

3.5 Computing Sequence
1) Assume the distribution of shear deformation $\{x^S\}$.
2) Calculate the restoring force $\{Q^W\}$ shared by the braced wall.
   (a) Repeat procedures (b)-(g), from the top to the bottom story.
   (b) Get T_k^A, T_k^B and Q_k^L, Q_k^R from the constitutive relationships (assumption 3 and 4).
   (c) Get Q_k^B, N_k^B from Eqs.(4) and (6).
   (d) Assume Q_k^W.
   (e) Get M_k^W from $M_k^W = M_{k+1}^W + W_k \cdot x_k + (Q_k^W \cdot h_k + Q_{k+1}^W \cdot h_{k+1})/2$
   (f) Get N_k^L, N_k^R from equations (5) and (7).
   (g) If Eq.(3) is not satisfied, go back to step (d).
3) Get $\{x^B\}$ and $\{x\}$ from Eqs.(8)-(10).
4) Get $\{Q^F\}$ from the constitutive relation for the open frame (assumption 6).
5) If $\{Q\} = \{Q^W\} + \{Q^F\}$ does not satisfy Eq.(2), improve $\{x^S\}$ by Eq.(11) and go back to step 2).

4 Computed Results and Discussions

4.1 Examples for Analysis

Dynamic response analyses of 8 and 16 story cantilever braced walls in Tables 1 and 2 are conducted for El Centro 1940 NS accelerogram.

Vertical springs with spring constant 5000 ton/cm (49000 kN/cm) are assumed under the border column bases, so as to simulate the elongation of piles, and basic properties of these frames are as follows;

$W = 100$ ton(980 kN):   story weight
$L = 7.5$ m: wall span $h = 3.75$ m: story height
$d = 40$ cm: effective depth of border columns
$x_0 = 2$ cm: elastic limit displacement of the open frame
$n_E = 4$:   non dimensional Euler load of brace, the ratio of the Euler load of the brace to the limit axial force.

Ishiyama's proposal (1986) for vertical distribution of the design earthquake load is adopted, and base shear coefficients C_B are adjusted in order to unify the fundamental periods of the system as 1 sec. for 8 story frames and 2 sec. for 16 story frames.

Input level $a = 0.35 - 1.4$ is defined as the ratio of the maximum ground acceleration and the yield level acceleration (the ratio of design base shear and the building mass).

The wall share q, the ratio of the restoring force shared by the braced wall to the total design load is set uniform in all stories, and each components are designed as follows;

Border columns
1) Maximum compression N_C of the compression side border column is determined from the design earthquake load.
2) Limit axial force $N_O = N_C/p$ and the cross sectional area A are determined for given axial force ratio p, the ratio of the maximum design load to N_O.
3) The full plastic moment and the moment of inertia are determined for given effective depth of sandwich section d as $M_O = N_O d/2$ and $I = A d^2 / 4$.

Braces
Braces are designed by one of the following two methods.
(a) Method-1
1) Horizontal strength shared by braces is set to be equal to the design earthquake load Q_W of the braced wall [solid lines in Fig.1(c),(f)].
2) Compression strength of braces are determined by the formulae proposed by Wakabayashi et al. (1977).
(b) Method-2
1) Horizontal forces shared by border columns under design earthquake load are obtained from their axial forces and the yield condition (Fig.4).
2) Braces are designed to resist the difference between the design load Q_W and the horizontal resistance of border columns [dashed lines Q_b in Fig.1(c),(f)].

Table 1   8 story Frames   Table 2   16 story Frames

No.	Design Method	C_B	Options
1	1	0.305	basic type
	2	0.362	
2	1	0.365	$p = 1.0$
	2	0.398	
3	1	0.241	$p = 0.6$
	2	0.335	
4	1	0.227	$r = 2$
	2	0.271	

No.	Design Method	C_B	Options
5	1	0.279	basic type
	2	0.306	
6	1	0.338	$p = 1.0$
	2	0.352	
7	1	0.220	$p = 0.6$
	2	0.260	
8	1	0.193	$r = 4, T = 1.8$
9	1	0.190	$r = 2, T = 1.6$

For basic type frames with no options, the wall share q, the axial force ratio p, the fundamental period T and the wall aspect ratio r are set as follows.

$$q = 1.0 \qquad T = \begin{cases} 1.0 \text{ sec.} \\ 2.0 \text{ sec.} \end{cases}, \quad r = \begin{cases} 4.0 & \text{(8 story frames)} \\ 8.0 & \text{(16 story frames)} \end{cases}$$
$$p = 0.8$$

Fig.8 shows the distribution of design shear force shared by braces. Total design load of the braced wall is carried by braces in Method-1, and in Method-2, horizontal forces shared by braces are reduced in lower stories, for large aspect ratio r and small axial force ratio p.

4.2 8 story frames

Computed results for 8 story frames are shown in Fig.9. The effect of design method of braces for example frame No.1 is shown in Fig.9(a). In the figure, "Story drift" corresponds to the maximum story drift, "Shear deformation" to the maximum shear deflection between stories and "Axial force" to the ratio of maximum compression of border columns to their limit axial force.

Most impressive fact is that Method-1 causes the maximum compression of the border column exceed the design value, 80% of the limit axial force, for the input level a larger than 0.5. Although, Method-2 does not cause the excess of column axial forces, even for $a = 1.4$. Figs.9(b) and (c) show the comparison of design method for the case $p = 0.6$ and 1.0 (example frames No.2 and 3), the design values of the axial force ratio of border columns. Method-1 causes the excess of the axial force of border columns for large input level. The smaller is the design value of the axial force ratio p, the larger is the excess ratio. Buckling of border columns may occur in the early stage for $p = 1$, in the actual behavior, although buckling of border columns are ignored in this analysis. On the other hand, Method-2 does not cause any excess of column axial forces in both cases.

The maximum compression of border columns in the earthquake excitation depends upon the aspect ratio of the wall. The wall aspect ratios for examples 1-3 are relatively large value, $r = 4$. Results for $r = 2$ (example 4) is shown in Fig.9(d). For small aspect ratio, axial force variations of border columns are not so large as for large aspect ratio, and even Method-1 does not cause the serious

Fig. 9 Examples No.1-4

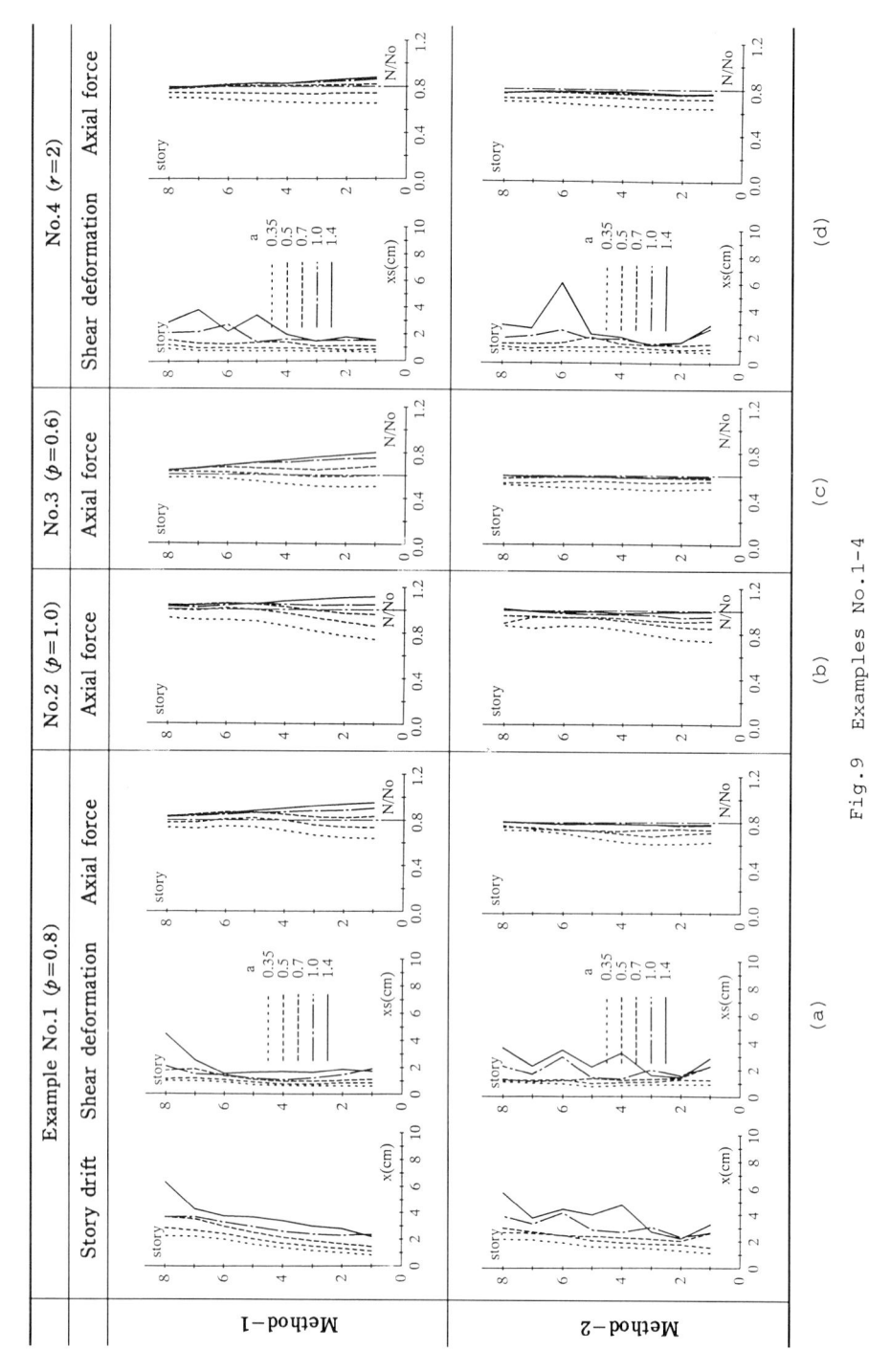

Fig.10 Examples No.5-7

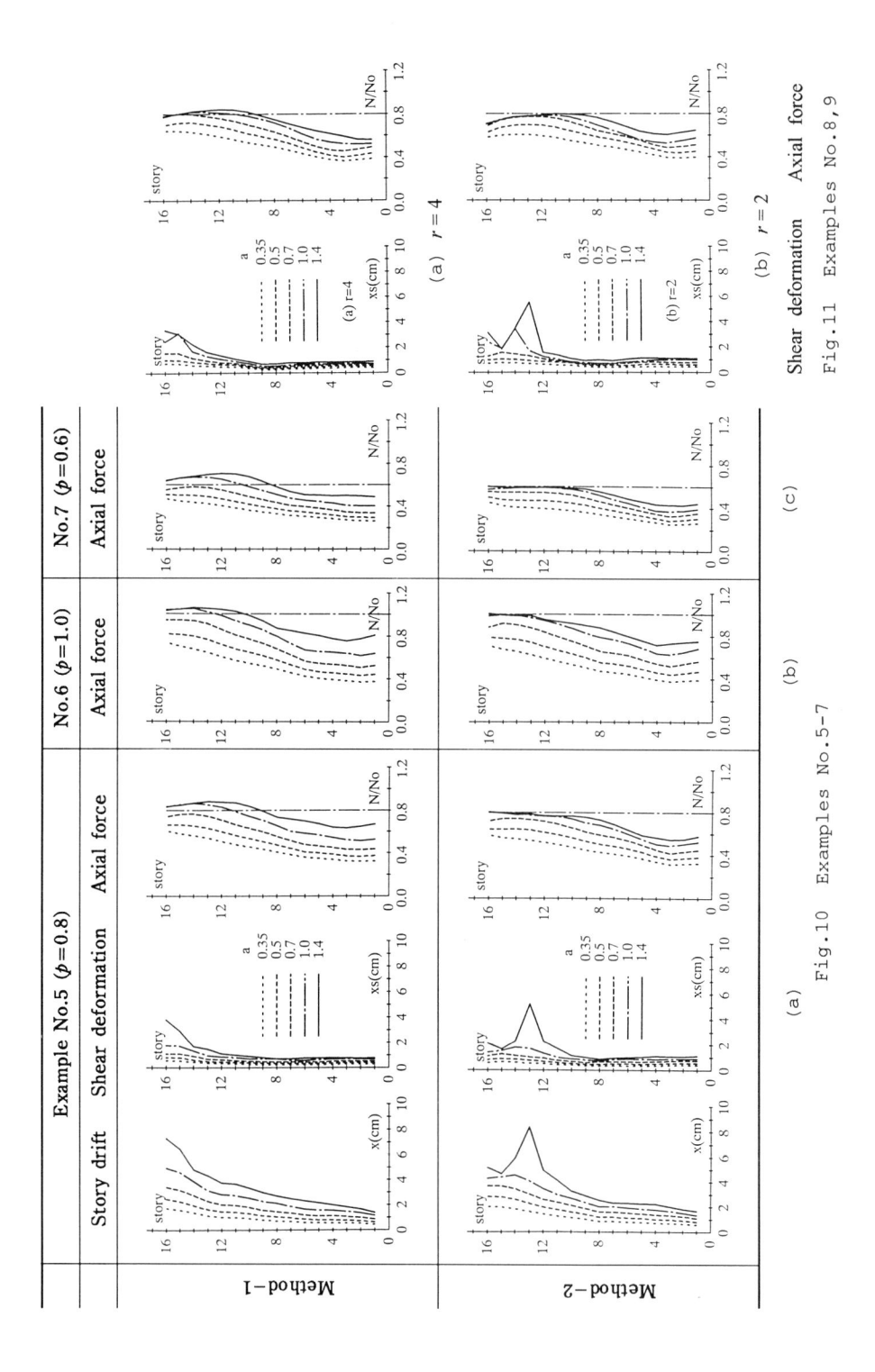

Fig.11 Examples No.8,9

damages of border columns.

4.3 16 story frames

Computed results for 16 story frames are shown in Figs.10 and 11. Fig.10 corresponds to Fig.9(a)-(c) for 8 story frames and the fundamental properties are similar to those of 8 story frames. But in the 16 story frame the axial force ratio of border columns of upper stories is larger than that of lower stories, and the excess of axial force by Method-1 occurs only in upper stories. This means that the de-sign overturning moment is rather overestimated in lower stories. Method-2 does not cause any excess of column axial forces as for 8 story frames. For examples 5-7, the wall aspect ratios are very large value $r=8$, and they are not suitable models for actual buildings. Fig.11 shows the results for $r=4$ and $r=2$ (examples 8 and 9). As the vertical distribution of column axial forces depends upon the wall aspect ratio, the maximum response of the column axial force does not exceed the design value for $r=2$, and for $r=4$ the excess is rather small.

5 Conclusions

1) In the slender cantilever type braced wall, the compression side border columns may be subjected to too large compression due to the excessive resistance of tension side columns, which may cause the local failure.
2) Excess resistance of border columns become large for lower stories, so relatively weak upper stories may be subjected to large deflection.
3) It is possible to control the maximum compression of border columns at the seismic excitation not to exceed the design value, by the reduction of brace strength of lower stories.
4) Uniform distribution of horizontal resistance of the wall is realized by the reduction of brace strength of lower stories.

6 References

Ishiyama, Y. (1986) Distribution of Lateral Seismic Forces along the Height of a Building, BRI Research Paper No.120, Building Research Institute, Ministry of Construction, Japan.
Shibata, M. (1988) Hysteretic Behavior of Multi-Story Braced Frame, Proc. 9WCEE, Vol.IV, 249-254.
Shibata, M. and Wakabayashi, M. (1985) Hysteretic Behavior of Columns in Multi-Story Braced Frame, Trans. AIJ, 353, 13-20. (in Japanese)
Wakabayashi, M., Nakamura, T., Shibata, M., Yoshida, N. and Masuda,H. (1977) Hysteretic Behavior of Steel Braces Subjected to Horizontal Load Due to Earthquake, Proc. 6WCEE, Vol.3, 3188-3194.

PART TWO
LOCAL DUCTILITY

11 LOCAL DUCTILITY IN STEEL STRUCTURES SUBJECTED TO EARTHQUAKE LOADING

A. S. ELNASHAI
Department of Civil Engineering, Imperial College, London, UK

Abstract

The paper reviews the current status of research and development in seismic performance of steel structures based on a hierarchical framework linking local to global ductility, with special reference to criteria for local ductility supply. Work on section curvature ductility evaluation based on local buckling criteria is discussed and related to member plastic hinge length and deflection ductility. Recent work on connections; rigid, semi-rigid and composite, is highlighted and its potential in supplying high overall energy dissipation is discussed. The paper concludes with suggestions regarding future developments that would enhance the use of steel structures in seismic design.
Keywords: Local Ductility, Curvature, Rotation and Displacement Ductility, Local Buckling, Connections, Hysteretic Behaviour, Capacity Design.

1 Introduction

The aim of seismic design is to arrive at a structural configuration with an acceptable level of safety at an affordable cost. At variance to conventional design to resist primary loads, seismic forces, which are considered as an accidental condition, are resisted through the imposition of a level of damage on the structure. Therefore, controlled energy dissipation, in a damaged state, in a structural system is essential for its survival under severe earthquake loading scenarios. Consequently, failure mode control is arguably the most important and fundamental requirement for safe seismic performance of structures.

Whilst structures have been designed for many years to resist seismic forces by a 'direct design' approach, recently, the philosophy of 'capacity design' has become accepted as the most rational basis for the control of the response and failure of structures. The difference between direct and capacity design philosophies is demonstrated with reference to Figure 1.

Members A-D, B-D and C-D are subjected to seismic actions $(M, N, V)_{A-D}$, $(M, N, V)_{B-D}$ and $(M, N, V)_{C-D}$ as shown. In direct design, each member is sized to resist the applied actions, using code design expressions. In capacity design, it is first decided which members are to yield first and whether the other members are to yield subsequently or remain elastic. Herein, it is assumed that member C-D is required to contribute to the development of the desired failure mode, hence it is considered a 'dissipative' zone. Members A-D and B-D should remain elastic. The capacity design procedure is initiated by dimensioning member C-D for the applied actions $(M, N, V)_{C-D}$. The overstrength of member C-D (ϕ_{C-D}) is then estimated, taking into account, at least, (i) actual yield strength, (ii) increase in cross-sectional area and (iii) strain-

Behaviour of Steel Structures in Seismic Areas. Edited by F. M. Mazzolani and V. Gioncu. Published in 1995 by E & FN Spon, 2-6 Boundary Row, London SE1 8HN. ISBN: 0 419 19890 3.

hardening. The action effects are then recalculated, elastically, to reflect the equilibrium condition at the instant of the realisation of the strength (including the parameter ϕ_{C-D}) of member C-D, leading to the stress resultants (M', N', V')$_{A-D}$ and (M', N', V')$_{B-D}$, which are employed in the design of the two 'non-dissipative' members A-D and B-D.

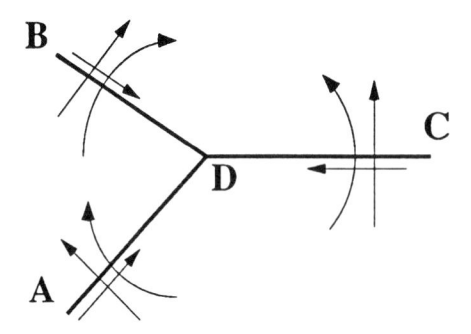

Figure 1: Seismic Design Forces of a Sub-assembledge

It is clear, however, that despite the sound philosophy on which capacity design is based, its implementation is by no means straightforward. For example, in addition to the possible sources of overstrength mentioned above (which exclude strain-rate effects), there are large uncertainties in our ability to estimate the various limit states, especially the ultimate situation. The controvercies regarding local buckling, rupture and low-cycle fatigue criteria have not been adequately resolved. Hence, unquantifiably-conservative estimates of overstrength factors are utilised. Moreover, it is obvious that at the equilibrium state corresponding to the overstrength of the dissipative zone, the action distribution will not be a factored version of the elastic distribution. Hence, in principle, the design framework presented above is only applicable alongside inelastic large displacement analysis procedures.

The role of local ductility in seismic design may be visualised by considering the hierarchical perspective of structural response shown in Figure 2. Local buckling only is listed, to the exclusion of fatigue and fracture, since it is considered to be significantly more important. It is only through an understanding of the inter-relation between local and global ductility that the response modification factors (behaviour factors) can be optimised, thus enabling the economic design of structures with a recognisable level of safety.

Local ductility in steel structures 135

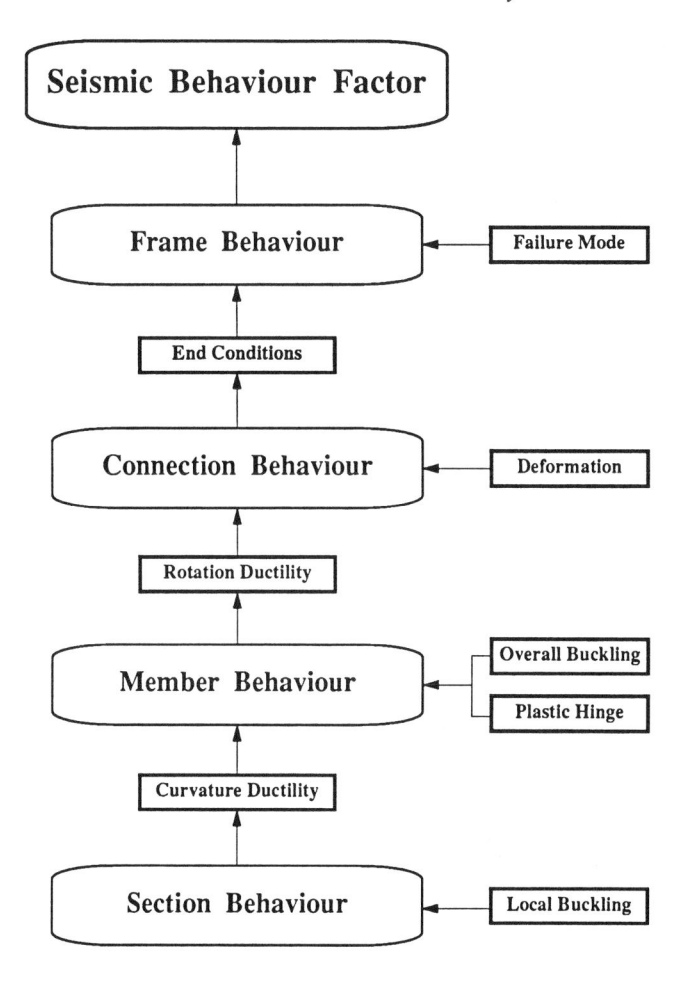

```
┌─────────────────────────────┐
│  Seismic Behaviour Factor   │
└─────────────────────────────┘
              ▲
              │
┌─────────────────────────────┐      ┌──────────────┐
│      Frame Behaviour        │◄─────│ Failure Mode │
└─────────────────────────────┘      └──────────────┘
              ▲
              │
      ┌───────────────┐
      │ End Conditions│
      └───────────────┘
              ▲
              │
┌─────────────────────────────┐      ┌──────────────┐
│    Connection Behaviour     │◄─────│ Deformation  │
└─────────────────────────────┘      └──────────────┘
              ▲
              │
      ┌──────────────────┐
      │ Rotation Ductility│
      └──────────────────┘
              ▲
              │
┌─────────────────────────────┐      ┌──────────────────┐
│      Member Behaviour       │◄─────│ Overall Buckling │
└─────────────────────────────┘      ├──────────────────┤
                                     │  Plastic Hinge   │
                                     └──────────────────┘
              ▲
              │
      ┌──────────────────┐
      │ Curvature Ductility│
      └──────────────────┘
              ▲
              │
┌─────────────────────────────┐      ┌──────────────┐
│     Section Behaviour       │◄─────│ Local Buckling│
└─────────────────────────────┘      └──────────────┘
```

Figure 2: Hierarchical Framework for Structural Response

It is customary in seismic design to ensure that shear stresses do not contribute greatly to member response, with the exception of shear yielding in short links of eccentrically-braced frames. Therefore, the following discussion on the relationship between local and member ductility is restricted to flexural response. In Figure 3 (Broderick, 1994), the deformed shape, curvature distribution and bending moment diagram lead to the definition of the plastic hinge length, hence to the deflection. Therefore, longer plastic hinges and/or higher section curvature ductilities lead to increased member, and hence structure, deflection ductilities.

136 *Elnashai*

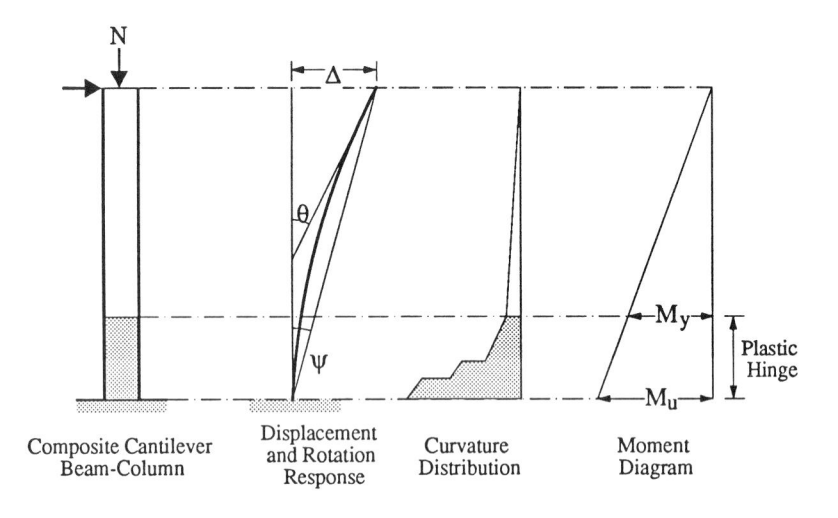

Figure 3: Relationship between Section and Member Ductilities

In summary, the material characteristics affect section curvature ductility, which, from a local buckling viewpoint, is also a function of the section dimensions. The rate and extent of the spread of plasticity from one section to the next affects the plastic hinge length, hence the member ductility for a given curvature ductility. The response of connections dictates the location of points of contraflexure, hence affects the overall buckling of members and plastic hinge initiation. Finally, all the above, alongside the redistribution potential, affects the overall ductility, hence the energy dissipation capacity of the entire system. This material-section-member-connection-frame approach is the only rational basis for controlled seismic design.

Hereafter, some of the significant contributions to the evaluation of local ductility supply in steel and composite steel/concrete structures are reviewed and the relevance of the work to the evaluation of behaviour factors is highlighted. This is not a comprehensive literature review or a state-of-the-art report, rather a selective assessment of some more recent developments.

2 Section Curvature Ductility

The curvature ductility supply of a steel section is defined as the ratio between the curvature at yield and that at the attainment of an ultimate limit state, as given below:

$$\mu_\varphi = \frac{\varphi_{ultimate}}{\varphi_{yield}} \qquad (1)$$

As such, it is affected by the definition of the limit states as well as by the structural response characteristics of the material and the section configuration. It is not intended herein to provide a critical discussion of the limit states and their effect, but rather attention will be focussed on material and section characteristics which affect the evaluation of curvature ductility.

2.1 Material properties

The material deformational and strength parameters have an effect on curvature ductility due to their effect on the yield and ultimate limit states. Under dynamic loading, the yield strength is affected by the rate of loading, hence a dynamic yield strength should be considered. In the work of Manzocchi (1991), the results of various experimental studies, summarised in Soroushian and Choi (1987), were used to derive a continuous model accounting for the instantanious strain-rate effect. Some analyses were undertaken, and preliminary conclusions pointed towards mild variations in local seismic response, with significant changes on the overall structural level. The model by Manzocchi was used by Izzuddin and Elnashai (1992) and Elnashai and Izzuddin (1993) to analyse offshore platforms and steel frames, respectively, subjected to blast and earthquake loading. In this case, the overall response of the selected structure was almost unaffected, for earthquake loading. However, the buckling sequence was changed when strain-rate was taken into account. This may be significant in braced frames, where buckling of the brace is an important seismic design limit state. Furthermore, it can be argued that the effect of strain-rate on the yield strength should be taken into consideration even in static analysis, where buckling may take place.

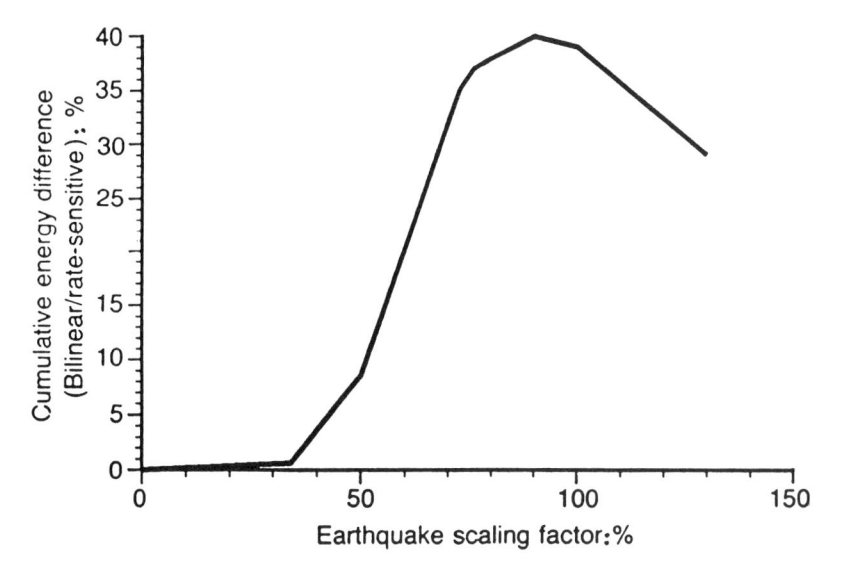

Figure 4: Effect of Strain-rate on Steel Frame Response

Steel manufacturers guarantee minimum yield strengths, but do not specify maximum values. This leads to very large variations in actual yield strength in comparison with the design value (Manzocchi, Chryssanthopoulos and Elnashai (1992); Mazzolani Mele and Piluso (1993)). An increase in the yield strength, in conjunction with a fixed ultimate limit state criterion which is independent of yield, leads directly to a reduction in curvature ductility. It is therefore of critical importance, due to the randomness in the properties of the material response parameters, to include an element of probabilistic analysis in the assessment of the ductility supply of steel sections. Manzocchi, Chryssanthopoulos and Elnashai (1994) derived contour plots relating the ductility supply of steel flexural members to the dispersion in the yield strength and the strain-hardening modulus for a given probability of exceedence, a sample of which is shown in Figure 5. Such a concept should be extended to cover sub-assembledges and frames with various configurations.

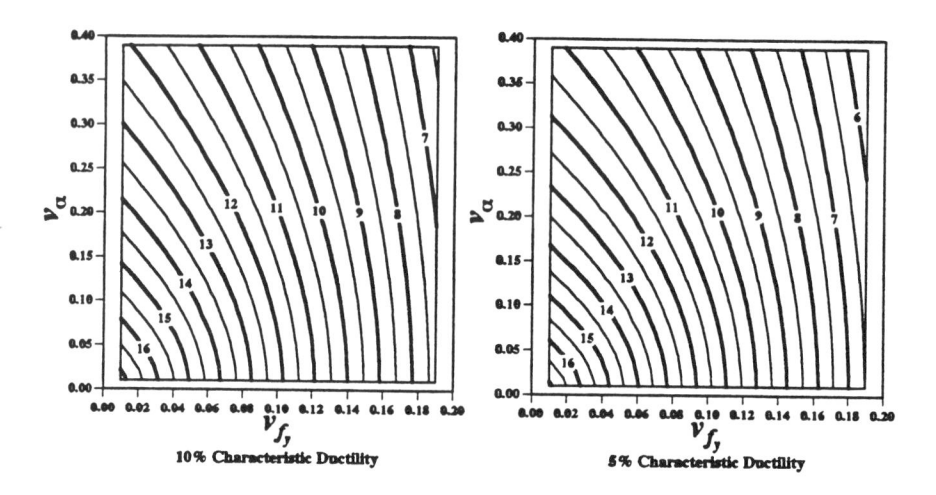

Figure 5: Contours of Ductility Supply of a Steel Cantilever Member

Brittle fracture under repeated loading was studied by Kuwamura and Akiyama (1994). They observed that local ductility and notch toughness were greatly reduced due to press-braking of the square sections studied. However, no quantitative measure for accounting for this effect was presented. Balio and Perotti (1987) derived a model for seismic analysis of steel members taking into account local buckling, low-cycle fatigue and fracture. The model was compared to a set of tests on steel cantilevers, and very good correlation was achieved. Whereas the effect of local buckling on curvature ductility needs no further proof, less cases have been observed where fatigue and fracture are governing considerations. It is, nonetheless, important to continue to develop such models and to calibrate analysis tools to well-controlled test data.

2.2 Section properties

At section level, one of the most important limit states is that relating to local buckling under repeated cyclic loading. Due to the nature of earthquake loading, local buckling criteria derived under monotonic loading give rather optimistic estimates of the section curvature ductility, as observed by Takanashi et al (1975). In the latter reference, the ductility supply under monotonic and cyclic loading regimes were compared, showing that very large reductions in ductility ensue under severe cyclic loading. Tests cionducted by Lee and Lee (1994), showed that the cyclic deformation capacity may be up to 60%

lower than that under monotonic loading conditions. The effect of cyclic degradation was shown to be more serious in the presence of high axial loads.

The approach of Ballio and Perotti (1987) was employed, with modifications to account for recovery, by Elghazouli (1992) who used test results on cantilevers to calibrate the normalised critical buckling strain of a flange outstand $\varepsilon_{critical}$, given by:

$$\frac{\varepsilon_{critical}}{\varepsilon_{yield}} = \frac{k\pi^2 E}{12\sigma_{yield}(1-v^2)(b/t)^2} \qquad (2)$$

Where k is a buckling coefficient, E is Young's Modulus, b and t are the width and thickness, respectively of the plate element, and v is the Poisson's ratio. The resulting model, in conjunction with the unloading-reloading relationships proposed in the latter reference, was used to analyse cantilever members under earthquake loading with excellent comparisons with test results being achieved (Elnashai and Elghazouli, 1993).

Analytical solutions, such as that given by Kato (1989, 1990) require calibration to account for cumulative plastic straining. Using tests on mild steel stub columns, and modifying the results to account for the presence of a moment gradient, Kato derived the expression,

$$\frac{1}{s} = 0.689 + \frac{0.651}{\alpha_f} + \frac{0.0553}{\alpha_w} +/_- 0.0303 \qquad (3)$$

Where s is the critical stress ratio (σ_{cr}/σ_y), α_f and α_w represent the flange and web slendernesses, respectively. This expression applied in the derivation of flange-web buckling interaction formulae giving allowable slenderness ratios for a target ductility level.

A similar approach was used by Broderick (1994), who employed the concept of equivalent sections proposed by Kato to derive a set of expressions from which the rotation ductility of composite steel/concrete (partially encased) members may be determined. The analytical results were compared to tests conducted at Imperial College with a high degree of correlation being avhieved, as shown in Table 1.

Table 1: Comparison between Analytical and Experimental Results

TEST	EI (Nmm^2)	M_y (kNm)	Δ_y (mm)	M_u (kNm)	Δ_u (mm)	μ_Δ
IC02 - Exp.	2.62x10^{12}	61.1	11.2	89.8	66	4.9
- Anal.	2.85x10^{12}	80.8	13.6	90.4	73.2	4.4
ICA2 - Exp.	2.61x10^{12}	71.1	11.0	90.0	75	5.8
- Anal.	3.07x10^{12}	81.4	10.7	89.5	73.6	5.9
ICB2 - Exp.	2.93x10^{12}	78.4	10.8	118.7	60	5.5
- Anal.	4.35x10^{12}	113.3	10.5	119.3	51.5	4.9

In the context of evaluating response modification factors (q-factors), which is the ultimate objective of local ductility studies, estimating the curvature ductility of composite beams assumes a role of great significance. While much work has been performed on the rotation capacity of such beams under monotonic loading, both in positive and negative bending, (Johnson, 1970), little information exists on their response under earthquake loading. The review given by Broderick (1994) concluded that the rotation capacity of composite beams under negative and positive moments may be conservatively estimated from existing work, pending analysis and tests on cyclically-loaded members.

3 Inelasticity in Steel and Composite Members

The evaluation of rotational ductility requires knowledge of the curvature distribution within that part of the member which deforms inelastically, referred to as the plastic hinge. It is customery to assume either a constant, linear or piecewise linear distribution of curvature within the plastic hinge length to facilitate the evaluation of member rotations.

Lay and Galambos (1967) quantified the length of member over which the strain exceeds that at which the onset of strain-hardening occurs. This is given by:

$$\frac{L_b}{b} = \frac{\pi t}{4w}\left[\frac{t}{w}\left\{\frac{A_w}{A_f}\right\}^{0.25}\right]\left[\frac{7}{3}\left\{\frac{1+\nu}{h}+0.25\right\}\right]^{0.25} \quad (4)$$

Where A_w amd A_f are the areas of web and flange, respectively and h is the height of the section. Using the above equation, the rotation capacity may be obtained, based on the assumption that the rotations in the elastic and pre-strain hardening regimes are negligible. Further solutions for the length of plastic hinge, hence the rotation ductility, were given by Kemp (1987). On the other hand, Kato (1989, 1990) used the critical stress criterion described above to propose expressions for rotation ductility for three ranges of normalised applied axial stress ρ (ratio of applied axial stress to yield stress) with regard to the critical stress. This approach has been verified and used, with modifications by Manzocchi (1994) for steel members and by Broderick (1994) for composite members.

Gioncu et al (1994) stipulated a collapse mechanism for the web of a member and developed a computer program to evaluate the ultimate plastic rotation of steel memebrs. Their results compared well with experimental studies, and it was concluded that the code concept of ductility classes based on section characteristics is inadequate. The same statement was made in a paper by Gioncu and Mazzolani (1994) in the context of reviewing briefly available methods of rotation ductility evaluation. Moreover, the study by Aiello and Ombres (1994) demonstarted the effect of axial and shear forces on rotation ductility and concluded that such effects should be included in classifying steel memebrs for design purposes.

In the work of Vayas and Psycharis (1994), the method of local ductility definition of EC8 Part 1.3 Chapter 2 (Concrete) was adopted for steel memebrs, leading to curvature ductility limits for various EC3 member classification. These were used to develop member limit state rules for seismic design that account for the interdependence of web and flange buckling as well as effect of axial load.

Elnashai and Elghazouli (1993) conducted a parametric study, using advanced inelastic analysis procedures, on the effect of various parameters on seismic design quantities, including the plastic hinge length of composite (partially-encased) members. The effect of axial load, yield strength, strain-hardening slope, slenderness width-to-thickness ratio, stirrup spacing and concrete confinement factor on the plastic hinge length was studied. It was concluded that a normalised plastic hinge length of 0.2L should be used to conservatively calculate rotational ductility, whilst 0.4L should be used for ductile detailing purposes. Elghazouli (1992) plotted the experimentally-observed curvature ductility-rotation ductility relationship from a series of cyclic and pseudo-dynamic tests on composite memebrs conducted at Imperial College. These were compared with the assumptions of constant and linearly-varying curvature distributions in the plastic hinge zone. It was concluded that the constant curvature distribution, leading to the relationship,

$$\mu_\theta = 0.54 \, \mu_q \tag{5}$$

is unconservative in many cases, whilst that of a linearly-varying curvature distribution, expressed as,

$$\mu_\theta = 0.28 \, \mu_q \tag{6}$$

is conservative for all cases where a realistic confinement factor is used. It is, however, noteworthy that the analytical results on which Elghazouli based his recommendation excluded shear deformations. Therefore, the constant curvature distribution may not be as unconservative as might be concluded from the above discussion.

The ductility of composite beams was recently studied by Pacurar (1994), who tested two beams and compared test results with analytical evaluations. The rotation ductilities obtained (under increasing vertical load in a two-point simple beam test) was on average about eleven. It is not clear, though, how can these results apply to earthquake loading, where the vertical loads are constant whilst the transverse load varies.

Subassembledges comprising composite beams and columns were tested by Plumier at al (1994), who observed that failure occured in the beams and not the columns. Little deterioration of the composite panel zone was observed. The conclusion on beam failure, satisfying seismic design requirements, is, however, questionable, since no slab was considered in the tests.

A substantial amount of work on member ductility was undertaken by Ballio and his co-workers. A review of these studies; experimental and analytical, is given by Ballio and Castiglioni (1994).

4 Connection Behaviour

Connections in steel frames form a critical component of the structure, from both economic and behavioural points of view. Fully-welded connections are expensive and require highly sophisticated welding procedures and skilled labour. The deformation and strength of connections not only affects directly the local ductility supply but also has a marked effect on the behaviour of the connected members, especially as fully-welded, so called rigid, connections may be full or partial strength, depending on the design actions and the capacity provided. Since even full strength rigid connections are deformable, and pinned connections have some definable stiffness and strength, all connections may be viewed as semi-rigid.

Figure 6, Elnashai and Elghazouli (1994), gives a broad classification of connection stiffness and strength. In this figure, Type A represents a fully-welded connection with no account taken of shear panel deformation, type B is similar to A, but with due account taken of shear panel stiffness and strength, type C is a semi-rigid configuration (top, seat angles and web angles) and type D is a flexible connection (two web angles only).

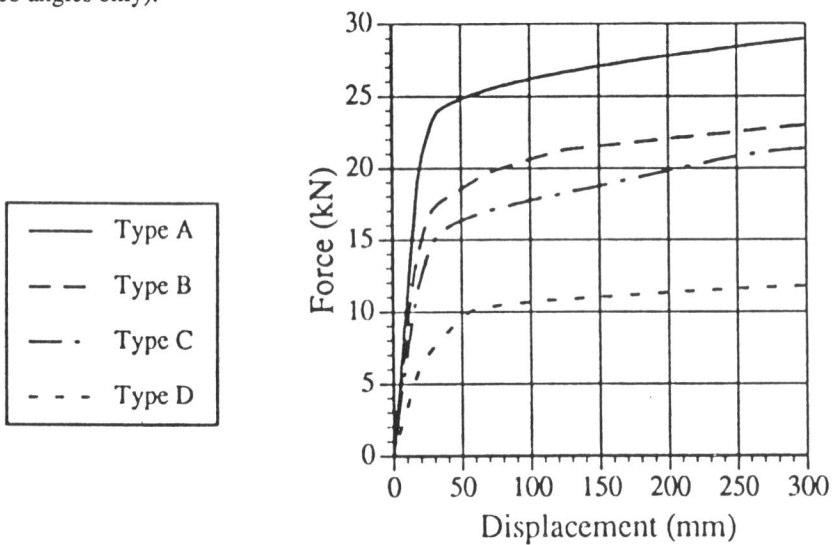

Figure 6: Classification of Connection Stiffness and Strength

Early tests on the effect of connection type on seismic response (Nander and Astaneh, 1989) showed that the rigidly connected frame is by no means the optimum solution, and that semi-rigid frames have an important role to play in seismic design. The study by Parra Rosales (1991) confirmed this observation by comparing the seismic response of four frames of various configurations under the effect of two earthquakes. It was concluded that semi-rigid connections, whilst exhibiting higher flexibility than rigid connections, attract lower inertial loads leading to displacements similar to, or even lower than, those experienced by rigid frames. An experimental study of two storey steel frames was undertaken by Takanashi et al (1992), as part of a collaborative programme between the Institute of Industrial Science, University of Tokyo and Imperial College, London. The test results clearly indicated that the semi-rigid frame design is feasible, even in areas of high seismicity, since it may produce lower inter-storey drifts than those of the rigid frame solution. As the control of these drifts is commonly the most critical seismic design criterion for moment resisting frames, this feature can lead to increased economy in material weight as well as in fabrication and erection costs.

Analytical models for connection response have mostly relied upon the representation of the moment-rotation behaviour using global (super-element) approaches. A component-based model, where only the geometric and strength properties of components are required, was derived by Madas and Elnashai (1992). The model, shown in Figure 7, was verified by comparisons with tests conducted by Ballio at al (1987) and the test series mentioned above (Takanashi et al, 1992). It was then used to conduct parametric studies on various types of connection (Elnashai and Elghazouli, 1994).

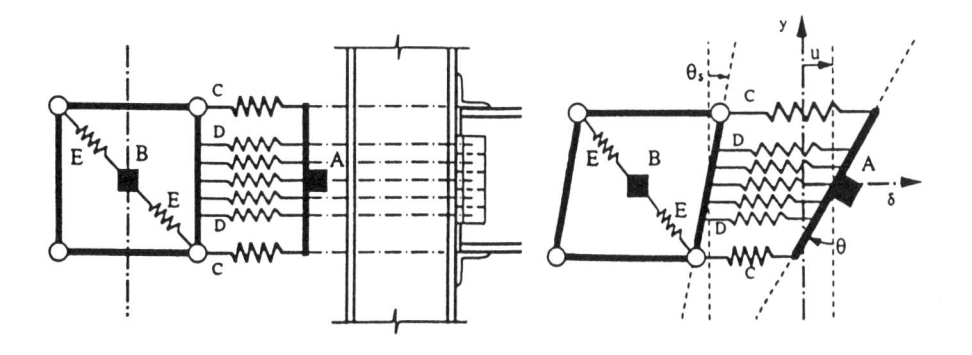

Figure 7: Component-based Connection Model

As mentioned above, accurate connection modelling is not only important with regard to connection deformation and strength, but also in the context of its effect on the members framing into it. The analysis undertaken by Elnashai and Elghazouli (1994) was used to plot the variation in plastic hinge length (normalised by the storey height) versus storey displacement of a two storey frame as a function of the connection type, as shown in Figure 8. It follows that the global ductility of a steel structure will be inaccurately evaluated if the response of the connection, even when it is fully welded, is not realistically represented.

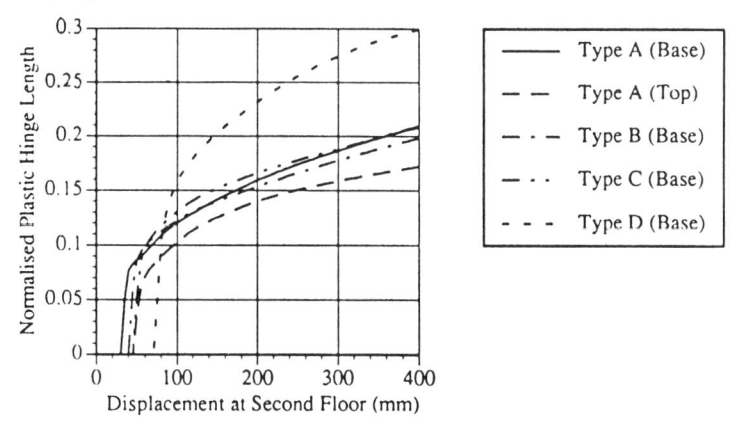

Figure 8: Normalised Plastic Hinge Length for Various Connection Types

Two types of welded beam-column connections were studied by Dimoiu and Dan (1994) using simplified procedures and 3D finite element analysis, but no ductility evaluation was undertaken, since the analysis did not proceed to total failure. Welded beam-column connections with slender panel zones were tested by Pasternak and Vayas

(1994). From the thirteen tests undertaken, it was concluded that thin-walled panel zone conecctions exhibit three resistance mechanisms; panel shear, tension field and frame action in the panel and its surrounding members. It was observed that frame action increases the energy absorbtion capacity but may cause higher susceptibility to low cycle fatigue.

In the context of testing connections, significant contributions have been made by Balio et al (1986, take from Plumiers paper in JCSR) and by Plumier (1994), amongst others. Such studies are essential for the further development of analytical models and their calibration. Composite connections were studied by Pradhan and Bouwkamp (1994), who tested fully-welded and bolted connections with composite panel zones. They concluded that bolted connections may be adequate, although the particular geometries tested indicated that the welded connections exhibit higher shear panel yielding, hence energy dissipation.

Interesting studies on special connections have been undertaken by several researchers. Semi-rigid ductile links between RC structural walls (replacing RC coupling beams) were studied by Campione et al (1994). The system developed showed high ductility and behaviour factors (q) of 3.4 to 4.0 for forces and 6.3 to 7.9 for displacements. Dunai et al (1994) studied experimentally steel-to-concrete end plate connections under cyclic loading, and concluded that rapid deterioration in stiffness and strength may occur, leading to an effective hinge. Finally, composite link beams in eccentrically-braced frames were studied by Selleri and Spadaccini (1994). The tests conducted, up to displacement ductilities of 8 to 12, clearly indicated that composite link beams are superior to their steel counterparts, with up to 60% increase in ductility.

5 Closing Remarks

Failure mode control and the evaluation of response modification factors, which are essential pre-requisites for safe and economic seismic design, require the accurate evaluation of local ductility supply. Material characteristics and section limit states are required for the evaluation of member ductility. The connection response is an essential ingredient, by virtue of its direct significance and its effect on the behaviour (end conditions) of the connected members. Significant developments have been undertaken in the context of local ductility evaluation, some of which are given above. However, much work is still needed, to reduce the uncertainty inherent in seismic design, at least from the supply prespective. Some of the many pressing issues emanating from the above discussion are given below.

Material Level: Brittle fracture; low cycle fatigue; variable amplitude strain-rate sensitivity; constitutive modelling of high strength steel.

Section Level: Interdependent flange/web buckling; local buckling recovery under reversed loading, influence of concrete slab on steel joist response.

Member Level: cyclic lateral torsional buckling; degrading interactive moment-thrust-shear limit surfaces; effective length as a function of connection stiffness, strength and ductility.

Connection Level: Variable amplitude cyclic moment-rotation relationships in 2D and 3D; criteria for ductility supply in partial strength connections; role of effective slab width on connection performance.

Developments towards the ends identified above will result in more rigorous failure mode control and will minimise earthquake design forces through the use of increased behaviour factors. Hence, the ultimate objective of building safe and economic structures may be realised.

6 Acknowledgement

This paper includes information from several research projects, past and present, conducted at Imperial College, London, mainly funded by the UK Science and Engineering Research Council. The contributions made to these projects by several researchers, especially Drs. A.Y.Elghazouli and B.M.Broderick, as well as Mr. G.M.Manzocchi, are gratefully acknowledged.

7 References

Aiello, M.A. and Ombres, L., (1994), Influence of cyclic behaviour on the local ductility of steel members, International Workshop on Behaviour of Steel Structures in Seismic Areas, 26 June to 1 July, Timisoara, Romania.

Ballio, G., Calado, L., De Martino, A., Faella, C. and Mazzolani, F., (1987), Cyclic behaviour of steel beam-column joints, Experimental research, Estratto dalla Rivista Costruzioni Metalliche, Vol.2, pp.3-24.

Ballio, G. and Castiglioni, C., (1993), Seismic behaviour of steel sections, Journal of Constructional Steel Research, Vol.29, Nos. 1-3 (Special Issue on Seismic Performance of Steel Structures), pp.21-54.

Ballio, G. and Perotti, F., (1987), Cyclic behaviour of axially loaded members; numerical simulation and experimental verification, Journal of Constructional Steel Research, Vol.7, pp.3-41.

Broderick, B.M., (1994), Seismic Testing, Analysis and Design of Composite Frames, PhD Thesis, ESEE Section, Imperial College, University of London, January.

Campione, G., Colajanni, P. and Scibilia, N., (1994), Behaviour of concrete walls coupled by ductile links with semi-rigid connections, International Workshop on Behaviour of Steel Structures in Seismic Areas, 26 June to 1 July, Timisoara, Romania.

Dimoiu, I and Dan, S., (1994), On the ductility of two welded steel beam column connections, International Workshop on Behaviour of Steel Structures in Seismic Areas, 26 June to 1 July, Timisoara, Romania.

Dunai, L., Ohtani, Y. and Fukumoto, Y., (1994), Cyclic behaviour of steel-to-concrete end-plate connections, International Workshop on Behaviour of Steel Structures in Seismic Areas, 26 June to 1 July, Timisoara, Romania.

Elghazouli, A.Y., (1992), Earthquake Resistance of Composite Beam-columns, PhD Thesis, ESEE Section, Imperial College, University of London, January.

Elnashai, A.S. and Elghazouli, A.Y., (1993), Performance of composite steel-concrete members under earthquake loading, Pat I: Analytcial model, Part II: Parametric study and design considerations, Journal of Earthquake Engineering and Structural Dynamics, Vol. 22, pp.315-368.

Elnashai, A.S. and Elghazouli, A.Y., (1994), Seismic behaviour of semi-rigid steel frames, Journal of Constructional Steel Research, Vol.29, Nos. 1-3 (Special Issue on Seismic Performance of Steel Structures), pp.149-174.

Elnashai, A.S. and Izzuddin, B.A., (1993), Sources of uncertainty and future research requirements in seismic analysis of structures, Nuclear Engineering, Vol.32, No.4, pp.213-220.

Elnashai, A.S. and Izzuddin, B.A., (1992), Optimal nonlinear dynamic analysis of steel jacket structures, Offshore and Polar Engineering Conference ISOPE '92, San Francisco, USA.

Gioncu, V., Mateescu, A. and Iuhas, A., (1994), Contribution to the study of plastic rotational capacity of I steel sections, International Workshop on Behaviour of Steel Structures in Seismic Areas, 26 June to 1 July, Timisoara, Romania.

Gioncu, V. and Mazzolani, F.M., (1994), Alternative methods for assessing local ductility, International Workshop on Behaviour of Steel Structures in Seismic Areas, 26 June to 1 July, Timisoara, Romania.

Izzuddin, B.A. and Elnashai, A.S., (1993), Influence of rate-sensitivity on the response of steel frames, 2nd European Conference on Structural Dynamics Eurodyn '93, Trondheim, Norway.

Johnson, R.P., (1970), Research on steel-concrete composite beams, American Society of Civil Engineers, Vol.96, No.ST3, pp.445-459.

Kato, B., (1989), Rotation capacity of H-section members as determined by local buckling, Journal of Constructional Steel Research, Vol. 13, pp 95-109.

Kato, B., (1990), Deformation capacity of steel structures, Journal of Constructional Steel Research, Vol.17, pp 33-94.

Kemp, A., (1987), Quantifying ductility in continuous composite beams, Composite Construction in Steel and Concrete, Proceedings of Engineering Foundation Conference, Henniker, New Hampshire, pp. 107-121.

Kuwamura, H. and Akiyama, H., (1994), Brittle fracture ubder repeated high stresses, Journal of Constructional Steel Research, Vol.29, Nos. 1-3 (Special Issue on Seismic Performance of Steel Structures), pp.5-20.

Lay, M.G. and Galambos, T.V., (1967), Inelastic beams under moment gradient, American Society of Civil Engineers, Vol.93, No.ST1, pp.381-399.

Lee, G.C. and Lee, E.T., (1994), Local buckling of steel sections under cyclic loading, Journal of Constructional Steel Research, Vol.29, Nos. 1-3 (Special Issue on Seismic Performance of Steel Structures), pp.55-70.

Madas, P.J and Elnashai, A.S., 1992, A component-based model for the response of beam-column connections. 10th World Conference on Earthquake Engineering, Madrid, Spain, July.

Manzocchi, G.M.E., (1991), The Effect of Strain Rate on Steel Structures, M.Sc. Dissertation, ESEE Section, Imperial College, University of London, August.

Manzocchi, G.M.E., Chryssanthopoulos, M.K.. and Elnashai, A.S, (1992), Statistical analysis of steel tensile test data and implications on seismic design criteria, ESEE Research Report 92-7, Imperial College.

Manzocchi, G.M.E., Chryssanthopoulos, M.K. and Elnashai, A.S., (1994), Reliability based limits on member ductility, ESEE Research Report 94-1, Imperial College.

Mazzolani, F.M., Mele, E. and Piluso, V., (1993), Statistical characterization of constructional steels for structural ductility control, Costruzioni Metalliche, No. 2 pp.89-101

Nander, M.N. and Astaneh, A., (1989), Experimental studies of a single storey steel frame with fixed, semi-rigid and flexible connections, Earthquake Engineering Research Centre, Berkeley report no. UCB/EERC 89-5, August.

Pasternak, H. and Vayas, I., (1994), Cyclic behaviour of thin-walled welded knee joints, International Workshop on Behaviour of Steel Structures in Seismic Areas, 26 June to 1 July, Timisoara, Romania.

Pacurar, V., (1994), Ductility of steel-concrete mixed section beams, International Workshop on Behaviour of Steel Structures in Seismic Areas, 26 June to 1 July, Timisoara, Romania.

Parra Rosales, J.G., 1991, Seismic resistance of of steel frames with semi-rigid connections, MSc dissertation, ESEE Section, Imperial College, August.

Plumier, A., Abed, A. and Tiliouine, B., (1994), Increase of buckling resistance and ductility of H sections by encased concrete, International Workshop on Behaviour of Steel Structures in Seismic Areas, 26 June to 1 July, Timisoara, Romania.

Pradhan, A.M. and Bouwkamp, J., (1994), Structural performance aspects on cyclic behaviour of the composite beam-column joints, International Workshop on Behaviour of Steel Structures in Seismic Areas, 26 June to 1 July, Timisoara, Romania.

Selleri, F. and Spadaccini, O., (1994), Design of dissipative composite steel-concrete links, International Workshop on Behaviour of Steel Structures in Seismic Areas, 26 June to 1 July, Timisoara, Romania.

Soroushian, P. and Choi, K.B., (1987), Steel mechanical properties at different strain rates, J. Struct. Eng., ASCE, Vol.113, No.4, pp. 663-672

Takanashi, K., Elnashai, A.S., Elghazouli, A.Y. and Ohi, K., (1992), Experimental behaviour of steel and composite frames under static and dynamic loading, Engineering Seismology and Earthquake Engineering, Imperial College report ESEE 92-10 (jointly with the Institute of Industrial Science), November.

Takanashi, K., Udagawa, K. and Tanaka, H., (1973), Failure of steel beams due to lateral buckling under repeated loads. IABSE Symposium on Resistance and Ultimate Deformability of Structures Acted Upon by Well Defined Repeated Loads, Lisbon, pp.163-169.

Vayas, I and Psycharis, I, (1994), Local cyclic behaviour of steel members, International Workshop on Behaviour of Steel Structures in Seismic Areas, 26 June to 1 July 1994, Timisoara, Romania.

12 ON THE DUCTILITY OF TWO WELDED STEEL BEAM-COLUMN CONNECTIONS

I. DIMOIU and S. DAN
Civil Engineering Faculty, Timisoara, Romania

Abstract

One type of the large used beam column connection is made by the cleats applied on the beam flanges and welded to the column flange. Another cleats or equal angles are used to joint the beam web to the column flange. This type of steel beam - column connection is placed at a distance of 0.2 beam length from the column. A small cantilever is welded at the column in the workshop. This fact ensures a hold in check of a good quality of the beam - column connection. A joint between the cantilever and beam is made at the site. This type at joint is largely used as a screwed joint and as well as, a welded one. An actual analytic method is performed. Using the plane cross section hypothesis and the yielding range of the construction steel, a procedure has been stated for the secant elastic modulus. The space beam column is modeled by rectangular finite elements. Assuming a load level which causes yielding of some finite elements, the secant elastic modulus can be computed. The displacement of some nodes prescribed of the analytic models had been noticed. The displacements differ with respect to the behaviour range: elastic and elasto - plastic. The curve M $- \Phi$ had been drawn.

Keyords: Beam-Column Connection, Finite Element, Elastic Modulus, Secant.

1 Introduction

The beam column connection, most frequently used at romanian steel structures is presented by the drawings of Fig.1. it is called direct beam column connection. A short cantilever is welded to the column in the manufacturing process at the workshop. A high quality welded structural joint is ensured. The beam is shaped so the flange gap and web gap to be in the different cross section. The beam is resting on the columns short cantilevers during the laying time. The lifting tool could be put out of work after laying. The cleats settled on both face of the flange realize a symmetrical stress run - off, shown in Fig.1b. An asymmetrical stress run - off is

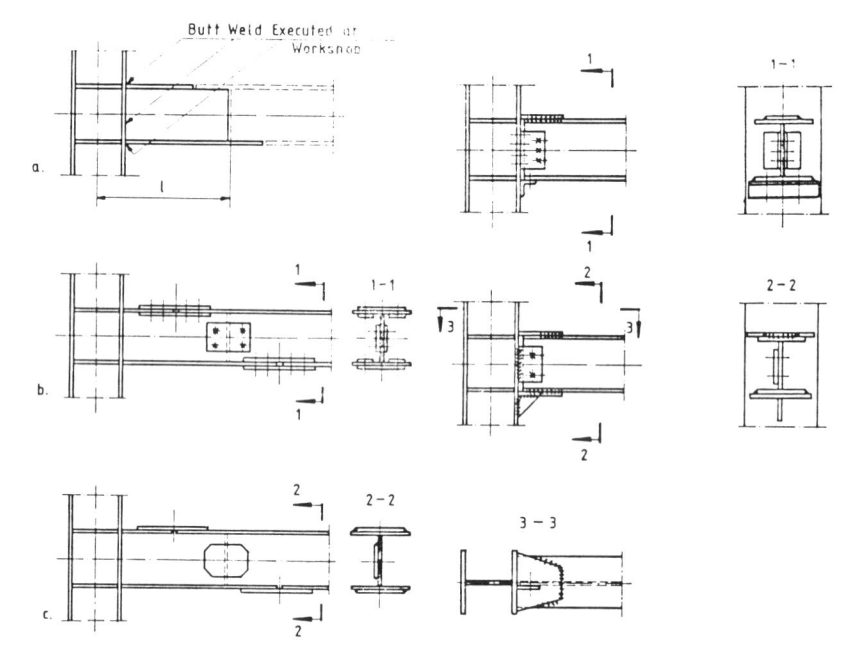

FIG. 1 INDIRECT BEAM-COLUMN JOINT   FIG. 2 DIRECT BEAM-COLUMN JOINT

presented by the joint of Fig. 1c. The gaps are covered by one sided cleats, only.

The drawings shown in the Fig.2. is another current beam – column connection type used in romanian steel structures. The beam end is attached to the column flange by the help of welded equal angles or welded cleats. The Fig.2a is showing a symmetrical stress run – off, while the Fig.2b is illustrating an asymmetrical one. This type of beam column joint is further called indirect connection.

The above presented sorts of beam – column connection are of a material discontinuity. The discontinuity is placed at an out of strong stress zone for the direct connection usually at a 0.2 span distance. For the indirect connection, the material discontinuity is placed just at the column flange face. This is a strong stress zone.

The yielding phenomenum, of the direct beam column connection is met in the butts which attach the beam flanges to the column flange. In another words, the yielding starts in the cross beam section. It is double T shaped cross section and is an adjacent beam section to the flange column Fig.3a.

For the indirect beam column connection, one can say that the yielding phenomenum is met in a beam cross section out at the joining zone. The added cleats or added angles are designed to bearing capacity condition of the butt weld: or fillet weld. The limit strength of the weld is lower than

Ductility of welded beam-column connections **151**

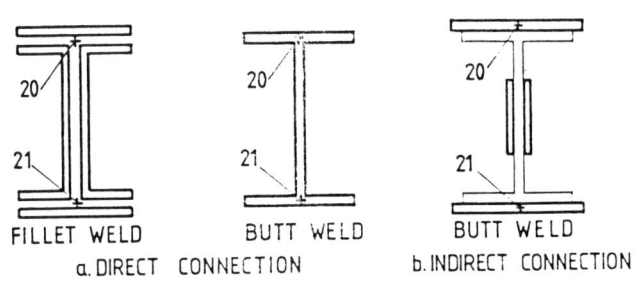

FIG. 3 CROSS SECTION OF WELDING MATERIAL
a. DIRECT CONNECTION — FILLET WELD / BUTT WELD
b. INDIRECT CONNECTION — BUTT WELD

limit strength of base material so the adding elements (cleats, angles) provide a cross section with a greater bearing capacity than of the current beam one. Only the material of the fillet weld or butt weld gives a bearing capacity greater or equal to the bearing capacity at the beam Fig.3b. Bu consequence the yielding is met in the welds.

2 Mathematical model

Consider the double T cross section shown in the Fig.4. The hatched aria of the flanges, indicates an yielded region. The external bending moment M is balanced by the moment of the cross section stresses about the neutral axis.

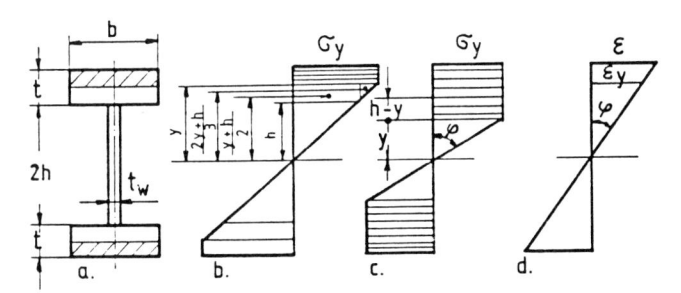

FIG.4 PARTLY PLASTIFIED DOUBLE T CROSS SECTION

$$M = M_{fp} + M_{fe} + M_w \qquad (1)$$

where M_{fp} is the moment of stresses form the yielded area of flanges, M_{fe} is the moment of stresses from the unyielded area of flanges and M_w is the moment of stresses from the web area.

Using the notations of the Fig.4b we have:

$$M_{fp} = b(t+h-y)\sigma_y(y+(t+h-y)/2) \qquad (2)$$

$$M_{fc} = b(y-h)\ (\sigma_y h/y)2(h+y)/2 + b/2(y-h)/y \cdot \sigma_y(y-h)2(h+2y)/3 \qquad (3)$$

$$M_w = 1/2\ t_w h \cdot h/y \cdot \sigma_y\ 2 \cdot 2/3h \qquad (4)$$

Performing the following notations:

$$B = t_w/b\ ;\quad C = M/b(h+t)^2\sigma_y;\quad H = h/(t+h) \qquad (5\text{-}7)$$

$$T = t/(h+t);\quad Z = y/(t+h) \qquad (8\text{-}9)$$

the equation (1) can be written

$$Z^3 - 3(1-C)Z + 2(1-B)H^3 = 0 \qquad (10)$$

There is only one real root of the equation (10). It is of form:

$$Z_1 = 2(1-C)^{1/2}\cos \alpha/3 \qquad (11)$$

where $\cos \alpha = (B-1)H^3/(1-C)^{3/2}$ \qquad (12)

If the external bending moment is increased the web fibres exceed the elastical range Fig.4b. and the above mentioned equilibrium equation (1) takes the form:

$$Z^2 = 3\left[\ C/B\ +\ T/B \cdot (2H+T)\ -\ H^2\ \right] \qquad (10')$$

Keeping the fundamental hypothesis of plane cross section, the magnitude of the rotation angle is expressed by:

$$\text{tg}\Phi = 1 \cdot \epsilon_y/y = \epsilon_y/Z(t+h) \qquad (13)$$

The strain of the extreme fibre is given by:

$$\epsilon = \epsilon_y/Z \qquad (14)$$

Finally, starting from the Young's modulus definition

$E = \text{tg}\alpha_y = \sigma_y/\epsilon_y$ a secant modulus can be defined as:

$$E = tg\alpha = \sigma_y/\epsilon = \sigma_y z/\epsilon = Z.E_y \qquad (15)$$

The upper half of cross section is stretched, the lower half of cross section is compressed. In the midle of the web, two adjacent fibres couldn't be one stretched, the other compressed.

Taking into account the real phenomenum of the entire plastified cross section the influence of the share force is neglected in the previous theoretical development.

3 Numerical example

The following dimensions belong to a double T cross section: $b = 90$ mm; $h = 100$ mm; $t = 20$ mm; $t_w = 14$ mm. The results noted in Table 1 are refered to the yielding phenomenum analysed by thye previous analitycal method.

Table 1   Plastic Moment and Rotation Angle

Place of yielding	Y mm	M kNm	E kN/mm²	ϵ x10⁻³	Φ x10⁵
Extreme flange fibre	120	106.0	196.46	1.22	1.10
Into the flange	112.2	117.4	122.74	1.95	1.63
Into the web	48.5	128.6	84.96	2.82	2.35

The ratio between the values of strain in the hardening range and yielding range about [1], [2] is $\epsilon_s/\epsilon_y = 10\text{-}20$. On the actual example, for an web elastic zone of $2y = 2 \times 48.5 = 99$ mm breadth, the similar strain ratio for the flange extreme fibre is $\epsilon_s/\epsilon = 0.00282/0.00114 = 2.47$. The hardining modulus is not still attained. The Fig.5 shows the relationship between bending moment and the angle of rotation.

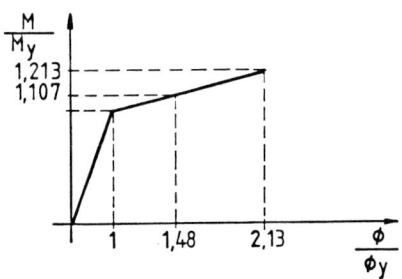

FIG. 5 M — ϕ DIAGRAM

The Fig.8d is showing a detailed zone of the flange joint. It is refered to two rows of rib finite elements and two rows of flange finite elements. They are neighbour to the column flange. A stress peak emphasized in this picture. It corresponds to the common point at the beam flange and column flange, on the symmetry axis. This poinht is noted by 20 on the upper beam flange and by 21 on the lower beam flange Fig.3.

Starting from the above mentioned pictures, one can say, the yielding phenomenum begins from these points. The finite elements of the beam flanges, located on the symmetry axis and adjacent to the collumn flange are firstly plastified. Then, others finite elements of the beam flange attaine yielding stage. The latter are symmetrically towards the flange edges; and adjacent to the column flange. During yielding of these finite elements, beam web finite elements are involved. Firstly, the web element neighbouring the beam flanges and adjacent to the column flange attaine the plastic stage.

The development of the phenomenun has been studied in a steps sequence. Every time the applied force was incrementaly increased. When a finite element attained yielding stress, a new elastic modulus has been provided to it for the next step. The stress magnitude is computed by the formula:

$$\sigma = \pm S_{min,max} / 1.t \pm 6 M_{min,max}/1.t^2 \qquad (16)$$

where t is the thickness of the finite element and S_{max}, S_{min}, M_{max}, M_{min} are forces on unit length. They are provided by the computer program.

The rotation angle of the cross section has been determined by the help of the displacements of node 20 and 21. They are common nodes for the column flange and beam flange. Knowing the displacement of the above mentioned nodes the angle of rotation is given by:

$$\Phi \approx tg\, \Phi = (\Delta u(x) + \Delta(y))/2h \qquad (17)$$

where $\Delta u(x) = u_{20}(x) - u_{21}(x)$ \qquad (18)

$\Delta u(y) = u_{20}(y) - u_{21}(y)$ \qquad (19)

Let M_y, Φ_y, the external bending moment and the rotation angle respectively at the start of yielding stage at the nodes 20 and 21. Let M, Φ, the external current bending moment and the rotation angle respectively. The latter are refered to the some beam cross section marked by the nodes 20 and 21.

The pairs of ratio values M/M_y, Φ/Φ_y have been obtained.

4 Approximate elastic – plastic analysis

4.1 Presentation of models

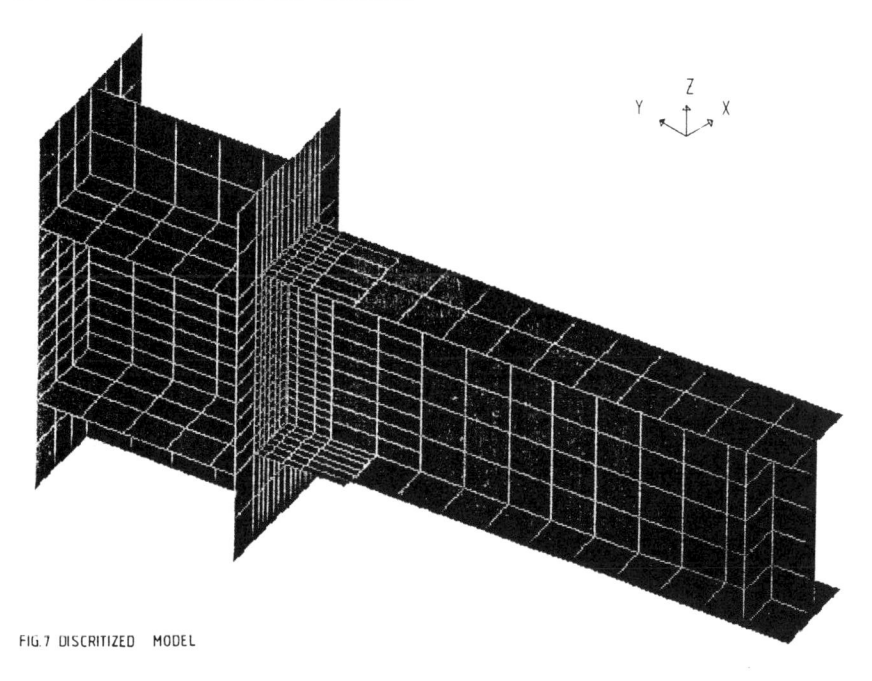

FIG.7 DISCRITIZED MODEL

Two construction steel models had been made. They are at 1:1 scale and are refered to the privious types of beam column connection. The geometry and dimensions of testing models are shown in the Fig.6.

Every considered model has been discretized in plane finite elements Fig.7. Some of the plane elements act in membrane stage, other in bending stage. The critical zones of the joint are discretized in great detail. The finite elements are of 2 mm up to 10 mm breadth.

A vertical lood is applied at the end at the short cantilever.

4.2 Direct beam – column connection

The drawings of pictures table noted by Fig.8 are illustrating the stress isolines. The stress isolines of the whole assembly are shown in the Fig.8a. A more detailed sight is presented in the Fig.8b and Fig. 8c. The stress isolines of the both beam web and column web are shown in the Fig.8b. The stress isolines of two beam flanges are shown in the Fig.8c, they are prolongede into the web column ribs.

FIG. 6 CONSTRUCTED MODELS

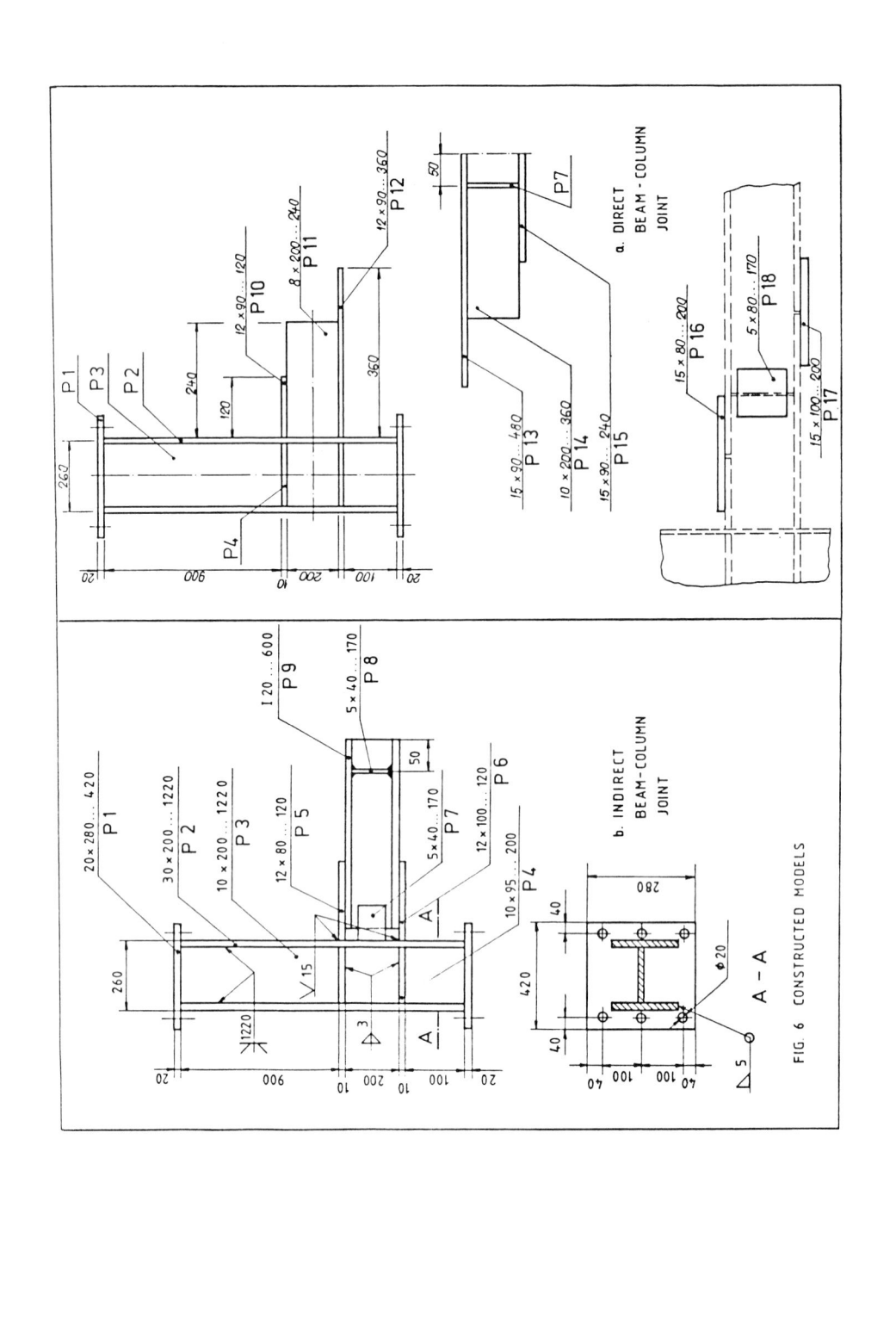

a. DIRECT BEAM-COLUMN JOINT

b. INDIRECT BEAM-COLUMN JOINT

They are in a relation ship expressed by the graph of Fig.9a.

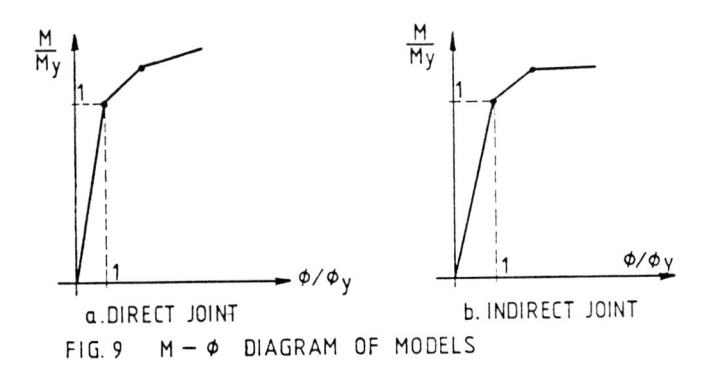

a. DIRECT JOINT

b. INDIRECT JOINT

FIG. 9   M – Φ DIAGRAM OF MODELS

4.3   Indirect beam – column connection

The yielding phenomenum is firstly met in the cross section shown in Fig 3b. It starts at the same node 20 and 21. The extending of the plastic zone is similarly to the direct beam column connection. A different things can be mentioned with respect to the yielding of web elements. They are involved after the yielding of the whole flange of the cross section.

The relationship between the applied bending moment and the rotation angle of the beam cross section, is shown by the graph of Fig.9b.

5   Conclusions

Every joint considered here, present a critical cross section. This section is neighboured to the column flange face. It is made of added material from welding in the case of butt welds. In the case of fillet welds the critical cross section is made of base material.

The yielded flanges and the yielded cleats are having the main contribution in the plastification phenomenum at direct joint and indirect joint respectively. A little contribution in plastification is due to the webs in the both case. That is the reason of a small shape of the ending part of the M – Φ curves.

The both analytical methods presented here are approximately. They ilustrate the main contribution of the flange or the horizontal cleats to the plastification.

a. STRESS ISOLINES ON THE WHOLE MODEL

b. STRESS ISOLINES ON THE TWO WEBS

c. STRESS ISOLINES ON THE BEAM FLANGE AND COLUMN RIB

d. DETAILED STRESS ISOLINES ON THE BEAM FLANGE AND COLUMN RIB

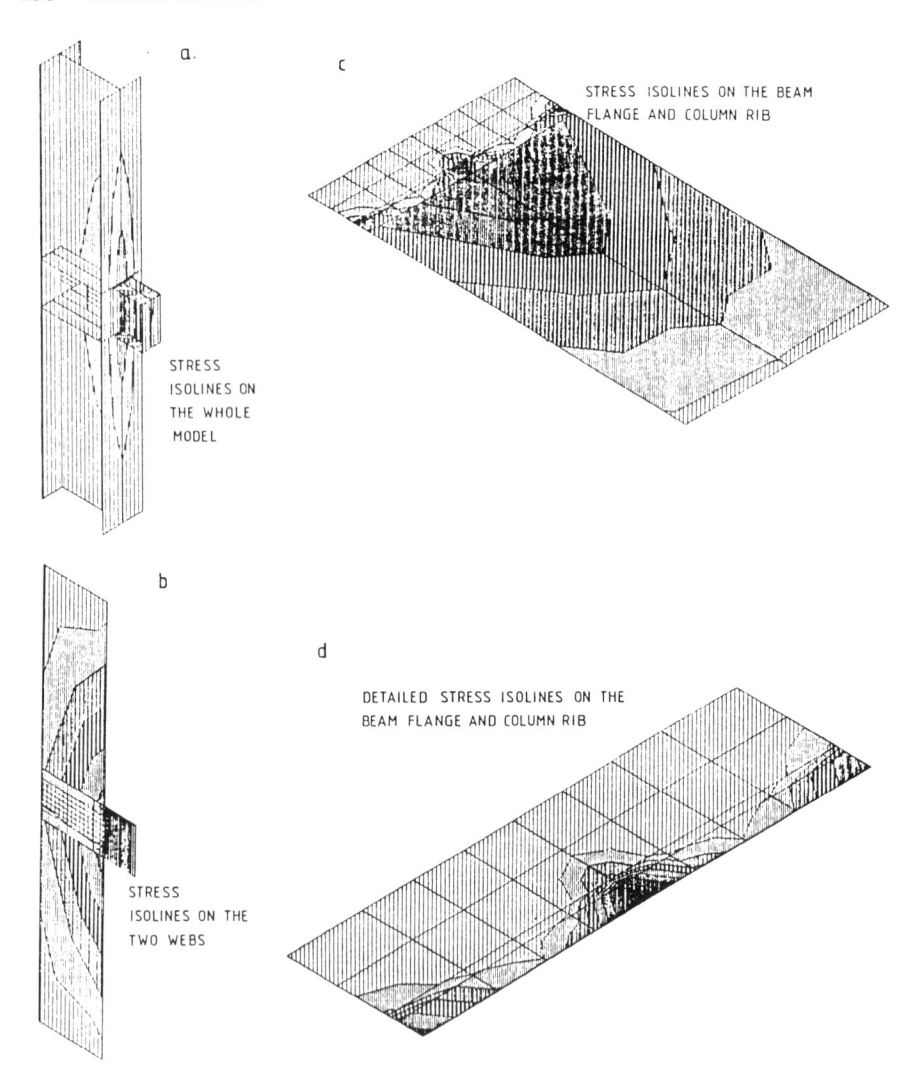

FIG. 8 STRESS ISOLINES

References

1 Neal B.G. The Plastic Methods of Structural Analysis.1956.
2 Fintel M. Handbook of Concrete Engineering Van Nostrand Reinhold Co.1974.
3 Wilson L.E, Habibulah A., SAP 90. Users Manual.

13 CYCLIC BEHAVIOUR OF STEEL-TO-CONCRETE END-PLATE CONNECTIONS

L. DUNAI
Technical University of Budapest, Budapest, Hungary
Y. OHTANI
Kobe University, Kobe, Japan
Y. FUKUMOTO
Osaka University, Osaka, Japan

Abstract
Experimental study is completed on the cyclic behaviour of typical steel-to-concrete end-plate connections under constant axial and cyclic bending loading conditions. Main focus of the research is on the rotational stiffness of the mixed connection and its cyclic deterioration. Bending deformation of the end-plate -- as a dominant source of the joint nonlinear behaviour -- is measured and investigated. Cyclic characteristics of the moment-rotation response of the joints are analyzed. Fast deterioration of the rotational stiffness is observed under cyclic loading.
Keywords: Mixed Connection, Experimental Study, Cyclic Loading, Stiffness Deterioration.

1 Introduction

End-plate type structural solutions for mixed connections are typically used as steel beam-to-reinforced concrete or composite column joints, and steel or composite column bases (Nishimura 1993, Wald 1993). There are also proposals to use steel-to-concrete end-plate connections in composite bridges as beam-to-beam joints (Ohtani and Fukumoto 1991). Moment end-plate mixed connection transfers the tension by bolts and/or studs, reinforcing bars and the compression by the bearing of the end-plate and concrete surfaces. Shear force is transferred by shear transmission elements (e.g., headed studs) and/or by the friction between the end-plate and concrete. The structural response of the connection is an interaction between the bolted/studded end-plate and the concrete/reinforced concrete behaviour. The local deformation of the flexible end-plate and the interacting contact problem result a highly nonlinear structural behaviour of the joint region.
  Experimentally verified formulas -- based on the equivalent rigid plate approach -- are available for the full-strength design of mixed connections (Wald 1993). Connections that are designed this way are considered as rigid. There is a lack of knowledge, however on the real rigidity and cyclic behaviour

due to the small number of available test data. Research results on "fixed" column bases call the attention on the role of rotational stiffness and its cyclic deterioration (Akiyama 1985, Astaneh et al. 1992, Nakashima et al. 1989, Penserini and Colson 1989). The influence of base fixity on the frame behaviour under cyclic horizontal effect was analyzed by Fukumoto et al. (1981).

In the current research a fundamental experimental program is designed to analyze the axial force-bending moment interaction and the cyclic loading effects in the behaviour of steel-to-concrete end-plate connections. The essential behavioural components are identified in the case of different structural solutions. Cyclic characteristics of moment-rotation responses are determined and compared. This paper reports on the details of the testing program and the observed cyclic behaviour.

2 Experimental program

2.1 Test specimens

Specimens of three typical structural arrangements are used in the test program, as shown in Fig. 1:

SP-1: flush end-plate connection,
SP-2: extended end-plate connection with bolts on the outer side of the flange,
SP-3: extended end-plate connection with bolts on both sides of the flange.

SP-1 and SP-2 are partial strength connections, while SP-3 is a full strength connection.

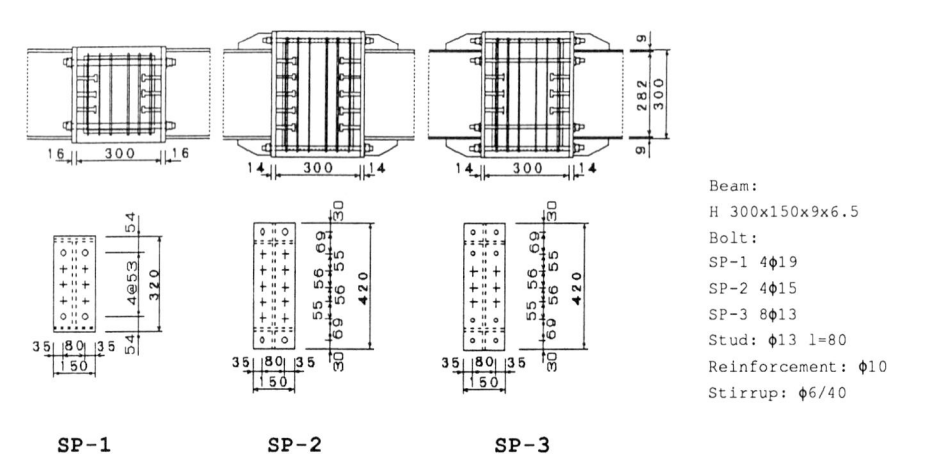

Beam:
H 300x150x9x6.5
Bolt:
SP-1 4ϕ19
SP-2 4ϕ15
SP-3 8ϕ13
Stud: ϕ13 l=80
Reinforcement: ϕ10
Stirrup: ϕ6/40

Fig. 1 Test specimens

The end-plate/bolt detail is designed as "intermediate plate" (Astaneh et al. 1992), in which there is a strong interaction between the plate bending and bolts elongation. The headed stud/reinforcement detail is applied to analyze its supporting effect on the end-plate bending and its contribution to the load bearing capacity.

The material of the steel member and end-plate is steel grade of SS 400 of JASBC (1987), with average yield stress of 295 MPa and tensile stress of 440 MPa. The yield and plastic moments of the steel section are 142 kNm and 160 kNm, respectively. The bolts are prestressing bars of grade SBPR 80/105 of JASBC (1987), with average yield stress of 850 MPa and tensile stress of 1100 MPa. The concrete is normal weight with cylinder mean strength of 47 MPa.

2.2 Test arrangement

The symmetrical steel-concrete-steel connection detail is placed in the center of a steel beam with a span of 2500 mm, as shown in Fig. 2. Cyclic bending moment is applied on the joint by the vertical actuator and rigid loading beam system in a four-point-bending arrangement. The cyclic vertical loading is supported by the pin and the two-way-roller support on the other end of the beam. The axial compression load is applied by the horizontal actuator.

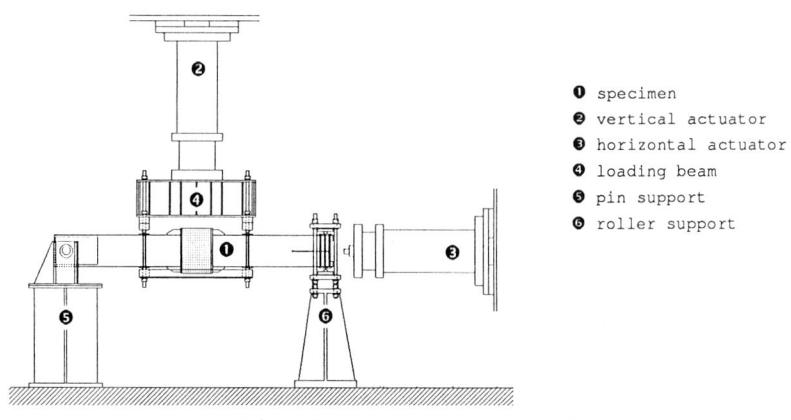

Fig. 2   Test arrangement

The measurement system focuses on the local connection behaviour. The end-plate deformation and the rotation components of the connection are measured by relative displacement measurement devices, as shown in Fig. 3. The interpretation and derivation of the rotation of the connection are illustrated in Fig. 3, too. The strain distributions in the steel beam, end-plate, bolts, studs and reinforcements are measured by strain gauges. Displacement transducers are applied to measure the vertical and horizontal deflections of the whole specimen and supports.

$\Theta = (\Delta u_a - \Delta u_b)/h$

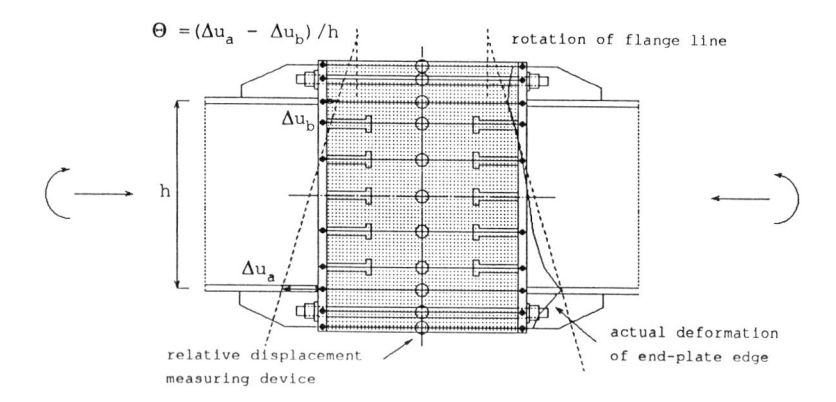

Fig. 3 Measuring and interpretation of rotation

2.3 Testing procedure

In the cyclic loading phase of the test program constant axial force is applied with an intensity of $P/P_y=0.075$, where P_y is the yield load of the steel section. The nominal elastic limit of the connection (M_y - yield moment, Θ_y - yield rotation) was determined by a 2D nonlinear analysis of the tension zone. Loading in the elastic range is controlled by the applied load: $M/M_y= 0.5, 0.75, 1.0$. As the elastic limit is reached, the predicted values are verified and adjusted. The inelastic cyclic loading is controlled by the rotation: $\Theta/\Theta_y= 1$ (2 cycles), 2(2), 3(1), 4(2), 5(1), 6(1), 8(1), 10(2). Note, that due to the concrete nonlinear behaviour and the manual controlling method, the above steps are partly modified during the loading process.

3 Experimental results and observations

3.1 Hysteretic moment-rotation curves

Typical elastic M-Θ curve of SP-3 can be seen in Fig. 4. The initial stiffness is high due to the pretensioning of the axial compression force. The nonlinear part of the curve is a transfer zone between the initial and P=0 stiffness.

The inelastic hysteretic M-Θ curves for the three specimens are shown in Figs. 5-7, respectively. The measured strengths of the connections are $M_u/M_{pl}= 0.66, 0.80$ and 1.13, respectively, where M_{pl} is the plastic moment of the steel section. The initial rotational stiffness is $K_0= 20, 28$, and 65 kNm/mrad, respectively. The maximum range of rotations is -25 to 25 mradians, which means about $\Delta u= \pm 15$ mm relative displacement differences between the flange lines of the connection. It can be seen from the curves that the basic trends of cyclic behaviour are similar for the three specimens.

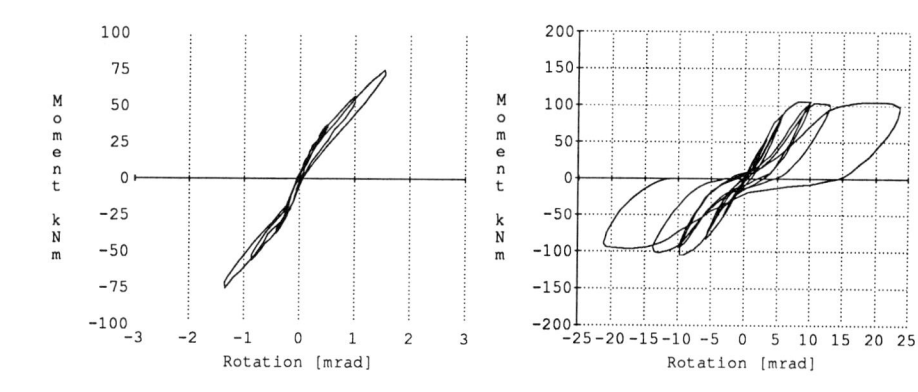

Fig. 4 SP-3: M-Θ /elastic

Fig. 5 SP-1: M-Θ /inelastic

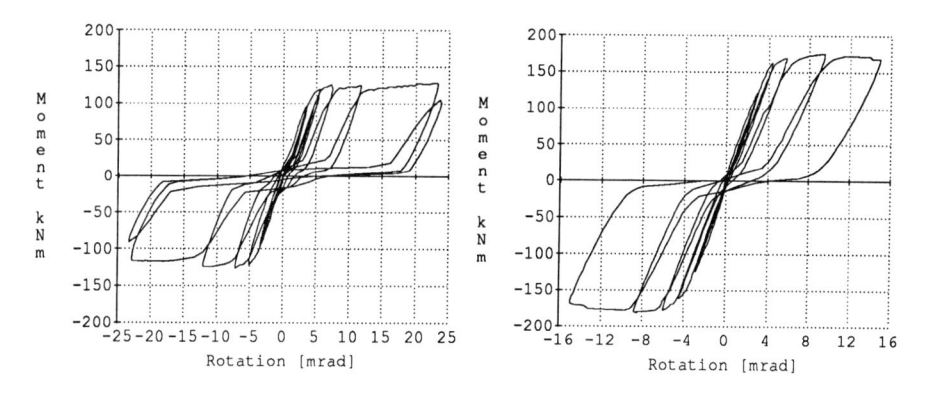

Fig. 6 SP-2: M-Θ /inelastic

Fig. 7 SP-3: M-Θ /inelastic

3.2 Features of cyclic response

The cyclic performance of the joints can be illustrated by the hysteretic M-Θ curves and the end-plate deformations. Fig. 8 shows the deformation of end-plate's edge of specimen SP-3, for two typical cycles: $\Theta/\Theta_y \approx 3$ and 8, respectively. The behaviour can be summarized as follows:

Elastic range: tension cracks develop in the concrete at the line of the studs' heads. Since the bolts are not anchored in the concrete, these cracks are not considered as significant cyclic behavioural aspects.

$\Theta/\Theta_y \approx$ 1-3: local plastic zones appear and spread in the tension zone of the connection due to the bending of the end-plate, flange and web elements. As a result of the residual plastic deformations, gaps appear between the end-plate and concrete and the end-plate and bolt head. The gaps create contact problems under reversal loading what indicates the degradation of the stiffness.

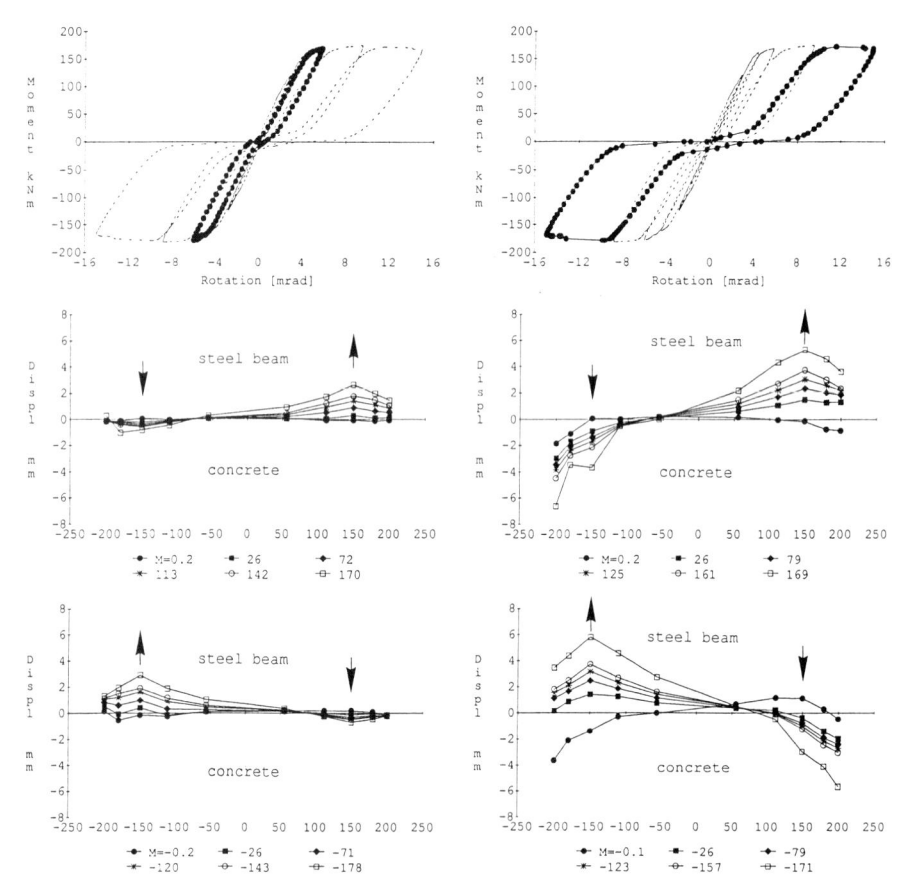

Fig. 8   SP-3: end-plate deformation under cyclic loading

$\Theta/\Theta_y \approx 3$: shell concrete crushes in the compression zone of the connection. As a result of it, the support of the outer part of the end-plate is vanishing. It indicates large deformation of the end-plate in the compression zone and significantly increases the connection gaps, defined before. The ultimate moment capacity of the connection is reached.

$\Theta/\Theta_y \approx 3-10$: the connection behaves practically as a "hinge" in the small range of the bending moment. The large gaps indicate rigid body rotation with a significantly reduced rigidity. The increasing effect of the axial compression force on the rotational stiffness is experienced in the "hinge-range" of the behaviour. After large rotations the strength of the connection is recovered.

$\Theta/\Theta_y \approx 10$: global failure of the connection occurred due to the interaction of crushing of the core concrete and the extended plasticity in the end-plate and bolts.

4 Analysis of experimental data

4.1 Control parameters

The measured experimental data is analyzed according to the recommendations of ECCS (1986). In the lack of monotonic test the limit of the elastic range (M_y) is predicted analytically and verified during the cyclic testing procedure. The elastic limit is defined as the starting of spread of plasticity in the tension field of the connection (end-plate, flange and web). The measured P=0 initial stiffness (K_o) is used to calculate the nominal elastic rotation (Θ_y). Due to the symmetrical arrangement the derived initial values are applied for both positive and negative half cycles.

$$M_y = M_y^+ = M_y^- \quad \Theta_y = \Theta_y^+ = \Theta_y^- \quad K_o = K_o^+ = K_o^- = M_y/\Theta_y \tag{1}$$

Main characteristics of the cyclic behaviour are the ductility, stiffness and strength degradation, and the cumulated absorbed energy. The pertinent control parameters - full ductility (ψ_i), rigidity (ξ_i), resistance (ε_i) and absorbed energy (η_i) ratios of (+) and (-) halves of inelastic cycle #i - are defined by assuming as a reference the perfect elasto-plastic behaviour, according to Eqs. 2-5 and Fig.

$$\psi_i^+ = \Delta\Theta_i^+ / (\Theta_i^+ + \Theta_i^- + \Theta_y) \quad \psi_i^- = \Delta\Theta_i^- / (\Theta_i^- + \Theta_i^+ + \Theta_y) \tag{2}$$

$$\xi_i^+ = tg\alpha_i^+ / K_o \quad \xi_i^- = tg\alpha_i^- / K_o \tag{3}$$

$$\varepsilon_i^+ = M_i^+ / M_y \quad \varepsilon_i^- = M_i^- / M_y \tag{4}$$

$$\eta_i^+ = A_i^+ / (\Theta_i^+ + \Theta_i^- - 2\Theta_y) M_y \quad \eta_i^- = A_i^- / (\Theta_i^- + \Theta_i^+ - 2\Theta_y) M_y \tag{5}$$

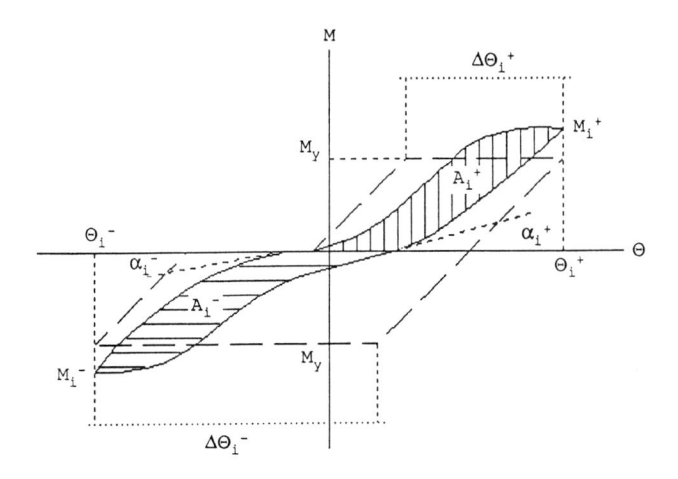

Fig. 9 Control parameters of inelastic cycle #i

4.2 Cyclic characteristics

Control parameters of the cyclic behaviour are calculated from the hysteretic M-Θ curves of the connections. Figs. 10-12 show the pertinent ratio - to - partial ductility (Θ/Θ_y) diagrams for the positive range loading of each specimen, respectively. Note, that the ratios -- derived from the negative half cycles -- exhibit similar features.

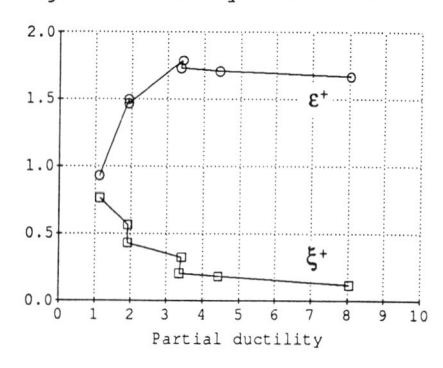

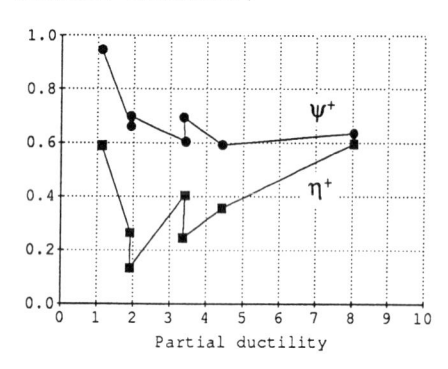

Fig. 10   SP-1: M-Θ cyclic characteristics

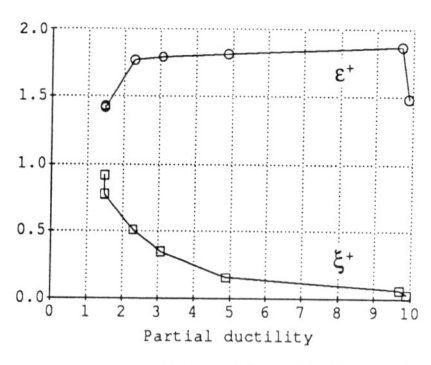

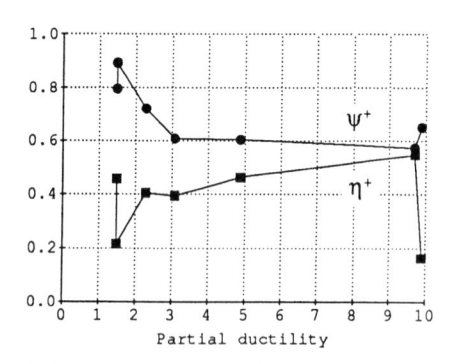

Fig. 11   SP-2: M-Θ cyclic characteristics

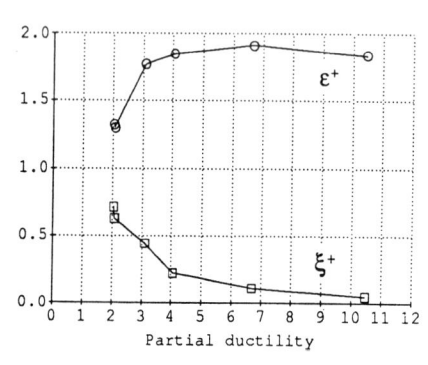

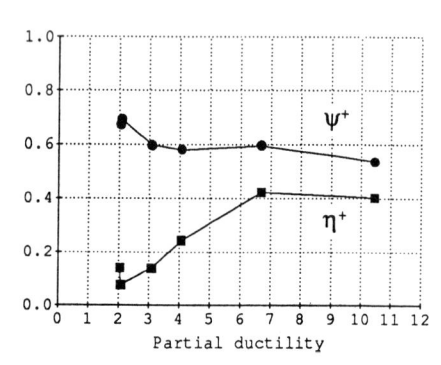

Fig. 12   SP-3: M-Θ cyclic characteristics

Rigidity and resistance ratios:
Similar trends are shown in the diagrams of rigidity and resistance ratios for all specimens. Significant increasing of the resistance is followed by fast decreasing of the rigidity. The stiffness degradation is the largest in the $\Theta/\Theta_y \approx 1-3$ range of partial ductility. In this zone the concrete in the compression part is practically undamaged. The stiffness drops back to about 0.30-0.35 in the case of SP-1 and SP-2, and 0.45 in the case of SP-3. In the same region the strength is increasing up to about 1.75. In the range of $\Theta/\Theta_y \approx 3-10$ the stiffness degrades gradually to about 0.1 value while the strength is practically unchanged.

Full ductility ratios:
A sharp decreasing of full ductility ratio is experienced in the range of $\Theta/\Theta_y \approx 1-3$ due to appearing of connection gaps. In the following cycles it is practically constant around 0.6 value for all the specimens.

Absorbed energy ratios:
The energy absorption is not significant in the low ductility range (only about 10-15% of the total cumulated energy is absorbed up to $\Theta/\Theta_y=3$), that is why the ratio has no main importance here. The absorbed energy and its ratio increase gradually to about 0.5-0.6 values in the larger range of ductility. The total absorbed energy at the level of absolute rotation $\Theta=15$ mrad equal to 1.4, 1.6 and 1.9 kJ for the three specimens, respectively.

5 Conclusion

Fundamental cyclic loading experimental study is completed on three steel-to-concrete end-plate connections emphasizing the rotational rigidity. The general features of cyclic behaviour are found to be similar in the case of different structural solutions. Fast degradation of rotational stiffness is observed parallel with the gradual increasing of the strength in the low region of ductility due to the developing of connection gaps. The rigidity is reduced practically to the "hinge" range after the crushing of the concrete.

Acknowledgments

The research is conducted under the support of Grant-in-Aid for Scientific Research of The Japanese Ministry of Education, Science and Culture. It is performed as a part of the first author's participation in a Postdoctoral Fellowship program of the Japan Society for the Promotion of Science.

The authors wish to express their thanks to R. Nishiyama, K. Mihara and K. Isshiki of Osaka University for their assistance in the experimental program.

References

Akiyama, H. (1985) Seismic resistant design of steel frame column bases. Gihodosuppan, Tokyo (in Japanese).

Astaneh, A., Bergsma, G. and Shen, J. H. (1992) Behavior and design of base plates for gravity, wind and seismic loads, in Proceedings of National Steel Construction Conference, American Institute of Steel Constructions.

ECCS (1986) Recommended testing procedure for assessing the behaviour of structural steel elements under cyclic loads. Technical Committee 1, TWG 1.3 - Seismic Design, No. 45

Fukumoto, Y., Itoh, Y. and Katsuya M. (1981) Theoretical and experimental studies on in-plane strength of towers of the Meikonishi-Ohashi cable-stayed bridge. Nagoya University Civil Engineering Research Report No. 8101, Nagoya University, Nagoya (in Japanese).

JASBC Manual (1987) Manual for the design of steel structures in Japan. JIS, Japan (in Japanese).

Nakashima, S., Suzuki, T. and Igarashi, S. (1989) Behaviour of full scale exposed type steel square tubular column bases under lateral loading. in Proceedings of IABSE Symposium, Part A, Helsinki, pp. 148-152.

Nishimura, Y. (1993) State-of-the-art report on composite and mixed structures in Japan. GBRC, 18, 2, 3-11. (in Japanese)

Ohtani, Y. and Fukumoto, Y. (1991) Experimental study of beam-to-beam joint of continuous composite bridge. Hanshin Expressway Corporation, Disaster Science Research Institute, Research Report (in Japanese).

Penserini, P. and Colson, A. (1989) Ultimate limit strength of column base connection. J. Constructional Steel Research, 14, 301-320.

Wald, F. (1993) Column base connections; a comprehensive state-of-the-art review. COST C1 Project, CIPE 3510 PL 20 143, Technical University of Prague, Prague.

14 CONTRIBUTIONS TO THE STUDY OF PLASTIC ROTATIONAL CAPACITY OF I-STEEL SECTIONS

V. GIONCU, G. MATEESCU and A. IUHAS
Building Research Institute, INCERC, Timisoara, Romania

Abstract
The paper deals with the determining of the plastic rotational capacity of I steel sections. This rotation is limited by the local flange and web buckling. The method of predicting the moment-rotation curve of a locally buckled element is developed using a collapse mechanism formed by plastic zones and yielding lines. A specialized program DUCTROT-93 is developed for determining the ultimate plastic rotation. The comparison of theoretical values obtained using this program, theoretical values determinated with FEM procedure, and experimental results shows a good correspondence.
Keywords:Ductility, Flange width-thickness ratio, Local buckling, Plastic buckling, Plastic rotation, Plastic rotation capacity, Rigid-plastic mechanism, Web width - thickness ratio.

1 Introduction

Ductil steel moment resisting frames have been used in seismic design due to their good performance. This good behaviour during a major earthquake is attributed to their excellent capacity to dissipate seismic energy, based on the rotation ductility of beams and beam-columns. The steel structures are generally ductile and a great inelastic capacity is possible to be reached. But this good ductility can be impaired by the occurrence of local plastic buckling of the constituent plates of the cross section. So, the available ductility of steel members is limited by the problem of local instability of compressed flanges and webs. The limitations of the slenderness of the cross section (flange and web b/t ratios) are prescribed in specifications: ECCS; TWG 1.3 (1986), EUROCODE 3 (1993), AIJ-LRFDSS (1990), P 1oo (1992) etc., but the values which sort the ductility classes are very doubtful .
    Attempts for the calculation of the rotation capacity

Behaviour of Steel Structures in Seismic Areas. Edited by F. M. Mazzolani and V. Gioncu.
Published in 1995 by E & FN Spon, 2-6 Boundary Row, London SE1 8HN. ISBN: 0 419 19890 3.

were performed by Kuhlman (1987, 1989), Spangemacher (1991) using FEM procedure, by Mazzolani-Piluso (1992, 1993) and Kato (1989, 1990) by integrating the moment-curvature relationship, Climenhaga-Johnson (1972), Lucky-Adams (1969), Ivanyi (1979), Kuhlman (1989), Gioncu et al. (1989) based on the collapse mechanisme, and Vayas-Psycharis (1990) using the effective width.

Between these methods it seems to be the best procedure the using of the collapse plastic mechanism coming from the experimental evidence. Thus in the following this method is used to determine the plastic rotation capacity of I steel sections.

2 Standard beam

Usually the standard beam for determining the plastic rotation capacity is considered the centrally loaded beam (fig.1a), because the behaviour of this beam can be compared to the one of the cantilever beam. The dimensions of the cross-section are presented in Fig.1b, while the moment diagram and plastic rotation are shown in Fig.1c. The considered stress-strain curve has the form presented in Fig.1d.

Due to moment gradient the yielded region is limited extent. The moment at which the flange yielding occurs is M_{pf}, less than the plastic moment M_p ($M_{pf}=0.90-0.94 M_p$). When the moment at a section reaches M_{ps}, there will be a jump in flange strain from yield to strain-hardening, Lay-Galambos (1967). Due this jump, if the loading is increasing, the plastic moment at the middle of beam is increasing to M_o and the yielding region is extended to (Fig.1c):

$$l_p = (1 - \frac{M_{pf}}{M_o}) l \qquad (1)$$

In spite of this extension of yielding region, the plastic rotation is concentrated at the middle of the beam (Fig.1c).

A very important feature of the beam behaviour is the moment-rotation curve, shown in Fig.2. Four characteristic points are presented. The first refers to the reach of flange yielding M_{pf}, second to the full plastic moment M_p, the third to the maxim moment M_{max}, attained due to plastic buckling in the hardening region and the last one, corresponding to the plastic moment in the lowering post-buckling curve. For each these moments a corresponding rotation is obtained. Due to the fact after the reach of the last

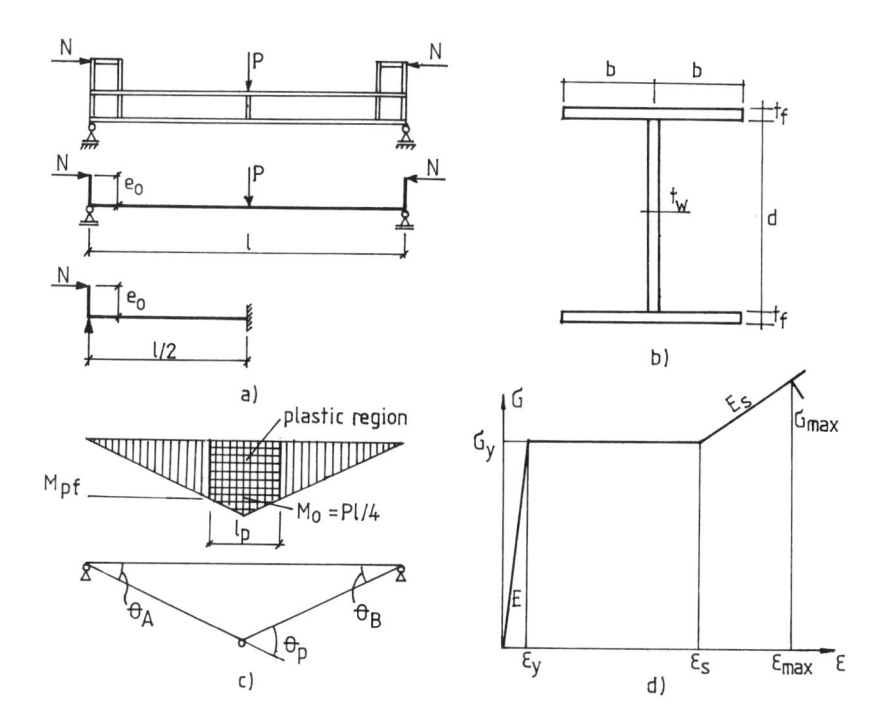

Fig.1. Characteristics of standard beam.

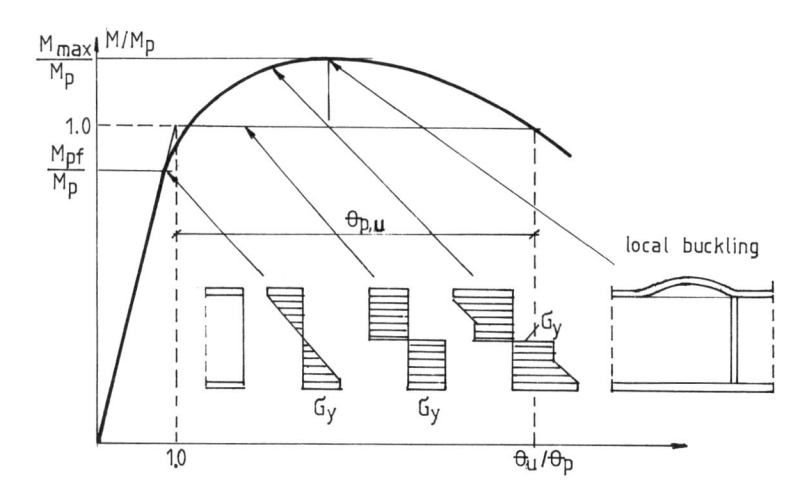

Fig.2. Moment-rotation curve.

moment an important degradation of the beam rigidity is produced; conventionally is defined this rotation as the last one, θ_u. To measure the deformation capacity non-dimensionally, the rotation capacity R_u is usually calculated as:

$$R_u = \frac{\theta_{p,u}}{\theta_p} = \frac{\theta_u - \theta_p}{\theta_p} = \frac{\theta_u}{\theta_p} - 1 \qquad (2)$$

where $\theta_{p,u}$ is the <u>ultimate plastic rotation</u> and θ_p the rotation corresponding to first plastic hinge.

The paper suggests a procedure for determining the rotation capacity and the ultimate plastic rotation, using the plastic post-buckling behaviour of the standard beam.

3 Collapse mechanism

A model for the collapse mechanism composed by plastic zones and yielding lines is used, adapted to the form of real experimental buckling shapes. Fig.3 shows a real collapse shape, while the Fig.4 presents the proposed collapse mechanism, which is a modified one used by Climenhaga-Johnson (1972) and Ivanyi (1979). One can see that due to the presence of middle stiffners, the collapse mechanism is asymmetrically, but for high beams (d>3b) the difference from a symmetrical one is not very important.

The beam rotates about the point O, whose location is fixed by δd, which is considered as a constant in the deformation of the mechanism. It is possible to minimize the M-θ relationship and to obtain the corresponding parameter δ. It has found that this minimum results for $\delta = 0.85 - 1.00$. As the rotation occurs, the zone below O remains in plane and simple tensil plastic zones AAAA and BOB are produced. Above the point O the web buckles due to formation of yielding lines OE, OC, CE, EE and plastic zone ECE. Buckling of the upper flange occurs by compression and rotation of plastic zone EFEF and rotation of yielding lines DD and EF. The length of flange buckling is βb, where β is obtained using the research works of Lay (1965) which has been determined this length from plastic condition. This buckling length cannot exceed the length of the yielding region, obtained from equation (1). So, the parameter β is determined as the minimum of the values:

Fig.3. Experimental collapse mechanism

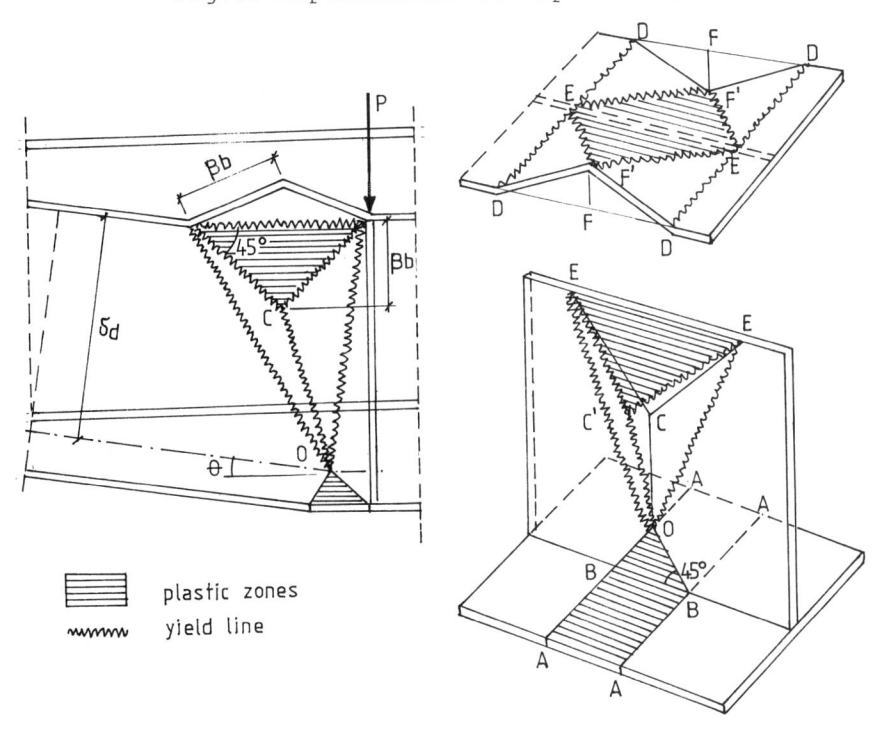

Fig.4. Proposed collapse mechanism.

$$\min \begin{cases} \beta = \frac{1}{4}(1 - \frac{M_{pf}}{M_{max}})\frac{1}{b} \\ \beta = 1.154(\frac{t_f}{t_w})^{\frac{3}{4}}(\frac{d}{b})^{\frac{1}{4}} \end{cases} \quad (3a,b)$$

This flange buckling length is a modification of the collapse mechanism used by Climenhaga, in which $\beta = 1$, hypothesis which is not confirmed by experimental tests, which show that $\beta > 1$. In comparison with the relationships developed by Ivanyi (1979) some simplifications are admitted, considering that the rotations of the yielding lines are small, so that sin x=x. An other modification of the collapse mechanism used by Climenhaga and Ivanyi refers to the considering different yield stresses for flanges and web, by introducing the parameter $\varrho = \sigma_{yw}/\sigma_{yf}$.

The works absorbed in each of deforming elements for plastic zones and yielding lines are:

$$W_z = V\int_\varepsilon \sigma\, d\varepsilon \quad ; \quad W_1 = 1\int_\theta M\, d\theta \quad (4a,b)$$

where V is the volum of plastic zones and l the length of yielding lines. After some tedious algebra the total work absorbed by the buckling beam W_b is

$$\frac{W_b}{bd^2\sigma_{yf}} = (A + \frac{B}{\theta^{\frac{1}{2}}})\theta \quad (5)$$

where

$$A = \frac{t_f}{b}\frac{b}{d}(2-\delta + 2\frac{t_f}{b}\frac{b}{d}) + \frac{1}{2}\frac{t_w}{d}\left[\frac{d}{b}(1-\delta)^2 + \beta\delta\right] \quad (6a,b)$$

$$B = \frac{1}{2}(\frac{\delta d}{\beta b})^{\frac{1}{2}}\left\{\varrho(\frac{t_w}{d})^2\left[(2\delta-1)+\frac{d}{b}(\delta+\frac{\beta^2}{\delta}+\beta\delta)+\frac{\beta b}{\delta d}\right]+6(\frac{t_f}{b}\cdot\frac{b}{d})^2\right\}$$

The work W_p produced by external forces P and N (Fig.1a) is:

$$\frac{W_p}{bd^2\sigma_{yf}} = \frac{C}{m_p m_q}\frac{M}{M_p}\theta \quad (7)$$

where

$$C = M_p\frac{1}{1-n_p\lambda^2} + \frac{d}{e}(\frac{e_o}{d}+\delta-\frac{1}{2}) \quad (8)$$

$$n_p = \frac{N}{N_p} \; ; \quad q_p = \frac{Q}{Q_b} \; ; \quad \lambda = (\frac{N_p}{N_{cr}})^{\frac{1}{2}} \; ; \quad e = \frac{M}{N} \qquad (10a\text{-}d)$$

$$m_p = \begin{cases} 1 & \text{if } 0 \leq n_p < 0.15 \\ 1.18(1-np) & \text{if } n_p \geq 0.15 \end{cases} \qquad (11)$$

$$m_q = \begin{cases} 1 & \text{if } 0 \leq q_p < 0.5 \\ 1 - \frac{d^2 t_w \sigma_{yw}}{M_p \cdot m_p} (q_p - \frac{1}{2}) & \text{if } q_p \geq 0.5 \end{cases} \qquad (12)$$

$$N_p = 4bt_f \sigma_{yf} + dt_w \sigma_{yw} \; ; \quad Q_p = \frac{dt_w \sigma_{yw}}{3^{\frac{1}{2}}} \; ; \qquad (13a,b)$$

$$N_{cr} = \frac{\pi^2 EI}{(\mu l)^2} \qquad (14)$$

Equating the absorbed work of beam and the one produced by external forces leads to a form of M-θ relationship:

$$\frac{1}{m_p m_q} \frac{M}{M_p} = \frac{A}{C} + \frac{B}{C} \frac{1}{\theta^{1/2}} \qquad (15)$$

The coeffients A, B and C depend on the geometrical and mechanical parameters of the beam, and the parameters β and δ, which define the form of collapse mechanism. The first parameter is obtained from the equations (3a,b) and the second one from the minimum of equation (15). This relationship represents a hyperbola related to the rotation of the plastic hinge.

4. DUCTROT-93 program

Based on the studied collapse mechanism it was found available to develop a numerical analysis for different I cross-sections in the aim to determine the rotation capacity R_u and the ultimate plastic rotation, θ_{pu}. A numerical program DUCTROT 93 (ductility of rotation) was elaborated for solving the equation (15) and representing the M-θ curve. One of the studied cases is illustrated in Fig.5.

The maximum values for moment is determined using the ultimate stresses given by Mazzolani-Piluso (1992, 1993)

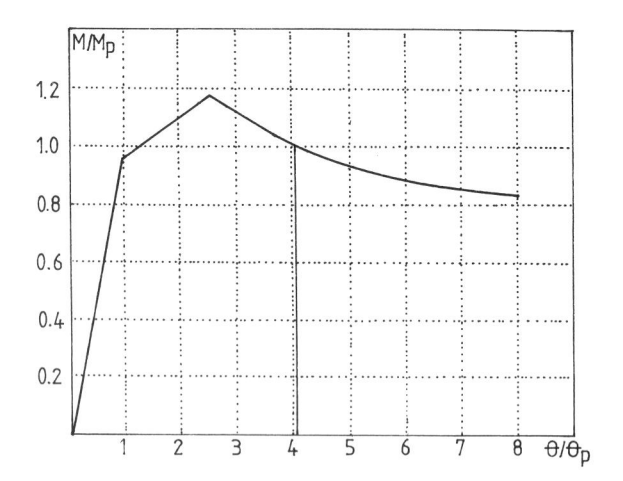

Fig.5. Typical moment-curvature (DUCTROT 93)

and Kato (1989, 1990). The ultimate plastic rotation can be found by intersecting the curve (15) with an horizontal line $M/m_p m_q M_p = 1$. The elastic rotation of the beam used for the determining the rotation capacity, is the sum of end elastic rotations:

$$\theta_p = \alpha \frac{m_p m_q M_p l}{EI} \qquad (16)$$

For standard beam $\alpha = 0.5$; for other beams this value can be determined from the real elastic rotation diagram.

The DUCTROT-93 program (Petcu-Gioncu, 1993) is an interactiv one, having the possibility to change the initial data and to elaborate tables with the ultimate plastic rotation, depending on geometrical and mechanical parameters.

5. Comparison with other theoretical results and experimental test values

In the aim to verify the values obtained with the proposed collapse mechanism, the theoretical results obtained by Spangemaher (1991) using FEM procedure are presented in Fig.6. The correspondence of the values obtained for ultimate plastic rotation and rotation capacity with DUCTROT-93 and FEM is quite good. Other five curves were compared, obtaining the same good results. Same differences exist only for small and high values of rotation capacity ($R_u < 4$, $R_u > 12$).

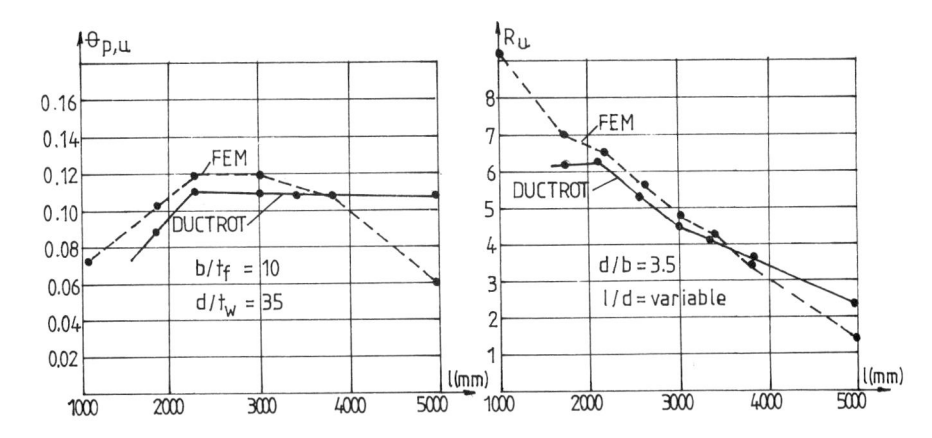

Fig.6. Comparison with FEM procedure.

The comparison with the experimental results obtained by Lukey-Adams (1969) (9 tests) and Kuhlman (1989) (15 tests) shows the same good correspondence. Moment-rotation curves obtained by two programs, ABACUS and PROFIL, experimental values determined by Kuhlman and the same curve determined with the DUCTROT-93 program are plotted in Fig.7a. The "scatter area" of the ratios between theoretical and experimental values is given in the Fig.7b. A quite good correspondence exists in the domain $4 < R_u < 12$, as in the case of comparison with FEM procedure.

In exchange of these very good confirmation, the comparison with the experimental results obtained by Spangemaher (1991) is not so good, probably due to the fact that for the cross-section with $d \simeq 2b$, the collapse mechanism is very asymmetrically, differing from the studied symmetrical one (see section 3). This shortcoming of the method shows, that for the future it is required some theoretical studies on the asymmetrical collapse mechanism.

6. Influence of axial forces and eccentricities

Using the DUCTROT-93 program, the influence of axial forces and eccentricities on the ultimate plastic rotation and rotation capacity is presented in Fig.8. One can see that the presence of the axial forces produce a very important decreasing of the rotation capacity especially in the field $n_p < 0.15$; for $n_p > 0.15$ the modification of values is unimportant, but at the very reduced level. The influence of eccentricities is different depending on the eccentricity sign. For positive one, the additional moment having the same effect as the moment due to

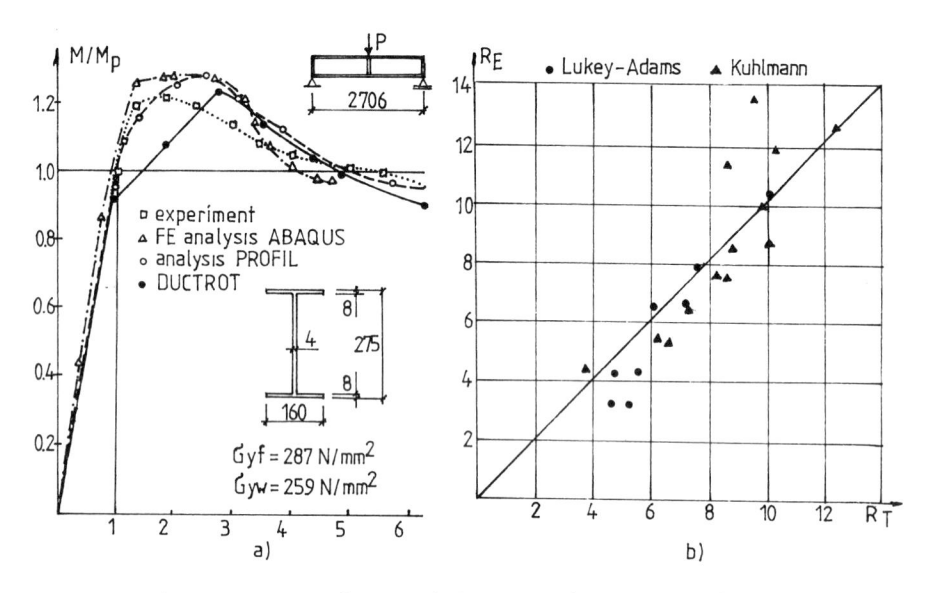

Fig.7. Comparison with experimental values

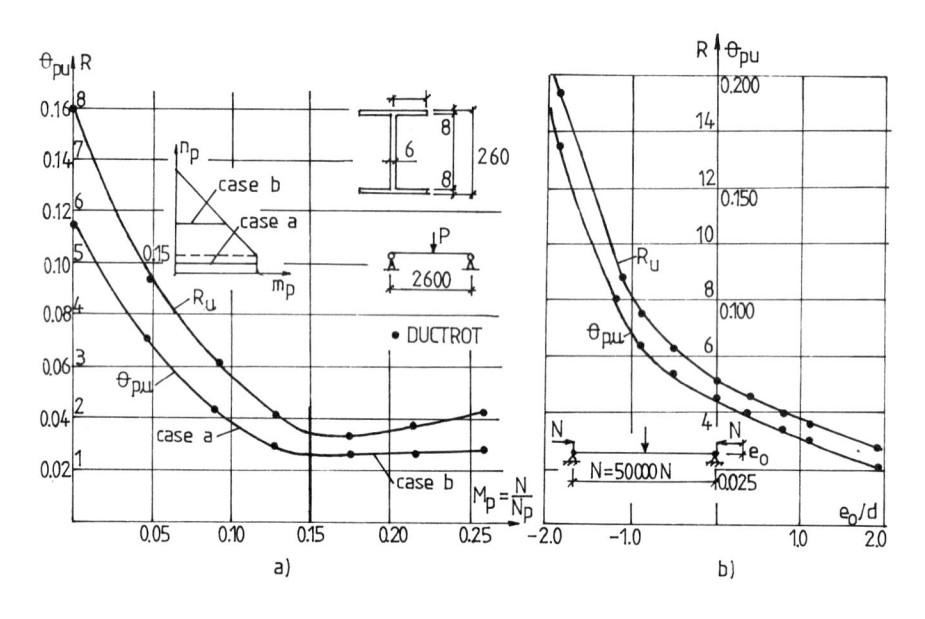

Fig.8. Effects of axial forces and eccentricities

centrally load, the capacity of rotation decreases, while for the negative sign, the increase is very important.
Unfortunately, theoretical and experimental results concerning the influence of axial forces on rotation capacity are not available. Thus, it is not possible to verify the accuracy of DUCTROT-93 results.

7. Proposed method

Due to the fact that we have shown (Gioncu et al. 1994) that is preferable to use the ultimate plastic rotation instead of the rotation capacity, a relationship for determining this beam or beam-column characteristic is proposed as:

$$\theta_{p,uN} = \alpha_N \alpha_e \theta_{p,u} \qquad (17)$$

where α_N and α_e are coefficients which introduce the influence of axial forces and of end eccentricities, while $\theta_{p,u}$ is the ultimate rotation of the beam, which can be determined from Tables, depending on geometrical parameters b/t_f, d/t_w, d/b, l/d, and mechanical ones, σ_{yf}, σ_{yw}. For instance, Table 1, determined using DUCTROT-93, gives the possibility to determine the ultimate plastic rotation of

Table 1. Ultimate plastic rotation.

$l/d = 30$		d/t_w						
d/b	b/t_f	30	40	50	60	70	80	90
3	6	0.4551	0.3297	0.3590	0.5866	0.1640	0.0822	0.0732
	7	0.4003	0.2852	0.2343	0.2760	0.1252	0.0637	0.0572
	8	0.3564	0.2560	0.1962	0.1790	0.1019	0.0526	0.0482
	9	0.3203	0.2323	0.1784	0.1459	0.0859	0.0455	0.0429
	10	0.2903	0.2123	0.1638	0.1317	0.0736	0.0396	0.0412
	11	0.2650	0.1951	0.1515	0.1221	0.0618	0.0344	0.0473
4	6	0.5340	0.4002	0.3152	0.2646	0.2743	0.2500	0.0736
	7	0.4532	0.3443	0.2733	0.2243	0.1941	0.1959	0.0663
	8	0.3922	0.3004	0.2407	0.1987	0.1680	0.1484	0.0610
	9	0.3448	0.2650	0.2142	0.1779	0.1509	0.1304	0.0543
	10	0.3071	0.2367	0.1921	0.1606	0.1369	0.1186	0.0455
	11	0.2768	0.2133	0.1738	0.1459	0.1250	0.1086	0.0369
5	6	0.5643	0.4439	0.3641	0.3061	0.2627	0.2331	0.2324
	7	0.4699	0.3713	0.3075	0.2607	0.2251	0.1971	0.1753
	8	0.4015	0.3173	0.2638	0.2254	0.1958	0.1723	0.1534
	9	0.3503	0.2760	0.2299	0.1973	0.1725	0.1525	0.1361
	10	0.3107	0.2438	0.2031	0.1747	0.1532	0.1361	0.1220
	11	0.2794	0.2180	0.1814	0.1563	0.1375	0.1225	0.1102
6	6	0.5725	0.4617	0.3904	0.3377	0.2965	0.2632	0.2363
	7	0.4735	0.3806	0.3223	0.2807	0.2480	0.2217	0.1998
	8	0.4034	0.3220	0.2726	0.2380	0.2114	0.1899	0.1721
	9	0.3516	0.2786	0.2350	0.2052	0.1829	0.1649	0.1501
	10	0.3124	0.2450	0.2061	0.1798	0.1604	0.1449	0.1323
	11	0.2815	0.2188	0.1833	0.1596	0.1423	0.1288	0.1177

a beam. Analogous Tables are obtained for the entire area of geometrical and mechanical parameters.

Concerning coefficients α_N, α_e, due to scarcity of experimental values, further research works are necessary, to express an approximate relationship.

Due to the fact that the present estimation of rotation capacity is based on monotonic loading, a very important question refers to the using of these results for the cyclic loading. The experimental tests show that in the last case some deteriorations are produced in post-buckling behaviour due to some cracks observed in the flange - web connection zones (Ballio - Calado, 1986). Thus, a reduced value for ultimate plastic rotation is proposed:

$$\theta_{p,u\ cycl} = 0.9\ \theta_{p,u\ mon} \qquad (18)$$

8. Conclusions

The using of collapse mechanism allows to determine the ultimate plastic rotation and rotation capacity with an accuracy confirmed by theoretical and experimental research works. The results show that the provisions for ductility classes, given in codes, are not sufficiently, because the rotation capacity depends on more factors than those considered in these codes.

9. References

AIJ (1990) **Standard for Limit State Design of Steel Structures.**
Ballio, G., Calado, L. (1986) Steel bent sections under cyclic loads. Experimental and numerical approaches. **Constr. Met.**, 1, 1-23.
Climenhaga, J.J., Johnson, R.P. (1972) Moment rotation curves for locally buckling beams. **J.Str.Div.**, ST6, 1239-1254.
ECCS, TWG 1.3 (1986) **European Recommendations for Steel Structures in Seismic Zones.**
Eurocode 3 (1993), **Design of Steel Structures.**
Gioncu, V., Mateescu, G., Orasteanu, S. (1989) Theoretical and experimental research regarding the ductility of welded I sections subjected to bending, in **Proceedings of SSRC Int. Symp. on Stability**, Beijing.
Gioncu, V., Mateescu, G., Iuhas, A. (1994) Contributions to the study of the ductility of unbraced frames in **Proceedings of STESSA '94,** Timisoara.
Ivanyi, M. (1979) Moment-rotation characteristics of locally buckling beams. **Periodica Polytechnica**, Civil Engineering, no. 3-4, 217-23o.

Kato, B. (1989) Rotation capacity of H-section members as determined by local buckling. **J.Construct. Steel Research** 13, 95-109.
Kato, B. (1990) Deformation capacity of steel structures. **J.Construct. Steel Research,** 17, 33-94.
Kuhlman,V., Roik, K. (1987) Rechnerische Ermittlung der Rotationskapazität biegebeanspruchter I- Profile, **Stahlbau,** Heft 11, 321-327.
Kuhlman, V. (1989) Definition of flange slenderness limits on the basis of rotation capacity values, **J.Construct. Steel Research,** 14, 21-40.
Lay, M. G. (1965) Flange local buckling in wide-flange shapes. **J. Struct. Div.,** ST6, 95-116.
Lay, M.G., Galambos, T.V. (1967) Inelastic beams under moment gradient, **J.Struct. Div.** ST1, 381-399.
Lukey, A.F., Adams, P.F. (1969) Rotation capacity of beams under moment gradient, **J.Struct. Div.,** ST6, 1173-1188.
Mazzolani, F.M., Piluso, V. (1992), Evaluation of the rotation capacity of steel beams and beam-columns, in **Proceedings of 1 st State of the Art Workshop COST C1,** Strasbourg.
Mazzolani, F.M., Piluso, V. (1993), Member behavioural classes of steel beams and beam-columns, in **Proceedings of C.T.A. Congress,** Viareggio, 405-416.
Petcu, D., Gioncu, V. (1993), DUCTROT-93: Plastic rotation capacity of H-section steel beams and beam-columns, **Computer Program,** INCERC Timisoara.
P-100 (1992), **Antiseismic Design Code for Civil, Agricultural and Industrial Buildings** (in Romanian).
Spangemacher, R. (1991), Zum Rotationsnachweis von Stahlkonstructionen, die nach dem Traglastverfahren berechnet werden. **Dissertation,** Technischen Hochschule Aachen.
Vayas, I., Psycharis, I.N. (1990) Behaviour of thin-walled steel elements under monotonic and cyclic loading, in **Structural Dynamic** (eds Kratzig et al), Balkema, Rotterdam, 579-583.

15 ALTERNATIVE METHODS FOR ASSESSING LOCAL DUCTILITY

V. GIONCU
Building Research Institute, INCERC, Timisoara, Romania
F. M. MAZZOLANI
Istituto di Tecnica delle Costruzioni, University of Naples, Naples, Italy

Abstract
The problem of the evaluation of the rotation capacity has been analysed by critically comparing the existing methods. In particular, the concept of member behavioural classes has emphasized as a suitable proposal to substitute the usual cross-section classification in the assessment of local ductility.
Keywords: Local Ductility, Rotation Capacity, Local Buckling.

1. Introduction

The plastic behaviour of a structure depends upon the amount of moment redistribution. The attainment of the predicted collapse load is strictly related not only to the hinge position where sections reach the full plastic moment, but also to the inelastic rotation which other hinges can have elsewhere. Hence, plastic hinges require a certain amount of ductility in addition to their strength requirement; the rotation capacity is a measure of this ductility. Rotation capacity usually is defined as that one connected to the deformation which a given cross-sectional shape can accept at the plastic moment without premature failure occuring.
In limit design of structures, it is postulated that plastic hinges have a sufficient rotation capacity. Therefore, it is clear that the cross-section of members have to satisfy precise geometrical requisites in order to allow for plastic deformations until the collapse mechanism of the structure is reached without loosing its load carrying capacity.

The rotation capacity of steel members is undermined by the occurrence of local buckling in the plate elements which constitute the member cross-section and, if torsional restraints are not provided, by the occurrence of lateral torsional buckling.

2. General remarks

Regarding the **local ductility**, all the modern codes define the concept of cross-section behavioural classes which depends only on the width-to-thickness ratios of flanges and of web. These proposals contain some shortcomings (Mazzolani and Piluso,1993a):
a) Independent limitations of these ratios are unreasonable, because, obviously, the flange is restrained by the web and the web is restrained by the flange.
   Only the Japanese code takes into account this interaction for H - shaped sections.
b) The local ductility depends not only on the width-to-thickness ratios, but also on the ratio between width of flange and web, member length, moment gradient, level of axial load, eccentricity of axial load, steel quality, flexural-torsional buckling, etc. As a consequence of such additional factors, it seems that the concept of cross-section behavioural classes should be substituted by the concept of member behavioural classes (Mazzolani and Piluso,1993b). Only the Japanese code partially considers this concept, by using in the classification also the slenderness ratio of beams.
c) As far as the values of b/t ratios in the different ductility classes are concerned, we can observe that this subdivision does not correspond to the actual behaviour of beam and beam-columns which is a continuous one and the given discrete values of b/t ratio seem to be very arbitrary.

The local ductility is also influenced by the low-cycle fatigue, which reduces the ductility, causing premature failure.

3. Local ductility evaluation

In the plastic classical theory it is assumed that in a plastic hinge of a bent member the plastic moment remains constant as far as the rotation increases. This assumption neglects two very important aspects:
a) the additional moment capacity due to the strain-hardening effects;
b) the possibility of moment reduction due to the local buckling of flanges and/or web.

Neglecting the first aspect in the structural design is on the safe side, but the omission of the second can produce an uncontrolled plastic redistribution, followed by a very important increasing of lateral displacement. Because this situation must be avoided, a method for eliminating the achievement of a very strong degradation field is required. The codes use the concept of cross-section behavioural classes to provide a sufficient plastic rotation capacity for a well controlled redistribution. This is a qualitative method, in which the actual behaviour is not able to be determined.
A more accurate method is to determine the rotation capacity, by using the actual behavioural M-θ curve.

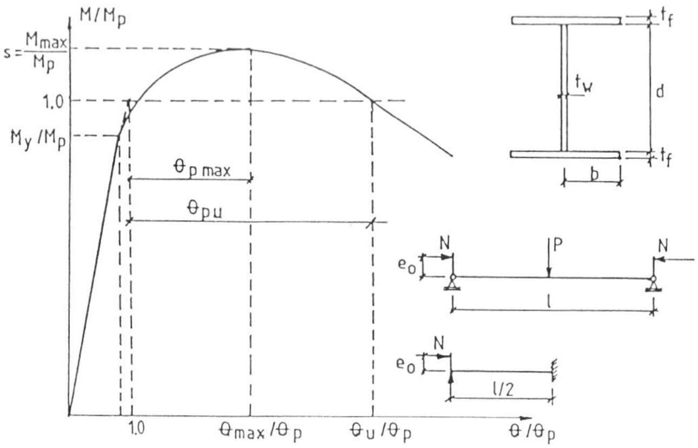

Fig.1. Moment-rotation curve.

On this base, two different approaches can be followed (Fig.1):
- The rotation capacity is defined by considering the stable part of the M-θ curve:

$$R_{max} = \frac{\theta_{p.max}}{\theta_p} = \frac{\theta_{max} - \theta_p}{\theta_p} = \frac{\theta_{max}}{\theta_p} - 1 \qquad (1)$$

being θ_p the rotation corresponding to the full plastic moment M_p and θ_{max} the plastic rotation corresponding to the maximum value of the moment M_{max}.
This way has been followed by Kemp (1985), Kato (1989 and 1990), Mazzolani and Piluso (1993b).
- The rotation capacity is defined by considering also the unstable branch of the M-θ curve up to the value θ_u, which corresponds to first yielding moment M_y in the lowering curve:

$$R_u = \frac{\theta_{p.u}}{\theta_p} = \frac{\theta_u - \theta_p}{\theta_p} = \frac{\theta_u}{\theta_p} - 1 \qquad (2)$$

This way has been followed by Climenhaga-Johnson (1972), Lucky-Adams (1969), Ivanyi (1979), Kuhlman and Roik (1986,1987 and 1989), Spangemacher (1991), Gioncu et al. (1994), Vayas and Psycharis (1990).
Fig.2 shows a comparison of R_{max} and R_u (Kuhlman, 1986).

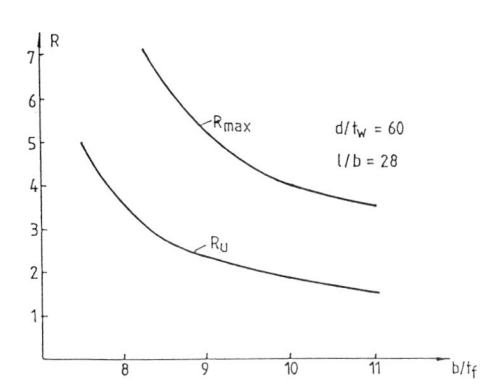

Fig.2. Comparison between R_{max} and R_u.

From the operative point of view, the calculation of the rotation capacity of steel members can be obtained by different methods (Mazzolani and Piluso, 1992):
- **Theoretical methods** based on the use of FEM for evaluation of the moment versus rotation relationship (Kuhlman, 1986).
- **Semi-empirical methods** in which the theoretical method is based on the experimental evidence. These methods are developed by integrating the moment-curvature relationship untill the maximum value of the moment experimentally determined; in this way expressions for practical calculations of rotation capacity are provided (Mazzolani and Piluso, 1992).
A second way is based on the interpretation of the collapse mechanism coming from the experimental evidence. This methodology provides values of the maximum plastic rotation θ_u (Gioncu et al., 1994).
The use of the effective width method can be also followed (Vayas and Psycharis, 1990).
- **Empirical methods** are based upon the statistical analysis of experimental data of full-scale member tests and provide practical relationships for determining the rotation capacity (Akiyama, 1980; Spangemacher, 1991).
Two major questions arise from the above definitions. The first question is to state when it is better to use R_{max} (Eq.1) or R_u (Eq.2) for the rotation capacity. The second refers to the problem if it should be preferable to use the plastic rotation θ_{max} or θ_u instead of rotation capacity R_{max} or R_u.

The answer to the first question is connected to the shape of the bilinear M-θ curve used in the structural analysis. If it is a strain-hardening post-yielding curve (Fig.3a), one must use the first definition of rotation capacity and the limit situation is defined by the maximum moment M_{max}. If a stress-strain curve with horizontal plateau is considered (Fig.3b), the second definition of rotation capacity can be used.

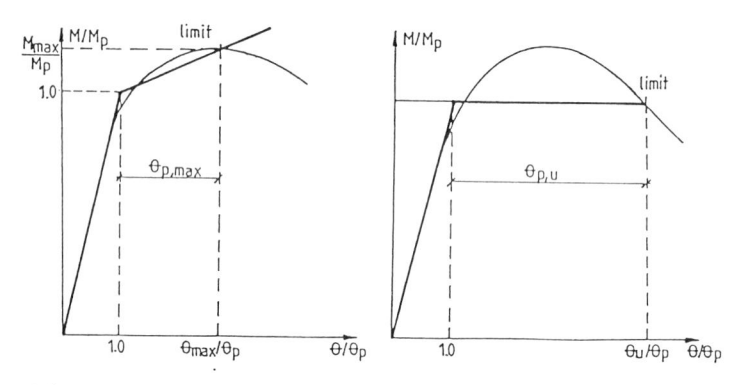

Fig.3. Cases for using θ_{max} or θ_u.

The answer to the second question is more difficult. For the evaluation of rotation capacity, a centrally loaded beam is commonly adopted as test specimen. The advantage of using this scheme consists on the possibility to simulate the behaviour of a cantilever member. The plastic rotations at the collapse state, θ_{max} or θ_u, are values which depend, in principle, on member properties, while the definition of rotation θ_p corresponds to the formation of plastic hinge. For the monotonic loading, the definitions of θ_p and R_{max} or R_u respectively can be done without difficulties. But for seismic load, many problems arise in a complex structure, because the first yield and the ultime state do not always arise at the same time and in the same section. For practical purposes, it can be easier to consider the plastic rotation and its limits, $\theta_{p,max}$ or $\theta_{p,u}$, instead of the rotation capacity R_{max} or R_u (Fig.3).

Fig.4 shows the influence of the ratios b/t_f, d/t_w, b/d, l/d, on the plastic rotation, obtained with the program DUCTROT-93 (Pectu and Gioncu, 1993), for determining the limit of plastic rotation, $\theta_{p,u}$. It is easy to observe that the provisions of the codes are very poor in comparison with the more complex actual behaviour.

Experimental works for determining the rotation capacity of H-section beams are made by Lukey-Adams (1969), Kuhlman and Roik (1986,1987 and 1989), Kemp (1985), Spangemacher (1991), Gioncu et al. (1989).

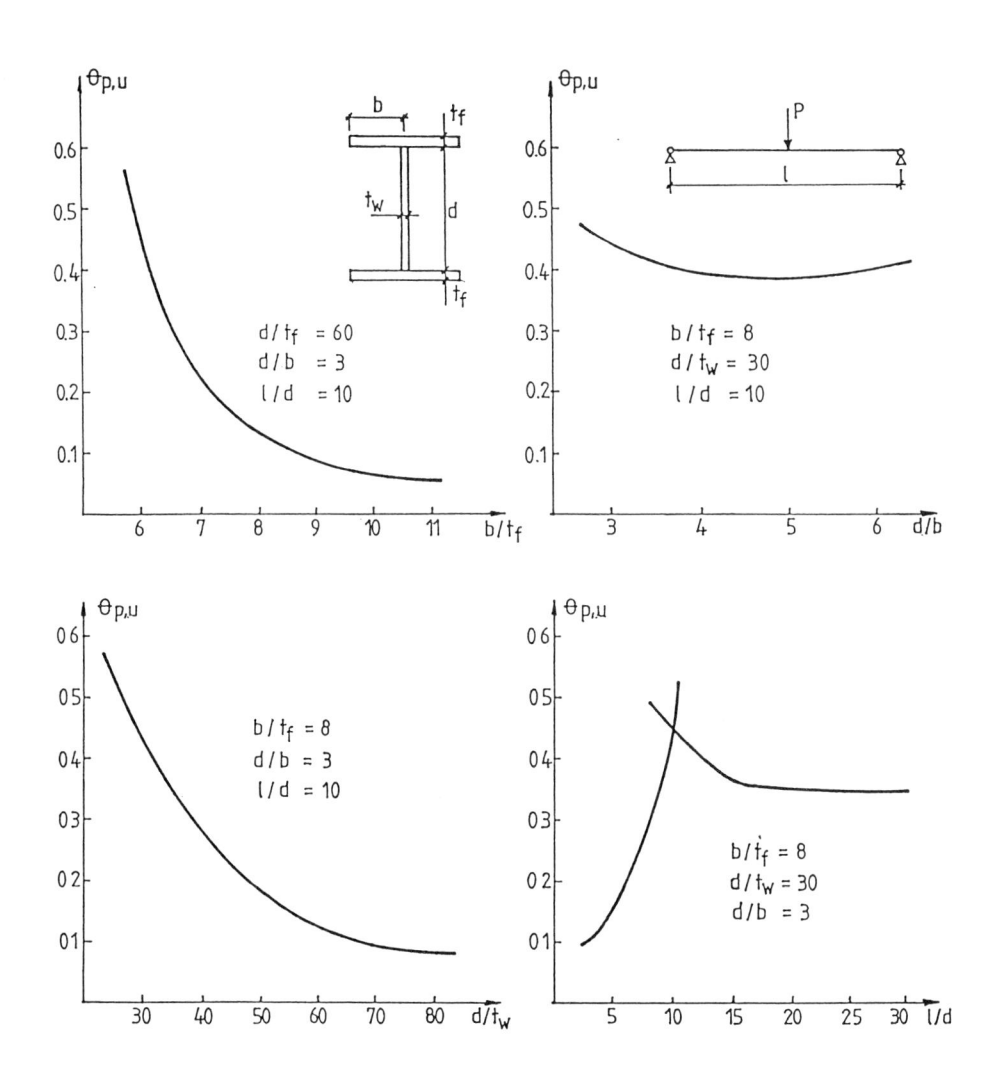

Fig.4. Influence of geometrical parameters on the plastic rotation.

Unfortunately, research results concerning the rotation capacity of the beam-columns are now a day very few, despite the influence of the axial forces very significantly reduces the capacity of plastic rotation. In Fig.5 two cases are presented: the rotation capacities θ_{max} and θ_u in presence of the axial forces after Mazzolani and Piluso (1993b) and Gioncu (1994). We can observe that the influence of axial forces on the maximum plastic rotation θ_u is more important than for the plastic rotation θ_{max}.

The comparisons between the Mazzolani and Piluso method for determining R_{max} rotation capacity and the Gioncu method for obtaining R_u rotation capacity, with some experimental avaiable values are presented in Fig.6. Taking into account the complexity of the phenomenon, the approximation obtained in both methods can be considered satisfactory.

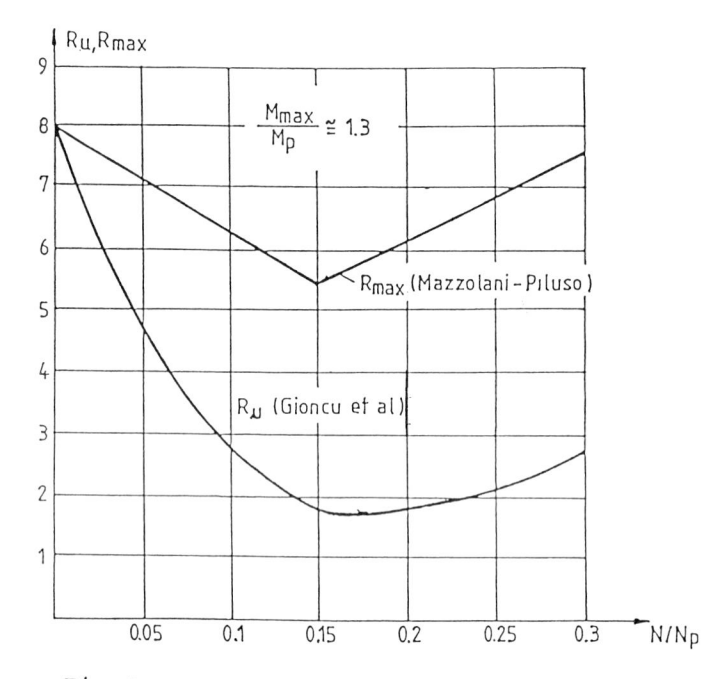

Fig.5. Influence of axial forces.

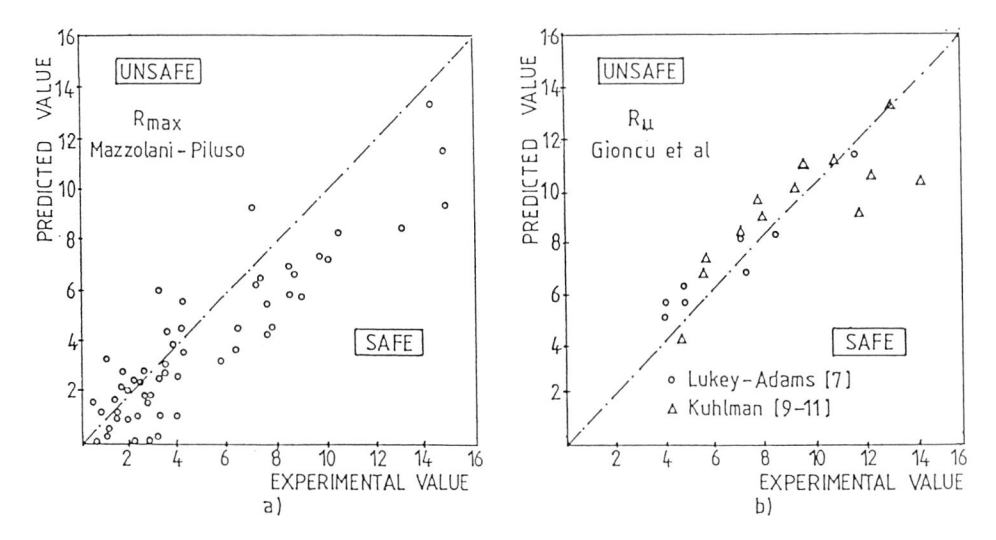

Fig.6. Comparison between theoretical and experimental values.

4. Conclusions

In conclusion, two very important aspects must be underlined:
- the cross-section behavioural classes, used in codes for determining the rotation capacity, should be substitued by the concept of member behavioural classes (Mazzolani and Piluso, 1993b);
- for the practical purposes it is easier to use the plastic rotation limitation instead of the rotation capacity;
- as an alternative for ductility classes, the actual values of plastic rotation limit can be determined using the relationship (Gioncu et al., 1994):

$$\theta_p = \alpha_N \alpha_e \bar{\theta}_p \tag{1}$$

where $\bar{\theta}_p$ is the rotation capacity limit of a beam, related to the ratios: b/t_f, d/t_w, b/d, l/d,
α_N and α_e, coefficients which introduce the influence of axial forces and of end eccentricity, respectively.

5. References

Akiyama, H. (1980) **Eartquake Resistant Limit State Design for Buildings**, University of Tokio Press.

Climenhaga, J.J. and Johnson, R.P. (1972) Moment-rotation curves for locally buckling beams, in **J. of the Struct. Div.**, vol.98, ST6, p.1239-1254.

Gioncu, V. Mateescu, G. and Iuhas, A. (1994) Contributions to the study of plastic rotational capacity of I steel sections in **Proceedings of STESSA'94**, Timisoara.

Gioncu, V. Mateescu, G. and Orasteanu, S. (1989) Theoretical and experimental research regarding the ductility of welded I sections subjected to bending, in **Proceedings of SSRC Int. Symp. on Stability**, Beijing.

Ivanyi, M. (1979) Moment rotation characteristic of locally buckling beams, **Technical University Budapest**, vol.23, 217-230.

Kemp, A.R. (1985) Interaction of plastic local and lateral buckling, in **Struct. Eng.**, vol.111, n.10, 2181-2196.

Kato, B. (1989) Rotation capacity of H-section members as determined by local buckling, in **J. Constr. Steel Research**, 13, p.95-109.

Kato, B. (1990) Deformation capacity of steel structures, in **J. Constr. Steel Research**, 17, p.33-94.

Kuhlman, U. (1986) Rotations kapazitat biegebeansprucher I-Profile unter Beriicksichtigung des plastischen Beulens, in **Technical Reports Institute fur Konstruction Ingenieurbau Ruhr** - Universitat Bochum, Mit.86-5.

Kuhlman, U. (1989) Definition of flange slenderness limits on the basis of rotation capacity values, in **J. Construct. Steel Research**, 14, 21-40.

Lukey, A.F. Adams, P.F. (1969) Rotation capacity of beams under moment gradient, in **J. of the Struct. Div.**, vol.95, ST6, June, 1173-1188.

Mazzolani, F.M. and Piluso, V. (1992) Evaluation of the rotation capacity of steel beams and beam-columns, in **Proceedings of 1st State of the Art Workshop COST C1**, Strasbourg.

Mazzolani, F.M. and Piluso, V. (1993a) ECCS Manual on "Design of Steel Structures in Seismic Zones, **ECCS document**.

Mazzolani, F.M. and Piluso, V. (1993b) Member behavioural classes of steel beams and beam-columns, in **Proceedings of C.T.A. Congress**, Viareggio, 405-416.

Pectu, D. and Gioncu, V. (1993) DUCTROT-93: Plastic rotation capacity of H-section steel beams and beam-columns, **Computer Programme INCERC**.

Roik, K. and Kuhlman, U. (1987) Rechnerische Ermittlung der Rotationskapazitat biegebeanspruchter I-Profile, in **Stahlbau**, Heft 11, 321-327.

Spangemacher, R. (1991) Zum Rotationsnachweis von Stahlkonstruktionen, die nach dem traglastverfahren berechnet werden, **Dissertation**, Technischen Hochschule Aachen.

Vayas, I. and Psycharis, I.N. (1990) Behaviour of thin-walled steel elements under monotonic and cyclic loading, in **Structural Dynamic**, (eds Kratzig et al.), Balkema, Rotterdam, 579-583.

16 INFLUENCE OF CYCLIC ACTIONS ON THE LOCAL DUCTILITY OF STEEL MEMBERS

M. A. AIELLO and L. OMBRES
Faculty of Engineering, University of Lecce, Lecce, Italy

Abstract

In this work the influence of strain limitation produced from shear and axial forces on the cyclic behaviour of steel structural members subjected to bending moment, shear and axial force, is examined.
By the definition of cyclic moment-curvature-axial force diagrams of critical cross-sections, the moment-rotation diagrams of members, modelled as rigid structures in which geometrical and mechanical properties are lumped in the above-mentioned sections, are determined.
At last, taking into account the reduction of resistance and rigidity caused from the cyclic actions, the ductility values of structural members are examined varying the number of cycles.
Keywords: Local Ductility, Cyclic Actions, Members.

1 Introduction

The ductility, that is the capacity of a structural system to undergo high plastic strains, is a fundamental requirement of the structural design in seismic areas. In presence of seismic forces, in fact, ductile structural members are able to dissipate notable energy amounts sensibly reducing damages that these forces can produce.
 Hence, the structural design cannot neglect the ductility; it can be referred to the overall structure (overall ductility) or to individual structural members (local ductility); moreover, the design with the ductility check allows, to address the static behaviour of structures inside the safety and effective limits.
 In seismic areas, the use of steel structures imposes the ductility check during the structural design; the Codes, in fact, define the behaviour factors q to evaluate seismic forces; they represent a measurement of the overall ductility of structures and depend both on the steel ductility and on the local ductility of individual structural members.
 The evaluation of the local ductility is very important

in order to define the overall behaviour of structures because a possible local deficiency can produce local crisis able to reduce the resistance of the overall structural system.

The local ductility is generally evaluated by moment-curvature diagrams with reference to cross-sections of beams or by moment-rotation diagrams considering individual structural members.

As a consequence, parameters influencing the local ductility value are the same that influence the strain state of structures, that is instability phenomena, constitutive laws of materials, the state of stresses and, at last, bond and load conditions of structural members.

Instability phenomena, relative to the flanges and the web of steel members, sensibly influence the local ductility because they can prevent the evolution in the elastic-plastic field of stresses and strains in some cross-sections reducing the ductility of structural members in which these sections are placed.

Besides, an adequate choice of the cross-section of members is necessary to avoid or to reduce the instability phenomena; for this reason can be useful the use of cross-sections of classes 1 and 2 geometrically defined from Eurocode EC3.

Anyway, this solution is not sufficient to guarantee acceptable values of the local ductility because effects furnished from other parameters, as the state of the stress acting on the structural members, are also influential.

Generally, in seismic areas structural members are subjected to a combined state of stress of bending moment, axial and shear forces which are of the same order of magnitude; then any one is not negligible to respect the other ones. The presence of bending moment, shear and axial forces, produces the collapse of the cross-section in corrispondence of a strain state more limited when both axial and shear forces increase ; as a consequence the local ductility depends from the combined state of stresses.

Seismic actions produce a decay of the resistance and the rigidity of structural members that corresponds to a reduction both of the strain capacity and of the ductility.This situation has been examined to avoid a noteable reduction of the initial ductility value of structures because of seismic actions and, moreover, to save the dissipative capacity of structures that, as known, is a very important property to prevent seismic effects.

On the basis of these considerations, in the present paper, the influence of cyclic actions on the ductility value of structural members with I-shaped cross-sections subjected to bending moment, axial and shear forces is analysed.

In particular, it is defined moment-curvature-axial force diagrams, varying the number of cycles of load on the basis of the strain limitation caused from the shear and axial forces values; afterwards, these diagrams are used to define moment-rotation diagrams of structural members and, at last, it is possible to determine the local ductility for each member giving prominence to the influence of the load cycles damage.

Numerical applications allow to put in evidence the influence of parameters that define the problem and furnish useful design informations.

2 Problem formulation

The ductility is evaluated by the ratio $\mu = \alpha_u/\alpha_e$ where α_u and α_e are values of a significant parameter of movement (displacement or rotation) of the cross-sections corresponding to the collapse and the elastic limit respectively. When we consider the cross-section of steel structures, the local ductility is evaluated as $\mu_\chi = \chi_u/\chi_e$ where χ_u and χ_e are the ultimate and the elastic value of curvature respectively, while, with reference to the structural member we have $\mu_\theta = \theta_u/\theta_e$ being θ_u and θ_e ultimate and elastic value of end rotations respectively.

The μ_χ and μ_θ values are not equals but related each other; from a design viewpoint the μ_θ value is more significant because it depends also from bound and load conditions and it takes into account instability phenomena, both local and overall. Anyway, the μ_θ value depends from the knowledge of moment-curvature diagrams of cross-sections of structures and, as consequence, from the steel constitutive law.

Bending moment-curvature-axial force diagrams, relative to I-shaped steel cross-sections in presence of monotonic loads, are defined analytically (Aiello et al.1992) examining all collapse mechanisms that can verify considering also local instability phenomena(web buckling).

In this paper we define a numerical procedure to determine the moment-curvature-axial force diagrams varying the number of cycles of load; using these diagrams we have determined moment-rotation diagrams and local ductility for structural members most used in steel constructions as cantilever beams and continuous beams.

Results are obtained using the steel constitutive law of elastic-plastic-hardening type; this law, is not extremely correct because, in presence of seismic forces, it is necessary to consider constitutive laws in which the structural damage produced from seismic actions is taken into account (Popov 1980, Ballio and Calado 1986, Castiglioni and Di Palma 1988). Anyway, this hypothesis sensibly simplifies the analytical procedure and it allows to obtain informations more significant as regard to the structural behaviour of steel members.

The aim of the paper, in fact, is to put in evidence the effectiveness of the value of the local ductility obtained considering structural members respect to the value relative to cross-sections of members. In this way, it is possible to take into account all parameters which define the structural behaviour of steel members that is not possible to consider with reference to the cross-section ductility value. As a consequence it is possible to define, more correctly,the member behavioural classes as proposed in (Mazzolani and Piluso 1993), respect to the cross-sectional classes of EC3 and EC8.

3 Moment-curvature diagrams of I-shaped steel cross-sections

Moment curvature diagrams in presence of axial and shear forces are defined on the basis of the stresses and strains variation on the cross-sections; the analysis is carried out using a constitutive law of elastic-plastic-hardening type.In order to semplify analytical relationships, we consider the reduced modulus $E_r = 2EE_h/(E+E_h)$.

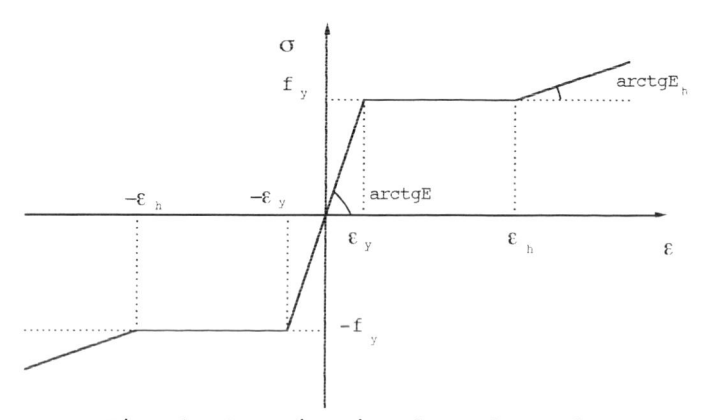

Fig. 1. Constitutive law of steel.

3.1 Monotonic load
Analytical relationships of the moment-curvature-axial force diagram for an I-shaped cross-section, are shown in (Aiello et al.1992). In the Fig. 2 the diagram obtained for the HE200M profile is shown; we can see that the presence of axial and shear forces, sensibly reduce the resistance and the strain of the cross-section. Particularly, when the non dimensional shear force exceeds the 0.3 value and the non dimensional axial force exceeds the 0.2 value, the resistance of the cross-section is remarkably reduced.

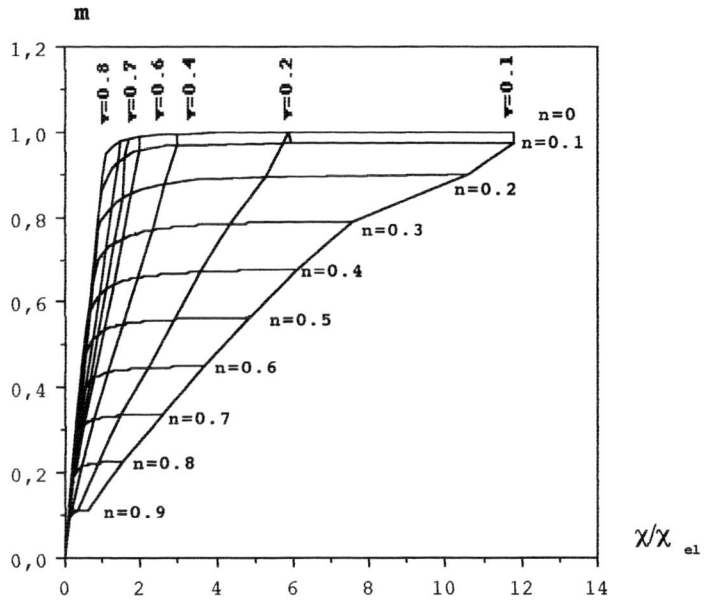

Fig. 2. Moment - curvature diagrams for the HE200M profile.

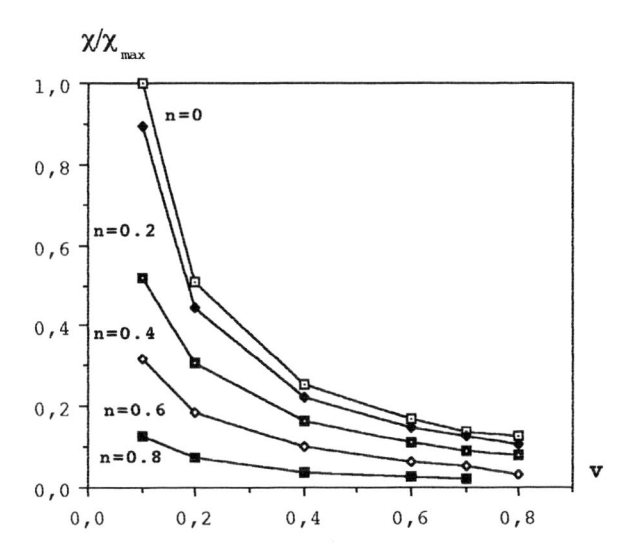

Fig.3. Diagrams χ/χ_{max} versus v.

In Fig. 3 is shown the χ/χ_{max}- n -v diagrams obtained using the m-χ-n-v diagrams where χ and χ_{max} are the ultimate curvature values corresponding to the collapse of the cross-section; we can observe the influence of axial and shear forces on the value of the ductility of the cross-section.

3.2 Cyclic loads

The definition of moment-curvature diagrams of cross-sections subjected to cyclic loads, is obtained using a numerical method based on the subdivision of the cross-section in strips.

The definition of cyclic diagrams must be carried out starting from a state of stresses and strains in the cross-section assigned on the basis of shear and axial forces values.

The problem, therefore, is conditioned from the stresses values and the cyclic behaviour of the cross-section is ruled from the strain limitation induced from axial and shear forces.

The same situation is verified when the cross-section is subjected to instability phenomena that produce a limitation of strain values.

In the Fig. 4 is shown the m- χ-n diagram relative to the HE200B profile considering a number of cycles equal to N=10.

It is possible to put in evidence that, increasing the number of cycles, we obtain a reduction both of bending moment and of curvature values. This result is confirmed from EC8 prescriptions which limit non dimensional shear and

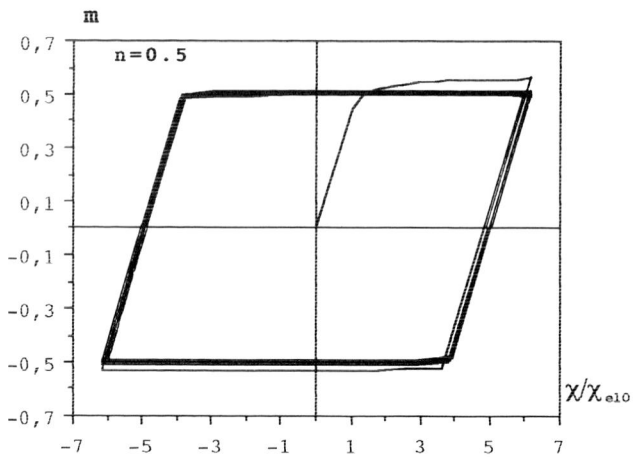

Fig.4. Non-dimensional moment-curvature diagrams for the HE200B profile (initial value of shear: v=0.2).

axial forces to 0.5 and 0.15 values respectively in
corrispondence to the cross-sections in which it is possible
the formation of plastic hinges.

4 Moment-rotation diagrams of steel members

Moment-rotation diagrams of steel members are defined using
moment-curvature values of cross-sections; as known, in
fact, rotation values are obtained by integration of
curvatures along axis of structural members.

Moreover, in presence of cyclic loads, the analytical
procedure is laborious and, anyway, very difficult; for this
reason, it is preferable to use simplified patterns that
allow to obtain immediately and with a good approximation,
the solution.

In this paper, it is considered a model in which the
structure is formed by rigid bars connected in
corrispondence of cross-sections in which all the
geometrical and mechanical properties of bars are lumped. In
particular the cross-sections above mentioned are the
critical cross-sections of the structure in which it is
possible the formation of plastic hinges.

The rotation values of bars and structures, therefore,
depend from the state of stress of critical cross-sections
and from the limit of strains corresponding to values of
shear and axial forces or to local instability phenomena.

Using this model, known the moment-curvature diagram of
the critical cross-section and taking into account the
strain limitation, the solution is immediate for a
statically determined member.

With reference to statically indetermined members,
generally the limit strain in each critical cross-section is
different;therefore the evaluation of the rotation value,
especially the ultimate value, using the proposed model, is
often difficult. In these situations, the use of simplified
procedures is not correct and the solution must be obtained
following the general procedure founded on the curvatures
integration.

5 Numerical applications

Numerical applications are carried out with reference to I-
shaped commercial steel members.

In the Fig.5 are shown the cyclic moment-rotation
diagrams relative to a centrally loaded beam ; results can
be compared to that one of a cantilever beam that represents
the unit of rigid frames configuration commonly used as
resistant steel structures in seismic areas.

The start of cyclic behaviour corresponds to the
curvature limit value defined on the basis of the shear

force values that, combined with axial force and bending moment, produces the collapse on the critical cross-section.

The variation of the local ductility of the beam, evaluated with reference to values relative to the total

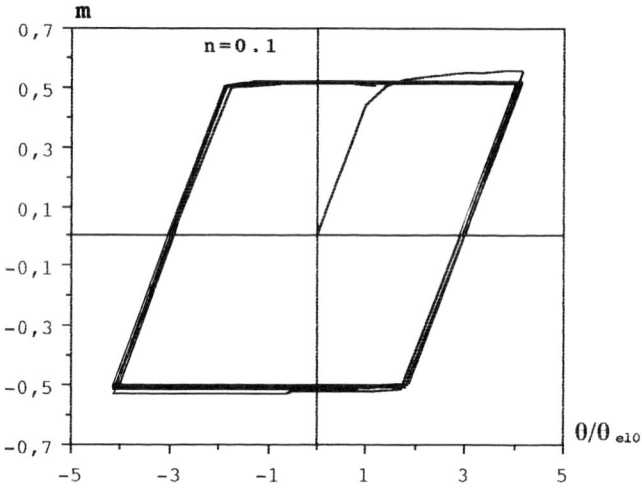

Fig. 5. Non-dimensional cyclic moment-rotation diagram for a centrally loaded beam (initial value of shear: v=0.3) [cross-section HE200B: b/t_f=13.33; b/t_w=22.22; h/t_w=22.22].

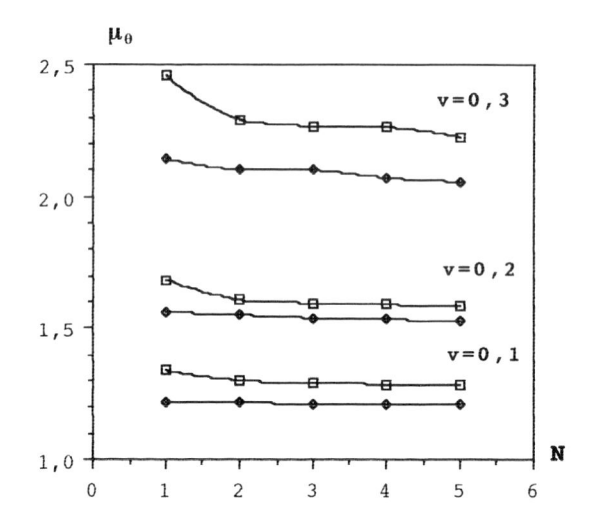

Fig. 6. Diagrams local ductility versus cycles number (centrally loaded beam HE200B profile) [⊞ n=0.5, ⟡ n=0.1].

inelastic excurtion for a half-cycle, is shown in the Fig.6 varying the number of cycles, the axial and shear forces values.

It can be observed that, the ductility value decreases when the cycles increase; in particular, for high values of axial forces the variation is more sensible in correspondence of initial cycles, afterwards, it becomes little.

In the Fig.7 are shown the non-dimensional moment-rotation diagrams relative to a cantilever beam with cross-section for which b/t_w= 17.86, b/t_f= 11.76 and h/t_w= 35.71.

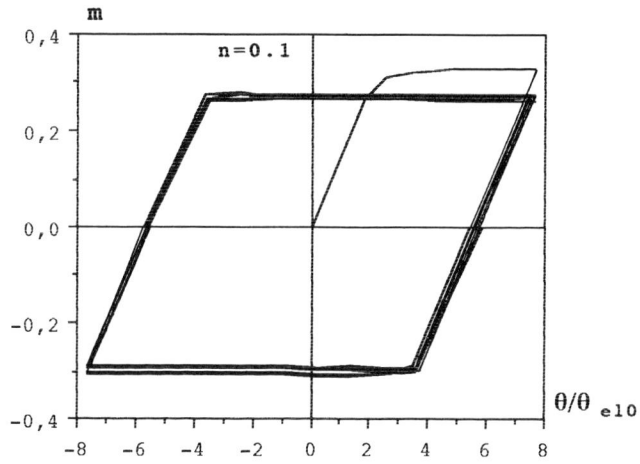

Fig. 7. Non-dimensional moment-rotation diagrams versus number of cycles (initial value of shear: v=0.3).

6 Concluding remarks

The work allows to put in evidence the influence of the strain limitation, caused from shear and axial forces acting on the critical cross-sections of steel members, on the local ductility value in presence of cyclic loads. This situation frequently happens in steel structural elements made in seismic areas; the strain limitation, produced also from instability phenomena of local or overall type can reduce the deformation capacity of members and, as a consequence, the local ductility value.

The deformation capacity of steel members depends not only on the geometrical properties of the cross-section but also on the axial and shear forces values, load and boundary conditions of members; for this reason it is necessary to evaluate the local ductility value considering the structural member in such a way as to take into account all parameters that influence the ductility. As a consequence it can be observed that the local ductility can not defined on the basis of the simple cross-sectional classification of EC3 but it is more correct to define a member behavioural classification that allows to obtain a more significant value of the local ductility.

Results obtained in this work show the dangereous effects produced from shear and axial forces on the resistance and rigidity of steel members, especially in presence of cyclic actions, and they confirm the opportunity to limit values of shear and axial forces as prevue from the Eurocode, EC8.

7 References

Aiello, M.A. La Tegola, A. and Ombres, L. (1992) Coupled instability of thin-walled members under combined moment, axial and shear force, **Proceedings of CIMS '92, First International Specialty Conference on Coupled Instabilities in Metal Structures**, Timisoara, Romania.

Ballio, G. and Castiglioni, C.A.(1993) Le costruzioni metalliche in zona sismica: un criterio di progetto basato sull'accumulazione del danno, **Atti del convegno C.T.A.**, Viareggio, 99-109.

Ballio, G. and Calado, L.(1987) Steel bent sections under cyclic loads. Experimental and numerical approaches, **Costruzioni metalliche**, 3, 1-23.

Ballio, G. Calado, L. Leoni, F. and Perotti, F.(1986) Numerical simulation of the cyclic behaviour of steel subassemblages, **Costruzioni metalliche**, 5, 268-295.

Castiglioni, C.A.(1992) Analisi numerica e sperimentale del danneggiamento di elementi strutturali in acciaio soggetti ad azioni sismiche, **Ingegneria sismica**, anno IX,1,9-15.

Castiglioni, C.A. and Di Palma, N.(1988) Membrature in acciaio soggette a carichi ciclici: modellazione numerica e confronti sperimentali, **Costruzioni metalliche**,6, 288 -312

Commission of the European Communities(1992) Eurocode n.3.

Commission of the European Communities(1992) Eurocode n.8.

Degenkolb, H.J.(1970!), Design of earthquake-resistant structures: steel frame structures, in **Earthquake Engineering**(coordinating editor R. Wiegel), Prentice Hall, Inc., England Cliffs, N.Y., pp.425-447.

Mazzolani, F.M. and Piluso, V.(1993) Member behavioural classes of steel beams and beam-columns, **Proceedings of the C.T.A. Conference**, Viareggio, 405-416.

Popov, E.P.(1980) Seismic behaviour of structural subassemblages. **Journal of Structural Division**, ASCE, 7, 1451-1474.

Pozzati, P.(1987) **Teoria e Tecnica delle strutture**, Vol. 3*, Ed. UTET, Torino.

17 DUCTILITY OF STEEL-CONCRETE MIXED SECTION BEAMS

V. PACURAR

1. General considerations

The violent earthquakes of recent years set into evidence the essential role of plastic deformations in the behavior of load carrying structures. To make certain that failure will take place within normal sections having an adequate ductility, the mixed steel-concrete elements should have such a structure that the shearing forces would be carried in optimum conditions, thus creating some plastic yielding joints.

The European standards for reinforced and prestressed concrete members calculation 1 accept a characteristic curve - b of parabolic - rectangular shape for concrete (fig.1).

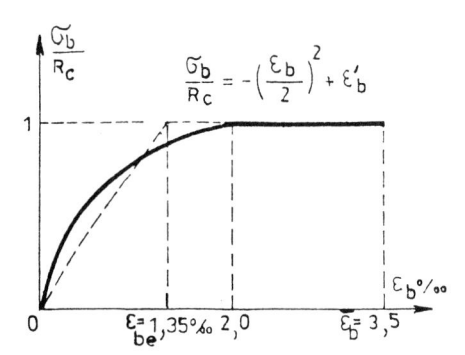

Fig.1 The stress-strain calcul diagram for concrete loaded to compresion

This shape has been also accepted for the calculation of mixed sections 3. For steel, the European standards 2 for steel member calculation as well as for mixed sections 3, accept the simplefied calculating shape of the membrane- (fig.2).

The plastic deformations of concrete (fig.1) is relatively small $b_p = b - b_e = 2, 1-3\%$ while the plastic deformations of steel are much higher, they could get to 5...8%.

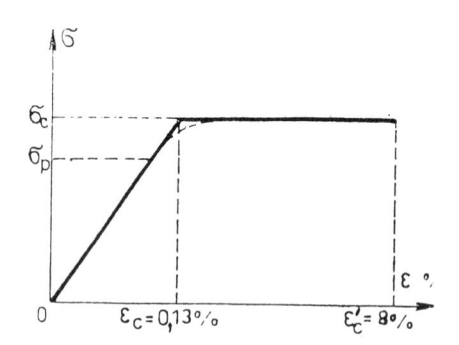

Fig.2 The stress-strain calcul diagram for steel OL37

The deformation calculation (rotations curvatures, sags) within plastic state, is performed by taking into considerations the folloing suppositions (hypotheses).

a) The stress-strain relationships, both for steel beam and reinforced concrete slab are non-linear. For the steel beam the simplified diagram of fig.2 is accepted, the specific creep strain (ε_c) marking the yield point is given by relation:

$$\epsilon_c = \frac{\sigma_c}{E_0}$$

For concrete within compressed area the parabola rectangle shaped chacterristic curve is admitted (fig.1).

b) The fracture of elements is produced by reaching creep strain within the compressed area of the metallic section.

c) If the neutral axis is located within the concrete slab or the slab is situated within the tensioned zone, the role of the tensioned concrete is not taken in consideration in carrying the load.

The ductility factor is one of the parameters characterising the behavior of a structure element or material, in plastic (post-elastic) range, and represents the ratio between the maximum deformation on collapsing and the deformation corresponding to the yield point.

If we consider the rotation of the section, the ductility factor can be determined with relation:

$$\Delta = \frac{\Theta_r}{\Theta_c}$$

where θ_r represent the rotation on fracture, and θ_c represent rotation on reaching yiel point of steel.

2 Parameters Influencing the Ductility Factor

From the investigations carried out on steel-concrete mixed sections element, we got to the conclusion that the ductility factor is directly or undirectly influenced by a great number of parameters. These can be groupe as:

a) Paramerters generally influencing the ductility of members and structures:
   - the mechanical and deformation properties of materials (steel and concrete);
   - type of loading: bending, central force bending, etc;
   - the manner the loads are applied: statically, dynamically, oscillating, alternating etc;
   - quality of craftsmanship.

b) Parameters specific to steel-concrete mixed section member:
   - type of joints and degree of cooperation;
   - manner of arranging the metallic structures and mixed section as a whole, including the manner of reinforcing the concrete slab.
   - the manner the section fails and which can occour by the formation of a plastic hinge due to web value or by losing general stability;
   - the way the structure elements are joint together.
Recent investigation on elements of mixed sections have set into evidence the difference between the steel-concrete mixed section behavior and the behavior of steel elements and concrete elements as well.
   The reinforced concrete increase local and general stability of metallic beams securing their failure by creating plastic hinges.

3 Experimental Program

The tests were carried out on two types of steel-concrete mixed beams G_1 and G_2.

3.1 Beams G_1 (fig.3)
Beams G_1 are made of welded OL37 steel beams, elastic Bc30 lap type monolithic reinforced concrete head and plate connectors being made with steel OB37 cross-ties and bors.
   As concerning the plastic deformation capacity the metal beam falls in class .
   Like steel-concrete mixed beam, since the plastic neutral axis is situated inside the metal beam web, this falls in 2nd class, the concrete slab heaving its ultimate

strain much lower than that of the steel.(3)
The beams were tested this way:
-an element G_1 - 1 was subjected to static loads showly applied, in stage of 30 kN, until failure occurred;
-an element G_1 - 2 was subjected to 500,000 loading & unloading cycles between long term loading (P_d) and working load (P_E) and then it was subjected to a steady static load until failure.
  The following measurements were taken during testing :
-specific deformatins over the beam height in more sections;
-slippages between metallic beam and reinforced concrete slab , the value of sags in several point;
  The values of long term, working, calculating and breaking loadings are given in table 1.

Fig.3 Beams G_1

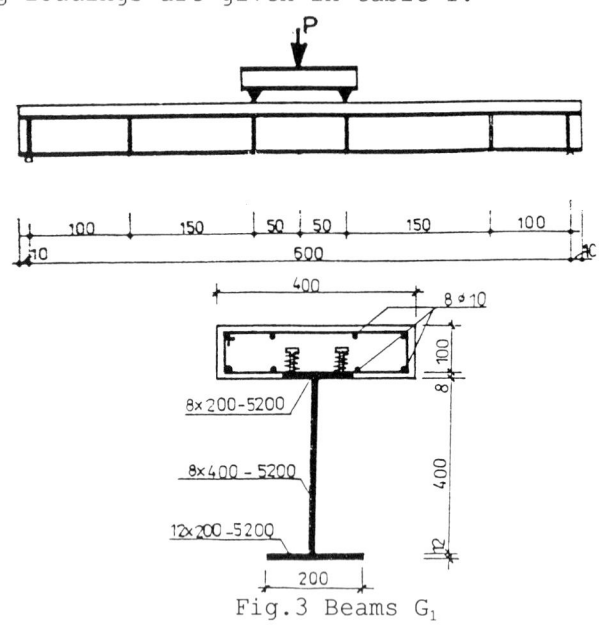

Table 1. The values of loads

	Long-term		Working		Calculating		Break			
							Calculated		Experimental	
	P_{ld}	M_{ld}	P_E	M_E	P_c	M_c	P_r^c	M_r^c	P_r^{exp}	M_r^{exp}
	kN	kNm	kN	kNm	kN	kNm	kN	kNm	kN	kNm
G1-1	135	168,7	180	225	275	344	324	405	402	502,5
G1-2									390	487,5
G2	70	105	100	150	140	210	185,5	278,2	175	265,2

subjected to 500,000 loading-unloading cycles (fig.4)

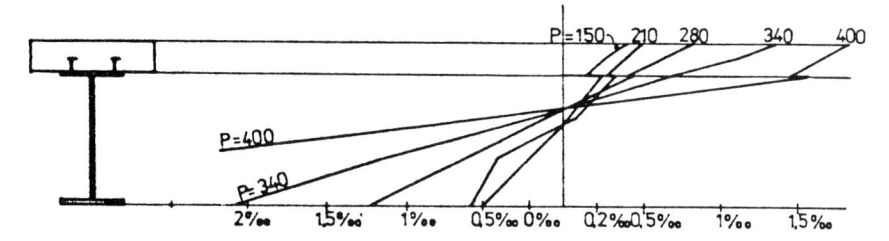

Fig.6 The variation of deformation over the beam height at midspan for differnd loading stages

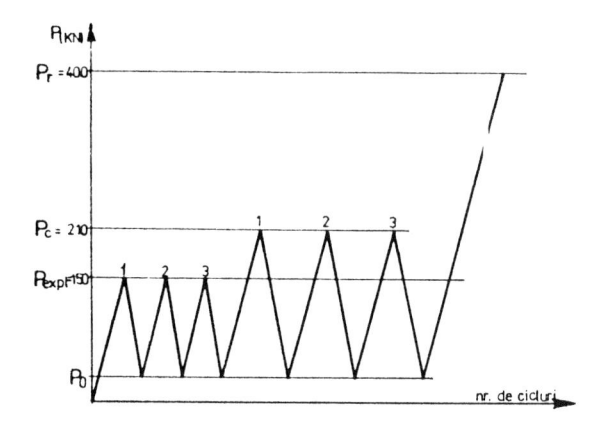

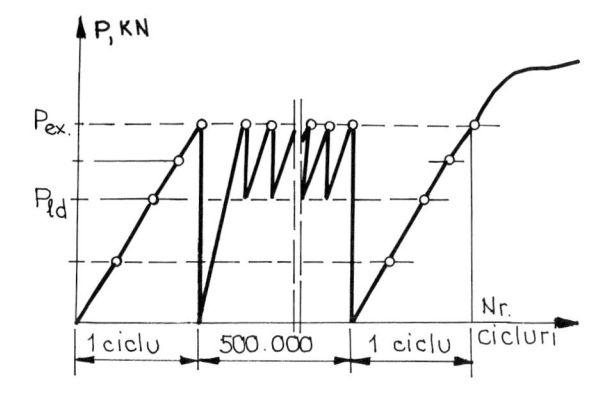

Fig.4 Loads application

Beam G_1-1-1 was loaded in stages of 30 kN according to diagram of figure 4.a. From force-flexure diagram (fig.5) follows an elastic behavior of the mixed beam until a value of 0.70-0.75 P_r of the load.

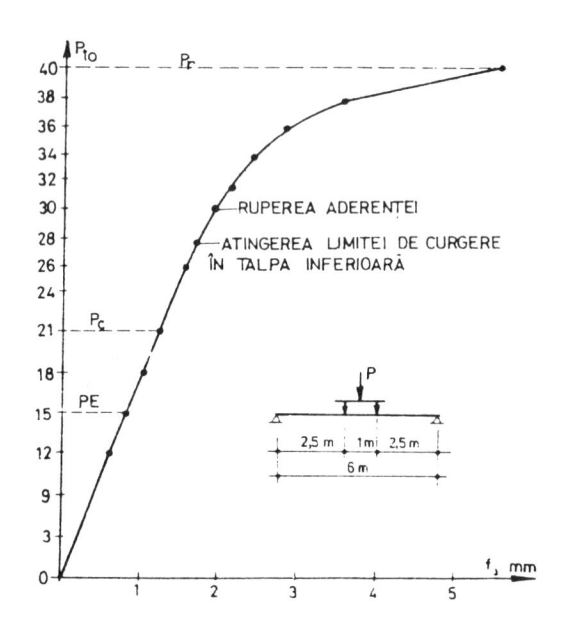

Fig5. Force-flexure diagram

Beyond this value when the failure of connection was produced, the force-sag relationship was no longer linear due to the inelastic deformations and slippages occuring between the metal beam and concrete slab.

The failure of the member occured in the central area by the fact that the lower flange of the metal beam began to slip, the concrete slab being crushed over a lenght of 40 cm.

Prior to failure, the maximum slips between the metal beam and concrete slab measured at ends were 0.57 mm and 0.63 mm.

The variation of deformation over the beam height at midspan was presented in figure 6, for different loding stages.

It has been noted that the neutral axis is all the time situated within the metal beam web. Beam G-1-2 was

ranging from $P_d=135$ kN and $P_E=180$ kN.
The loading frequency was 40-50 oscillations per minute. No esential modifications have been recorded with beam behavior after 560,000 loading cycles, the flexures increase at 55% with respect to the first loading cycle, and the creep was only local having very low values 0.05 mm.

After loading cycles were concluded the beam was loaded up to yield point in the same way as beam G-1-1. Beam G-1-2 broke down identically to beam G-1-1, the ultimate strenght being about 3% lower.

3.2 Beam G_2 (fig.7)
The specimen G_2 is made up of the two I30 steel OL37, profile, being wind braced at the lower part and the concrete slab made up of 19 Bc30 prefab reinforced concrete slabs. The metal beams were rigidity connected to the reinforced concrete beam (fig.7). On testing this member we examined the way the loading hypotheses were observed with designing and testing road bridges.

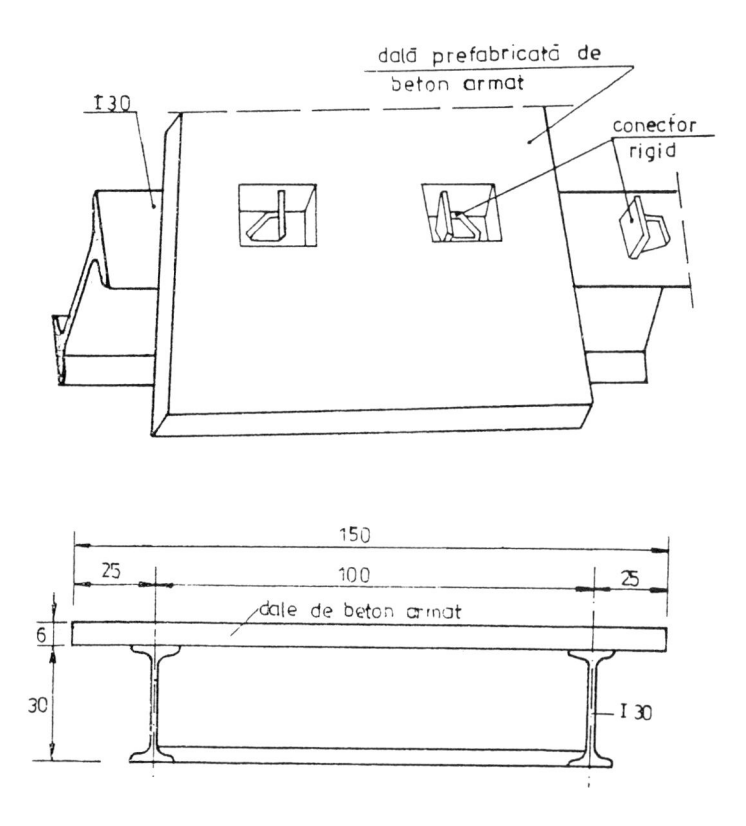

Fig.7 Beam G_2

The loding stage were so designed that they would simulate long term loading, working and calculating loading.

The specific deformations were measured with this element, as well, durring loading cycles, in three zones over the beam height, the slips between metal beam and rinforced concrete slab and the value of flexure were measured in several points, as well due to the rigid connectors, only local slips of value longer than 0.05 mm were recorded.

The failure of the member took place by crushing the concrete within prefab web plate joints being accompanied by severe plastic deformation of metal beams. The value of the experimental moment of rupture was 5% lower that the calculated one.

4 Estimation of Beam Ductility.

The values of ductility factor determined with relations (2) are given in table 1, together with the experimental values of ductility factor resulting as a ratio of breaking torque (θ_r) and creep (θ_c), measured within the central area of beams G_1 and G_2.

The creep limit deformation considered within the extreme grains of mettalic beams was 1.30%, and the specific limit deformations of concrete, taken into consideration for ductility factor assessement were b_e=1.35% and b_r=3.5% (see table 2)

Table 2 Designing Experimental values of Ductility Factors

Member		Designing values			Experimental values		
		θ_c	θ_r	$\Delta = \dfrac{\theta_r}{\theta_c}$	θ_c^{exp}	θ_r^{exp}	$\Delta^{exp} = \dfrac{\theta_r^{exp}}{\theta_c^{exp}}$
G_1	a	$\dfrac{2,65\%o}{h}$	$\dfrac{25,5\%o}{h}$	9,62	$\dfrac{2,60\%o}{h}$	$\dfrac{31,98\%o}{h}$	12,3
	b				$\dfrac{2,31\%o}{h}$	$\dfrac{24,95\%o}{h}$	10,8
G_2	a	$\dfrac{2,65\%o}{h}$	$\dfrac{28,2\%o}{h}$		$\dfrac{2,65\%o}{h}$	$\dfrac{31,53\%o}{h}$	11,9
	b				$\dfrac{2,61\%o}{h}$	$\dfrac{31,58\%o}{h}$	12,1

The design limit deformation obtained in mettalic beam G_1 was 25.5% and 28.2% in G_2, respectively.

The experimental values of specific deformations over the height of the section were measured by means of the removable Huggenberger deformeter on the metallic beam and by means of a dial extensometer on concrete.

The limit shrinkage measured on concrete was 4.1-4.3% having a mean value of 4.2% with beam G_2, respectively 4.15 with G_1.

The limit elongation of the lower grains of metallic beams measured until the stage the beam load was 29.38% (and 22.64%) with beams G_1 and 22.64% with the pair of G_2 beams.

The fracture occurred when the metallic beam started to field, followed then by concrete failure in compressed section.

Analysing the experimental and design values of ductility factors contained in table 2 we found them quite close and that the accepted design hypotheses are close to actual values.

5 Conclusions

The experimental investigations carried out on beams of mixed steel-concrete sections proved them to have a remarcable behavior in ultimate state. The experimental values of ductility factor were higher than the calculated values with all beams under loading.

The analysis of beams ductility of mixed steel-concrete sections sets into evidence the extremaly positive behavior of such types of sections under circumstances of their utilisation in arhieving high strength structuresof building situated in seismic areas.

Although the compressed flange of beams G_1 has a flexibility $\frac{b_1}{t} = \frac{96}{8} = 12$ higher than the admissible limit value, 10, for the beams of class 2-a 2, its value did not turn out on yielding because it was fixed to the concrete beam with connectors.

The use of mixed sections results in an increase of safety degree and local stability of structural members, alone side with improvement of their behavior in ultimate load stage.

6 References

1 EUROCODE 2: "Calcul des structures en béton" Partie 1-1: Règles générales et règles pour bâtiments. 1992

2 EUROCODE 3: "Calcul des structures en acier" Partie 1-1: Règles générales et règles pour bâtiments. 1992

3 EUROCODE 4: "Desing of composite steel and concrete structures. Part 1-1: General rules and rules for buildings.

4 Păcuraru V., Tertea I., Oneţ T., Viorel Gabriela: Efectul încărcării repetate asupra comportării grinzilor de pod cu secţiune compusă oţel-beton. Buletinul Ştiinţific al Institutului Politehnic Cluj-Napoca, nr.28-1985, pag.17-28.

18 INCREASE OF BUCKLING RESISTANCE AND DUCTILITY OF H-SECTIONS BY ENCASED CONCRETE

A. PLUMIER and A. ABED
University of Liège, Liège, Belgium
B. TILIOUINE
Ecole Polytechnique d'Alger, Algiers, Algeria

Abstract
Concrete encased between the flanges of H beams positively influences their behaviour under cyclic loading. Recent research results are explained and analysed to evaluate this effect.
Keywords : Composites Structures, Encased Concrete, Cyclic Behaviour, Local Buckling, Rotation Ductility, Seismic Design.

1 Introduction

The study of seismic behaviour of the composite steel-concrete structures is of a special interest in regions of high seismicity. Works in this field was started in Japan and in North America. Recent works in the same field has been done also in Europe (Office for Official Publications of the European Community 1992, Abed and Plumier 1993, Plumier and Schleich 1993, Elnashai et al. 1991). However, significant gaps in the knowledge remain, to which corresponds a lack in design rules for composite steel-concrete structures. Consequently, to fill some of the gaps mentioned above, researchs on seismic behaviour of composite structures have been realised, with an active participation of the University of Liège.Tests have been carried out on composite beam-column specimens under cyclic loads, aiming at a better knowledge of the connection behaviour in general and of the influence of concrete with respect to the behaviour of pure steel components.

This presentation essentially deals with the influence of the concrete encased between the flanges of H sections on their behaviour in the non linear range. It is mainly based on the results of research by Abed and Plumier (1993) and to some results of research work by the Office for Official Publications of the European Community (1992). As the work by Abed and Plumier (1993) is unpublished, we first summarise its content.

Behaviour of Steel Structures in Seismic Areas. Edited by F. M. Mazzolani and V. Gioncu.
Published in 1995 by E & FN Spon, 2-6 Boundary Row, London SE1 8HN. ISBN: 0 419 19890 3.

2 Test set-up

The tests were performed on the test set-up for cyclic testing on structure available at the University of Liège (Fig. 1). The tested specimens are hinged on top and bottom of the column and at the end of the steel beam extender. The displacements are produced by means of 2 actuators in line, so that the available range of displacements around the origin is + 400 mm, -400 mm. Maximum applied force is 1000 kN. Forces are recorded by means of an external load cell installed at the end of the beam extension.

3 Tested specimens

The specimens are made of a HE 300 B column and a HE 260 A beam. Concrete is encased between the flanges of the beams and the columns. There are two types of beam-column connection : bolted connection B and welded connection W. The dimensions and the reinforcement of the concrete as well as the cross section of typical tested specimen are shown in Fig. 2. For each type of connection, three different values of web thickness t_w in the column panel zone have been realised : t_w = 11(original thickness), 7 and 15 mm. Each type of specimen has been tested according to 2 testing procedures. A total of 12 composite specimens are tested.

4 Testing procedure

The loading programme, given by two displacement histories, is shown at Fig. 3. The first displacement history (Fig. 3.a) complies with the ECCS testing procedure (ECCS 1986) , slightly modified according to the results of a previous research on composite sections (Office for Official Publi. of the Eur. Com. 1992). A real change in the ECCS procedure was also introduced. It consists in a cycle with great displacement amplitude introduced just after the first three equal cycles in the plastic range. The second displacement history (Fig. 3.b) consists in a regular displacement increase at every step. These displacement histories are based on the yield displacement e_y and on the plastic displacement Δe of an estimate of the maximum practical ultimate displacement (e_u). The value of e_y was evaluated on the basis of the plastic rotation observed for similar HE 260 A beams in a previous research (Office for Official Publi. of the Eur. Com. 1992).

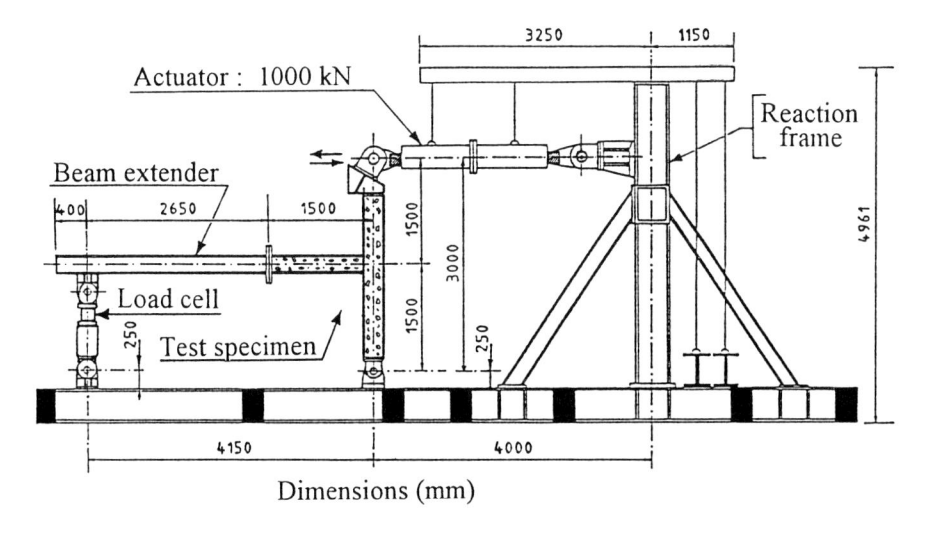

Fig. 1. Tested specimens in test set-up.

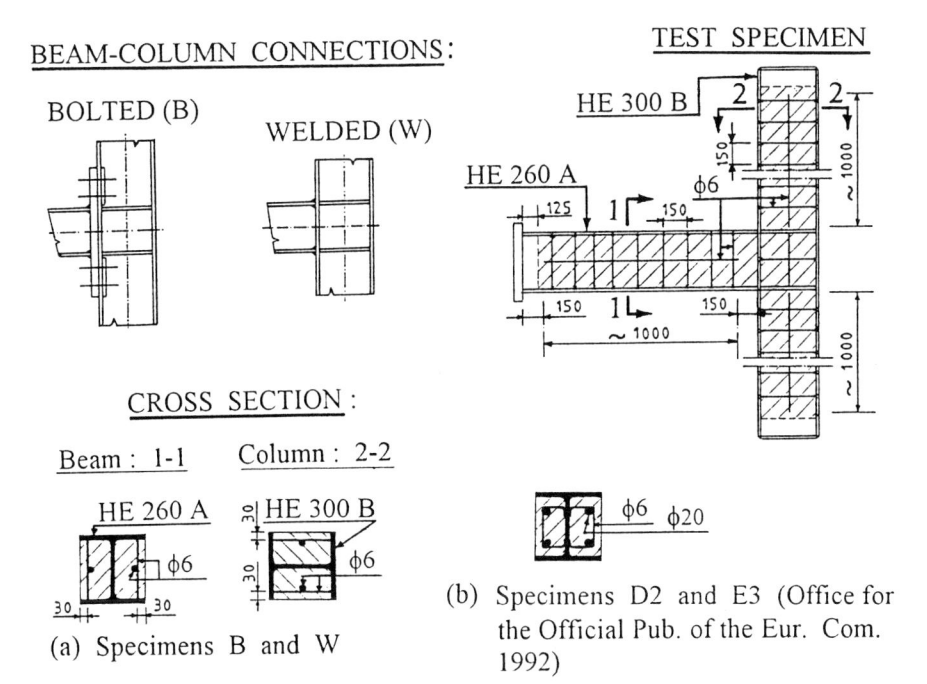

BEAM-COLUMN CONNECTIONS:
BOLTED (B)   WELDED (W)

TEST SPECIMEN

CROSS SECTION:
Beam: 1-1   Column: 2-2
HE 260 A   HE 300 B

(a) Specimens B and W

(b) Specimens D2 and E3 (Office for the Official Pub. of the Eur. Com. 1992)

Fig. 2. Dimensions and reinforcement of the concrete of the tested specimens.

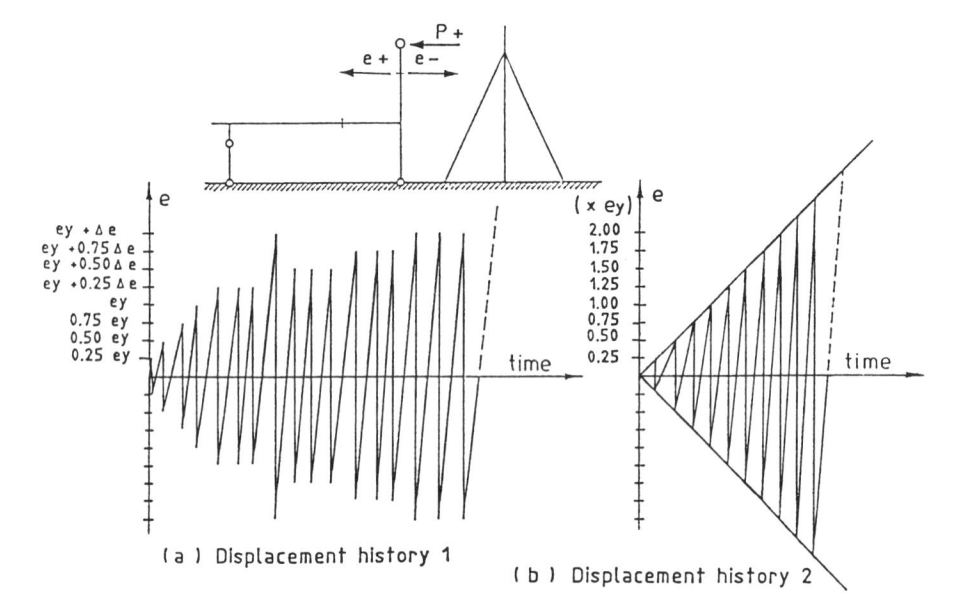

Fig. 3. Displacement histories.

5 Test results

Fig. 4 shows moment-total rotation relationships of two tested specimens, in which bending moment and total rotation are those of the beam. Some observations can be made :

- for each one of the 2 types of connection, the tested specimen has, before failure, practically the same behaviour under both history of displacement, the envelope curve of the hysteresis curve being almost identical for the two cases ;
- all the observed yield mechanisms take place in the beam, in one of the following way : plastic bending of the beam, involving the welds to the column or end plate and final flange buckling (every specimen) ;
- the changes in the thickness of the web of the column play a relatively minor rule and only some very slight cracks are observed in the concrete filling the panel zone ; they correspond to negligible rotation of the beam. This is due to the locking of the panel zone by a somewhat over resistant concrete (fck = 53 N/mm^2).
- the concrete encased between the beam flanges clearly opposes to buckling towards the inner part of the beam (Fig. 5). When there is no early event in the welds, the buckles in the beam flanges push out partly or totally the concrete infillment and finally crack.

(a) Specimens B ; displacement history 1

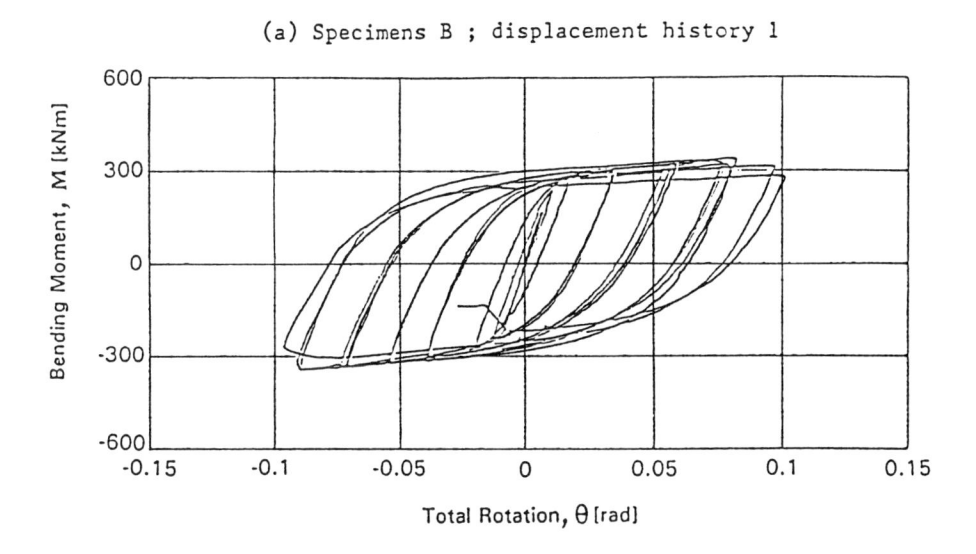

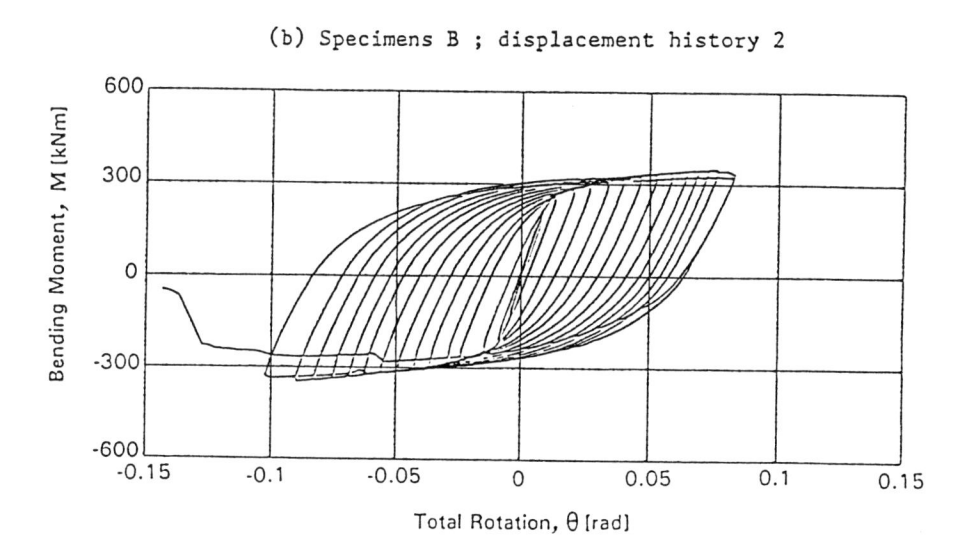

(b) Specimens B ; displacement history 2

Fig. 4. Moment - total rotation hysteresis curves.

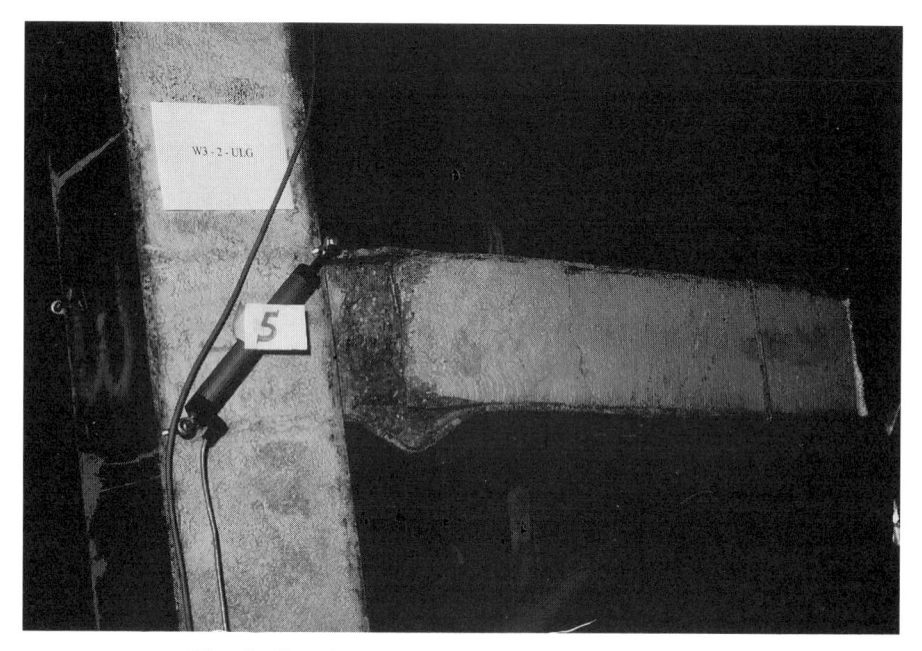

Fig. 5. Specimens W - Failure mechanism.

6 Analysis of the influence of encased concrete

An evaluation of test results is illustrated at Figs. 6 to 13. This evaluation refers to the research programme discussed above and to that by the Office for Official Publi. of the Eur. Com. (1992), in which similar steel sections have been tested. In that work, however, the encased concrete and its reinforcement were different (Fig. 2.b). The specimens tested in the research work by the Office for Official Publi. of the Eur. Com. (1992) are referred to by the symbol D for the bolted specimens and by the symbol E for the welded ones. The index associated to D and E have the following meaning :

> D2 : composite similar to B
> E2 : bare steel beam, welded to a composite column ; E2 is the reference for all the comparison between bare steel and composite beams
> E3 : composite similar to W.

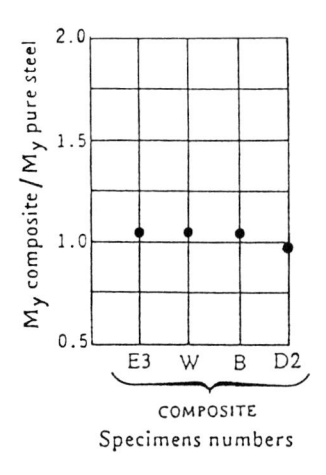

Fig. 6. Normalised plastic moment.

Fig. 7. Normalised elastic stiffness.

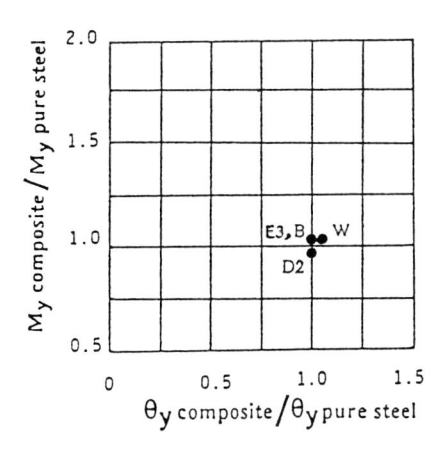

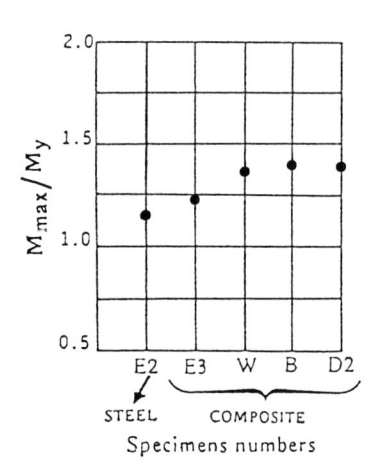

Fig. 8. Normalised plastic moment and plastic rotation.

Fig. 9. Normalised maximum Bending moment.

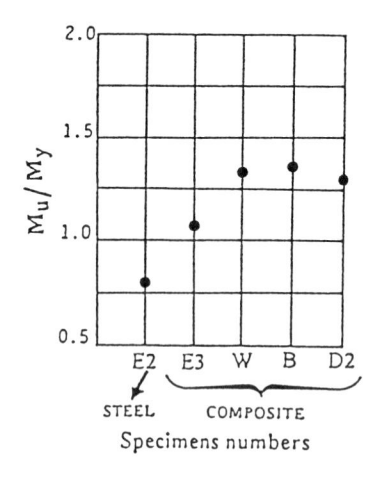

Fig. 10. Normalised ultimate bending moment.

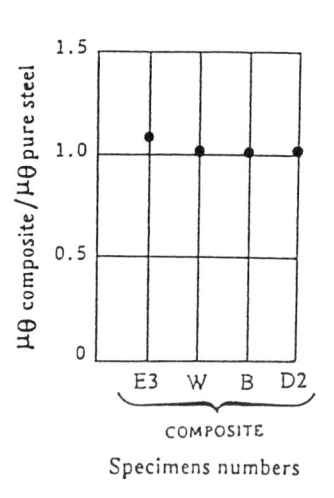

Fig. 11. Normalised maximum rotation ductility.

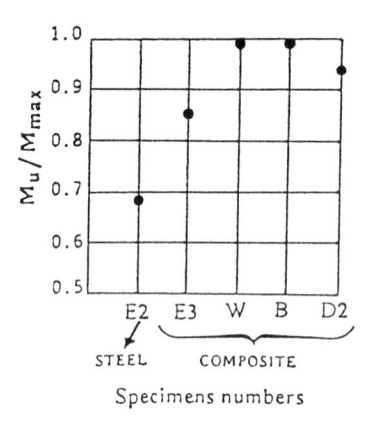

Fig. 12. Maximum moment deterioration.

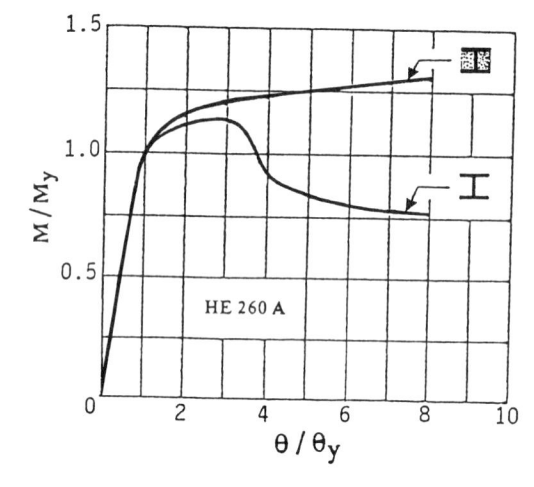

Fig. 13. Envelope of hysteresis curves for steel and composite beams.

The test results of the research by Abed and Plumier (1993) are the mean value for each type of specimen B or W. These results are compared with those of the research work by the Office for Official Publi. of the Eur. Com. (1992).

Fig. 6 shows the plastic moment M_y of each specimen normalised by the plastic moment of the steel section. There is almost no influence of the encased concrete on M_y. This is due to the position of the concrete close to the neutral axis of the beam, to the lack of connection between the concrete in tension and the beam and to a relatively weak link able to transmit shear between steel and concrete.

Fig. 7 shows the elastic stiffness K_e normalised by the elastic stiffness of the steel section. The increase of the elastic stiffness by the composite effect is about 25 % for the specimen D2 and 16 % for E3, while there is no increase for B and W. This is due to a very low longitudinal reinforcement of the concrete in these specimens and a totally ineffective shear transfer.

The normalised plastic moment and the plastic rotation θ_y normalised by the plastic rotation of the steel section are shown at Fig. 8. The value of the normalised plastic rotation is almost equal to 1.0 for all the specimens. It means that the θ_y value is unique : $\theta_y \cong 10^{-2}$ rad. $\cong 0.6°$.

The maximum load-carrying capacity, measured in terms of maximum moment M_{max} reached in a specimen normalised by its plastic moment (M_y), is given at Fig. 9. The value of the maximum load-carrying capacity of the composite specimens (B, D2, W, E3) ranges around a value of about 35 % higher than the plastic moment, while this value is about 14 % for the bare steel specimen E2. This difference can clearly be attributed to the restraint to buckling brought by the concrete.

Fig. 10 shows the ultimate moment M_u of the specimens normalised by their plastic moment. M_u corresponds to the maximum total rotation θ_u reached without failure. The normalised ultimate moment in the composite specimens is 70 % higher than that in the steel specimens.

Fig. 11 shows the maximum rotation ductility μ_θ normalised by the maximum rotation ductility of the steel section. It is close to 1.0 for all the types of specimens. This means that the θ_u value is unique, at least for the histories of displacement involved : $\theta_u \cong 8 \cdot 10^{-2}$ rad.

Deterioration effects, causing a reduction in the load-carrying capacity can be seen at Figs. 10, 12 and 13. It is measured by the ratio of the ultimate moment M_u previously defined to the maximum moment M_{max} (Fig. 12). The reduction in load-carrying capacity in the great displacement range is quite weak for the composite specimens. For the steel specimens the reduction is substantial :

$\frac{M_u}{M_{max}} = 0.68$ for specimen E2. Again, this can be explained by the restraint to buckling brought by the concrete encased between the flange of the H section. In steel specimens, the local buckling is not prevented, begins earlier than in the composite specimens and progresses during the loading cycles, causing then a progressive deterioration of the load-carrying capacity of the tested element. This phenomenon is largely delayed when concrete is present.

This positive effect of concrete can also be translated in terms of ductility. If ductility is defined considering a limited loss in bearing capacity of the bended element, we can also conclude that :

$$\mu_{composite} > \mu_{steel}$$

For instance, if we admit a 0 % loss on M_y, we have, on the mean :
$$\mu_{composite} = 8, \text{ while } \mu_{steel} = 4 \text{ (Fig.13)}.$$

7 Practical conclusions for seismic design

All the results given above can be synthesised in the envelope curves given at Fig. 13. If a structure made of H sections with encased concrete is designed, we suggest for its design against seismic actions :

- the stiffness of the elements is similar to that of steel elements ;
- $M_y$composite = 1.1 $M_y$steel, with higher ductility, for HEA section.

8 References

Abed, A. and Plumier, A. (1993) Reserach on beam-column connections. **Inter. Rep. Eur. Assoc. Struct. Mech. Labos. WG3,** Test programme in Liège, August.

Elnashai, A.S. Takanashi, K. Elghazouli, A.Y. Dowling, P.J. (1991) Experimental behaviour of partially encased composite beam-columns under cyclic and dynamic loads. **Struct. and Build. Board,** Pap. 9714.

European Convention for Constructional Steelwork (1986) Recommended testing procedure for assessing the behaviour of structural steel elements under cyclic loads. **ECCS - TWG 1.3,** Rep. 45.

Office for Official Publications of the European Community (1992) Seismic resistance of composite structures. **Rep. EUR 14428 EN,** Luxembourg.

Plumier, A. and Schleich, J.B. (1993) Seismic resistance of steel and composite frame structures. **J. Constr. Steel Res.,** 27, pp. 159-176.

19 STRUCTURAL PERFORMANCE ASPECTS ON CYCLIC BEHAVIOR OF THE COMPOSITE BEAM-COLUMN JOINTS

A. M. PRADHAN and J. G. BOUWKAMP
Institute of Steel Construction, Technical University of Darmstadt, Germany

Abstract
Pseudodynamic tests have been performed to study the cyclic behavior of full scale composite beam-column (B/C) joints with fully welded and high-strength bolted connections. The nonlinear joint behavior as affected by the interactive plastification of the beam-end and the column shear-panel zone are presented, covering both the effects of the type of connection (welded and bolted) and shear-panel thickness (t_w) as well as the contribution of the concrete.

Test results indicate a basic influence of the beam-end and the shear-panel zones on the local ductility. However, with the column remaining elastic, the joints studied showed basically identical combined ductile characteristics. This illustrates that also in composite construction the joint design can be developed utilizing the shear-panel distortions (plastification) in order to achieve adequate structural ductility. This potential behavior could prevent early connection failure and allow optimum energy dissipation.

In as far as shear-panel plastification can be accepted for low-rise frame structures, the results of an earthquake study of a typical EC-code designed composite building frame, with different shear-panel zone characteristics, has been presented. In this study the joint algorithms have been formulated reflecting the analytically assessed joint characteristics as defined by earlier correlative experimental/numerical studies.

Keywords: Beam-column Joint, Composite, Cyclic – Pseudodynamic, Earthquake Response

1. Introduction

The cyclic behavior of composite joints are influenced by:

a) type of connection between beam and column,
b) relative strength and stiffness characteristics of beam and shear-panel zone (column remaining elastic) and
c) the relative contribution of steel and concrete within the shear-panel zone.

In order to investigate the cyclic behavior of composite beam-column (B/C) joints as affected by the interaction between the basically brittle filled-in concrete and ductile steel both experimental and analytical studies have been performed. These studies, supported by the CEC and administered by the JRC, Ispra, are part of a multi-university project carried out at the Universities of Athens (NTU), Darmstadt (THD) and Liége (ULG). The experimental investigation at Darmstadt covered a study of 24 full scale joint specimens with 6 different joint designs. For comparison, 4 additional full scale steel beam-column joint specimens have been tested. In this paper only the results of a selected number of tests have been presented to illustrate the basic findings of this study. Specifically, the relative ductility demand between the beam and the shear-panel zone as well as the steel/concrete interaction effect on the joint behavior is addressed.

2. Test Specimens and Setup

The test setup and the basic welded and bolted joint specimens are shown in fig. 1.

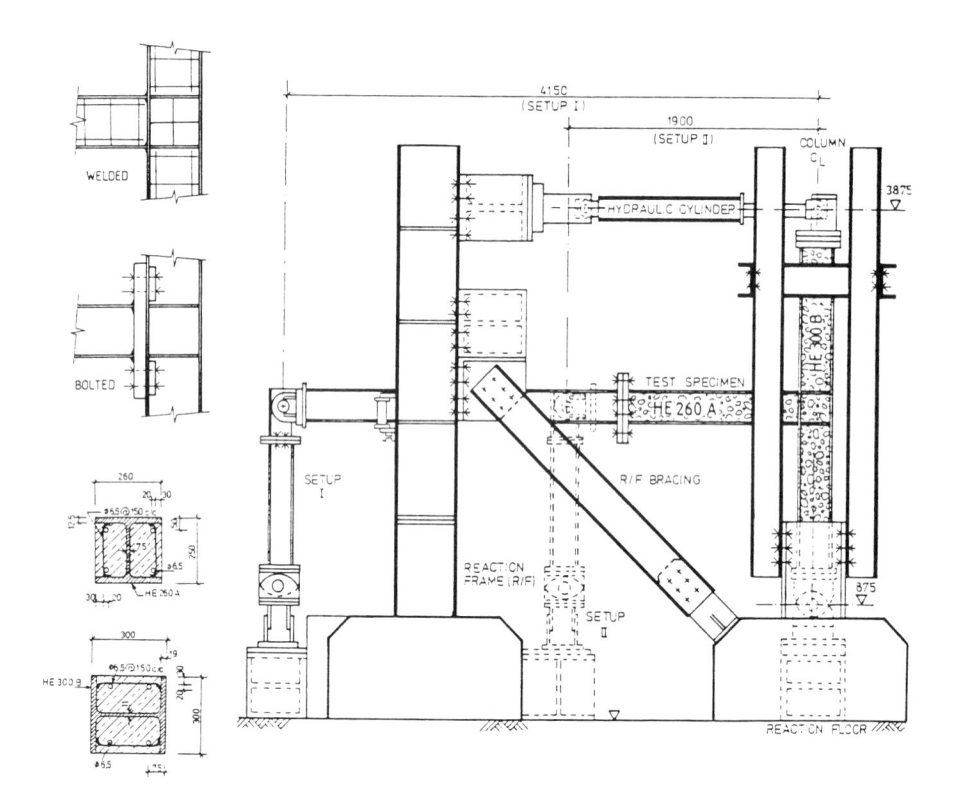

Figure 1: Test specimens and setup

The composite beam and column sections are designed as standard rolled steel sections (St-37-2 HE 260 A and HE 300 B, respectively) with filled-in reinforced concrete (Constantinescu et al. 1992). The reinforcement in the panel-zone consists of two crossed $\phi 6.5$ mm bars and the reinforcement in the remaining beam and column sections consists of $\phi 6.5$ mm @ 15 cm c.c. in both directions (fig. 1 sections). Steel coupon tests showed yield strength values for the beam and column sections as 290 and 370 N/mm² respectively. The concrete cube strengths was observed as 24 N/mm².

The design of the specimens covered the following basic variations:

a) Connection type: Welded and bolted (fig. 1),
b) Concrete contribution: Bare steel and steel/concrete composite,
c) Shear-panel web thickness (t_w): 7, 11 and 15 mm.

In general, the bolted connections, including the beam-end plate and column washer-plates, were designed to remain elastic for a moment of 1.2 times the nominal plastic beam-moment.

The tests were performed in a specially designed reaction frame anchored to the laboratory tie down slab. A displacement controlled horizontal force was applied at the top of the beam-column specimen by means of a hydraulic actuator. The 4 bare-steel and 2 composite specimen were tested under quasi-statically controlled conditions. However, the remaining composite specimen were tested pseudo-dynamically with prescribed harmonic, steadily increasing, base-input excitations. In this phase, a lumped mass located at the top of the specimen had been assumed in the PSD SDOF system model, in order to permit, for certain specimens, a comparison with real-time shaking table tests performed at Athens (NTU).

3. Structural Performance

3.1 Welded versus Bolted Joints

The effect of the connection type is evaluated by comparing the results of the following 3 sets:

a) composite welded and bolted joints ($t_w = 7$ mm),
b) composite welded and bolted joints ($t_w = 11$ mm) and
c) bare-steel welded and bolted joints ($t_w = 11$ mm).

In this study, invariably, the inelastic shear-panel deformations were larger in the welded joints than in the corresponding bolted joints (fig. 2b, beam-moment versus shear panel distortion). This behavior is due to the different shear stress levels in the column panel zone resulting from the shear-panel forces caused by the beam-moment transfer. The welded connection has typically a smaller moment lever-arm than the bolted connection and thus higher shear stress distortions. Results indicate that for design purposes, the bolted connection layout can be designed to effectively reduce column shear-panel stresses and avoid using doubler plates. Although smaller shear-panel deformations placed a higher ductility demand on the beam section (fig. 2c, beam-moment versus beam-end rotation), the overall displacement response characteristic of the joint specimen were not significantly affected by the connection type (fig. 2a, horizontal actuator force versus displacement). Test data further indicated a slightly lower elastic beam stiffness for bolted joints as a result of a small, linearly increasing, separation between beam end-plate and column-flange.

The welded and bolted composite joint specimens exhibited invariably prior to failure an outward buckling of the beam half-flange without a reduction of the beam moment resistance. This favourable behavior resulted from a direct-contact compressive load transfer between concrete and steel in the beam compression zone. Ultimate failure may result either from increased buckling distortions (including crushing of the concrete in the beam section) or failure of the welds between the beam and column (or end plates).

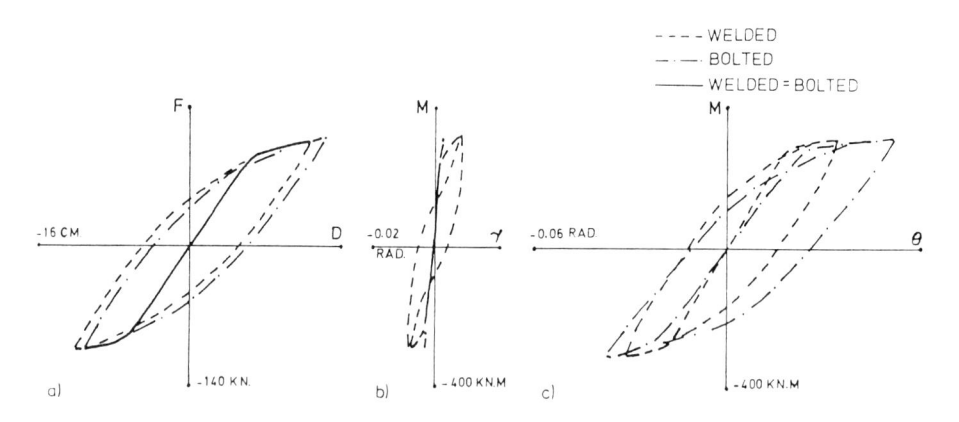

Figure 2: Welded and bolted beam-column composite joints ($t_w = 7$ mm)

3.2 Concrete Contribution

The effect of the concrete can be observed clearly from the shear-panel force – distortion relationship ($Q - \gamma$) (fig. 3b), showing a significantly reduced distortions "γ" for the composite panel zone. This effect results consequently in reduced beam rotations for the steel joint as compared to the composite joints (fig. 3c). The overall actuator-load – displacement relationship ($F - D$) (fig. 3a) provides the total specimen response, showing relatively similar stiffness characteristics for the two joints, with the composite joint having, as expected, a larger force resistance. Although, for the beams, the concrete prevents an early inward buckling of the beam flanges, the concrete has little effect on the moment capacity as shown in fig. 3c, presenting the moment – rotation relationship ($M - \theta$) of the beam.

These observations indicate that for small to moderate earthquakes, the concrete can be efficiently utilized to provide strength and stiffness of the shear-panel zone and only limited energy dissipation (ductility concentrated in the beam sections). However, for larger earthquakes the concrete in the shear panel zone may be crushed, resulting in a significant energy dissipation in the column-web steel, thereby contributing to the overall ductile resistant capacity of the structure.

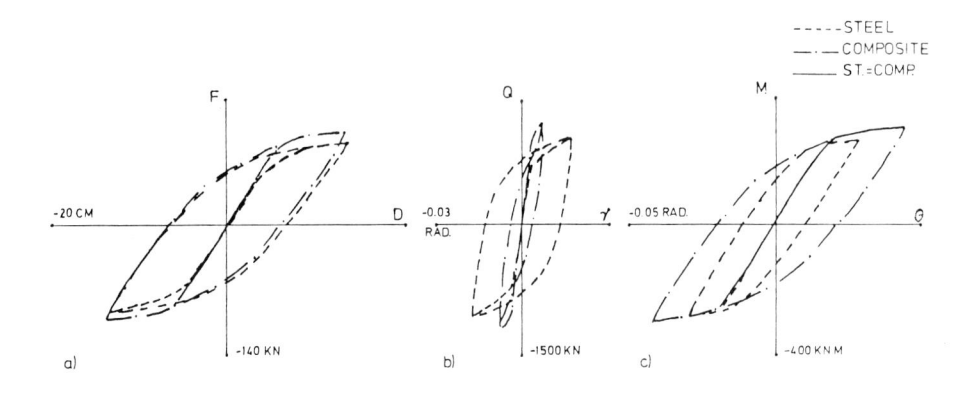

Figure 3: Bare-steel and composite beam-column joints ($t_w = 11$ mm)

3.3 Shear-panel Thickness

The effect of the shear-panel thickness "t_w" has been studied by testing welded composite joint specimens with column-web thicknesses of 7, 11 and 15 mm in the shear-panel zone. For the 7 mm test specimen, the column-web in the shear-panel zone had been machined down from a nominal thickness of 11 mm to a thickness of 7 mm (prior to introducing the column stiffeners). The 15 mm specimen was formed by adding a 4 mm doubler plate in the shear-panel zone. Varying the shear-panel thickness "t_w" not only affects the overall shear resistance in the panel zone but also changes the relative contribution of the steel and concrete in this area.

In general the test results indicated, as expected, larger shear-panel distortions in the nonlinear range for the specimen with a 7 mm steel-web thickness as compared to those with either 11 or 15 mm web thicknesses. In the first case the concrete contributed to the shear resistance while in the latter two cases this contribution had basically no effect in comparison with steel-web resistance and did not show an appreciable difference between the 11 and 15 mm web thicknesses. This basic behavior has been illustrated by the shear-panel force – distortion relationships ($Q - \gamma$) for specimens with web thicknesses of 7 and 11 mm (fig. 4b), showing a larger distortion for the 7 mm web-thickness specimen. The contributing effect of the concrete in the shear-panel for the 7 mm web-thickness specimen is also illustrated by the pinching effect in the force – displacement response envelope shown in fig. 4a and the Q – γ hysteretic response curve shown in fig. 4b. Furthermore, it can be noted that the overall maximum force – displacement characteristics in both instances is basically identical (fig. 4a) although the ductility in the beam for the 7 mm web-thickness specimen is less than for 11 mm case (fig. 4c). Consequently, the total ductility could be achieved by energy dissipation in both of the two dissipating elements (beam and shear-panel zone).

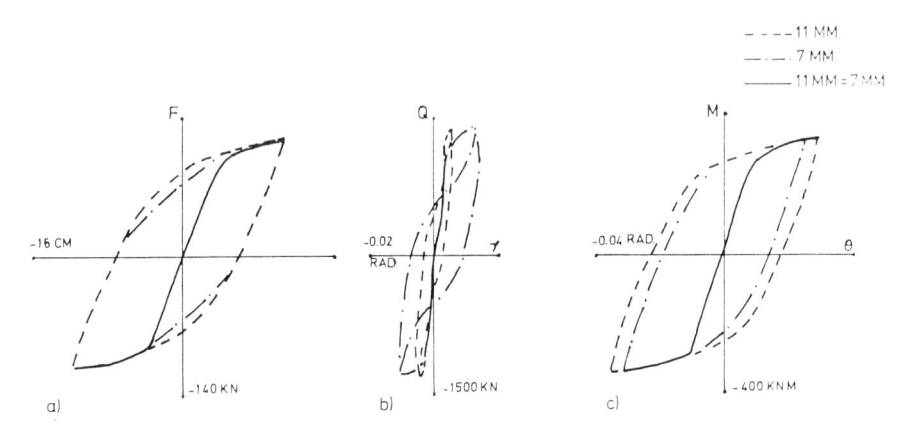

Figure 4: Beam-column composite joints ($t_w = 11$ and 7 mm)

4. Nonlinear Dynamic Analyses and Experimental Correlations

Nonlinear dynamic analyses have been carried out to capture the experimentally observed dynamic behavior of the composite beam-column joint using not only the same input base excitation, mass and damping but also the same integration time-step as in the corresponding PSD tests. Based on the material properties, strength and stiffness characteristic for the joint components could be derived, showing good correlation with the basic test base-curves. These data have been used to formulate the specimen characteristics for efficient nonlinear analyses.

The test structure is modelled as an assemblage consisting of nonlinear beam-column elements and two nonlinear rotational connection springs reflecting the shear – distortion behavior of both the steel section and concrete in the column shear-panel zone (fig 5a). For the beam-column elements bilinear hysteretic models have been considered. For the shear-panel zone, the steel "spring", including the characteristics of both the steel column-web and the column-flange bending effect (Krawinkler & Mohasseb 1987), has been modelled bilinearly with hysteretic inelastic unloading behavior (fig. 5b). The concrete "spring" is modelled as a bilinear "elasto-platic with gap" hysteretic spring (fig. 5c).

With experimental results in the elastic range showing an only partial contribution of the concrete to the composite action in beams and columns, an effective contribution of only 40% of the concrete strength and 25% of the concrete stiffness has been used in formulating the strength and stiffness properties of the beam-column elements.

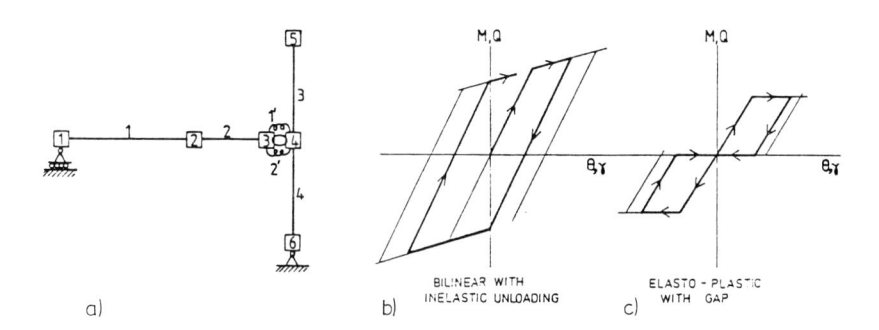

Figure 5: Structure and hysteretic models

In general, the correlative dynamic analyses showed a satisfactory agreement with the observed overall and detailed test behavior. As an example, fig.6 presents the analytical and experimental results for a welded composite joint specimen with a web thickness of 7 mm. Since in all analyses the same beam and column strength proterties have been used, a deviation from the experimental yield level of less than 5% has been observed. This effect, together with deviations between the assumed stiffness of the nonlinear branch of the beam model (4% of the elastic stiffness) and the experimentally observed stiffness in that range, leads to certain variations of the observed and analytically predicted nonlinear behavior. However, whereas these differences were minimal, the nonlinear dynamic analyses were considered to provide sufficiently accurate results for an efficient earthquake response prediction analyses. These observations are illustrated by the results shown in figs. 6b&c showing the shear-force – shear-distortion response $(Q - \gamma)$ of the column web and the moment – rotation relationship $(M - \theta)$ at the beam-column junction. Fig. 6a supports the claim that the numerical model selected give sufficiently accurate results for a predictive response analysis.

5. Case study

In order to assess the basic effect of the composite action in the shear-panel zone under a given earthquake ground motion, a single frame of a 4 story (3.5 m story height) building with composite ductile moment resisting frames has been studied. The 3-bay frames, spaced at 6 m o.c., having a total height of 14 meters (fig. 7), were designed for a maximum EC-8 stipulated earthquake zone (Bouwkamp & Fehling 1992).

For the nonlinear dynamic response analyses the ground motion time history of the 1986 Kalamata earthquake has been used. To assess the influence of the composite column-panel zone action, the following 3 joint-design cases have been studied:

a) Case 1 – composite frame with rigid shear-panel zone (reflecting the doubler plate as designed),
b) Case 2 – composite frame with only steel shear-panel zone (concrete and doubler plate omitted), and
c) Case 3 – composite frame with composite shear-panel zone (doubler plate omitted).

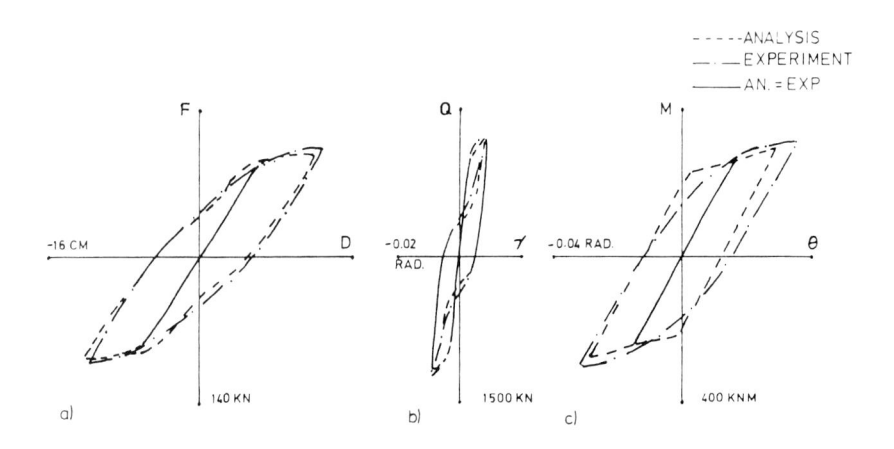

Figure 6: Numerical – Experimental Correlation

The nonlinear dynamic analyses have been performed with the different joint models described in accordance with the strength and stiffness characteristics noted before in this joint study. The member stiffnesses were modelled assuming full composite contribution of the filled-in concrete and without regarding the composite slab action. The analyses were performed using the computer program DRAIN-2DX (Prakash & Powell 1992). The results have shown that in Case 1, yielding in the ground- and first-floor interior columns and first-floor beams did occur after about 1 second of the earthquake. In Case 2, with bare-steel shear-panels without doubler plates, yielding took place in the shear-panel zones only starting within 1 second of the earthquake. In Case 3, first yielding was observed at the first-floor exterior-column shear-panel after about 1 second; subsequent yielding occured in the ground-floor columns and first-floor beams as well as in the shear-panels at the first-floor level. The first-story drift (in percentage of the story-height) versus time as presented in fig. 8 shows the relative performance of the 3 Case studies. Together with fig. 9, the results indicate that at no time the story drift limitation of 1.5% was exceeded. Basically it may be concluded that the composite joint solution (Case 3) could well be adopted as a design alternative, thereby effectively omitting the doubler plates. In fact, during the joint tests the cyclic response has indicated that drift levels of 7% could be acommodated without loss of hysteretic stable joint behavior.

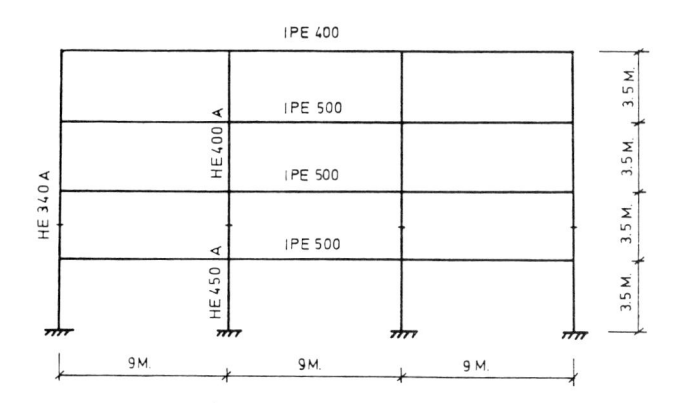

Figure 7: Case study – composite frame

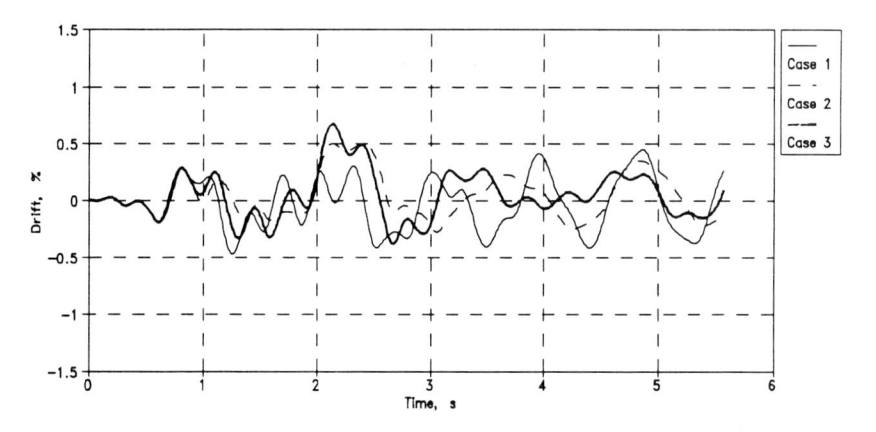

Figure 8: 1st Story drift time history

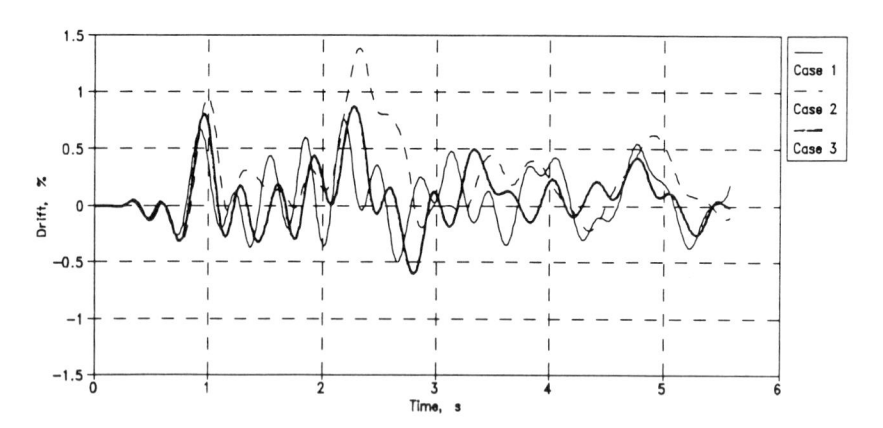

Figure 9: 4th Story drift time history

6. Conclusion

The test studies indicated that because of the joint geometry selected, welded beam-column joints would provide a larger shear-panel energy dissipation potential than joints with HS-bolted beam-end plates. On the other hand, bolted connections can be layed out such that shear-panel yielding will be prevented and plastification will be forced into the beams. In both types of joints, the use of concrete in the shear-panel column-web zone provides sufficient resistance to allow omitting the use of doubler plates if necessary. In general, the use of composite concrete offers the potential to achieve balanced ductility characteristics for beam and shear-panel zones. The appropriateness of such a design philosophy has to be assessed against the potential post-earthquake damage intensities which can be expected in the shear-panel zone.

7. References

Bouwkamp, J.G. and Fehling E. (1992) Comparison of R/C and composite frames for earthquake resistant design. in **Proc. of the 10th World conference on Earthquake Engineering**, 19-24 July 1992, Spain, pp. 6:3455-3460.

Constantinescu, D.R. Van Kann, J. Pradhan, A.M. Ashadi, H.W. and Bouwkamp, J.G. (1992) Pseudo-dynamic tests on composite steel/concrete joints. in **Proc. of the 10th World conference on Earthquake Engineering**, 19-24 July 1992, Spain, pp. 6:3139-3144.

Krawinkler, H. and Mohasseb, S. (1987) Effects of panel zone deformations on seismic response. **J. Construct. Steel Research**, 8, 233-250.

Prakash, V. and Powell, G.H. (1992) **DRAIN-2DX – User Guide.** Department of Civil Engineering, University of California, Berkeley, California.

20 LOCAL CYCLIC BEHAVIOUR OF STEEL MEMBERS

I. VAYAS and I. PSYCHARIS
National Technical University of Athens, Athens, Greece

Abstract
The seismic design of steel building structures relies upon the capability of structural members to dissipate part of the input seismic energy through inelastic action. To that end member cross sections are classified into classes with different energy absorption capacity. The present work reviews the Eurocode 8 provisions on Classification, makes proposals for the formulation of material independent criteria complying with the requirements of the seismic design and derives application rules for symmetrical I-sections. The rules refer to slenderness limitations of the web and the flanges of the cross section that are mutually dependent. The influence of a normal force on the section is studied. For the various cross sectional classes the energy dissipation capacity is finally given.
Keywords: Steel Members, Classification, Criteria, Codes, Cyclic Behaviour.

1 Introduction

In a steel framed structure members are conventionally designed for strength, i.e. the verification refers to their resistance that shall be greater than the internal moments and forces acting upon them. However there are cases where ductility is also a governing criterion for element design. Ductility is required for structures in seismic areas where a portion of the input seismic energy shall be dissipated by inelastic action of the structural members. Another example where ductility shall be provided is when a plastic design of the frame is envisaged, where plastic hinges form in the members and a plastic redistribution of forces and moments between members takes place.

In order to determine the ductility of an element, its behaviour beyond the attainment of its maximum strength shall be studied. This study shall be extended for a seismic design considered by the present work, to cyclic loading conditions that the element is affected during a strong earthquake.

The scope of the present paper in to review the Eurocode 8 (1993)-provisions regarding local ductility of steel members and to make proposals in respect to the compliance criteria and the application rules included in the Code for their improvement. The proposals are based on analytical investigations for both monotonic and cyclic loading.

2 Classification of cross sections according to Eurocode 8

In Eurocode 8, as in most modern seismic Codes, inelastic acceleration spectra are introduced by application of a qlobal behaviour factor q to the corresponding elastic spectra. The behaviour factors q take into account the fact that part of the input seismic energy will be dissipated through inelastic action within the structure. In order to secure the assumed value of the behaviour factor, and therefore the capacity of the frame to dissipate energy, specific measures in respect to the global and local ductility shall be taken. The requirement for local ductility in a moment resisting frame that is considered here, refers to the ability of the steel members (beams and columns) to sustain various reversed deformation cycles iwthout significant stregth and stiffness degradation. In Eurocode 8, 3 classes of members are introduced, each referring to a different ductility level and leading to a different value of the behaviour factor q. As a criterion for the classification serves the form of the moment-rotation (M-θ) curve of the member (Fig.1) when subjected to bending. However the classification refers to cross sections rather than members, so that Fig.1 should actually refer to a moment-curvature (M-k) curve of the cross section and not to a M-θ curve of the member.

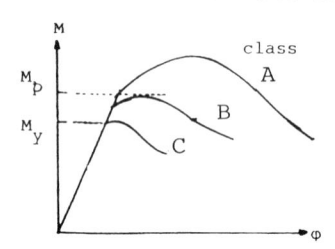

Fig.1. Moment-rotation curres of steel members

The criteria for the 3 classes of cross sections according to EC8 are the following:
Class A cross sections are able to develop their plastic resistance moments and possess adequate rotational capacity.
Class B cross sections develop the plastic resistance moment but don't possess enough ductility at that moment

Class C cross sections develop the moment at which yielding at the extreme fibre starts, but can't reach the plastic moment.
The most important factor of the behaviour of the cross section in respect to strength and ductility is the slenderness of its compressed plated parts (web, flange), since local buckling results in a reduction of strength and ductility. Therefore the classification leads to the determination of limiting slendercenses of the plated parts. For that reason according to the rules of EC8, in a first step the cross section is decomposed into its individual plated elements. For each isolated-panel different limits for its relative slenderness $\bar{\lambda}_p$ are assigred to every class. This is equivalent to limiting the b/εt ratio of the panel since the following relationship is valid

$$\bar{\lambda}_p = \sqrt{\frac{f_y}{\sigma_{cr}}} = \frac{1}{28,4(k_\sigma)^{1/2}} \frac{b}{\varepsilon t} \tag{1}$$

where σ_{cr} = critical buckling stress
f_y = yield stress
b, t = width, thickness of the panel
k_σ = buckling coefficient
$\varepsilon = (235/f_y)^{1/2}$ with fy in [MPa]

In that manner each panel is assigned to a specific class and the cross section is classified according to the class of its most slender panel. The application rules of this classification procedure for a symmetrical I-cross section are shown in Fig.2a. Fig.2b shows the relevant application rules of Eurocode 3(1993).

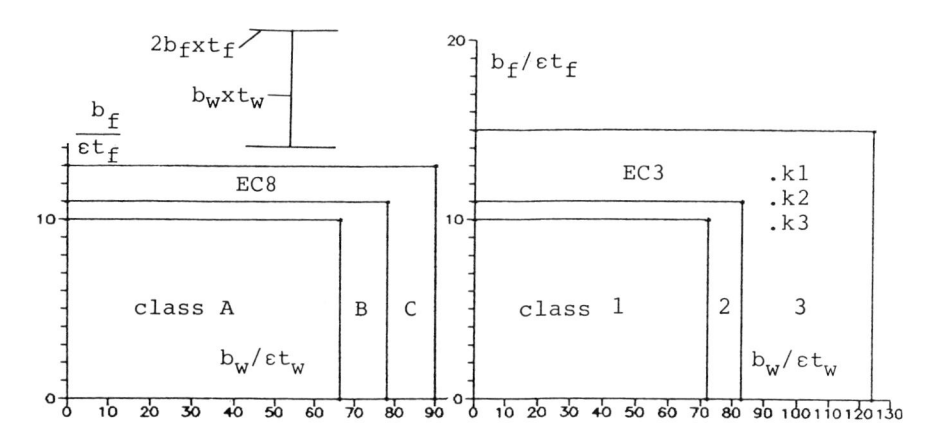

Fig.2a. Cross section classification according to EC8(a) and EC3 (b).

Following observations may be made in respect to the results of Fig.2:
a) Although the criteria for the classification according to EC8 and EC3 are as previously formulated the same, they unexpectedely lead to different application rules.
b) The slenderness limits for the flanges are in both Codes the same for class A(class 1 in EC3) and B(2) cross sections, but differ to approximately 15% for class C(3) sections.
c) The slenderness limits for the webs are reduced in EC8 in respect to EC3. But while this reluction is about 10% for class A(1) and B(2) cross sections it increases to 30% for class C(3) sections.
d) A cross section, as e.g.k, is classified in class C of its web is of class C independently of its flange which may be of class A, B or C (see sections k_1, k_2 or k_3 in Fig.2).

In the following it will be tried to formulate general classification criteria in compliance with the stated requirements of the seismic design and to derive relevant application rules for the case of symmetrical I-sections.

3 Proposed classification criteria

As explained before, the requirements concerning classification differ between a seismic design and a design for non seismic actions. In the latter, strength and ductility shall be supplied for analysis purposes, while in the former case a definite energy dissipation capacity of the member shall be secured. Consequently the relevant compliance criteria should also be different.

The criteria set up in EC3 seem not to need a modification since they are consistent to their implication in design. For a seismic design it is proposed here to adopt the relevant criteria that are proposed by the part of EC8 that deals with reinforced concrete structures and to adapt them as appropriate to steel framed structures.

Considering an M-k curve of a cross section (Fig.3) the curvature ductility ratio is defined as the ratio

$$\mu_k = \frac{k_u}{k_y} \qquad (2)$$

Where k_u = post-critical curvature at the level of the design resistance moment M_{Rd} and
k_y = corresponding curvature in the ascending branch of the M-K curve at the same moment level

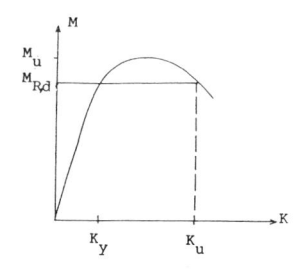

Fig.3. Definition of curvature ductility ratio

It is presently proposed to apply the following criteria for the classification of cross sections:
- Class A cross sections shall have a ductility ratio of $\mu_k=15$
- Class B cross sections shall have a ductility ratio of $\mu_k=6$
- Class C cross sections shall have a ductility ratio of $\mu_k=3$

The design resistance moment M_{Rd} is equal to

$$M_{Rd} = \frac{M_{p\ell}}{\gamma_m} \tag{3}$$

where $M_{p\ell}$ =plastic resistance moment of the cross sections and
γ_m =1,10 partial safety factor for the resistances.

For concrete structures the same definitions apply except that $M_{Rd}=0,85\ M_u$, since the partial safety factor for reinforcing steel-whose yielding is supposed to coincide with the limit state-is equal to $\gamma_s=1,15$. The proposed criteria are therefore, as the relevant requirements, independent on the building material (concrete,steel).

The established classification criteria allow for the deviration of relevant application rules. This will be done subsequently for I cross sections.

4 Application rules for I-beams

In order to derive application rules satisfying the stated criteria by analytical means, a method is needed that permits the description of the element behaviour in the descending branch of its resistance. Such a strain-oriented method is developed by the authors. For its detailed description reference in made to Vayas, Psycharis (1992 and 1993).

For various I-sections, the M-K diagrams were determined by this method.

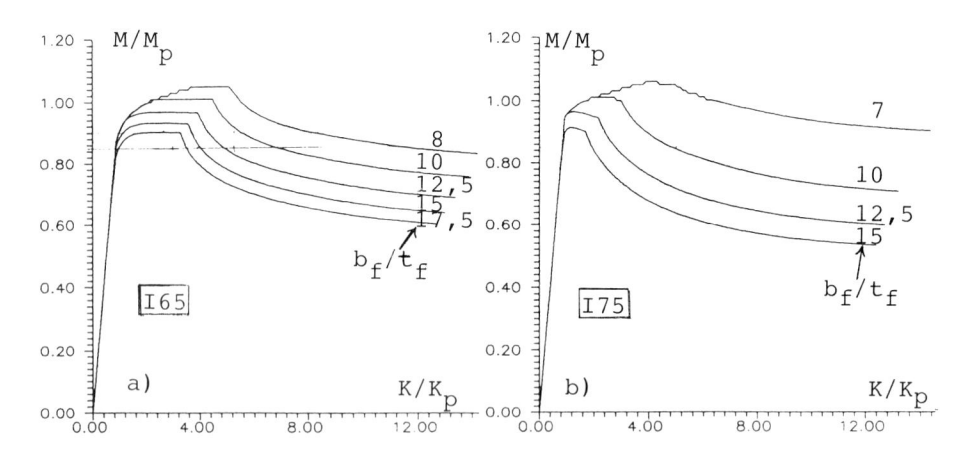

Fig.4.Moment-curvature diagrams for I cross sections.

The parameters which were varied were the web and flange slenderness. All cross sections studied here are designated by two figures. The first is equal to the web-and the second to the flange-slenderness. Consequently a section that is characterised as I 30/20 has a web slenderness of $b_w/t_w=30$ and a flange slenderness of $b_f/t_f=20$. The material is supposed to be Fe 360 for which $\varepsilon=1$ in eq(1). Strain hardening of steel was taken into account with a value of the tangent modulus E_T equal to $E/50$, where E is the modulus of elasticity. Some results are shown in Fig.4,their evaluation in respect to classification of the cross sections in shown in Fig.5. As expected, the web and the flange slendernesses are not independent variables, since a compact web has a stiffening effect on the flange, while a slender web may prematuraly buckle and lead to the loss of strength of the entire section.

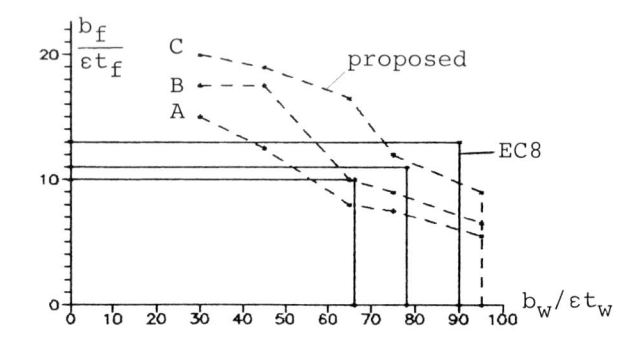

Fig.5. Proposed slenderness limits for I-beams

5 Application rules for I-columns

In order to derive application rules for I-columns, moment-curvature diagrams of I-sections for various levels of the applied normal force were defermined. The level of the applied axial force is expressed by the parameter.

$$n = \frac{N}{N_p} \tag{4}$$

where N = applied axial force
        N_p = plastic normal force of the cross section.

The influence of the axial force on the behaviour of the cross section is shown in Fig.6. Both the moment resistance and the ductility are reducedby the presence of that force.

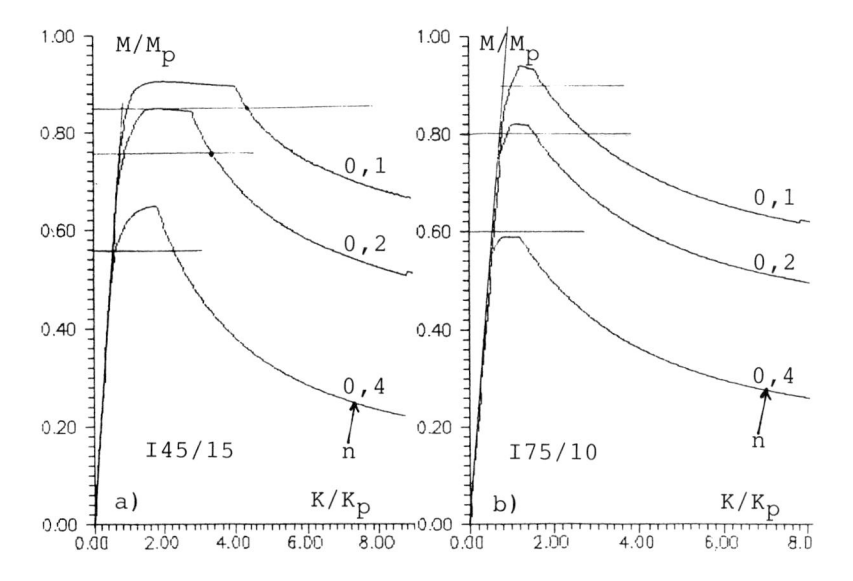

Fig.6. Moment-curvature diagrams of I-setions for various axial force.

If ductility ratios are to be determined, Mpl shall be modified in eq(3) to allow for the normal force. Adopting the relevant expression of EC3 for I-sections, the design resistance moment is given by

$$M_{Rd} = \frac{M_{p\ell}}{\gamma_m} \frac{(1-n)}{1-0,5\alpha} \tag{5}$$

where $\alpha = \dfrac{A_w}{A}$ = ratio of the web to the overall area $\leq 0,5$
n   according to eq. (4)

Some results in the form of the ratio $\mu_k/\mu_{k,n=0}$ where μ_k= ductility ratio at the presence of a normal force n=-η and $\mu_{k,\eta=0}$=ductility ratio for η=0 in dependence of the normal force η are presented in Fig.7.

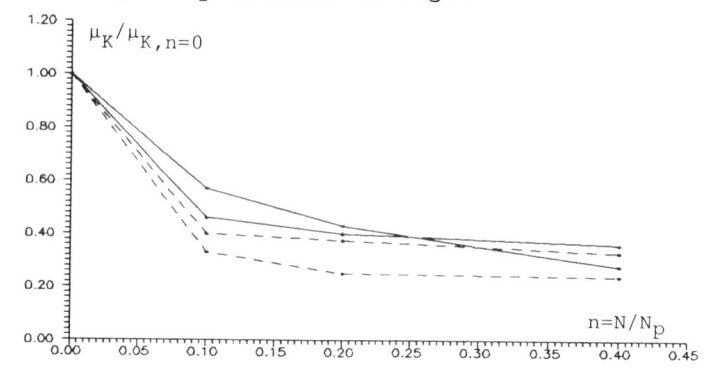

Fig.7.Ductility ratios in relation to an applied normal force

According to the provisions of EC8, the presence of an axial force in the cross section is taken into account for the web classification only, due to the change of the stress distribution along the element, while the slenderness limits for the flanges remain unaffected. The relevant slenderness limitations according to the provisions of EC8 and the evaluation of the present method are given for two levels of the axial force in Fig.8.

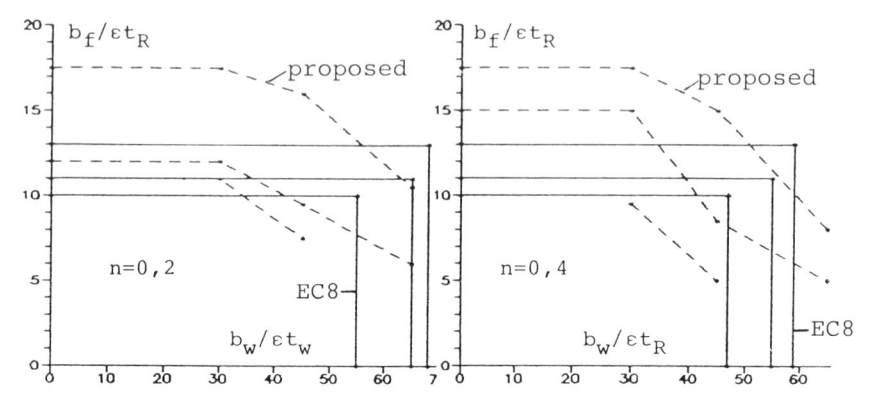

Fig.8.Slenderness limitations according to EC8 and the present method for various axial forces

6 Cyclic behaviour

So far the examination of the behaviour of cross sections was restricted to monotonic loading conditions. However as it was stated before, it is necessary to study the cyclic behaviour of the various classes of cross sections in order to verify the correctness of the classification criteria in respect to the stated requirements. This will be done by application of the strain-oriented method as formulated by Vayas, Psycharis (1990)

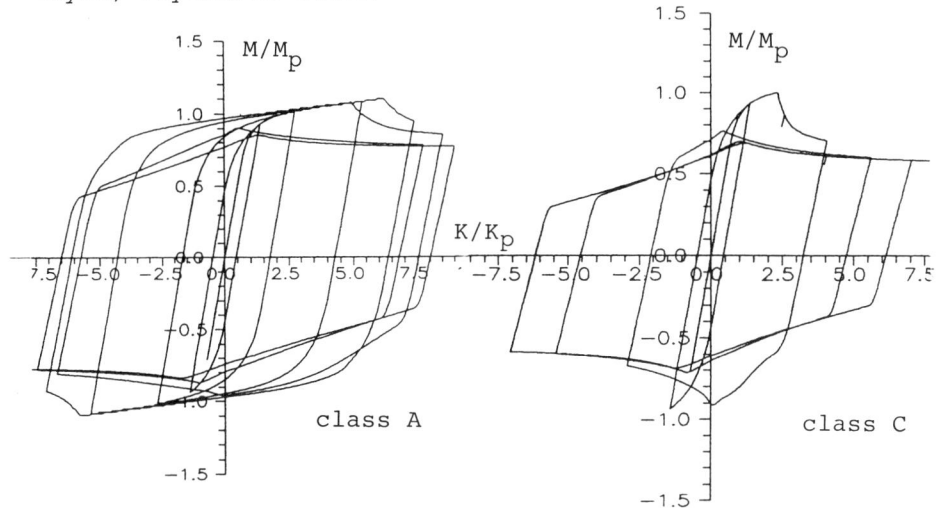

Fig.9. Behaviour of cross sections with variable slenderness under cyclic loading.

for cyclic loading. Some results are shown in Fig.9. From cyclic curves as presented in Fig.9, the energy dissipation capacity of the section may be determined. To that end, the absorbed energy ratios vs. the full ductility ratios as defined by the ECCS-Recommandations (1986) are evaluated. The former parameter is the area of an actual hysteresis loop A divided by the corresponding area of a linear elastic-perfectly plastic material and is determined by the expression

$$\text{Absorbed energy ratio} = \frac{A}{M_p(k^+ - k^-)}$$

The later parameter is defined as

Ductility ratio = $\dfrac{k^+ - k^-}{2k_p}$

It shall be observed that this parameter has a different meaning than the corresponding one introduced in section 3 and shown in Fig.3. The results of the evaluation are shown in Fig.10.

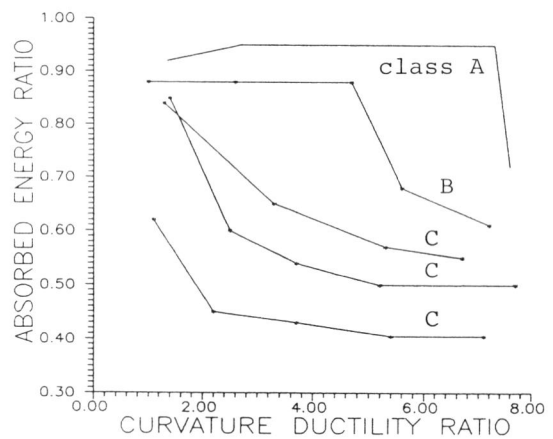

Fig.10. Energy dissipation capacity of various cross sections.

Obviously the absorbed energy ratio decreases with increasing slenderness and ductility ratios. The hysteresis loops for class A cross sections are stable over the entire range of ductility ratios. The same happens to class B cross sections up to a ductility ratio of 5, whereas the energy disipation capacity of class C cross sections is limited. Slender cross sections are even lessable to dissipate energy.

7 Conclusions

Following conclusions may be drawn:

1) It is possible to define criteria for the classification of cross sections that are independend on the structural material (steel, concrete etc).
2) These criteria may be formulated in terms of required curvature ductility ratios at the level of the design resistance moments.
3) The application of these criteria to I-sections lead to slenderness limitations for the web and the flanges that are mutually dependent.

4) The presence of a normal force drastically reduces the ductility of the cross section, showing the necessity of an appropriate normal force limitation for columns in a seismic design.
5) The proposed criteria show a good correspondance to the relevant requirements concerning the cyclic behaviour of the cross section.

8 Acknowledgements

The writers wish to thank Mrs Gratiela Mateescu from INCERC, Timisoara, Romania who offered during her stay in Athens useful assistance in the preparation of the paper.

9 References

Eurocode 3 (1993) Design of Steel Structures. CEN, European Committee for Standardiration, ENV.
Eurocode 8 (1988). Structures in Seismic regions. Design, Commission of the European Communities.
Vayas, I.and Psycharis,I.(1990). Behaviour of thin-walled steel elements under monotonic and cyclic loading, in Structural Dynamics (eds. W.B.Kratzig et al.), Balkema, Rotterdam.
Vayas, I. and Psycharis, I. (1992). Dehnungsorientierte Formulierung der Methode der wirksamen Breite, Stahlbau, 61, 275-283.
Vayas, I. and Psycharis, I. (1993). Ein dehnungsorientiertes Verfahren zur Ermittlung der Duktilitat ron Tragern aus I-Profilen, Stahlbau, 63, 333-341.
European Convention for Constructional Steelwork (1986). Recommended Testing Procedure for Assessing the Behaviour of Structural Steel Elements under Cyclic Loads.

21 CYCLIC BEHAVIOUR OF THIN-WALLED WELDED KNEE JOINTS

H. PASTERNAK
Technical University of Cottbus, Cottbus, Germany
I. VAYAS
National Technical University of Athens, Athens, Greece

Abstract
Thirteen experiments are performed to investigate the cyclic behaviour of beam-to-column joints of steel frames with slender joint panels. The performance of the joints in respect to strength, rigidity, ductility and low cycle fatigue are examined. Three different load carrying mechanisms can be identified. They are the shear action of the web panel, the tension field action, and the frame action of the surrounding elements. The various characteristics of these actions as determined by the experiments are discussed. It is concluded that a good performance can be achieved by properly selecting the dimensions of the web panel, the surrounding flanges and their relative proportions.
Keywords: Knee Joints, Cyclic Behaviour

1 Introduction

Welded plate girders are widely used as elements of moment resisting frames, especially in low-rise industrial buildings. The main reasons for this development are:

The use of automatic welding procedures that allows for a fully automatic production comparable to the production of hot-rolled profiles,
the possibility to produce variable cross sections along the beam length as to fit best to the stress conditions, saving thus steel and
the possibility of the steel fabricators to enlarge their turnover by transferring work from the steel producers to themselves.

Normally a high bending resistance due to the large spans is required, resulting in the application of deep beams with compact flanges, that supply this resistant to the most extend. The webs are mainly expected to resist the applied shear and are usually slender in order not to unduly increase the weight of the girder. Plate girders with slender webs are most economically designed in the postcritical region, since they exhibit a considerable postbuckling shear resistance due to the formation of tension fields. Relevant design rules are well established for the span region between supports (e.g. Eurocode 3 (1992)). The design of the joints is still governed by the critical buckling load of the joint panels, requiring thus a costly stiffening of the joint region.

Experimental and theoretical investigations on such joints subjected to monotonic loading were performed recently which allowed for the study of their behaviour and the development of relevant design rules (Scheer et al (1992)). The investigations showed that the joints may be designed similar as in the span region in the postcritical range, since tension fields are developing here too.

For seismically induced loads, the behaviour of the joints in respect to strength and ductility are of great importance. For this reason a series of cyclic tests on 13 joints was conducted. The hysteretic energy dissipation capabilities of the joints for various levels of ductility were determined, and the mechanisms of failure were identified. The results of the tests provide reference data for the development of design rules for such joints when applied in moment resisting steel frames.

2 Description of the test specimens

All specimens were scaled down to the same size as those that have been investigated under static loading (Scheer et al (1992)). The geometric properties of the specimens with the notation of Fig.1 are summarized in Table 1. The main parameters were the aspect ratio of the joint panel and the various thickness ratios between the web and the flanges.

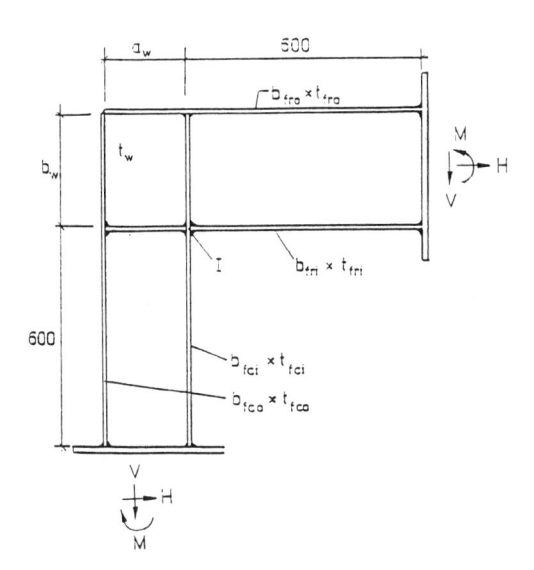

Fig.1. Geometry of the joint

The first six specimen have a very slender web panel, the next six have a web of moderate slenderness and the last one a stronger web. The aspect ratio of the joint panel was varied between 0,5 and 1,0, the thickness ratio of the web to the flanges varied between 1,25 and 5,0 and the thickness ratio between the flanges of the members of the joint varied between 0,5 and 1,0.

For the fabrication of the specimen, the two joint members including the end plates were welded together. Subsequently the joint panels were fitted into this "frame" and welded to the surrounding elements. All welds were carried out as one-sided fillet welds as it is usually done by many steel fabricators.

The material properties of the specimen have been determined by coupon tensile tests as prescribed by the relevant German standards (Table 2). A full description of the tests is given by Scheer et al (1993).

Table 1. Specimen dimensions in [mm]

1	2	3	4	5	6	7	8	9	10	11	12	13
No. of specimen	Specimen	a_w	b_w	t_w	$b_{f,r,o}$	$t_{f,r,o}$	$b_{f,r,i}$	$t_{f,r,i}$	$b_{f,c,o}$	$t_{f,c,o}$	$b_{f,c,i}$	$t_{f,c,i}$
1	AZ-05-10-2	200	200	1.0	150	5.1	as columns 6 and 7					
2	AZ-10-10-1	200	200	1.0	150	9.7	as columns 6 and 7					
3	CZ-10-10-H	240	300	1.0	150	9.7	as columns 6 and 7					
4	CZ-10-10-Q	300	240	1.0	150	9.7	as columns 6 and 7					
5	DZ-05-10-H	200	300	1.0	150	4.9	as columns 6 and 7		150	9.7	as columns 10 and 11	
6	DZ-05-10-Q	300	200	1.0	150	9.7	as columns 6 and 7		150	4.9	as columns 10 and 11	
7	B-03-20-1	301	301	2.0	150	3.0	150	7.6	150	3.0	150	7.6
8	B-03-20-2	301	299	2.0	149	5.1	150	7.5	149	3.0	150	7.5
9	B-03-20-3	301	300	2.0	90	5.0	150	7.5	149	3.0	150	7.5
10	F-05-20-1	150	299	2.0	149	3.0	150	7.5	149	5.0	150	7.5
11	C-03-20-1	240	300	2.0	150	3.0	150	7.5	150	3.0	150	7.5
12	C-05-20-1	240	300	2.0	149	5.2	150	7.6	149	5.2	150	7.6
13	B-05-40-1	300	300	3.9	150	5.0	150	7.6	150	5.0	150	7.6

Table 2. Material properties in [MN/m^2]

	Specimen	1 flange $f_{r,o}$		2 flange $f_{c,o}$		3 web	
		f_y	f_u	f_y	f_u	f_y	f_u
1	AZ-05-10-2	304	412	as col. 1		230	311
2	AZ-10-10-1	278	411	as col. 1		230	311
3	CZ-10-10-H	275	408	as col. 1		226	309
4	CZ-10-10-C	283	412	as col. 1		226	309
5	DZ-05-10-H	332	415	270	408	226	309
6	DZ-05-10-C	270	408	332	415	230	311
7	B-03-20-1	232	337	as col. 1		319	392
8	B-03-20-2	324	446	232	337	319	392
9	B-03-20-3	324	446	232	337	319	392
10	F-05-20-1	232	337	324	446	319	392
11	C-03-20-1	232	337	232	337	319	392
12	C-05-20-1	324	446	324	446	319	392
13	B-05-40-1	324	446	324	446	246	391

3 Tests set up and instrumentation

The test joints are connected to strong extension arms as shown in Fig.2. This allows for the multiple use of parts of the test arrangement and the achievement of a large lever arm for the realisation of the joint moment, keeping the applied force relatively small. One of the extension arms is fixed to the foundation, the other one is supported on rollers. This movable arm is also supported through roller bearings against lifting, so that its out of plane displacements are completely restricted. The friction of the rollers is negligible. The load is applied by a hydraulic actuator to the two arms.

Its position between the arms is variable, permitting the variation of the applied normal and shear forces in relation to the applied moment as shown in Fig.1.

Initially the lengths g_1 to g_3, w_1 to w_3 and g_1 to g_3 were measured, permitting thus the determination of the overall geometry of the test set up. The measurements during the tests included the applied force, the distance w_3 and therefore the joint rotation and the lengths of the two panel zone diagonals.

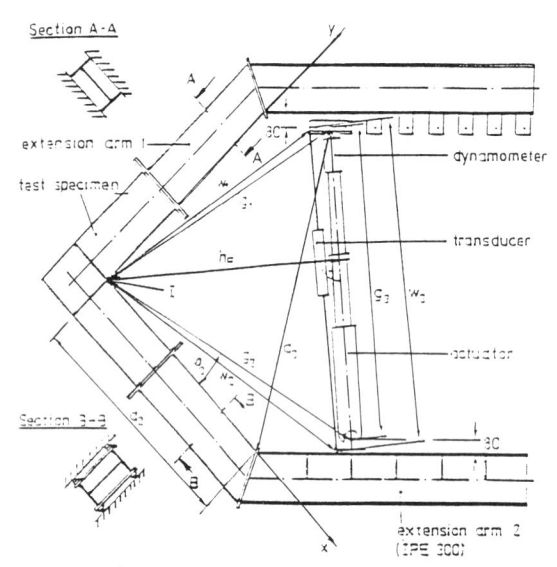

Fig.2. Test arrangement

4 Testing procedure

The loading sequence followed the ECCS "Recommended Testing Procedure for Assessing the Behaviour of Structural Steel Elements under Cyclic Loads" (1986).

The displacements of the hydraulic actuator as measured by the length w_3, were selected as the reference deformation values v for the testing control procedure. These displacements represent the joint panel rotation as practically the only appearing deformations for the given geometry of the test arrangement and the test specimens.

Being v_y^+ and v_y^- theyield displacements under positive and negative bending moments, the load cycles were selected according to the above Recommendations as following (Fig.3):

one cycle in the $\{n \cdot v_y^+/4, n \cdot v_y^-/4\}$ interval n=1,...,4.
three cycles in the $\{2n \cdot v_y^+, 2n \cdot v_y^-\}$ interval n=1,2,...

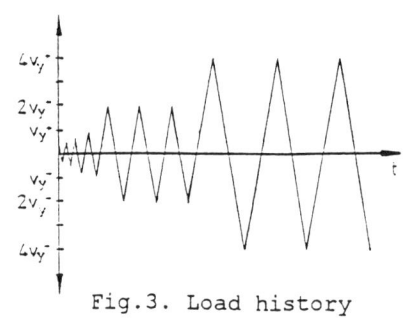

Fig.3. Load history

This procedure was continued until failure of the specimen, with was considered to be reached when the material of the web panel tore due to low cycle fatigue. The test was however continued beyond that level, since the carrying capacity of the specimens was not exhausted. In that range only one cycle for each ductility level was applied.

The yield displacements (rotations) v^+_y and v^-_y are defined as those, at which the sloping line at the origin of the moment-displacement (rotation) curve intersects with the tangent of this curve, having a slope equal to 1/10 of the first line. Since the actual yield displacements of the specimens were not known in advance, the short testing procedure of the ECCS Recommendations was followed, i.e. the first cycles were run till the evaluation of the yield displacements with estimated values.

The results are represented in the form of moment-rotation curves. Both quantities are referred to the point I (Figs. 1 and 2), since the joint "opens" and closes" around this corner. For each cycle, the following parameters characterizing the joint behaviour in comparison to a perfect elasto-plastic behaviour are computed (Fig.4).

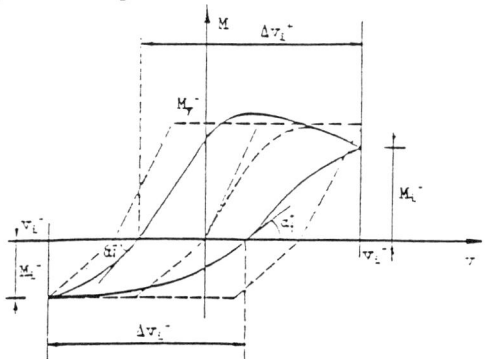

Fig. 4. Characteristic value of the moment-displacement curve

Partial ductility: $\mu^-_{oi} = v^+_i/v^+_y$
Full ductility: $\mu^+_i = \Delta v^+_i/v^+_y$
Full ductility ratios: $\psi^+_i = \Delta v^+_i/(v^+_i+(v^-_i-v^-_y))$
Resistance ratios: $\varepsilon_i = M^+_i/M^+_y$
Rigidity ratios: $\xi^+_i = \tan a^+_i/\tan a^-_y$
Absorbed energy ratios: $n^+_i = A^+_i/(v^+_i+v^-_i-v^+_y-v^-_y) \cdot M^+_y$

The corresponding negative values of the above mentioned parameters are evaluated, similar to these positive values, by interchanging the sings. The characteristic values of these parameters for a group of 3 cycles of the same ductility level are:

for positive deformations, the minimum values out of the three and for negative deformatins, the maximum values out of the three.

5 Results

The most important data of the 13 specimens are presented in Table 3. The linear buckling moment of the web panel was determined under the assumption of simply supported edges. The yield moments have been determined conventionally, following the procedure of the ECCS Recommendations as explained before.

Table 3. Evaluation of joint moments [kNm]

No	Specimen	$\bar{\lambda}$	M_{cr}	M_{yw}	M_u^+	M_y^+	M_u^-	M_y^-
1	AZ-05-10-2	1.75	1.74	5.31	7.3	5.6	-6.9	-5.6
2	AZ-10-10-1	1.75	1.74	5.31	12.9	5.8	-12.1	-5.6
3	CZ-10-10-H	2.26	1.84	9.39	17.5	8.8	-16.2	-8.3
4	CZ-10-10-Q	2.26	1.84	9.39	15.4	10.0	-14.4	-9.0
5	DZ-05-10-H	2.00	1.99	7.83	12.1	3.1	-10.2	-7.0
6	DZ-05-10-Q	2.00	1.99	7.97	12.1	9.6	-9.9	-6.3
7	B-03-20-1	1.55	13.91	33.37	25.6	25.6	-20.3	-20.8
8	B-03-20-2	1.54	13.93	33.15	24.5	24.5	-22.4	-18.9
9	B-03-20-3	1.55	13.92	33.26	25.6	24.0	-20.5	-20.5
10	F-05-20-1	0.94	$> M_{yw}$	16.52	18.9	17.4	-17.4	-17.4
11	C-03-20-1	1.34	14.71	26.52	22.2	20.2	-20.2	-20.2
12	C-05-20-1	1.34	14.71	26.52	24.0	24.0	-24.0	-24.0
13	B-05-40-1	0.70	$> M_{yw}$	49.85	54.5	49.1	-48.0	-47.8

Fig.5 presents the general response of specimen DZ 05-10-H. Its aspect ratio is 0.67, whereas the flanges are 5 times thicker than the web. The first web buckle appeared at ductility 4. The change of the tension field's direction was accompanied by a loss of stiffness. It occurred at the time of change of the load sign. The stiffness increased again when the loading curve intersected the original envelope curve for the first time. Subsequently the moment-rotation graph was directed towards the extreme point of the envelope curve of the previous cycle and afterwards along the sceleton curve. At larger ductilities a strength deterioration occured and in any new cycle the strength was reduced as compared to the strength that was reached at the previous cycle for the same ductility level.

The absorbed energy ratios droped linearly from an initial value of 1,5 to 0,6 at ductility 12 and remained constant up to the end of the test. The general behaviour of the specimens under cycling loading, as observed during the tests, is depending on the behaviour and the relative proportions of the web

is exhausted tension fields develop along the diagonals (Fig. 6). The anchorage of the tension fields is provided by the flanges. Accordingly, the second moment carrying mechanism of the joint, the tension field action, is dependent on the relative strength of the web panel to the flanges. The final load carrying mechanism of the joint is the frame action that is supplied by the surrounding frame of the joint panel. This action is progressing up to the development of a kinematic mechanism in the frame. Its magnitude is evidently dependent on the strength of the frame itself. regarding the rigidity of the joint, it may be observed, that it is primarily provided by the shear action of the web panel, prior to its buckling, and afterwards by the tension field action. The joint behaves then as an eccentrically braced frame. The contribution of the frame action to rigidity is as expected small.

Regarding the energy absorbtion capacity of the joint, it may be observed that joints with a strong frame action have excellent characteristics. On the contrary, joints with weak flanges and very slender webs were not able to absorb large amounts of energy. The change of direction of the tension field was accompanied by a stiffness and energy absorption capacity loss, which could not be counter balanced by the frame action.

Regarding the low cycle fatigue strength of the specimens as expressed by the appearance of the first crack, it may be observed, that it was highest for joints with strong flanges and lowest for joints with weak flanges and slender webs. In general four cracks appeared in each specimen at the intersection of the two opposite tension fields. The cracks didn't lead to an overall collapse of the joint. The stress could be redistributed within the panel due to the ductility of the steel. The reported cyclic behaviour of the joint is therefore expected in cases where mild steel is used.

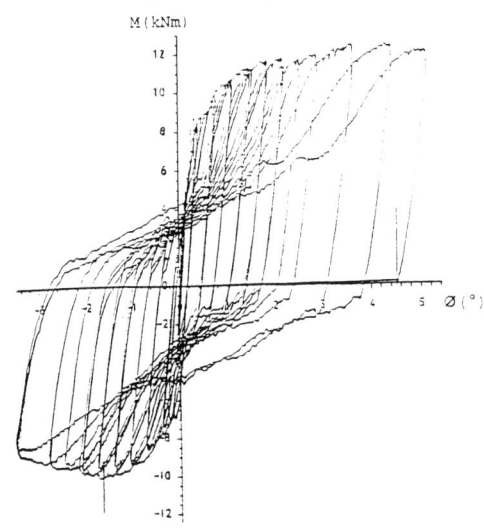

Fig.5. Moment-rotation response for specimen DZ-05-10-H

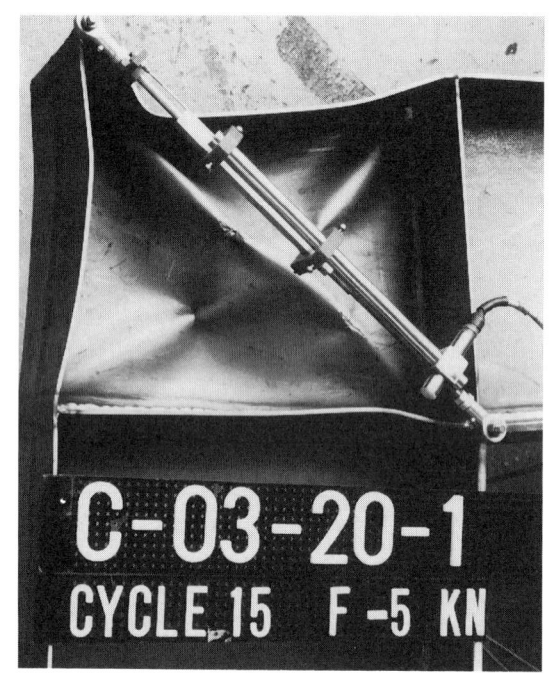

Fig.6. Specimen under positive bending moment

6 References

Scheer, J., Pasternak, H., and Schween, T. (1992): "Ermittlung der Traglast von geschweissten Rahmenecken mit Hilfe von Zugfeldmodellen", TU Braunschweig, Inst. f. Stahlbau, **Report No. 6202/2**

Scheer, J., Pasternak, H., Schween, T., and Vayas, I. (1993): "Experimental Investigation of Beam to Column Joints with Slender Webs under Cyclic Loading", TU Braunschweig, Inst. f. Stahlbau, **Report No. 6317**

PART THREE
GLOBAL DUCTILITY

22 AN OVERVIEW OF GLOBAL DUCTILITY

H. AKIYAMA
University of Tokyo, Tokyo, Japan

Abstract
The inverse of q-factor in EC8 and the D_s-value in Japanese building code are equivalent and signify a strength reduction factor ascribed to the energy absorption capacity of a structure. Despite that these strength reduction factors have been deemed to be an empirical one, the logical framework to estimate the strength reduction factor must be established for the sound development of the earthquake resistant design based on the energy concept. In this paper, an outlined feature of the strength reduction factor is presented and comments on papers presented at the Workshop are added in the light of theoretical derivation of the strength reduction factor.

Keywords: Energy concept, q-factor, D_s-value, cumulative inelastic deformation ratio, inelastic deformation ratio, damage concentration, P - δ effect, global ductility, local ductility.

1 Introduction

It has been widely recognized that the inelastic deformation together with the strength is very important for a structure to resist to earthquakes. Thus, the inelastic energy absorption capacity can be the primary index of structural resistance against earthquakes.

On the other hand, the loading effect of earthquakes on structures can be most concisely grasped by means of the energy input. At present it is possible to obtain realistic behaviors of a structure under an earthquake through numerical response analyses by constructing a suitable structural model and applying minute restoring force characteristics for each structural element. Such a direct analytical measure, however, can not be a single tool for us to understand the consequence between the seismic input and the structural behavior, because the fluctuation of seismic phenomena and the complexity of structural systems yield too much information to make us allow a simple derivation of the strength reduction factor.

For this reason, we must construct a logical framework parallel to the direct numerical approach.

This framework must serve as a firm ground for design judgement and a storage of useful information obtained by numerical analyses.

To be logical, the following conditions must be satisfied.

1) the energy balance between the total energy input exerted by an earthquake and   the absorbed energy by a structure is kept.

2) A multi-story structure can be decomposed into story frames. Then, the total energy absorption made by the structure must be the sum of the energy absorption made by story frames.

3) The only well-defined loading condition is monotonic loading and the hysteretic behavior under seismic motion must be related to the monotonic behavior.

The concept of strength reduction factor is only effective as far as the structure is prevented from the collapse by applying it under the possible maximum level of earthquake (design earthquake). Therefore, the loading effect of the possible maximum level of earthquake must be clearly defined. To do this, the introduction of the spectrum of the energy input made by an earthquake is indispensable.

The load reduction factor is defined as follows.

$$D_s = \frac{R}{R_e} \tag{1}$$

where D_s: the D_s -value in Japanese building code
R : the required strength of the structure to resist to the design earthquake
R_e: the required strength of the structure deprived of the inelastic deformation capacity to resist to the design earthquake

It is obvious that the q-factor in EC8 is related the D_s -value as follows.

$$q = \frac{1}{D_s} = \frac{R_e}{R} \tag{2}$$

It is important to note that the loading effect of the design earthquake must be kept constant in obtaining both R and R_e

In this paper, first, a loading process of derivation of the load reduction factor is presented based on the energy concept, taking moment-resistant frames as an example. Since in moment-resistant frames the shear deformation is predominant, the overall flexural deformation is ignored.

Next, contents of papers presented at the Workshop are introduced and comments are added referring to the general aspect of the global ductility.

2 Energy Input due to Earthquakes

The loading effect of earthquakes on structures can be grasped most concisely by means of the energy input. The energy input can be expressed in a form of energy spectrum as follows(Akiyama, 1985).

- The Fourier spectrum of accelerogram of a seismic motion coincides with the V_E - T relationship as to the same seismic motion. V_E is the equivalent velocity obtained through the following conversion from the total energy input to the undamped one-mass vibrational system with the natural period of T.

$$V_E = \sqrt{\frac{2E}{M}} \tag{3}$$

where E : the total energy input to the system
M : the mass of the system

- The V_E - T relationship defined by Eq.(2) is termed as the energy spectrum.

The energy spectrum for the damped system or inelastic system can be obtained by smoothing (or averaging) the energy spectrum for the undamped elastic system. The extent of smoothing increases proportionally to the extent of nonlinearity of the system.

- The energy spectra for the inelastic systems can be represented by the energy spectra for the elastic damped system with 10% of critical damping (h = 0.1). In this sense it can be easily understood that the total energy input made by an earthquake mainly depends on

the total mass, strength distribution, stiffness distribution, mass distribution and type of restoring force characteristics of the structure.
• Since the structure is softened by undergoing inelastic deformation, the substantial period of vibration is lengthened according to the extent of plastification. The substantial period of vibration, which should be applied to the energy spectra in place of T_o, has been found to be

$$T_e = \sqrt{\frac{T_o^2 + T_o T_m + T_m^2}{3}} \qquad (4)$$

where T_e : the substantial period of vibration
T_o : the natural period in the elastic range
T_m : the longest period of vibration experienced under an earthquake

As far as the restoring force characteristics of the structure are clearly defined, T_m can be easily related to the maximum deformation of structure.

Compared to the conventional acceleration response spectrum and the velocity response spectrum which can be directly applied only to elastic systems, the energy spectrum has decisive advantages as following.

• The energy spectrum can be applied both to elastic and inelastic systems.
• The energy spectrum can be represented by a single curve which corresponds to the energy spectrum for the elastic system with $h = 0.1$.
• Two major indices of structural damage are expressed in terms of cumulative inelastic deformation and maximum deformation. The cumulative inelastic deformation can be directly related to the total energy input and it is not difficult to find a relationship between the maximum deformation and the cumulative deformation through numerical response analyses.
• On the basis of energy spectrum, the earthquake resistant design can be clearly formulated to be Control of Energy Distribution under a Constant Energy Input.

The major problems met in the route of earthquake resistant design are:

• How to get the energy spectra for specific ground condition ?
• How to distribute the input energy over a structure ?

The first problem belongs to the field of seismology. The mechanism as to the generation of seismic motions are considered as follows.

When the strain energy stored in the bedrock in the hypocentral area reaches a certain limit value, the rocks collapse, and the released strain energy propagates as the vibrational energy through bedrocks.
The seismic motion which has arrived at the deep bedrock beneath the site of construction is transmitted to the surface of ground, being amplified through rocks and soils. Seismic motions on the surface of ground also involve some components of surface waves.

An example of the energy spectra is shown in Fig.1.

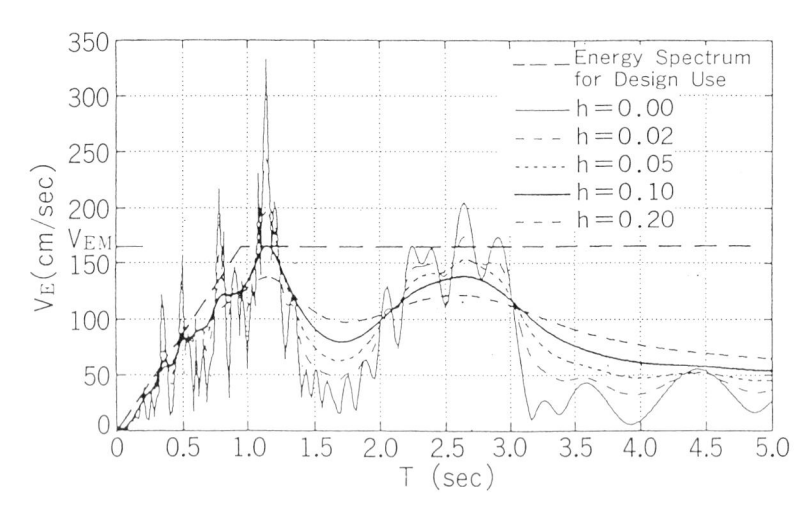

Fig. 1 Energy Spectrum of Hachinohe Record

The used seismic record is of the E W - component of Hachinohe record of the Tokachi -Oki earthquake (1986). Smoothing effect of damping is clearly seen in the diagram. The energy spectrum with h = 0.1 which should be applied to the structural design is a smooth curve and can be enveloped by a bilinear relationship shown by two broken lines. These lines are expressed as follows

for $T \leq T_G$ (shorter period range)

$$V_E = \frac{V_{EM} T}{T_G} \quad (5)$$

for $T > T_G$ (longer period range)

$$V_E = V_{EM} \quad (6)$$

where T_G : the transient period which separates the range of period
         V_{EM} : the level of V_E in the longer period range

Considering the elongation of substantial period of vibration due to the plastification of structure, the energy spectrum for design use must be modified by introducing a magnification factor a_s as follows.

$$V_E = \frac{a_s V_{EM} T}{T_G} \quad (7)$$

where T : the fundamental natural period in elastic range

The value of a_s depends on the restoring force characteristics and the extent of plastification of the structure and lies in the rage of 1.2 to 1.5 for the practical structures.

When the structure of ground is well known, the energy spectra on the surface of ground can

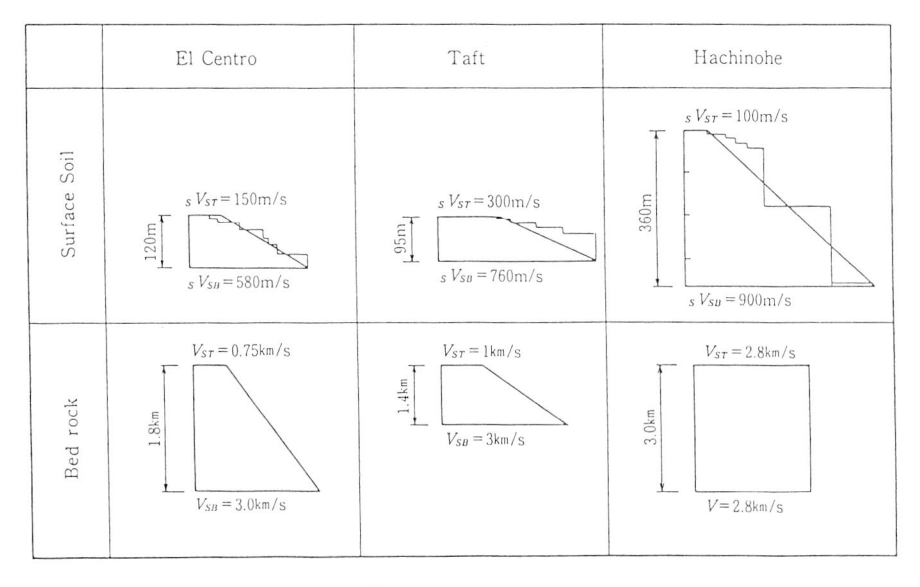

An overview of global ductility 257

2 (a) Assumed V_s - profile in Sites of Strong Record

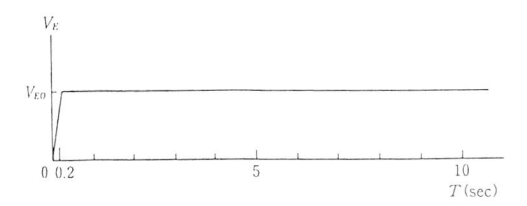

2 (b) Energy Spectrum for Ground Motions in Base Rock

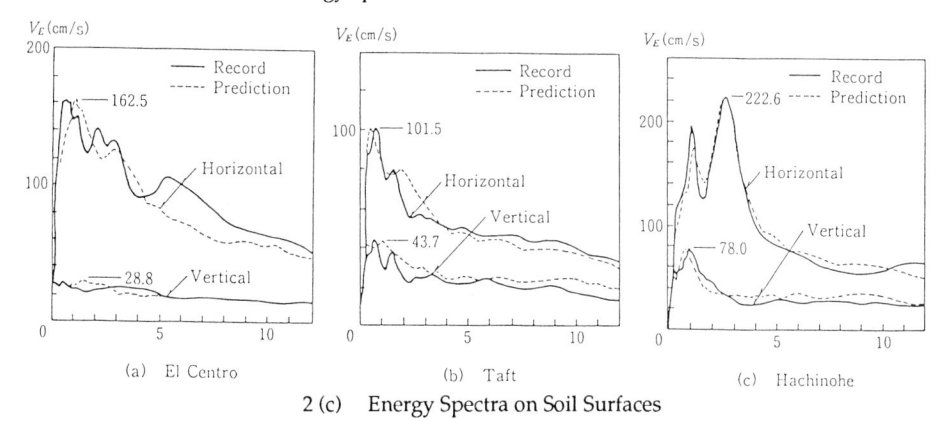

(a) El Centro   (b) Taft   (c) Hachinohe

2 (c) Energy Spectra on Soil Surfaces

Fig. 2 Prediction of Energy Spectra on the Surface of Ground

be predicted by applying one-dimensional wave propagation theory.

In Fig, 2 (a), the profiles of shear wave velocity of surface soils in each site are shown by fine lines and are appoximated by linear profiles by solid lines together with the profiles of shear wave velocity of bedrocks. A synthetic ground motion which possesses a flat energy spectrum shown in Fig. 2 (b) was generated and applied to the bottom of bedrocks. The energy spectra of the surface ground motions calculated through the wave propagation theory are compared with the energy spectra of observed ground motions in Fig. 2 (c). As is seen in Fig. 2 (c), the position of peaks and valleys in energy spectra can be predicted provided that the ground structure is known. However, the application of a constant level of energy input can be permitted for the following reasons.

- Generally, it is impracticable to predict the local ground conditions. Thus, it must be recognized that the position of peaks and valleys of energy spectra is changeable.
- The application of constant level of V_E does not yield an over-conservative structural design.

3 Derivation of strength Reduction Factor

The equilibrium of energy can be written as

$$W_e + W_h + W_p = E \tag{8}$$

where W_e : the elastic vibrational energy
W_h : the energy absorbed by damping
W_p : the cumulative strain energy

The total energy input E is a very stable amount irrespective of structural behavior.

On the other hand, the distribution of energy over a structure depends on the structural type, and mechanical properties of structural components. The structural damage corresponds to W_p.

In order to know the distribution of energy, numerical response analyses are indispensable. By summarizing the results of numerical analyses, it is possible to construct a simplified and conceptual design method based on the energy spectra.

W_p consists of cumulative inelastic strain energy in every story W_{pi}. Thus,

$$W_p = \sum_{i=1}^{N} W_{pi} \tag{9}$$

where N : the number of story

W_p and W_{pi} can be regarded as structural damage. Each story of a shear type multi-story structures is considered to be composed of a stiff element and a flexible element. The flexible element has a small stiffness and remains elastic, whereas the stiff element has a large stiffness and behaves inelastically. The relation between the shear resistance and the story

An overview of global ductility 259

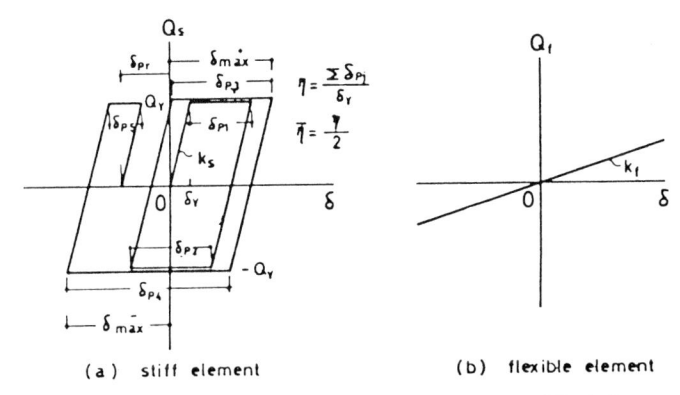

Fig. 3 Restoring Force Characteristics of Each Story

displacement is depicted in Fig. 3 where the elastic-perfectly plastic restoring force characteristics of the stiff element is assumed. Assuming that the spring constant of the stiff element k_s, is sufficiently larger than that of the flexible element k_f and the contribution of energy absorption of the flexible element can be neglected, then, the damage of the first story of a building is written as

$$W_{p1} = \frac{M g^2 T^2}{4 \pi^2} \times \frac{2 \alpha_1^2 \overline{\eta}_1}{\kappa_1} \tag{10}$$

where α_1 : the yield shear force coefficient of the first story ($Q_{Y1} / M g$)
Q_{Y1} : the yield shear force of the first story
$\overline{\eta}_1$: the averaged cumulative inelastic deformation ratio of the first story (= cumulative inelastic deformation δ_{p1} divided by two times of yield displacement δ_{Y1})
δ_{Y1} : the yield displacement (see Fig. 3 (a))
$\kappa_1 = k_1 / (4\pi^2 M / T^2)$
g : the acceleration of gravity
k_1 : the spring constant of the first story

The total damage of a structure, W_p can be formally related to W_{p1} as

$$W_p = \gamma_1 W_{p1} \tag{11}$$

In the shear-type multi-story structures, it has been made clear that γ_1 is expressed by the following formula.

$$\gamma_1 = 1 + \sum_{j \neq 1} s_j (p_j / p_1)^{-n} \tag{12}$$

where $p_j = \dfrac{\alpha_j}{\alpha_1 \overline{\alpha}_j}$ $s_j = \left(\sum_{i=j}^{N} m_i / M \right)^2 \overline{\alpha}_j^2 (k_1 / k_j)$

$\overline{\alpha_j}$: the optimum yield shear force coefficient distribution
α_i / α_1 : the actual yield shear force coefficient distribution
m_i : the mass of ith floor
k_i : the spring constant of i t h story

p_j means a deviation of the actual yield shear force distribution from the optimum yield shear force distribution under which the damage of every story $\overline{\eta}_j$ is equalized, and is termed the damage concentration factor. n is termed the damage concentration index. When n becomes sufficiently large, γ_1 becomes unity. It means that a sheer damage concentration takes place in the first story. When n is nullified, a most preferable damage distribution is realized. Practically, the value of n ranges between 2.0 and 12.0. Weak-column type of structures are very susceptible of damage concentration, and the n-value for them should be 12.0. In weak-beam structures, the damage concentration is considerably mitigated due to the elastic action of columns, and the n-value can be reduced to 6.0. In Fig. 4 (c), a generalized form of weak-beam type structure is shown. The presence of a vertically extending elastic column is essential to this type of structure. The elastic column by itself is not required to withstand any seismic forces. While ordinary frames pin-connected to the elastic column absorbs inelastically seismic energy, the elastic column plays a role of damage distributor. By applying this type of structure, the n-value can be reduced to 2.0. The damage concentration is also governed by the value of p_j. To simply estimate the damage concentration in the first story, an unified value may be applied as p_j as follows.

$$p_1 = 1.0 \qquad p_{j \neq 1} = p_d \qquad (13)$$

- possible place of plastic hinge formation
- real hinge

(a) weak-column type $n = 12.0$
(b) weak-beam type $n = 6.0$
(c) generalized form of weak-beam type $2.0 < n < 12.0$

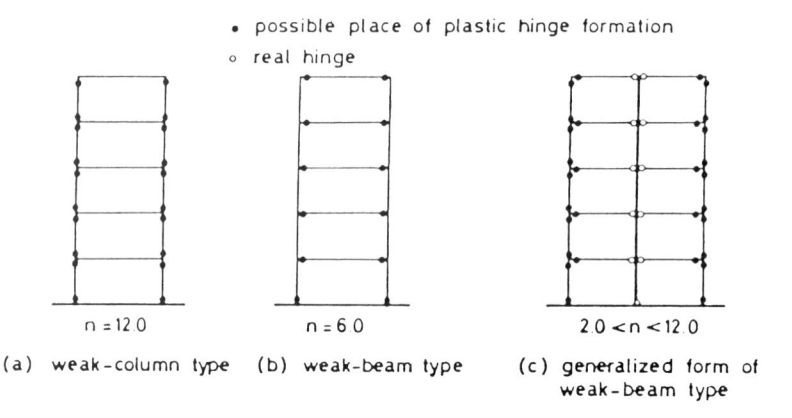

Fig. 4   Various Structural Types

Eq. (13) signifies that the strength gap is assumed between the first story and the other stories. It is impossible to make the yield shear force distribution of an actual multi-story building agree completely with the optimum distribution. The reasons are easily found in the scatter in mechanical properties of material and the rearrangement of geometrical shapes of structural members for the purpose of simplification in fabrication process. Taking account of such a situation. The following value is proposed as a minimum strength gap to be taken into account in the design procedure.

$$p_d = 1.185 - 0.0014\ N \qquad p_d \geq 1.1 \tag{14}$$

The elastic vibrational energy is expressed as follows.

$$W_e = \frac{M\ g^2\ T^2}{4\ \pi^2} \times \frac{\alpha_1^2}{2} \tag{15}$$

Taking account of damping, Eq. (8) is reduced to

$$W_p + W_e = E \times \frac{1}{(1 + 3h + 1.2\sqrt{h})^2} \tag{16}$$

where h = damping constant

Substituting Eqs. (15) and (11) into Eq.(10), the following formula is obtained.

$$\alpha_1 = \frac{\alpha_e}{\sqrt{1 + 4\frac{\gamma_1\ \overline{\eta}_1}{\kappa_1}}} \qquad \alpha_e = \frac{2\ \pi\ V_E}{T\ g}\ \frac{1}{1 + 3h + 1.2\sqrt{h}} \tag{17}$$

Eq. (17) is rewritten as

$$\alpha_1\ (T) = D_s\left(\overline{\eta}_1\right) \alpha_e\ (T) \tag{18}$$

where $\alpha_e\ (T)$: the required minimum yield shear force coefficient for the elastic system with the fundamental natural period T
$\alpha_1\ (T)$: the required minimum yield force coefficient of the first story for the inelastic system with T
$D_s\left(\overline{\eta}_1\right)$: the reduction factor for the yield shear force coefficient, which depends on $\overline{\eta}_1$

The distribution of masses is assumed to be uniform. The yield deformation of every story is also assumed constant. Then, the stiffness distribution k_i / k_1 becomes equal to the strength distribution Q_{Yi} / Q_{Y1} under the optimum yield shear force coefficient distribution.

The optimum yield shear force coefficient distribution $\overline{\alpha}_i$ is given by the following formula.

$$\overline{\alpha}_i = f\left(\frac{i-1}{N}\right) \tag{19}$$

for $x > 0.2,\ f(x) = 1 + 1.5927x - 11.8519x^2 + 42.583x^3 - 59.48x^4 + 30.16x^5$
for $x \leq 0.2,\ f(x) = 1 + 0.5x$

Using the above-mentioned parameters, γ_1 in Eq.(12) and κ_1 are calculated and approximated by the following relations.

$$\gamma_1 = 1 + 0.64(N-1)p_d^{-n} \tag{20}$$

$$\kappa_1 = 0.48 + 0.52\,N \tag{21}$$

Then, the D_s - values is written as

$$D_s = \frac{1}{\sqrt{1 + \dfrac{4\{1 + 0.64(N-1)p_d^{-n}\}\bar{\eta}_1}{0.48 + 0.52\,N}}} \tag{22}$$

Fig. 5 shows the relationship between the D_s - values and the number of story N for specific values of $\bar{\eta}_1$. The difference of D_s - values is caused by the difference of the damage concentration index n which governs the damage distribution in multi-story buildings. D_s - values inevitably increase with the increase of N due to the effect of damage concentration.

The goal of earthquake resistant design can be summarized as follows.
 1) to minimize α_1
 2) to minimize the maximum story displacement δ_{max}
 3) to minimize the residual story displacement δ_{pr} (see Fig. 3 (a))

To attain the first item, the following two measures are practicable.
 1) to increase the deformation capacity $\bar{\eta}_1$
 2) to reduce the damage concentration index n

The former is realized by applying mild steel to the stiff element. When the structural members are carefully selected so as to avoid structural instability such as local buckling and lateral buckling, it is not impossible to attain the value of $\bar{\eta}_1$ greater than 100. The later is realized by applying the weak-beam type structure or more general damage dispersing systems as shown in Fig. 4 (c). High-strength steels can be most effectively used as a flexible element and a vertical damage distributor.

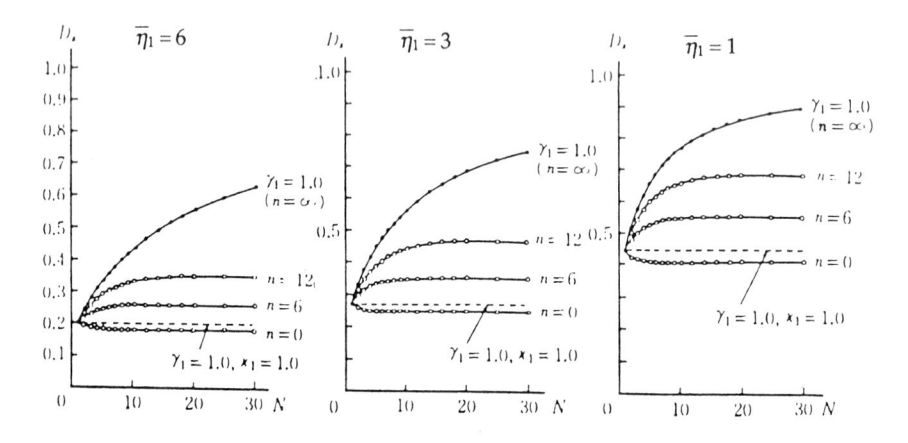

Fig. 5 D_s - Values for Multi-Story Frames

To discuss the maximum story displacement, the inelastic deformation ratio, μ is introduced as follows.

$$\bar{\mu} = \left(\bar{\delta}_{max} - \delta_Y\right)/\delta_Y \tag{23}$$

where $\bar{\delta}_{max}$: the average value of the maximum story displacement in the positive and negative directions.

The residual story displacement, δ_{pr} is equal to the difference between the cumulative inelastic deformations of positive and negative directions as seen in Fig. 3 (a). To reduce δ_{pr} and $\bar{\mu}$, the most effective measure is the application of "the flexible-stiff mixed structure". Only a slight participation of the flexible element enables to nullify δ_{pr} and to reduce $\bar{\mu}$ remarkably as is seen in the following empirical relations.

$$\text{for} \quad k_f/k_s = 0, \quad \bar{\mu} = \frac{\eta}{2} \quad \text{for} \quad k_f/k_s > 0.03, \quad \bar{\mu} = \frac{\eta}{4} \quad \text{to} \quad \bar{\mu} = \frac{\eta}{6} \tag{24}$$

The stiff elements are the source of energy absorption, whereas flexible elements restrain effectively development of excessive deformations and one-sided deformations. Again, high-strength steel can be very suitable materials to form the flexible element.

4 Relationship between Local Ductility and Global Ductility

The global ductility of ith story of a structure is expressed by the cumulative inelastic deformation ratio $\bar{\eta}_i$ and the inelastic deformation ratio $\bar{\mu}_i$. Since $\bar{\mu}_i$ can be related to $\bar{\eta}_i$ according to the structural characteristics as is expressed in Eq. (24), it is essentially necessary to estimate $\bar{\eta}_i$. A multi-story frame can be decomposed into story-frames, and $\bar{\eta}_i$ of ith story frame can be related to the load-deformation curve of the story-frame under the monotonic loading. The analytical method to be applied to the story-frame under the monotonic loading has been well established.

4.1 Decomposition of a frame into story frames

When the ith story of a frame, as is shown in Fig. 6 (a), is on the brim of yielding under an earthquake, the upper $(i+1)$th story and the lower $(i-1)$th story rest in almost the same

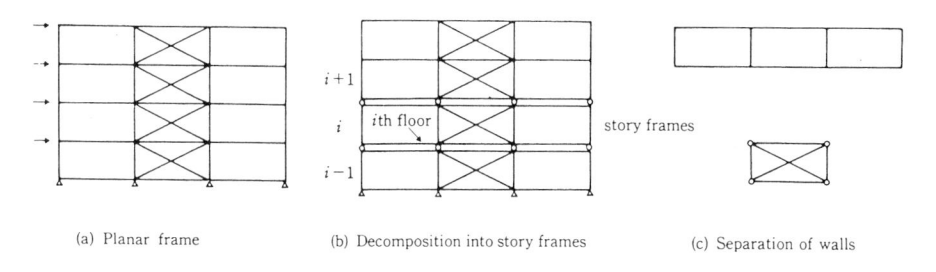

(a) Planar frame   (b) Decomposition into story frames   (c) Separation of walls

Fig. 6   Decomposition of a Multi-Story Frame into Story Frames

stress state. Beams and beam-to-column connections on the ith floor carry stresses induced by shear forces in the ith and (i - 1)th stories. In such a state, it can be considered that approximately half of the stresses in the beam and beam-to-column connection are stresses induced by the story shear force in the ith story. In this context, beam and beam-to-column connections may be decomposed to form story frames as shown in Fig. 6 (b). The story frames are assumed to be connected to each other with pinned supports, as shown in the figure. Moreover, the story frame is decomposed into the pure moment-resistant frame and the pure shear wall. The pure moment frame is a portion eliminated from the walls. The pure shear wall is formed by wall elements enclosed by a pin-connected rectangular frame consisting of rigid members. The enclosing frame, which is unable to resist horizontal forces by itself, is installed for the convenience of calculating strength and deformation.

The distribution of story shear forces is assumed to be proportional to the yield-shear force distribution prescribed by the optimum yield-shear force coefficient distribution, $\overline{\alpha}_i$. Denoting this standard story shear force distribution associated with $\overline{\alpha}_i$ by $\overline{Q}_i$, $\overline{Q}_i$ can be described by

$$\overline{Q}_i = \frac{\left(\sum_{j=i}^{N} m_j\right) \overline{\alpha}_i}{M} \qquad (25)$$

Since the upper beam in the ith story resists the moment, $_bM_i$, which is produced by the ith story shear force and the moment, $_bM_{i+1}$, which is produced by the (i + 1)th story shear force, it is natural to assume that the upper beam in the ith story should be decomposed so as to be proportional to $_bM_i$, and $_bM_{i+1}$. Assuming the shear span of a column to be one-half of the column length, the dividing ratio for the observed beam, d_i, should be obtained as follows

$$d_i = \frac{h_i \overline{Q}_i}{h_i \overline{Q}_i + h_{i+1} \overline{Q}_{i+1}} \qquad (26)$$

where h_i, h_{i+1} : the story heights of the ith and (i + 1)th stories.

Denoting structural properties of the observed beam, including strength and stiffness, by S_o, structural properties thus decomposed become

$$S_i = S_o d_i$$
$$S_{i+1} = S_o (1 - d_i) \qquad (27)$$

where S_i, S_{i+1} : structural properties of decomposed beams in the ith and (i + 1)th stories

The above discussion can be applied in exactly the same manner to the decomposition of panel zones in beam-to-column connections.

4.2 Transformation of elastic-plastic systems into elastic-perfectly plastics system

Generally speaking, the horizontal force-resistant element of a one-story rigid steel frame may be considered to be composed of an elastic element which remains elastic and an inelastic element which undergoes inelastic deformations. The restoring-force characteristics of an inelastic element under monotonic loading can be expressed as a tri-linear relation, as shown in Fig. 7 (b), where the Q - δ curves under positive and negative loadings can be considered

An overview of global ductility   265

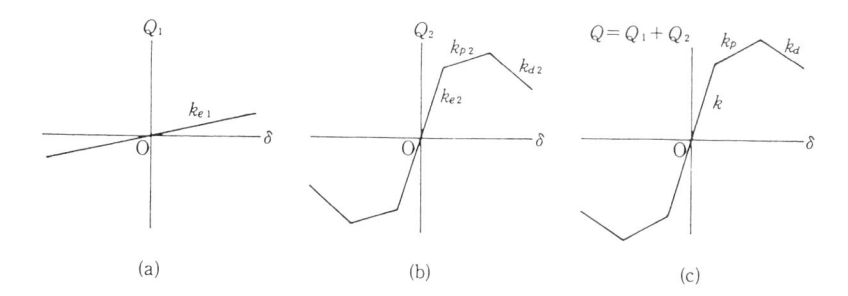

Fig. 7   General Restoring-Force Characteristics

to be identical to one another. Fig. 7 (a) indicates the load-deformation function of an elastic element. Combining the behaviors of the elastic element and the inelastic element, the load-deformation relation for a story under monotonic loading is obtained, as shown in Fig. 7 (c). k, k_p, and k_d - denoting rigidities in elastic, inelastic and degrading domains respectively - are expressed in the following relations referred to in Figs. 7 (a) and (b).

$$k = k_{e1} + k_{e2}$$

$$k_p = k_{p2} + k_{e1} \tag{28}$$

$$k_d = k_{d2} + k_{e1}$$

The $P - \delta$ effect is produced by vertical loading and horizontal story displacement. It is an elastic effect with a negative slope, as expressed in the following equation, and may be included in the stiffness of the element k_{e1}

$$Q_{p\delta} = -\frac{W \delta}{H} \tag{29}$$

where   W : the total vertical load applied to the story
         H : the height of the story

The $Q - \delta$ relationship shown in Fig. 7 (c) is shown again in Fig. 8 (a), neglecting elastic deformations. In this figure, nondimensional notations are used as follows.

$$\tau = Q / Q_Y, \quad \eta = \delta / \delta_Y,$$

$$\tau_m = Q_{max} / Q_Y, \quad k'_p = k_p / k_e, \quad k'_d = k_d / k_e$$

When Bauschinger effect does not appear on the load-deflection relationship, the following fact can be observed between the load-deformation diagram under arbitrary changing deformations and that under the monotonic loading.

> The inelastic part of a load-deformation curve under a repeated loading consists of curve-segment which corresponds to each loading cycle. Curve-segments connected sequentially in one loading domain coincide with the inelastic load-deformation curve under the monotonic loading.

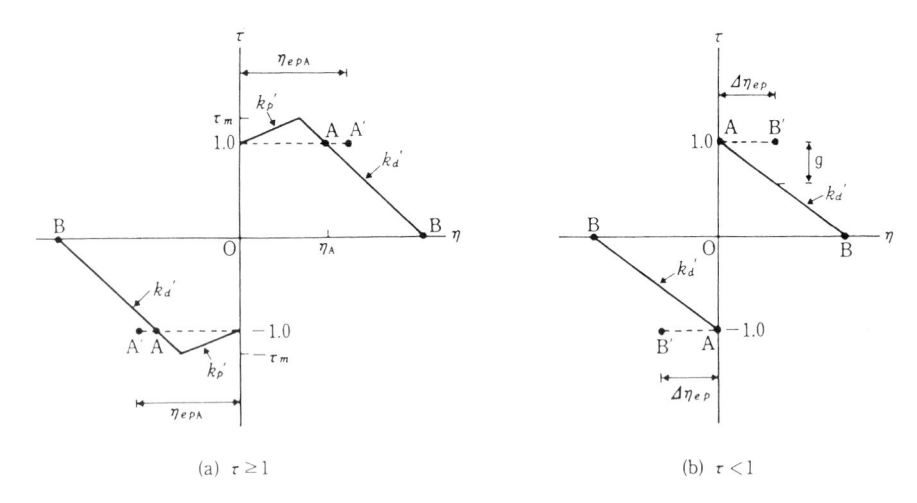

Fig. 8 Transformation of General Systems into Elastic-Perfectly Plastic Systems

(a) $\tau \geq 1$

(b) $\tau < 1$

Thus, the inelastic load-deformation curve under the monotonic loading can be considered to express the relationship between the strength and the cumulative inelastic deformation in one loading domain.

In the same manner, the elastic-perfectly plastic restoring force characteristics can be depicted as is shown by a broken line in Fig. 8 (a)

The collapse state of the story is indicated by the point B where the resisting force of the story is totally lost. Up to the point A where the resisting force returns to the level of Q_Y, the cumulative plastic deformations can be assumed to take place in the same manner as in the case of the elastic-perfectly plastic system. Thus, the state A with η_A in the elastic-plastic system can be transformed to the state A' with η_{epA} in the elastic-perfectly plastic system by considering the equivalence of the absorbed energy in both systems, Beyond the point A, the degradation of strength accelerates one-sided development of inelastic deformation in either the positive or negative direction.

The transformation of this degraded part is expressed by $\Delta\eta_{ep}$ as shown in Fig. 8 (b).

$\Delta\eta_{ep}$ denotes the cumulative inelastic deformation ratio in one direction in the elastic-perfectly plastic system, produced by the energy input, which makes the story reached at point A move to point B. Then, the cumulative inelastic deformation ratio in one direction in the elastic-perfectly plastic system produced by the input energy which causes the collapse of the observed story in the elastic-plastic system is written as

$$_u\overline{\eta}_{ep} = \eta_{epA} + \Delta\eta_{ep} \tag{30}$$

As a final result, the average cumulative inelastic deformation ratio of the elastic perfectly plastic system, which corresponds to the collapse state of a general-elastic-plastic system can be expressed by

$$_u\overline{\eta}_{ep} = \frac{(\tau_m^2 - 1)}{2}\left(\frac{1}{k_p'} + \frac{1}{|k_d'|}\right) + \frac{g_d}{|k_d'|} \tag{31}$$

where
$$g_d = \frac{n\, s_k\, p_k^{-n} + 4\sum_{j=1}^{N} s_j\, p_j^{-n}}{4(n+2)\sum_{j=1}^{N} s_j\, p_j^{-n}}$$

Generally, Bauschinger effect may appear in the path which leads to point - A, and increases the energy absorption capacity. By considering Bauschinger effect, Eq. (31) can be modified as follows

$$\overline{u\eta}_{ep} = \frac{a_B\,(\tau_m^2 - 1)}{2}\left(\frac{1}{k'_p} + \frac{1}{|k'_d|}\right) + \frac{g_d}{|k'_d|} \tag{32}$$

The amplification factor due to Baushinger effect a_B lies in the range of 1.0 to 2.5.

5 Comments on Contributions at the Workshop

1) Statistical Evaluation of the Behavior Factor for Steel Frames by Calderoni, Ghersi and Rauso, 1994

In this paper, the evaluation of the behavior factor q by means of a statistical procedure, based on the analysis of the dynamic inelastic responses of moment resisting multi-story frame designed with different q-values to a set of accelerograms is made. The paper points out the inconsistency of the commonly used definition of q-factor as the ratio $a_{g,u} / a_{g,y}$, mainly owed to the fact that the peak ground acceleration which corresponds to the reaching of the ultimate plastic rotation in a section $a_{g,u}$ is related to the overstrength of all sections while the peak ground acceleration which corresponds to the yielding of the most stressed section $a_{g,y}$ is strongly affected by the behavior of a single section, leading to the wide scatter of q-values owing to the characteristics of each accelerogram.

Four different types of moment-resisting frames are proportioned by means of modal analysis using the average spectrum of applied thirty accelerograms reduced by different q-values and the inelastic dynamic response analysis is made to individuate the maximum value of q which enables the structure to limit to an acceptable value the probability of structural failure. The results of analysis made clear the important role of the spread of overstrength of sections in a structure in determining the q-factor.

The proposed methodology is quite orthodox on the reason that the method provide us with the direct approach to check the structural safety under the design earthquakes.

Not only the extension of analyses to a large number of frames but also the construction of a logical frame work to synthesize the results of analysis are strongly expected to enable us to assess the values of the behavior factor on a firm base.

2) Contribution to the Study of the Ductility of Unbraced Frames by Gioncu, Matteescu and Iuhas, 1994

In this paper, a procedure for determining the q-factor for various types of moment-resisting frames is presented based on the nonlinear response analysis. The different parameters, as accelerogram type, natural vibration period, axial forces, story number, degradation in restoring force characteristics are taken into account.

The applied definition of q-factor is given as

268 Akiyama

$$q = \frac{a_u}{a_p} \qquad (33)$$

where a_u : the peak ground acceleration causing the collapse of the structure
        a_p : the peak ground acceleration causing the attainment of the first plastic hinge

The results of analysis were illustrated by the θ - q diagrams, where θ is the hinge rotation at the observed point. Those result are realistic and implicative indicating extremely complex features of the q - factor influenced by applied parameters. The q - factor is the structural behavior factor and should be removed of the fluctuation on of seismic input. Most of the results presented in the paper inevitably includes the influence of the fluctuation of seismic input and this makes the results complicated.

The definition of Eq. (33) is same as that given by Eq. (2) as far as the loading effect of the seismic input is proportional to the value of q. As is shown in Fig. 1, the loading effect depends on the magnitude of damping, or in other words on the extent of plastification of structure. Thus, the peak ground acceleration can not be a suitable index of the seismic input, as is pointed out by the first paper . To eliminate the influence of fluctuation of seismic input, Eq. (33) should be replaced by

$$q = \frac{V_{Eu}}{V_{Ep}} \qquad (34)$$

where V_{Eu} : the energy input converted into the velocity causing the collapse of the structure
        V_{Ep} : the energy input converted into the velocity causing the attainment of the first plastic hinge

3) **A Method for the Evaluation of the Behavior Factor for Steel Regular and Irregular Buildings by Vayas, Syrmakezis and Sophocleous, 1994**

A method for the determination of the behavior factor q for steel moment-resisting frames is presented.

The basic assumption is made that the maximum deflection at the representative point of a structure does not depends on the strength of the structure under an earthquake excitation and is identical to the elastic response of the structure. Then, the q -factor can be defined as

$$q = \frac{\delta_u}{\delta_y} \qquad (35)$$

where δ_u : the representative deflection in the state where the local ductility is exhausted under the monotonic loading
        δ_y : the representative deflection at the elastic limit of structure under monotonic loading

To be more exact, the definition of q given by Eq. (35) must be supported by another assumption that the local ductility can be uniquely determined in terms of the maximum deflection of structure irrespective of type of loading .

The first assumption is examined by applying Eq. (17).

For the single degree of freedom system, the following relationship holds.

$$\frac{\alpha}{\alpha_e} = \frac{1}{\sqrt{1+4\overline{\eta}}}$$

$$\overline{\mu} = \frac{\overline{\eta}}{a} \qquad (36)$$

$$\frac{\overline{\delta}}{\delta_e} = \frac{\alpha}{\alpha_e}\left(1+\overline{\mu}\right)$$

where
α : the required strength for the inelastic system to resist to an earthquake
α_e : the strength for the inelastic system to resist to the same earthquake
$\overline{\delta}$: the maximum deflection of the inelastic system averaged in positive and negative directions
δ_e : the elastic limit deflection of the system
a : a constant which relates $\overline{\eta}$ to $\overline{\mu}$

From Eq. (36), the following relation is obtained.

$$\frac{\overline{\delta}}{\delta_e} = \frac{(4a-1)\left(\frac{\alpha}{\alpha_e}\right)+\left(\frac{\alpha_e}{\alpha}\right)}{4a} \qquad (37)$$

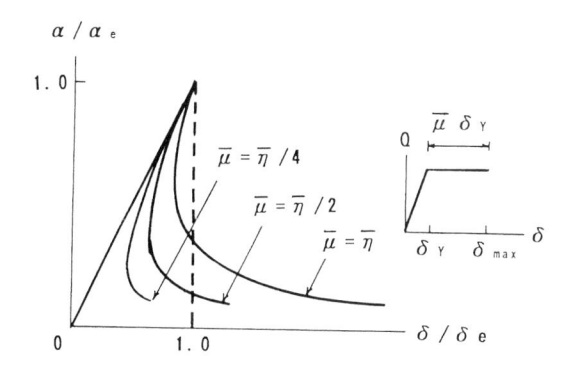

Fig. 9 Constancy of the Maximum Deflection Response

In Fig. 9, the $\alpha/\alpha_e - \overline{\delta}/\delta_Y$ relation is shown for various values of a. The first assumption is depicted by a broken line. As is seen in the figure, the assumption may lead to a conservative side of estimate of the maximum deflection estimate for the elastic-perfectly plastic systems with the value of a greater than 2.0. The proposed method is simple and consistent approach as far as the basic assumption is valid. However, some modification is required in order to apply this method to the degrading systems in which one-sided development of deflection is liable to occur and the multi-story frames in which the reference point to calculate δ_u is not uniquely determined.

4) On the Seismic Resistance of Light Gange Steel Frames by Calderoni, Martino, Ghersi and Landolfo, 1994

In this paper the possibility of using the light gage steel members in dissipative zones of seismic structure is examined through the numerical response analyses on single story one bay frames which are characterized by the geometric and mechanical cyclic deterioration. The applied method is similar to that applied in the first paper, and the definition of q described by Eq.(33) is also introduced. The results of analyses are very significant and are summarized as follows

- The dynamic response of frames with degrading characteristics is quite similar to that of correspondent ideal elasto-plastic system.
- The q - factor which can be used for such degrading type of frames is in the range of 2 - 3. This low value however does not agree with EC8 provisions that impose fully elastic design of q = 1 for such type of structures.

In the AIJ Recommendation for the light weight steel structures (AIJ, 1985), the applied values of D_s are in the rage of 0.3 to 0.52 which are equivalent to the q - factors of 2.0 to 3.3.

In these structures characterized by degrading characteristics, the dynamic instability limit seems to be a very clear criterion of the collapse of structure.

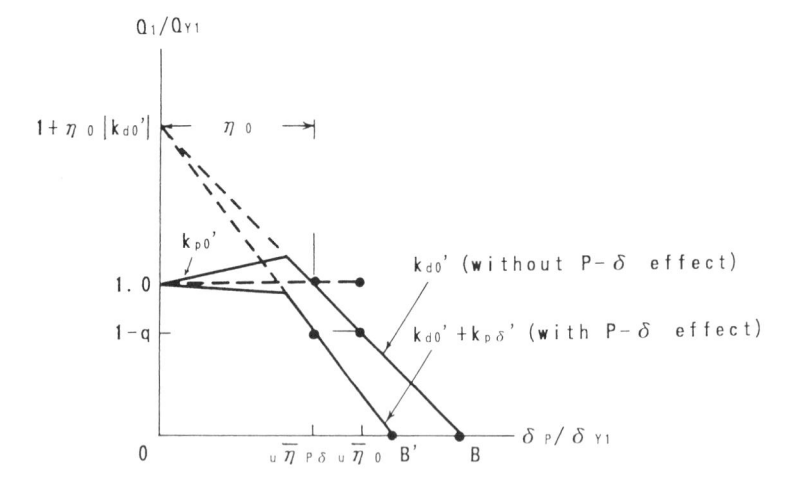

Fig. 10   P - δ Effect in Deformation Capacity

In Fig. 10, a simplified load-deflection curve of a degrading system is shown, neglecting the elastic part of it. The dynamic instability limits correspond to the points B and B'. The load-deflection curves can be approximately expressed by two line segments; one is the elastic-perfectly plastic relation as is indicated by the broken line pallalel to the abscissa and another is a degrading part with a slope of k'_{do} for the system without P - δ effect or $k'_{do} + k'_{p\delta}$ for the system with P - δ effect. $k'_{p\delta}$ is the ratio of the negative slope to the slope of the elastic range and is expressed as

$$k'_{p\delta} = \frac{\Phi}{q} = \frac{\theta_e}{\alpha} \tag{38}$$

An overview of global ductility 271

where Φ : the coefficient defined by EC8
θ_e : the elastic limit story inclination in radian
α : the yield shear coefficient of story

According to Eq.(32), the collapse state is transplanted on the elastic-plastic relationship with the deformation capacity $_u\eta$ as follows.

For the system without P - δ effect,

$$_u\eta_o = \frac{g_d + \eta_o |k'_{do}|}{|k'_{do}|} \tag{39}$$

For the system with P - δ effect

$$_u\eta_{p\delta} = \frac{g_d + \eta_o |k'_d|}{|k'_{do}| + |k'_{p\delta}|} \tag{40}$$

where k'_{do} : the nondimensionalized slope in degrading range
g_d : the limit degrading ratio given in Eq. (31)

The q - factor for the one-story frame is obtained from Eq. (22), taking N = 1.

$$q = \sqrt{1 + 4\overline{\eta}} \div \sqrt{4\overline{\eta}} \tag{41}$$

Then, the q - factor influenced by P - δ effect is obtained as

$$q_{p\delta} = q \sqrt{\frac{|k'_{do}|}{|k'_{do}| + |k'_{p\delta}|}} \tag{42}$$

where the q - factor for the case without P - δ effect
the q - factor for the case with P - δ effect

The frame - A in the paper is characterized by the following parameters,

$$k'_{do} = -0.12, \quad \overline{\eta}_o = 1.90$$

Taking N = 1 and n = 12, the values of g_d and q for the frame - A are obtained as

$$g_d = 0.286, \quad q = 4.26$$

The upper - bound value of $|k'_{p\delta}|$ is around 0.1. Substituting $|k'_{do}| = 0.12$ and $|k'_{p\delta}| = 0.1$ into Eq. (42), $q_{p\delta}$ is found to be 2.87.
The obtained value well agrees with the conclusion found in the paper, and the importance

of the $P - \Delta$ effect together with the mechanical deterioration is confirmed.

5) A New method to Design Steel Frames Failing in Grobal Mode by Mazzolani and Piluso, 1994

In this paper, a method for designing moment resisting frame in global mode is presented. The global mode is the complete weak-beam type of collapse mechanism with all columns remaining elastic except for the plastic hinge formation at the column bases. The beam section properties are assumed to be known quantities in order to resist vertical loads.

The unknowns of the design problem are the column sections. The kinematic theorem of the plastic collapse is applied. The column sections are so determined as to satisfy the condition that the kinematically admissible multiplier of the horizontal forces corresponding the global mechanism (real mechanism to be attained)has to be the smallest among all kinematically admissible multipliers.

The method developed in a previous work by the authors is extended to include the $P - \Delta$ effect. Three types of mechanism with degrees of freedom of n_s (n_s being the number of stories) are assumed.

The similarity of the virtual displacement vectors in both horizontal and vertical direction enables the simple derivation of the multiplier of the horizontal forces influenced by the $P - \Delta$ effect.

The proposed design procedure stated in a clearest closed form can be applied straight-forwardly. The reliability of the method to attain the global mode has been fully examined by the authors.

One of the matters of concern is the sensitivity of the column sections to the scatter of the vector of horizontal forces $\{F\}$, because the horizontal force distribution may change in the practical situation during earthquakes.

6) Resistance Deterioration during Earthquakes Compiled in Earthquake Response Test Database on Steel Frames by Ohi and Takanashi, 1994

In this paper, thirty-four pseudo-dynamic tests on single-story steel moment-resisting frames are presented as a database. This elaborate set of various test records can afford an invaluable realistic information next to that of actual damage caused by earthquakes.

The author claims that the presented database can be used for education of structural engineers and students, verification of analytical model for response prediction, and consensus making about limit states.

The classification of structural damage is made with the emphasis on the deterioration of lateral resistance as follows

   A : no deterioration   for $\gamma = 0$
   B : early deterioration   for $0.1 > \gamma > 0$
   C : considerable deterioration for $1.0 > \gamma > 0.1$
   D : complete collapse   for $\gamma = 1.0$

γ is defined to be the amount of resistance deterioration divided by the maximum lateral resistance.

In Table 1 in the paper, a brief summary of the database is given. In order to make a precise estimate of the q - factor, the information given in the Table is not sufficient. However a rough estimate of the damage level is possible based on the nondimensionalized slope in the degrading part and the strength level of each model given in the Table.

The load deflection curve in the inelastic range is assumed to be a linear one as is indicated in Fig. 11.

Then, the q -factor is obtained by applying Eq. (41) as follows

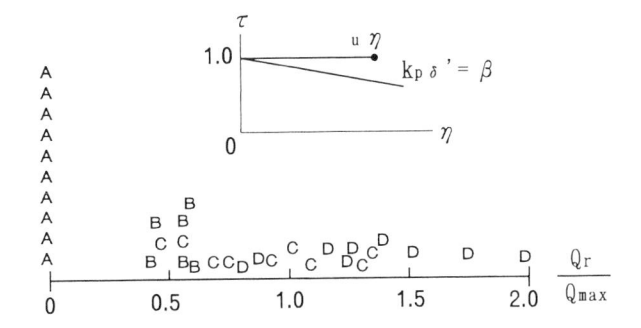

Fig. 11   Correspondence of Damage Level to Q_r/Q_{max}

$$q = \sqrt{1 + \frac{4 g_d}{|\beta|}} \qquad (43)$$

where β : the nondimentionalized negative slope relative to the slope in the elastic range

The required strength to prevent the total collapse of the structure Q_r is expressed as

$$Q_r = \frac{Q_e}{q} \qquad (44)$$

where Q_e : the required strength for the elastic system under the same earthquake

The preserved maximum strength level Q_{max} / Q_e is given in the Table.
Then, Q_r / Q_{max} is expressed as

$$\frac{Q_r}{Q_{max}} = \left(\frac{Q_e}{Q_{max}}\right) / q \qquad (45)$$

The damage level of tested frames are indicated on the scale of Q_r / Q_{max} in Fig. 11.
It is clearly seen that the damage level well corresponds to the preserved strength level relative to the required strength level estimated by means of the q - factor. Thus, the database is very useful to confirm the accuracy of the q - factor.

7) Remarks on Behaviour of Concentrically and Eccentrically Braced Steel Frames by Mazzolani, Georgescu and Astaneh, 1994

In this paper, very important remarks on the behavior of braced frames are presented.
It is pointed out that the concentrically braced frames have fundamental drawbacks in the energy dissipating capacity and the strength keeping capacity and should be basically designed on the basis of the elastic design.
To overcome the deficiency of CBF, two measures are discussed;
   1) Using an assemblage of two or more rigid CBF connected by weak plastic beams, a large range of structural systems balancing stiffness and ductility could be obtained.
   2) Eccentrically braced frames with links can dissipate a large amount of energy.
In these structures, all the other structural component outside the beam-ends or the links

must be designed to remain essentially elastic. To satisfy this condition and avoid undesirable failure mechanism, the analytical approach proposed in the fifth paper is mentioned to be an effective measure for attaining the global mechanism in braced frames.

As for the energy absorption capacity, the concentrically braced frame should be considered to be not so poor.

In Fig. 12, an example of diagonally braced frames is shown.

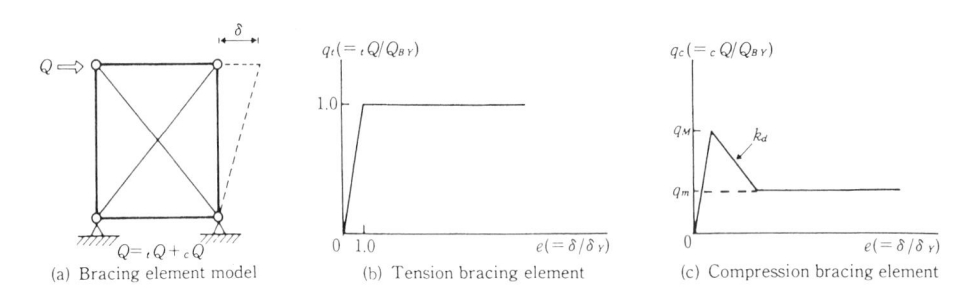

Fig. 12　Deformation Characteristics of Diagonal Bracings

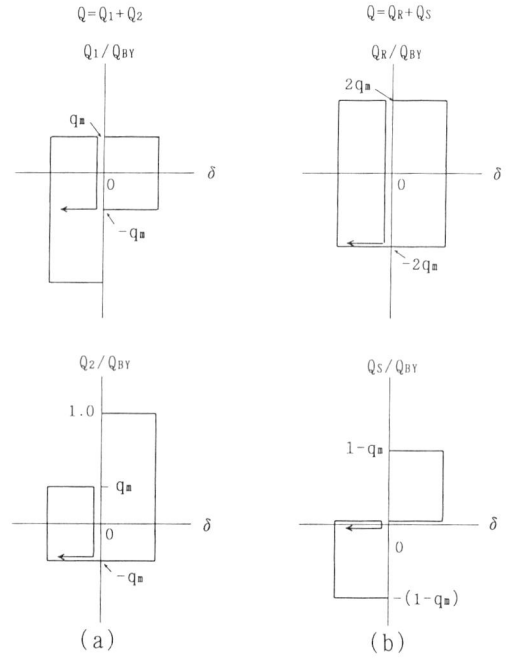

Fig. 13　Decomposition of Restoring-Force Characteristics

$_tQ$ and $_cQ$ are the lateral resistances of the tensile bracing and the compressive bracing respectively.

Q_{BY} is the resistance of the tensile bracing at the elastic limit. e is the nondimensionalized lateral deflection relative to the yield deflection. The simplified load-deflection curve of the compressive bracing shown in Fig. 12 (c) is characterized by the buckling of the bracing. Neglecting the degrading part, the curve can be approximated by a bilinear curve with the level of plateau of q_m. Then, the simplified load-deflection curve with neglect of elastic deflection under the repeated loading is obtained as is shown in Fig. 13 (a).

The total horizontal resistance, consisting of two components shown in Fig. 13 (a), can be decomposed into elastic perfectly plastic restoring force characteristics and the slip types as shown in Fig. 13 (b). For this combined systems with the elastic-perfectly plastic element and the slip element, it has been found the following relationship.

$$\bar{\mu} = \frac{\bar{\eta}}{4} \tag{46}$$

It stands for two cycles of hysteresis loop with the amplitude of $\bar{\mu}$.

Then, the energy absorption capacity of the braced frame W_{pB} is written as

$$W_{pB} = \left[4\, q_m \cdot 4\,\bar{\mu} + (1 - q_m) \cdot 2\,\bar{\mu} \right] Q_{BY} \tag{47}$$

On the other hand, the energy absorption capacity of the moment resisting frame equipped with the elastic-perfectly plastic restoring force characteristics and the strength level of $2\,W_{BY}$ is expressed as follows.

$$W_M = 8\, Q_{BY}\, \bar{\mu} \tag{48}$$

The ratio of the q - factor of the braced system to that of the moment resisting frame is obtained as follows, referring to Eq. (41)

$$\frac{q_B}{q_M} = \sqrt{\frac{7\, q_m + 1}{4}} \tag{49}$$

where q_B : the q - factor for braced frames
       q_M : the q - factor for moment resisting frame

q_m can be represented by the following value within the range of $\lambda_e < 2.5$.

$$q_m = 0.2 \tag{50}$$

λ_e is defined by

$$\lambda_e = \sqrt{\varepsilon_Y}\, \lambda$$

where ε_Y : the yield strain of bracing
       λ : the slenderness of bracing

Substituting $q_m = 0.2$, q_B / q_M is obtained as

$$\frac{q_B}{q_M} = 0.77 \tag{51}$$

As far as the q - factor of the single-story frame is concerned, the concentrically braced frame is not so defective compared to the moment-resisting frame.

Nevertheless, the multi-story concentrically braced frames is liable to be subjected to severe damage concentration due to partial mechanism and the value of n for them to be applied in Eq. (22) should be less than 12.0.

8) Behavior of 'K' Braced Frames at Load Seismic Type by Georgescu, and Gosa, 1994

In this paper, the analytical method for the $P - \Delta$ diagrams of K-braced frames is presented, using the model with as initial imperfection and a plastic hinge in the middle of steel bar.

In all plastic fields the deformation of the bar is kinematic, keeping the plastic hinge a plastic bending moment reduced by the axial force.

The results of analysis were compared with experimental results of several types of K-braced frames. The tests results have shown that K-shaped braced steel frames possesses a certain capacity to dissipate energy under seismic loading and have confirmed the hypothesis of the mathematical model.

In order to obtain a good behavior, a slenderness of the diagonals $\lambda < 100$ and the use of thick gusset plates with local strengthening are recommended.

Obtained results are consistent with the applied value of q_m in Eq(50), and the possibility of further application of eccentrically braced frames is suggested.

6. Conclusion

An overview of the global ductility was presented and contributions at the Workshop were summarized and added with comments for discussions and common understanding of the strength reduction factor.

Major points are summarized as follows
1) The development of computational tools has made it possible to obtain structural responses for any type of structures. Nevertheless, it seems more necessary to construct a logical framework for understanding the global ductility, parallel to the direct numerical analysis.
2) Although the numerical approach may be realistic and practical, it leads to too versatile structural behavior to make a reasonable estimate of the q - factor.
3) In estimating the q - factor, the loading effect of earthquakes and the structural behavior must be clearly separated.
4) The separation of the loading effect of earthquakes can be made by introducing the energy spectra in place of the conventional peak acceleration spectra, because the total energy input into inelastically stressed structures is very stable, only dominated by the total mass and the fundamental period of structures.
5) In applying direct numerical approach the intensity of an earthquake should be scaled by the level of the total energy input in place of the peak ground acceleration of it, since the energy spectrum of an actual earthquake record may be greatly influenced by the extent of plastification of a structure and thus, is not proportional to the peak ground acceleration.
6) A logical frame work for the q - factor can be constructed on the basis of the balance between the energy input and the dissipated energy of a structure.
7) The global ductility of structures mainly depends on the manner of damage distribution (concentration), while the local ductility depends on the mechanical structural behavior.
8) In order to derive the q-factor based on the numerical analysis, the assumption on the constancy of the maximum deflection response must be a strong rule of thumb. However, the assumption must be examined in the case of multi-story frames, and the degrading systems.

9) To control the damage distribution, a sense of the capacity design is very effective. In order to realize a desired yielding mechanism, the kinematic theorem of plastic collapse can afford a simple and reliable tool covering the P-δ effect.
10) The P-δ effect caused by the gravity loading and the sidesway deflection is of the primary importance in the estimate of q-factor in the case of multi-story frames and the degrading systems such as light weight steel frames.
11) The global ductility must be recognized on the synthesis of a wide range of experienced structural damage. The database on the responses of structures up to their collapse state must be openly exchanged among researchers to arrive at consensus on the limit state of the structure.
12) The braced frames can be utilized for the purpose of realizing preferable global mechanism of frames by keeping them elastic.
13) The energy absorbing capacity of the concentrically braced frames is widely recognized to be far poorer than that of the moment-resisting frames, but it should not be disregarded in estimating the q-factor realistically.

7 References

Akiyama, H. (1985) : **Earthquake-Resistant Limit-State Design for Buildings,** University of Tokyo Press, 1985

Architectural Institute of Japan (1985) : **Recommendations for the Design and Fabrication of Light Weight Steel Structures**

Akiyama, H. and Kitamura, H. (1992) : Design Energy Spectra for Specific Ground Conditions, **International Workshop on Recent Developments in Base-Isolation Techniques for Buildings, Tokyo, April.**

Calderoni, B., Ghersi, A. and Rauso, D. (1994) : Statistical Evaluation of the Behavior Factor for Steel Frames, **STESSA '94**

Gioncu, V., Mateescu, G. and Iuhas, A. (1994) : Contribution to the Study on the Ductility of Unbraced Frames, **STESSA '94**

Vayas, I., Syrmakezis, C. and Sophocleous, A. (1994) : A Method for the Evaluation of the Behavior Factor for Steel Regular and Irregular Buildings, **STESSA '94**

Calderoni, B., DeMartino, A., Ghersi, A. and Landolfo, R. (1994) : On the Seismic Resistance of Light Gauge Steel Frames, **STESSA '94**

Mazzolani, F.M. and Piluso (1994) : A new Method to Design Steel Frames Failing in Global Mode including $P = \Delta$ effects, **STESSA '94**

Ohi, K. and Takanashi, K. (1994) : Resistance Deterioration during Earthquakes Compiled in Earthquake Response Test Database on Steel Frames, **STESSA '94**

Mazzolani, F.M., Georgescu, D. and Astaneh, A. (1994) : Remarks on Behaviour of Concentrically and Eccentrically Braced Steel Frames, **STESSA '94**

Georgescu, D. and Gosa, O. (1994) : Behavior of 'K' Braced Frames at Load Seismic Type, **STESSA '94**

23 STATISTICAL EVALUATION OF THE BEHAVIOUR FACTOR FOR STEEL FRAMES

B. CALDERONI and D. RAUSO
Istituto di Tecnica delle Costruzioni, University of Naples, Naples, Italy
A. GHERSI
University of Catania, Italy

Abstract

The evaluation of the behaviour factor q by means of a statistical procedure, based on the analysis of the dynamic inelastic response of schemes designed with different q-values to a set of accelerograms, is here proposed. The problems connected to the method (definition of the seismic input, design of frames, criteria for evaluation of q) are discussed. The application of the procedure to some steel frames points out some preliminary conclusions and outlines future directions for research.

Keywords: Dynamic response, Inelastic behaviour, Design of frames, Behaviour factor, Overstrength.

1 Introduction

As it is well known, every seismic code bases its prescriptions on the assumption that, during severe earthquakes, any designed structure will be able to dissipate a large part of the energy input through plastic deformations. The energy dissipation capacity of the structure is taken into account by performing a linear analysis based on a reduced spectrum (normalized design spectrum) the ordinates of which are significantly lower than the ones of the elastic response spectrum. The European seismic code (Eurocode 8 - EC8) accomplishes such reduction by means of the behaviour factor q, which is defined as "an approximation of the ratio of the seismic forces, the structure would experience if its response were completely elastic, to the minimum seismic forces that may be used in design with a conventional linear model still ensuring a satisfactory response of the structure". Values of the behaviour factor are given, depending on many characteristics of the structure, like material, ductility level of the section, structural scheme, regularity. No indication is yet provided to allow the designer to calculate by himself correct values of q in specific situations.

Behaviour of Steel Structures in Seismic Areas. Edited by F. M. Mazzolani and V. Gioncu. Published in 1995 by E & FN Spon, 2-6 Boundary Row, London SE1 8HN. ISBN: 0 419 19890 3.

Many studies have been up to now carried out to correlate values of behaviour factor to structural characteristics. Veletsos and Newmark (1960) proposed to evaluate the response spectrum for elastic-perfectly plastic single degree of freedom systems (with $T > 0.5$ s) dividing the corresponding ordinates of the elastic response spectrum by the value of the ductility factor. More recently, Mahin and Bertero (1981) tested the reliability of such criterion, showing that the inelastic response is particularly sensitive to the actual seismic input. Nevertheless, many simplified methods for the evaluation of q in complex structures were proposed on this assumption, based on the evaluation of the horizontal displacement and of the global ductility by means of dynamic or static elastoplastic analyses (Setti, 1985, Cosenza et al., 1986). A different approach proposed by Kato and Akiyama (1982), based on the energy concept, assesses the safety of a structure by comparing its energy dissipating capacity with the earthquake input energy.

A comparative presentation of these and other simplified methods, performed by Guerra et al. (1990), showed an extremely wide dispersion in the values of q obtained for a set of regular steel frames; q-values were furthermore strongly different from the values provided by the dynamic inelastic response of the same schemes to three artificial accelerograms. Calderoni et al. (1991, 1993) analysed the non-linear response of r.c. and steel frames, designed in accord to EC8 prescriptions, to some historical accelerograms scaled with increasing peak ground acceleration (PGA), so as to individuate the values of PGA corresponding to the yielding of the most stressed section ($a_{g,y}$) and to the collapse ($a_{g,u}$), conventionally defined as the reaching of the ultimate plastic rotation in a section. The authors pointed out the inconsistency of the commonly used definition of q-factor as the ratio $a_{g,u}/a_{g,y}$, mainly owed to the fact that $a_{g,u}$ is related to the overstrength of all sections while $a_{g,y}$ is strongly affected by the behaviour of a single section (the most stressed one); alternative definitions of q provided interesting values, which were however strictly related to the characteristics of each accelerogram and thus unable to settle an unique safe value of the behaviour factor.

The careful overview of the technical literature, here briefly recalled, leads to the conclusion that, in spite of the many interesting ideas and methods proposed, a lot of work must still be done to sufficiently assess the values of the behaviour factor. A consistent procedure to evaluate q, i.e. the statistical analysis of the response of structural schemes to homogeneous sets of accelerograms, was usually considered cumbersome and discarded for this reason. This approach appears now to be possible, thanks to the progressive enhancement in the available computing capacity. The present paper aims therefore at examining the problems connected to it (definition of the seismic input, design of frames, criteria for the evaluation of q) so as to draw the most correct way to obtain statistically significant values of the behaviour factor.

2 Definition of seismic input and response spectra

For a proper design of a structure in a specific site, it would be necessary to have two statistically significant sets of accelerograms, corresponding to historical earthquakes with a short and a long return period respectively (the first one to check the serviceability limit state of the structure, the second one to grant it against collapse). For research purpose, artificially generated accelerograms are often used, because they can correspond to a given elastic response spectrum and it is thus easy to obtain a large set of homogeneous accelerograms. However, the uncertainty on the reliability of artificial-quakes energy contents led us to prefer historical events. We therefore examined the Enea-Enel data base which collects more than one thousand of recordings of the earthquakes occurred in Italy during the last twenty years. Most of them were discarded, because of their weakness (PGA< 0.05 g). The other ones were divided into sets, basing on site soil, magnitudo, PGA value, shape of response spectrum, energy spectrum. The number of recordings with high PGA was not sufficient to create a statistically significant set of strong earthquakes. For this reason we finally selected only one set of 30 accelerograms (tab. 1) which presented an average value of PGA equal to 0.16 g. The elastic response spectra of the selected accelerograms were statistically analysed, evaluating at each period T the mean (plotted in the average response spectrum, fig. 1), the standard deviation, the coefficient of variation and the frequency distribution. The shape of the mean spectrum is very similar to that of the elastic spectrum proposed by EC8 for soil site A; the coefficient of variation often attains values greater than 0.5, showing a large dispersion of samples, in spite of our efforts for homogeneity; the frequency distribution proves to be approximately lognormal. Basing on the above set, we defined the set of strong earthquakes, scaling the actually recorded ground accelerations by a factor 2.2 so as to get a mean PGA equal to 0.35 g, and the set of weak earthquakes, reducing the strong ones by a factor 4, i.e. multiplying the actual values by 0.55. Some problems may arise in consequence of this procedure, because the energy contents of actual weak and strong earthquakes are usually not proportional to their PGA.

Table 1 - Reference number of the selected accelerograms in the Enea-Enel data base

m032.ew	m032.ns	m038.ns	m143.ew	m143.ns	m152.ew
m152.ns	m153.ew	m156.ew	m156.ns	m168.ew	m168.ns
m169.ew	m169.ns	m177.ew	m301.ew	m301.ns	m302.ew
m302.ns	m350.ew	m621.ew	m621.ns	m627.ew	m627.ns
m636.ew	m636.ns	m643.ew	m643.ns	m644.ew	m644.ns

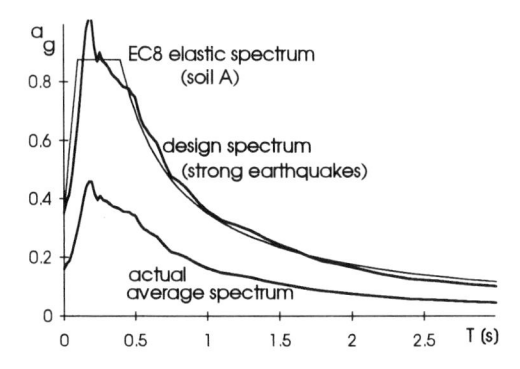

Fig. 1 - average elastic response spectrum of the selected accelerograms and design spectrum used in the analysis

Aim of future works will be the extension of the inquiry to wider seismic data bases, so as to overpass the present lack of strong motions; it must yet be noted that this is probably not the major problem and, moreover, that the main purpose of the paper is to assess a procedure, which may (or probably must) be used with different sets of accelerograms.

3 Design of steel frames

In order to evaluate the behaviour factor for specific structural schemes and given design criteria, many schemes (defined in geometry and vertical loads) must be proportioned by means of modal analysis using the elastic response spectrum reduced by different q-values; their inelastic dynamic response to the selected earthquakes must then be checked, so as to individuate the maximum values of q able to limit to an acceptably low value the probability of structural failure. The elastic spectrum must obviously correspond to the seismic input; the average spectrum, which covers about 50% of samples, has been used in the present work, according to EC8 prescriptions.

3.1 Second order effects

While proportioning a steel frame (which is a very deformable structural scheme) the second order geometrical effects play a very important role. Eurocode 3, Design of steel structures, checks the sensitivity of a frame to global buckling by means of the coefficient θ, defined at each storey (in the case of linear elastic analysis) as

$$\vartheta = \frac{\delta}{h}\frac{V}{H}$$

where δ is the horizontal relative displacement

and   h   is the storey height
       H   is the total horizontal reaction at the bottom of the storey
       V   is the total vertical reaction at the bottom of the storey.

Eurocode 8 asks to multiply the above value by a coefficient q_d to take into account the differences between the conventional elastic analysis and the actual inelastic response to strong motions; coherently to the criterion of ductility factor given by Newmark, q_d may be assumed coincident to q. The second order effects may be approximately accounted by multiplying the effect of horizontal loads by the coefficient $1/(1-\theta)$. The structure is not allowable when $\theta > 0.3$. It must be noted that, in the case of seismic design, the ratio δ/H do not depend upon the value of q; θ is therefore linearly related to q_d.

The numerical analyses performed pointed out that the limit value of θ often corresponds to q-values ranging from 4 to 6. The consequence of it are shown in fig. 2. The weight of the structure necessary to satisfy strength checks, without any limit to displacement, is drawn in function of q (curve a for first order analysis and b for second order analysis); the weight necessary to satisfy deformative limits, without any strength control, is also drawn (curve c). It is evident the value of q above which deformative limits become predominant. It must be noted that in this case the structure is oversized in respect of the design actions and the displacement itself is overestimated. We therefore believe more proper to assume the coefficient q_d as the ratio q/ρ being ρ the overstrength factor M_{Rd}/M_{Sd}, and we suggest to modify in this way the prescriptions given by EC8. The weight of the structure thus evaluated is nearly constant with q, as shown by curve d.

The statistical analysis of the dynamic inelastic response of steel frames, proposed in this paper, may be used to clarify the two possibilities which arise from the above considerations.

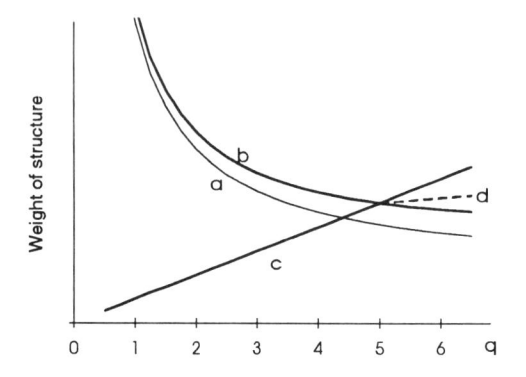

Fig.2 - weight of structural elements as a function of q

If the dynamic second order effect will be proved to be correlated to q in an approximately linear proportion, as assumed by the ductility factor criterion, no relevant savings may be obtained in design by using q-factors greater than a value which in some examined cases seems to be about 4. On the contrary, the value of behaviour factor $q=6$ presently proposed by EC8 for regular frames may be considered significant only if the second order effects will be less sensitive to the increase of q.

3.2 Design of comparable frames with different q-values

A common way to estimate the value of behaviour factor for a structural scheme is to analyse its response to seismic events scaled with increasing PGA. Although useful, the values so obtained do not correspond to the correct definition of q and do not allow to properly judge the second order effects. On the other side, some problems arise when proportioning a given scheme with different values of the behaviour factor. First of all, the necessity of using commercial sections leads to a remarkable oversizing of the structure, accentuated by the use of the capacity design criterion which asks to proportion columns basing on the resistant capacity of the beams; the actual strength of the frame is thus not exactly correspondent to the value of q. Secondly, the use of a greater behaviour factor leads to a structure more deformable, which presents an higher period T; the behaviour of the different frames is thus strongly affected by the difference in period. To overcome these problems, we fictitiously changed the strength of the material, assuming an unique value of f_y for beams and a different one for columns, so as to exactly attain yielding in the most stressed beam and column. All other sections are obviously oversized, as it always happens in actual steel frames. Further investigations will be later on carried out to evaluate the influence of such "spread" overstrength; our first numerical applications make it very relevant, inducing to warn about the self-sizing potentiality given by some computer programs, which might use iterative procedure to strongly reduce it.

3.3 Analysed schemes

A consistent statistical approach would require the analysis of a large number of schemes. Presently, in order to test the proposed procedure, we analysed only four frames, named A, B, C and D (fig. 3), each one proportioned with three different values of the behaviour factor (3, 4 and 5); the value $q=6$ was never adopted, because of the above discussed deformability limits. Cross sections and masses were defined for each frame so as to obtain structures with the same period, $T=1.5$ s. The first scheme, with the same sections but with slightly different masses, had already been examined in a previous paper (Calderoni et al., 1993) and will be later on named frame A'.

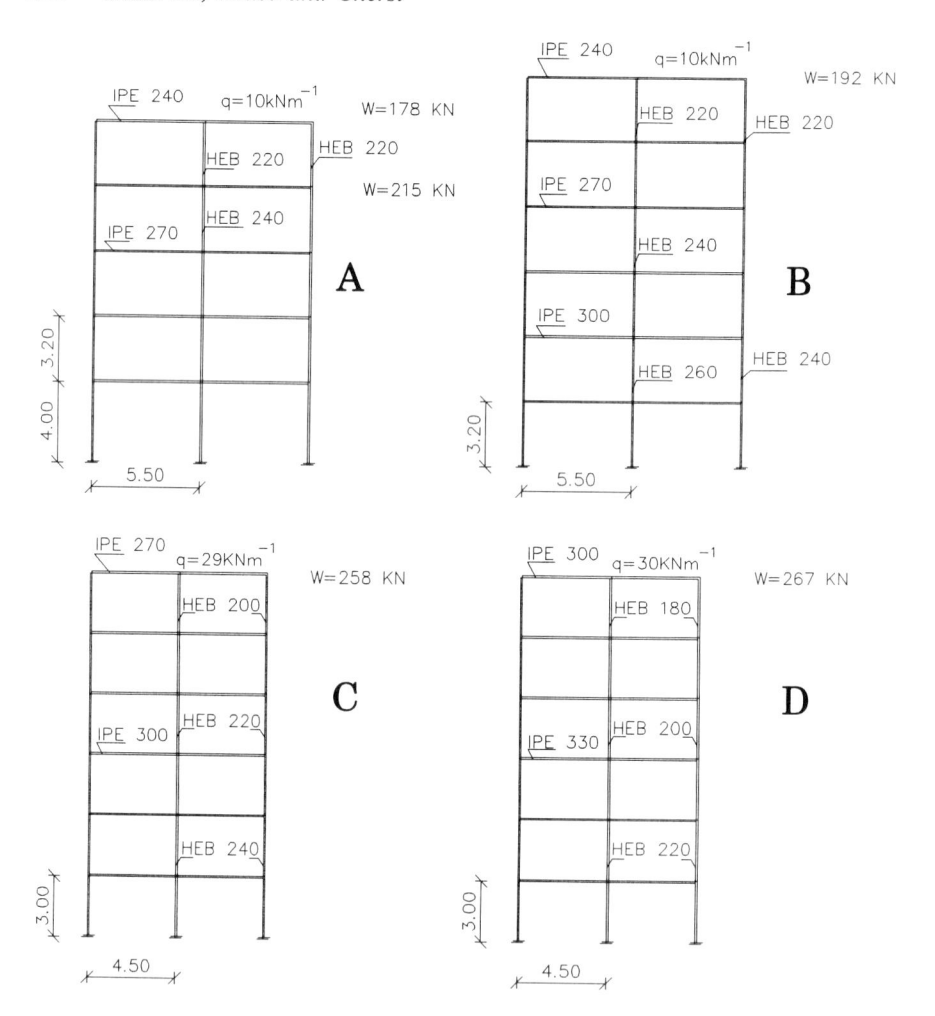

	frame A		frame B		frame C		frame D	
q	f_y [MPa]		f_y [MPa]		f_y [MPa]		f_y [MPa]	
	beams	columns	beams	columns	beams	columns	beams	columns
5	140.4	113.7	127.4	123.4	180.7	196.6	141.9	247.1
4	154.0	123.9	139.3	130.8	195.0	206.9	154.2	264.7
3	177.6	141.7	160.3	143.8	219.5	224.5	175.2	294.7

Fig. 3 - loads, geometrical and mechanical characteristics of the examined schemes

4 Dynamic analysis

The numerical analysis has been performed by means of the well-known code DRAIN-2D (Kanaan and Powell, 1973), which integrates step-by-step the differential equations of motion corresponding to a given seismic input, taking into account the possibility of the yielding of the end of the members. The ultimate rotation of plastic hinges has been evaluated by means of empirical and semi-empirical methods (Mitani and Makino, 1980, Mazzolani and Piluso, 1992); the preliminary numerical tests showed that the differences between the values provided by the two methods are not relevant on the statistical evaluation of the failure probability. Each set of analyses consists in the evaluation of the dynamic inelastic response of the scheme under examination to the 30 selected accelerograms. The results are then statistically analysed, pointing out:
- percentage of cases in which the yielding or the failure has been detected at least in one section (in beams or in columns);
- percentage of yielding and failure for each section;
- mean value of the ratio $\varphi_{pl}/\varphi_{pl,u}$ of plastic over ultimate rotation, for each section;
- mean value of the plastic level Σ, defined as the sum of the ratios $\varphi_{pl}/\varphi_{pl,u}$ extended to all beams or columns;
- mean value of the horizontal displacement of each floor.

Just as a starting point, a series of elaborations has been carried out on the frame A', which had been designed using EC8 spectrum for soil B and ground acceleration 0.25 g and numerically tested (Calderoni et al., 1993) by means of two historical Italian seismic recordings (Bagnoli Irpino and Tolmezzo) scaled with increasing PGA. The same procedure (many sets of statistical analyses, in which the selected accelerograms are scaled by increasing factors) has been used, although the results may be only partially significant because of the slight differences between the design spectrum used and the mean spectrum of the selected earthquakes. The statistical analysis shows a 10% probability of yielding at a_g=0.10 g and a 50% at a_g=0.25 g (comparable to the values $a_{g,y}$=0.08 g obtained for Bagnoli Irpino and $a_{g,y}$=0.18 g for Tolmezzo); the probability of failure is 10% at a_g=0.35 g and 20% at a_g=0.50 g (while $a_{g,u}$ was 0.51 g for Bagnoli Irpino and 1.02 g for Tolmezzo). The distribution of sections with high probability of yielding is analogous to those provided by the deterministic analysis. A failure probability above 10% has been detected at the ends of the beams of first and second level and at the bottom of the central column (fig. 4).

The procedure we propose in order to define the value of behaviour factor is much more simple, because it only requires the evaluation of the dynamic response of some frames, designed on a single scheme with different q-factors, to the above defined set of strong motions. The analyses thus performed point out first of all the substantial effectiveness of the capacity design criterion. In all frames, the probability of yielding is very high (nearly 100%) in the beams, excluded those of the upper storey, while it is significant only at the bottom of the external columns and at the lower storeys of the central one (fig. 4). As a consequence, the plastic level Σ of the beams is many times greater than that of the columns. Nevertheless, the behaviour of frame A is clearly different from that showed by frames B, C and D.

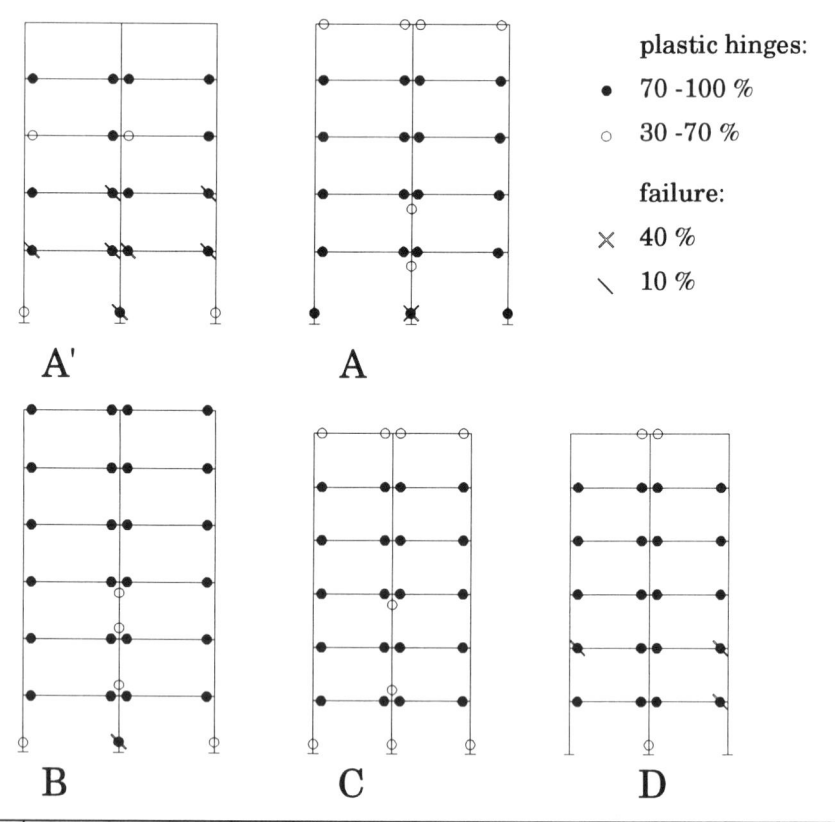

q	frame A plastic level Σ		frame B plastic level Σ		frame C plastic level Σ		frame D plastic level Σ	
	beams	columns	beams	columns	beams	columns	beams	columns
5	1.52	0.53	3.19	0.32	2.62	0.24	2.51	0.17
4	1.37	0.51	2.64	0.44	2.17	0.18	2.32	0.11
3	1.07	0.30	1.81	0.17	1.81	0.25	1.79	0.23

Fig. 4 - plastic hinges (when $q=5$) and plastic level of the examined frames

The failure probability of the first frame is high: about 40% when it is proportioned with $q=5$, 30% when $q=4$, 25% when $q=3$. The collapse always corresponds to the reaching of the ultimate plastic rotation at the bottom section of the central column; the probability of failure of other sections of columns or beams are extremely low. Such condition is confirmed by the values of Σ: the plastic level of beams is in fact only three times that of columns, and this last one is given by the yielding of only few sections.

On the contrary, the other frames have a probability of collapse of about 10% when they are designed with $q=5$ and this value only slightly decreases with q. Many beams and few columns present approximately the same low probability of failure. The plastic level of beams is now about ten times greater than that of columns.

From these results, judging allowable a small possibility of failure (10-15%), we may conclude that it is possible to assume $q=5$ in the schemes B, C and D but only $q=3$ in the case of scheme A.

To further investigate on such difference, the statistical analysis has been repeated using the set of weak earthquakes. Also in this case the behaviour is different: the probability of beam yielding is in any case remarkable (up to 40% when $q=5$), but no plastic hinge occurs in the columns of frame B, C and D while the bottom of the central column of frame A has the same probability to yield than the beams. A retrospective examination of the design calculations was very useful to enlighten the case. In the scheme A, the use of an unique section at the four bottom levels of the central column, together with the greater height of the first storey, made the bottom section of it to be the most stressed one. In the other frames the maximum stress was instead reached in the intermediate storeys. For this reason, that section had no overstrength in frame A ($\rho=1$), while in the other schemes it was $\rho=1.3 \div 1.5$, leading to a different behaviour in spite of the same careful respect of the code prescriptions.

5 Conclusions

The proposed procedure appears able to analyse easily and effectively the dynamic behaviour of structures, putting in evidence any lack which may reduce their capability to resist to strong motions. The extension of the investigation to a large number of schemes will be able to assess on a strong base the values of the behaviour factor and to check the previsions of simplified methods. The examined cases point out some preliminary conclusions and future lines of research:
- the value of the behaviour factor for steel frames is strongly conditioned by the deformability and by the global stability problems; for this reason, in spite of the actual ductility of such structures the value $q=6$ might non be easily utilizable;

- the spread overstrength of the sections plays a very relevant role, which has to be better clarified and quantified in future;
- the design criteria, although basic and commonly valid, might in some cases not be sufficient to grant the required inelastic behaviour.

The authors thank ing. Zila Rinaldi who collaborated to the numerical analysis during the preparation of her graduation thesis.

References

Calderoni, B., Ghersi, A., Landolfo, R. and Mazzolani, F.M. (1991) Esame comparativo del comportamento non lineare di telai sismoresistenti. **Atti del 5° Convegno nazionale "L'ingegneria sismica in Italia"**, Palermo, 285-298.

Calderoni, B., Landolfo, R., Lenza, P. and Mazzolani, F.M. (1993) Comparative analysis of seismic elastoplastic behaviour of reinforced concrete and steel frames. **Proc. 10th World Conference on Earthquake Engineering**, Madrid, vol.7, 4025-4030.

Cosenza, E., De Luca, A., Faella, C. and Mazzolani, F.M. (1986) On a simple evaluation of structural coefficients in steel structures. **Proc. 8th European Conference on Earthquake Engineering**, Lisbon, vol.3, 6.1/41-48.

Guerra, C.A., Mazzolani, F.M. and Piluso, V. (1990) Evaluation of the q-factor in steel framed structures: state-of-art. **Ingegneria sismica**, 2, 42-63.

Kanaan, A.E. and Powell, G.H. (1973) DRAIN-2D, a general purpose computer program for dynamic analysis of inelastic plane structures. **University of California**, Berkeley, Report no.UBC/EERC-73/6.

Kato, B. and Akiyama, H. (1982) Seismic design of steel buildings. **Journal of the Structural Division**, ASCE, vol.108, ST8, 1709-1721.

Mahin, S.A. and Bertero, V. (1981) An evaluation of inelastic seismic design spectra. **Journal of the Structural Division**, ASCE, vol.107, ST9, 1777-1795.

Mazzolani, F.M. and Piluso, V. (1992) Evaluation of the rotation capacity of steel beams and beam-columns. **Proc. 1st Cost C1 Workshop**, Strasbourg.

Mitani, I. and Makino, M. (1980) Post local buckling behaviour and plastic rotation capacity of steel beam-columns. **Proc. 7th World Conference on Earthquake Engineering**, Istanbul. vol.6, 493-500.

Setti, P. (1985) A method to compute the behaviour factor for constructions in seismic zones. **Costruzioni Metalliche**, 3, 128-139.

Veletsos, A.S. and Newmark, N.M. (1960) Effect of inelastic behaviour on the response of simple systems to earthquake motions. **Proc. 2nd World Conference on Earthquake Engineering**, Tokio, vol.II, 895-912.

24 CONTRIBUTION TO THE STUDY OF THE DUCTILITY OF UNBRACED FRAMES

V. GIONCU, G. MATEESCU and A. IUHAS
Building Research Institute, INCERC, Timisoara, Romania

Abstract
A rational procedure for the determining the q-factor for one and multi-story frames is presented based on the non - linear dynamic (time-history) analysis, using DRAIN-2D computer program. The collapse load is determined corresponding to the attainment of ultimate plastic rotation in beams or columns, obtained through of DUCTROT-93 computer program. The different parameters, as accelerogram type, natural vibration period, axial forces, story number, degradation etc., are analized. For the interaction q-θ some simple relationship are proposed depending on the position of plastic hinge in structure.
Keywords: Global ductility, Structural behaviour factor, Accelogram, Natural vibration period, Ultimate plastic rotation.

1 Introduction

The key of design approach of a structure in the seismic zone is the controlled dissipation of the seismic energy input through plastic deformation. The value of these deformations must be limited according the local and global ductility, so that, even serious damages are produced, the collapse of structure be prevent.
   With a view to consider this condition, in the codes the design forces are derived from a linear elastic design response spectrum, reduced by means of a factor q, namely structural behaviour factor which takes into account the dissipative capacity of the structure. The values of this factor depend on the type of accelerogram acting at the base of structure, the natural vibration period, the plastic redistribution capacity, the type of collapse, the local ductility, the type of connections etc (Gioncu, Mazzolani, 1994). Due the difficulties to estimate the influence of these factors, the codified values are determined as a function of the structural type only. Thus, the generic q-factor given in codes is not able to determine the actual behaviour of a structure.

The state-of-art of q-factor evaluation in steel framed structures was presented by Guerra et al (1990). In order to compare the methods, seven frames with different configurations are examined. A great scatter in the q-factors values are observed, so it was impossibly to recommend the best method. In this situation, the single reliable method is the non-linear dynamic analysis, called also the time-history analysis, which is using a real accelerogram approximating the seismic conditions of the site. The frame collapse must be defined as the reaching of the ultimate plastic rotation at the end of the most important member of the frame.

In the paper, using DRAIN-2D and DUCTROT-93 computer programs, a series of inelastic analyses were performed to evaluate the q-factor and the possibility to obtain a simple method for the evaluation of this factor.

2 Selected structures

Some typical structures were selected for the analysis. The geometry was chosen to be typical for an actual unbraced frame. The result of this selection is presented in Fig.1: three types of one story frames (A-type with hinged connections, B-type with rigid connections, C-type mill - building with traveling crane bridge) and multi-story frames (D-type with one span and 2 stories and E-type with 2 spans and 2-6 stories). For one story frames, the variables were the accelerogram type, the span and height

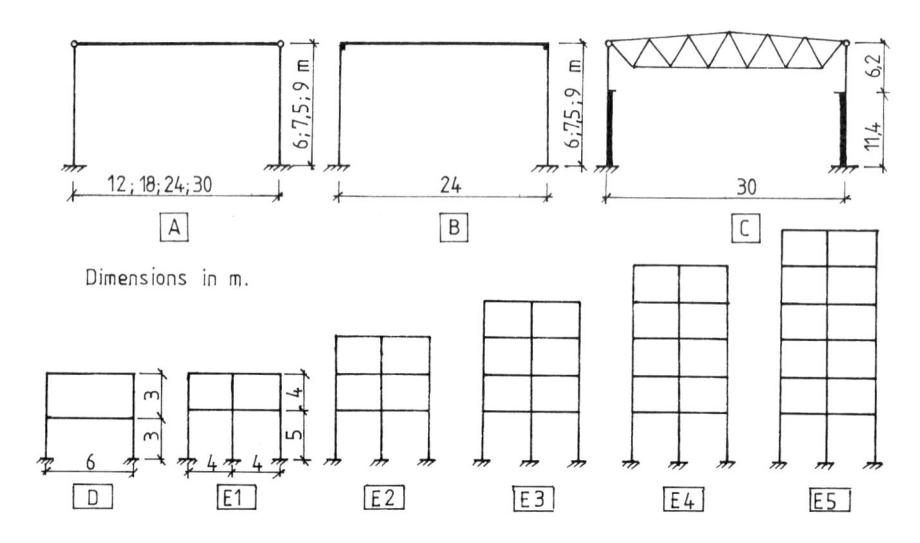

Fig.1. Geometry of frames

dimensions, column rigidity etc., while for multi-story frames, the story number and the degradation in rigidity after local buckling.

3. Computer model

The DRAIN-2D computer program (Powell, 1973) was used to analize the different frame configurations. The adopted method is based on the general definition of the behaviour factor q as

$$q = \frac{a_u}{a_p} \tag{1}$$

were a_u is the peak ground acceleration causing the collapse of the structure and a_p is the same peak causing the attainment of the first plastic hinge. Thus, the first step in using the computer program is the determination of the acceleration which produced the first plastic hinge

$$a_p = \lambda_e a_o \tag{2}$$

were a_o is the basic peak ground acceleration and λ_e, a numerical multiplier. After the determination of λ_e, the structure is analysed for

$$a = n \lambda_e a_o \qquad n = 2,3,\ldots \tag{3}$$

till the collapse of structure is attained.

$$a_u = \lambda_u a_o \tag{4}$$

Thus, the ultimate behaviour factor is

$$q_u = \frac{\lambda_u}{\lambda_e} \tag{5}$$

For defining the failure criteria, a different alternative can use the lateral displacements or the rotation of plastic hinges. Because for short free periods the use of displacements give no results, it is preferable to use the rotations. In this case the definition of ultimate plastic rotations is required. For this purpose, the DUCTROT-93 computer program (Gioncu et al., 1994, Gioncu-Petcu, 1993) is used. Thus, the adopted method performs the plotting the curve $q-\theta_p$ by means of Drain-2D computer program, and the limitation of this curve to the ultimate plastic rotation, θ_{pu}, determined through DUCTROT-93.

4. Computed results

A series of analysis were performed with the aim to obtain informations concerning the influence of different factors. Only some of these are presented in the following.

4.1. Effect of different acceleration records

The acceleration record has a considerable impact on the predicted response. In Fig.2 two records are plotted, one for EL CENTRO (with short time period) and VRANCEA-77 (with long time period). The two acceleration records produce dramatically different results in the plastic behaviour in spite of the practically same values for λ_e. It can see that the q_μ-factor for ELCENTRO is more than twice the one obtained for VRANCEA. Thus, for earthquake with long time period the structure, converted in a mechanism due to the plastic hinges, allows more ample deformations than those produced by the earthquake with short time period. It is very clearly that each country must assess its own q-factors, related to its accelerogram characteristics.

4.2. Effect of different cross-sections

The increase of the columns rigidity leads to an increase of the frame deformability, and in spite of the greater ultimate plastic rotation, the q_μ-factors diminish (Fig.3). This effect is due to the fact that the increase of the seismic forces are not counter-balanced by the increase of the geometrical characteristics of the cross-section.

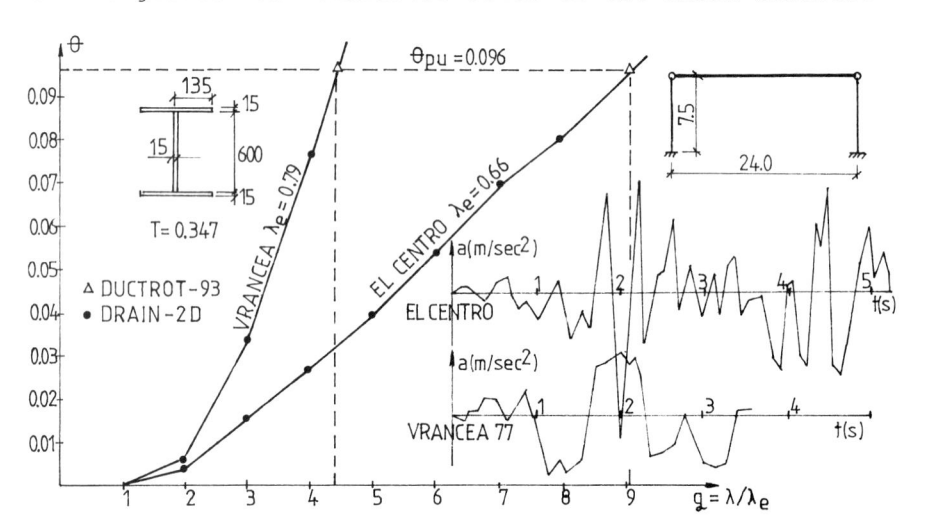

Fig.2. Effect of different acceleration records.

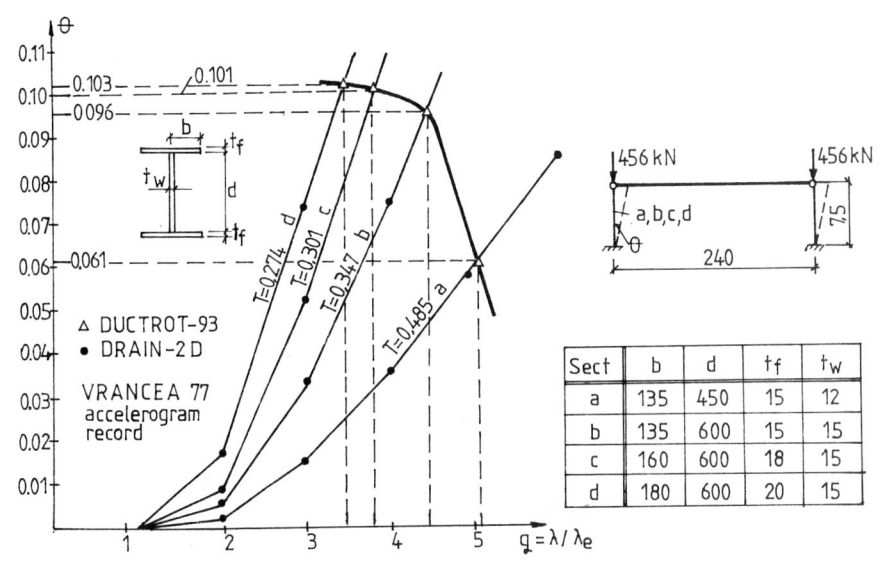

Fig.3. Effect of cross-section increase.

This is a very good subject for optimization studies.

4.3. Effect of natural vibration period

It is very well-known that the seismic forces are depending on the natural vibration period. The question refers in what manner behaves a structure transformed in a mechanism (without eigenvalues) under inertial forces which depend on structural response. From Fig.3 one can see that plastic rotations increase with the diminishing of natural vibration period. But this increasing is limited (Fig.4) to a period around about 0.2 sec., after which the plastic rotation values decrease. It is very important to underline that these values do not correspond to the one refering to the maximum elastic spectrum, which is larger.

4.4. Effect of axial forces

Fig.5 shows the influence of axial forces in columns, for one story frames with different spans. Naturally, the parameter λ_e for the first plastic hinge is increasing with the diminishing of axial forces, but the behaviour after the development of plastic mechanism is very different; the plastic rotations have larger amplitudes for low values of axial forces than the corresponding ones for high values. This fact is due to the level of accelerations when the structure becomes a plastic mechanism, greater for the first case than for the second.

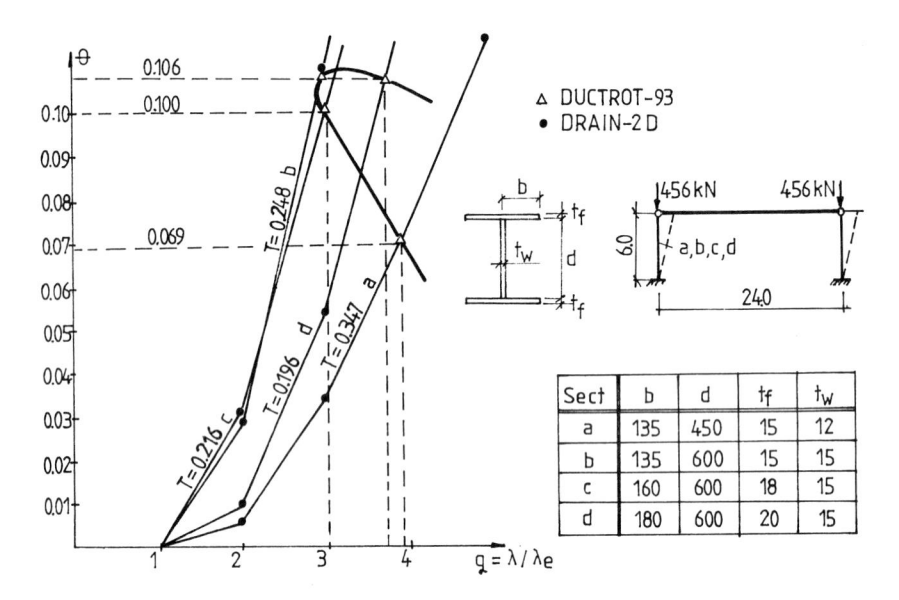

Fig.4. Effect of natural vibration period.

4.5. Effect of steel grade
The diagrams of Fig.6 confirm the above mentioned observation that the behaviour after the development of plastic mechanism depends on the acceleration level. For low values of yield stresses this level is more reduced than for the high values. So for high steel grade, Fe 510, the q_u-factor has lower values than for Fe 360 steel.

4.6. Effect of connections
The series of A-type frames with hinged beam-column connections [between beams and columns] are compared with the series of B-type frames with rigid connections. The results give some very contradictory conclusions because for A-6; B-6 frames the increase of q_u-factor is very important, while for the other frames, the effects are more reduced. These results are due to fact that for rigid connections the natural vibration periode decreases and therefore the inertial forces are increasing; in the same time primary moments and eccentricities of axial forces are produced, which decrease the ultimate plastic rotation. On the other hand, the acceleration level at which the structure becomes mechanism plays a very important role. Thus, the effects of connections are very difficult to be analysed with simple judgments, due to very contradictory factors which can induce misconceptions in the designing.

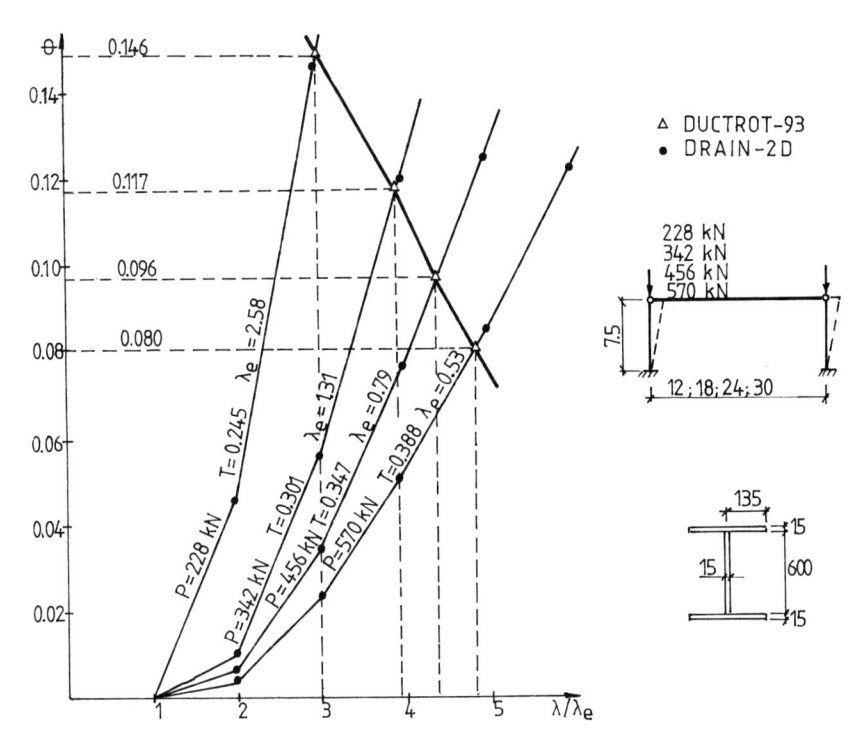

Fig.5. Effect of axial forces.

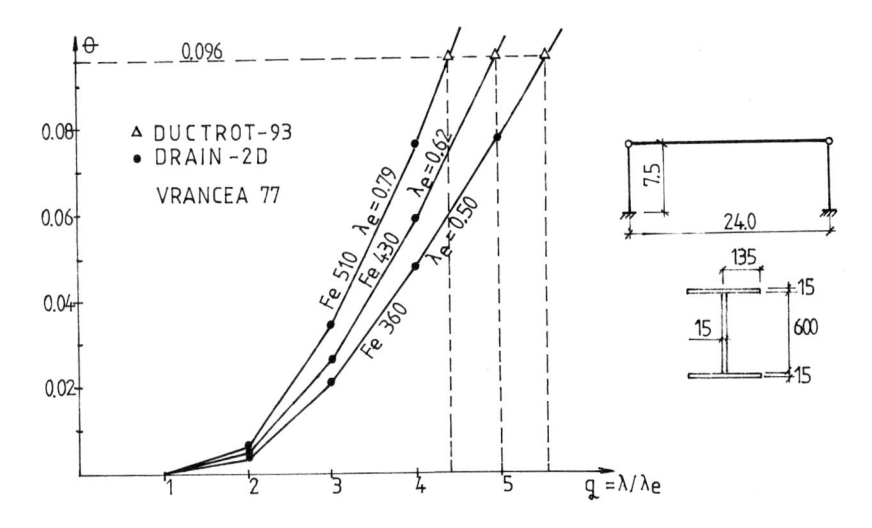

Fig.6. Effect of steel quality

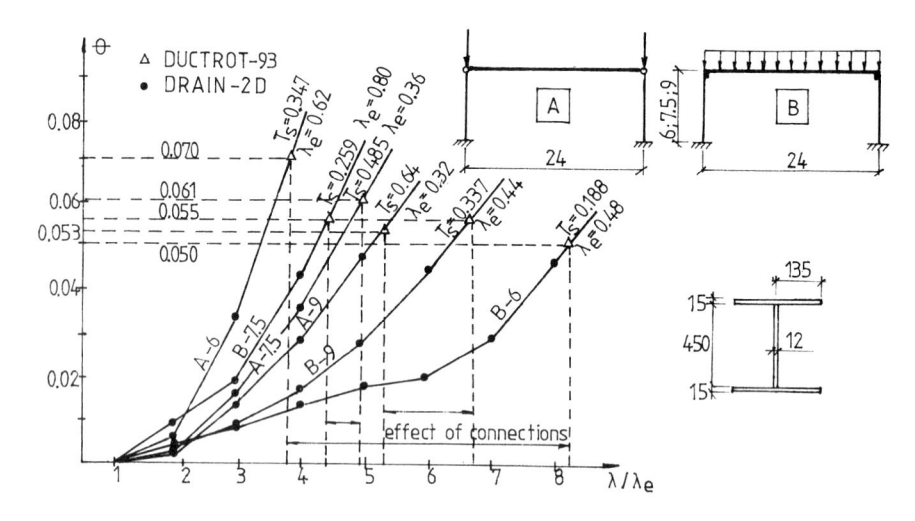

Fig.7. Effect of connections.

4.7 Behaviour of stepped columns

For mill-buildings with traveling crane bridge (Fig.8) the lower part of the column is sized for higher forces than those from seismic action. Thus only the upper part of the column is in danger to be involved in the collapse mechanism. Fig.8 shows that the effect is in opposite sense as for the unstepped column (Fig.3) the one with reduced

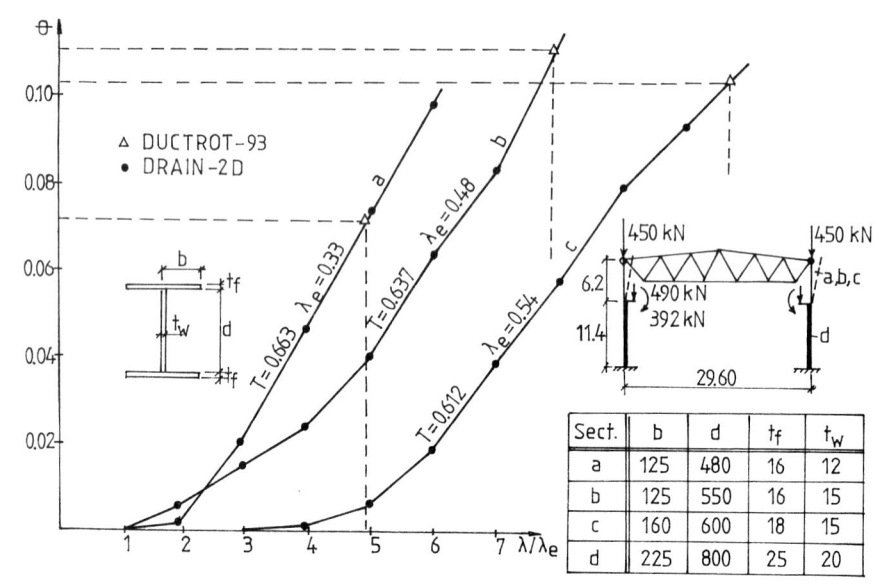

Fig.8. Stepped columns.

Sect.	b	d	t_f	t_w
a	125	480	16	12
b	125	550	16	15
c	160	600	18	15
d	225	800	25	20

section (a) has a greater deformations than the others (b and c). This effect is due to the fact that the movement of the structure is imposed by the whole column, not only by the upper part of the frame.

4.8 Effect of story number

Five multi-story frames were analysed and presented in Fig.9. The first plastic rotations for beams and central columns have been followed. One can see that the first plastic hinge is produced in beams and the second one in the middle columns, at the base. The ultimate plastic rotation is always attained in the beams. A very important observation refers to the form of q-θ curves: for beams,

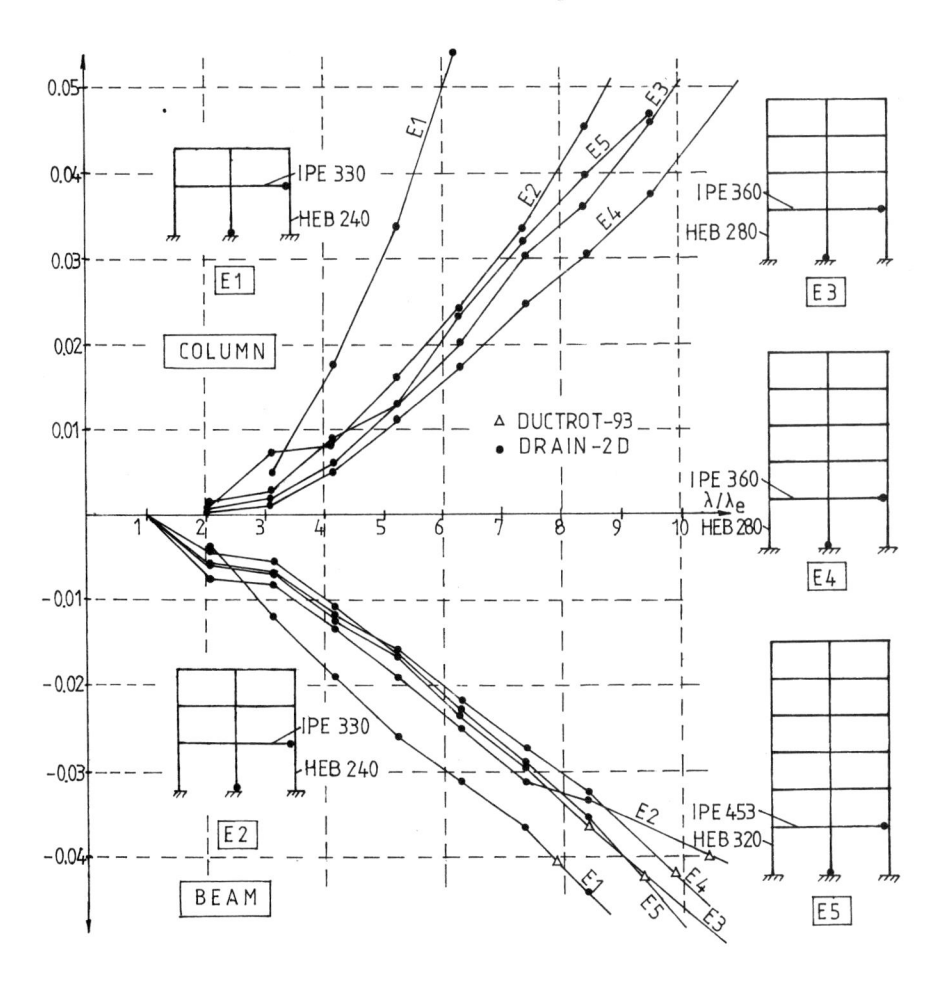

Fig.9. Effect of story number.

these curves are practically linearly, while for columns, in the presence of axial forces, the curves are nonlinearly.

4.9 Consideration of supplimentary rotation capacity

One could ask if the consideration of ultimate plastic rotation is not to drastically, because the sections continue the rotations after the attainment of these values. In Fig.10, the q_u-factors are determined for rotation capacities R_u, $1.1R_u$ and $1.2R_u$. One can see that the rotations after reaching the rotation capacity, R_u, are increasing dramatically, so the modification of q_u-factor is insignificantly.

4.10 Effect of degradation

Fig.10 shows the consequence of a degradation in the curve M-θ due to the low-cycle fatigue, premature buckling or cracks in welding zones. A comparison with a section without degradation shows a good correspondence before the mechanism, but after this structural changing, the differences are very important, the increases of plastic rotation

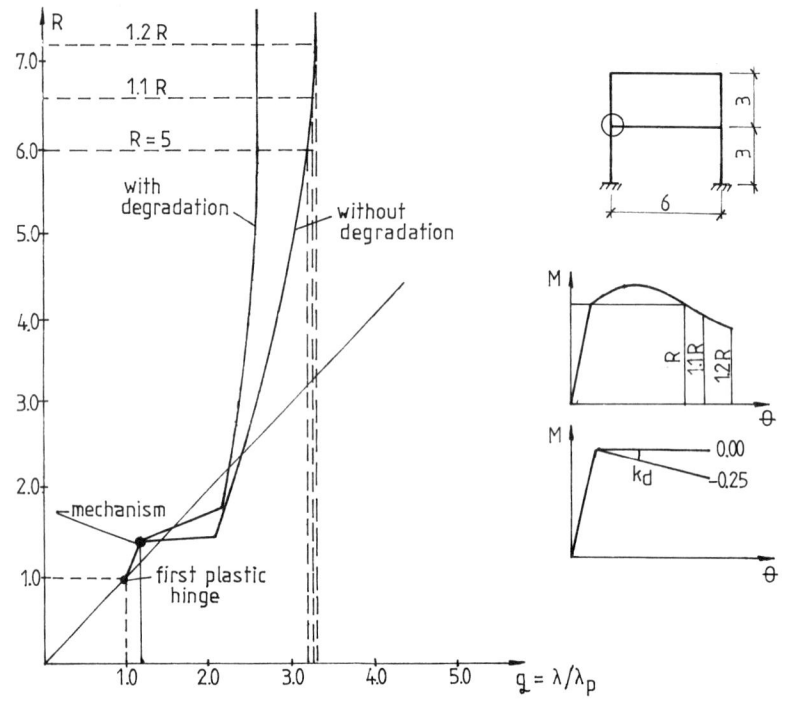

Fig.10. Effects of supplementary rotation capacity and degradation

are dramatically for the case with degradation. So, the
situations which can produce some degradations must be
avoided.

5. Conclusions

Based on these numerical tests it is possible to understand
the behaviour of a structure, converted into a plastic
mechanism, during a major seismic action. The major factor
of this behaviour is the level of the acceleration at which
this transformation is produced. There are also other
factors which influence this behaviour: natural vibration
period, plastic moment value, axial forces etc. It seems
that for the plastic hinge in beams, the q-θ relationship
is linearly, while for the columns, the variation is
parabolically. Some supplementary studies are required
for a complete control of the unbraced frames behaviour.

6. References

Gioncu, V., Mateescu, G., Iuhas, A.(1994), Contributions
  to the study of plastic rotational capacity of I steel
  sections, in **Proceedings of STESSA,94,** Timisoara.
Gioncu, V., Mazzolani, F.M. (1994), On the evaluation of
  the global ductility of steel structures: Proposals of
  codification, in **Proceedings of STESSA,94,** Timisoara.
Gioncu, V., Petcu, D. (1993), DUCTROT-93: Plastic rotation
  capacity of H-section steel beams and beam-columns.
  Computer Program, INCERC Timisoara.
Guerra, C.A., Mazzolani, F.M., Piluso, V.(1990), Evaluation
  of the q-factor in steel framed structures: state-of-
  -art,**Ingegneria sismica,** VII,2, 42-63.
Powell, G.H.(1973), **DRAIN-2D User's Guide,** University of
  California, Berkely.

25 A NEW METHOD TO DESIGN STEEL FRAMES FAILING IN GLOBAL MODE INCLUDING P-Δ EFFECTS

F. M. MAZZOLANI
Istituto di Tecnica delle Costruzioni, University of Naples, Naples, Italy
V. PILUSO
Department of Civil Engineering University of Salerno, Salerno, Italy

Abstract
In this paper a method for designing moment resisting frames failing in global mode is presented. It is based on the application of the kinematic theorem of the plastic collapse. The beam section properties are assumed to be known quantities, as they are designed in order to resist vertical loads. The unknowns of the design problem are the column sections. They are determined by means of design conditions expressing that the kinematically admissible multiplier of the horizontal forces corresponding to the global mechanism has to be the smallest among all kinematically admissible multipliers.

The method in its basic format has been already developed in a previous work, but herein is improved, from the theoretical point of view, by including in the algorithm the influence of second order effects, which can be particularly significant in the seismic design of steel frames.

Keywords: Limit Design, Collapse Mechanisms, Second Order Effects

1 Introduction

The simple design criteria, suggested by the modern seismic codes, do not always lead to frames failing in global mode. For this reason, a more sophisticated design procedure, assuring the development of a collapse mechanism of global type, has been proposed by Mazzolani and Piluso for MR-frames (1993a, 1993b). The reliability of this method has been verified on a great number of structural schemes, leading in all cases to the fulfilment of the design object. Therefore, it has allowed to cover an important gap in the design tools for seismic resistant steel frames.

The theory of limit design has been mainly used in order to compute the collapse multiplier of a given structure. In case of frames, the most known methods for reaching this purpose are the elementary mechanism combination method (Neal and Symonds method) and the moment distribution method (Horne method). The moment distribution method has been also applied in order to search structural solutions leading to the minimum weight.

The structural design oriented to the control of the failure mode is a relatively recent problem arised from seismic design needs, which up-to-now has been mainly faced through simplified rules provided in seismic codes. These rules are not derived by means of the limit design approach and in most cases they do not allow to attain the requested failure mechanism.

Starting from these considerations, a new design method has been proposed (Mazzolani and Piluso, 1993a and 1993b) for the control of the failure mode of seismic resistant steel frames. It is based on the observation that the collapse mechanisms of frames under horizontal forces can be considered belonging to three main typologies (Fig.1). The collapse mechanism of the global type is a particular case of the type 2 mechanism. As a consequence, the control of the failure mode can be performed through the analysis of $3n_S$ mechanisms (being n_S the number of storeys). We assume that the beam sections are already designed in order to resist vertical loads. So, we state that the values of the plastic section modulus of columns have to be defined in such a way that the kinematically admissible multiplier of the horizontal forces corresponding to the global mechanism is less than the ones corresponding to the other $3n_S-1$ kinematically admissibile mechanisms. It means that, according to the upper bound theorem, the above stated multiplier is the true collapse multiplier and, therefore, the true collapse mechanism is represented by the global failure mode.

In the original method (Mazzolani and Piluso, 1993a and 1993b) the second order effects, which are not included in the limit analysis, have been indirectly covered by means of an approximated method based on the use of correction factors. In this paper, from the theoretical point of view, a significant improvement is obtained by accounting for second order effects by means of a procedure based on the equilibrium curves of the considered mechanisms.

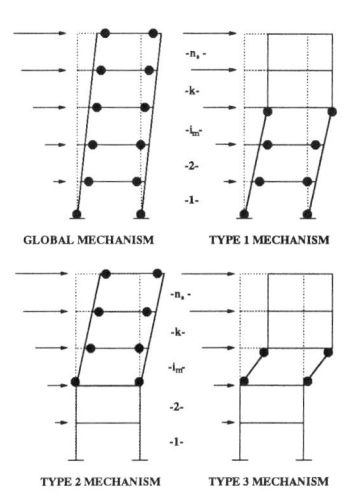

Fig.1
The analysed collapse typologies

2 Mechanism equilibrium curves

As already pointed out, the collapse mechanisms of moment resisting frames under seismic horizontal forces can be considered belonging to three main typologies (Fig.1). The collapse mechanism of the global type is a particular case of type 2 mechanism.

The influence of the collapse mechanism on the slope of the softening branch of the behavioural curve α–δ can be analysed by means of the evaluation of the equilibrium curves of the different mechanism typologies.

The vector of the design horizontal forces $\{F\}$ is given by:

$$\{F\}^T = \{F_1, F_2, \ldots, F_k, \ldots, F_{n_s}\} \quad (1)$$

being F_k the horizontal force applied to the kth storey and n_s the number of storeys.
The vector of the storey heights $\{h\}$ is given by:

$$\{h\}^T = \{h_1, h_2, \ldots, h_k, \ldots, h_{n_s}\} \quad (2)$$

being h_k the height of the kth storey.
The vector of the storey vertical loads $\{V\}$ is given by:

$$\{V\}^T = \{V_1, V_2, \ldots, V_k, \ldots, V_{n_s}\} \quad (3)$$

being V_k the total vertical load acting at the kth storey. In the following, it is assumed that vertical loads are concentrated at the nodes of the frame.

The equilibrium curve of the generic collapse mechanism can be obtained by means of the principle of virtual works:

$$\alpha \{F\}^T \{du\} + \{V\}^T \{dv\} = \{M_p\}^T \{d\theta\} \quad (4)$$

being:
- α the multiplier of the horizontal forces;
- $\{du\}$ the vector of the virtual horizontal displacements corresponding to the given mechanism;
- $\{dv\}$ the vector of the virtual vertical displacements corresponding to the given mechanism;

- { $d\theta$ } the vector of the virtual rotations of the plastic hinges of the given mechanism;
- { M_p } the vector of the plastic moments corresponding to the plastic hinges of the given mechanism.

With reference to Fig.2, it can be observed that the horizontal displacement of the k-th storey involved in the generic mechanism is given by $u_k = r_k \sin\theta$, being r_k the distance of the kth storey from the center of rotation C and θ the angle of rotation.

The top sway displacement is given by $\delta = H_o \sin\theta$, being H_o the sum of the interstorey heights of the storeys involved by the generic mechanism.

The relation between vertical and horizontal virtual displacements is given by $dv_k = du_k \sin\theta$.

It shows that, as the ratio dv_k/du_k is independent from the considered storey, vertical and horizontal virtual displacement vectors have the same shape.

In fact, the virtual horizontal displacements are given by $du_k = r_k\, d\theta$, where r_k defines the shape of the virtual horizontal displacement vector, while the virtual vertical displacements are given by $dv_k = (\delta / H_o)\, r_k\, d\theta$ and, therefore, they have the same shape r_k of the horizontal ones.

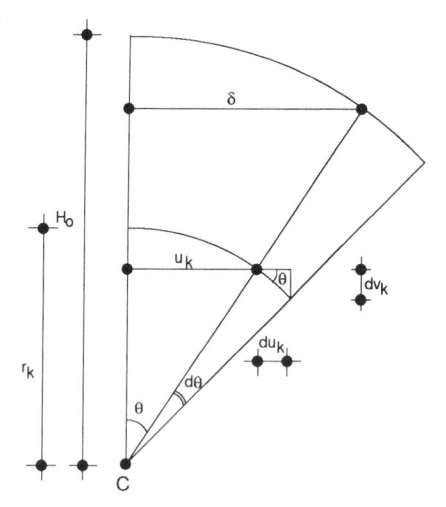

Fig.2
Evaluation of vertical displacements

It can be concluded that, denoting with $\{s\}$ the shape vector of the horizontal displacements, it results $\{du\} = \{s\}\, d\theta$ and $\{dv\} = (\delta / H_o)\, \{s\}\, d\theta$ while $\{d\theta\} = \{I\} d\theta$, because all plastic hinges are subjected to the same virtual rotation ($\{I\}$ is the unity vector).

As a consequence, equation (4) provides the following expression for the equilibrium curve of the generic collapse mechanism:

$$\alpha = \frac{\{M_p\}^T \{I\} - \{V\}^T \{s\} \dfrac{\delta}{H_o}}{\{F\}^T \{s\}} \quad (5)$$

which can be written as $\alpha = \alpha_o - \gamma_s\, \delta$, being α_o the kinematically admissible multiplier of the horizontal forces provided, for the generic collapse mechanism, by the limit design theory under the assumption of rigid-plastic behaviour of the material, while γ_s is the absolute value of the slope of the softening branch of the α–δ curve. They are given by:

$$\alpha_o = \frac{\{M_p\}^T \{I\}}{\{F\}^T \{s\}} \qquad \gamma_s = \frac{\{V\}^T \{s\} \dfrac{1}{H_o}}{\{F\}^T \{s\}} \quad (6)$$

In the following, the previous expressions (6) will be explicitated for the different collapse mechanism typologies. The following notation is herein adopted:
- n_s is the number of storeys;
- n_c is the number of columns;
- n_b is the number of bays;
- k is the storey index;
- i is the column index;

- j is the bay index;
- i_m is the mechanism index;
- $M_{c,ik}$ is the plastic moment, reduced for the presence of the axial internal force, of the ith column of the kth storey;
- $M_{b,jk}$ is the plastic moment of the jth beam of the kth storey.
- $\alpha_o^{(g)}$ and $\gamma_s^{(g)}$ are, respectively, the kinematically admissible multiplier of the horizontal forces (rigid-plastic theory) and the slope of the softening branch of the α–δ curve, corresponding to the global type mechanism;
- $\alpha_{i_m}^{(t)}$ and $\gamma_{s_{i_m}}^{(t)}$ have the same meaning of the previous symbols, but they are referred to the i_mth mechanism of the tth typology (t=1,2,3).

In case of global type mechanism (Fig.1), the shape vector of the horizontal displacements is given by $\{s\} = \{h\}$, while $H_o = h_{n_s}$ because all storeys participate to the mechanism.

Therefore, the kinematically admissible multiplier and the corresponding slope of the softening branch are given by:

$$\alpha_o^{(g)} = \frac{\sum_{i=1}^{n_c} M_{c,i1} + \sum_{k=1}^{n_s}\left(\sum_{j=1}^{n_b} 2 M_{b,jk}\right)}{\sum_{k=1}^{n_s} F_k\, h_k} \quad ; \quad \gamma_s^{(g)} = \frac{\frac{1}{h_{n_s}}\sum_{k=1}^{n_s} V_k\, h_k}{\sum_{k=1}^{n_s} F_k\, h_k} \quad (7)$$

With reference to the i_mth mechanism of type 1 (Fig.1), the shape vector of the horizontal displacements can be written as $\{s\}^T = \{h_1, h_2, h_3, \ldots, h_{i_m}, h_{i_m}, h_{i_m}\}$, where the first element equal to h_{i_m} corresponds to the i_mth component.

Moreover, in this case, it results $H_o = h_{i_m}$; therefore, the kinematically admissible multiplier corresponding to the i_mth mechanism of type 1 is given by:

$$\alpha_{i_m}^{(1)} = \frac{\sum_{i=1}^{n_c} M_{c,i1} + \sum_{k=1}^{i_m-1}\left(\sum_{j=1}^{n_b} 2 M_{b,jk}\right) + \sum_{i=1}^{n_c} M_{c,ii_m}}{\sum_{k=1}^{i_m} F_k\, h_k + h_{i_m}\sum_{k=i_m+1}^{n_s} F_k} \quad (8)$$

while the corresponding slope of the softening branch of the α–δ curve is given by:

$$\gamma_{s_{i_m}}^{(1)} = \frac{\frac{1}{h_{i_m}}\left(\sum_{k=1}^{i_m} V_k\, h_k + h_{i_m}\sum_{k=i_m+1}^{n_s} V_k\right)}{\sum_{k=1}^{i_m} F_k\, h_k + h_{i_m}\sum_{k=i_m+1}^{n_s} F_k} \quad (9)$$

With reference to the i_mth mechanism of type 2 (Fig.1), the shape vector of the horizontal displacements can be written as:

$$\{s\}^T = \{0, 0, 0, \ldots, 0, h_{i_m} - h_{i_m-1}, h_{i_m+1} - h_{i_m-1}, \ldots, h_{n_s} - h_{i_m-1}\} \quad (10)$$

Moreover, it results $H_o = h_{n_s} - h_{i_m-1}$; as a consequence, the kinematically admissible multiplier corresponding to the i_mth mechanism of the type 2 and the corresponding slope of the softening branch of the α–δ curve are given by:

$$\alpha_{i_m}^{(2)} = \frac{\sum\limits_{i=1}^{n_c} M_{c,ii_m} + \sum\limits_{k=i_m}^{n_s}\left(\sum\limits_{j=1}^{n_b} 2 M_{b,jk}\right)}{\sum\limits_{k=i_m}^{n_s} F_k (h_k - h_{i_m-1})} \quad ; \quad \gamma_{s_{i_m}}^{(2)} = \frac{\dfrac{1}{h_{n_s} - h_{i_m-1}} \sum\limits_{k=i_m}^{n_s} V_k (h_k - h_{i_m-1})}{\sum\limits_{k=i_m}^{n_s} F_k (h_k - h_{i_m-1})} \quad (11)$$

Finally, with reference to the i_mth mechanism of type 3 (Fig.1), the shape vector of the horizontal displacements can be written as:

$$\{ s \}^T = \{ 0, 0, .. 0, 1, 1, 1, ..., 1 \} (h_{i_m} - h_{i_m-1}) \quad (12)$$

where the first term different from zero is the i_mth one.
Moreover, it results $H_o = h_{i_m} - h_{i_m-1}$; therefore, the kinematically admissible multiplier of the i_mth mechanism of type 3 and the corresponding slope of the softening branch of the α–δ curve are given by

$$\alpha_{i_m}^{(3)} = \frac{2 \sum\limits_{i=1}^{n_c} M_{c,ii_m}}{\left(h_{i_m} - h_{i_m-1}\right) \sum\limits_{k=i_m}^{n_s} F_k} \quad ; \quad \gamma_{s_{i_m}}^{(3)} = \frac{\sum\limits_{k=i_m}^{n_s} V_k}{\left(h_{i_m} - h_{i_m-1}\right) \sum\limits_{k=i_m}^{n_s} F_k} \quad (13)$$

3 Failure mode control

3.1 Design conditions

In order to design a frame failing in global mode, the cross-sections of columns have to be dimensioned in such a way that, according to the upper bound theorem, the kinematically admissible horizontal force multiplier corresponding to the global type mechanism is the minimum among all kinematically admissible multipliers. As a consequence, the following design conditions have to be imposed:

$$\alpha_o^{(g)} - \gamma_{s_{i_m}}^{(g)} \delta \leq \frac{\alpha_{i_m}^{(t)} - \gamma_{s_{i_m}}^{(t)} \delta}{v_{i_m}^{(t)}} \quad i_m = 1,2,3,....,n_s \quad t = 1,2,3 \quad (14)$$

which for $\delta = 0$ are coincident with the ones given in the original method (Mazzolani and Piluso, 1993a and 1993b)

Therefore, there are $3n_s$ design conditions to be satisfied in the case of a frame having n_s storeys. These conditions, which derive directly from the application of the upper bound theorem, will be integrated by conditions related to technological limitations.

The coefficients $v_{i_m}^{(t)}$ (with $i_m = 1,2,3,....,n_s$ and $t = 1,2,3$) are herein introduced only for comparison with the original procedure, where they were used in order to take into account, in an approximate way, the influence of second order effects which are neglected in limit design. These coefficients had to be chosen in order to include in the design process the different influences that the second order effects play in case of different collapse mechanisms. For practical purposes the value $v_{i_m}^{(t)} = 1.15$ was suggested with the exception $v_1^{(2)} = 1$, because the corresponding mechanism is coincident with the global one (Mazzolani and Piluso, 1993a and 1993b). In the revised procedure herein proposed the coefficients $v_{i_m}^{(t)}$ are not necessary, because second order effects are directly taken into account through the equilibrium curves of the considered mechanisms. As a consequence, in the following it has to be assumed $v_{i_m}^{(t)} = 1$ ($i_m = 1,2,3,.....,n_s$; $t = 1,2,3$). Obviously, a fundamental role is now played by the displacement δ, which has to correspond to the one leading to the complete development of the collapse mechanism.

3.2 Conditions to avoid type 1 mechanisms

The n_s conditions expressed by relation (14) for $t=1$ can be explicitated in a convenient form. For this scope, a constant term and two known discrete functions are introduced:

$$\theta_1 = \sum_{k=1}^{n_s}\left(\sum_{j=1}^{n_b} 2 M_{b,jk}\right) \; ; \; \xi_{i_m} = \sum_{k=1}^{i_m-1}\left(\sum_{j=1}^{n_b} 2 M_{b,jk}\right) \; ; \; \lambda_{i_m} = \frac{\sum_{k=1}^{n_s} F_k h_k}{\sum_{k=1}^{i_m} F_k h_k + h_{i_m} \sum_{k=i_m+1}^{n_s} F_k} \tag{15}$$

The above quantities are known terms, because the plastic moments of beams $M_{b,jk}$ are known, as the beams have known sections due to the fact that they are designed in order to resist vertical loads, and the horizontal forces F_k are assigned.

Moreover, with respect to the original method (Mazzolani and Piluso, 1993a and 1993b), additional parameters related to the influence of second order effects have to be defined. They are:

$$\Delta_{i_m}^{(1)} = \frac{1}{h n_s} \sum_{k=1}^{n_s} V_k h_k - \frac{\lambda_{i_m}}{h_{i_m}}\left(\sum_{k=1}^{i_m} V_k h_k + h_{i_m} \sum_{k=i_m+1}^{n_s} V_k\right) \tag{16}$$

In addition, it is convenient to introduce the following parameter:

$$\rho_{i_m}^{(1)} = \frac{\sum_{i=1}^{n_c} M_{c,i i_m}}{\sum_{i=1}^{n_c} M_{c,i1}} \tag{17}$$

which is the ratio between the sum of the reduced plastic moments of the i_mth storey columns and the same sum corresponding to the first storey columns.

By means of the above parameters, the i_mth condition to be satisfied in order to avoid type 1 collapse mechanisms can be written in the following form:

$$\rho_{i_m}^{(1)} \geq \frac{(1-\lambda_{i_m}) \sum_{i=1}^{n_c} M_{c,i1} + \theta_1 - \lambda_{i_m} \xi_{i_m} - \Delta_{i_m}^{(1)} \delta}{\lambda_{i_m} \sum_{i=1}^{n_c} M_{c,i1}} \tag{18}$$

which has to be applied for $i_m=1,2,3,......n_s$.

3.3 Conditions to avoid type 2 mechanisms

The n_s conditions obtained from (14) for $t=2$ can be explicitated in a convenient form by introducing a new series of parameters:

$$\theta_{i_m} = \sum_{k=i_m}^{n_s} \left(\sum_{j=1}^{n_b} 2 M_{b,jk} \right) \quad ; \quad \gamma_{i_m} = \frac{\sum_{k=1}^{n_s} F_k h_k}{\sum_{k=i_m}^{n_s} F_k (h_k - h_{i_m-1})} \tag{19}$$

where θ_{i_m} is still a known function, because plastic moments of beams are already established, being the beams designed in order to resist vertical loads. Moreover, it is useful to note that the constant θ_1 (the first one of Eqs.15)) of the previous Section is the value of the function θ_{i_m} for $i_m=1$. In addition, γ_{i_m} is a known discrete function of the mechanism index i_m.

Furthermore, with respect to the original method (Mazzolani and Piluso, 1993a and 1993b), additional parameters related to the influence of second order effects are requested:

$$\Delta_{i_m}^{(2)} = \frac{1}{h_{n_s}} \sum_{k=1}^{n_s} V_k h_k - \frac{\gamma_{i_m}}{h_{n_s} - h_{i_m-1}} \sum_{k=i_m}^{n_s} V_k (h_k - h_{i_m-1}) \tag{20}$$

Following the method adopted in the previous section, a new series of parameters is introduced, such as:

$$\rho_{i_m}^{(2)} = \frac{\sum_{i=1}^{n_c} M_{c,ii_m}}{\sum_{i=1}^{n_c} M_{c,i1}} \tag{21}$$

By means of the above parameters, the i_mth condition to be satisfied in order to avoid type 2 collapse mechanisms can be written in the following form:

$$\rho_{i_m}^{(2)} \geq \frac{\sum_{i=1}^{n_c} M_{c,i1} + \theta_1 - \gamma_{i_m} \theta_{i_m} - \Delta_{i_m}^{(2)} \delta}{\gamma_{i_m} \sum_{i=1}^{n_c} M_{c,i1}} \tag{22}$$

3.4 Conditions to avoid type 3 mechanisms

A new known discrete function of the mechanism index i_m (β_{i_m}) and a final series of parameters ($\rho_{i_m}^{(3)}$) are now introduced. They are given by:

$$\beta_{i_m} = \frac{\sum_{k=1}^{n_s} F_k h_k}{\sum_{k=i_m}^{n_s} F_k \Delta h_{i_m}} \quad ; \quad \rho_{i_m}^{(3)} = \frac{\sum_{i=1}^{n_c} M_{c,ii_m}}{\sum_{i=1}^{n_c} M_{c,i1}}$$ (23)

Furthermore, with respect to the original method (Mazzolani and Piluso, 1993a and 1993b), other parameters related to the influence of second order effects have to be introduced. They are:

$$\Delta_{i_m}^{(3)} = \frac{1}{h n_s} \sum_{k=1}^{n_s} V_k h_k - \beta_{i_m} \sum_{k=i_m}^{n_s} V_k$$ (24)

The n_s conditions, provided by relation (14) for $t=3$, which have to be satisfied in order to avoid type 3 collapse mechanisms, can be written in the following form:

$$\rho_{i_m}^{(3)} \geq \frac{\sum_{i=1}^{n_c} M_{c,i1} + \theta_1 - \Delta_{i_m}^{(3)} \delta}{2 \beta_{i_m} \sum_{i=1}^{n_c} M_{c,i1}}$$ (25)

3.5 Technological conditions

According to the above formulations, the $3 n_s$ design conditions have been derived directly from the application of the upper bound theorem. In particular, for each storey there are three design conditions to be satisfied because three collapse mechanism typologies have been considered. As these design conditions have to be contemporary satisfied for each storey, the ratio:

$$\rho_{i_m} = \frac{\sum_{i=1}^{n_c} M_{c,ii_m}}{\sum_{i=1}^{n_c} M_{c,i1}}$$ (26)

between the sum of the reduced plastic moments of columns of the i_mth storey and the same sum corresponding to the first storey columns allows to satisfy the above design conditions if the following relation is verified:

$$\rho_{i_m} = \max \left\{ \rho_{i_m}^{(1)}, \rho_{i_m}^{(2)}, \rho_{i_m}^{(3)} \right\}$$ (27)

As the section of columns can only decrease along the height of the frame, the values of ρ_{i_m} (with $i_m=1,2,....,n_s$) obtained by means of the conditions derived through the application of the upper bound theorem have to be modified in order to satisfy the following technological limitation:

$$\rho_1 \geq \rho_2 \geq \rho_3 \geq .... \geq \rho_{n_s}$$ (28)

3.6 Evaluation of the axial load in the columns at the collapse state

If the sum of the reduced plastic moment of columns of the first storey is specified, then the previously explained design conditions allow the definition, through the ratios ρ_k ($k=1,2,....n_s$), of the same sum corresponding to the kth storey, which guarantees that failure dos not occur according to mechanisms belonging to the three examined typologies. In order to define the plastic section modulus of the columns, the evaluation of the axial load in the columns at the collapse state is requested.

The evaluation of the column axial forces is particularly easy, because they can be derived taking into account that the shear forces in the beams, at the collapse state, are due to the vertical loads and to the fact that plastic hinges are located at both ends of beams, so the plastic moments are there applied. The sum of these shear forces trasmitted by the beams at each storey, above the considered one, provides the axial forces in the columns of the considered storey.

4 Design algorithm

It has been pointed out that the upper bound theorem allows the assessment of a condition for avoiding each undesired collapse mechanism, by considering the ratio between the sum of the reduced plastic moments of the k-th storey column and the same sum corresponding to the first storey. As three different collapse mechanism typologies have been considered, there are $3n_S$ design conditions to be satisfied, which are provided by relations (18),(22) and (25). These design conditions have to be integrated by the technological condition (28). The above mentioned relations can be used in order to design a frame failing in global mode and, therefore, having an ultimate multiplier of the horizontal forces expressed by the first one of equations (7). The algorithm to solve this problem is now presented. The following steps have to be performed:

a. Selection of the maximum displacement up to which it is desired to assure that the collapse mechanism cannot be different from the global one; this displacement has to be the ultimate displacement and therefore it can be evaluated as $\delta = \theta_{p_u} h_{n_s}$, being θ_{p_u} the ultimate value of the beam plastic rotation which can be evaluated starting from the beam plastic rotation capacity. The computation of the beam plastic rotation capacity can be performed by means of a method proposed by the same Authors (Mazzolani and Piluso, 1992).

b. Computation of the storey functions θ_{im}, β_{im}, γ_{im}, λ_{im} and ξ_{im}, which are provided by the equations (15), (19) and the first one of Eqs.(23).

c. Computation of the parameters $\Delta_{im}^{(t)}$, related to the influence of second order effects, given by equations (16), (20) and (24).

d. Computation of the slopes $\gamma_{S_{im}}^{(t)}$ of the equilibrium curves of the considered mechanisms, as provided by equation (9) and the second ones of equations (11) and (13).

e. Computation, through the first one of equations (7), of a tentative value α_t of the ultimate multiplier corresponding to the global mechanism by imposing that the reduced plastic moment of the first storey columns is not less than the plastic moment of the beams.

f. Computation of the limit values $\rho_{im}^{(1)}$, $\rho_{im}^{(2)}$ and $\rho_{im}^{(3)}$ provided by equations (18), (22) and (25), respectively.

g. Computation of the value of ρ_{im} which allows to avoid failure modes corresponding to the three examined collapse mechanism typologies:

$$\rho_{im} = \max \left\{ \rho_{im}^{(1)}, \rho_{im}^{(2)}, \rho_{im}^{(3)} \right\} \tag{29}$$

h. Modification of the computed values of ρ_{im} in order to satisfy the following technological condition:

$$\rho_1 \geq \rho_2 \geq \rho_3 \geq \ldots \geq \rho_{n_s} \tag{30}$$

i. Computation of the corresponding kinematically admissible multipliers $\alpha_{i_m}^{(1)}$, $\alpha_{i_m}^{(2)}$ and $\alpha_{i_m}^{(3)}$ provided by equation (8) and the first one of equations (11) and (13), respectively.
l. Computation of the ultimate multiplier as the minimum among all the kinematically admissible multipliers, but including the influence of second order effects for the selected displacement:

$$\alpha_u = \min\left\{\alpha_{i_m}^{(t)} - \gamma_{s_{i_m}}^{(t)}\delta \quad ; \quad \text{with} \quad i_m = 1, 2, 3, \ldots n_s \quad \text{and} \quad t = 1,2,3\right\} \tag{31}$$

m. If the condition:

$$\left|\alpha_u - \alpha_t\right| > tolerance \tag{32}$$

is verified, then the value of $\sum_{i=1}^{n_c} M_{c,i1}$ corresponding to $\alpha_u = \alpha_o^{(g)} - \gamma_s^{(g)}\delta$ has to be computed and the procedure has to be repeated starting from step **f**. In the opposite case, it can be assumed $\alpha^{(g)} = \alpha_o^{(g)} - \gamma_s^{(g)}\delta = \alpha_u$ and the column sections can be derived according to the following steps.
n. Computation of the axial force in the columns at the collapse state.
o. Computation, for each storey, of the sum of the reduced plastic moments by means of equation (26). The reduced plastic moment of each column can be now obtained by assuming that each column provides the same contribution to the above mentioned sum.
p. Definition of the section of the *j*th column of the *k*th storey, by assuming that the point (N_{jk}, $M_{p,jk}$) belongs to the yielding surface.

The described design method has been applied to a wide series of structural schemes, obtained by varying the number of bays from 2 to 6 and the number of storeys from 2 to 8. The design results have been verified by means of static inelastic analyses including second order effects. Structures failing in global mode have been obtained in all the examined cases, pointing out the reliability of the design method.

5 References

Horne, M.R., and Morris, L.J. (1973): «Optimum Design of Multi-Storey Rigid Frames», Chapter 14 of **«Optimum Structural Design - Theory and Application»**, edited by R.H. Gallagher and O.C. Zienkiewicz, Wiley.
Horne, M.R. and Morris, L.J. (1981): «**Plastic Design of Low-Rise Frames**», Constrado, Collins Professional and Technical Books, London.
Mazzolani, F.M. and Piluso, V. (1992): «Evaluation of the Rotation Capacity of Steel Beams and Beam-Columns», **1st COST C1 Workshop**, Strasbourg, 28-30 October.
Mazzolani, F.M. and Piluso, V. (1993a): «Failure Mode and Ductility Control of Seismic Resistant MR-Frames», Giornate Italiane della Costruzione in Acciaio, **Italian Conference on Steel Construction**, Viareggio, 24-27 Ottobre.
Mazzolani, F.M. and Piluso, V. (1993b): «Dimensionamento a collasso dei telai sismo-resistenti in acciaio», **VI Convegno Nazionale, L'Ingegneria Sismica in Italia**, Perugia, 13-15 Ottobre.

Acknowledgements

This work has been partially supported by Research Grants CNR.93.02287.CT07 and CNR.GNDT.93.02471.PF54

26 REMARKS ON BEHAVIOUR OF CONCENTRICALLY AND ECCENTRICALLY BRACED STEEL FRAMES

F. M. MAZZOLANI
Istituto di Tecnica delle Costruzioni, University of Naples, Naples, Italy
D. GEORGESCU
Civil Engineering Institute, Bucharest, Romania
A. ASTANEH-ASL
Department of Civil Engineering, University of California at Berkeley, USA

Abstract
Some remarks on the behaviour of concentrically and eccentrically braced framed during a high intensity earthquake are presented in this paper. These remarks are based on recent theoretical and experimental works, as well as on the personal experience of the authors.
Keywords: Steel Frames, Bracings, Ductility, Seismic Behaviour.

1. Introduction

It is now generally accepted that the structures in seismic areas shall be designed to satisfy two fundamental requirements. First, the structures must withstand minor earthquakes, with no structural damage and with only limited non-structural damage (Mazzolani et al.,1994). Therefore, under frequent earthquake the structure should possess sufficient strength and stiffness in order to behave elastically and to keep the lateral displacements in the limit of allowable interstory drift. Second, under an unfrequent severe earthquake, significant damages, both in the structural and non-structural elements are acceptable, but the collapse must be avoided and the risk of the loss of life is to be limited. These requirements are usually met by providing the structure with a significant energy dissipation capacity during large inelastic deformations.

Behaviour of Steel Structures in Seismic Areas. Edited by F. M. Mazzolani and V. Gioncu.
Published in 1995 by E & FN Spon, 2-6 Boundary Row, London SE1 8HN. ISBN: 0 419 19890 3.

As a result, the seismic design of a structure requires a balance between strength, stiffness and energy dissipation. Moment resisting frames, MRF, provide a satisfactory strength and possess an eccellent ductility, but tend to behave too flexibly and so, especially for high-rise buildings, MRF becomes uneconomical in developping the designed stiffness required by the drift condition.

The recent January 17, 1994 Northridge Los Angeles earthquake indicated that some of the steel moment frames developed fracture through their connections. Even though it is early to declare the exact cause, it appears that welded moment frames did not behave as ductile as expected.

Concentrically braced frames, CBF, are capable of easily ensuring both the required strength and stiffness, but provide a poor ductility.

Eccentrically braced frames, EBF, attempt to combine the strength and the stiffness of CBF with the ductility of MRF. It is, therefore, generally accepted that the ductile behaviour under the severe earthquake motions and the energy dissipation capacity of a structure during a large excursion in the plastic range is the main goal in seismic design, especially with regard to braced frame systems.

Some remarks concerning this aspect are presented in the paper.

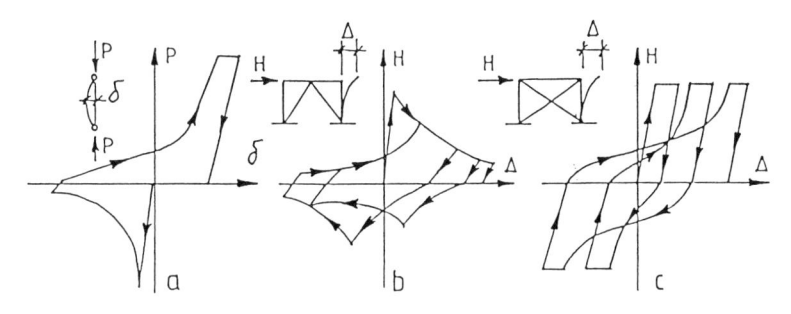

Fig.1. Typical hysteresis loops.

2. Evolution of concentrically braced frames

In designing CBF in seismic regions two tendences have been followed in the last years:
- to keep the structure in the elastic range as much as possible;
- to improve the behaviour of the structure and bracings in the plastic range.

The first tendency could be fulfilled by reducing the behaviour factor q. With this purpose, in the last 20 years significant theoretical and experimental works investigating the behaviour of CBF have analysed (Popov, 1988) by emphasizing:

- the deterioration of the cyclic inelastic behaviour of compressed bar (Fig.1,a);
- the V braced frames (Fig.1,b) with compression braces which deteriorate both strength and stiffness;
- the braced frames (Fig.1,c) dissipating energy in tension which practically deteriorate only stiffness (Fig.1,c).

As a result, in reducing the behaviour factor q all the recent codes take into account the fact that V and X braced frames behave differently. Table 1 shows the ratio q_{BF}/q_{MRF} between the behaviour factor q_{BF} of braced frames, related to the behaviour factor q_{MRF} of the moment resisting frames MRF.

TABLE 1

Code	q_{BF}/q_{MRF}		
	CBF		EBF
	X-bracing	V-bracing	
American UBC 1991 SEAOC 1990	0.666	0.666	0,833
European EUROCODE 8 1993	0.666	0.333	1,0
American ATC 1978	0.666	0.666	-
Japanese JBSL 1981	0,857	0,750	-

In order to increase the safety of the V braced frames, the recent American code (AISC, 1990), in addition to the reduction of the behaviour factor q, introduces two special conditions:
- the design force of V brace members shall be at least 1,5 times the required force coming from calculations;
- the compression design strength of a bracing member in axial compression is reduced by the factor 0,8.

In terms of behaviour factor that means to replace the value q given in the codes by a reduced q equal to:

$$q_{red} = \frac{0,8}{1,5} q = 0,533 q \qquad (1)$$

By introducing in Eq.(1) the value q=2, recommended by EUROCODE 8 [4] for V braced frames, one obtains q_{red}=0,533x2=1,06, i.e the structure shall remain in elastic range even under the action of a severe earthquake motion.

The second tendency is represented by a number of proposals to improve the post-elastic behaviour of braced frames. Starting from the idea that since in a bolted construction (Fig.2,a) some slip in the connections can occur under severe earthquakes, a damping system can be easily obtained by using slotted holes (Fig.2,b).

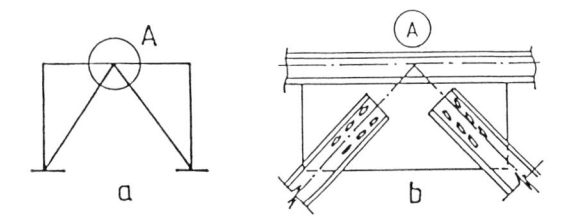

Fig.2. Slotted hole system.

More refined proposals (Vulcano and Mazza, 1993) consists of introducing in the joint different kinds of devices able to dissipate energy by allowing some displacements in the joints.
With regard to the above considerations we can observe:

(1) To keep in the elastic range a structure during severe earthquake motions could be an unreasonable structural solution at least for two reasons:
- it is uneconomical to satisfy the required strength;
- it becomes very difficult to avoid undesiderable plastic deformations of the ground.

Both inconveniences discussed above result from the high level of the horizontal forces to be considered when a structure behaves elastically under a severe earthquake. For a building located in a region with high seismic activity (A=0,35), behaving elastically (q=1) and assuming R(T)=2,5 the shear force F_b, according to EUROCODE 8, becomes:

$$F_b = \frac{AR(T)}{q}\Sigma W_j = 0,875\Sigma W_j \qquad (2)$$

i.e, the structure would withstand an horizontal force F_b equal to 87,5% of its total gravitational load ΣW_j, which, from an economical point of view, is very costly. As a result, to provide a CBF with a certain energy dissipation capacity becomes a real and pressing necessity.

(2) Taking into consideration the concept, generally recognized now, which located the energy dissipation zones out of the joints, all the solutions with slipping joints could be questioned by some designers.
In order to overcome the above difficulties some new structural CBF systems dissipating energy in bending could be conceived (see Fig.3).
The system of Fig.3,a is based on the "strong column - weak beam" concept, largely used in EBF (Fig.3,c), applied for the CBF in the form "rigid braced frames - weak beam" (Fig.3,b), a form similar to a concrete structure with rigid diafragms

and elastic connecting beams (Fig.3,d). The structural system of Fig.3,e is an extension of the "strong column-weak beam" concept, largely used in MRF system (Fig.3,f and g).

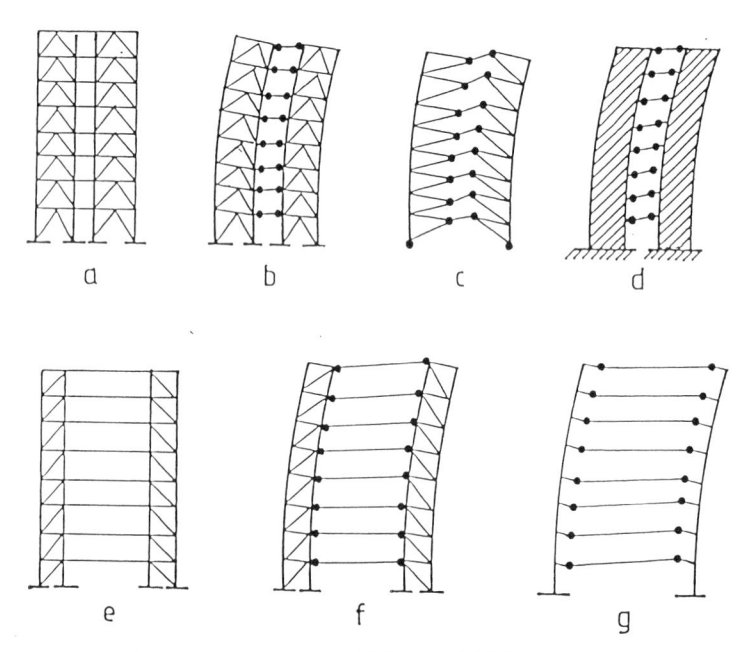

Fig.3. Some new CBF systems.

Applications of both structural system (Fig.3,a and e) have been done in Fig.4a and b, by using structural steel Fe510. The non linear dynamic analysis was performed taking into account the accelerogram A (Vrancea N-S Mars 1977) and using a DRAIN-2D program, processed by Cretu (1994), as it has been done by Mazzolani et al. (1994).
Three design situations were considered, corresponding to different limit states:
- accelerogram "0,5A", corresponding to a return period of about 10 years;
- accelerogram "A", corresponding to a return period of 50 years;
- accelerogram "1,5A", considered as illustrative for an earthquake with a return period of about 500 years.

In comparison with MRF (Fig.3,g) and EBF (Fig.3,c), which dissipate energy "inside" the system, the proposed structures for CBF dissipate energy "outside" the bracing system (Fig.3,b and f).

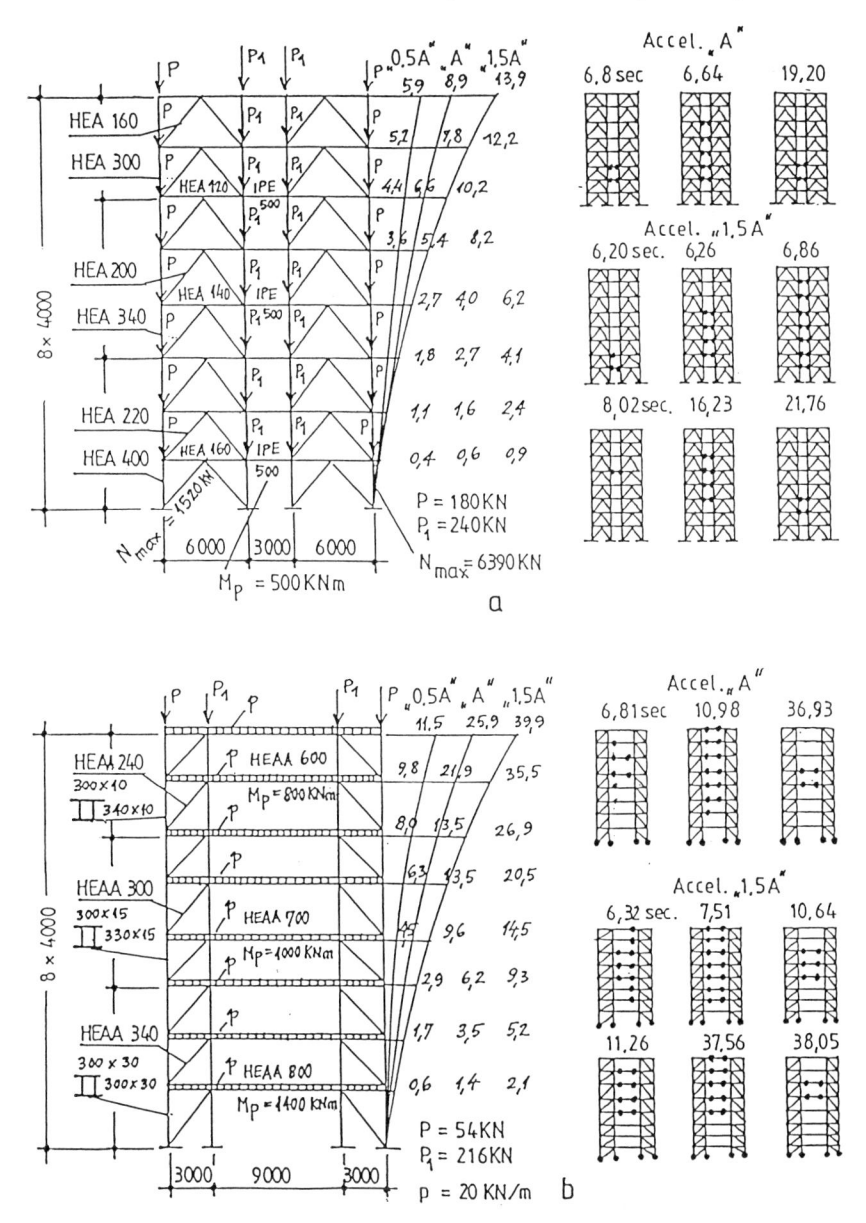

Concentrically and eccentrically braced frames 315

Fig.4. Numerical results.

The structure in Fig.4,a behaves similarly to an EBF structure, but with a quite moderate excursion in the plastic range. Such a type of structure could be characterized by:

(i) A very large stiffness even during severe earthquake motions, diminishing the risk of damage of non-structural elements. EUROCODE 8 allows a drift equal to 0,004h (0,004x400=1,6cm), corresponding to a moderate earthquake, requirement which is satisfyied by the structure of Fig.4,a even for a quite severe earthquake motion (in case of accelerogram A the maximum drift at the 4th storey is equal to 4,0-2,7=1,3cm).

(ii) In spite of its moderate capacity to dissipate energy, the structural system of Fig.4,a represents a significant improvement of the "classical" solution of Fig.5, in which the intermediate beams are simply pin-ended. The maximum displacement at the top diminishes from 29,7 cm to 13,9 cm and the maximum axial forces in the columns and braces at the base of the building diminish from 9230 KN to 6390 KN and from 2280 KN to 1520 KN, respectively (Fig.4,a and Fig.5).

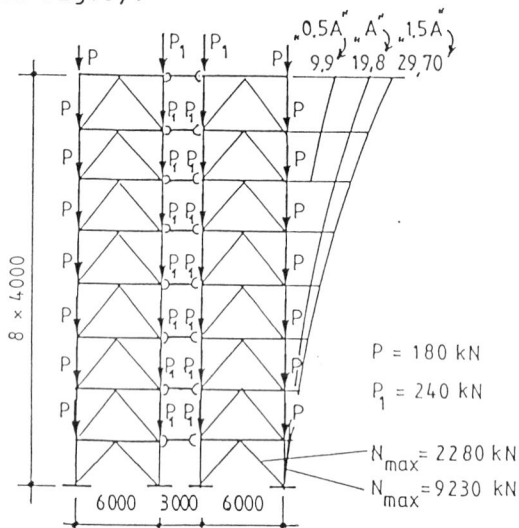

Fig.5. Classical solution with pin-ended beams.

The structure of Fig.4,b behaves similarly to the structure of a MRF system and could be characterized by:

(i) A quite large transversal deformability under severe earthquakes (in comparison with the structure in Fig.4,a the maximum top displacement increases from 13,9cm to 39,9cm in case of accelerogram 1.5A).

(ii) A good stiffness under moderate earthquakes (the maximum drift 9,8-8,0=1,8cm quite equal to the allowable drift 1.6cm in case of accelerogram 0,5A).

From the above results, we find the possibility to generalize this concept and to define a new typology of structures ranging from classical CBF to MRF, with different values of stiffness and ductility. By varying the distance b between the central columns B and C of Fig.6 we can obtain structural systems with different seismic properties.

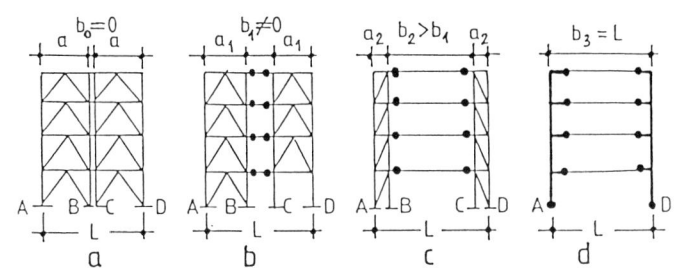

Fig.6. Structural systems obtained by varying the distance b.

The case of $b_0=0$ corresponds to a classical rigid CBF system (Fig.6a) characterized by great stiffness S_0 and a very poor structural ductility D_0. By providing a certain distance between the central columns B and C ($b_1 \neq 0$) the system of Fig.6,b provides good stiffness $S_1 > S_0$ and moderate ductility $D_1 > D_0$. When $b_2 > b_1$ as in Fig.6,c the stiffness decreases ($S_2 < S_1$) and the ductility increases ($D_2 > D_1$). Finally, when $b_3 = L$ the MRF system of Fig.6,d is obtained, having moderate stiffness $S_3 < S_2$ and better ductility $D_3 > D_2$. Summing up, from the above commentary with regard to the evolution of the structural systems of Fig.6, we can underline:
(i) A rational improvement of the poor ductility of a CBF can be acheived not so much by ductilyzising the system itself (as in Fig.2), but in looking for systems of two or more rigid CBF subsystems, connected by weak beams (as in Figs.6,b and 6,c), which introduce ductility in the structure.

(ii) The generation of structural systems as proposed in Fig.6 offers the possibility to obtain a very large range of structures, laying between a very rigid CBF (Fig.6a) up to a very ductile MRF (Fig.6,d).
A similar evolution can be acheived starting from a classical one bay CBF system (Fig.7).
As a result, in obtaining a system with rationally balanced stiffness and ductility, both EBF and CBF subsystems could be used.

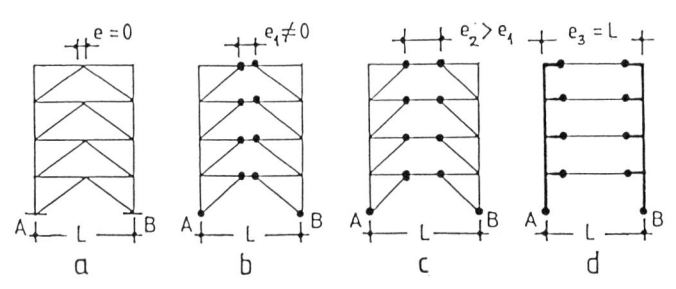

Fig.7. Structural systems obtained by varying the distance e.

3. Design suggestions for eccentrically braced frames

The recent American code (AISC, 1990) synthesizes the main conclusions resulting from a large number of theoretical and experimental works, which have been carried out in the last 15 years on the EBF behaviour (Popov, 1988). The provisions included in this code are largely based on:

(i) Plastic hinges dissipating energy should be located only in the links (Figs.8a and 8b), since all the other structural components outside the links must be designed to remain essentially elastic under the maximum forces that can be generated by the fully yielded and strain hardened links.

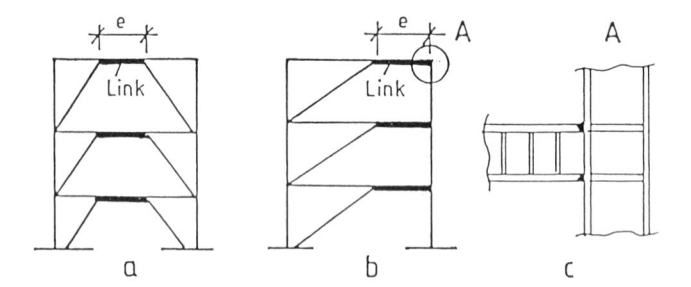

Fig.8. Types of links.

(ii) In order to provide lateral stiffness comparable to that of CBF and to offer high energy dissipation capacity during large inelastic deformations, short links, defined by $e \leq 1,6 M_p/V_p$ (Fig.8,a and 8,b), are to be preferred. Next to columns (Fig.8,b) it is particulary recommended to use only short links having the flanges connected to the columns by means of complete penetration welded joints (Fig.8,c).

With regard to these requirements the following remarks can be made:
(1) The first requirement (i) results from the "capacity design" concept. There are two important provisions in the American code (AISC) intended to prevent the occurance of plastic hinges outside the links:
- the axial column strength must be checked under the application of an amplified earthquake load $E'=3E/q$;
- all bracings and beams outside the links shall be sized for 1,5 times the axial forces and moments determined by using the design base shear force F_b.

With regard to these conditions it has to be noticed that:

(a) The provisions included in the American code (AISC) are largely based on the assumption that the ultimate limit state corresponds to a global mechanism.
In designing some EBF structures D.Georgescu(1993) observed that always, when the ultimate limit state is not governed by a global mechanism, the axial forces and moments in the bracing and columns could be greater than those resulting by applying the American code provisions, i.e the intented concept to keep all the structural members outside the links in the elastic range could be no longer satisfied.

(b) Astaneh et al. (1991) performed a study concerning the seismic behaviour of a 49 storey steel building in San Francisco under the action of the Loma Prieta earthquake of October 17, 1989. The lateral resisting system of the building consists of five MRF in the longitudinal direction and a mix of four MRF and four EBF in the transverse direction. The DRAIN 2 D program was used to perform the inelastic dynamic analyses in a step by step procedure. With regard to the post elastic behaviour of the EBF the dynamic analysis put in evidence the formation of plastic hinges in the columns of the 41 st floor and the formation of plastic hinges in beams not only in the links but also outside the links. Such plastic hinges are not desirable, since they can cause instability of the beams on braces, affecting the capacity of the structure. In addition, some brace members in lower floors buckled when structure was subjected to severe earthquakes.

(c) Mazzolani and Piluso (1993) proposed a rational approach to analyse the post elastic behaviour of MRF structure in order to avoid undesirable failure modes. This approach can be easily applied to EBF systems (Fig.9). The proposed method is based on the limit design concept in the kinematic upper bound form and its originallity consists on the fact that not only the global mechanism is taked into account, but partial mechanisms can also be considered (Fig.9).

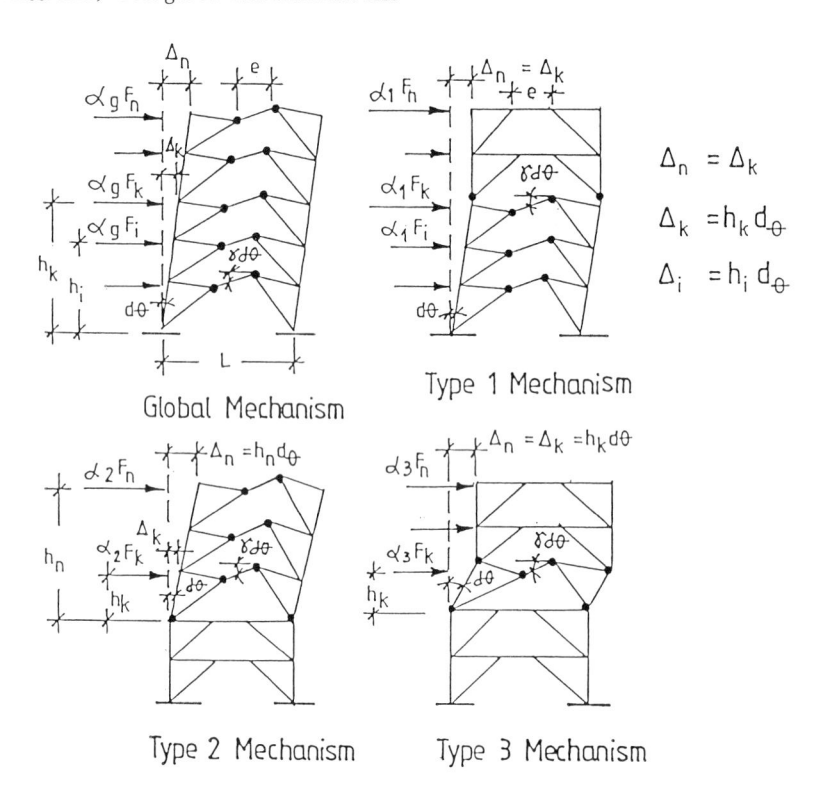

Fig.9. Different collapse mechanisms.

The kinematic equation of the equality between the external dL_e and internal dL_i virtual works is written (Fig.9):

$$dL_e = dL_i \qquad (3)$$

Three main typologies of possible collapse mechanisms are identified, the global failure mode being a particular case of the type 2. As a result, using Eq.3, four kinematically admissible multipliers α can be found, i.e α(g) corresponding to global mechanism and α(1), α(2), α(3), corresponding to 1,2,3 mechanisms respectively (Fig.9).

In order to design a frame failing in a global mode, the following design conditions should be satisfied:

$$\alpha^{(g)} < \alpha^{(1)}$$

$$\alpha^{(g)} < \alpha^{(2)} \quad (4)$$

$$\alpha^{(g)} < \alpha^{(3)}$$

(d) Summing up the above remarks, it results that the simple application of the provisions included in the American code could not be enough in sizing the structural elements of EBF in order to ensure a satisfactory plastic behaviour during high intensity earthquake motions. Therefore, it is of a great interest to control very carefully the dynamic response of the structure in post elasic range. The MAZZOLANI-PILUSO upper bound model could be an alternative design method for avoiding many of the difficulties arising from the step by step analysis.

(2) The second requirements (ii) in the American code is a consequence of the experimental results which put in evidence that the links in shear possesses a greater rotation capacity than the ones in bending.
In comparing the links in shear of figures 8,a and 8,b we can observe that:
- from a theoretical point of view both behave equally in shear;
- from a design point of view the beam-to-column joint of Fig.8,a, which is permitted to be designed as a pin, is much more easily built in the field compared to complete penetration welded joints (Fig.8,c). As a result, whenever possible, the structural solution in Fig.8,a is to be prefered.

4. Conclusive remarks

A new structural system to improve the ductility of CBF is proposed in this paper. Using an assemblage of two or more rigid CBF connected by weak plastic beams, a large range of structural systems balancing stiffness and ductility could be obtained.

In order to provide a satisfactory behaviour of the EBF systems during a high intensity earthquake motions, the dynamic post elastic response of the structure must be very carefully controlled. To fulfil this requirement the MAZZOLANI-PILUSO model (1993) could be an alternative method in sizing the structural elements, avoiding many of the difficulties arising from the step by step analysis.

5. References

A.I.S.C. (1990) Seismic Provisions for Structural Steel Buildings-Load and Resistance Factor Design.

Astaneh, A. Chen, C.C. and Bonowitz, D. (1991) Studies of a 49-Story Instrumented Steel Structure Shaken During the Loma Prieta Earthquake, in **Proceedings of the Second Conference of Tall Buildings in Seismic Regions**, Los Angeles.

Cretu, D. and Muscalu, L. (in preparation) Plot and Post Processor, DR 386-P.C. Version DRAIN-2D.

EUROCODE 8 (1993) Earthquake Resistant Design of structures.

Georgescu, D. (1993) Recent Developments of Theoretical and Experimental Results Seismic Resistant on Steel Structures, in **Proceedings of 14th C.T.A. Congress**, Viareggio.

Mazzolani, F.M. Georgescu, D. and Astaneh, A. (1994) Security Levels in Seismic Design, in **Proceedings of STESSA'94**, Timisoara.

Mazzolani, F.M. and Piluso, V. (1993) Failure Mode and Ductility Control of Seismic Resistant MR - Frames, in **Proceedings of 14th C.T.A. Congress**, Viareggio.

Popov, E.P. (1988) Eccentrically Braced Frames: Coming of Age, in **Proceedings of the 50th Regional Conference**, Los Angeles.

Vulcano, A. and Mazza, F. (1993) Seismic Performances of Steel Braced Frames with Damped Bracing System, in **Proceedings of C.T.A. Congress**, Viareggio.

27 RESISTANCE DETERIORATION DURING EARTHQUAKES COMPILED IN EARTHQUAKE RESPONSE TEST DATABASE ON STEEL FRAMES

K. OHI and K. TAKANASHI
Institute of Industrial Science, University of Tokyo, Tokyo, Japan

Abstract
Thirty-four pseudo-dynamic tests on single-story steel moment resisting frames have been collected and compiled in the database, SCARLET, which stands for 'System for Computer-Aided Researches on Limit states using Earthquake response Test data.' This paper outlines this database and makes a discussion about the classification of structural damage with the emphasis on the deterioration of lateral resistance.
Keywords: Pseudo-dynamic Test, Database, Resistance Deterioration, Human Judgment, Damage Classification.

1 Introduction

Recently, earthquake response simulations based on actual structural behaviors have been performed by many researchers through shake table tests and on-line pseudo-dynamic tests. Then many records of inelastic response and failure process are being collected. For whatever special purpose each simulation was intended, systematic processing over a set of various test records is expected to provide another fruitful perspective about earthquake resistant design. Such a database can be used for 1) education of structural engineers and students, 2) verification of analytical model for response prediction, 3) consensus making about limit states, and so on.

2 Contents of SCARLET

Thirty-four single-degree-of-freedom (SDOF) response tests in the database are carried out on steel moment resisting frames subjected to unidirectional earthquake motions. Excitations are the scaled N-S component recorded at El Centro in 1940 except for the test code 01. Tested structural models are as follows:
 Test code 01: A portal frame composed of rigid beam and rectangular cross-section columns cut from hot-rolled mild steel plate (Takanashi and Ohi 1984); the scaled E-W component recorded at Hachinohe Harbor in 1968 is used as the excitation.
 Test code 02 through 09: Response tests on braced frames composed of members cut from hot-rolled mild steel plates (Ohi and Takanashi 1987). The shapes of hysteresis loops are greatly different from those of moment resisting frames, and then not discussed in this paper.

Behaviour of Steel Structures in Seismic Areas. Edited by F. M. Mazzolani and V. Gioncu.
Published in 1995 by E & FN Spon, 2-6 Boundary Row, London SE1 8HN. ISBN: 0 419 19890 3.

Table 1 SDOF inelastic responses of steel moment resisting frames compiled in SCARLET

Code	T [1]	N/N_Y [2]	β_{pd} [3]	β [4]	Q_{max}/Q_Y [5]	τ_{max} [6]	μ [7]	γ [8]	Q_{max}/Q_e [9]	Damage level
1	0.65	0.02	−0.01	−0.03	1.06	0.083	3.2	0.06	0.27	B
10	0.51	0.00	0.00	−0.02	1.19	0.032	3.5	0.02	0.22	B
11	0.31	0.00	0.00	0.00	1.12	0.019	2.1	0.00	0.21	A
12	0.31	0.00	0.00	−0.02	1.23	0.064	7.0	0.11	0.16	C
13	0.80	0.00	0.00	0.00	1.14	0.036	2.1	0.00	0.37	A
14	0.80	0.40	−0.19	−0.05	1.30	0.034	3.3	0.12	0.41	C
15	0.80	0.40	−0.21	−0.05	1.32	0.040	3.6	0.10	0.35	C
16	0.80	0.00	0.00	0.00	1.11	0.028	1.9	0.00	0.37	A
17	0.80	0.40	−0.16	−0.04	1.19	0.026	2.8	0.07	0.40	B
18	0.80	0.40	−0.15	−0.12	1.41	0.054	6.4	0.65	0.29	C
19	0.80	0.00	0.00	0.00	1.12	0.032	2.0	0.00	0.43	A
20	0.80	0.40	−0.16	0.00	1.14	0.019	1.9	0.00	0.44	A
21	0.80	0.40	−0.16	−0.21	1.18	0.051	5.1	1.00	0.26	D
22	0.80	0.00	0.00	0.00	1.10	0.022	1.5	0.00	0.61	A
23	0.80	0.00	0.00	0.00	1.17	0.031	2.1	0.00	0.42	A
24	0.80	0.40	−0.16	−0.37	1.04	0.033	3.4	1.00	0.56	D
25	0.80	0.00	0.00	0.00	1.09	0.036	2.3	0.00	0.31	A
26	0.80	0.40	−0.15	0.00	1.19	0.020	1.9	0.00	0.45	A
27	0.80	0.40	−0.15	−0.20	1.33	0.066	6.5	1.00	0.31	D
28	0.80	0.00	0.00	0.00	1.26	0.032	2.7	0.00	0.32	A
29	0.80	0.40	−0.13	−0.09	1.08	0.017	1.7	0.05	0.46	B
30	0.80	0.40	−0.13	−0.44	1.07	0.039	3.8	1.00	0.42	D
31	0.80	0.40	−0.13	−0.42	1.06	0.039	3.8	1.00	0.37	D
32	0.80	0.00	0.00	−0.24	1.00	0.032	2.2	0.28	0.31	C
33	0.80	0.20	−0.07	−0.51	1.03	0.026	2.2	0.57	0.53	C
34	0.80	0.20	−0.07	−0.42	1.03	0.041	3.4	1.00	0.45	D
35	0.50	0.10	−0.25	−0.04	1.07	0.089	2.4	0.06	0.31	B
36	0.50	0.10	−0.25	−0.19	1.31	0.255	6.9	1.00	0.21	D
37	0.50	0.10	−0.28	−0.15	1.01	0.089	2.0	0.15	0.33	C
38	0.50	0.10	−0.28	−0.22	1.39	0.254	5.2	1.00	0.20	D
39	0.50	0.10	−0.13	−0.03	1.01	0.054	2.2	0.04	0.35	B
40	0.50	0.10	−0.13	−0.09	1.07	0.064	2.6	0.11	0.25	C
41	0.50	0.10	−0.18	−0.06	1.22	0.131	4.6	0.24	0.30	C
42	0.50	0.10	−0.18	−0.11	1.22	0.082	2.9	0.17	0.23	C

[1] T: Natural period in seconds.
[2] N/N_Y: Constant axial load in the column by its axial yield capacity.
[3] β_{pd}: Simple-plastic slope considering P−Δ effect by elastic slope.
[4] β: Averaged slope in plastic range measured from the test curve.
[5] Q_{max}/Q_Y: Maximum lateral resistance by initial yield resistance.
[6] τ_{max}: Peak drift angle in radian.
[7] μ: Response ductility factor.
[8] γ: Amount of resistance deterioration by maximum lateral resistance.
[9] Q_{max}/Q_e: Maximum lateral resistance by elastic response shear.
Damage level: A: No deterioration ($\gamma=0.0$), B: Early deterioration ($\gamma<0.1$),
  C: Considerable deterioration ($0.1<\gamma<1.0$),
  D: Complete collapse ($\gamma=1.0$).

Test code 10 through 12: Portal frames composed of rigid columns with pinned feet and hot-rolled mild steel H-shaped beams (Ohi and Takanashi 1988).

Test code 13 through 24: Portal frames composed of rigid beams and H-shaped columns welded from low-yield-ratio high-strength steel plates (Takanashi, Ohi, Meng and Chen 1992).

Test code 25 through 34: Portal frames composed of rigid beams and square-box columns welded from low-yield-ratio high-strength steel plates (Meng, Ohi and Takanashi 1992).

Test code 35 through 42: Cantilever H-shaped columns welded from low-yield-ratio high-strength steel or mild steel plates (Kuwamura, Suzuki, Ohi and Meng 1989).

3 Modeling error study using SCARLET

A pair of the restoring force and the response displacement compiled in the database gives a sample of hysteresis curve. On the other hand, a number of ordinary cyclic loading tests were reported on steel members and frames in the past literature. Of course, the hysteresis curves compiled in SCARLET can be used to build or check a mathematical model as done with ordinary cyclic test data. The hysteresis curves in SCARLET, however, are particularly advantageous in the verification of mathematical models for dynamic response analysis, because the restoring force and the displacement always satisfy the equation of motion under a certain earthquake.

Consider a test hysteresis curve of a braced frame as shown in Fig. 1(a). To simulate this curve, a certain hysteresis model is chosen, and the behavior of the model under the same displacement history is found to be as shown in Fig. 1(b). Most of engineers feel satisfaction with this model, because it seems to express basic features of the test hysteresis curve. Such a checking could be done even if the test hysteresis were the result from an ordinary cyclic loading test.

(a) Test Hysteresis Curve

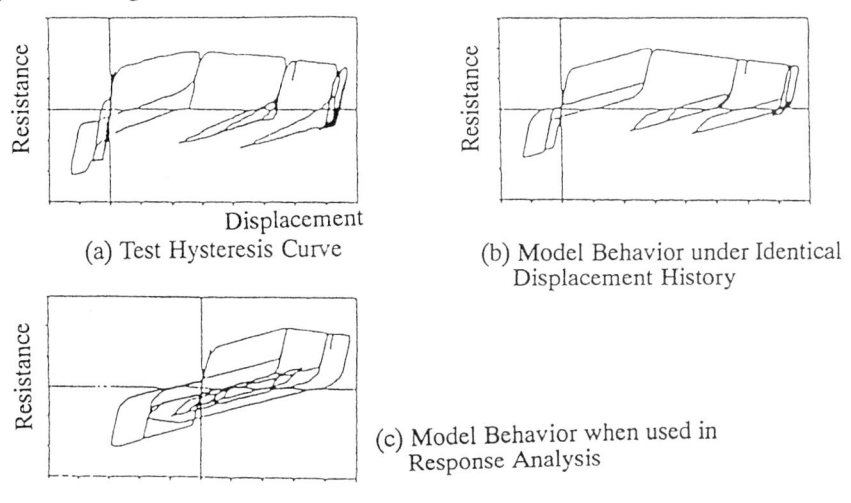

(b) Model Behavior under Identical Displacement History

(c) Model Behavior when used in Response Analysis

Fig.1 Modeling Error Study using *SCARLET*

However, when the test hysteresis is a SCARLET hysteresis, a further checking can be done. The response hysteresis simulated by the model under the same ground motion is shown in Fig. 1(c). Some of engineers do not think this result is good. Thus, a good model in a static sense is not always a good model in the prediction of dynamic inelastic response. SCARLET hysteresis data can be used to check the validity of a mathematical model when used in the dynamic response analysis.

Bilinear hysteresis is sometimes used as the simplified model that exhibits the basic hysteresis behavior of a steel moment frame. The SDOF responses of unbraced frames in SCARLET are compared with those predicted by bilinear hysteresis, and shown in Figs. 2(a) through 2(c). As for the values of the initial stiffness and the yield resistance in the bilinear hysteresis, those read from the test curve are assigned in the analysis. By adjusting the value of the second slope of the bilinear hysteresis in a bit-by-bit manner, the response prediction error of the peak displacement is suppressed within 20 percent of the test results. The model parameters determined in such a way are also used as they are for the prediction of other response quantities. It can be seen that larger errors are induced in the prediction of the permanent sets after earthquakes as shown in Fig. 2(b), while the prediction errors of the energy absorption are suppressed as shown in Fig. 2(c).

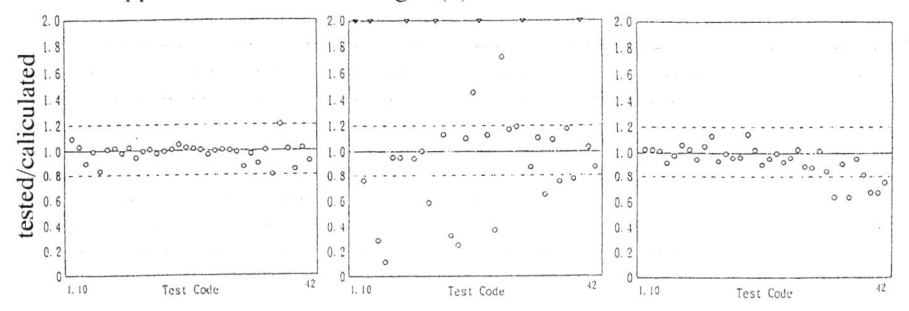

(a) Peak Displacement (b) Permanent Set after Earthquake (c) Energy Absorption

Fig.2 Response Prediction Error with Bilinear Hysteresis Model

4 Human judgments about acceptable damage

Magnitude or degree of structural damage after a severe earthquake may be assessed and evaluated from various factors such as peak deformation or permanent set, occurrence and growth of local buckling or fracture, occurrence of overall structural instability, and so on. In this paper, the discussion is limited to the factor that is directly measured from the hysteresis loops.

Several hysteresis loops compiled in SCARLET are shown to 91 individuals, and the inquiry is made to them whether each case is acceptable or not, if it is the response of an ordinary steel office building to a destructive earthquake like the 1923 Earthquake in Tokyo. Ninety-one individuals are classified into the following four groups:

Researchers of structural engineering including university professors: 23
Structural engineers engaged in design of buildings: 19
Graduate students whose major is structural engineering: 21
Adults who are non-experts but interested in earthquake disasters: 28

Percentage of 'unaccepted damage' is plotted for each group in Figs. 3(a) and 3(b) to two kinds of response quantities, the ratio of strength deterioration to the maximum strength, denoted by γ, and the peak drift angle. It is found that the answers from the first two groups, assumed to be experts, have a similar tendency, and that the γ-factor has a priority for this judgment, when the peak drift angle is different in order from the γ-factor.

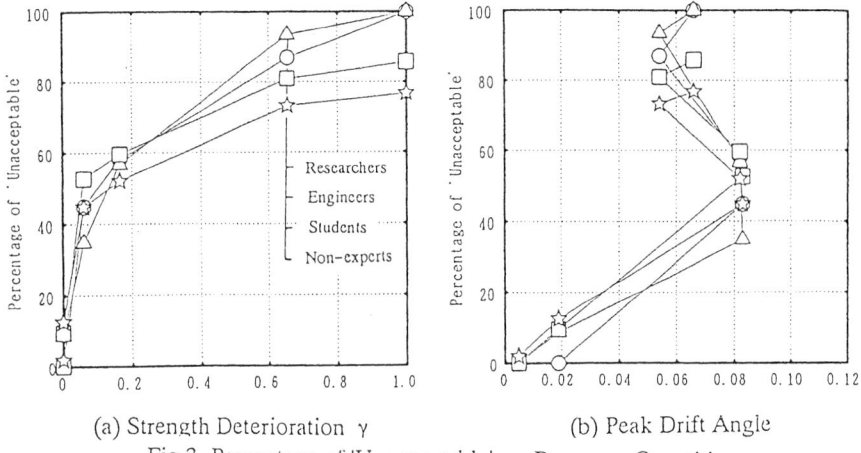

(a) Strength Deterioration γ (b) Peak Drift Angle

Fig.3 Percentage of 'Unacceptable' vs. Response Quantities

5 Damage classification by resistance deterioration

According to the value of the γ-factor, the test runs are classified into the following four degrees of structural damage:

A: 'No Deterioration'   for $\gamma = 0$ and inelastic response
B: 'Early Deterioration'   for $0.1 > \gamma > 0$
C: 'Considerable Deterioration' for $1.0 > \gamma > 0.1$
D: 'Complete Collapse'   for $\gamma = 1.0$

Examples of hysteresis loops classified into the four classes are shown in Figs. 4(a) through 4(d). According to the previous section, about 50 percent of the experts would think the hysteresis curve unacceptable if it is classified into 'C: Considerable Deterioration' or 'D: Complete Collapse.' Most of such unacceptable hysteresis curves have plastic deformations that are accumulated to one direction, and then the curves are very similar to the monotonic test curves. Therefore, if such a skeleton curve is given, the γ-factor can be determined from the peak displacement.

When the skeleton curve is simplified as a bilinear curve as shown in Fig. 5(a), the γ-factor and the response ductility factor, μ, are related with:

$$\mu = 1 + \gamma/|\beta| \qquad (1)$$

where β: the ratio of the second slope in the bilinear curve to the initial elastic slope

Eq. (1) can generate boundaries between two of the damage degrees in $\beta-\mu$ coordinate plane. First, the slope of the first-order mechanism line denoted

(a) No Deterioration

(b) Early Deterioration

(c) Considerable Deterioration

(d) Complete Collapse

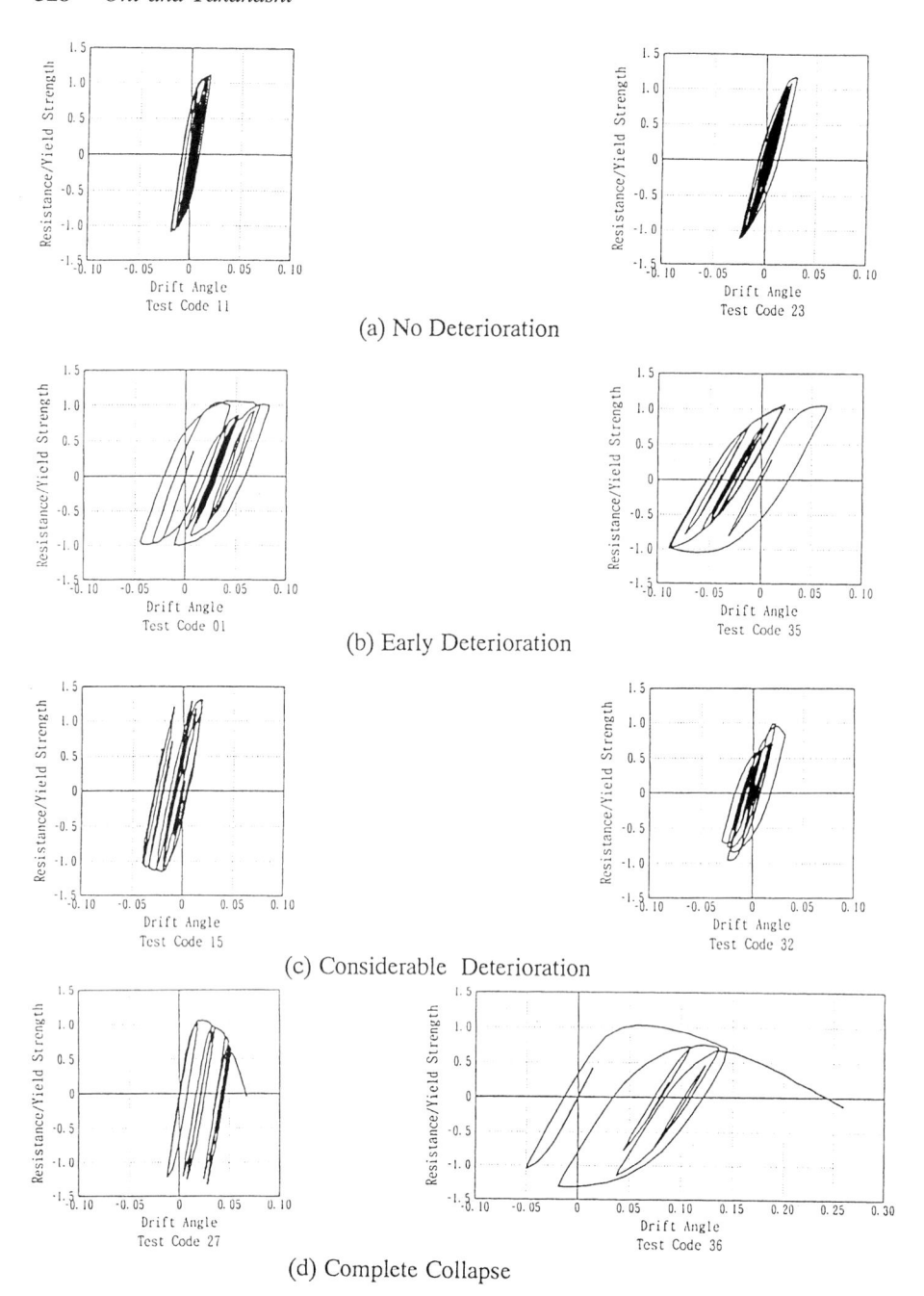

Fig.4 Classification of Structural Damage

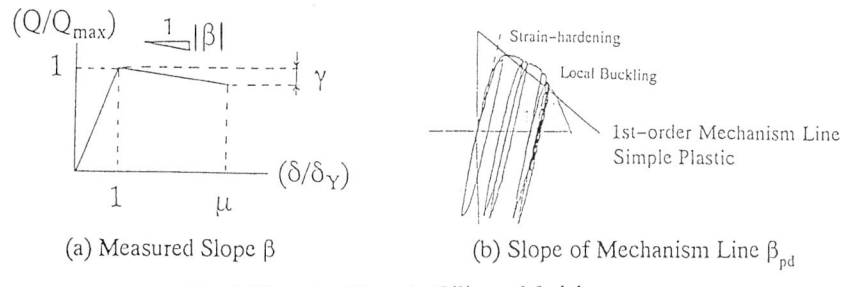

Fig.5 Negative Slope in Bilinear Model
(a) Measured Slope β
(b) Slope of Mechanism Line β_{pd}

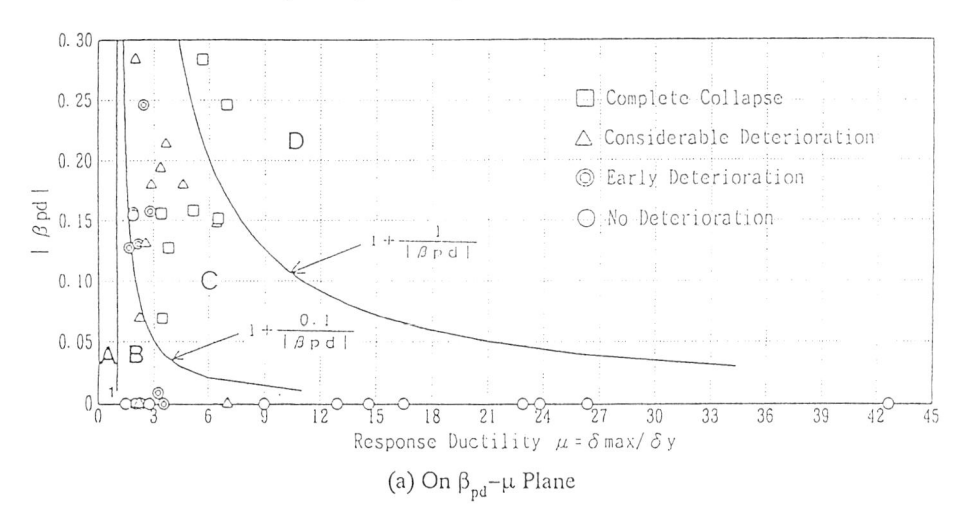

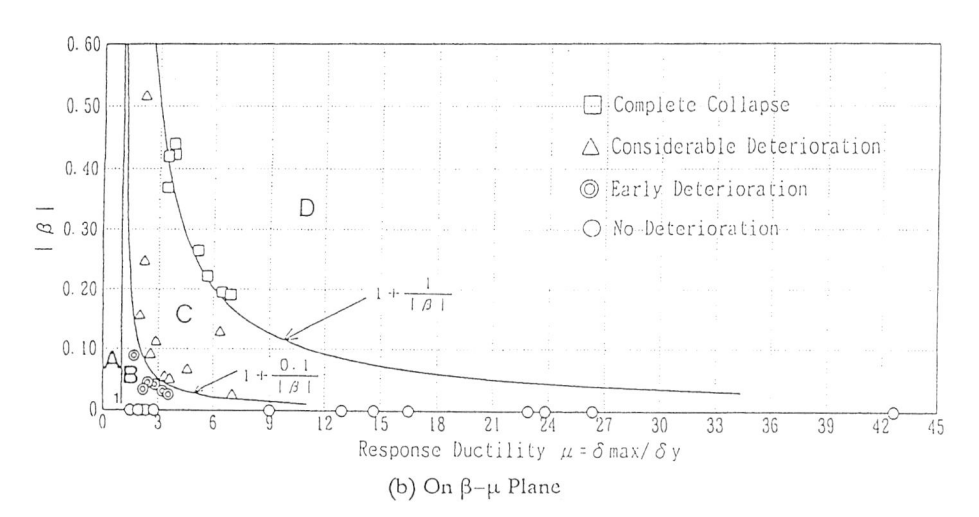

(a) On β_{pd}-μ Plane

(b) On β-μ Plane

Fig.6 Zoning of Damage Classification on β or β_{pd} - μ Plane

by β_{pd} is used for the value of β in Eq. (1), and the results are shown in Fig. 6(a). The damage zones predicted do not well agree with the actual classification, that is, some of the cases classified into 'D: Complete Collapse' or 'B: Early Deterioration' are placed in the 'C: Considerable Deterioration' zone predicted. This reason is that the slope of the first-order mechanism line does not always represent the negative slope observed in actual hysteresis loops, which is affected by strain-hardening and local buckling as well. For the comparison, the average slope after yield until the final reduced resistance is read from the actual hysteresis loops so that it satisfies the conditions shown in Fig. 5(a) exactly. It is natural that the boundaries of damage degrees are absolutely expressed by Eq. (1) as shown in Fig. 6(b). From this discussion, a careful estimation of the change of negative slope over the whole plastic range of skeleton curve (not simply as a straight line) is needed to assess the possible damage degree of a certain steel frame to severe earthquakes.

6 Correlation with linear-elastic response

As described before, the peak displacement SDOF responses of steel moment frame would be approximately predicted even by use of a simple hysteresis rule such as bilinear, but this would be possible if the model parameters could be adjusted to the optimal values. Especially with the presence of negative slope after yielding, numerically predicted responses are very sensitive to the model parameters. Instead of carrying out numerical response analysis, the peak displacement of inelastic system is sometimes predicted from the linear-elastic response as proposed in the past literature. In this section, the hypothesis that identical amounts of strain energy are exerted into inelastic and linear-elastic systems (Veletsos and Newmark 1960) is adopted to predict the damage degrees in the term of γ-factor.

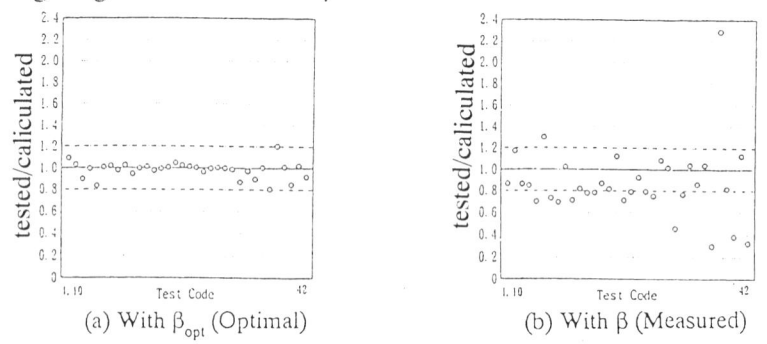

(a) With β_{opt} (Optimal) (b) With β (Measured)

Fig.7 Prediction Error of Displacement Responses

From the hypothesis of identical strain energy shown in Fig. 8, we obtain:

$$0.5\,(Q_c^2/K) = 0.5\,(Q_{max}^2/K)\,\{\,1 + (2-\gamma)\gamma/\,|\beta|\,\} \qquad (2)$$

where Q_c: linear-elastic response shear force
  Q_{max}: maximum strength
  K: elastic stiffness

then $(Q_{max}/Q_c) = [\,|\beta|\,/\,\{\,|\beta|\,+ (2-\gamma)\gamma\,\}\,]^{1/2}$ (3)

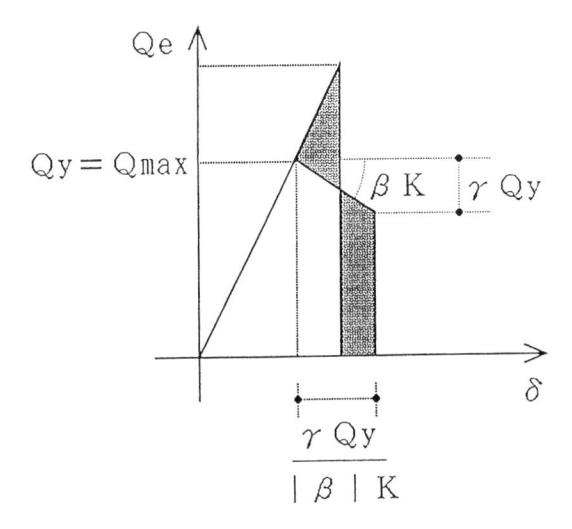

Fig.8 Hypothesis of Identical Strain Energy

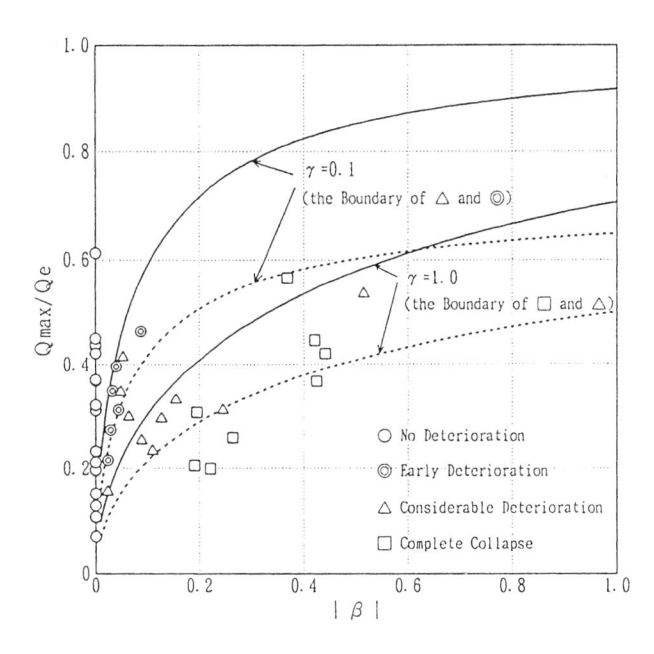

Fig.9 Prediction of Damage Degree from Linear-Elastic Response Shear

The boundaries of the damage degrees predicted by Eq. (3) are plotted on (Q_{max}/Q_c)-β coordinate plane as shown in Fig. 9. These boundaries have a tendency to overestimate the actual damage. The reason of such a conservative result is that a few cyclic reversals are experienced during the deterioration process, and they dissipate a considerable amount of energy. If we double the terms of inelastic strain energy in Eq. (3) to compensate such an extra amount of energy dissipation, the dotted boundaries are obtained as shown in Fig. 9.

7 Concluding remarks

A database of earthquake response tests on steel frame models, SCARLET, is outlined, and the damage degrees observed in thirty-four SDOF tests on steel frames are classified into four damage levels according to the amount of the resistance deterioration.

(1) More than 50 percent of forty-two Japanese experts in structural engineering regard the class 'C: Considerable Deterioration' as unacceptable structural damage in ordinary steel buildings to a destructive earthquake. That corresponds to more than 10 percent decrease of the maximum lateral resistance.

(2) If the average negative slope after yielding is estimated until it reaches the remaining strength, the amount of strength deterioration can be related through that slope with the maximum deformation. To evaluate such a negative slope, P-Δ effect, strain-hardening, and local buckling shall be carefully considered.

(3) The hypothesis of identical strain energy with linear-elastic response provides a conservative prediction of damage degree, because it ignores the energy dissipation due to cyclic reversals. Some of the test results show that the structure can dissipate almost twice of such a monotonic energy by cyclic reversals during an earthquake.

8 References

Kuwamura, H., Suzuki, T., Ohi, K. and Meng, L. (1989) Pseudo-Dynamic Failure Test of Small-Scale Columns of Different Steel Grades. **J. Struct. Engrg.** Architectural Inst. Japan (AIJ), Vol. 35B.

Meng, L., Ohi, K. and Takanashi, K. (1992) A Simplified Model of Steel Structural Members with Strength Deterioration Used for Earthquake Response Analysis. **J. Struct. Constr. Engrg.** AIJ, No. 437.

Ohi, K. and Takanashi, K. (1987) Hysteresis Loops Observed in Earthquake Response Tests on Steel Frame Models. **Bull. Earthquake Resistant Struct.** Research Center, Inst. Indu. Sci., No. 20.

Ohi, K. and Takanashi, K. (1988) An Improvement of On-line Computer Test Control Method. **Proc. 9WCEE**, Tokyo-Kyoto.

Takanashi, K. and Ohi, K. (1984) A Correlation Study in Pseudo-Dynamic On-line Tests vs. Shaking Table Tests. **1984 Annual Technical Session and Meeting of SSRC.**

Takanashi, K., Ohi, K., Meng, L. and Chen, Y (1992) Collapse Simulation of Steel Frames with Local Buckling. **Proc. 10WCEE**, Madrid.

Veletsos, A. and Newmark, N.M. (1960) Effect of Inelastic Behavior on Response of Simple System to Earthquake Motions. **Proc. 2WCEE**.

28 ON THE SEISMIC RESISTANCE OF LIGHT GAUGE STEEL FRAMES

B. CALDERONI, A. DE MARTINO and
R. LANDOLFO
Istituto di Tecnica delle Costruzioni, University of Naples, Naples, Italy
A. GHERSI
University of Catania, Italy

Abstract

As a first step of a wider research, which has the aim to examine the possibility of using thin gauge steel members in dissipative zones of seismic structures, the dynamic behaviour of three single storey one bay frames is analysed in this paper. The portal frames are built up with a cold-formed steel beam, and columns with the member cross section of different slenderness. The horizontal load vs. displacement curve is defined both on the basis of existing experimental results and by theoretical simulations, taking into account the geometrical and mechanical cyclic deterioration. The dynamic response of the structures, obtained by using a large number of historical italian accelerograms, is compared to the analogous results based on an ideal elasto-plastic s.d.o.f. system.

Keywords: Dynamic behaviour, Cold-formed steel sections.

1 Introduction

The technical improvement in the production and working of steel during the second half of this century led to a wide use of thin gauge sections mainly obtained by cold-forming steel sheets (Walker, 1975). In seismic countries, like Italy, the use of cold-formed members has not yet reached the same economical importance of hot rolled traditional steel profiles. A reason for this is that the seismic protection of buildings requires structural systems able to undergo large local deformations in the plastic range. This is easily achieved by means of the hot rolled members (plastic or compact cross section), while the thin-walled ones show a limited ductile behaviour, since influenced by the local buckling of the compressed elements of the section.

As it is well known, the elastic post-local-buckling behaviour of cold-formed thin gauge section is characterized by the increase of the strength as the deflection grownws up, together with a decrease in stiffness. The inelastic behaviour of thin gauge sections has been the object of many investigations, mainly devoted to the definition of specific limit conditions (yielding moment, ultimate moment) more than to the evaluation of the whole moment-curvature relationship. However, both experimental tests (De Martino et al., 1990) and numerical models (De Martino et al., 1992) show that a thin walled section reaches the maximum strength in the early stages of the plastic range and presents after it a more or less remarkable loss in load carrying capacity.

The behaviour of a cold-formed frame is remarkably non-linear, because of the progressive decrease in stiffness (due to local buckling) as load and deflection increase. Its strength may be evaluated by means of numerical models which assume a reduced effective rigidity varying along the member length depending upon the stress magnitude (Blandford et al., 1984). Tests performed in Japan on beams and simple frames (Ono,

1984, Ono and Suzuki, 1986) have shown that yield strength and deformation capacity of these structures can reach significant values. In spite of this, the European seismic code (EC8, 1988) and the ECCS Recommendations (ECCS, 1988) allow the use of thin gauge sections only in non-dissipative zones, neglecting their ductility even in the most favourable cases. A deeper examination of the problem is therefore necessary in order to avoid such a severe and probably unjustified penalization of cold-formed members (Calderoni et al., 1993b).

A careful investigation will require a large number of tests. The experimental way is obviously the best one to obtain a clear interpretation of the local and the overall behaviour of members or structures. However, useful informations may be provided by numerical models, which allow a wide parametric analysis of the problem. In previous studies has been developed:
- a general analytical expression of moment-curvature relationship, in which a suitable definition of the involved parameters gives the possibility to generate different shapes of curves (Calderoni et al., 1993a);
- a numerical model of the inelastic F-δ curve of cold-formed simple frames in order to obtain it directly from the moment-curvature relationship of the member sections (Calderoni et al., 1993b).

The use of these two procedures allows to investigate on the behaviour of thin gauge frames in a wide range of size and shape of sections.

2 The frame model behaviour

The behaviour of a single storey one bay frame under horizontal monotonic loads is well interpreted by the horizontal load versus horizontal displacement relationship $(F\text{-}\delta)$ of the upper nodes. In the case of a portal frame, built up using thin-walled members for columns and beam, the load-displacement curve $(F\text{-}\delta)$ shows a linear elastic branch followed by a short non linear one, up to the maximum value of the horizontal load (F_{lim}). Therefore, for increasing displacement, the curve continues with a softening branch, which slope is strictly related to the local slenderness of the member cross sections. Generally speaking, the maximum frame bearing capacity, in term of horizontal loads, is not reached with fully plastic moment in the joints, but it depends upon the achieving the local buckling conditions, as this is showed in the local moment-curvature curve of the specific member cross section. The shape of the F-δ curve of the frame changes as the slope of the softening branch of moment-curvature curve of end cross sections of the beam or of the column varies. The curve represents the static behaviour of the frame under monotonic horizontal actions, indipendently from the second order effect of vertical loads (geometrical deterioration).

In order to perform the dynamic analysis of the frame under specific seismic input, we need, moreover, to define the cyclic behaviour of the structure, taking into account the stiffness deterioration for repeated action, and consequently for cyclic repetitions of plastic displacement. By assuming a simplified F-δ curve, constituted by a bi-linear increasing branch, a linear softening branch and an horizontal costant one for the residual bearing capacity (fig.1), cyclic deterioration rules have been stated:
- if horizontal displacement δ is smaller than δ_{lim}, in positive or negative field, the reverse behaviour (unloading and loading in reverse field) remains elastic with the same stiffness of the loading branch, even if there is a residual deformation (in the case of $\delta > \delta_{el}$);
- if horizontal displacement δ is greater than the positive (negative) value of δ_{lim}, in the following unloading condition the negative (positive) value of F_{lim} is reduced and the relative limit elastic displacement is increased; the ratios between F_{lim} and F_{el} and between δ_{lim} and δ_{el} remain costant. So we get a reduction both in strenght and stiffness,

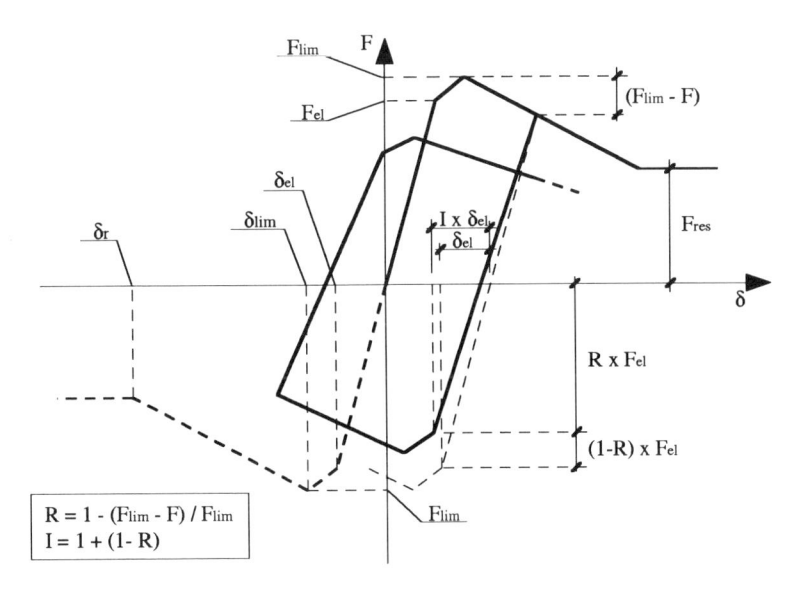

Fig.1

as greater as the plastic excursion is, by means of a linear correlation, as showed in fig.1.

The validity of the above mentioned semplified model to interpret the monotonic and the cyclic behaviour of the frame, has been verified on the basis of experimental results (Ono, 1984, Ono and Suzuki, 1986) and theoretical assumptions (Takanashi et al., 1992).

3 Frame selection

The most suitable way to define the load-displacement law for a frame under monotonic actions should be the experimental one, by performing real scale tests. But, since the object of this paper is just to explore the possibility to use the thin walled members in seismic areas, it is justified to use also a fully theoric approach: the F-δ curve of the frame is defined numerically on the basis of the moment-curvature relationship (M-χ) of the member cross section, obtained by others numerical simulations, as provided in a previous study (De Martino et al., 1992).

We have also followed a third way: it equally consists in a numerical determination of the frame F-δ curve, but based, this time, on a moment-curvature (M-χ) of the member cross section obtained experimentally.

By following the previously referred approachs, the present paper analyzes three one storey one bay portal frames, that are characterized by different F-δ laws:

- FRAME A: the columns are compact section members, while the beam is a thin-walled rectangular hollow section member; the frame F-δ curve is obtained numerically on the basis of the theoretical elasto-plastic moment-curvature (M-χ) law for the column and of the simulated one for the beam; the assumed curve, the load condition and the mechanical and geometrical characteristics are showed in fig.2;
- FRAME B: this frame has been chosen among those tested by Ono e Suzuki (1986); the beam and the column are built up by using member composed by a double cold formed

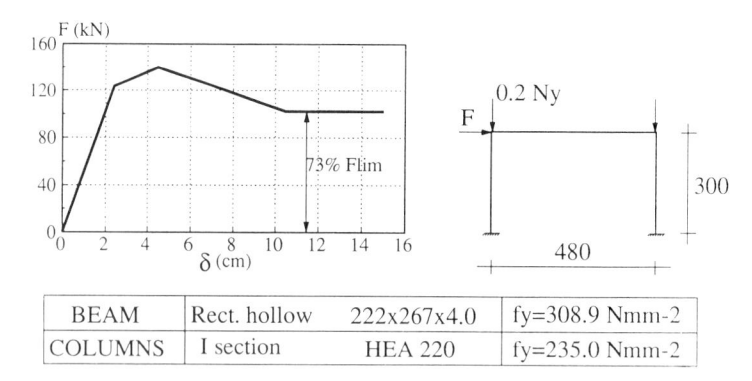

| BEAM | Rect. hollow | 222x267x4.0 | fy=308.9 Nmm-2 |
| COLUMNS | I section | HEA 220 | fy=235.0 Nmm-2 |

Fig.2. FRAME A

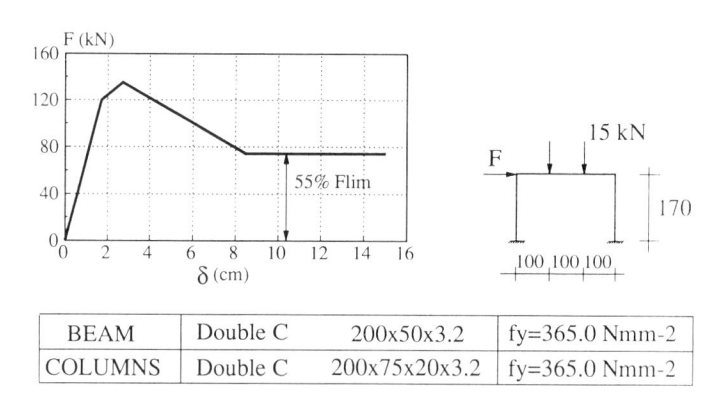

| BEAM | Double C | 200x50x3.2 | fy=365.0 Nmm-2 |
| COLUMNS | Double C | 200x75x20x3.2 | fy=365.0 Nmm-2 |

Fig.3. FRAME B

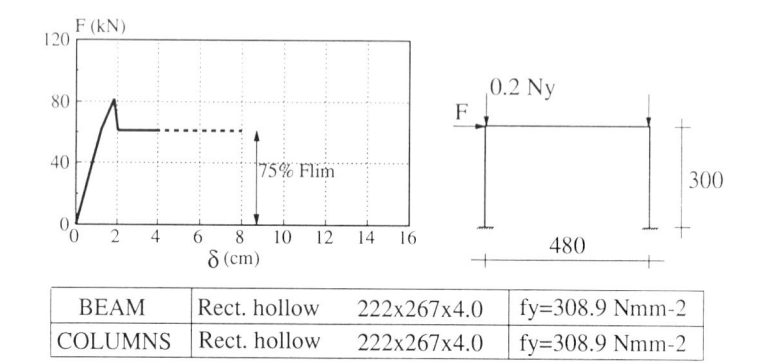

| BEAM | Rect. hollow | 222x267x4.0 | fy=308.9 Nmm-2 |
| COLUMNS | Rect. hollow | 222x267x4.0 | fy=308.9 Nmm-2 |

Fig.4. FRAME C

channel; the F-δ curve, based on experimental results, is showed in fig.3;
- FRAME C: the columns are made by thin-walled rectangular hollow section member; its M-χ law has been experimentally obtained by Watanabe et al. (1992); the beam has the same cross section of the column, but shows a different moment-curvature law, for the absence of the axial load, herein numerically simulated (fig.4).

The comparison, of the F-δ curves, allows to clearly underline the different behaviour of the frames under horizontal loads, in terms both of strenght and deformability. If we not consider the values of the bearing capacity (F_{lim}), the differences appear meaningfull expecially in the softening branch. It may be noticed that the slope of this branch increases substantially going from the frame A to the frame C. The values of the residual bearing capacity are also different, varying from 0.55 F_{lim} to 0.75 F_{lim}.

4 Dynamic response of frames under seismic input

By considering the horizontal displacement of the upper joints as the unique unknown parameter, a portal frame can be modelled as a s.d.o.f. system, characterized by a non-linear degrading F-δ law.

The authors, for each one of the considered frames, have performed a step by step dynamic analysis taking into account the simplified mechanical deterioration rules mentioned in Section 2. As input load conditions thirty real accelerograms recorded during recent italian earthquakes have been used. These accelerograms have been selected from the ENEA-ENEL data base in order to get a statistically significant set (Calderoni et al., 1994): the average elastic response spectrum (50% of probability to be exceeded) obtained from these selected earthquakes is very similar to that provided by Eurocode 8 for the soil site type A, scaled to a PGA equal to 0.15g (value set by actual codes for the third category of seismic areas).

The step by step dynamic analysis has been performed by taking into account also the degrading effects of the vertical loads (P-δ effect), depending on the value of the ratio mg/h, where m is the system mass, g the gravity acceleration and h the column height.

The value of the mass is $m = F_{lim} / \beta$, being β the EC8 design acceleration, evaluated for PGA=0.15g, a behaviour factor $q=1$ and the appropriate response coefficient $R(T)$.

	FRAME A		FRAME B		FRAME C	
	Flim = 140000 N m = 63830 kg T = 0.70 s R(T) = 1.43		Flim = 135000 N m = 43760 kg T = 0.49 s R(T) = 2.02		Flim = 81500 N m = 23450 kg T = 0.43 s R(T) = 2.33	
q	θ	A.C.	θ	A.C.	θ	A.C.
1	0.0405	1.000	0.0358	1.000	0.0153	1.000
2	0.0810	1.915	0.0716	1.925	0.0306	1.968
3	0.1215	2.746	0.1073	2.777	0.0459	2.908
4	0.1620	3.493	0.1431	3.555	0.0612	3.815
5	0.2025	4.155	0.1787	4.260	0.0765	4.690
6	0.2430	4.734	0.2144	4.890	0.0918	5.535

Table 1. Frames analysis data

The second order effects have been considered, as in a real design procedure, by means of the coefficient $\theta=(mg/h) \times (q\delta/F)$, where δ/F, in this case, is the lateral stiffness K of the structure.

Each frame has been analysed by varying the behaviour factor q from 1 to 6. Since the ultimate bearing capacity F_{lim} is fixed, the structure can withstand, with increase of q factor, a stronger design earthquake. Therefore, to obtain the different values of q, we have magnified the thirty selected accelerograms by means of a coefficient (A.C.), as reported in table 1.

In this way 540 numerical elaborations (given by three frames, six q factor and thirty accelerograms) have been performed. The analyses have been carried out also with regard to an ideal elasto-plastic s.d.o.f. system, with stiffness, mass and plastic strenght equal to those of the actual frames.

The s.d.o.f. dynamic equilibrium equation has been integrated by means of the iterative Newmark method. For each analyzed scheme, the time history of displacement, velocity and acceleration of the upper node of the system have been obtained.

5 Analysis of the results

The large amount of results have been statistically elaborated. We assumed that each frame withstands an earthquake if the maximum displacement δ_{max} obtained by the dynamic analyses is lower than a prefixed allowable value δ_{all}. In fig.5 the percentage of the withstanded earthquakes are reported versus q factor. In each diagram the two curves are referred respectively to the actual degrading frame (DF) and to the ideal elasto-plastic one (EP), with regard to a specific value of the available ductility $\mu=\delta_{all}/\delta_{el}$. With reference to $\mu=\infty$, the reported values denote the number of earthquakes in which the dynamic instability are not reached.

It can be noted that the differences between EP and DF are very slight as regard to frame A and frame B, at all level of μ. On the contrary there are more significant scatters in the case of frame C, that increase with μ. Similar observations can be done on fig.6, where the mean values of ductility demand $Y=\delta_{max}/\delta_{el}$ versus q factor are showed. The curve of the 84% probability values (mean plus standard deviation) is also drawn, but only for DF system. At the bottom of this diagram, for each value of q, the number of earthquakes withstanded by the degrading frame (as percentage) is reported: the mean values of Y, obviously, has been evaluated only on the results corresponding to these last earthquakes.

Useful remarks can get from fig.7, in which the curves for the DF system, separately reported in fig.5, are all compared among them and with the curve corresponding to EP for $\mu=\infty$; the probability for a thin-walled portal to withstand an earthquake is widely influenced by the q factor, the available ductility μ and the shape of the F-δ curve.

Since we have chosen a design elastic spectrum practically conform to the average elastic spectrum of the used accelerograms, and the frames are designed according to EC8 without any overstrenght, we expect at least a probability of 50% for all the analyzed cases. On the contrary it can be seen that only the ideal elastic-plastic system without any ductility limit exhibits a 50% percentage for all examined values of q factor. For the actual frames, $\mu=\infty$ provides a q factor lower than 6 for frame A, lower than 5 for frame B and lower than 4 for frame C, according with the slope of the softening branch in the F-δ curve, that increases by moving from frame A to frame C.

So it seems that, in a degrading structural system, characterized with cyclic reduction in both strenght and stiffness, the slope of the softening branch in the F-δ curve is more important than the value of the residual bearing capacity. In fact frame C, which presents an hard slope of the decreasing branch with an high residual strenght (75% F_{lim}), has

Light gauge steel frames **339**

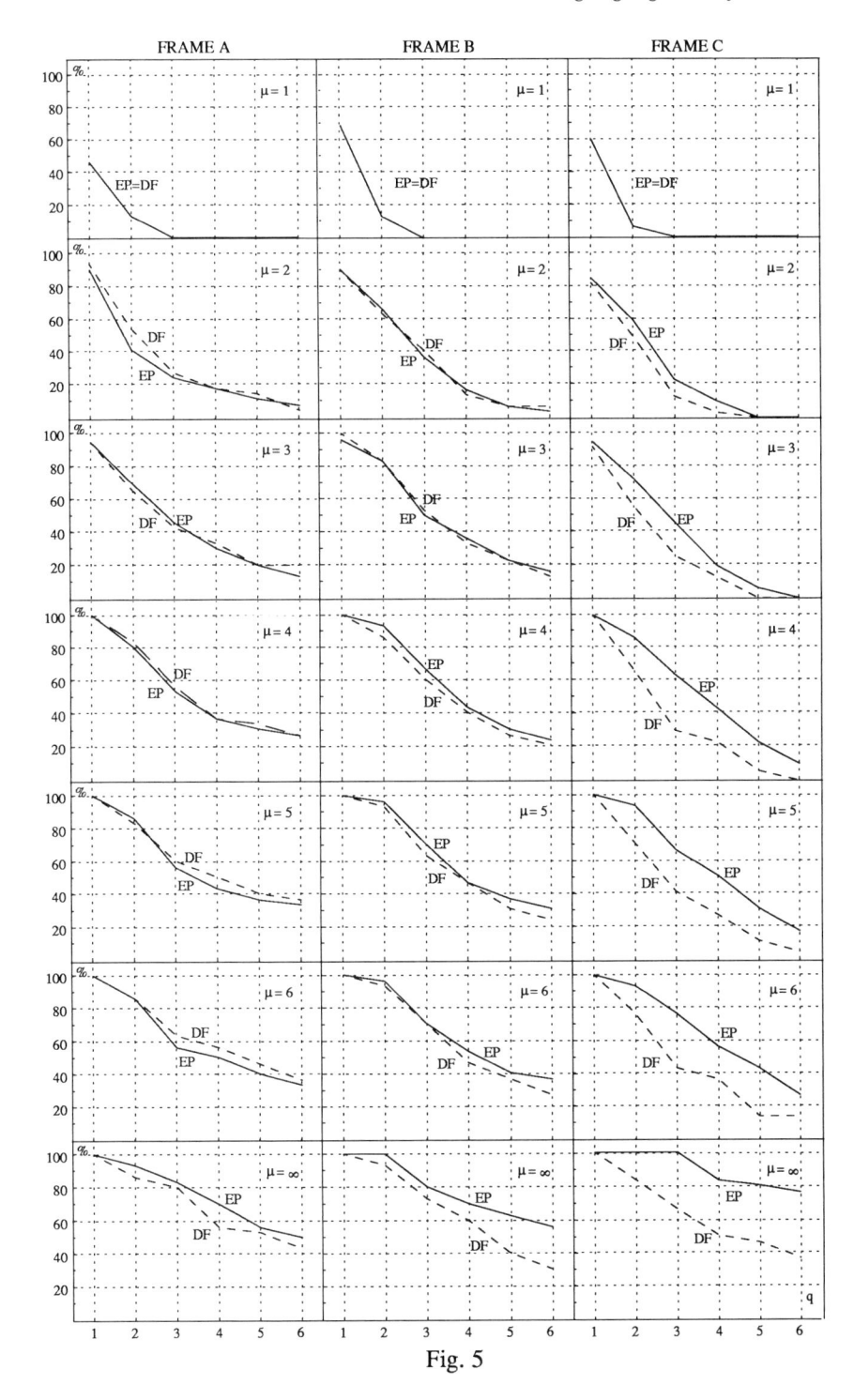

Fig. 5

showed the dynamic behaviour mostly different than the behaviour of corrispondent EP system and, certainly, the worst among those under examination.

It can be also said that a thin-walled member frame, showing a softening branch with a moderate slope (like frame A and frame B), has a dynamic response to real earthquakes quite similar to an ideal elasto-plastic structure. On the contrary it is more sensitive to the instability effects of vertical loads, because of the cyclic decrease of stiffness during the seismic event.

By means of the obtained results it is confirmed that one of the most important parameters, to judge the reliability of a thin-walled structure with respect of seismic loads and to assess a safe value of q factor, is the available ductility. If we assume as realistic value $\mu=3$, it can be seen that the maximum allowable q varies in the range 2-3 for all examined frames, where the minimum value corresponds to frame C. Requiring, restrictively, an higher probability level (e.g. 84%) of withstanding earthquakes, the q factor must be reduced in the range 1.5-2.

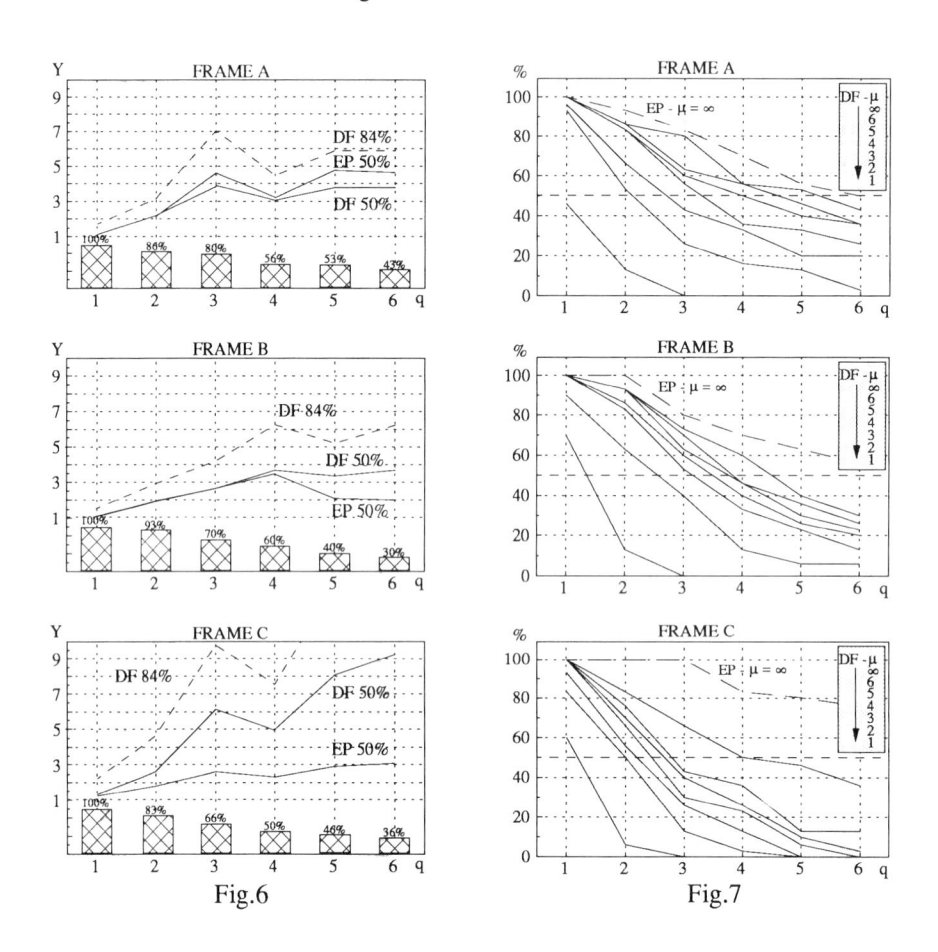

Fig.6

Fig.7

	FRAME A				FRAME B				FRAME C			
	Flim = 140000 N				Flim = 135000 N				Flim = 81500 N			
q	m (kg)	θ	T (s)	R	m (kg)	θ	T (s)	R	m (kg)	θ	T (s])	R
1	63827	0.0405	0.70	1.43	43760	0.0358	0.49	2.02	23444	0.0153	0.43	2.33
1.5	121894	0.1160	0.97	1.04	84981	0.1042	0.69	1.45	49277	0.0483	0.62	1.61
2	170371	0.2162	1.14	0.88	121100	0.1980	0.82	1.22	78015	0.1019	0.78	1.28
2.5	201034	0.3189	1.24	0.81	145360	0.2971	0.90	1.11	104119	0.1700	0.91	1.10

Table 2. Characteristics of the further analyzed frames.

6 Further analyses on the frames

To evaluate the influence, on the obtained results, of the procedure used to get different values for q factor and to perform a more realistic design process, another set of elaborations has been executed.

In the previous cases we magnified the accelerograms without changing the system mass. This time, instead, we have fixed the PGA to the effective values of the selected earthquakes (that provide an average PGA=0.15g), as we are designing the frames in a moderate seismicity zone. Since F_{lim} is also fixed, as the q factor increases, the frames are able to support a greater seismic mass, because the design acceleration is reduced. This is to say, in an actual design situation, that a designer can place, in a given building, a different number of resisting frames in relation to the q factor. In this way for each value of q we get a new frame, characterized by a specific mass, period and instability coefficient θ, as indicated in table 2. It can be underlined that the maximum value of q has been limited to 2.5. This is due to the two following reasons:
- the previous results confirms that q factor greater than 2.5-3 does not seem to be realistic for such type of structures;
- the increase of the mass makes greater the instability coefficient θ; so, for frame A and frame B, q values greater than 2.5 lead to θ values greater than 0.3, that is the limit stated by EC8.

These new frames are analyzed in the same way of the previous ones, by perfoming analogous dynamic analyses. The statistically elaborated results are reported in figures 8 and 9.

The results discussed in Section 6 are practically confirmed by these further analyses. It can be seen that, for frame A and frame B, a q factor equal to 2.5 gives a probability to whitstand the selected earthquakes at least of 50% with μ=3. For frame C the value of q can become non plus than 2, to have the same probability. In the same way, the ductility demand results quite similar for DF and EP systems, with regards to frame A and B, while, for frame C, shows greater scatters.

Substantial independence of the percentage data from μ, when it becomes greater than 4, can be noted in fig.8. This fact, together with the difficulties in the design procedure to get a q factor greater than 2.5, confirms that the instability problems, related with low elastic initial stiffness, are perhaps more important than ductility for such type of structures.

7 Conclusion

The wide dynamic analyses, performed on three different thin-walled member simple frames subjected to a large number of historical earthquakes, have confirmed that a degrading frame has a dynamic response quite similar to that of a corrispondent ideal elasto-plastic system (which represents a structure with compact-section members). Only when the slope of the softening branch become very hard, some remarkable differences can be noted.

This fact means that the design procedure and the choice of an appropriate behaviour factor is related especially to the shape of the F-δ curve, the lateral stiffness and the available ductility. In particular, if this last parameter is set to 3, it seems that the maximum value of q factor, which can be used, is in the range 2-3, at least for the analyzed cases. This low value however does not agree with EC8 provisions, that impose fully elastic design (q=1) for such type of structures with slender members (class IV of EC3).

All the way a more reliable assessment of the problems requires further reaserches both in theoretical and experimental fields. In particular it will be useful to study in deep the actual ductility of these structures, the post local buckling behaviour of the sections, the influence of the low cycles plastic effects, the role of the vertical loads with regards to the residual bearing capacity and the technological problems of the connection joints.

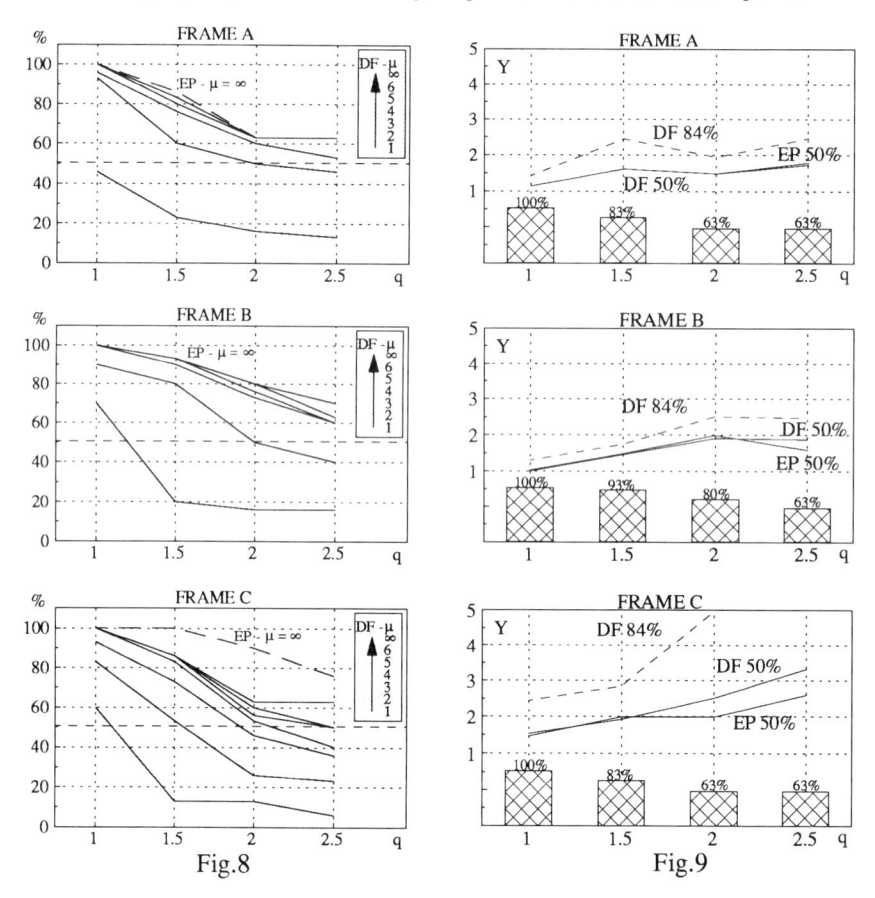

Fig.8

Fig.9

8 References

Blandford, G.E. Wang, S.T. and Wang, N.T. (1984) Geometric Non-linear Dynamic Analysis of Locally Buckled Frames, in the **Proceedings of the 7th International Specialty Conference on Cold-Formed Steel Structures**, St. Louis, Missouri.

Calderoni, B. De Martino, A. Ghersi, A. and Landolfo, R. (1993) Analisi del comportamento di strutture in acciaio con aste in parete sottile sotto carichi sismici, in the **Proceedings of the 6th Congress L'Ingegneria Sismica in Italia**, Perugia, pp. 755-764.

Calderoni, B. De Martino, A. Ghersi, A. and Sveldezza, M. (1993) Flexural behaviour of thin gauge members: portal frames under horizontal loads, in the **Proceedings of the 9th Congress C.T.A.**, Viareggio, pp. 209-220.

De Martino, A. Ghersi, A. Landolfo, R. and Mazzolani F.M. (1992) Calibration of a bending model for cold-formed sections, in the **Proceedings of the 11th International Specialty Conference on Cold-Formed Steel Structures**, St. Louis, Missouri.

De Martino, A. De Martino, F.P. Ghersi, A. and Mazzolani, F.M. (1990) Il comportamento flessionale dei profili sottili sagomati a freddo: indagine sperimentale. **Acciaio**, n.12/190.

Eurocode n.3 (1993) **Design of Steel Structures, Part 1.3, Cold-Formed Thin Gauge Members and Sheetings.**

Eurocode n.8 (1988) **Structures in seismic regions** - Design - Part 1. General and building.

European Convention for Constructional Steelwork (1988) **Recommendations for Steel Structures in Seismic Zones.**

Ono, T. (1984) Earthquake Resistance of Light-Gauge Steel Frames, in the **Proceedings of the 7th International Specialty Conference on Cold-Formed Steel Structures**, St. Louis, Missouri, pp.189-202.

Ono, T. and Suzuki, T. (1986) Inelastic Behaviour and Earthquake-Resistance Design Method for Thin-Walled Metal Structures, in the **Proceedings of the IABSE Coll. on Thin-Walled Metal Structures in Building** Stockholm, pp.115-122.

Takanashi, K. Ohi, K.I. and Chen, Y. (1992) Collapse Simulation of Steel Frames with Local Buckling, in the **Proceedings of the 10th World Conference Earthquake Engineering,** Madrid, pp.4481-4484.

Walker, A.C. (1975) **Design and Analysis of Cold-Formed Sections.** International Textbook Company Limited, London.

Watanabe, E. Sugiura, K. and Yamaguchi, T. (1992) Strength and ductility of thin-walled beam-columns with low yield steel and round corners, in the **Proceedings of the 10th World Conference Earthquake Engineering,** Madrid, pp.2897-2902.

29 A METHOD FOR THE EVALUATION OF THE BEHAVIOUR FACTOR FOR STEEL REGULAR AND IRREGULAR BUILDINGS

I. VAYAS, C. SYRMAKEZIS and A. SOPHOCLEOUS
National Technical University of Athens, Athens, Greece

Abstract
A method for the determination of the behaviour factor q for steel framed structures is presented. It is based on elastoplastic second order analysis of the frame. The definition of the limit state is such that the local ductility of structural members is exhausted. 3 local ductility levels are introduced for which different values of the q-factor may be determined. The infuence of various irregularities on the value of the q-factor has been studied. The studies refer to stiffness and strength reductions, to the geometric configuration of the frame and the capacity design criteria.
Keywords: Steel frames, Seismic Design, Behaviour Factors, Irregularities.

1 Introduction

It is widely known that according to the requirements of most modern Seismic Codes, a conventionally building structure shall be designed such as to ensure:
a) that structural damages are limited in the event of a moderate earthquake which has a large probability of occurence during its design life and
b) that the structure shall not collapse during a strong earthquake with a small probability of occurence.
Consequently, inelastic deformations of parts of the structure are allowed for under a severe earthquake which implies that part of the imput seismic energy may be dissipated through inelastic action. This results in a significant reduction of the seismic induced forces that would develop if the structure behaved elastically. This reduction takes in most Codes the form of a behaviour factor, denoted in Eurocode 8 (1993) as q, to be applied to the forces that are derived from an elastic response spectrum.
  The Scope of this paper is to present a method for the determination of the behaviour factor for unbraced plane frames and to establish criteria referring to local and

global member ductilities, frame irregularities and other design rules that have an influence on the value of the behaviour factor.

2 Proposed method for evaluating behaviour factors

There is a variety of methods concerning the evaluation of the behaviour factor for a framed structure. They range from inelastic dynamic analysis of the complete frame for various accelerograms, to approximate dynamic analyses of equivalent elasto-plastic single degree of freedom systems or methods based on the energy approach. A detailed description of selected methods is given by Guerra et al.(1990) where reference in made to. The present method is based on the fact that for a multi-storey unbraced frame, inelastic deformations are due to the formation of plastic hinges in its members. The amount of energy dissipation is equal to

$$E_p = \sum M_p \theta_{p\ell}$$

where M_p is the plastic moment at a member cross section when a plastic hinge is formed,
$\theta_{p\ell}$ is the plastic rotation of that plastic hinge and the summation refers to all plastic hinges (Fig.1)

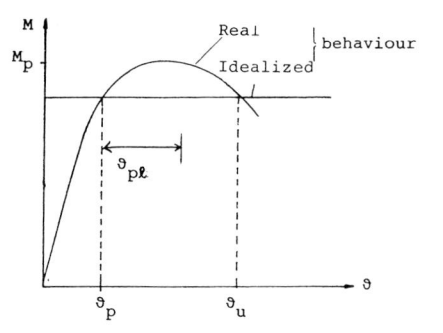

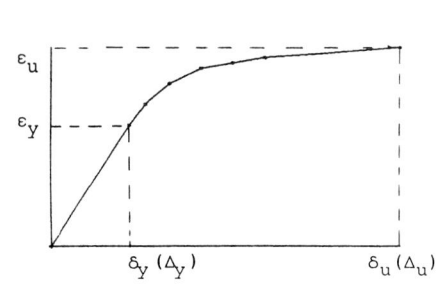

Fig.1.Moment-rotation diagrammes of a beam

Fig.2.Horizontal load-deflection curves

The value of the behaviour factor is though directly related to the number and distribution of the plastic hinges, since this factor is expressing the ability of the structure to dissipate energy. The above referred quantities may be determined by elastoplastic analysis as described subsequently.

Of a multi-storey building frame its topology, the cross sections and material prorerties of its members and the gravity loads W existing at the time of the seismic event are considered to be known. The base shear of the frame is given by

$$F_d = \varepsilon \sum W \qquad (1)$$

where ε is the seismic coefficient and

$\sum W$ is the sum of the quasi-permanent gravity loads. The distribution of F_d along the height of the building is assumed to follow the fundamental mode of the frame. This assumption is valid if the flexibility of the frame is limited to such an extend, that the contribution of higher modes of vibration is negligible. For other cases, the method is still valid in an approximative way considering the overall degree of accuracy.

Initially a frame analysis is performed by considering the gravity loads as acting alone. Subsequently the horizontal seismic forces are applied and the frame is analysed according to the second order plastic hinge theory. By considering constant gravity loads the base shear is increased stepwise. This corresponds according to eq.(1) to an increase of the seismic coefficient ε. This loading pattern represents the actual loading conditions that arise during a seismic event, where the seismic induced loads vary with the time while the gravity loads remain constant. The seismic coefficient ε is increased at each step to such an extend that a new plastic hinge is formed. After the formation of a plastic hinge at the end of a member its plastic rotation starts and is increasing by subsequent loading. The limit state of the frame is defined as the state at which the plastic rotation of a plastic hinge exceeds its permissible value. The relevant criterion may be expresed as

$$\text{act } \theta_{p\ell} = perm\ \theta_{p\ell} \qquad (2)$$

where act $\theta_{p\ell}$ = actual plastic hinge rotation and
perm $\theta_{p\ell}$ = permissible rotation.
Usually the limit state of a frame is considered to be reached either at the attainance of a complete plastic mechanism or at the loss of stability. Both states can be however controlled by expression (2).

When the loading procedure is completed and the limit state is attained, the load-deformation curves may be drawn. The deformations refer to both the top storey horizontal displacements or the interstorey drifts Δ_i. The loads are represented by the values of the seismic coefficient ε.

The yield values of ε_y, δ_{yi} and Δ_{yi} are defined as those values that correspond to the first application of the horizontal load since at that load level the first plastic hinge is formed. The relevant values at the limit state are denoted by ε_u, δ_{ui} and Δ_{ui} (Fig.2).

The overstrength of the frame is expressed by the relation

$$\Omega = \frac{\varepsilon_u}{\varepsilon_y} \qquad (3)$$

the ductility ratios in terms of displacements by

$$\mu_{\delta i} = \frac{\delta_{ui}}{\delta_{yi}} \quad \text{or} \quad \mu_{\delta rel} = \frac{\Delta_{ui}}{\Delta_{yi}} \qquad (4)$$

In a single-storey building the two ratios μ_δ and $\mu_{\delta rel}$ are identical. For multi-storey buildings they differ from storey to storey. The displacement ductility ratio of the frame is then a weighted average of these ratios for the different storeys. In the present work the displacement ductility ratio of the frame is taken equal to μ_δ since the displacements of the top storey represent in a way the entity of the frame deformations.

The behaviour factor of the building is then given by the relation

$$q = \mu_\delta \qquad (5)$$

It shall be noted that according to the present methodology only monotonic loading is considered. In reality however the seismic loads act in a cyclic way imposing thus additional ductility requirements to the structural members of the frame, since ductility shall be provided not only for monotonic but also for cyclic loading. This shall be taken into account when the permissible values of the plastic rotations are evaluated. For such a procedure reference is made to Vayas, Psycharis (1994). The above mentioned procedure supplies lower bounds of q since it does not account for energy dissipation through damping. This influence is generally taken into account by considering elastic response spectra for a definite value a damping.

3 Effects of local member ductility

Acording to EC8 (1993) 3 classes of cross sections with various strength and ductility levels are defined. As a criterion for the classification serves the slenderness of their compressed plated parts (webs, flanges). The classes are characterized as A, B and C and correspond to high, medium and low ductility. Different values of the behaviour factor as given by eq.(6) are assigned to those classes:

Class A $\quad 5<q<6$ (6a)
Class B $\quad 4<q<5$ (6b)
Class C $\quad 2<q<4$ (6c)

The influence of these local ductility levels on the value of the behaviour factor will be studied subsequently. In order to do this, the permissible values of $\theta_{p\ell}$ for the 3 classes shall be estimated.

The ductility ratio of a beam in terms of rotation is defined as the ratio (Fig.1).

$$\mu_\theta = \frac{\theta_u}{\theta_p} \qquad (7)$$

After the formation of a plastic hinge the continuity between the joint and the end of a member is interrupted. The difference

$$\theta_{p\ell} = \theta - \theta_p \qquad (8)$$

is expressing the plastic rotation of the member which is the relative rotation between its end and the joint. Since the value of θ_p and thus the inclination of the M-θ curve is dependent on the overall geometric configuration it is advisable to make the verification of the plastic hinge rotation in terms of the absolute value of the plastic rotation, as indicated by eq.(2) rather than in terms of ductility ratios μ_θ, since θ_p includes the total joint rotation before the formation of the plastic hinge.

According to the background information provided by EC3 (1989) and EC8 (1988), the cross section classification was done by making use of experimental and theoretical results on simply supported, tranversely loaded beams. From the maximum moment-end rotation curves of these investigations, the resistance and ductility of the beams were derived. There is no clear relationship between the class of cross section and the relevant ductility ratio μ_θ, because there are in addition to the characteristics of the cross section other parameters influencing μ_θ (moment gradient, plastic hinge length etc). However from all available information, the following values of μ_θ in respect to the class of the cross section may be approximatively used:

Class A $\quad \mu_\theta^* = 8$ \hfill (9á)

Class B $\quad \mu_\theta^* = 4$ \hfill (9b)

Class C $\quad \mu_\theta^* = 2$ \hfill (9c)

From these values of μ_θ^*, the ductility ratios in terms of curvature may be approximatively determined according to

$$\mu_k = \left(\frac{\mu_\theta^* - 1}{2}\right)\frac{\ell}{\ell_p} \qquad (10)$$

The values of μ_K are, in contrast to μ_θ^*, a function of the cross section characteristics only and not of the loading and restrain conditions of the system.

For members that are part of a framed structure it shall be considered that the end conditions correspond to a fixed rather than to a simply supported situation. For a fixed beam the relation between μ_K and μ_θ may be approximatively written as

$$\mu_k = \frac{\mu_\theta - 1}{6} \frac{\ell}{\ell_p} \qquad (11)$$

From eqs.(7) to (11) and the assumption that the plastic hinge length ℓ_p is approximately equal to $0,10\ell$ the following values of the permissible plastic rotation at any plastic hinge may be derived:

Class A: perm $\theta_{p\ell} = 2,1\ k_p.\ell$ (12a)
Class B: perm $\theta_{p\ell} = 0,9\ k_p.\ell$ (12b)
Class C: perm $\theta_{p\ell} = 0,3\ k_p.\ell$ (12c)

The application of the proposed method will be shown on calibration frames whose behaviour factors have been determined by other authors (Fig.3).

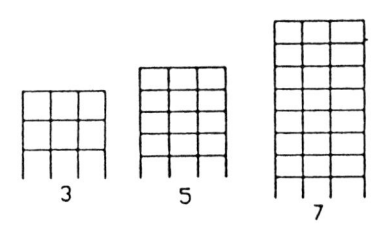

FRAME	L	H	BEAMS	COLUMNS
3	400	400	IPE330	HE240B
5	400	300	IPE360	HE280B
7	400 400 500	300	IPE450	HE360B

L=Beam spans in cm
H=Storey heights in cm

Distributed vertical load 52KN/m

Fig.3. Calibration frames

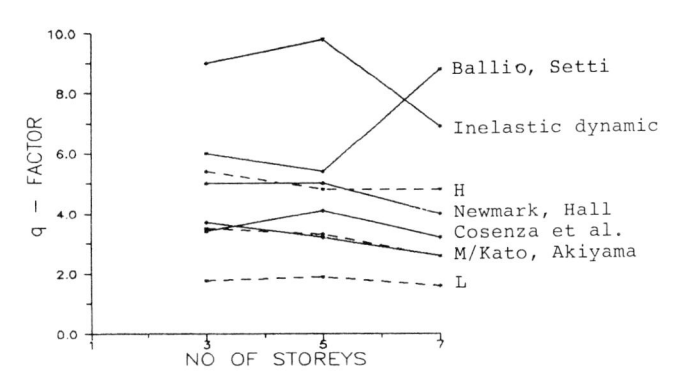

Fig.4. Behaviour factors for the frames

The values of the q-factors determined by various methods as presented by Guerva et. al.(1990) and the present method for the 3 local ductility levels are shown in Fig.4. The results indicate that for class A and B cross sections eqs.(6a) and (6b) are valid but that eq(6c) must be changed to

    Class C: q<2

4 Effects of Stiffness distribution

Frames are considered according to EC8 as regular if for the stiffnesses of the vertical elements

$$0.8 < \frac{x_i}{x_{i+1}} < 1.2 \quad \text{and} \tag{13a}$$

$$0.9 < \frac{x_i}{\bar{x}_i} < 1.1 \tag{13b}$$

where $\bar{x}_i$ is the mean value of the stiffness at each storey i. In

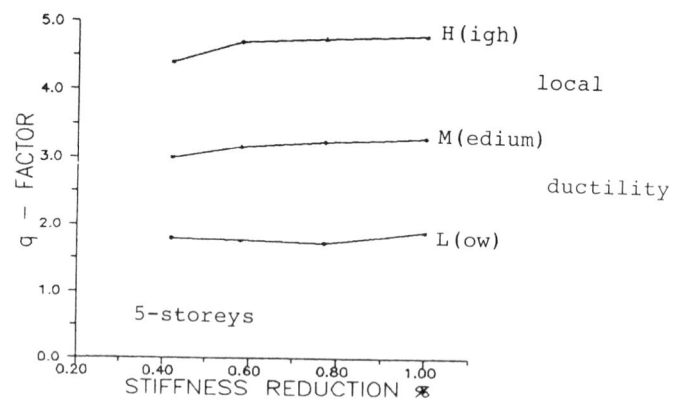

Fig.5. Effects of levels of stiffness reduction (1st floor)

order to verity the validity of this rule a number of frames have been analysed where the stiffnesses at specific storeys of the frames of Fig.3 have been reduced. Fig.5 presents the q-factors of the 5-storey building in terms of the stiffness reduction of the 1st floor. It may be seen that the stiffness reduction does not influence the behaviour factor as strong as the EC8-rules predict. In the 7-storey frame of Fig.3, the stiffnesses of the 1st, 3rd and 5th floor were reduced by 50%, in order to verity if it is important at which floor the reduction takes place. The results are shown in Fig. 6. The level of the soft storey seems to play a secondary role in the value of the behaviour factor.

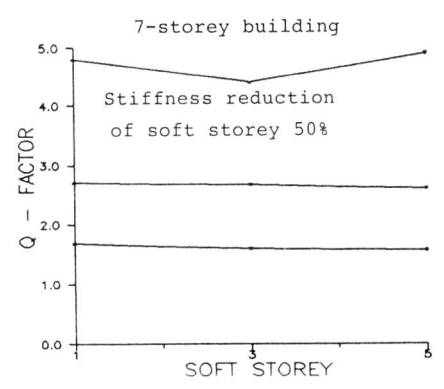

Fig.6. Effects of stiffness reduction at various storeys

Fig.7 shows the q-factors for the 3 buildings of Fig.3 compared to the corresponding values of the same frames with a 50% stiffness reduction of the 1st floor. The differences seem to be very small.

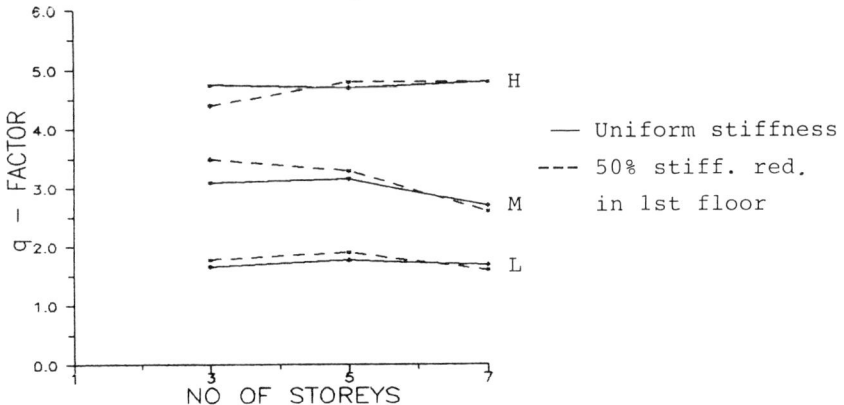

Fig.7. Effects of stiffness reduction for variour buildings

The above results, although limited in number, indicate that the stiffness distribution is not a decisive factor for the q-factor.

5 Effects of strenngth distribution

For the distribution of strength along the height of the building similar expressions as (13) are proposed by EC8. Xi is then referred to "strength" instead of "stiffness". For the 5-storey building of Fig.3 both strength and stiffness of the ground floor were reduced by 30%, and 50% correspondingly. The results are shown in Fig.8. The q-factor for ductility level H is drastically reduced although that the column resistance moments at the joints are still larger than the corresponding ones of the beams. It seems that the q-factor of a frame is greately reduced by a combination of a soft with a weak storey.

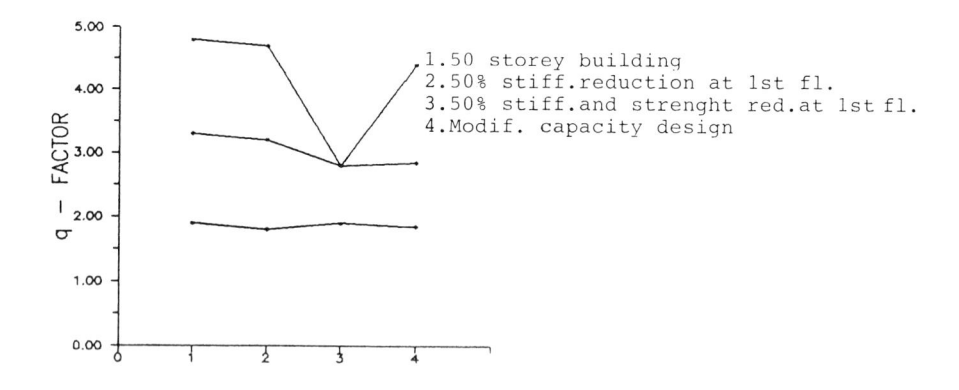

```
          5.00
                                    1.50 storey building
          4.00                      2.50% stiff.reduction at 1st fl.
                                    3.50% stiff.and strenght red.at 1st fl.
    FACTOR                          4.Modif. capacity design
          3.00

    | 2.00
    q

          1.00

          0.00
               0    1    2    3    4
```

Fig.8. Effects of strength variations

6 Capacity design criteria

The capacity design criterion of EC8 is defined such that at each frame joint the sum of the column resistance moments shall be larger that the beam resistance moments. In order to possibly relax this rule, the 5-storey building of Fig.3 was analysed considering in the intermediate span of the 1st floor a beam of IPE 500 instead IPE 450. In that manner the capacity design criterion is not fullfilled in the two central joints of the 1st storey. However the sum of the column resistance moments of all 4 joints of the 1st storey is larger than the resistance moments of all beams of this storey. The q-factor as shown in Fig. 8 is little affected as compared to the initial frame, indicating that it is possible to relax the very strict capacity design criterion of EC8.

From all the above presented results it is evident that the q-factors for only the high local ductility level are influenced by possible non-regularities. For the low local ductilility level the frame behaviour is quasi elastic resulting in low values of q, that are not sensitive to other boundary conditions.

7 Effects of geometry

Deviations of the frame geometry from the orthogonal form constitute irregularities according to EC8. In order to study the geometric effects, various frames with setbacks in the vertical were studied. The relevant q-factors are shown in Fig.9. It may be seen that the q-factors for frames with setbacks are indeed reduced for the high ductility level, so that a corresponding reduction in the Code provisions seems to be justified.

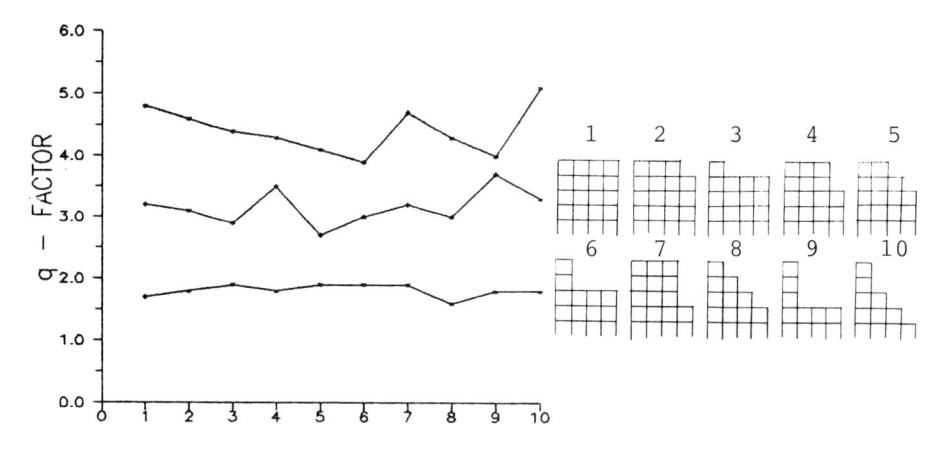

Fig.9. q-factors for frames with setbacks

8 Conclusions

From the results of the proposed method for the determination of the q-factor for unbraced plane frames the following conclusions may be drawn:
a) The q-factor strongly depends on the local ductility of the frame members.
b) Stiffness reductions are not largely influencing the values of q.
c) Stiffness and strength reductions simultaneously present at the same storeys may strongly reduce the q-factor.
d) The capacity design criterion as proposed by EC8 may be relaxed. A possible relaxation could be such that the sum of the column resistance moments shall be greater than the sum of the beam resistance moments at the storey level instead of the joint level.
e) For low local ductility levels the values of q are low and insensitive to other irregularity conditions since the frame behaviour is quasi elastic.

The above presented results are initial results of an ongoing research programme sponsored by the Greek Earthquake Planning and Protection Organisation whose finacial support is greately aknowledged. More conclusive evidence will be drawn after the completion of the programme.

References

Background document No 5.02 of Eurocode 3.(1989) Commission of the European Communities.
Eurocode 8 (1993). Structures in Seismic regions. Design. Commission of the European Communities.

Background Documents for Eurocode 8, Part 1 (1988). Vol 2-Design rules (1988). Commission of the European communities.

Guerra, C.A.,Mazzolani, F.M.,Piluso, V. (1990). Evaluation of the q-factor in steel framed structures: State of the art, Ingegneria sismica, 2, 42-63.

Vayas, I. and Psycharis, I. (1994). Local cyclic behaviour of steel members. Intern. Workshop STESSA, Timisoara, Champman and Hall.

PART FOUR
EXPERIMENTAL METHODOLOGY

30 EXPERIMENTAL METHODOLOGIES: GENERAL REPORT

D. DIACONU and A.-C. DIACONU
INCERC-IASSY, Romania

ABSTRACT
This report include two parts:
    1-st PART - Main methods , procedures and experimental techniques used for monothonic, cyclic-reversal, pseudo-dynamic and dynamic-seismic tests are synthetically presented, for:
    A) Materials
    B) Elements, connections, subassemblages
    C) Structural models, portions of structures and structures.
    2-nd PART - A synthesis regarding the contributions brought by the Experimental Methodology section papers is presented.

PART 1

A. MATERIALS

1. General aspects
The materials used in resistant structures of buildings subjected to dynamic-seismic loading, are characterised by their strength, deformability and energy dissipation properties in both elastic and inelastic range of behaviour, which are different from their known static or dynamic (fatigue) properties. This properties are mostly dependent on both the dynamic excitation type and complex characteristics of the materials. Specific concepts and testing methods are needed for this purpose.
In order to establish the phisico-mecanical properties of building materials, the following types of methods are generally used :
    - methods based on testing installations providing monothonic increasing and decreasing loading aimed to determine complete σ-ϵ relationship;
    - methods based on dynamic facilities (pulsators) which subject the material to a large number of loading cycles (fatigue) of high frequences and constant or variable excitation level, the material beeing initially loaded at service stress level (σ_o) and then overloaded under cyclic stress providing the yelding;
    - methods using dynamic facilities (pulsators) operating at low

Behaviour of Steel Structures in Seismic Areas. Edited by F. M. Mazzolani and V. Gioncu.
Published in 1995 by E & FN Spon, 2-6 Boundary Row, London SE1 8HN. ISBN: 0 419 19890 3.

frequencies (up to 15 Hz) under a limited number of cycles (from 5 to 10), with the maximum stress near proportionality or yelding limit of the material. The tests are performed at a constant referece stress level (σ_0) initial loading of specimens, representing the level of permanent loads.

2. Low-cycle fatigue concept

The low-cycle fatigue phenomena of the material were less investigated then the dynamic fatigue, entering only in the last decade in the systematic concern of the researchers.

The fatigue under a reduced number of loading cycles of the material is related to the material strength behaviour under seismic action, modeled as a dynamic action having a certain number of loading cycles of specific kinematic parameters.

The low-cycle fatigue strength represents the maximum stress that the material is able to resist without the destruction of the selected specimen for a given type of stress (tensile, compressive, torsion, etc.) under a prescribed number of loading cycles simulating the seismic action (as regarding the frequency, duration, number of reversals, amplitudes, etc.).

The fatigue resistance to a small number of cycles (R_{onrc}) is a function of both intrinsic characteristics of material and the loading history simulating the seismic action and its kinematic characteristics.

Is necessary to specify that the main types of "time-histories" (monothonic increasing, sinusoidal, constant cyclic stress, constant or gradually increasing cyclic deformation), the intensity and rate of loading used in various investigations on the behaviour of building materials under a reduced number of uniaxial or multiaxial loading cycles are intending to represent the seismic action.

Analysis of some actual severe earthquake records have revealed that the number of repeated cycles of relatively large intensity is less 50...100.

Although during the last 10 years, the concern for this phenomenon has increased continuosly, definite general conclusions and recomandation for fatigue strength of materials under a reduced number of cycles have not however been emphasized and each research laboratory used specific experimental method or metodology.

An alternative to determine the materials strength under a reduced number of cycles is presented as an interesting example in the aseismic design of structures. This results from the tension (σ) and the logarithm of the number of cycles (lg N) relationship.

The following variation diagrams σ-lg N have been identified :
   a) liniar diagram, given as follows :

$$\sigma_N = (a - b*lgN)*R_{us} \qquad <daN/cm^2> \qquad (1)$$

in wich : R_{us} is the ultimate static strength of the material under a certain loading,
   a,b are empirical coefficients depending on material and type of loading,

b) nonlinear diagram of the form (2) or (3) :

$$\sigma_N = C_v * R_{us} * (1 - \frac{lgN}{lgN_0} * (1 - \frac{1}{a-b*\rho})) \qquad <daN/cm^2> \qquad (2)$$

The relation (2) can be used to evaluate the materials strength under any number of repeated cycles and any coefficient of symmetry (ρ). No is the reference number of cycles and Cv is the coefficient of variation.
Other investigators have proposed the following relation for the material strength under reduced number of loading cycles :

$$\sigma_N = R_{us} * (\alpha - \beta * lgN) \qquad <daN/cm^2> \qquad (3)$$

in wich : $\alpha = R_{10}/R_{us}$ is the ratio of the strength for a single cycle

(N = 1) to the ultimate static strength R_{us};
$\beta = (R_{01}-R_{10})/(R_{us}*lgNo)$

The value of resistance at reduced number of cycles (R_{onrc}) is generally above the static resistance limit (R_{us}) and bellow the shock resistance (R_{o1}) depending on both the loading rate and the deformability characteristics of the materials. Thus, the more casant is the material, the smaller should be the difference between R_{onrc} and R_{o1}. A similar effect is observed with regard to the influence of the rate of loading.

3. Dynamic deformability of materials

The determination of the dynamic modulus of elasticity can be performed mainly by means of:
- distructive testing methods (involving the characteristic σ-ϵ curve of a material subjected to a dynamic or seismic action),
- nondistructive testing methods such as : resonance methods, ultrasonic impulse methods, complex modulus method and mechanical impedance method.

The analysis and interpretation of experimental data regarding the characteristic relationship σ-ϵ under a reduced number of loading cycles, may be based on :
a) envelope characteristic curves for which various mathematical relations have been proposed;
b) loading-unloading characteristic curves under a reduced number of cycles.

The methods used for determination of damping are differentieted according to the damping characteristic factor. Generally, two experimental techniques are used in either free vibration regime or forced vibration regime at resonance.
These methods take the form of :
a) logarithmic decrement method;
b) half power points method and

c) method based on the evaluation of energy dissipated per cycle.

4. Remarks

Experimental results obtained in Romania, regarding some phisico-mechanical characteristics for OL-37 steel, loaded under a reduced number of cycles (N), with coefficient of simmetry $\rho=0$ and the frequency of 100...300 cycles/min, provided the following relations:

$$E_d = (1.04 \div 1.26) E_{st}$$

$$\sigma_N = R_{us} * (1.34 - 0.131 * lgN) \qquad <daN/cm^2>$$

In spite of all existing difficulties it is necessary to elaborate an experimental methodology recognized in all countries, in order to determine strength, modulus of elasticity and the loss coefficients for different type of steels subjected to a reduced number of loading cycles over an initially service stress level (σ_o) or a zero initial stress state.

B. TESTS ON ELEMENTS, CONNECTIONS AND SUBASSEMBLAGES

1. General aspects

Experimental investigations pursue the collection of data needed to cover some goals regarding models and computations :
- clarification of some computation relations and design methods to be used in practical applications;
- verification of strength and deformability capacities for resistance qualification or comparison;
- validation of certain strengthening solutions.

The majority of experimental works are refering to the behaviour of elements, joints and subassemblages under bending moments, shear forces, axial forces and torsion considered as acting separatelly or in various combinations as well as to the assessment of the strength and deformation capacities available at certain damage ranges. Another goal in experimental works refer to the application of various strengthening procedures.

Apart from the availability of some strain-stress hysteretic curves for sections, individual members and structural subassemblages, the experimental works involving dynamic-seismic actions have as objectives to put in evidence the bond characteristics of various elements including their connections.

The problems arised in undertaking any experimental program of research on dynamically loaded elements, connections and subassemblages are :
- selection of representative member, joint or subassemblage within the structural system under consideration;
- restoring the geometric-mechanical conditions of stress continuity at marginal sections for the tested specimen to satisfy the

requirements regarding the real stress state within the structure;
- establishing the dynamic similitude criteria when the expeimental conditions do not permit a full test scale .
The experimental process can be conducted into two main manners :
   a) according to force calibration, with free deformations;
   b) according to prescribed deformation calibration, with corresponding resulted forces.
For the time being the method can be applied only to static monothonic increasing, static alternant and pseudo-dynamic loading tests.
For dynamic-seismic loading tests, the amplitude and frequency calibration of the excitation force P(t) is necessary. Positive or negative alternate actions (with regard to the stress) or certain variations in the positive zones only, may result (see fig. 1).
To accomplish the mentioned requirements, special facilities have been constructed both at INCERC-Iassy and at INCERC-Bucharest to allow experimental investigations on elements, joints, planar or spatial subassemblages at full scale or modeled at relatively large scale.
The INCERC branch of Iassy has an universal testing stands for elements (fig.2) connections (fig.3) and planar or spatial subassemblages (fig.4a,b).The subassemblage testing facilities have the following characteristics :
* the loading apparatus (fig.4a)
   - maximum value of P_i = 5000 daN
   - maximum value of N = 100000 daN
   - maximum value of H(t)= 12000 daN, which can be applied monothonic statically or alternate dynamically, whith a constant or variable frequency of 0...5 Hz. The loads P_i and N are maintained constant during a test while H(t) can be applied along any direction.
* the loading apparatus (fig.4b)
   - maximum value of P_i = 25000 daN
   - maximum value of N = 200000 daN
   - maximum value of H(t) = 30000 daN, which can be applied monothonically, alternate statically or alternate dynamically with constant or variable frequency of 0...30 Hz.
In the near future the loading system will include seismic and random loading capabilities.
It should be mentioned that both P_i and N are maintained constant during a test by means of a hydraulic compensation system .

2. Remarks

The experimental works performed on elements, joints and planar or spatial subassemblages, provide useful data to point out and to rise knowledges regarding some specific phenomenas. The quantification and generalization of these data into computing and testing models, offer the possibility to the design engineers to develop computation procedures to evaluate the behaviour of the structural assembly at various loading stages.
Before conducting such experimental works, additional parametric studies and various behaviour models are needed to identify the zones of interest and the characteristic sections.
One goal of the experimentation on individual elements, connections

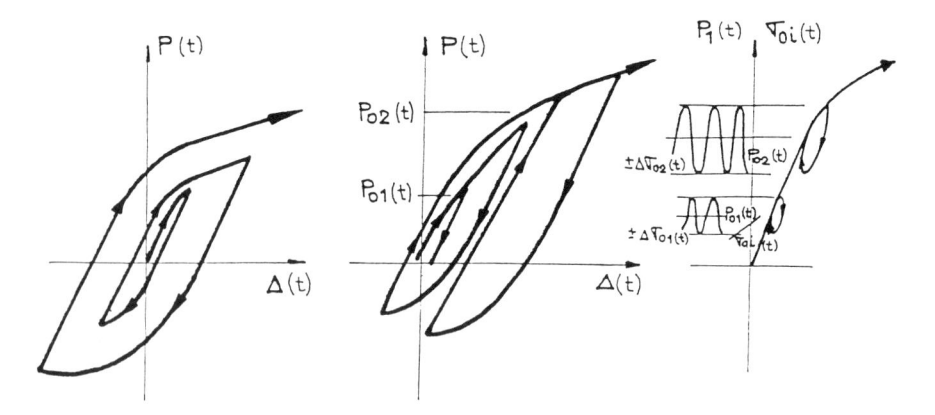

Fig. 1. Dynamic-seismic loading types.

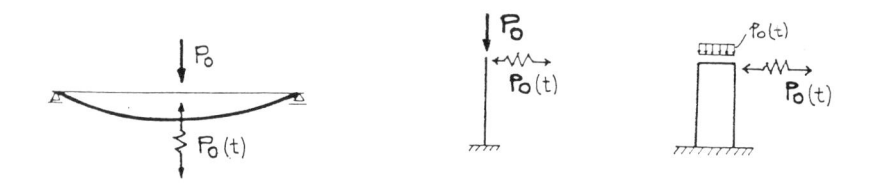

Fig. 2. Testing stands for elements.

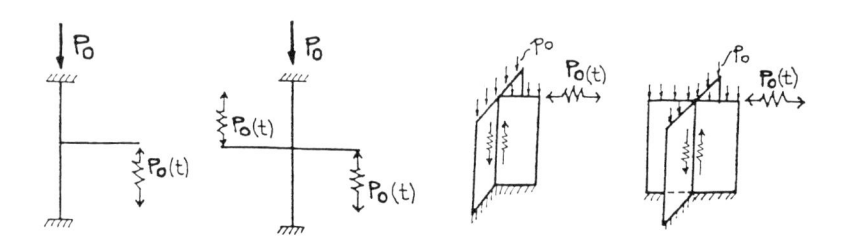

Fig. 3. Testing stands for connections.

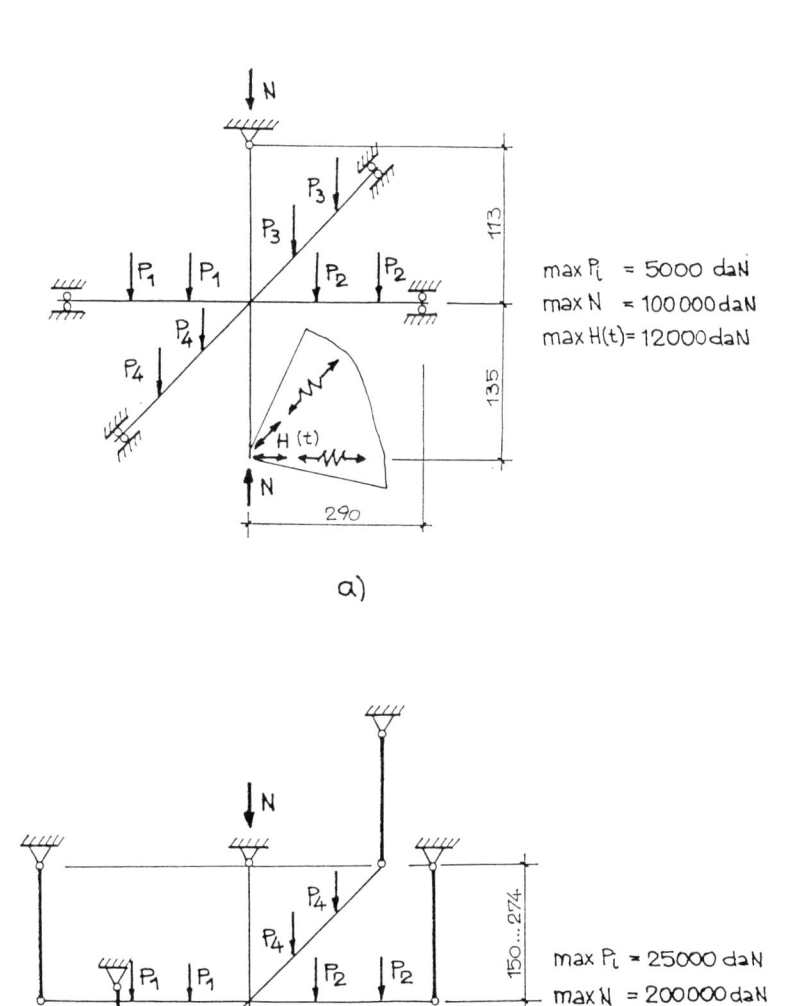

Fig. 4. Testing stands for subassemblages.

and subassemblages is to filtrate and to optimize the solutions, leading to efficient structural assemblies.

In this context the harmonization of testing methods is necessary to be undertaken, and in the meantime, those methods should be correlated to the existent experimental facilities.

C. TESTS ON STRUCTURAL MODELS, PORTIONS OF STRUCTURES AND STRUCTURES

1. General aspects

The experimental works involving dynamic-seismic loding, are generally performed on various structural assemblies and fragments of civil engineering structures (buildings, factories, agricultural and engineering constructions).

The structural assemblies can be tested in situ or in specialized laboratoriess equipped with suitable dynamic simulators (shaking tables of various capacities).

The shaking table experiments involve a complex testing methodology and a suitable setup in equipment and personnel.

The main objective of such an experiment is to determine the characteristic components of the global seismic response related to the structural type under consideration, the behaviour stage and the energetic level of the programed table motion.

The elements characterizing the global seismic response comprise :

- variation of dynamic characteristics (periods, modal shapes and damping ratios) in various behaviour stages up to structure failure,

- modification of lateral stiffness matrix coefficients in various stages up to failure, including the problems of simmetry and matrix conformation;

- development of hysteretic curves ($M-\Phi$, $F-\Delta$ and so on) for critical sectins and structural system as a whole;

- assessment of acceleration distribution both along the structure height and within the horizontal plane, distribution of lateral drifts and lateral displacements as well as variation of dynamic stresses within characteristic sections;

- evaluation of average curvatures at selected sections of both vertical and horizontal members, as well as strains;

- evaluation of relative rotations between walls and floors;

- study of the dynamic of floors including their deformability in a horizontal plane to ascertain the role of the horizontal components in the distribution of the seismic forces to vertical structural elements;

- to obtain experimental data on the interaction of structural elements of different stiffnesses as well as the interaction between structural and nonstructural elements;

- to study the role of infrastructure upon seismic stability and resistance;

- to investigate the soil-structure interaction problems in earthquake engineering which require cooperation of different disciplines;

- to identify the mechanism of seismic energy dissipation and failure mechanism of a structure.

The difficulties arised in shaking table tests refer mainly to establish the test methodology. Two testing methods are developed in connection with the use of the shaking tables:

a) A method involving the dynamic seismic load application with gradually increasing intensities using different compositions of actioning programs (sine sweep, sine beat, continous sine near the resonance of tested model, actual earthquake records or artificial earthquakes selected according to a certain criteria, narrow and large band white noise and other dynamic excitations).

The dynamic tests are supplemented with static tests under lateral horizontal forces applied at different stages of stiffness degradation to obtain the lateral flexibility matrix coefficients and their variation during experimental program.

b) A method involving the dynamic-seismic load application directly at high intensity levels (design level or damage level). This method could lead to structure failure during a single test run and could waste a series of experimental information that could not be recorded. The procedure is employed especially in qualification tests of technological equipment.

The difficulties encountered in such experiments are :
- selection of the experimental model with or without considering the influence of foundation soil;
- to establish the similitude requirements ;
- calibration and fidelity of driving forces according to the programed dynamic action or recomended site spectra;
- the distribution and calibration of whole system related to structural response (data aquisition, equilibration, amplification, plotting, etc.);
- storage and processing of experimental data by means of specific computer programs (the INCERC-Iassy system can perform simultanously the recording of 100 data chanels in dynamic-seismic range);
- theoretical correlations, investigations and generalizations;
(A special attention is given to the problems concerning some computations scenarios or the methods for decondensation or condensation of the models which imply many complex tasks and require careful correlation with theoretical studies.)
- error checking and identification of error sources (due to consideration of mass, damping, stiffness, loads, as well as due to calibration of actioning and response chain, including data processing and interpretation);
- to identify and evaluate the whole assembly of problems regarding the dynamic interaction between structural and nonstructural components, structure and foundation soil at various loading intensities and to establish appropriate hysteresis models.

In situ tests under dynamic-seismic loading refer especially to the etermination of dynamic characteristics (frequencies, modal shapes and damping values) at various damage stages, used as opportunities to evaluate directly the construction quality.

These tests involve:
- the choice of magnitude for the structural response and the manner of testing (microtermors, vibration generator, pull-back tests, controlled detonations in surrounding ground, explosions or real earthquake motions);
- the distribution of the response measurement points, whithin the

structures;
- the location of transducers, equilibration, recording and storage chain and the processing of the obtained data;
- some admissibility criterias to correlate the analysis of experimental data in order to quantify specific phenomenas for certain type of structures and loading, are require. Those criterias should be based mainly on comparing the valuess of experimental results with admissible limit values, also related to the energetic level of the excitation.
The admissibility limits are to be particularized in function of constituent materials, structural configuration, sensibility of the structural system to dynamic-seismic loading and used testing methodology.

2. Remarks

In this context the development of a new set of guidelines is proposed to harmonize the methods and the procedures used in conjunction with seismic behaviour (from monothonic, alternate cyclic or dynamic-seismic loading) for materials, members, junctions, structures or their models and technilogical equipment.

PART 2

CONTRIBUTIONS OF EXPERIMENTAL INVESTIGATIONS SYNTHESIS

With respect to the needs and globaly presented methodologies in the first part of this report, the experimental works registred in the present scientific meeting deals with a large number of specific problems having distinct implications in pointig out the earthquake motion complex phenomenas using elaborated experimental investigations.
The aim of the first paper by E. Cosenza, A. de Martino and G. Manfredi is to deepen the existing testing procedures for steel elements subjected to quasi-static, pseudo-dynamic and dynamic-seismic loading.
The use of more refined design methods, marked also by EC 3 is related with suitable experimental techniques regarding characteristic phenomenas quantification of the material, elements, joints, connections and structure.
These determined the formed European Committees (RILEM-1990 and ECCS-1983) to work for standardization of the tests on the materials and structural elements, having the final aim to provide guides for experimental testing, appropriate for each type of loading.
Regarding these, the paper present the spread of the experimental methods presented in IX and X WCEE and an interesting setup cost to loading equipment analysis, as well as the quasi- static testing procedure detailed presentation. Also, a series of testing classifications on the base of loading shape, the loading rate, number of cycles and the aim (comparison, qualification, modelling).
To improve the pseudo-dynamic tests, O.S. Bursi and P.B. Shing investigate the effect of the experimental measurement errors, especially for displacements as well as their use in a predictor-

multi-corrector algorythm for loading. It is also deepen the influence of α-Operator-Splitting and OSM algorithms in numerical computations for pseudo-dynamic tests.

Distribution and influence of the errors are related to the angular frequency of system and an admissibility criteria formed as a ratio between achieved and admitted residual displacements.

The achievement of beam-to-column connections have a very important role for the structure safety under earthquake ground motions.

Based on ECCS proposals and on the previous papers of Ballio and Mazzolani, L. Calado and J. Ferreira investigate the behaviour of steel beam-to-column angle connections under increasing cyclic lateral loading.

Four different beam-to-column angle connections were tested under step increasing displacements with three repetitions in elastic range and two repetitions in plastic range. The results comprised constitutive curves as well as data regarding connection ductility, resistance of connection type, their stiffness, the absorbed energy and maximum load level in each cycle.

Finally some conclusions were drawn regarding the influence of the angle thickness as well as the number and the position of bolts.

F. Wald, I. Simek and J. Seifert also investigate a problem of connection but for column base. Several types of connections were analyzed with the aim to optimize thickness and size of the plate as well as the diameter, the length, the position in plate and the number of bolts.

To deepen the tension and compression zones phenomenas a constant vertical loading and an horizontal cyclic loading are used. The above mentioned variants in connection with N/M ratio are used to complete EC 3.

If element and connection tests give the possibility to investigate some local phenomenas, then the structural assembly tests also permit to determine the characteristics of global seismic response and location of specific critical zones. This kind of problem it is investigated in A. Kakaliagos, G. Verzeletti, G. Magonette and A.V. Pinto paper.

The behaviour of two full scale three-story one-bay steel moment resisting frame selected from a structural assembly designed according to EC 8 and tested pseudo-dynamically at various loading intensities (0.5, 1.0, 1.5 from Kalamata eathquake motion and from an artificial generated accelerogram) till failure is described.

The synthesis of gained results leads to some aspects regarding the failure (in beam-to-column connection and joint panels): the shear yielding of the joint panels, crack accurance at the first story beam-flange to column-flange, a good behaviour for columns and girders according to EC 8 and an unsatisfying capacity design for joint panels and connections.

Therefore, the current EC 8 provisions regarding the joint panel and connections must be improved. Future investigations are required for a delicate problem regarding the balance of plastic rotation which can be developed in girder, joint panel and connections.

The behaviour of steel framed structures with cladding corrugated sheet panels under seismic motions approach is a very complex problem from the computational point of view as well as the technical achievment.

Regarding this problem, R. Landolfo and F.M. Mazzolani paper investigate the shear behaviour of corrugated sheet panels with connections. This work is part of a general research programm regarding the effect of cladding panels on the behaviour of the steel framed structures.

The aim is to asses some design rules for this kind of structures based on finite element analysis using plane stress elements as well as a suitable experimental programm for input data.

The performed tests, made with a suitable setup for shear loadings have quantified the stiffness of the connecting system for subpanel-to-subpanel (aluminium rivets) and panel-to-frame (self-drilling screws) in both horizontal directions.

For a realistic input data the phisico-mechanical characteristics of the panels were determined (continous flat sheep-isotropic f.e.; corrugated sheets-orthotropic f.e.).

Five elastic characteristics (E_1, G, μ_1) to define the orthotropic thin plate were used in the numerical simulations.

The analysis and experiments of the cladding panels subjected to shear loading revealed new informations regarding deformability and strength capacity related to the computing algorithm.

The analytical and experimental gained informations can be used to a general computation methodology for sandwich panels-structure interaction.

In situ experimental measurements for the dynamic characteristics reveal data regarding seismic loading level calibration for future earthquake ground motions.

In this context, A. la Tagola and W. Mera paper present the measurements of the dynamic characteristics for some steel structures from Ecuador, using natural ambient vibration excitation and with emphasis on the soil influence.

The quantified limits of the expected errors for computed and measured first period, with or without soil influence, are related to the building mass /soil mass ratio. Some comparison data regarding Italian seismic code and EC 8 are also given.

It is important to emphasize that the period of vibration of the structures is strongly influenced by the energy level of the excitation (ambient vibrations, wind, earthquake ground motion) and the stage of stiffness degradation of the structural members considered together.

A. Acatrinei and M. Rotaru present the main characteristics of the seismic response regarding four models of a single story steel factory building subjected to seismic excitation on the 140tf shacking table from INCERC-Iassy first on transversal then on the longitudinal direction,up to failure.

The structural type consists of hinged beams, welded beams (rigid joint), braced in "X", "V" and "K"-links evidentiating comparative data concerning the structural configuration, the elements designing, the efficiency of the constructive means and main comparative elements of the seismic response.

The gained experimental data were used in order to base the specific provisions of the Romanian aseismic code P 100/92.

In order to minimize the horizontal level drifts several devices are used. In G. Palamaru, C. Mihai and M. Rotaru work experimental researches on five models of dampers using regressive stiffness are presented. The specimens are made of steel X-shaped plates and can be used in bracing systems for framed structures.

Based on the analytical and theoretical investigations a design methodology for these dampers was established.

* * *

Both the complexity of the problems approached in the first part of this report and the pointed out contributions of the authors mentioned in the second part of the same work emphasize the idea that the large number of theoretical and experimental investigations (regarding materials, elements, connections, experimental models, portions of structures and structures)require an elaboration of experimental testing guidelines under various mechanical excitations of seismic type. In order to minimize the investigation cost-prices a national or zonal corroborated research effort must be accoplished in this field.

APPENDIX I

CHARACTERISTICS OF SEISMIC SIMULATORS

BUILDING RESEARCH CENTER OF IASSY

Nr. crt.	INSTALLATION	USEFUL DIMENSIONS (M)	FREQUENCY RANGE (Hz)	DYNAMIC PERFORMANCE (g)	NOTE
1.	S - 15 TF	3.2 x 3.2	0.50 ... 35	1.5	IN USE
2.	S - 140 TF	10.0 x 10.0	0.25 ... 12	1.0	IN USE
3.	I - 1000 TF	14.4 x 14.4	0.00 ... 30	3.4	NOT YET IN USE
4.	I - 200 TF	12.0 x 12.0	0.00 ... 35	3.1	NOT YET IN USE
5.	I - 30 TF	6.0 x 7.5	0.00 ... 50	5.2	NOT YET IN USE
6.	I - 5 TF	3.5 x 3.5	0.00 ... 50	11.5	NOT YET IN USE
7.	I - SBS	12.0 x 12.0	1.00 ... 80	—	NOT YET IN USE

BUILDING RESEARCH CENTER OF BUCHAREST

8.	IB1	6.0 x 6.0	1.00 ... 30	5.7	NOT YET IN USE
9.	IB2	3.0 x 3.0	1.00 ... 40	5.5	NOT YET IN USE
10.	IB3	2.0 x 3.0	1.00 ... 33	2.0	IN USE
11.	Reaction wall	18.0 x 24.0	0.00 ... 5	—	IN USE

31 ERROR EFFECTS IN PREDICTOR–CORRECTOR ALGORITHMS FOR PSEUDODYNAMIC TESTS

O. S. BURSI
Department of Structural Mechanics and Design Automation,
University of Trento, Trento, Italy
P. B. SHING
Department of Civil, Environmental and Architectural Engineering,
University of Colorado, Boulder, USA

Abstract
The error-propagation characteristics of Operator-Splitting-based algorithms with numerical dissipation and force correction are investigated in this paper. These algorithms are shown to suppress the experimental error amplification but they tend to degrade the accuracy of solutions. Hence, an explicit predictor-multi-corrector algorithm is investigated. The algorithm shows superior performance and can be effectively adopted for pseudodynamic test computations in many circumstances.
Keywords: Transient Algorithm, Experimental Error Growth, Numerical dissipation.

1 Introduction

Experimental errors associated with displacement control and displacement and force measurements are often introduced into a numerical computation in pseudodynamic tests. Even though the experimental feedback errors introduced in each step are small, cumulative errors in the numerical results can be significant because of the large number of time steps generally involved. Furthermore, these errors tend to excite the spurious higher-mode response in multiple-degree-of-freedom systems as shown by Shing and Mahin (1987).

Different approaches have been proposed for suppressing the experimental error growth associated with pseudodynamic test computations. Nakashima and Kato (1987) incorporated in a transient algorithm a force correction, namely the *I-Modification*, based on the initial stiffness of a structure and the difference between the commanded and measured displacements. Shing et al. (1991) introduced into the α-method developed by Hilber et al. (1977) a correction procedure for residual errors. Later, it was shown that this correction procedure incorporated with the constant-average-acceleration method, denoted as S algorithm, can effectively eliminate the spurious higher-mode response (Shing et al. (1993)). Recently, Nakashima et al., (1993) controlled the experimental error growth by means of the dissipative properties of the α-*Operator-Splitting*, referred to as the $\alpha - OS$ algorithm.

Additional error-propagation analyses, numerical simulations and experimental studies have been carried out to obtain a further comprehension of these numerical procedures (Comberscure and Pegon (1994)). To address this issue, the error-propagation characteristics of the $\alpha - OS$ algorithm and the Operator Splitting procedure coupled with the *I-Modification*, denoted as OSM algorithm, are investigated in this paper. Theoretical analyses and numerical simulations show that the $\alpha - OS$ and OSM algorithms can suppress the experimental error amplification. However, both procedures tend to introduce significant amounts of frequency error and artificial damping.

Behaviour of Steel Structures in Seismic Areas. Edited by F. M. Mazzolani and V. Gioncu.
Published in 1995 by E & FN Spon, 2-6 Boundary Row, London SE1 8HN. ISBN: 0 419 19890 3.

Finally, it is proposed to start the modified-Newton iterations of the S procedure with an explicit predictor. The resulting predictor-multi-corrector algorithm enhances the accuracy of numerical results and can be effectively utilized for pseudodynamic test computations in many circumstances.

2 The $\alpha - OS$ algorithm for non-linear problems

The recently developed $\alpha - OS$ algorithm is an explicit predictor-corrector algorithm which naturally arises by combining the Operator-Splitting technique and the α-method (Nakashima et al. (1993)). The procedure adopts a modified time-discrete equation of motion:

$$Ma_{i+1} + (1+\alpha)Cv_{i+1} - \alpha Cv_i + (1+\alpha)(Kd_{i+1} + \tilde{r}_{i+1} - K\tilde{d}_{i+1})$$
$$-\alpha(Kd_i + \tilde{r}_i - K\tilde{d}_i) = (1+\alpha)f_{i+1} - \alpha f_i \qquad (1)$$

and the numerical approximations of Newmark,

$$d_{i+1} = d_i + \Delta t v_i + \Delta t^2 \left[\left(\frac{1}{2} - \beta\right)a_i + \beta a_{i+1}\right], \quad v_{i+1} = v_i + \Delta t\left[(1-\gamma)a_i + \gamma a_{i+1}\right] \quad (2)$$

in which M and C are the mass and damping matrices of the structure and K is a predictor stiffness matrix. $\tilde{d}_{i+1}$ and $\tilde{r}_{i+1}$ are vectors of Newmark's predictor displacements and restoring forces, respectively at time equal to $(i+1)\Delta t$, where Δt is the integration time step. d_i, v_i and a_i are vectors of nodal displacements, velocities and accelerations and f_i is the external force excitation.

A necessary condition for the unconditional stability of the algorithm for a non-linear system is that the operator $\delta K_{i+1} = K - K_{i+\frac{1}{2}}$ is positive semi-definite, where $K_{i+\frac{1}{2}}$ is defined as the average stiffness matrix between the time steps $i\Delta t$ and $(i+1)\Delta t$ (Plesha and Belytschko (1985)). When the system is linearized and $\alpha \in [-\frac{1}{3}, 0], \gamma = \frac{(1-2\alpha)}{2}$ and $\beta = \frac{(1-\alpha)^2}{4}$, the α-OS algorithm inherits both the unconditional stability and the low mode second-order accuracy of the α-method (Hilber et al. (1977)). Like the α-method, it maintains favourable energy-dissipation properties in the higher part of the eigenfrequency spectrum.

The α-OS algorithm is implemented for pseudodynamic tests as follows.

Predictor phase

$$d_{i+1}^{(p)} = \tilde{d}_{i+1} = d_i + \Delta t v_i + \Delta t^2\left[\left(\frac{1}{2} - \beta\right)a_i\right] \qquad (3)$$

Corrector phase

$$K^*(d_{i+1}^{(c)} - d_{i+1}^{(p)}) = -(1+\alpha)r_{i+1}^{(p)} + \bar{P}_{i+1} \qquad (4)$$

$$K^* = \bar{K} + (1+\alpha)K, \quad \bar{K} = \frac{1}{\Delta t^2 \beta}M + \frac{(1+\alpha)\gamma}{\Delta t \beta}C \qquad (5)$$

$$\bar{P}_{i+1} = (1+\alpha)f_{i+1} - \alpha f_i - Cv_i - (1+\alpha)(1-\gamma)\Delta tCa_i + \alpha(Kd_i + \tilde{r}_i - K\tilde{d}_i) \qquad (6)$$

Updating phase

$$a_{i+i} = \frac{1}{(\Delta t)^2 \beta}(d_{i+i}^{(c)} - d_{i+1}^{(p)}), \quad v_{i+i} = v_i + \Delta t\left[(1-\gamma)a_i + \gamma a_{i+i}\right] \qquad (7)$$

In this implementation, vector $d_{i+1}^{(c)}$ is chosen as the primary unknown of Eq. (4). $r_{i+1}^{(p)}$ defines the restoring force vector corresponding to the predictor displacement $d_{i+1}^{(p)}$.

3 Error-propagation analysis of $\alpha - OS$ and OSM algorithms

The error-propagation analysis summarized here is based on the method developed by Shing and Mahin (1987). In this method, one can examine an undamped linearly elastic single-degree-of-freedom system which represents the most severe condition. Eqs. (1)-(2) can be expressed in the following one-step recursive form

$$x_{i+1} = Ax_i + (1+\alpha)Lf_{i+1} - \alpha L f_i \tag{8}$$

where the amplification matrix A and the load operator L are defined respectively as,

$$A = \frac{1}{D}\begin{bmatrix} 1 & 1 & \left(\frac{1}{2}-\beta\right)(1+(1+\alpha)\beta\frac{K}{K^a}\Omega^2) \\ +(1+\alpha)\beta\frac{K}{K^a}\Omega^2 & +(1+\alpha)\beta\frac{K}{K^a}\Omega^2 & +\alpha\beta^2\Omega^2 \\ -\beta\Omega^2 & -(1+\alpha)\beta\Omega^2 & +\beta\Omega^2\left(\beta-\frac{1}{2}(1+\alpha)\right) \\ & 1 & (1-\gamma)(1+(1+\alpha)\beta\frac{K}{K^a}\Omega^2) \\ -\gamma\Omega^2 & +(1+\alpha)\beta\frac{K}{K^a}\Omega^2 & +\alpha\beta\gamma\Omega^2 \\ & -(1+\alpha)\gamma\Omega^2 & +\gamma\Omega^2\left(\beta-\frac{1}{2}(1+\alpha)\right) \\ -\Omega^2 & -(1+\alpha)\Omega^2 & \alpha\beta\frac{K}{K^a}\Omega^2 + \Omega^2\left(\beta-\frac{1}{2}(1+\alpha)\right) \end{bmatrix} \tag{9}$$

$$L = \frac{\beta\Omega^2}{K^a D}\begin{Bmatrix} 1 \\ \frac{\gamma}{\beta} \\ \frac{1}{\beta} \end{Bmatrix} \tag{10}$$

with $D = 1 + (1+\alpha)\beta\frac{K}{K^a}\Omega^2$. $x_i = \{d_i, \Delta t v_i, \Delta t^2 a_i\}^T$ is a numerical vector, $\Omega = \omega^a \Delta t$ is the sampling frequency and $\omega^a = \sqrt{K^a/M}$ is the actual angular frequency of a structure. K represents a predictor stiffness which is assumed to be greater or equal to the actual stiffness K^a.

To eliminate numerical damping, one assumes that $\alpha = 0$, $\gamma = \frac{1}{2}$ and $\beta = \frac{1}{4}$ Hence, the $\alpha - OS$ algorithm is referred to as OS algorithm and Eq. (8) leads to the following analytical expression for the numerical solution of free vibration

$$d_n = exp(-\bar{\xi}\bar{\omega}\Delta t n)(c_1 \cos \bar{\omega}\Delta t n + c_2 \sin \bar{\omega}\Delta t n) \tag{11}$$

where

$$\bar{\xi} = 0 \quad \text{and} \quad \bar{\Omega} = \bar{\omega}\Delta t = \arctan\frac{\Omega\sqrt{4\left(1+\frac{1}{4}\frac{K}{K^a}\Omega^2\right)-\Omega^2}}{2\left(1+\frac{1}{4}\frac{K}{K^a}\Omega^2\right)-\Omega^2} \tag{12}$$

$\bar{\xi}$ represents the algorithmic damping ratio and $\bar{\Omega}$ is the distorted algorithmic frequency.

The frequency error $\frac{\Omega-\bar{\Omega}}{\Omega}$ associated with the ratio $\frac{K^a}{K}$ is plotted in Fig. 1. It can be observed that the actual frequencies can be shortened by a significant amount due to the splitting of the stiffness K^a. A negative value of α increases the amount of frequency distortion as shown by Nakashima et al. (1993). At $\frac{K^a}{K} = 1$ one gets $\bar{\Omega} = \arctan\frac{4\Omega}{4-\Omega^2}$.

This also represents the frequency distortion of the S algorithm (Shing et al. (1991)).

The schematic of the pseudodynamic test method and sources of experimental errors are shown in Fig. 2. The method is implemented with a dual displacement control by using displacement transducers external as well as internal to hydraulic actuators (Nakashima and Kato (1987)). Fig. 2 reports the displacement control error e^{dc}_{i+1} and both displacement and force measurement errors e^{dm}_{i+1} and e^{rm}_{i+1}. These control and measurement errors amount to the total feedback errors introduced in each step of a test (Shing and Mahin (1987)). However, the displacement and force measurement errors are substantially smaller than the actual displacement e^{dc}_{i+1} and force errors $K^a e^{dc}_{i+1}$ and are, therefore, neglected from now on. Hence, one obtains the following relations

$$(\tilde{d}^a_{i+1} - \tilde{d}^r_{i+1}) = e^{dc}_{i+1}, \quad \tilde{r}^a_{i+1} - \tilde{r}^r_{i+1} = K^a e^{dc}_{i+1} \tag{13}$$

It is important to note that $\tilde{d}^r_{i+1}$ and $\tilde{r}^r_{i+1}$ denote the exact numerical solutions evaluated from experimental feedback quantities $\tilde{d}^a_{i+1}$ and $\tilde{r}^a_{i+1}$, while $\tilde{d}_{i+1}$ and $\tilde{r}_{i+1}$ correspond to the true numerical solution in the absence of experimental errors.

Following the procedure of Shing and Mahin, (1987), one can express the cumulative error vector $\bar{\epsilon}_n = (\boldsymbol{x}^r_n - \boldsymbol{x}_n)$ as

$$\bar{\epsilon}_n = \sum_{i=1}^{n} \boldsymbol{A}^{n-i} \boldsymbol{L} g_i \tag{14}$$

in which

$$g_i = -K^a e^{dc}_{i+\alpha}, \quad e^{dc}_{i+\alpha} = e^{dc}_i + \alpha(e^{dc}_i - e^{dc}_{i-1}) \tag{15}$$

Eq. (14) represents the growth of cumulative errors in the α-OS algorithm due to stepwise errors e^{dc}_i.

To determine the growth of cumulative errors for the OSM algorithm, one has to consider the error correction on the restoring force, namely the *I-Modification*, as follows.

$$\tilde{r}^{co}_{i+1} = \tilde{r}^a_{i+1} - K(\tilde{d}^a_{i+1} - \tilde{d}^r_{i+1}) \tag{16}$$

Such a correction can be incorporated in the constant-average-acceleration method coupled with the Operator-Splitting technique. As a result, the cumulative error vector $\bar{\epsilon}_n = (\boldsymbol{x}^{co}_n - \boldsymbol{x}_n)$ can be expressed in the same form of Eq. (14) in which

$$g_i = (K - K^a) e^{dc}_i = (K - K^a)(\tilde{d}^a_i - \tilde{d}^r_i) \tag{17}$$

The influence of the displacement control error e^{dc}_{i+1} on the response of α-OS and OSM algorithms is analyzed according to the method of Shing et al. (1991). Following that development, both the time history of d_i in a simple harmonic response and g_i can be plotted in Fig. 3. Because g_i is equivalent to an excitation force which is in a specific phase relation with the displacement response, see Fig. 3, it has a negative damping effect for the $\alpha - OS$ algorithm and a positive damping effect for the OSM algorithm. The positive damping effect characterizes also the behavior of the S algorithm as was proved by Shing et al. (1991).

A preliminary comparison in terms of error-propagation among $\alpha - OS$, OSM and α-S algorithms is done by assuming $\alpha = 0$, ($\gamma = \frac{1}{2}$ and $\beta = \frac{1}{4}$). Then, the cumulative displacement error, $\bar{e}^d_n$, i.e. the first component of the vector $\bar{\epsilon}_n$ can be expressed as

$$e^d_n = D \sum_{i=1}^{n-1} e^{dc}_i \sin[\bar{\omega} \Delta t(n-i)] \tag{18}$$

Error effects in pseudodynamic tests 375

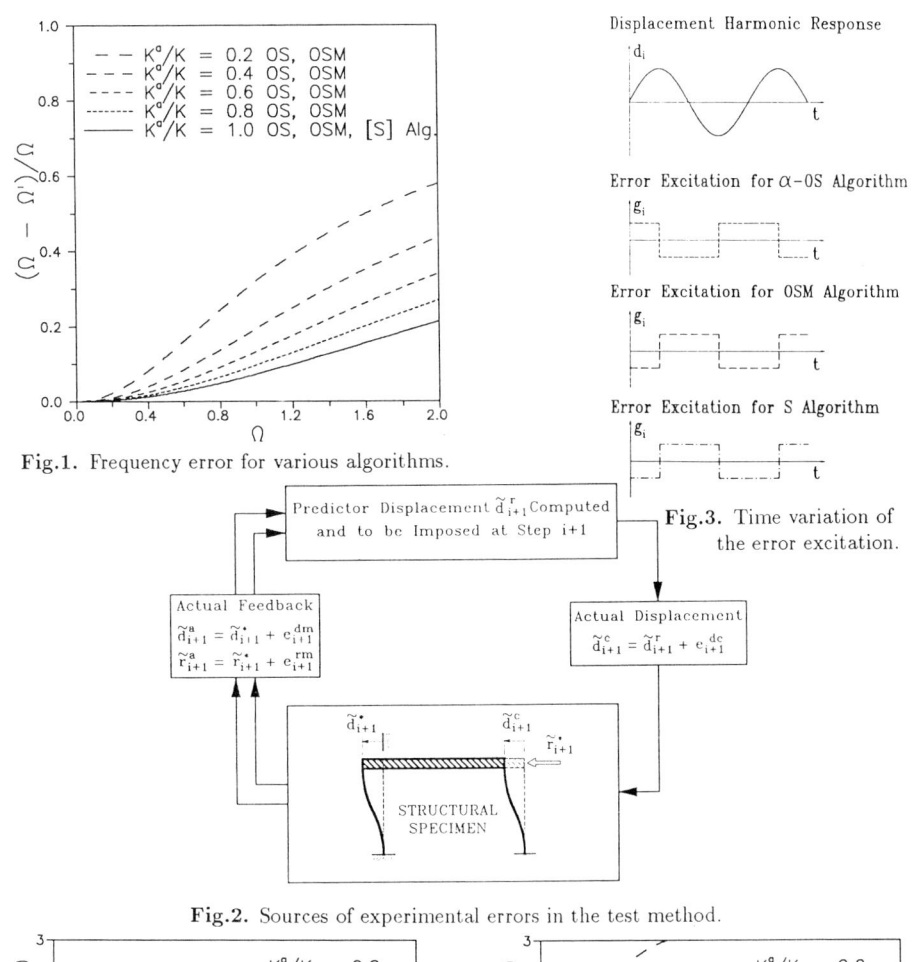

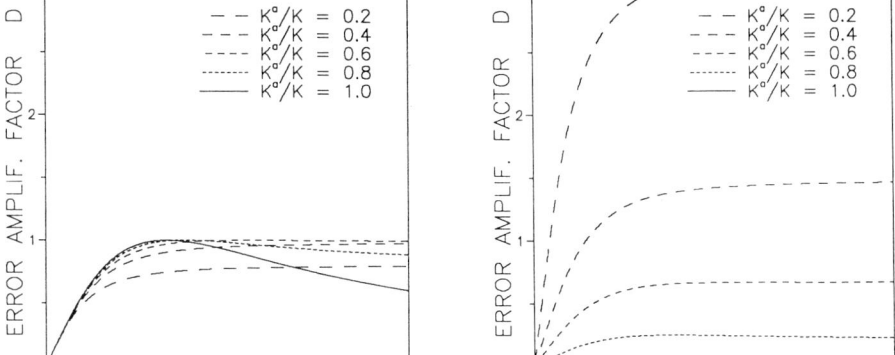

Fig.1. Frequency error for various algorithms.

Fig.3. Time variation of the error excitation.

Fig.2. Sources of experimental errors in the test method.

Fig.4. The ampl. factor for the OS algorithm. Fig.5. The ampl. factor for the OSM algorithm.

in which $\bar{\omega}$ is obtained from Eq. (12). The amplification factor depends on the algorithm in that

$$D = \frac{\Omega\sqrt{1 + \frac{1}{4}\frac{K}{K^a}\Omega^2 - \frac{1}{4}\Omega^2}}{\left(1 + \frac{1}{4}\frac{K}{K^a}\Omega^2\right)} \qquad (19)$$

for the OS algorithm and

$$D = \frac{\left(\frac{K}{K^a} - 1\right)\Omega\sqrt{1 + \frac{1}{4}\frac{K}{K^a}\Omega^2 - \frac{1}{4}\Omega^2}}{\left(1 + \frac{1}{4}\frac{K}{K^a}\Omega^2\right)} \qquad (20)$$

for the OSM algorithm. These factors are plotted in Fig. 4 and 5, respectively, as a function of Ω. Both D's are zero for $\Omega = 0$, while for large values of Ω $\lim_{\Omega \to \infty} D = 2\frac{K^a}{K}\sqrt{\left(\frac{K}{K^a} - 1\right)}$ for the OS algorithm and $\lim_{\Omega \to \infty} D = 2\left(1 - \frac{K^a}{K}\right)\sqrt{\left(\frac{K}{K^a} - 1\right)}$ for the OSM algorithm. Thus, both D's approach zero only if the $\frac{K^a}{K}$ ratio is one. The above results can be compared with the D factor which governs the overall convergence error of the S algorithm (Shing et al. (1991)). It is plotted in Fig. 6. In this case, if Ω is large then D approaches zero for any $\frac{K^a}{K}$ ratio. Furthermore, it is generally smaller in magnitude than the previous factors. It can be concluded that the S algorithm is far better than OS and OSM algorithms in terms of error amplification. In addition, it is better than the $\alpha - OS$ algorithm because the amount of numerical damping necessary to control the error growth degrades significantly the accuracy of the solution. To support the above conclusions numerical results which include numerical dissipation are shown in the next section.

4 Numerical simulations

In this section, the error-propagation characteristics of OS, $\alpha - OS$, OSM and S algorithms are compared by means of numerical experiments. The properties and the modal frequencies of the elastic two-storey shear frame considered in the simulations are shown in Table 1. The indexes 1 and 2 represent the lower and upper stories, respectively. Viscous damping is assumed to be zero and the integration time interval is selected to be 0.035 sec. for all cases. The stiffness factor SF which defines the predictor stiffness as $\boldsymbol{K} = \frac{1}{SF} \cdot \boldsymbol{K}^a$ is equal to 0.6. Only undershooting displacement control errors are assumed in the simulations. Their values are proportional to the maximum displacement of 0.36 in. developed at the lower storey. Furthermore, they are simply generated by introducing a constant undershoot in each step at both stories. The frame is subjected to the NS component of the 1940 El Centro ground motion, with its peak acceleration scaled to 0.045-g.

In the first sample of simulations, undershoots with magnitudes of 0.0022 in. and 0.0072 in. are considered. These errors correspond to about 0.6 and 2 per cent of the maximum displacement developed at the lower storey. The displacement-time histories of the top storey obtained with the OS algorithm are shown in Figs. 7(a)-(b). It can be noted that the algorithm cannot control the error amplification. When the $\alpha - OS$ algorithm is adopted with $\alpha = -\frac{1}{3}$, one gets the result shown in Fig. 8(a) for an error of 0.0072 in. The algorithm suppresses effectively the error growth but degrades the accuracy of the response due to a significant amount of frequency distortion. The corresponding displacement-time history determined with the OSM algorithm is shown in Fig. 8(b). The absence of numerical damping reduces the frequency distortion effects. However, the response is dominated

Error effects in pseudodynamic tests 377

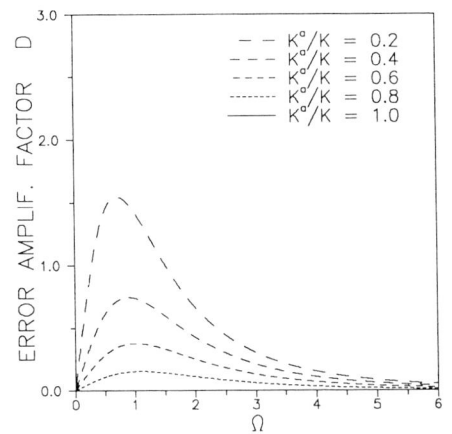

Fig.6. The ampl. factor for the S algorithm.

Table 1: Properties of the Shear Frame*.

Storey		Elastic Properties		
i	M $\left(\frac{kip-sec^2}{in}\right)$	K $\left(\frac{kip}{in}\right)$	ω_1 $\left(\frac{rad}{sec}\right)$	ω_2 $\left(\frac{rad}{sec}\right)$
1	0.010	8.0	11.2	35.7
2	0.020	4.0		

* : $1\,in = 0.0254\,m$; $1\,kip = 4.45\,kN$.

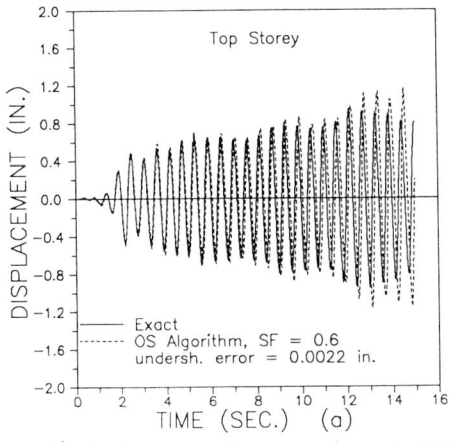

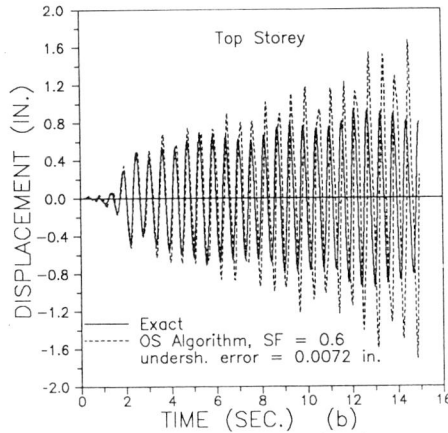

Fig.7. Responses of two-story frame to 1940 El Centro earthquake with undershooting errors.

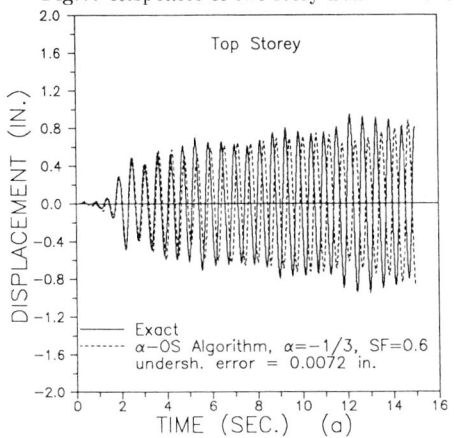

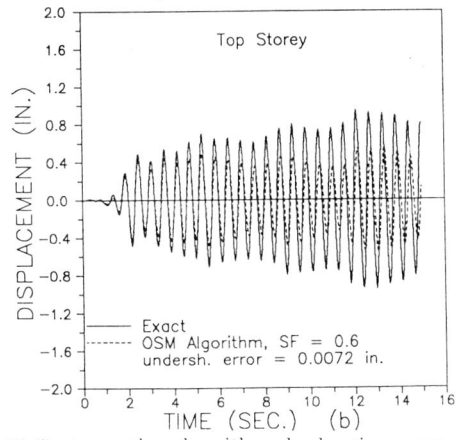

Fig.8. Responses of two-story frame to 1940 El Centro earthquake with undershooting errors.

by a large positive damping, as predicted above. In the second set of simulations, the S algorithm is considered. The displacement-time histories for error magnitudes of 2 and 6 per cent are shown in Fig. 9. Even with a large undershooting error of 0.022 in., the response accuracy both in terms of amplitude and frequency distortion is evident.

5 An explicit predictor-multi-corrector algorithm

The convergence of a modified-Newton iteration algorithm at each time step depends crucially on predictor values. At present, the S algorithm adopts in the predictor phase the value from the last converged solution. However, starting values for iteration can include the Newmark's predictor defined in Eq. (3), or, the corrector determined from Eq. (4). With these values, two explicit predictor-corrector procedures, namely the SP and the SC algorithms emanate from the S procedure, respectively. The implementation of the SP procedure with the α-method is described here.

Predictor phase

$$d_{i+1}^{(k)} = d_{i+1}^{(p)}, \quad r_{i+1}^{(k)} = r_{i+1}^{(p)} \quad (21)$$

Multi-corrector phase

$$K^* \Delta d_{i+1}^{(k)} = -R_{i+1}^{(k)} \quad (22)$$

$$-R_{i+1}^{(k)} = -\bar{K} d_{i+1}^{(k)} - (1+\alpha) r_{i+1}^{(k)} + \bar{K}\tilde{d}_{i+1} + \bar{F}_{i+1} \quad (23)$$

$$\bar{F}_{i+1} = (1+\alpha) f_{i+1} - \alpha f_i - C v_i - (1+\alpha)(1-\gamma) \Delta t C a_i + \alpha r_i \quad (24)$$

$$d_{i+1}^{(k+1)} = d_{i+1}^{(k)} + \Delta d_{i+1}^{(k)}, \quad r_{i+1}^{(k+1)} = r_{i+1}^{(k)} + \Delta r_{i+1}^{(k)} \quad (25)$$

Updating phase

$$d_{i+1} = d_{i+1}^{(l)} + \Delta d_{i+1}^{(l)}, \quad r_{i+1} = r_{i+1}^{(l)} + K \Delta d_{i+1}^{(l)} \quad (26)$$

$$a_{i+i} = \frac{1}{(\Delta t)^2 \beta}(d_{i+i} - \tilde{d}_{i+1}), \quad v_{i+i} = v_i + \Delta t \left[(1-\gamma) a_i + \gamma a_{i+i}\right] \quad (27)$$

$R_{i+1}^{(k)}$ represents the residual vector and k is an iteration index. To reduce the effect of stepwise residual errors on numerical results, a correction is introduced in the updating phase according to Shing et al. (1991). When experimental errors are small, both procedures lead to improved accuracy of the S algorithm, due to additional measurements of the restoring force. However, if undershooting errors are not negligible, the performance of the SC algorithm degrades dramatically. As a result, only the SP algorithm, described through Eqs. (21)-(27) and with $\alpha = 0$, ($\gamma = \frac{1}{2}$ and $\beta = \frac{1}{4}$) is examined from now on.

Both benefits and limits of the SP algorithm are shown through numerical experiments on the same elastic two-story shear frame. In the first sample of simulations, undershooting errors which correspond to about 2 and 6 per cent of the maximum displacement developed at the lower storey are considered. The displacement-time histories of the top storey are shown, for both cases, in Fig. 10. It can be noted that the SP algorithm performs well with the smaller error but damps down the response for larger errors.

In the second set of numerical simulations, the performances of S and SP algorithms are examined only with respect to convergence errors. In view of this, displacement tolerances of about 2 and 6 per cent of the maximum displacement developed at the lower storey are assumed. The displacement-time histories at the top storey determined with the S algorithm are shown in Fig. 11. A damping effect can be observed in the responses

Error effects in pseudodynamic tests 379

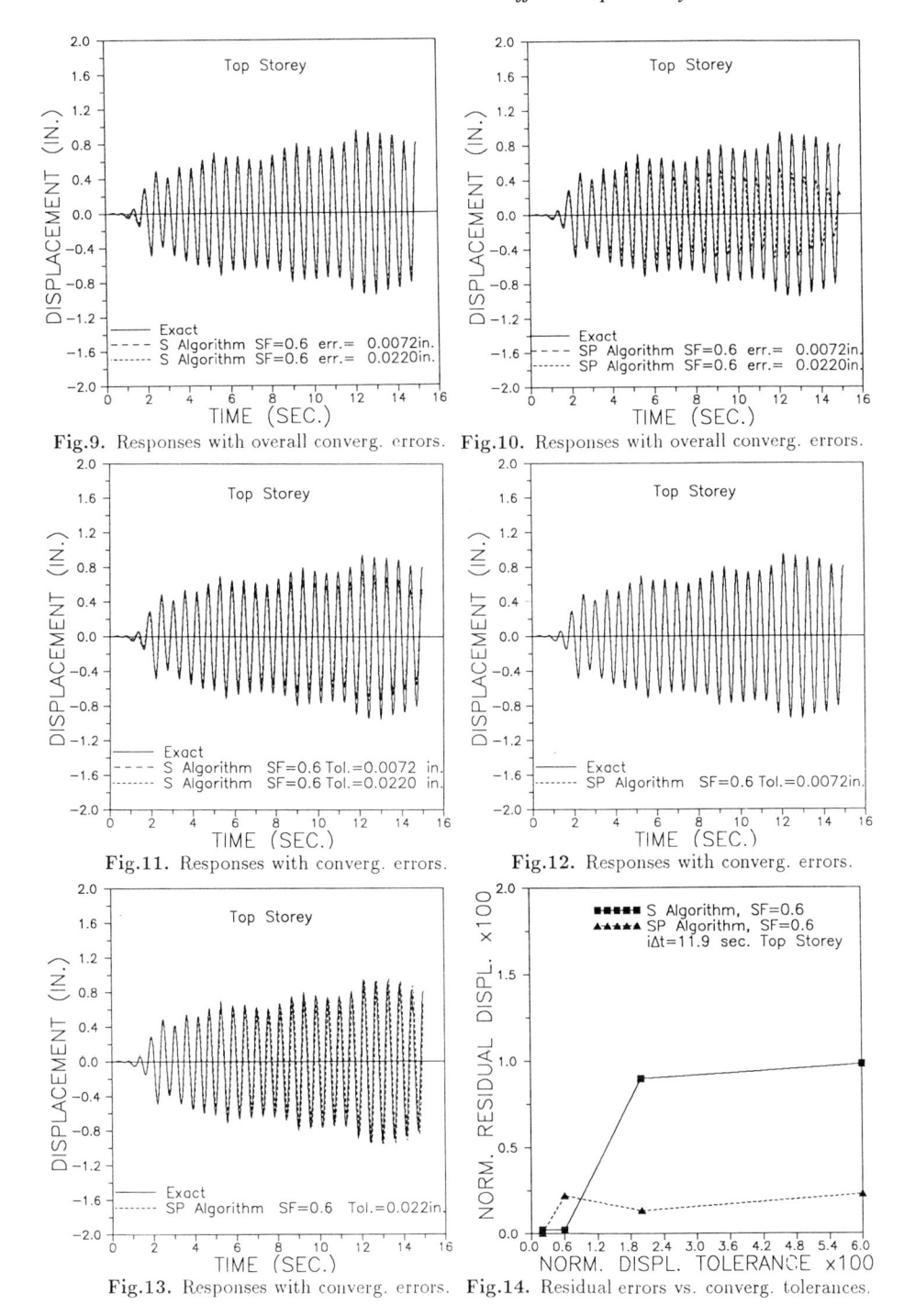

Fig.9. Responses with overall converg. errors.
Fig.10. Responses with overall converg. errors.
Fig.11. Responses with converg. errors.
Fig.12. Responses with converg. errors.
Fig.13. Responses with converg. errors.
Fig.14. Residual errors vs. converg. tolerances.

due to the approximate error correction. The corresponding results obtained with the SP algorithm are shown in Figs. 12 and 13, respectively. The increase of accuracy is evident because the Newmark's explicit predictor decreases the residual error in modified-Newton iterations. This trend can be observed in Fig. 14, where top storey residual displacements are plotted versus displacement tolerances both normalized with the maximum displacement developed at the lower storey.

6 Conclusions

The error-propagation characteristics of Operator-Splitting-based algorithms investigated herein can suppress the experimental error amplification but degrade the accuracy. The synthesis of the constant-average-acceleration method with a Newmark's explicit predictor results in a predictor-multi-corrector algorithm which can be effectively adopted for pseudodynamic test computations in many circumstances.

7 Acknowledgements

The study conducted here was sponsored by the C.N.R. under Grant No. 93.02186.CT07 and M.U.R.S.T. of Italy. However, opinions expressed in this paper are those of the writers, and do not necessarily reflect those of the sponsors.

8 References

Comberscure, D. and Pegon, P., (1994) Alpha-Operator Splitting Time Integration Technique for Pseudodynamic Testing, in **CEC, JRC.**, (draft) Ispra Italy.

Hilber, H.M., Hughes, T.J.R. and Taylor, R.L. (1977) Improved Numerical Dissipation for Time Integration Algorithms in Structural Dynamics. **Earthquake Eng. Struct. Dyn.**, 20, 283-292.

Nakashima, M. and Kato, H. (1987) Experimental Error Growth Behavior and Error Growth Control in On-Line Computer Test Control Method. **Research Paper, Building Research Ins., Minist. of Construction**, Japan, 123.

Nakashima, M., Akazawa, T., and Sakaguchi, O. (1993) Integration Method Capable of Controlling Experimental Error Growth in Substructure Pseudo Dynamic Test. **J. of Struct. Constr. Engng, AIJ**, 454, 61-71.

Plesha, M.E. and Belytschko, T. (1985) A Constitutive Operator Splitting Method for Nonlinear Transient Analysis. **Computer & Structures**, 20, 4, 767-777.

Shing, P.B. and Mahin, S.A. (1987) Cumulative Experimental Errors in Pseudodynamic Tests. **Earthquake Eng. Struct. Dyn.**, 15, 409-424.

Shing, P.B., Vannan, Mani T. and Carter, E. (1991) Implicit Time Integration for Pseudodynamic Tests **Earthquake Eng. Struct. Dyn.**, 20, 551-576.

Shing, P.B., Bursi, O.S., Radakoviz-Guzina, Z. and Vannan, M.T. (1993) Recent Developments in Pseudodynamic Testing based on an Implicit Time Integration Method, in **U.S.-Japan Seminar on Developments and Future Dimensions of Structural Testing Techniques**, Hawaii.

32 CYCLIC BEHAVIOUR OF STEEL BEAM-TO-COLUMN CONNECTIONS – AN EXPERIMENTAL RESEARCH

L. CALADO and J. FERREIRA
Department of Civil Engineering, Instituto Superior Técnico, Technical University of Lisbon, Portugal

Abstract
The behaviour of steel beam-to-column connections under cyclic loading is commonly assumed in the analysis and design of steel structures as fully rigid or ideally pinned. The real moment-rotation behaviour of these connections is actually between those two limit behaviours and depends on the type of connection (welded, bolted, with angles, with end plates, etc.) and on the connected beam and column.
  The scope of the research, that is presented in this paper, is the experimental assessment of the behaviour of semi-rigid steel beam-to-column angle connections under cyclic loading and its dependence on some of its details.
  Results of the experimental tests are presented, as well as the main conclusions.
Keywords: Beam-to-Column Connections, Cyclic Loading, Experimental Tests, Semi-rigid Behaviour, Steel Connections.

1 Introduction

The study of the behaviour of structural steel connections under cyclic loading is important as it allows to predict their response under seismic conditions, which is fundamental to assess strength and ductility of steel structures.
  Experimental tests on connections are necessary both to investigate their failure modes and to obtain data to verify the accuracy of analytical models. In order to allow the comparison between the results obtained in different laboratories and research centres [1-3], the ECCS [4] proposed unified testing procedures to test and evaluate the cyclic behaviour of structural steel elements. This recommendations had already been followed by Ballio et al [5], Zandonini [6] and Bernuzzi [7].
  Based on the previous work developed by Ballio et al [5], and Mazzolani et al [8], and following the ECCS recommendations, a experimental research programme was conducted in the new Laboratory for Structures and Strength of Materials (LERM) of the Department of Civil Engineering of the Technical University of Lisbon, on the cyclic behaviour of semi-rigid steel beam-to-column connections.
  Four full scale different beam-to-column angle connections were tested representing some of the current details in steel construction.
  The results of this test programme are presented as well as the main conclusions.

2 Test Apparatus

The experimental set-up used for the tests on the semi-rigid steel connections is shown in

Fig. 1. It consists mainly in: a) foundation, b) supporting girder, c) reaction wall, d) power jackscrew and e) lateral frame.

The power jackscrew (d), which displays a 1000 kN capacity and a 400 mm stroke, is attached to a specific frame, designed to accommodate the screw backward movement, which has been prestressed against the reaction wall (c).

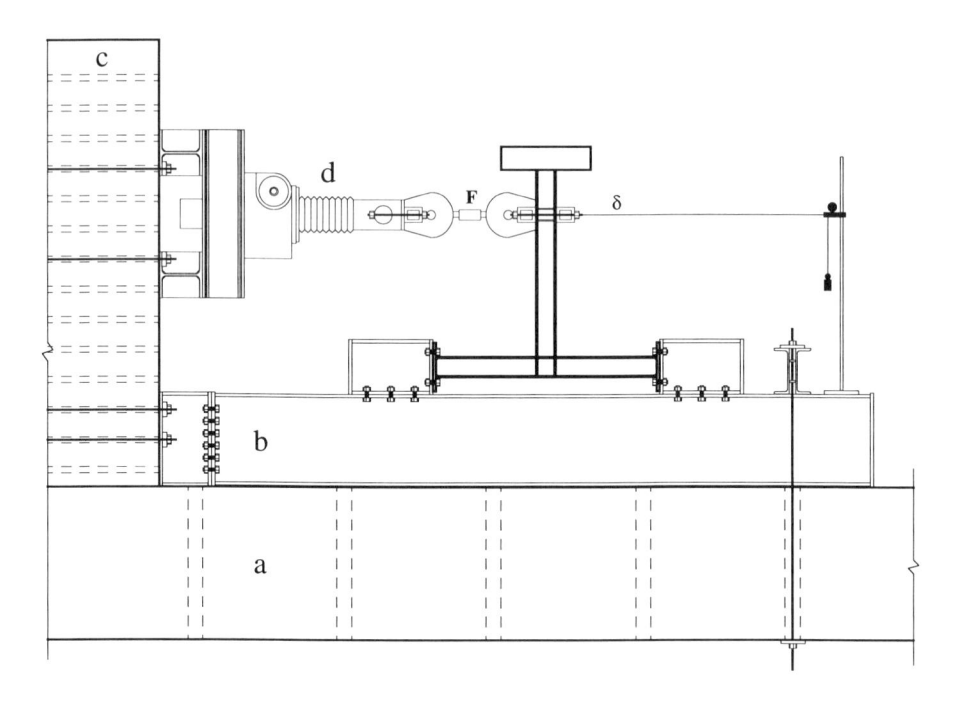

Fig. 1 - Test set-up

In the reaction wall there are pairs of holes with 195 mm interval between them to set the jack at different heights.

The specimen was connected to the supporting girder through two steel elements. The supporting girder was fastened to the reaction wall and to the foundation by prestressed bars.

The lateral frame (e) was designed to prevent specimens lateral displacements. This is achieved by means of four wheels attached to the frame, that contact the top head of the specimens, as shown in Fig. 2.

The forces F are measured in a load cell located between the power jackscrew and the specimen, while the top displacement δ, at the level of the applied force, is evaluated by a special device which consists of a transducer, a pulley, a wire and a weight. Whenever there is a displacement, the pulley imposes to the transducers a displacement that is 1/10 of the real displacement. With this device it is possible to measure large displacements using transducers with small amplitude.

An automatic testing technique was developed to allow the control of the power jackscrew, the displacement and all the transducers used to monitorize the specimen, to be done by computer .

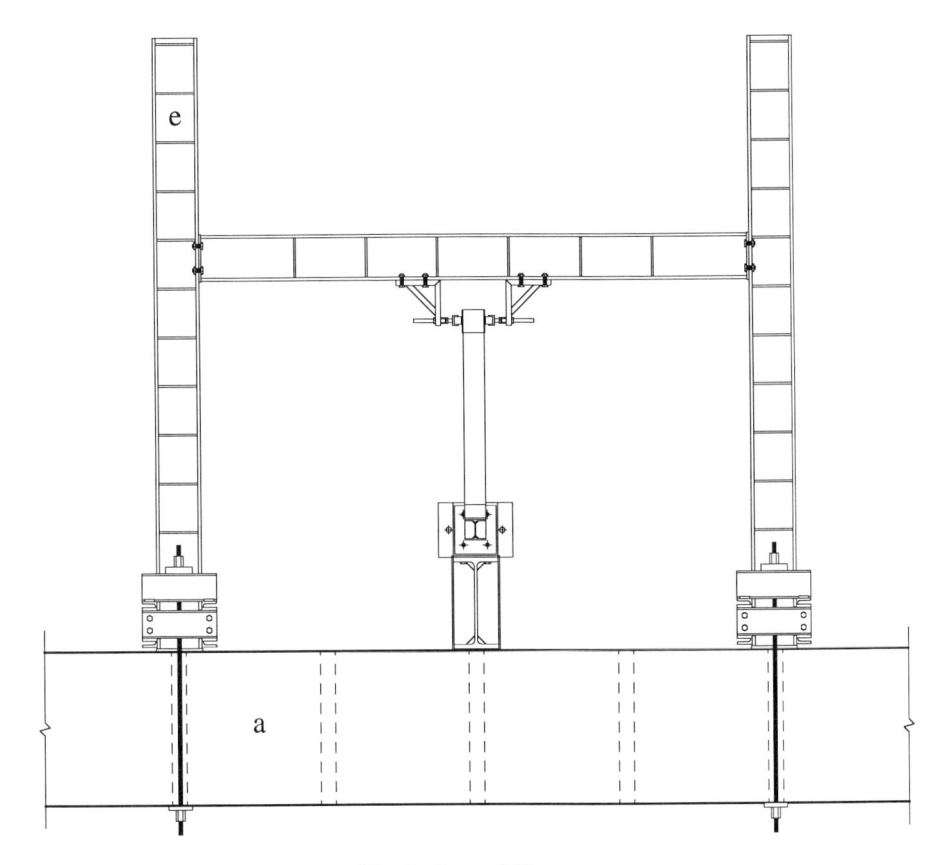

Fig. 2 - Lateral Frame

3 Experimental Programme

3.1 Tested Specimens

Four different angle connections named VPH1, VPH2, VPH3 and VPH4 were tested, which represent possible structural connections between beam and column in steel construction. Fig. 3 shows the geometry of the specimens while Fig. 4 shows the structural detailing of each connection. In order to impose cyclic displacement to the beam with the equipment shown in Fig. 1, the beam had to be the vertical element while the horizontal one was the column.

The VPH1 specimen is the reference connection, being the other specimens obtained by a variation of the elements of the connection, namely the number of bolts, the thickness of the angles and use of welding. VPH1 has two 100x100x10 angles with four shear and two tension M16 (8.8) bolts each.

Specimen VPH2 is obtained from VPH1 by adding two bolts in each connecting angle.

VPH3 connection differs from the reference one because it has thicker angles - 100x100x12 instead of 100x100x10.

In VPH4 specimen, the shear bolts in VPH1 were substituted by welding.

No axial load was imposed to the columns.

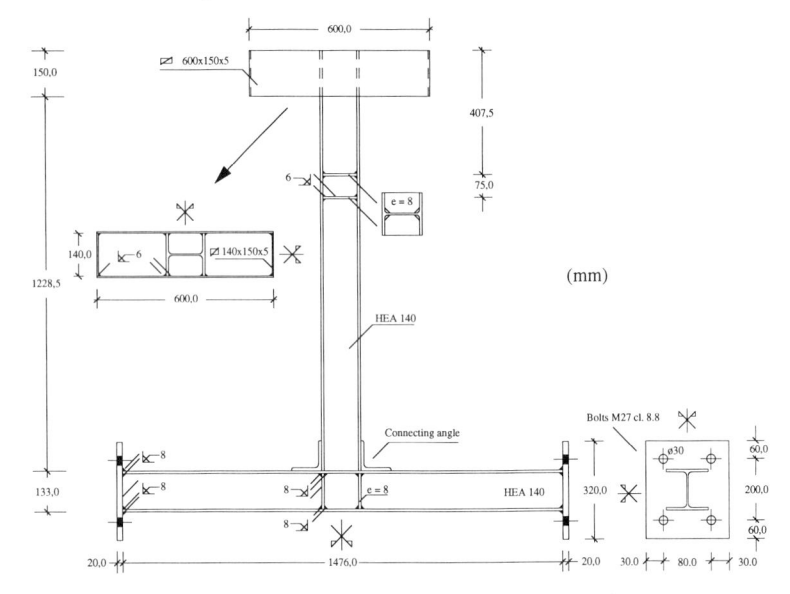

Fig. 3 - General geometry of VPH connections

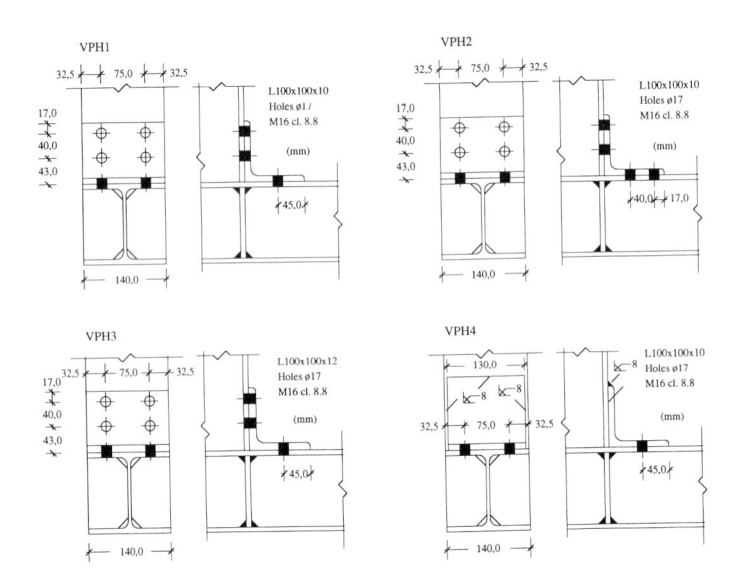

Fig. 4 - Detailing of tested connections

The steel characteristics were quantify through tensile tests on samples extracted from the beam and column web and flanges and from the angles. Some of these tests results are shown in Table 1.

Table 1 - Steel characteristics

	f_y [MPa]	f_{max} [MPa]	f_u [MPa]	ε_u [%]
web of HEA140	350	450	360	30
flange of HEA140	310	425	325	37
leg of L100x100x10	280	440	330	40
leg of L100x100x12	280	420	320	37

3.2 Transducers Positioning
The specimens were instrumented with electrical displacement transducers as shown in Fig. 5.

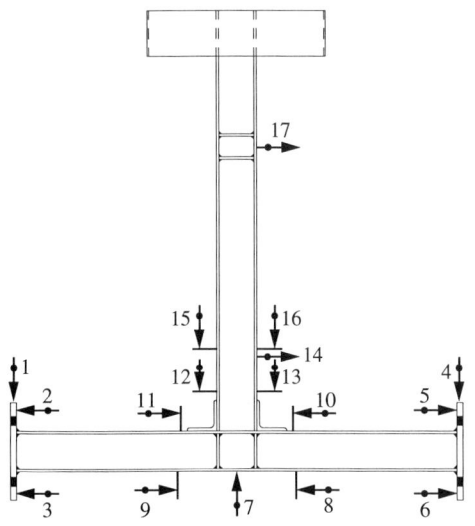

Fig. 5 - Transducers positioning

Transducers 1 to 6 monitor the displacement of the specimen supporting plates, while 7 gives the vertical displacement of the joint. Transducers 14 and 17 measure the horizontal displacement at the beam. The remaining transducers, as they are positioned in pairs with their axis parallel to the corresponding member axis, allow the assessment of the cross-section rotation. With this set of transducers it is possible to evaluate the relative beam-to-column rotation at the joint.

3.3 Displacement History
The ECCS recommendations for testing procedure [4] were followed for the assessment of the cyclic response of the connections. However, in the plastic range, only two repetitions for the same level of displacement were imposed, instead of the three recommended, because it is noticed that the third cycle doesn't provide a significant gain of information [1, 5-7]. The displacement history was comprised by the sequence of the following

cycles: i) one cycle for 1/4, 1/2, 3/4 and unity yield displacement, in both loading application directions; ii) two repeated cycles for each even multiple of the yield displacement until collapse is reached.

All specimens were submitted to the same displacement history in order to compare all the tests results. The yielding reference displacement was assessed by analytical calculation considering a rigid joint, so that the first yielding would occur at the beam flange at the joint cross-section.

4 Experimental Results

From the F-δ curve (Fig. 6 and Fig.8) some functions [4] concerning the cyclic behaviour in the plastic range were deduced for each specimen such as full ductility, relative resistance, relative rigidity and relative absorbed energy.

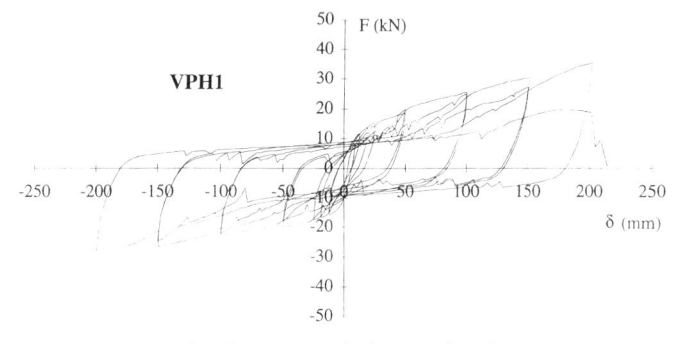

Fig. 6 - Hysteresis loops of VPH1

All specimens were tested until the effective failure took place, aiming to obtain the maximum possible information. In cyclic tests the failure of a specimen can be defined in terms of dissipated energy of the cycle, maximum force reached in the cycle or in terms of other measurable parameter. However, in this experimental research the failure of the specimen was considered when a connection element failed.

Some observations on the behaviour of the connections during the tests can be made.

VPH1 - This connection is characterised by growing slippage between bolted elements as the number of cycles increases. The increment of plastic deformations in both legs of the angles and in the column flange at the vicinity of the bolts was observed during the tests. Slip occurred mainly in the vertical plane between the beam flange and the vertical angle leg due to the ovalization of the holes in the flange of the beam and in the vertical leg of the angle. A vertical separation between the beam and the column was observed during the test, caused by the large plastic deformations of the angles and the local flexural deformations of the flanges of the column where bolts are connected. During the cyclic displacement with amplitude of 150 mm a horizontal crack started in the middle of each vertical leg of the angles when they were in tension. These cracks were growing with the increase of the displacement amplitude. During the second cycle of 200 mm the crack reached all the cross-section of the vertical leg of the angle, causing the failure of the connection (Fig. 7). No plastic deformation was observed in the central panel of the column.

VPH2 - The behaviour of this connection was generally similar to that of the VPH1 although the number of bolts in the horizontal leg of the angle have duplicated. However, some differences could be noted. The local flexural deformation of the column flange in the vicinity of the bolts was smaller in this connection, as well as the deformation in the horizontal leg of the angles. A crack also started in each of the vertical leg of the angles, until it reached all the cross-section during the first cycle of +200 mm.

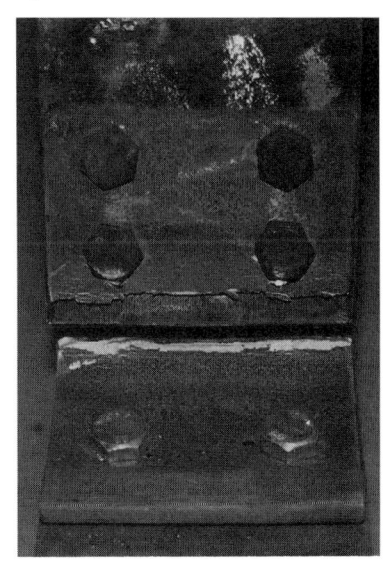

Fig. 7 - Failure mode of VPH1

VPH3 - This connection differs from VPH1 in the thickness of the angles; it has 12 mm instead of 10 mm of VPH1. It's behaviour was similar to the VPH1. The plastic deformation of the column flange in the vicinity of the bolts was greater than in VPH1 and VPH2, due to the increment of thickness of the angles. The failure of the connection occurred during the fourth cycle of +200 mm, caused by the crack in the vertical leg of the angle.

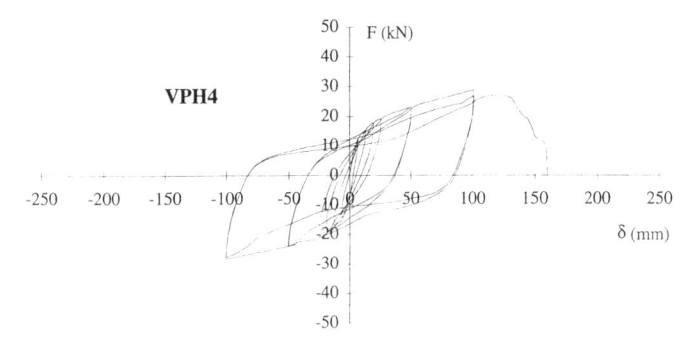

Fig. 8 - Hysteresis loops of VPH4

VPH4 - This connection showed a different behaviour when compared to the remaining ones. It was not observed plastic deformations and slip in the vertical legs of the angles.

However, local flexural deformation of the horizontal leg of the angle and of the flange of the column was observed during the test. The failure of the connection was also caused by cracking of the vertical leg during the first cycle of +150 mm of amplitude (Fig. 9).

Fig. 9 - Failure mode of VPH4

5 Conclusions

On the basis of the obtained results some qualitative conclusions are made on relations existing between the behaviour of the connections and the modifications introduced with respect to the base joint VPH1.

In Fig. 10 is shown a histogram comparing the maximum positive and negative force

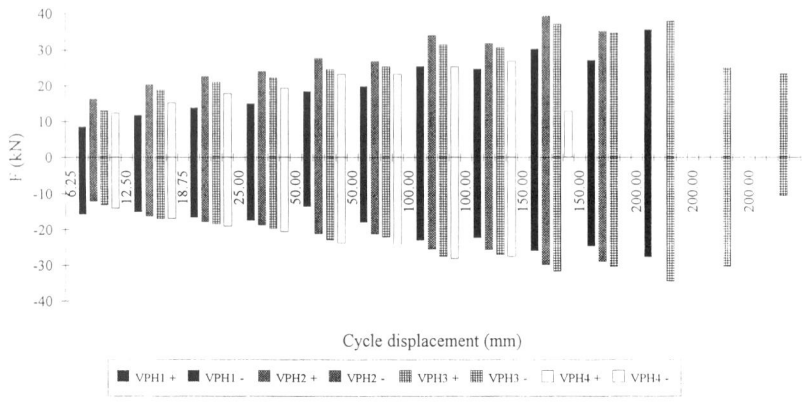

Fig. 10 - Load level for maximum displacement in each cycle

for each level of displacement and for each connection.

Joint VPH2 was obtained from the base joint by introducing two more bolts in each horizontal leg of the angles. With this modification the ductility of the connection is approximately the same but its resistance was improved. It could also be concluded that increasing the angles thickness, as in the specimen VPH3, a larger ductility was achieved in respect to the reference behaviour VPH1.

The increment of the number of bolts to reduce the deformation of the angles in VPH2 caused also an increase in the maximum forces. Providing welds instead of bolts to avoid slip, increase the resistance but decreased the stiffness of the connection and its ductility when compared to the base connection. The failure in all connections was due to the cracking of the angles cause by the large plastic deformations.

6 Acknowledgements

This experimental work was sponsored by the JNICT - Junta Nacional de Investigação Científica e Tecnológica through the research programme PBIC/C/CEG/1344/92 - Ligações em Estruturas Metálicas.

The authors wish to thank the collaboration of Prof. Ricardo Zandonini from University of Trento and Prof. Giulio Ballio and Prof. Carlo Castiglioni from Politecnico di Milano. The execution of the experimental tests had the participation and dedication of Ing. Jorge Proença from the Instituto Superior Técnico.

Siderurgia Nacional provided some of the steel elements used in this tests.

7 References

1. Popov, E. P. and Stephen, R. M. (1972) Cyclic loading of full-size steel connections. AISI Bulletin n. 2.
2. Astaneh, A. Nader, M. N. and Malik, L. (1989) Cyclic behaviour of double angle connections. ASCE, Journal of Structural Engineering, vol. 115, n.5.
3. Stelmack, T. W. Marley, M. J. and Gerstle, K. H. (1986) Analysis and tests of flexibly connected steel frames. ASCE, Journal of Structural Engineering, vol. 112, n°7.
4. ECCS, Technical Committee 1 - Structural Safety and Loadings, Technical Working Group 1.3 - Seismic Design (1986) Recommended testing procedure for assessing the behaviour of structural steel elements under cyclic loads.
5. Ballio, G. Calado, L. De Martino, A. Faella, C. and Mazzolani, F. M. (1987) Indagine sperimentale sul comportamento ciclico di nodi trave-colonna in acciaio. Costruzioni Metalliche n.2.
6. Bernuzzi, C. Zandonini, R. and Zanon, P. (1992) Semi-rigid steel connections under cyclic loads. First World Conference on Constructional Steel Design. Acapulco.
7. Bernuzzi, C. (1992) Cyclic response of semi-rigid steel joints. COST project C1 - Semi rigid behaviour.
8. De Martino, A. Faella, C. Mazzolani, F. M. (1984) Simulation of beam-to-column joint behaviour under cyclic loads, Costruzioni Metalliche n° 6.

33 EXPERIMENTAL TESTING FOR THE ANALYSIS OF THE CYCLIC BEHAVIOUR OF STEEL ELEMENTS: AN OVERVIEW OF THE EXISTING PROCEDURES

E. COSENZA and G. MANFREDI
Istituto di Tecnica delle Costruzioni, University of Naples, Naples, Italy
A. DE MARTINO
Istituto di Costruzioni, Faculty of Architecture, University of Naples, Naples, Italy

Abstract

The aim of this paper is to deepen the current position regarding testing structural elements, sub-structures and structures, for the purpose of assessing their resistance to seismic action. The goal is not to trace the state of the art of all relevant contributions on this matter; rather, it is to appreciate the present status of the subject in order to discuss some of the more common experimental procedures in the study of cyclic structural behaviour. The paper describes the planning of test processes on structural elements, with particular reference to the steel ones, based on the needs of the researchers and of the designers. Moreover, a discussion is presented on the scope of each experimental procedures to link the type of tests with a specific goal in seismic research.
Keywords: Experimental procedures, Cyclic behaviour, Earthquake engineering.

1 Introduction

The improvement of more sophisticated experimental techniques is related with the use of more refined design structural methods and procedures as confirmed by the new codes. With particular reference to steel material, we can refer, for instance, to:
• EC3, as much for the main body philosophy based on the use in computational analysis of the joint stiffness, maximum strength and ultimate rotation capacity, as, chapter n.8 "Design by testing" [EC3 (1992)], for the analytical-experimental procedures;
• EC3 Annex J for the joint classification, joint behaviour prevision and modelling (formulation closely based on experimental available data) [EC3 (1993)];

This need is showed by the activity of some European Committees which have worked for a standardization of monotonic and cyclic tests on materials [RILEM (1990)] and structural elements [ECCS (1983)]. In relation with this topics the RILEM Technical Committee 134 MJP has been formed 1992 among three different technical committees related with the tests on steel structures and elements with the subject *"Characterization of performance of metal joints"*. The aim of T.C. 134 MJP [De Martino et al. (1994)] is the standardization of monotonic and cyclic tests on structural joint to provide a guide for laboratories.

It is unquestionable that the job of this committee is at the current stage of the knowledge very hard. In fact, if on one hand very much has been done for the study, in the elastic and post-elastic behaviour, of structures not placed in seismic areas and some problems arise, even in this field, for instance in the evaluation or the prevision [EC3 (1993)] of joint behaviour, on the other hand, no standard monotonic test to determine moment-rotation curve of semi-rigid connection has been already proposed. As consequence study and research on experimental and theoretical procedures, in the field of structures under cyclic loading like seismic actions, are far from the conclusion. The evidence of this observation is demonstrated by the different theoretical, experimental or theoretical-experimental approaches of researchers in the study of aseismic behaviour of structures in the last decade. This paper develops some background topics for planning of test procedures in relation with the needs of the research and structural design. The authors will discuss these topics on the base of the most used testing processes referred in all relevant contribution presented in the last seismic design world conferences. Before going in details the authors, for clarity reasons on the terminology used in this paper below, present an overview on the different testing monotonic and cyclic procedures based on different possible testing classifications.

First classification divides the experimental procedures on the base of the shape of the loading branches in relation of the expected load condition:

- <u>monotonic test</u>: this kind of test is meaningful only for structures or structural details for which the expected load or displacement condition is characterized by the load always increasing until the collapse; a single test (no repetition is here considered for statistical consideration) can be performed to qualify as to model the structural element, on the base of a suited choice of the parameter to be measured; the test control parameter, applied to the specimen, will be indifferently the load or the displacement, if the elastic behaviour is to be investigated, the displacement if informations are needed until the collapse is reached.

- <u>cyclic test</u>: this kind of test is necessary for structures or structural details for which the expected load condition is characterized by a load history of a cyclic type as much with a known amplitude and distribution as with an unknown ones (for instance structure in seismic area).

A second classification divides the cyclic test on the base of the speed by which the load/displacement control parameter is applied to the specimen or to the structure:

- <u>dynamic test</u> is characterized by the speed of application of the control parameter; shaking machine or table are usually used to simulate dynamic condition for structure under dynamic events or seismic ones; this kind of test is generally a very expensive one and for this reason not much diffused.
- <u>pseudo-dynamic test</u> is characterized by the application of static forces in order to simulate the dynamic ones; this method provides realistic seismic simulation using equipment that is considerably less expensive than the shaking table one. In the pseudo-dynamic method, conventional time domain analysis procedures are combined with experimentally measured other parameters. The equations of motion for a discrete parameter model of the test specimen structural system are solved on line using a step-by-step numerical integration method. Inertial and damping forces are modelled analytically, while non linear structural restoring force characteristics are measured experimentally.
- <u>quasi-static test</u> is characterized by the applications of forces by means of electro-hydraulic actuators with very low speed; it is less able than the previous one to simulate dynamic or seismic condition; but it is generally also less expensive and for this reason the test apparatus more diffused in research laboratories; in the following this kind of test will be referred to.

A third classification classify the cyclic test on the base of the number of test performed on the specimen looking at the type of structural behaviour characterization to be achieved:

- <u>characterization of structural behaviour based on a single test</u>: the control parameter sequence history has to be as similar as possible to the action history expected on the specimen under test on the base of the known external action history; this kind of test is not able to model the behaviour of the structural element, but it is only able to verify the specimen under the probable action (<u>qualification test</u>) or to compare different structural elements under the same type of test condition (<u>comparison test</u>);

- <u>many cyclic tests</u>: only in this case the tests are able to provide the parameters needed to define a behavioural cyclic model of tested element (<u>modelling cyclic test</u>), which can be used in the modeling of the whole structure to perform numerical global evaluation under different load-displacement external histories.

2 Overview of testing procedures

The aim of this clause is to present an overview of existing in experimental methods and to give an evaluation of the spread of various procedures in seismic researches. For this purpose, the percentage of the quasi-static, pseudo-dynamic and dynamic tests presented during IX and X Word Earthquake Engineering Conference is shown in figure 1.

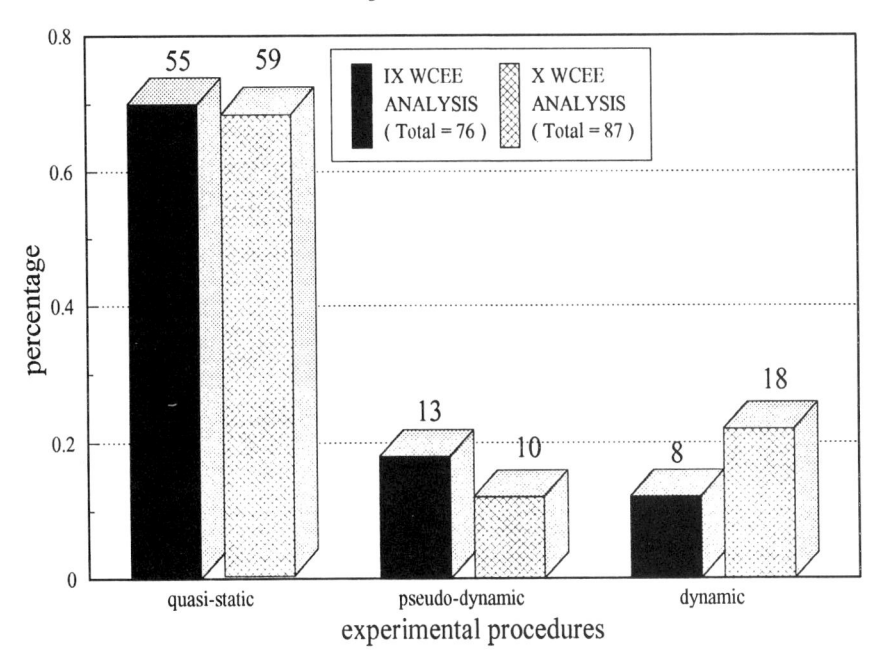

Fig. 1. Experimental procedures: percentage of quasi-static, pseudo-dynamic and dynamic tests presented during IX and X Word Conference.

The three types of testing procedures request a different cost in their employment and provide different results, as underlined in the previous clause. The quasi-static tests are the most common with a percentage greater than 60% and are mainly applied in the study of elements, like columns, beams and joints, or plane structures like shear-walls and frames.

The choice of the force or displacement history is a key issue in this type of test and this topic will be widely discussed in the following.

In the above mentioned conferences, the pseudo-dynamic type test has a percentage near 20% and it is applied to structures can be easy modelled with a few degrees of freedom system (cantilever, one or two storey frames, etc). This limitation is due to the characteristics of method that requires an actuator for each degree of freedom. The tested elements are mainly steel or r.c. ones and not masonry. This preponderance is related to the low reliability of pseudo-dynamic procedure in testing of specimens with distributed mass and brittle behaviour phenomena.

Also the dynamic tests have a percentage near 20 %. The examined structures are principally bi- or three-dimensional ones, generally in small scale, that cannot be tested with others methods. In fact only this peculiarity justify the use of this procedure that demands a very expensive experimental set-up (i.e. shaking tables).

The quasi-static tests are generally very simple; they don't require an high know-how and, therefore, are the most common ones. This smoothness is balanced by the necessity of a rational planning of loading process: in fact the choice of loading history (load/displacement) is connected with the aims of the research.

The quasi-static loading type tests generally used are summarized in fig.2. The simplest type is the linear one (a) that is characterized by an increase ΔX of the control parameter X for each cycle. The step type (b) and the ECCS type (c) also show a linear increase of control parameter but respectively with two or three cycles for each increment ΔX. This last loading process is proposed as standard test by ECCS for steel elements. The sinusoidal type (d) presents a sinusoidal variation of control parameter without repetitions and tries to simulate the shape of a typical accelerograms

The low cycle fatigue type (e) is characterized by a fixed imposed value of the control parameter $\overline{X}$ with n repetition up to failure; instead, the accumulation type (f) presents i steps of load/displacement $\overline{X}_i$ with n_i repetitions. Finally, the random type (g) is made applying a generic history $X(t)$ of the control parameter. The control parameter in quasi-static tests is generally displacement.

It is worth to underline that the quasi-static loading type tests are many and their definition is an open problem in the standardization of tests. The uncertainty in this choice is proved by the fig.3 where the loading patterns used in the research presented in last Earthquake Engineering Word Conference are summarized.

The majority of tests are of linear and of step function types with an high percentage of ECCS type too. Then the low cycle fatigue, the random and, finally, the accumulation type tests follow in decreasing order.

This trend is confirmed by the percentage related to only steel and steel-concrete composite structures. It is interesting to observe that the ECCS type tests is not particularly used for this group of structures, although it was proposed for the steel elements as standard procedure; on the other side, the low cycle fatigue type and the accumulation type are practically used only for this typology of structures.

Other interesting observations can be provided by the analysis of geographic distribution of experimental procedures analyzed in presents paper (fig.4). In fact the majority of tests is developed in Japan and in this country the linear type and the step function type are preferred. The ECCS type, also used in Japan, is more used in other countries as Europe and USA. The low cycle fatigue type and the accumulation type are practically utilised only in Japan.

3 Trends in testing procedures

The cyclic test can be classified, as above discussed, in three groups with regards to the aim: comparison tests, qualification tests and modelling tests.

The comparison tests have the objective of comparing the behaviour of the tested structural element with the behaviour of different others; the qualification tests aim at evaluating if the specimen is adequate to resist to standard real cyclic events; the modelling tests are devoted to the complete characterization of the cyclic behaviour of the specimen.

The dynamic and the pseudo-dynamic tests are customarily qualification ones and their results are not very suitable for developing analytical models , because they require the use of powerful analytical tools (i.e. nonlinear system identification).

The quasi-static tests have the advantage of an easy employment but the usefulness of results is strictly related with a good choice of loading patterns. In fact, by means of one test, it is possible to develop only a *comparison* or a *qualification* tests.

In the first case, it is not possible generally to detect synthetic informations on the behaviour of the specimen under real cyclic events, but on the contrary we have the advantage to compare the behaviour of the tested structural element with the behaviour of different others, being the cyclic sequence defined "a priori"; furthermore we have the possibility to create a database based on the results of homogeneous cyclic tests and to classify the structural behaviour by the same parameters measured during the test as the dissipated energy, stiffness, ductility functions etc.

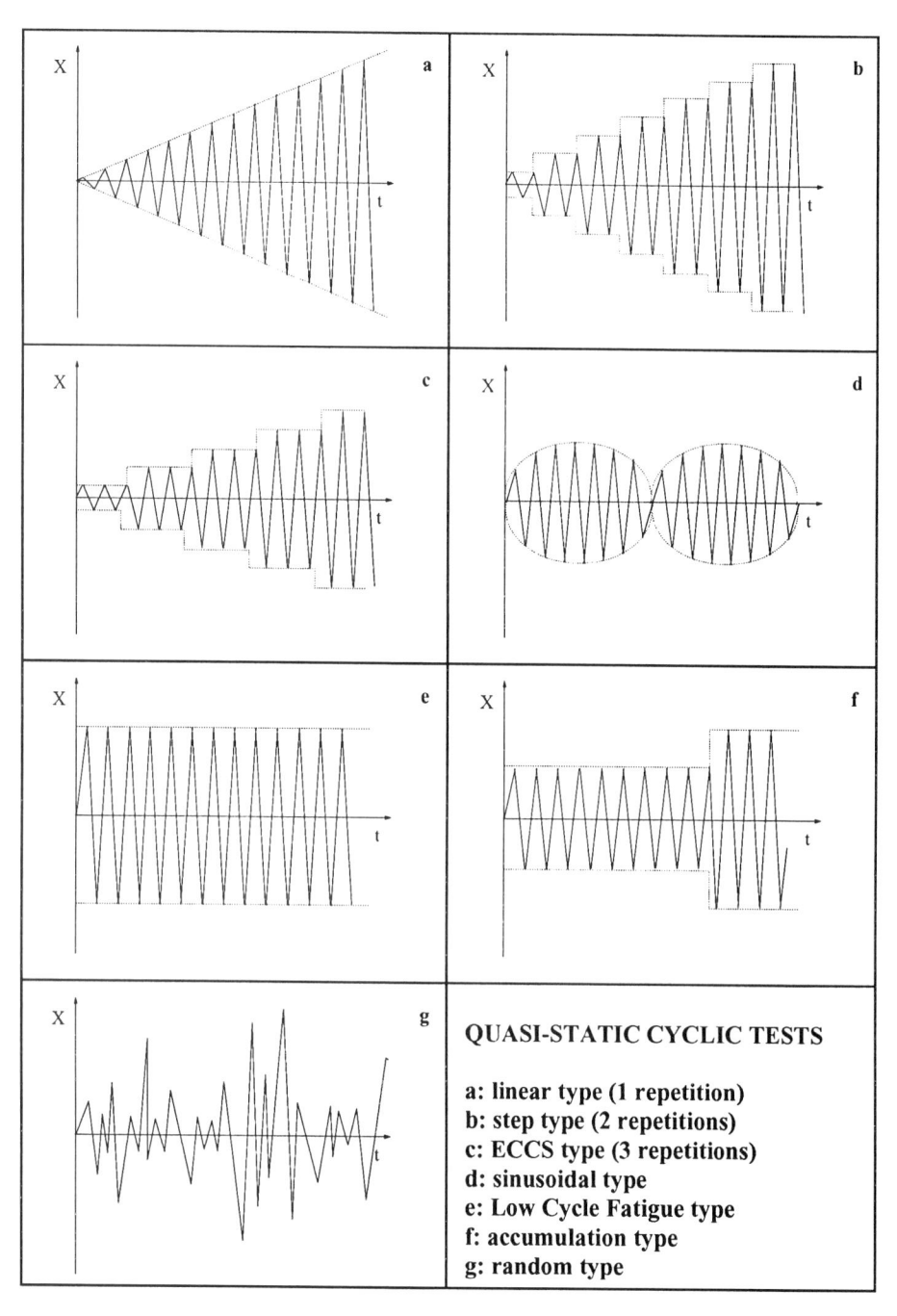

Fig. 2. Overview of quasi-static loading type tests.

QUASI-STATIC CYCLIC TESTS

a: linear type (1 repetition)
b: step type (2 repetitions)
c: ECCS type (3 repetitions)
d: sinusoidal type
e: Low Cycle Fatigue type
f: accumulation type
g: random type

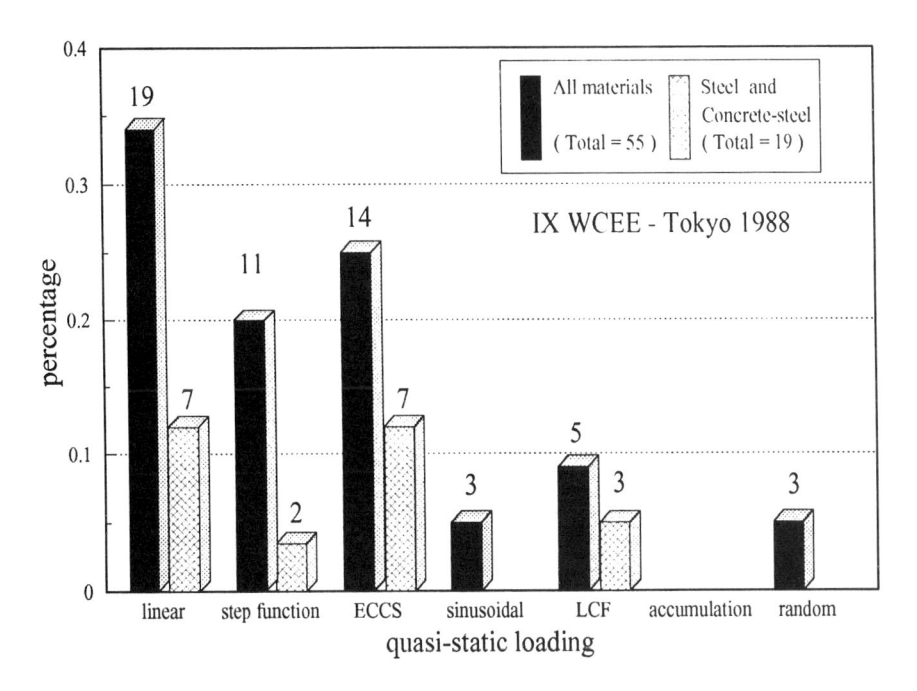

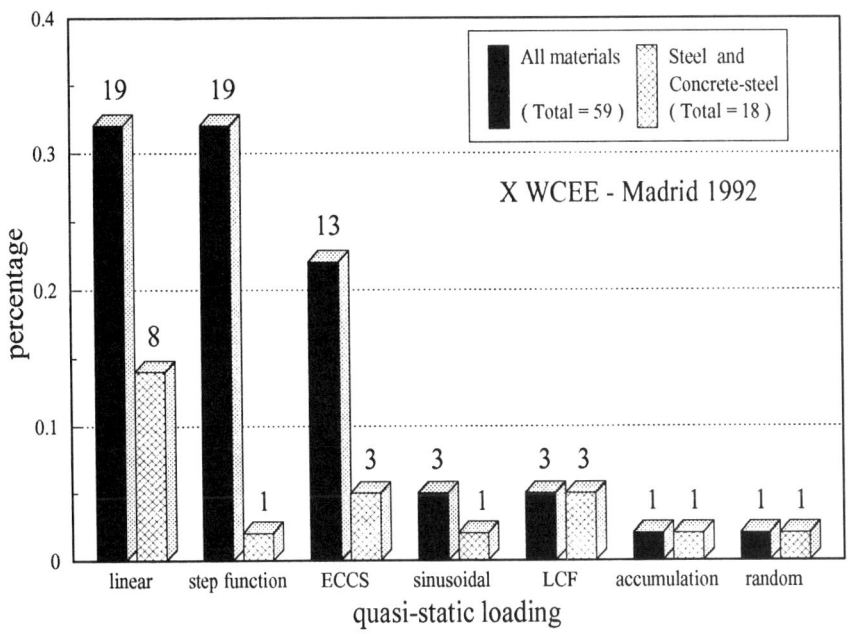

Fig.3. Percentage of quasi-static loading type tests presented during IX and X Word Conference.

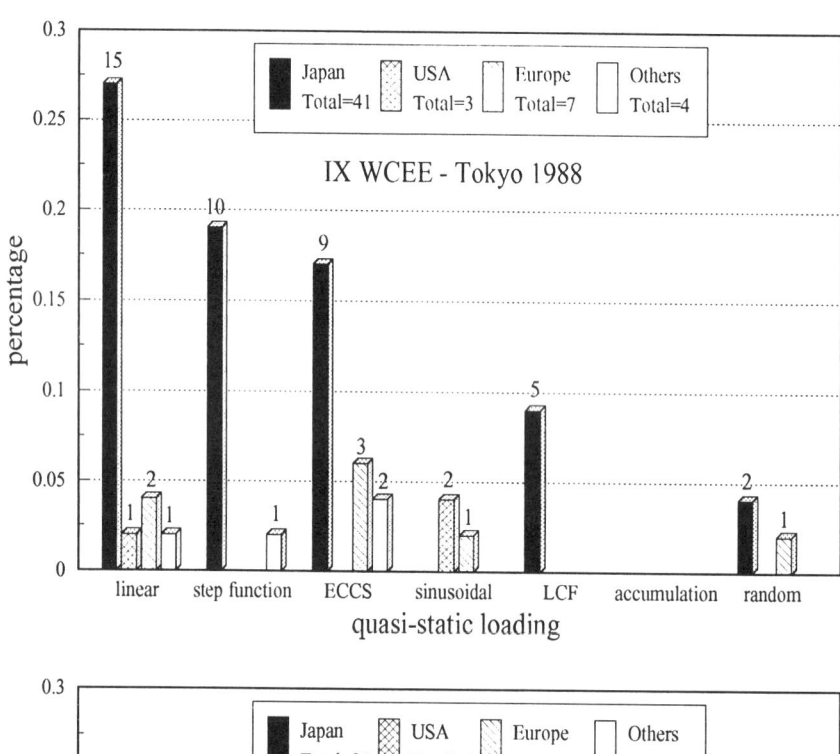

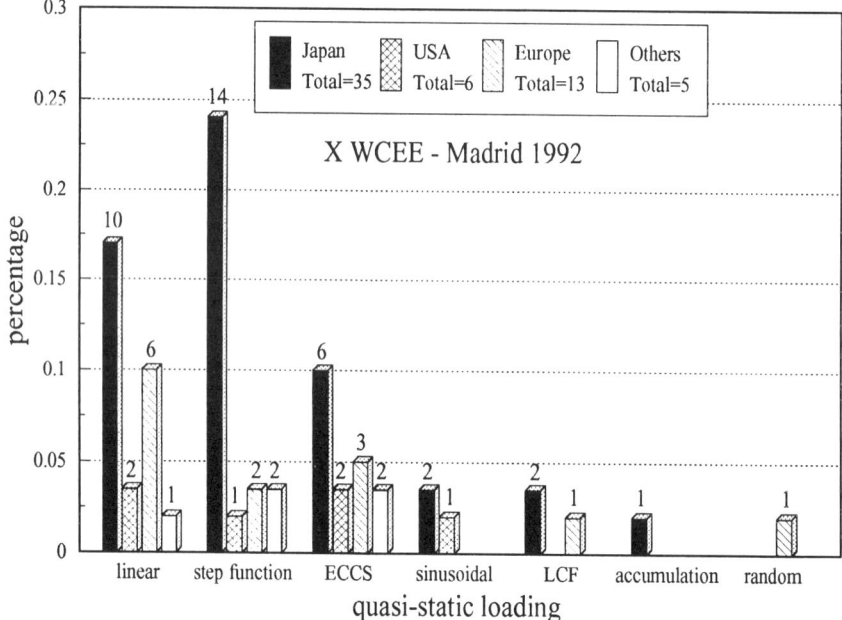

Fig.4. Geographic distribution of quasi-static loading type tests presented during IX and X Word Conference.

In the second case, is possible to arrange a random type procedure that can be developed by imposing a cyclic history related to the real working condition of the structural detail (like seismic one) and the aim of the test becomes that to establish if the specimen is able to bear this cyclic action; in other words, if it has been well designed to resist, on the safe side, to the expected cyclic event . By this test it is possible to verify if the specimen is suitable to resist to the possible cyclic event; if the specimen is adequate, and doesn't collapse during the test, it is also possible to get informations on the residual strength of the element by performing a monotonic test until the collapse, after the main cyclic test.

In case many equal test patterns are available it is feasible to develop *modelling tests* in order to evaluate the cyclic behaviour of the structural element so that different numerical analysis can be performed under any cyclic history.

The matter is closely related to the damage mechanic under strong intensity cyclic actions. The structural cyclic behaviour and the definition of the its leading parameters really still need more investigations have to be pursued by researchers; nevertheless some suggestions exist in literature that can help to deepen the problem.

It can firstly be observed that if the specimen doesn't show clear cyclic damage signs and the collapse is reached for a particular value of the displacement, then the monotonic test is able to define completely the model, and the below described tests are no more useful. For this reason it is suggest to perform initially a monotonic test and one of the above mentioned cyclic tests and then to decide on the base of results if the specimen need to be analyze with respect to its damage response.

In the case the structural damage needs to be modelled, this sequence of tests can be carried on:

a) monotonic test to collapse;

b) low cycle fatigue type tests (at least 3) of prefixed constant amplitude, which is defined as a percentage of the maximum displacement provided by the type a) test; in details, by defining with Δx_{max} the plastic displacement x_u-x_y, different percentage p_i of Δx_{max} can be chosen providing a generic amplitude named Δx_i. Then tests are carried on with constant amplitude Δx_i to find the number of cycles N_i which leads to failure.

c) accumulation type tests with imposed displacement $\Delta x_i = p_i \Delta x_{max}$ with a number of cycles n_i less than the one N_i which should lead to collapse the specimen, and residual cycles until failure with different amplitude.

These kind of tests can completely characterize the cyclic behaviour of the specimen. In fact the type "b" tests allow to evaluate the cyclic damage of the stiffness in the unloading branch as much as in loading one; on the contrary the type "c" tests allow to verify if a linear cumulative damage occurs; in other words, if the cyclic fuse is or not coincident with the monotonic curve. Moreover if we accept that the structural damage is supported by the plastic fatigue, that is true for many structural types, then it is possible, trough the above mentioned tests, to provide a detailed description of the cyclic behaviour of the model.

4 Concluding remarks

The definition of the tests procedures on structural elements is a central topic in the structural analysis and have to be related to the needs of the researchers and the designers. In this field, however, the procedures used by different researchers are very dissimilar as shown in this paper and this situation is a obstacle to the use of tests results. Therefore, it is necessary to discuss the introduction of standardized testing procedures in order to obtain comparable experimental data coming from different laboratories.

5 References

De Martino A. and Manfredi G. (1994), Activity of RILEM Technical Committee 134 MJP in the field of experimental tests on structural components, **V U.S. National Conference on Earthquake Engineering**, Chicago.
IX **Earthquake Engineering World Conference** (1988), Tokio, Japan.
X **Earthquake Engineering World Conference** (1992), Madrid, Spain.
ECCS (1983), **Recommended testing procedure for assessing the behaviour of structural steel elements under cyclic loads.**
Eurocode No.3, Part 1, Revised Annex J (1993), **Beam-to-column connections.**
Eurocode No.3 (1992), **Design of Steel Structures.**
RILEM T.C.83 (1990), Tension testing of metallic structural materials for determining stress-strain relations under monotonic and uniaxial tensile loading, **Materials and Structures**, 35-46.

34 PSD-TESTING OF A FULL-SCALE THREE-STOREY STEEL FRAME

A. KAKALIAGOS, G. VERZELETTI,
G. MAGONETTE and A. V. PINTO
European Laboratory for Structural Assessment, AMU, STI,
Joint Research Centre, Commission of the European Communities,
Ispra, Italy

Abstract

A full scale three-story one-bay steel moment resisting frame has been constructed and tested pseudo-dynamically at the European Laboratory for Structural Assessment (ELSA)-Reaction Wall Facility at Ispra. The frame connections and shear panel zones have been designed with Eurocode 8 (EC8). The objectives of this research were on one hand to study the seismic behavior of a realistic steel frame, conduct numerical seismic time-history frame analyses using available computer models and on the other hand to identify potential up-dating needs in the Eurocodes. The test results have demonstrated that the steel frame structure had flexible beam-to-column joints and connections. Under seismic loading, plastic hinges were formed in those locations and not in the girder-ends as anticipated in EC8. Both the study of the experimental and analytical results showed the necessity to size the joint shear panel zone using Allowable Stress Design (ASD) criteria.

Keywords: Pseudo-dynamic Testing, Steel Moment Resisting Frames, Seismic Behavior.

1 Introduction

Within the efforts of the Commission of the European Communities in developing common engineering standards in all member states, a continuous effort is focused in developing the EC8 for Earthquake Resistant Design. Full scale pseudo-dynamic testing (PSD) is a powerfull method to demonstrate the efficiency of the code requirements in fulfilling their objectives.

The first project at the ELSA-Reaction Wall was dealing with the PSD-testing of a full scale steel moment resisting frame designed with EC8. This project demonstrated the PSD-testing capabilities of the Ispra-Lab, Donea et al (1988), inasmuch as the steel moment resisting frame was the first significant structure to be tested in the ELSA-Reaction Wall Facility.

Behaviour of Steel Structures in Seismic Areas. Edited by F. M. Mazzolani and V. Gioncu.
Published in 1995 by E & FN Spon, 2-6 Boundary Row, London SE1 8HN. ISBN: 0 419 19890 3.

An extensive instrumentation system was developed capable of monitoring the structural deformations and rotations of the frame columns/girders and joints. Additionally, the internal force/moment distribution of the frame column/girder members was recorded. Such a procedure helped identifying the formation of plastic hinges in the steel skeleton during severe seismic excitations, whereby, such information was used to validate the EC8 design provisions and compare the experimental results with analytical methods developed, Kakaliagos (1994).

2 Test Structure and Specimen Design

With the pseudo-dynamic tests to be performed on a realistic steel moment resisting frame, the steel structure erected in the ELSA-Reaction-Wall-Lab was considered as a part of a complete building (Fig.1). It was decided to develop a test specimen which would represent a typical steel frame structure using standard rolled sections and realistic welded beam-to-column connections and joints. The structure consisted of two one-bay three story steel frames placed vertically in front of the Reaction Wall in the perpendicular direction. The steel frames were placed in parallel to each other at a distance of 5m. At each story level, the frames were connected to each other with reinforced concrete slabs by means of headed shear studs, thus forming composite slabs (Fig.1), Kakaliagos (1994).

The EC8 "design" spectrum (Fig.4) (response factor $\beta(T)=2.5$, ground acceleration factor a=0.3, soil condition factor S=1.0), the calculated steel frame fundamental period T=0.58sec (determined with rigid joints and connections), and a behavior factor q=6 were employed, to determine the base shear coefficient of (13%), and hence, evaluate the design lateral loads. The steel yield stress was 240 N/mm^2 (ST37 steel=A36 US).

The steel skeleton has been designed for dead/live and horizontal seismic loads implementing EC8-Capacity Design. Herein, for the seismic load combination of G+0.3P+E, the critical frame locations were checked against their plastic moment capacity, which considered shear interaction effects. In the latter load combination, G is the dead load, P the live load and E the horizontal EC8-seismic loads with the frame floor masses determined using 15% of the live load at each floor and 30% for the top floor.

The structure had welded beam-to-column joints and connections, without flange continuity plates in the joint. The unstiffened column flanges in the joint were sized to transfer the girder-flange tension/compression forces resulting from the girder plastic moment. Those connections/joints were assumed rigid for design purposes, Eurocode (1988). Implementing the strong-column weak-girder design concept, the welded connec-

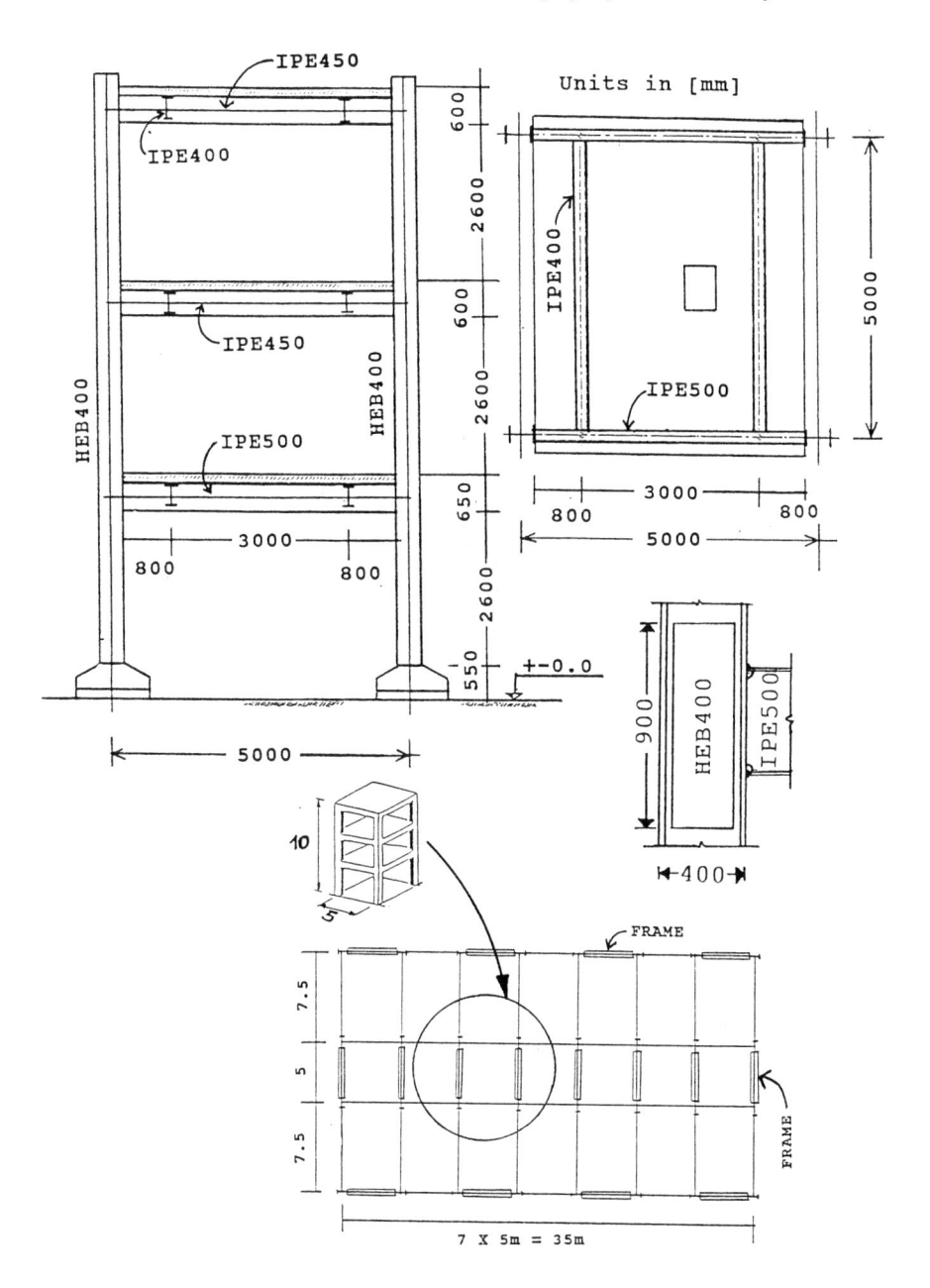

Fig.1 Full Scale Steel Frame Specimen

tions and shear panel zones were sized to resist the steel girder plastic moment. In addition, the girder/column steel sections were compact fulfilling EC8 requirements for high ductile hysteretic seismic performance (q=6).

3 Test Objectives

To assess the seismic behavior of a realistic steel frame structure (Fig.1) it was decided to use actual European Design Standards. These considerations resulted in a study aimed at investigating the seismic behavior of the welded beam-to-column connections and shear panel zones as well as the girder-flange to column-flange full penetration welds. In addition to the latter mentioned tasks, the seismic behavior of the girders and columns was also a subject of investigation.

The experimental campaign had also as a task to identify whether the frame's welded beam-to-column connections and joints (Fig.1), which in common civil engineering praxis are assumed to be rigid, are flexible during seismic excitations and can potentially affect the structure's seismic behavior. Previous static-cyclic tests on such types of connections by Kakaliagos and Bouwkamp (1993) have identified that these connections are flexible.

This experimental research focused on the seismic behavior of the steel skeleton, aiming to identify whether the EC8 code provisions could provide adequate resistance to horizontal earthquake ground motions and thus realize the EC8 Capacity Design concept for ductile steel moment resisting frame.

The whole experimental procedure was backuped with numerical frame nonlinear-seismic time-history analyses employing available computer models for the frame beam-to-column connections, shear panel zones, as well as for the column/girder members, Kakaliagos (1994).

4 Test Set-up and Instrumentation

In the PSD-Tests the structure masses were assumed to be lumped at the floor levels. Consequently, the structure's earthquake inertia horizontal loads had to be introduced in these locations. In principle, two 500KN double acting hydraulic actuators on each floor imposed the structure's horizontal inertia loads, which were reacted by the Reaction-Wall (Fig.2). Two optical displacement transducers on each floor, monitored the resulting floor horizontal displacements. These measurements were taken relative to an independent instrumentation frame. The displacement transducers were connected in parallel, monitoring equal horizontal displacements of the two steel frames.

This arrangement imposed a horizontal in-plane motion of the steel frame floor slabs (Fig.2). The output signal from these transducers was used directly in the PSD-algorithm.

In addition to the floor inertia load and associated horizontal floor displacement measurements, the internal frame column and girder moment, shear/axial load distribution was recorded by placing strain gages on the columns at locations near the frame joints where elastic column behavior was expected, Kakaliagos (1994).

Deformations were expected to appear in the beam-to-column connections and joints during seismic excitations, Kakaliagos (1993). Consequently, the joint deformation pattern shown in Fig.3 was anticipated. Herein, the joint rotation consisted of two rotational components: the connection rotation (ψ) and the joint shear rotation (γ). The first rotational component (ψ) appeared due to the local introduction of the concentrated girder flange force into the column web, while the second joint rotational component (γ) due to the shear deformation of the joint column web (Fig.3). Consequently, the group of LVDT's in the joint region aimed to yield the joint distortions and to identify individual (ψ) and (γ) rotations (Fig.3).

5 Steel Frame Testing

The experimental campaign was divided in two phases (Table 1). To present in detail the seismic steel frame response as identified in the PSD-tests, the paper describes this part of the experimental campaign only.

Table 1 Steel Frame Test Sequence

System	Phase I Identification	Free Vibration and Static Tests
Pseudo-Dynamic Tests	Phase II	Kalamata x 0.5- (KAL05) Artificial x 0.5- (ART05) Kalamata x 0.9- (KAL09) Artificial x 1.0- (ART10) Kalamata x 1.5- (KAL15) Artificial x-1.4-ART14)

The steel frame was subjected to both natural and artificially generated accelerograms. For this purpose the Kalamata-Greece (September 1976) earthquake motion and one artificially

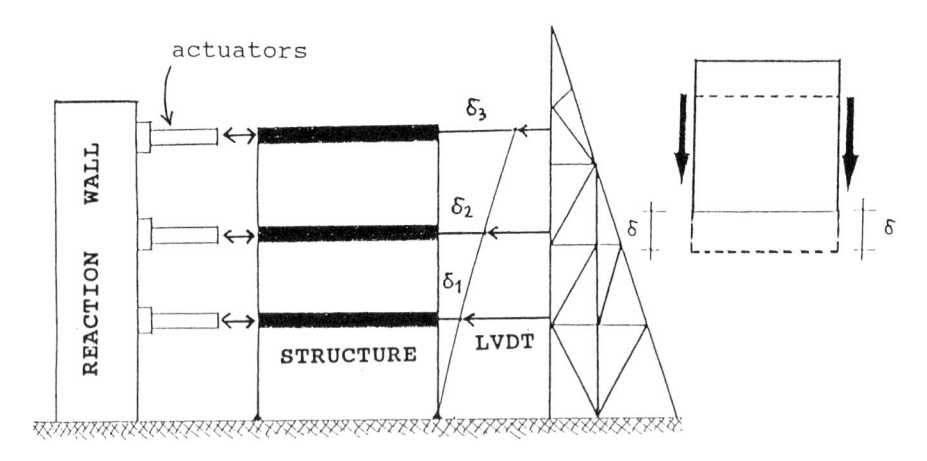

Fig.2 Pseudo Dynamic Test Setup

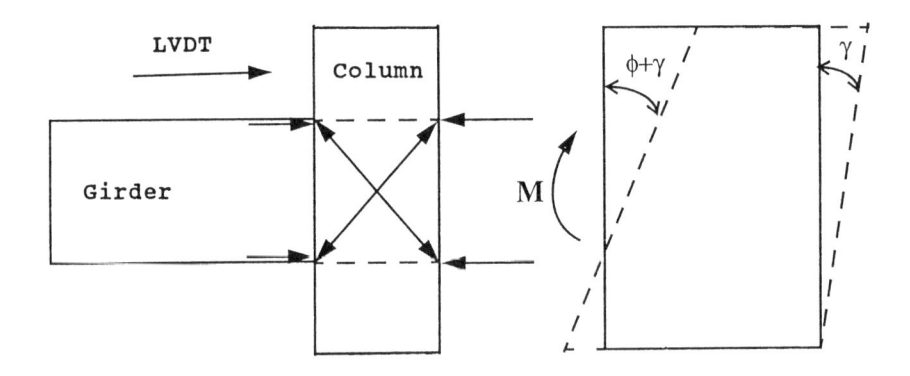

Fig.3 Beam-to-Column Rotation Measurement

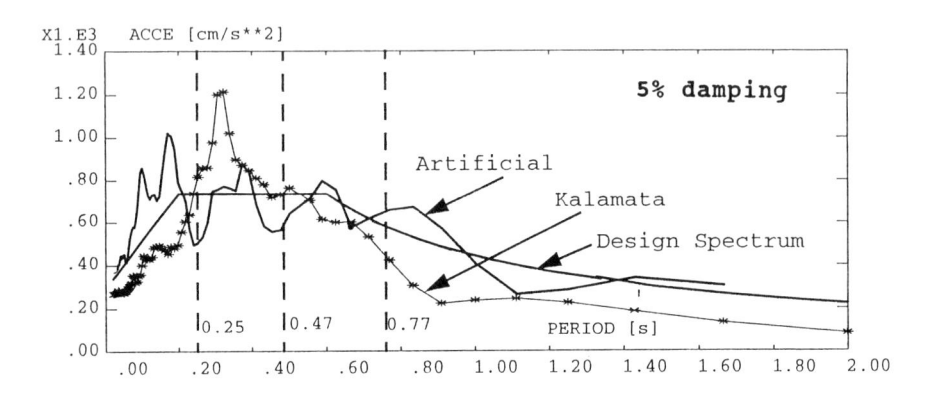

Fig.4 Earthquake Spectra and Steel Frame Periods

generated accelerogram were employed in the tests (Fig.4). Specifically, the latter accelerogram had similar spectral characteristics as considered in the frame design. Initially, the earthquake magnitude was scaled at a level insufficient to introduce plastic strains into the structure. Subsequently, the earthquake intensity was progressively increased, in order to introduce plasticity into the structure (Table 1).

6 Steel Frame Periods

The steel frame periods were obtained from free vibration tests. Herein, the frame's fundamental period was 0.25sec, with the frame floor mass, for the structure in the Lab being 7.3ton (Fig. 4). This result reflected stiffness contributions from flexible beam-to-column connections/joints. The connection/joint flexibility was verified in the PSD-tests. During all free vibration tests executed, a structural damping of 0.88 % was observed.

In the steel frame design, a fundamental period of 0.58sec was employed, reflecting rigid beam-to-column connections and joints. Herein, the frame floor mass from the complete building (total building floor mass/7=69.3ton) was employed (Fig.1). Using the information obtained from the free vibration tests (steel frame with flexible joints) and the floor mass of 69.3ton, a first mode period of 0.77sec was derived. For the tested steel frame a comparison of rigid/flexible joints yielded in an increase in the structure periods of 33% (0.77sec/0.58sec-Fig.4). Consequently, for the subsequent PSD-tests it was decided to simulate numerically in the PSD-algorithm a structure with a period of 0.47sec. This decision was taken in order to simulate a stiffer structure, realizing the fact that in the building considered (Fig.1) partition walls and elevator shafts influence the building's dynamic properties and hence would probably decrease the building's fundamental period of 0.77sec (Fig.4).

7 Experimental Results from Pseudo-Dynamic Tests

The recorded floor horizontal displacement time histories were used to identify elastic or plastic response. All three floor displacement time-histories were in phase, whereby the max. displacement appeared at the 3-floor (Fig. 6). A visual inspection of the latter time histories could not reveal higher mode participation. Such effect was confirmed by investigating the time histories of the floor inertia loads.

The girder-end moment, axial and shear forces were evaluated from joint equilibrium at the three floors, using the column mo-

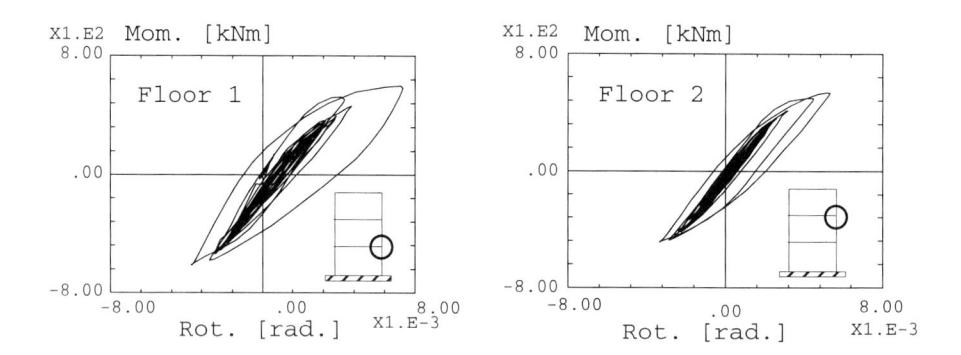

Fig.5 Hysteresis Diagrams-Girder moment at column face vs Joint Rotation($\psi+\gamma$)

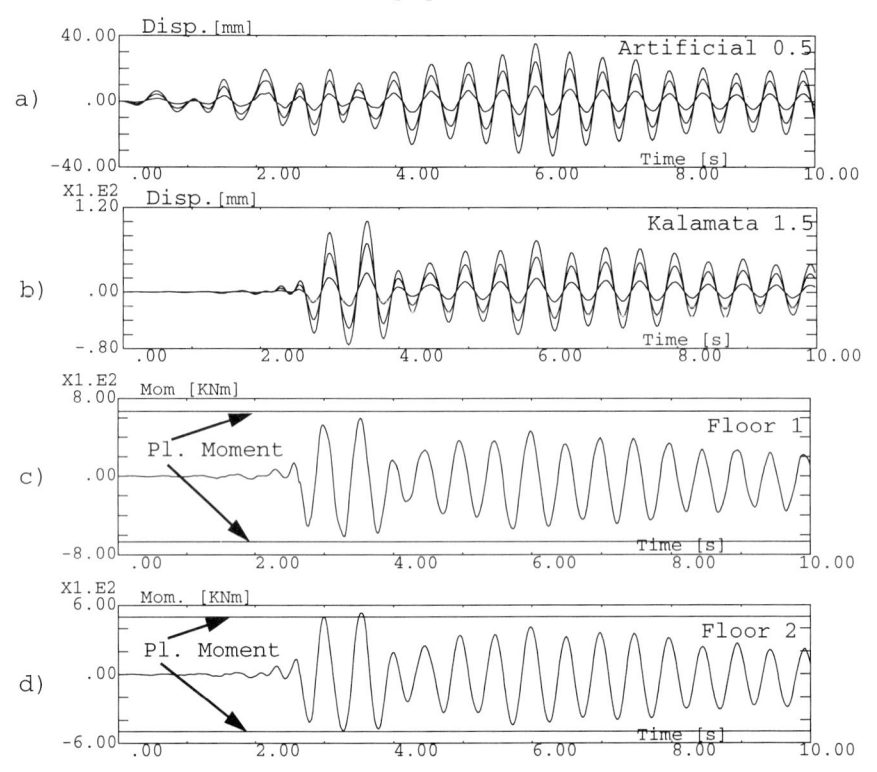

Fig.6 Floor horizontal displacements time-histories: (a) ART05, (b) KAL15 / Time histories of Girder moment at column face for KAL15 : (c) Floor1, (d) Floor 2

ments, shear and axial forces above and below the joint. Time histories of the latter quantities revealed that under all seismic excitations the frame girders and columns were elastic (Fig.6) while the seismic energy absorption took place in the beam-to-column connections and joints. (Fig. 5). The recorded bending moment distribution in the girders was antisymmetric. In addition, it was verified that shear and axial force did not affect the girder's plastic moment capacity.

The obtained girder-end moment (M)-(ψ+γ) rotation hysteresis diagrams (Fig.5) permitted a direct evaluation of the hysteretic joint behavior. Yielding first appeared in the column web in the area of girder-flange/column-flange junction. This effect was due to the local introduction of the beam-flange force into the column web. Subsequently, shear yielding appeared also in the panel zone. During KAL15 and ART14 seismic events, both connection (ψ) and panel zone (γ) were plastic (Fig.5). During this procedure, the columns and girders framing into the joints were elastic. The experimental data showed evidently that connection and joint rotation contributed in the frame's deformations.

Ultimately, during ART14 earthquake, cracks appeared in the first story beam-flange to column-flange full penetration welds. The rotation developed in the joint (about 8 mRad during KAL15 and ART 14 -Fig. 5), while the column and girder framing into the joint were elastic, led to a situation where the girder flange welds were subjected to bending.

8 Conclusions

From the PSD-tests the following conclusions were drawn:
1. EC8 Capacity Design is appropriate for sizing the frame's girders and columns in order to ensure seismic safety.
2. During seismic excitations, seismic energy absorption takes place in the beam-to-column connections and joints, while the adjacent girders remain elastic. Consequently, the strong-column weak-girder design concept cannot be fulfilled. To fulfil this latter objective, ASD design criteria for the joint shear panel zone must be employed.
3. The current EC8 provisions for sizing the thickness of the joint panel zone appear to require further review. Herein, there is a very delicate balance as to how much plastic rotation can be developed in the girders, and how much can be permitted to appear in the joint panel zones and connections. This effect is governed by the joint rotation (ψ+γ-Rotation-Fig.5). Under seismic loading the induced local curvature together with the longitudinal flange stress tend to cause cracks in the girder flange welds. ASD design for the joint panel zones can improve this situation.

9 Acknowledgements

The research presented in this paper was executed by the first author during a post-doctoral stay at the Joint Research Center, Ispra. He would like to thank especially the Commission of the European Communities for providing the financial support, and Prof. Jack Bouwkamp, from the Technical University of Darmstadt/Germany, for his critical suggestions in developing the test specimen and the test setup.

10 References

Donea J, Jones P.M, Magonette G, Verzeletti G. (1988) **The Pseudo-dynamic Test Method for Earthquake Engineering.** Commission of the European Communities, Joint Research Centre. Report EUR 12486 EN.

Eurocode 8 (1988) **Structures in seismic regions. Design Part 1. General and Building.** Report EUR 8849 EN. Commission of the European Communities

Kakaliagos A. (1994) **Pseudo-Dynamic Testing of a Full Scale Three-Story One-Bay Steel Moment-Resisting Frame. Experimental and Analytical Results.** Commission of the European Communities, Joint Research Centre. Report EUR-15605 EN.

Kakaliagos A, Bouwkamp J.G. (1993) Tests on Steel - Composite Beam-Column Connections. Strength and Stiffness Aspects. **Earthquake Spectra, Vol. 9, number 4.**

35 EXPERIMENTAL DYNAMIC MEASUREMENT OF TALL BUILDINGS WITH STEEL STRUCTURES

A. LA TEGOLA
Faculty of Engineering, University of Lecce, Lecce, Italy
W. MERA
Facultad de Ingeniería, Universidad Católica de Santiago de Guayaquil, Ecuador

Abstract
The seismic response of buildings with steel structures, theoretically evaluated in the structural design process, may have a noticeable difference with respect to the real structure. One cause may be that of the contribution of the non-structural elements, which although they are designed to perform a different function, in reality they have strength characteristics that can not be disregarded.
This incidence may not be ignored as compared to what can be observed in similar reinforced concrete structures.
Keywords: Seismic, Buildings, Vibrations.

1 Introduction

Steel structures require heavy additional elements which contribute to the hiding of electrical, plumbing or air-conditioning installations, protection for fire resistance, architectural cladding, etc. The quantity of such elements is a variable which is a function of the use given to the building. In some commercial buildings the vertical structure is additionally covered with architectural cladding that because of its dimensions, may influence noticeably the dynamic response of the building due to horizontal forces. This influence is noticed in the period of vibration of the structure.
  The main objective of this research is the direct measurement of the first period of vibration of steel buildings using the natural ambient vibration excitation. This is accomplished in four buildings of different use in the city of Guayaquil, Ecuador. Because the buildings are located on a very compressible soil, the natural period of vibration of the soil has also been measured.
  It has been demonstrated that in some cases when the period of vibration of the soil is very similar to the measured period of vibration of the building, there is the doubt that this measured period does not correspond to that of the building. Following there is some theoretical

considerations with reference to a two-degree of freedom system that models the building structure linked to the compressible soil, which shows that the measured period is in fact the period of the building structure.

2 A two degree of freedom model

With reference to the model in Fig. 1, the quantities k1, m1, k2, m2 represent the stiffness and the mass of the soil and the building respectively.

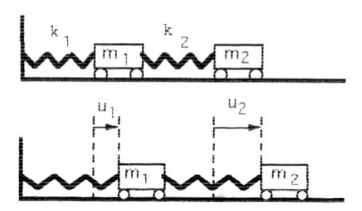

Fig.1. Two degree of freedom model.

The equation of dynamic equilibrium at a time t is as follows:

$$k_1 u_1 = -m_1 \ddot{u}_1 - m_2 \ddot{u}_2 \qquad (1)$$

$$k_2 (u_2 - u_1) = -m_2 \ddot{u}_2 \qquad (2)$$

Assuming for u_1 and u_2 the following solution:

$$u_1 = U_1 \sin(\omega t) \qquad (3)$$

$$u_2 = U_2 \sin(\omega t) \qquad (4)$$

It is obtained the homogeneous system:

$$(k_1 - m_1 \omega^2) U_1 - m_2 \omega^2 U_2 = 0 \qquad (5)$$

$$k_2 U_1 - (k_2 - m_2 \omega^2) U_2 = 0 \qquad (6)$$

In which when the determinant of the coefficients is equaled to zero,

$$\begin{vmatrix} (k_1 - m_1 \omega^2) & -m_2 \omega^2 \\ k_2 & -(k_2 - m_2 \omega^2) \end{vmatrix} = 0 \qquad (7)$$

From this it is obtained a second degree equation in ω^2 that allows to calculate the angular velocity and therefore the system's own frequency. For $m_2=0$ it is obtained $\omega^2 = k_1/m_1$. After naming this value ω_s^2 and dividing the first row by m_1, the determinant may be written as:

$$\begin{vmatrix} (\omega_s^2 - \omega^2) & -\dfrac{m_2}{m_1}\omega^2 \\ k_2 & -(k_2 - m_2\omega^2) \end{vmatrix} = 0 \qquad (8)$$

If it is assumed that the mass of the building is considerably less than the mass of the soil, this hypothesis being acceptable for buildings with steel structures, the two frequencies are uncoupled and therefore, ignoring the quantity m_2/m_1, it is obtained:

$$\omega^2 = \omega_s^2 \qquad (9)$$

$$\omega^2 = \dfrac{k_2}{m_2} \qquad (10)$$

that represents respectively the square of the angular velocity of the soil and of the building. Dividing the second equation by m_2 and naming $\omega_{st}^2 = k_2/m_2$ it is obtained:

$$\begin{vmatrix} (\omega_s^2 - \omega^2) & -\dfrac{m_2}{m_1}\omega^2 \\ \dfrac{k_2}{m_2} & -(\omega_{st}^2 - \omega^2) \end{vmatrix} = 0 \qquad (11)$$

And defining $(\omega_{12})^2 = \omega_{st}^2\, m_2/m_1$ leads to the equation:

$$\omega^2 = \dfrac{\omega_s^2 + \omega_{st}^2 + \omega_{12}^2}{2} \pm \dfrac{1}{2}\left((\omega_s^2 + \omega_{st}^2 + \omega_{12}^2)^2 - 4\omega_s^2\omega_{st}^2\right)^{\frac{1}{2}} \qquad (12)$$

With the variable $\mu = m_2/m_1$ and $s = \omega_s/\omega_{st}$ the following equation can be written:

$$\omega^2 = \omega_{st}^2\left(\dfrac{s^2 + 1 + \mu^2}{2} \pm \dfrac{1}{2}\left((s^2 + 1 + \mu^2)^2 - 4s^2\right)^{\frac{1}{2}}\right) \qquad (13)$$

From this relationship it is possible to calculate for different values of μ the value of the angular velocity ω. It can be easily verified that if the mass of the building is less than the mass of the soil, the error involved is not

relevant. It is evident that the measured period of vibration
of the building is larger than the case when the building is
assumed to be rigidly fixed to an undeformable soil.
If we assume the period of vibration of a certain soil to be
0.8 sec and the period of the structural mass to be 1.2 sec,
we obtain the following table as a function of μ:

Table 1.

μ	0.000	0.100	0.200	0.300	0.400	0.500	0.600	0.700	0.800	0.900	1.000
T1(sec)	0.800	0.797	0.788	0.774	0.757	0.737	0.716	0.694	0.673	0.651	0.630
T2(sec)	1.200	1.205	1.219	1.240	1.269	1.303	1.341	1.383	1.427	1.475	1.524

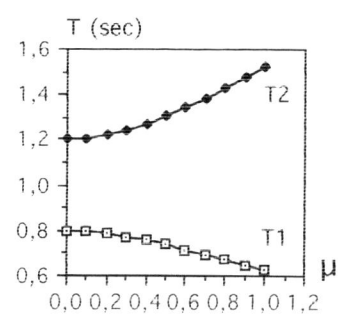

Fig.2. Periods of two freedom degree model

The error obtained is in the order of 27% only for m2=m1 and
therefore it is almost negligible.

3 Description of the equipment used

In order to determine the periods of vibration of four steel
buildings and different soils of Guayaquil, a Kinemetrics
equipment has been used, (the SSR-1), which consist of a
digital seismographic event recorder that records into solid-
state RAM, and three sensors (SS-1), which are Short Period
Seismometers that operate in vertical or horizontal mode and
register at the sampling rate of 100 sps.
   Selecting the appropriate parameters, the SSR-1 can record
continuosly the input comming fom the sensors, that is the
small displacements of the steel structures caused by the
ambient vibration (wind, traffic, machines, etc.) and also
caused by microtremors. The sensors are placed on top of the
buildings and oriented in the two orthogonal horizontal
directions of the structures (major and minor axis). Also, a

measurement is made on the base of the building to determine the natural period of vibration of the soil. The average recording time in each case is about three minutes. The recorded signals are stored in a computer for future use.

4 Description of the software used

After the measuring of the ambient vibration has been done, the records are retrieved from the SSR-1 unit using a Notebook computer and they are processed with an appropriate software for the calculation of the Fast Fourier Transform (FFT) of the signal. This mathematical procedure allows for the transform of the signal from the time domain to the frequency domain. Also the Power Spectral Density is determined which leads to the calculation of the natural frequency of vibration of the structure that has been measured, either a building or a soil. The inverse of the natural frequency is the natural period of vibration of such structure.

5 Experimental measurements

The experimental measurements were carried out in four buildings with steel structures which have different characteristics originated from the use of the buildings. The first building is a parking garage in which there is not present any important infrastructure. The second and third buildings are commercial banks with important infrastructure. The fourth building has a dual use: commercial activities and housing.

6 Results and conclusions

The measured periods of vibration along the two important directions on each of the buildings (major and minor axis) reveals a very important fact. In such buildings the walls or partitions made out of clay bricks, contribute to the stiffness of the structure, as well as the columns and the slabs made out of concrete. Therefore, it is concluded that in the deflection of the structure, the shear mode governs as compared to the bending mode. As a consequence of this, the number of levels of the structure does not influence the period of vibration.

In order to demonstrate this observation, a great number of measurements of reinforced concrete buildings have been carried out as part of another research project in the city of Guayaquil. In such buildings the columns have considerable dimensions and the deflection of the structure, specially for tall buildings, is governed by the bending mode, and the period of vibration increases almost linearly with the number

of levels, which also confirms some considerations of the
Ecuadorian Code which establishes that the period of
vibration (T) of a reinforced concrete structure may be
estimated to be aproximately equal to 10% of the number of
levels (N), or T = 0.10 N. In the following figure it is
plotted the Period of Vibration of Reinforced Concrete
buildings, as well as buildings with Steel Structures, along
the two main orthogonal directions, versus the number of
levels of such structures.
The difference between the values given by the Education Code
and the measured periods of vibration is as an average 30%,
always as a function of the number of levels, being the
measured periods smaller than the theoretical ones. This
results give similar differences with respect to those
obtained from the aproximate relationship from the Seismic
Italian Code:

$$T = 0,1 \frac{H}{B^{1/2}} \quad \text{with H,B in meters} \tag{14}$$

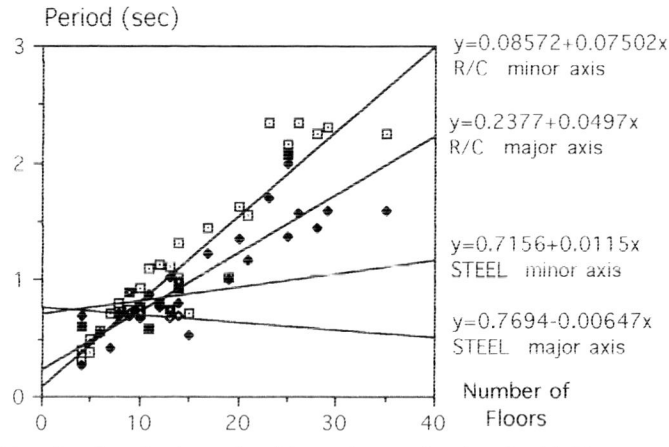

Fig.3. Periods of vibration of the buildings.

The results of the measurements of the periods of vibration
of the four buildings using the ambient vibration acting upon
the structures is presented in the following table 2,

Table 2

Building name	f1 Hz	f2 Hz	T1 sec	T2 sec	Number of Floors
Centro Park	1.12	1.41	0.89	0.71	9
Banco Internacional	1.30	1.41	0.77	0.71	10
MultiComercio	1.33	1.49	0.75	0.67	13
Seguros Cóndor	1.02	1.45	0.98	0.69	14

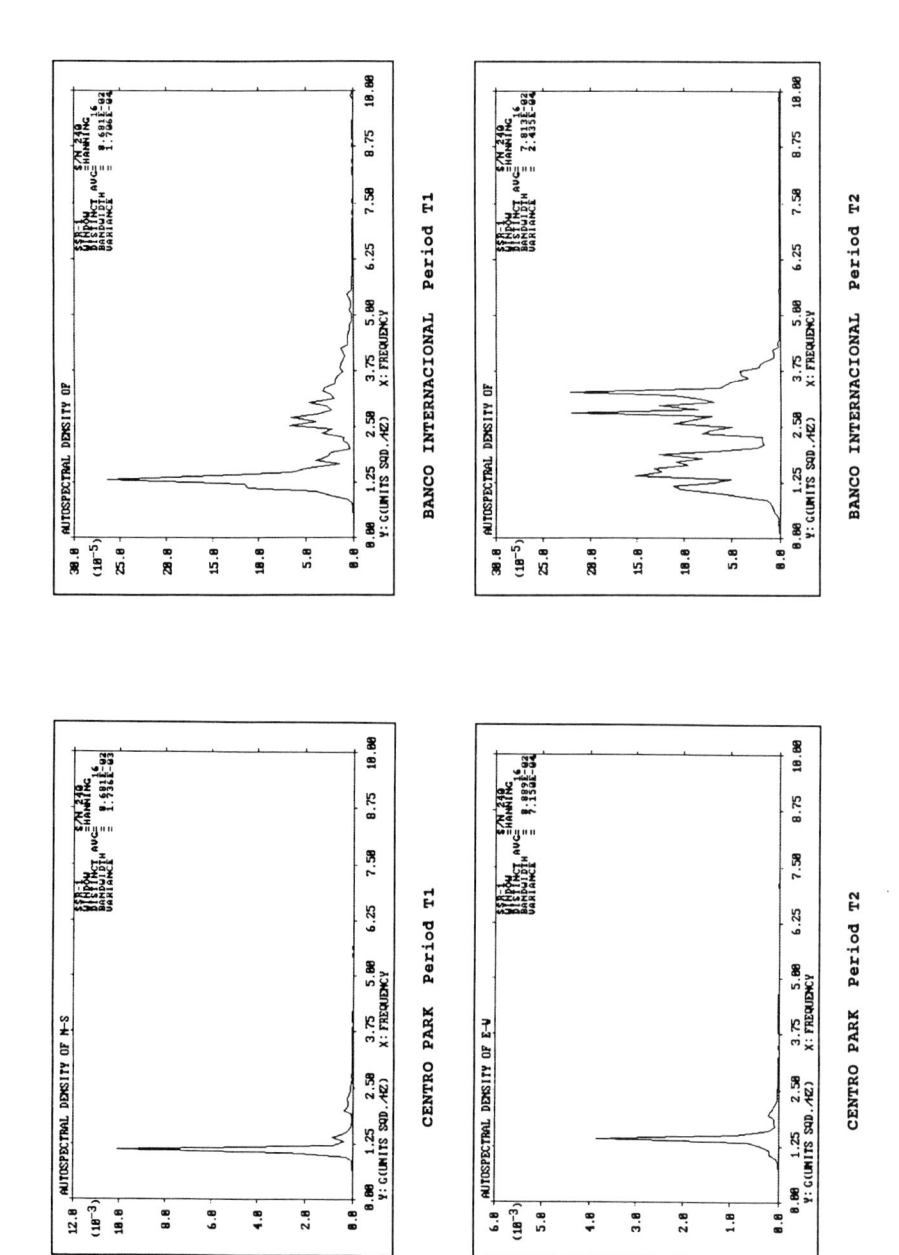

Fig. 4.a

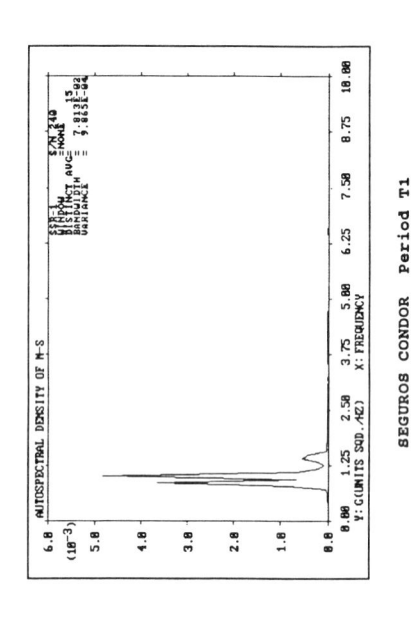

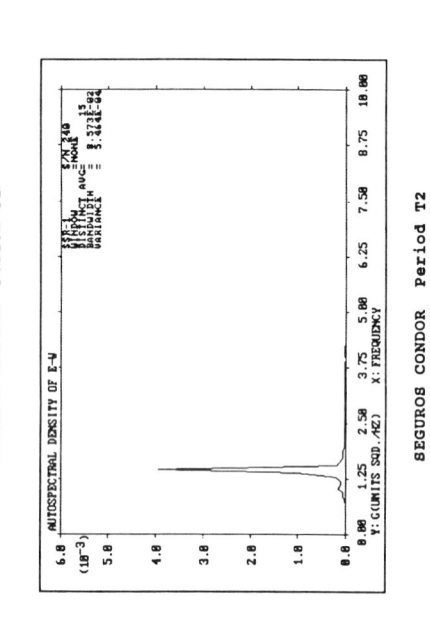

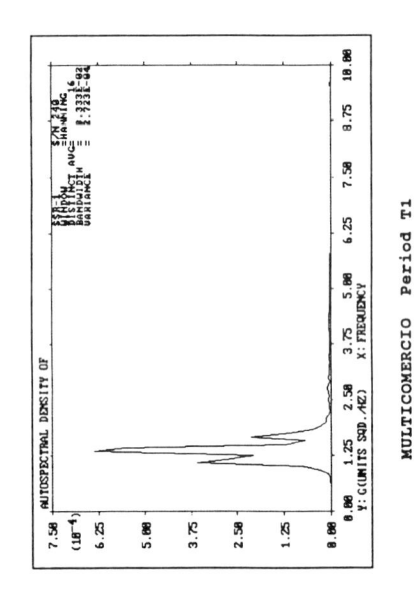

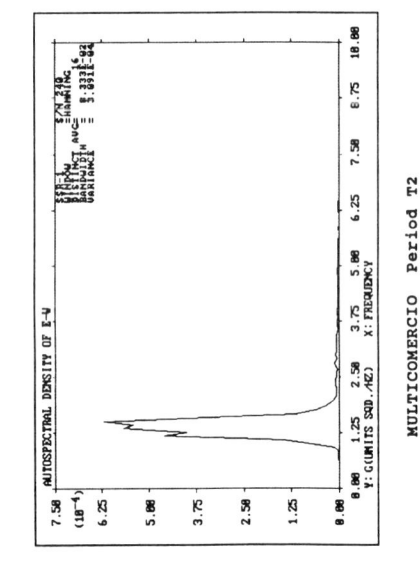

Fig. 4.b

where f1 and T1 are the frequency and period of vibration of the buildings in the major direction, and f2 and T2 are the frequency and period of vibration in the minor direction.

The graphs containing the Power Spectral Density of the signals generated by the four buildings in each direction of the structures are shown in Fig. 4.

A comparision between the periods of vibration of the buildings obtained from the measurements of the ambient vibration (T1, T2) and the periods of vibration obtained from the analytical study of the steel structures (T1a, T2a) gives the following results:

Table 3.

Building name	T1 sec	T2 sec	T1a sec	T2a sec	Number of floors
Centro Park	0.89	0.71	1.12	0.92	9
Banco Internacional	0.77	0.71	1.21	1.05	10
MultiComercio	0.75	0.67	1.49	1.25	13
Seguros Cóndor	0.98	0.69	1.54	1.38	14

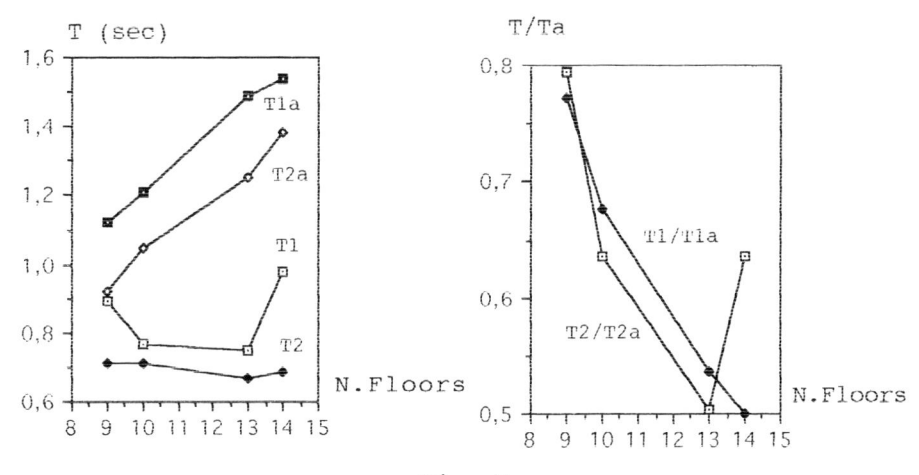

Fig. 5

The graph in fig. 5 shows the difference between the experimental and the theoretical results. It can be noticed that for Centro Park (a parking building), in which the superstructure is not considerable, gives a better correspondance between the calculated periods and the measured ones. The larger difference can be found in MultiComercio where the additional structural elements alter the measured structural response.

I any case, it can be said that the methodology used in the measurement of the periods of vibration of steel buildings in Guayaquil gives reliable results and that the theoretical periods of vibration are always larger than measured ones.

Because the real periods are always smaller than the theoretical periods with reference to the structural model adopted, one should expect that the real seismic forces that develop in the building are larger than those considered in the structural project and therefore a large amount of damage could occur for seismic accelerations smaller than those considered (according to the italian seismic code, the response coefficient increases from 10% to 20%; according to the EuroCode EC8, it increases from 30% to 40%).

One last important fact is that in steel buildings, being the floors much more rigid than the vertical elements, the shear deflection mode prevails and the periods of vibration are not influenced by the number of levels, as it is the case for buildings with reinforced concrete structures.

References

La Tegola, A.(1963) Stabilità dell'equilibrio e dinamica dei telai elastici multipiani. Nota I e II, **Costruzioni Metalliche n.1,2.**

Mera, W. (1991) Vulnerabilidad sísmica de Guayaquil: Determinación de los períodos de vibración de estructuras y de suelos usando la medición de la vibración ambiental; riesgo de resonancia. **Boletín de la Coordinación de Investigación de la Universidad Católica de Guayaquil n.5.**

36 SHEAR BEHAVIOUR OF STEEL CORRUGATED PANELS

R. LANDOLFO and F. M. MAZZOLANI
Istituto di Tecnica delle Costruzioni, University of Naples, Naples, Italy

Abstract

The first results of a general research program devoted to study the influence of cladding panels on the structural behaviour of framed structures are presented in this paper. The complete investigation analyses both experimental and theoretical aspects. With regard to numerical studies, the finite element analysis has been chosen as the suitable technique for evaluating both shear flexibility and strength of panels under monotonic loads. An appropriate model has been set-up, allowing to interpret the single panel behaviour taking into account both the panel geometry and its connecting system. The reliability of such models has been, therefore, deeply investigated and verified against existing experimental results.
Keywords: Seismic Behaviour, Stressed Skin Design, F.E. Analysis.

1 Introduction

It is well know that the interaction between cladding panels and framed structure represents an important aspect in evaluating the response of steel structure. The stiffening effect of panel, in fact, is always present and influence anyway the structural behaviour even if it is usually neglected due to the difficulties of taking it into account [Mazzolani and Sylos Labini (1984)]. In case of building in seismic areas, on the contrary, the cladding panels can give a significant contribution to the dynamic response of framed structure and they can suffer considerable damages in accordance to their absorption of earthquake energy [Mazzolani and Piluso (1990),(1991)]. The exact computation of this contribution becomes, therefore, a crucial aspect in order to know the actual bearing capacity of the "skin-frame" system [Davies and Bryan (1982)].

Starting from these considerations, a general research programme has been promoted to analyse the contributing effect of building panels on steel structure resistance to seismic and aeolian phenomena [Landolfo, Mazzolani and Piluso (1993)]. The research, supported by ECSC – Steel Research Program, and developed with the cooperation of italian consortium CREA (Massa), intends to investigate on the main aspects of the problem in order to asses some design rules which take into account the effect of panels on the whole behaviour of the steel structures.

This aim is pursued by means the following phases:

1. experimental analysis on corrugated sheet panels, linked together by different kinds of seam connections, submitted to monotonic and cyclic shear force;
2. f.e. analysis to model the behaviour of a single panel in terms of load-displacement relationships;
3. Full-scale tests on frame structures with bracing panels applying dynamic loads by means of suitable systems, namely shaking table or pseudo-dynamic methods;
4. numerical simulation of the dynamic behaviour of frame structures with collaborating panels by using an appropriate mathematical model which allows the interpretation of the results obtained in the previous phases as well as development of parametric analyses in order to reach some design guidelines.

The experimental activity is in progress. The sandwich panel typologies has been chosen because it represents the most widely used in practice for roofing and cladding. With regard to the connecting system , screws and glued strips have been considered.

As far as the numerical studies are concerned, finite element analysis of panels under monotonic shear force has been performed, according to Phase 2. In particular, an appropriate model has been set-up, allowing to interpret the single panel behaviour in terms of load-displacement relationships, taking into account both the panel geometry and its connecting system. In the following, the formulation of this model and its application to simulate the existing test results are shown.

2 Finite element model

The most advanced method to predict the effective shear response of diaphragms, as an alternative to the experimental full scale tests, is the finite element computer analysis. Test procedures have been standardised by the American Iron and Steel Institute (AISI), but their application is very expensive and time consuming, and the results applicable only to specific combinations of panels, supporting members and connectors.The essential feature of the finite element method is that a continuous body is divided into a large number of small elements arranged in such a way to approximate the actual geometry of the structure. The application of this method is particularly effective for shear diaphragms. In fact, the actual system consists on a large number of individual members, connected at discrete points; for this reason the finite element representation is more realistic than the one based on an equivalent continuum. Such system, therefore, can be idealized as a plane stress type problem in which each individual sub-panel becomes a finite element with nodes at point of interconnection with other sub-panel and supporting frame. Nevertheless, the definition of this model is quite complex and requires to pay particular attention for modelling the connection system and panel geometry [Nilson and Ammar (1974)]

Three different elementary structural systems have be modelled: the edge members, the panel sheets and the connections. Edge members at the perimeter of the diaphragm are represented as one-dimensional elements, able to resist axial loads as well as bending. If the panel is a continuous flat sheet, it can be modelled as an uniform, isotropic elastic plate loaded in its own plane and with elastic constants equal to that of the base material. On the contrary, for corrugated sheets – i.e. trapezoidal sheets – the behaviour of the panel is better interpreted by means of orthotropic plate with different extensional moduli in the two principal directions. In this case the corrugated sheet is modelled by an equivalent orthotropic thin plate of uniform thickness equal to that of the actual sheet. Five elastic constants are necessary to locate such a behaviour: two extensional moduli in the direction parallel (1) and perpendicular (2) to the corrugations, the effective shear modulus and two Poisson's coefficients in the same directions 1 and 2. All these elastic constants must be derived considering not only the properties of the base material, but also the geometry of the cross section. In particular the shear modulus is deeply influenced by the warping effect at the end of sheeting and depends on the type of edge fasteners, if this is made at discrete locations. With regard to the evaluation of the elastic constants, theoretical formulations are suggested in technical literature as an

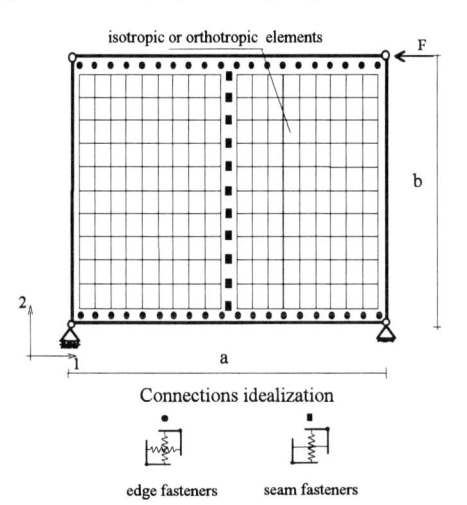

fig.1

alternative to experimental tests. The comparison between the results provided by this formulation and the ones obtained by means of finite element computation has shown that the theoretical approach leads to strongly underestimate the actual stiffness of the panel in transversal direction.

With regard to the connecting system, in both isotropic and orthotropic sheets, a regular discretizzation for the equivalent plate is adopted, in which the nodes are placed according to the position of the connectors. The displacements due to the connector slip play an important rule in the actual response of shear diaphragm. As a consequence particular attention for modelling these elements is required. In finite element model, the connections are idealized by means of two-directional springs, which interconnect the dual nodal points placed at any connector location. For spring elements a non-linear force-displacement relationship has to be assumed in order to account for the plastic redistribution of forces among various connectors. In the whole system a non-linear behaviour has been introduced for the connection system only according to previous experimental investigations, which have shown that this is the only important source of mechanical non-linearity [Atrek and Nilson (1980)].

The panel idealization for the application of the method is showed in Fig.1.

3 Influence of connecting system

In a shear diaphragm there are essentially two systems of connectors: seam fasteners between subpanels and panel-to-beam edge fasteners. A finite element analysis of "ad hoc" panel has been, therefore, performed -by using the advanced non-linear code Abaqus- in order to study the influence of the connecting system on the shear flexibility [De Matteis (1994)]. The sheet considered in the analyses is an isotropic elastic plate of uniform 5 mm thickness; the connectors consist of two-dimensional spring linkage and the edge members are rigid elements with pin-ended connections. The response of the spring elements has been characterised by assuming the experimental load-slip relationship showed in Fig.2, which is provide by Atrek and Nilson (1980). Several numerical applications have been carried out in order to calculate the shear flexibility of the diaphragm by summing the single component contribution (Fig.3). The first application is devoted to a single subpanel by considering a linear behaviour os spring elements. The shear strain of the plate (S), has been, preliminary, evaluated by assuming an infinite value for the elastic constants of every springs modelling the panel-to-beam connections. Then, the flexibility due to the slip in the fasteners in the two directions of the plane has been analyzed by means of three different steps. In the first one (S+O), an infinite stiffness value for the spring elements in longitudinal direction has been assumed. As regard to the springs in transverse direction, the initial stiffness of load-slip relationship showed in Fig.2 has been assigned ($k=12440$ Nmm^{-1}).

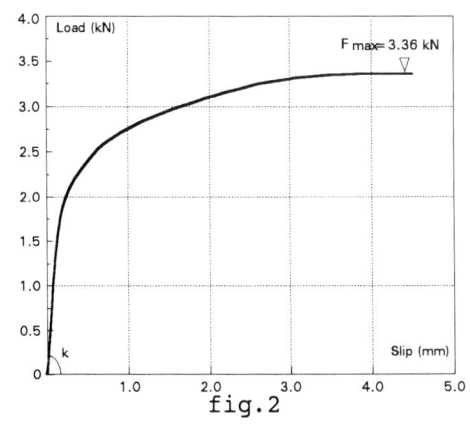

fig.2

In the second step (S+V), relative displacements between panel and beams have been allowed in longitudinal direction only by assigning a stiffness equal to 12440 Nmm^{-1} to the springs along this direction. Finally the spring elements with the same finite stiffness in both directions have been considered (S+O+V). The summary of obtained results is shown in Fig.3. In the histograms the panel flexibility have been normalized with respect to the value of the shear strain flexibility of the plate without connectors (S). In the same figure, also the results for the diaphragm formed by two subpanels are shown. In this case the seam fasteners in longitudinal direction have stiffness $k=12440$ Nmm^{-1}, while in the transverse direction an infinite value for k has been assumed to simulate the absence of relative displacements in this direction due to the presence of the constraint produced by the adjacent sheets. The comparison

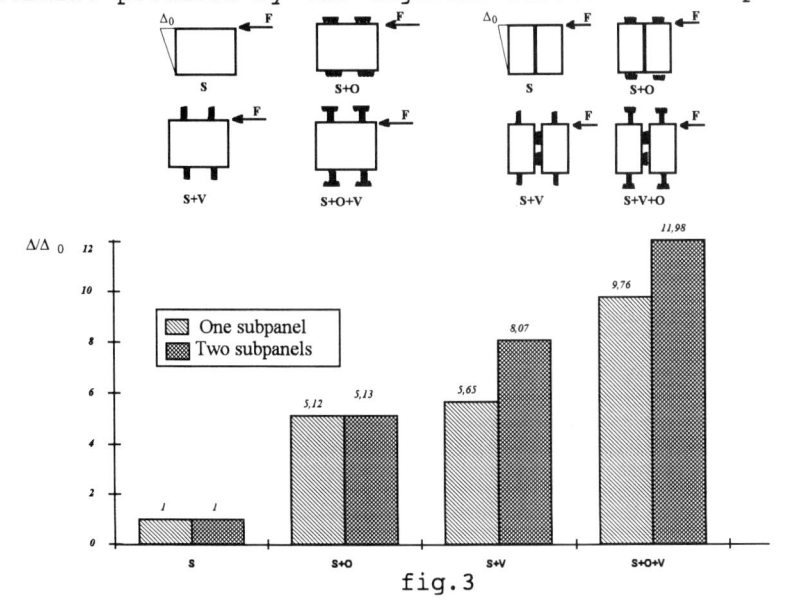

fig.3

between the two histograms allows to emphasise the flexibility contribution due to the seam fasteners.

The shear flexibility of the whole panel under the hypothesis of non-linear behaviour of the connections has been also investigated, by assuming the load-slip relationship showed in Fig.2. The comparison between linear and non-linear curves (Fig.4) allows to underline that the inelastic global response becomes significant starting from about 40% of failure load. In each case, the analyses have evidenced that the contribution of connecting system to the shear flexibility of the panel is always prevalent.

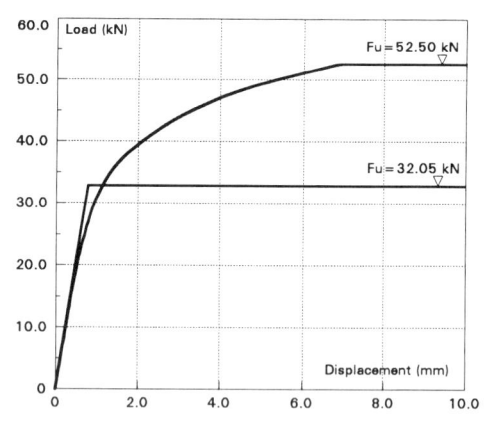

fig.4

The diaphragm failure load has been defined as the value of the external shear force corresponding to the collapse of the first spring connector. In the linear analysis, this occurs for an external shear load of 32.05 kN, which allows the ultimate strength in the corner panel-to-beam connector to be reached. On the contrary, when a non-linear approach is followed, a significant increment (up to 52.50 kN) in the diaphragm shear strength is observed, due to the plastic redistribution of stresses along connectors.

4 Comparison with experimental results

In order to calibrate the proposed model, it has been applied for simulating the monotonic shear behaviour of a panel tested at the University of Pisa [Sanpaolesi, Bolzi and Tacchi (1983)]. The examined prototype, shown in Fig.5, is composed by three symmetrical trapezoidal sheets, with cross section depth equal to 47 mm and 1.2 mm thickness, assembled by means of aluminium rivets of 6.3 mm in diameter at a pitch of 254 mm. The complete panel is connected to the upper and lower UPN 120 chords with Φ 6.3 mm self-drilling screw every 210 mm. The external frame is made of HEA 300 columns connected to IPE 240 beam by means of double-angle joints. The experimental shear force versus shear displacement curve under monotonic load shows a non-linear behaviour of the diaphragm (Fig.5). The collapse of the system is reached by means of rivet failure up to the

Shear behaviour of steel corrugated panels 427

Elementary sheet 1172 x 2643 mm

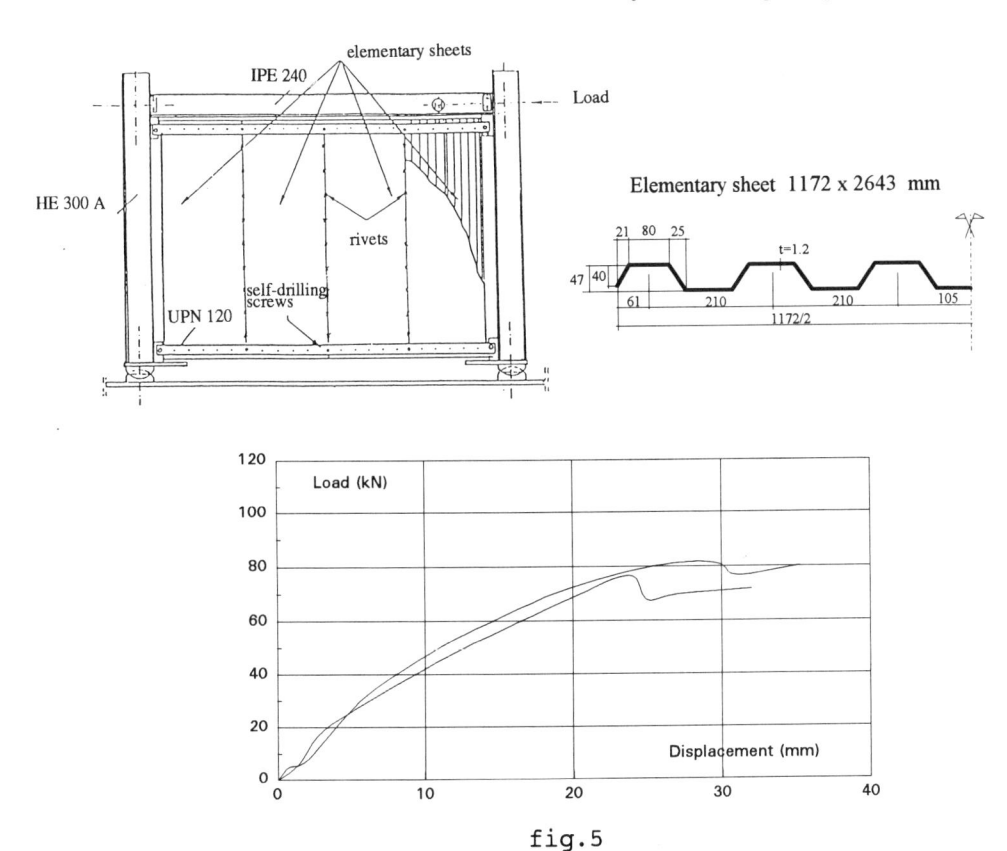

fig.5

G_{eff}	E_1	E_2	ν_{12}	ν_{21}
3500	963	266700	2.08E-3	0.576

Orthotropic specifications (N/mm²)

Self-drilling screw   Aluminium rivets

fig.6

full disconnection of each sheets with consequent loss of load bearing capacity.

The diaphragm idealization for finite element analysis is shown in Fig.6. The trapezoidal sheets have been represented by means of orthotropic plate of uniform thickness of 1.2 mm. Each elastic constant has been evaluated through a finite element analysis of the actual geometry of the sheet; the obtained values are summarised in the same figure.

The force-displacement relationship of the two systems of connection has been obtained by additional tensile tests carried out at Engineering Faculty of the University of Naples. The tests have concerned steel sheet elements with thickness equal to 1.2 mm fastened together by means of aluminium rivets, and fastened to beam elements by means of self-drilling screws. Both rivets and screws have diameter of 6.3 mm. The average experimental curve has been assumed to define the behaviour of the vertical spring elements of the seam fasteners (Fig.7a), while an infinite value of stiffness has been assigned to the orizontal spring elements.

In case of screwed connections, the tests have displayed that the failure occurs for bearing and tearing of the sheet. The experimental load-slip relationships obtained for three specimens are shown in Fig. 7b. In the same figure, also the average curve introduced in the numerical simulation to describe both orizzontal and vertical spring elements behaviour along the chords is reported.

The particular kind of beam-to-column connection used in the external supporting frame (see Fig.5) required the modification of the model by introducing two semi-rigid joints with appropriate values of stiffness.

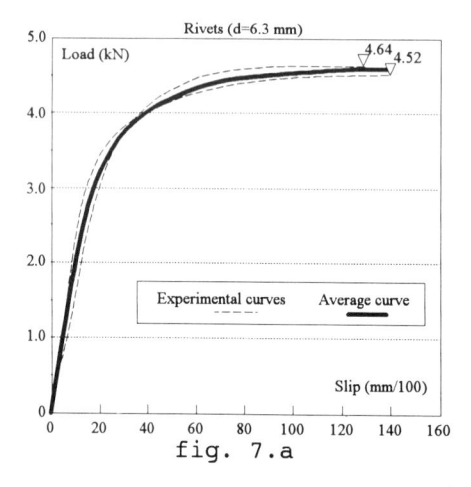

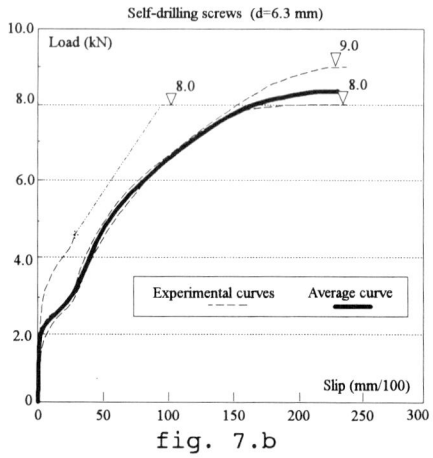

fig. 7.a

fig. 7.b

fig. 8

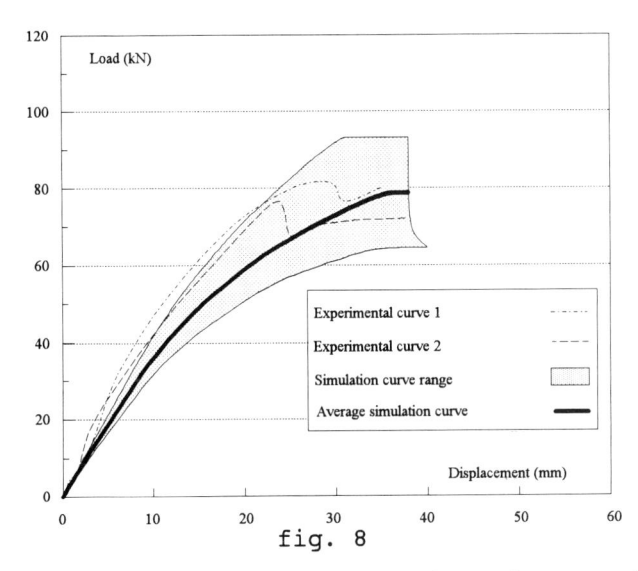

In order to overcome the uncertainty due to the correct evaluation of semi-rigid joint stiffness ($k\varphi$), a band criteria has been assumed. According to this assumption, the values of $k\varphi$ ranged from 0 to $2EI/L$ have been considered, being EI/L the bending stiffness of the upper beam (IPE 240).

The comparison between the range of simulated curves and the experimental data is shown in Fig.8.

The comparison between the averange curves of scatter band evidences that a good prediction of the strength of the panel.

For what concerns the flexibility, the main influencing parameter is the semirigidity of the beam-to-column joint. In particular, from the camparison of the slopes at the beginning of the curves it sems that the best correlation of the experimental results can be obtained by setting $k\varphi=2EI/L$, which has been considered as an upper bound value of the actual one.

5 Conclusions and further developments

The finite element analysis has given suitable results in evaluating both shear flexibility and strength of panels. The proposed model has allowed to simulate the shear behaviour of panels under monotonic loads, provided that the behavioural properties of each component are correctly introduced. These results represent the first step of a general research programme and they will be used as a base for the further developments. The improvement of this model

will follow different directives. First of all, it will be generalized in order to investigate the structural performance of sandwich panels. Subsequently the panel cyclic behaviour will be analyzed both theoretically and experimentally.
The testing program has been set up and typical sandwich panels used in the constructional practice have been selected.

6 References

Atrek, E. and Nilson, A.H. (1980) Nonlinear analysis of cold-formed steel shear diaphragms. **Journal of the Structural Division**, Vol.106, No. ST3, 693-710.

Davies, J.M. and Bryan, E.R. (1982) **Manual of Stressed Skin Diaphragm**. Granada Publishing Ltd., London.

De Matteis, G. (1994) **La modellazione del comportamento a taglio dei pannelli in lamiera grecata**. Graduation thesis, Institute of Tecnica delle Costruzioni, Engineering Faculty, University of Naples

Landolfo, R., Mazzolani, F.M. and Piluso, V. (1993) L'influenza dei pannelli di chiusura sul comportamento sismico dei telai in acciaio, in the **Proceedings of the 6th Congress L'Ingegneria Sismica in Italia**, Perugia, pp. 785-794.

Mazzolani, F.M. and Sylos Labini, F. (1984) Skin-frame interaction in seismic resistant steel structures. **Costruzioni Metalliche**, 4, 212-225.

Mazzolani, F.M. and Piluso, V. (1990) Skin-effect in pin-jointed steel structures. **Ingegneria Sismica**, 3, 30-47.

Mazzolani, F.M. and Piluso, V. (1991) Influence of panel connecting system on the dynamic response of structures composed by frames and collaborating cladding, in **Connections in steel structures: behaviour, strength and design** (eds R. Bjorhovde, A. Colson, G. Haaijer, J.W.B. Stark), Pittsburg, pp 344-353.

Nilson, A.H., and Ammar, A.R. (1974) Finite element analysis of metal deck shear diaphragms. **Journal of the Structural Division**, Vol.100, No. ST4, 711-726.

Sanpaolesi, L., Biolzi, L. and Tacchi, R. (1983) Indagine sperimentale sul contributo irrigidente di pannelli in lamiera grecata, in the **Proceedings of the 9th Congress C.T.A.**, Perugia, pp. 251-264.

37 THE TESTS OF THE COLUMN-BASE COMPONENTS

F. WALD, I. ŠIMEK and J. SEIFERT
Czech Technical University, Prague, Czech Republic

Abstract
This paper is a summary of an experimental part of a research in column base behaviour. The full scale tests and the component tests were performed for a typical column base with H shape columns and simple ubraced base plates. The results are used for numerical simulation and for development of an prediction analytical model based on assumption, that each component is possible to describe as independently.
Key words: Steel Structures, Column Base, Semi-rigid Connection, Tests, Component Method, Eurocode 3.

1 Introduction

The base plate is used as a simple and economic detail, to transfer the load from the steel column into the concrete foundation. The base plates should be fixed with two or four bolts, Fig. 1. Both of these details transfer a significant moment into the foundation. The first detail closely resembles the behaviour of a pinned connection while the second detail behaves more like a fixed joint. The behaviour of the column–base detail depends on the joint parameters but is best represented as a semi-rigid connection.

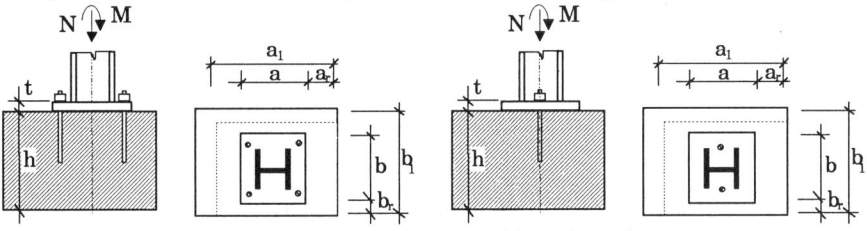

Fig. 1 Column base of typical building frame, bolts outside and inside of the cross-section.

Tab. 1 Main parameters of experiments.

Scheme	Test No.	Plate, mm Thickness	Plate, mm Size	Column	Bolt Diam., Mat.	Loading History	Loading Direction
	B6A	6	110x80	P10	–		F(v)
	B6B	6	110x80	P10	–		F(v)
	B10A	10	110x80	P10	–		F(v)
	B10B	10	110x80	P10	–		F(v)
	W1–200	10	300x200	P20	–		F(v)
	W2–300	10	300x300	P20	–		F(v)
	W3–400	10	300x400	P20	–		F(v)
	BK6	6	70x80	P10	2xM16, 5.6		F(v)
	BK8	8	70x80	P10	2xM16, 5.6		F(v)
	BK10	10	70x80	P10	2xM16, 5.6		F(v)
	W4–80	20	100x80	P20	2xM24, 5.6		F(v)
	W5–90	20	100x90	P20	2xM24, 5.6		F(v)
	W6–105	20	100x105	P20	2xM24, 5.6		F(v)
	BTA	10	220x220	H100	–		F(v)
	BTB	10	220x220	H100	–		F(v)+F(h)
	W8	20	300x220	H160	2xM24, 5.6		F(h)
	W9	20	300x220	H160	2xM24, 5.6		F(h)
	W7	20	300x220	H160	2xM24, 5.6		F(v)+F(h)
	W10	20	300x220	H160	2xM24, 5.6		F(v)+F(h)
	W11	20	300x220	H160	2xM24, 5.6		F(v)+F(h)
	W12	20	300x220	H160	2xM24, 5.6		F(v)+F(h)
	Š13	12	300x220	H160	2xM24, 5.6		F(v)+F(h)
	Š14	12	300x220	H160	2xM24, 5.6		F(v)+F(h)
	Š15	12	300x220	H160	2xM24, 5.6		F(v)+F(h)
	Š16	12	300x220	H160	2xM24, 5.6		F(v)+F(h)
	Š17	12	300x220	H160	2xM24, 5.6		F(v)+F(h)
	Š18	12	300x220	H160	2xM24, 5.6		F(v)+F(h)
	Š19	–	–	–	2xM24, 5.6		F(h)
	Š20	–	–	–	2xM24, 5.6		F(h)

Foundation 550x550x500 mm, C12/15; grout 20 mm, C20/25; steel S 235.

For a steel frame analysis, the column–base bending stiffness is as important as the beam, the column and the beam-to-column connections stiffness Wald at al. (1993b). A few existing experimental data concerning column bases from full scale frame tests, Robertson (1991), show that a significant column–base bending stiffness exists for most of the different cases, even for a simple column base Wald and Seifert , (1991).

The rigidity of column–base connections is primarily dependent on the moment-to-normal force interaction Akiyama (1985), Bealieu (1985), Thambiratnam and Paramasivan (1986), Nakashima (1986). The influence of the plate thickness, Targovski (1993), and of the bolt tree length, Astaneh at al. (1992), limit the behaviour under the pure moment loading. The history of loading could play a significant role in column base design. The presented review is focused on the column–base component tests results Seifert at al. (1993), Šimek at al. (1994) and on the results of the H shape columns.

2 Tests types

The characteristic parameters of the column–base tests are included in Table 1. The three tests types ware used to check the stiffness and resistance of the component on the tension side Fig. 2 of the compression side component Fig. 5 and of the component interaction in the full column base connection, Fig. 7.

2.1 Components in tension

The component tests ware designed to observe the separate behaviour of tension and compression part. The validity was checked on full scale tests.

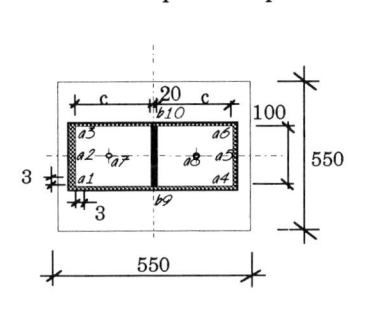

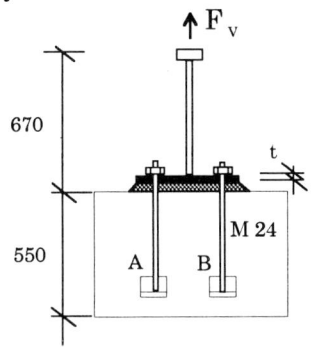

Fig. 2.   The tested stubs in tension.

The main observed problems on the tension side were the prying action Wald at al. (1994) and bolt elongation. Fig. 3 shows the deformation of a very thin base plate loaded in tension where the prying force occurred. The bolt force for a more realistic base plate thickness is drown against the applied load on one bolt, Fig. 4. The prying force are visible close to the anchor bolt collapse.

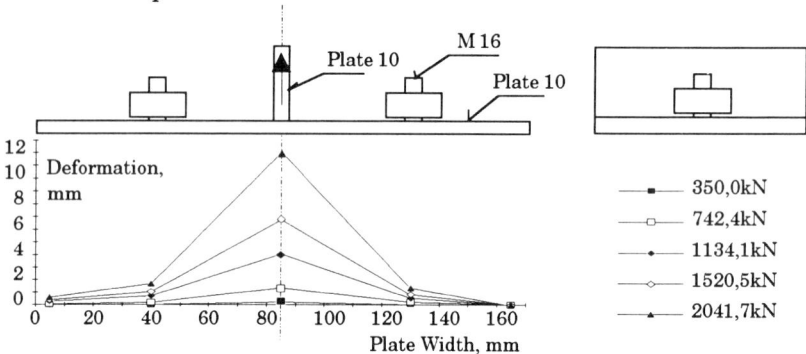

Fig. 2 The mean values of the deformation of the base plate of test #BT10.

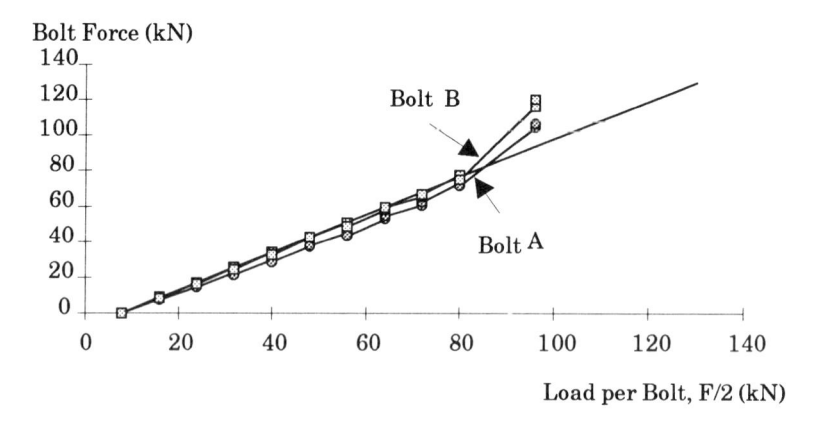

Fig. 4 The Bolt force load per bolt ratio, plate 20 mm, tests #W5-90.

2.2 Components in compression

The compression side is more important from the ultimate resistance modelling point of view. The Eurocode 3 (1992) advanced design model is based on resistance of the concrete concentrated strength and under the flexible plate. The new proposed model Wald at al. (1994b) for Eurocode 3 (updating) is more oriented to the plate behaviour, tests #BTA and #BTB. Fig. 5 shows the deformation on the base plate loaded in compression. Fig. 6 shows the complex 3D problem of the behaviour of the base plate under the H cross section loaded with eccentricity.

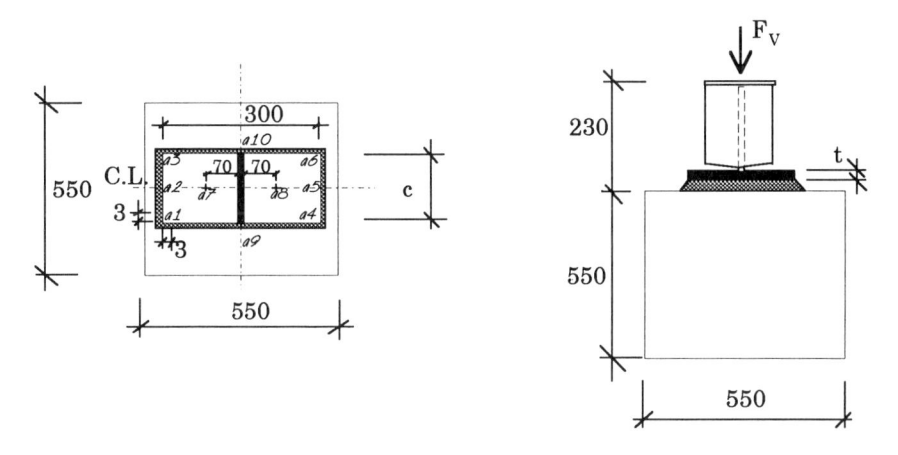

Fig. 5. The tested stubs in compression.

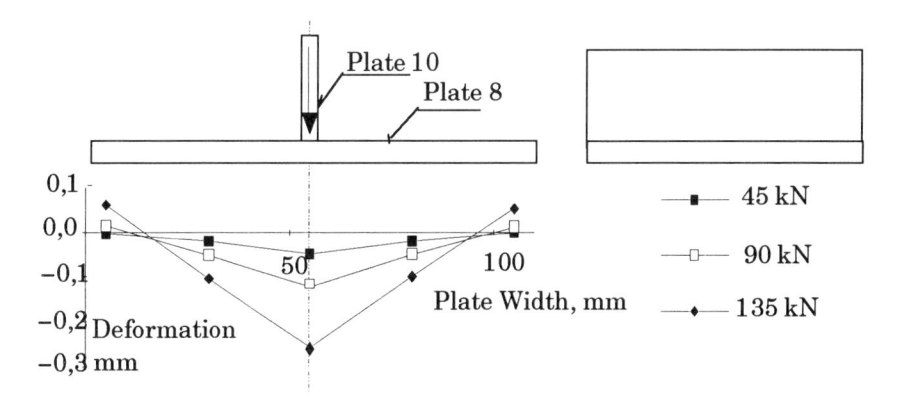

Fig. 6. The mean values of the deformation of the base plate of test #BC 8.

2.3 Column base

The column base of column with shape H were tested in compression, in compression with bending and in pure bending, The load was applied in different order, Tab. 1. with cycling of horizontal force close to the predicted ultimate value.

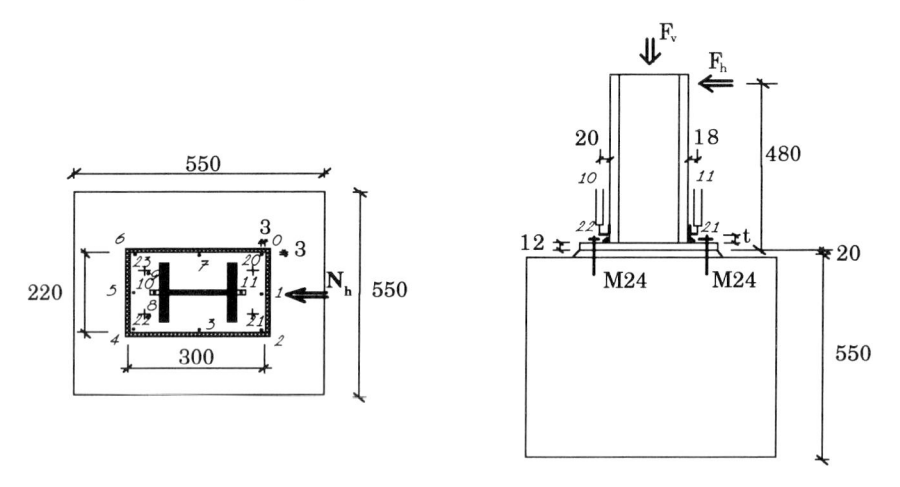

Fig. 7. The column base of full scale

3 Data bank

All the results of the performed tests have been adopted into the data bank, Wald (1993). The data bank is situated on the computer net and allows the public access and co-ordination of the data output of all the tests carried out on the topic. For the purpose of co-ordination of the data output of tests performed at different institutions are established the data sheets, with the all geometric, mechanical parameters and pure tests results. The

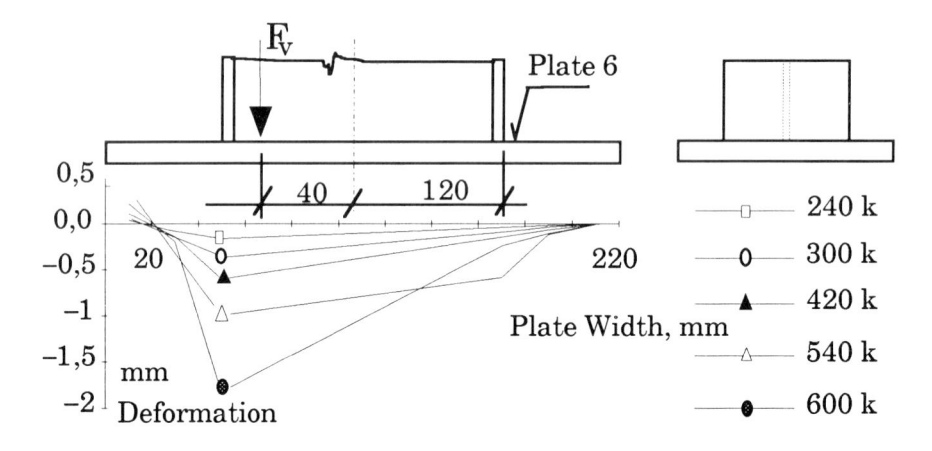

Fig. 8 The deformation of the base plate on the neutral axes, the test #BS 10.

data sheets are the data bank input and output pages produced by special program – interpreter. The data sheets are available in printed form as well. The limited necessary flexibility is included. The ratio of flexibility in data was not solved satisfactory till now. The data bank is also expected to lead to the establishment an acceptable component model of column base behaviour.

4 Conclusions

The column base connection tests depends to one of the most complex testing problems.
The already exist tests results streamed on locally used models will be able to use to help to establish a prediction resistance model based on component behaviour.
The limited number of stiffness tests with fully published moment rotation relation and including different column base connection detailing complicated analytical prediction of column/base stiffness.

5 Acknowledgement

This research has been conducted at the Department of Steel Structures as a part of the research project "Flexibility of Column Bases" which has been supported by Czech government grant COST C1. The sponsorship of Czech Convention for Constructional Steelwork is gratefully acknowledged.

6 References

ENV 1993-1-1, Eurocode 3: "Design of steel structures", (1992) Part 1.1: General rules and rules for buildings. European Committee for Standardisation (CEN), Brussels.

AKIYAMA H., (1985): *Seismic Design of Steel Column for Architecture*, in Japanese, Gibodoskupan, Tokyo.

ASTANEH A., BERGSMA G., SHEN J.H., (1992): *Behavior and Design of Base Plates for Gravity, Wind and Seismic Loads*, Proceedings AISC, National Steel Construction Coference, June, Las Vegas.

BEAULIEU D., PICARD A., (1985): *Cotribution des assemblages avec plaque d'assise a' la stabilite' des poteaux*, Construction Me'tallique, No.2, 1985.

NAKASHIMA S., SUZUKI T., IGARASHI S., (1989): *Behavior of Full Scale Exposed Type Steel Square Tubular Column Bases under Lateral Loading*, in IABSE Symposium, part A, Helsinki, pp. 148–152.

THAMBIRATNAM D.P., PARAMASIVAM P., (1986): *Base plates under Axial Loads and Moments*, Journal Structural Engineering, Vol. 112, No. 5, pp. 1166–1181.

YEE Y., MELCHERS R. E., (1986): *Moment–rotation curves for bolted connections*, Journal of Structural Engineering, Vol. 112, No. 3, pp. 615–635.

ROBERTSON A.P., (1991): *A study of base fixity effects on portal frame behaviour*, The Structural Engineer, Vol.68, No.2, pp. 17–24.

SEIFERT J., WALD F., BAŽANTOVÁ Z. (1993): "*Experiments with column bases*", in Czech "*Experimentální stanovení únosnosti a tuhosti patek sloupů*", ČVUT, Praha, 1993, p. 24.

ŠIMEK I., WALD F., BAŽANTOVÁ Z. (1994): "*Column–base Component Experiments*", in Czech *Experimentální stanovení tuhosti částí patek sloupů*, ČVUT, Praha, p. 28.

TARGOVSKI R., LAMBLIN D, GUELMENT G., (1993): *Baseplate Column Connection under Bending: Experimental and Numerical Study*, Journal of Construct. Steel Research, Vol. 27 , No. 1–3, pp. 37–54.

WALD F., SEIFERT J., (1991): *The Column–Base Stiffness Classification*, in Nordic Steel Colloquium, 9–13 September, Odense, pp. 413–421.

WALD F., OBATA M., MATSUURA S., GOTO Y., (1993a): "*Flexible Base Plates Behaviour Using FE Strip Analysis*", Acta Polytechnica 1, Vol. 33, No. 1, Prague, .p. 83–100.

WALD F., (1993): "*Column–Base Connections, A comprehensive state–of–the–art Review*", TU Budapest, Technical Report No. 12., Budapest.

WALD F., ŠIMEK I., BEZDĚK P., (1993b): "*Column–Base Connections*", Data Bank, COBADAT, COST C1, No.C2–WD/93–03, Liege.

WALD F., OBATA M., MATSUURA S., GOTO Y., (1994a): "*Prying Effect of Anchor Bolts*", JSPS, Nagoya.

WALD F., SOKOL Z., PERTOLD J., (1994b): "*Strength Design of Column–Base Connections, Model Verification*", ČVUT, Praha, p.76.

PART FIVE
CODIFICATION

38 CODIFICATION: GENERAL REPORT

A. PLUMIER
University of Liège, Liège, Belgium

1 Introduction

The main troubles undergone by structures, complete failure included, can be the result of :

1-a <u>conceptual</u> mistake, caused by the ignorance of some mechanical behaviour of the structure.
2-a mistake caused by the inadequate <u>modelling</u> of some aspects of the mechanical behaviour
3-an underestimation of the applied action.
4-a calculation mistake.

It is interesting in this year 1994 to look back to the past and recall the developments that drove us to the present situation, especially concerning the <u>concepts</u> set forward and the <u>modelling</u> of earthquake action and structural behaviour. This is done at Table 1, in which are given as a function of time :

some specific seismic events which were the origin of scientific studies later translated in code requirements ;
some scientific and technological achievements which brought an impulse to seismic design ;
the successive concepts, parameters and rules brought to surface as necessary to achieve successful seismic design.

The dates associated with the concepts and rules of Table 1 are approximate starting dates. They do not mean that the modelling of the corresponding theory is already at the best possible level. However, it is known that, in science, a big part of the gap in knowledge is filled very soon after the discovery of the concept.
Then come improvements and, eventually, perfection.

Behaviour of Steel Structures in Seismic Areas. Edited by F. M. Mazzolani and V. Gioncu.
Published in 1995 by E & FN Spon, 2–6 Boundary Row, London SE1 8HN. ISBN: 0 419 19890 3.

TABLE 1. HISTORY OF SEISMIC DESIGN.

Earthquake	Date	Sciences and Technology steps	Parameters considered
		Some constructional practices Ex. : historical structures in Japan, reconstruction of Lisbon after 1755.	
MESSINA (I)	1908		Horizontal force % weight
LONG BEACH (USA)	1933 1943	Accelerometer	Force depends on height (= period)
	1950's	Elastic structural dynamics-Response spectrum.	
AGADIR (MA) CARACAS (V) NIIGATA (J)	1960's	Computers Nuclear power plants. Dynamic linear Softwares	Time history analysis Elastic design response spectra Soil influence, regularity, torsion liquefaction, interest of ductility, approximate structural behaviour factors. Serviceability and ultimate limit state (OBE, SSE)
MANAGUA FRIULI (I) BUCAREST(RO)	1970's	Dynamic non linear softwares	Concept of capacity design Rules for local ductility of concrete.
IRPINIA (I) EL ASNAM (DZ) MEXICO SPITAK (USSR) LOMA PRIETA (USA)	1980's	Personal Computers Base isolation. Tests on connections Numerous parametric non linear analysis. Improved modelling of connections behaviour	Rules for local and global ductility of steel structures, overstrength rule for connections, weak beams strong columns concept, upper bound characteristics for materials. Computed behaviour factors and reduced earthquake spectra for design. Influence of topology, regularity.
LOS ANGELES (USA)	1990's	Tests on composite Cumulative damage in low cycle fatigue.	Rules for eccentric braced frames Rules for composite concrete structures.

Concerning earthquake resistant design, we can say, looking at the Historical Table, that a lot has been achieved during the latest 40 years.

All the recent codes, all over the world contain rules which cover the reality of structural behaviour and earthquake action well enough to consider that designers are now offered tools giving a fair chance to succeed in designing safe steel structures. This positive issue, to which many of us present in this workshop have contributed, is stressed by the globally very good behaviour of steel structures during recent earthquakes.

Having an effective design tool however does not mean that it is a very accurate and fully consistent tool. The development of design methods and damage assessment methods in earthquake engineering are still a wide research field and the content of codes can still be improved in many aspects. The papers presented in this session contribute to this improvement, which will remain for many years a continuous process.

2 Review of the session on codification

The presentation by ASTANEH covers one of the latest topics mentioned in our historical Table : composite steel concrete structures. It is clearly a type of structure for which test results and their analysis has been missing up to a very recent period. This is confirmed by the fact that composite steel concrete structures are the only type for which no rule is given in
Eurocode 8, because it has been considered (December 1993) that there was not enough scientific susbstance to support such a chapter ; the proposed chapter has been kept only as an informative annex. This might change with the paper by Professor ASTANEH who has been involved in conducting research on composite members and in designing composite structures. He was a member of the group that developed the first U.S. seismic code on that topic. His presentation of the results of the research and of the code is very complete.

DALBAN, IOAN and SPANU contribution explains new provisions of the Romanian Code for multi-storey braced frames and single storey structures. Some of these provisions are original or give more informations on detailing design than the Eurocode 8 or the U.S. Codes. In particular :

a) special requirements are given for gusset plates dimensions in term
of resistance and geometry, in order to sustain properly bracings in cyclic buckling and to allow buckling without unacceptable strains in the gusset.
b) detailed requirements are given for the links, beams and diagonals of eccentric braced frames.

It is also suggested that columns in frames should have a cruciform or box section, if one column is part of the earthquake resisting frames in the two principal directions.

Professor DALBAN and his colleagues can be congratulated for their efficiency in making out of Romanian Codes fully up to date documents.

DINCULESCU and GEORGESCU give principles and constructional solutions for the repair and strengthening of girders damaged by earthquakes.

The difficulty to give general rules for problems which are always particular is well known and the economic importance of maintaining the existing structures is very high.

The document contains many wise proposals. It stresses the necessary care about weldability of the existing steel, precautions required by welding and local heating in case of low weldability steel. Procedures to strengthen elements are described and the problem of provisional supports is dealt with. The importance of adding new restrains to unstable zones of beams is stressed.

This document is a sound basis to repair problems and it would be interesting to increase it by including some considerations to the following parameters :

-the belonging of the distorted zone to a dissipative zone of the structure or not;
-the necessary reevaluation of the whole repaired structures, in terms of capacity design ; the strengthening of some sections may bring unbearable effects to other sections or the connections in case of a next earthquake and be the cause of an early failure;
-the nature of the distorsion ; if the latter is, for instance, a local buckling of a flange, a repair realised only by a heating strengthening process is certainly not advisable;
- in some cases, like for instance the buckling of a sheared web, doing nothing might be accepted;
-bolted repair solutions should also be mentioned ; they can be very good in case of weldability or access problems ; the types and diameters to be chosen for repairs should be explained.

The presentation by GEORGESCU and BUGHEANU deals with some specific provisions concerning one storey industrial buildings.
It is explained how some simplified design rules have been established for that type of structures. These design approaches concern transverse frames, longitudinal frames and roof bracing systems.
While the methods concerning the first two types of earthquake resisting structures are based on classical structural analysis considerations, the design rules for roof bracing systems are quite specific and undoubtedly influenced by a real concern about the effectiveness of these horizontal diaphragms. It is difficult

with the explanations given to understand the rules or their basis. For instance, it is not clear why the torsion moment M is unaffected by the 1st assumption, stating that the acting force in a transverse frame is fully resisted by that frame : indeed this statement means that there is no diaphragm effect, so that the translational design force in each frame is certainly increased, but it also would mean that there is no real eccentricity of mass to consider, except a conventional one, so that the torsion moment should be reduced in comparison with the hypothesis of a rigid diaphragm.

GIONCU and MAZZOLANI have assessed the present situation concerning the structural behaviour factors and the general problem of global ductility of steel structures. They point out the various incongruencies in the present state of codification and propose some founded revisions.

Amongst the factors influencing the global ductility and presently unmentioned in codes are the types of accelerograms, the natural period of the structure, the plastic redistribution capacity, the type of collapse mechanism, the semi rigid connections and the local ductility.

GIONCU and MAZZOLANI point out several problems to be clarified for determining correct q factors :

-up to now, the failure criteria considered refer to the formation of a
global plastic mechanism, but the authors correctly mention that the rotation at plastic hinges, that means local ductility, can be the limit.
I would comment this saying that if codes include this parameter, the duration of the earthquakes, which is connected with the number of cycles and the damage accumulation, will also have to be defined in the code;
- for structures with small natural periods, the plastic rotation at hinges is the only limit and q factors should be determined on that basis;
- some countries, including Japan, accept weak-columns - strong beams design, because more economical, and provide structural behaviour factors for the corresponding storey mechanism ; this is not done in Eurocode 8.
-the design procedure for columns actually does not guarantee the achievement of hinges in the beams and there is a proposal to improve the situation;
- the definition of regular structures and the influence of setbacks is still an open subject and more rational parameters should solve that problem.

GIONCU and MAZZOLANI propose a rational approach to the choice of the member class. They also suggest that the real behaviour factor of a designed structure should be checked after design as being greater than the code ones, by means of simplified method to be established.

The paper by MAZZOLANI, GEORGESCU and ASTANEH contains a proposal of design method in which 3 limit states should be considered, instead of 2 in most recent codes.

The serviceability limit state correspond to moderate and frequent earthquakes (5-10 years return period) and should be withstood without any damage or discomfort.

The damageability limit state correspond to a quite large earthquake, with a high probability to be once sustained during the life of the structure (50 years return period). Minor damages and some inelastic behaviour are accepted in that case.

The level 3 earthquake (500 years return period) correspond to a collapse limit state.

This proposal can be considered as a logical refinement of the existing codes. Before becoming the common practice, it has to give convincing arguments that its approach really changes the design and it will have to overcome two already existing problems :

-a rational definition of the level 1, 2, 3 earthquakes;
-a rational definition of drift limits for level 1 and 2 earthquakes.

PACURAR and MOGA present recent tests results and analysis of the shear resistance of steel plate girders. They confirm that the design formula of Eurocode 3 give correct estimates. Further research are suggested. This idea can be supported because data on the cyclic behaviour of such elements are certainly welcome.

SANDI makes a keen review of earthquake resistant design and set forward a series of problem to which only approximate or even inconsistent solutions have been given.

His first remarks bear on earthquake action - SANDI stresses the need for guidelines defining the way to represent the 3 dimensional reality of earthquake motion. He also insists on better ways to represent local geological conditions and the randomness aspect of attenuation.

In phase with previous authors, he suggest that several intensity or acceleration values corresponding to several return periods should be specified in codes.

Other critical appraisals are made on design criteria. The first one is the necessity to link action data to prescribed limit states ; the second remark explains that local ductility demand should be prescribed. It is indeed very true that up to now if a structural part of a structure has to be qualified to become a dissipative part, no criteria are given on what should be proven by tests by this structural part.

Codes should also specify requirements aimed to better structural modelling in case non linear oscillations are analysed (soil structure interaction, non linear materials). A similar claim is made concerning accelerograms to be used in time-history analyses.

SANDI also explains several logical considerations about capacity design :

- capacity design should be accompanied by some suitable analysis aimed at the control of the risk of exceedance of available ductility;
- capacity design should be improved considering 3 dimensional action effects in structures;
- the limits to overpressure on the ground could be relaxed;
- qualitative criteria should be given to control the proper implementation of the capacity design strategy.

3 Conclusions

The Historical table given at the beginning of this General Report on Session 5.

Codification has shown that earthquake engineering has been a field of continuous developments since 40 years. It certainly has been a kind of a "last frontier" in civil engineering during the end of this century.

The presentations of today demonstrate this intense activity and show that still many problems have to be worked on to succeed in giving the accurate and safe approaches needed by designers and required by public safety.

At the present state of knowledge, I would suggest that the main existing rules to be improved and the main new rules to be defined are those of Table 2.

Table 2 : Future of codification.

Existing Rules to improve.

- Detailing of rigid connections.
- Design of composite structures.
- Design earthquakes and related limit state.
- Structures with multiple supports.
- Post design checks of real structural behaviour factor q.
- Soil structure interaction.
- Regularity.
- Improved q (light gage sections, ...).

Rules to be written for the first time.

- Semi rigid connections.
- q factors related to damage accumulation and earthquake duration.
- Qualification test and criteria on ultimate rotation.

39 SEISMIC DESIGN OF COMPOSITE STRUCTURES IN THE UNITED STATES

A. ASTANEH-ASL
University of California at Berkeley, USA

Abstract
A number of recent research, technology development and code-writing activities in the area of seismic design of composite building structures are discussed. The research projects include studies of composite columns, bracings, connections and connectors. The background of the development of the new, first U.S. code on seismic design of composite building structures is provided along with a discussion of the main provisions of the code.
Keywords: Composite Structures, Composite Columns, Composite Connections, Seismic Design

1 Introduction

The use of composite systems in highly seismic areas of the United States has been limited to the use of steel structures with composite floors and, recently, the use of steel structures with concrete-filled composite columns. Perhaps the main reason for not using the many advantages of composite systems in seismic areas has been the lack of seismic design codes. However, in less seismically active areas, a number of spectacular highrises have been designed and constructed using composite systems. Some examples discussed in the literature (Viest 1992) are the 75-story Texas Commerce Tower in Houston, the 57-story Three First National Plaza Tower in Chicago, the 73-story Dallas Main Center in Dallas, and Philadelphia's 65-story One Liberty Place.

These and other composite highrises built in the United States during the 1970s and 1980s are testimony to the development of innovative structural systems and construction methods. The main structural engineering innovation in these highrises has been the development of a composite tube system used as the exterior framing to resist wind loads along with gravity loads. In this type of construction, the steel core of columns is constructed several floors ahead and then the steel columns are encased in concrete to make them composite. These and other innovations have made the composite systems very effective, economical and fast to construct. However, the use of this efficient system in highly seismic areas such as California has been slow.

Considering the recent damage to connections of a number of steel moment frames and the failure of a number of reinforced concrete

Behaviour of Steel Structures in Seismic Areas. Edited by F. M. Mazzolani and V. Gioncu.
Published in 1995 by E & FN Spon, 2-6 Boundary Row, London SE1 8HN. ISBN: 0 419 19890 3.

structures during the January 1994 Northridge-Los Angeles earthquake, it appears that the use of steel-concrete composite systems could mitigate some of the vulnerabilities of steel and reinforced concrete structures.

Compared to other industrialized areas of the world, the practice of structural engineering in the United States is more restricted by legal issues and codes. As a result, it is very difficult if not impossible for a typical structural engineering firm to use a technology that is not explicitly included in the design codes. Therefore, a major step taken in recent years has been the development of the first U.S. code on seismic design of composite structures. The code (NEHRP 1994) was developed by the Building Seismic Safety Council of the National Hazard Mitigation Program. The development of the provisions for seismic design of composite structures was the result of more than 8 years of intensive work by a special task group with the author as one of the members. Other members are listed in the Acknowledgments.

A brief summary of highlights of the NEHRP code is provided in Section 3 of this paper and the reader is referred to the main document (NEHRP 1994) for detailed information.

Following is a summary of some of the recent research efforts in the United States in the field of seismic behavior of composite structures.

2 Summary of some recent research projects in the U.S.

2.1 Research on composite systems

One of the major research efforts in the United States on composite structures was the U.S.-Japan project on studying a six-story steel structure with composite floors (Hanson 1994). The project led to other studies on composite system such as the Lehigh University research (Lu et al. 1994).

2.2 Recent composite column studies

In recent years, an experimental and design-oriented research project has been conducted in the U.S. to study the behavior of composite columns. The project at Berkeley (Chen and Astaneh-Asl 1994) consisted of testing 17 full scale encased composite columns with high strength concrete. Because of concern for brittleness of high-strength concrete, current U.S. design codes do not permit the use of composite columns with concrete stronger than f'c of 55 MPa (8 ksi). The main objectives of the above-mentioned project were to study compressive behavior and post-buckling ductility of composite columns of high-strength concrete with f'c of 83 MPa (12 ksi).

Figure 1 shows force-displacement behavior of 5-m-long (15-ft) columns (Chen and Astaneh-Asl 1993). Tests and analyses indicated that the failure mode was, in general, relatively brittle local failure of the column due to buckling of the rebar, shattering of the concrete surrounding the rebar, and local buckling of the flange of steel rolled shape. Even though the column appeared to be appropriate for gravity load resistance, its use in high seismic areas would require closer placement of ties and more confinement of concrete.

The studies resulted in development of design recommendations for axially loaded high strength composite columns (Chen and Astaneh-Asl 1994).

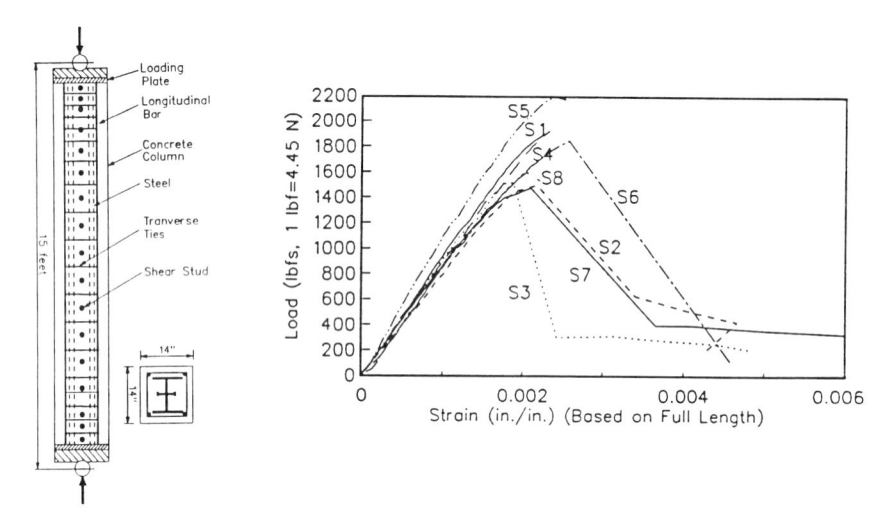

Fig. 1. High-strength composite column specimens and axial load test results (Chen and Astaneh-Asl 1993).

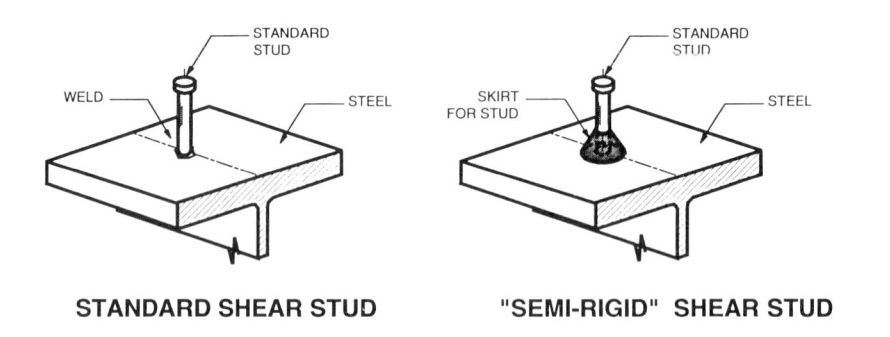

Fig. 2. Standard and semi-rigid shear stud (Astaneh-Asl 1993).

2.3 Seismic behavior of composite connectors

The most common type of connector in composite construction currently used in the United States is shear stud welded to steel member and embedded in concrete. The information on cyclic behavior of shear studs is very limited.

In 1990, A. Astaneh-Asl developed the idea of "semi-rigid" shear studs for seismic application (Astaneh-Asl 1993). Figure 2 shows a view of a standard and the new "semi-rigid" shear stud proposed by A. Astaneh-Asl. The only physical difference between a standard and semi-rigid shear stud is that in the semi-rigid shear stud, by placing a cone-shaped element called a "skirt" at the base of the stud, concrete is prevented from engulfing the whole length of the stud. As a result, a short length of the base of the stud is left free to act as a shear fuse, as shown in Figure 3. Such a seemingly minor difference between standard and semi-rigid shear studs causes significant changes in cyclic behavior of the stud and the composite system. Under cyclic shear, the semi-rigid stud behaves in a much more ductile manner and its strength and stiffness can be controlled to reduce the seismic forces in the structure.

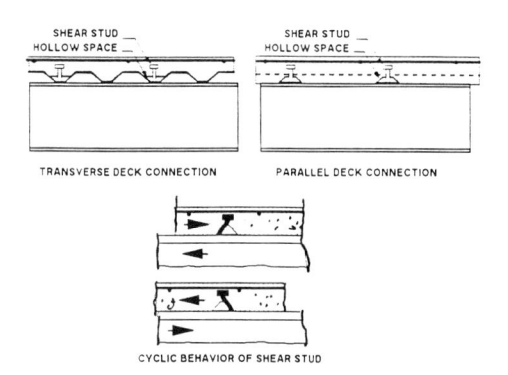

Fig. 3. Cyclic shear deformation of 'semi-rigid' shear studs.

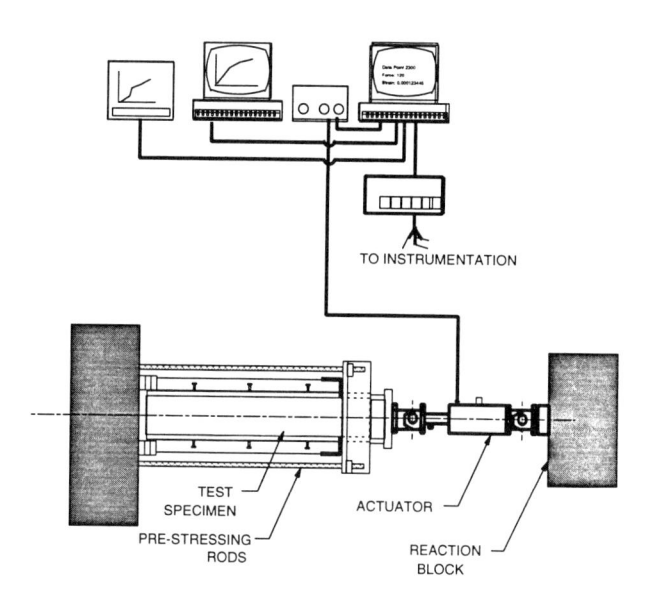

Fig. 4. Test specimens and test set-up for semi-rigid shear studs.

To test the potential of the proposed semi-rigid shear studs, four tests were conducted at the University of California at Berkeley. The test program, part of a larger project, consisted of constructing four push-through specimens (Figure 4) and subjecting the specimens to cyclic shear. Figure 5 shows a comparison of the cyclic shear behavior of the standard and "semi-rigid" shear stud (McMullin et al. 1993). The standard shear stud developed large stiffness and could tolerate only a few cycles of small shear displacements before the welds at the base of the stud fractured. This behavior, quite acceptable for non-seismic and small amplitude cyclic load, is not desirable in highly seismic areas where shear studs can be subjected to relatively large cyclic shear forces.

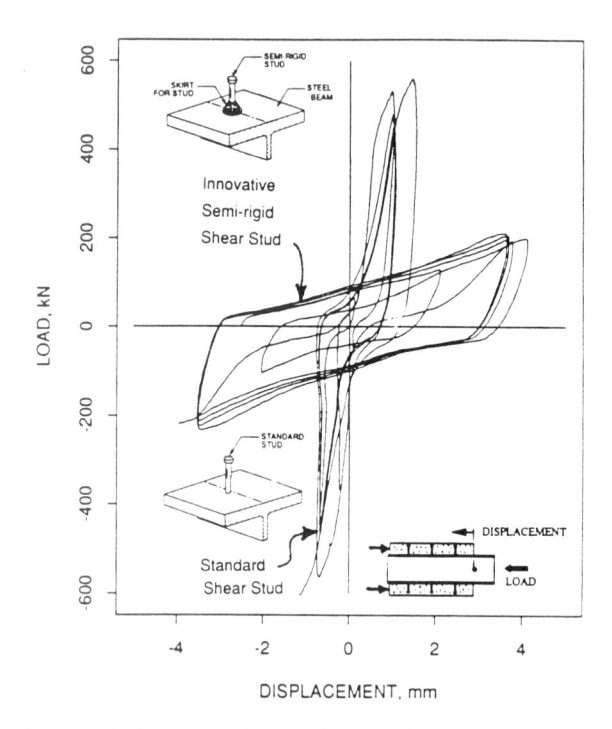

Fig. 5. Cyclic behavior of standard and semi-rigid shear studs.

Unlike the standard shear studs, the semi-rigid stud behaved in a very ductile manner for many cycles without fracturing the welds or crushing the concrete. Furthermore, in semi-rigid shear studs the stiffness and strength can easily be controlled by the designer by changing the height of the skirt cone at the base of the stud. In standard studs, the strength of the stud can be controlled by the designer; however, there is no way to control the stiffness independent of the strength.

Further research and development on semi-rigid shear studs is currently underway by A. Astaneh-Asl and his research associates to develop more data on the cyclic behavior of these potentially ductile elements and to formulate design procedures and guidelines. The semi-rigid shear studs can be used in many applications including composite floors, columns, walls, connections, piles and other elements where high cyclic ductility demand exists.

Initial results of an ongoing study to investigate effects of semi-rigid shear studs on the seismic behavior of composite systems indicated that, in contrast to standard studs, semi-rigid studs used between the floor slab and the floor beams generally increased period and damping and decreased seismic forces and drift values (Figure 6).

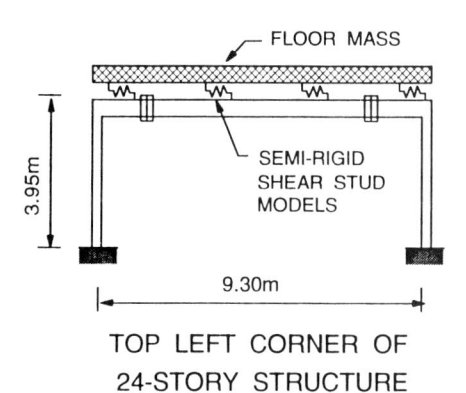

TOP LEFT CORNER OF 24-STORY STRUCTURE

RESPONSE OF THE FRAME SUBJECTED TO TAFT 1952 ULTIMATE LEVEL EARTHQUAKE

RESPONSE PARAMETER	STANDARD STUDS	SEMI-RIGID STUDS
Period, sec.	0.245	0.253
Lat. Disp., mm	26	21
Base Shear, kN	1522	1112
Max. Shear in Stud, kN	445	276

Fig. 6. Comparison of response of frame with standard and semi-rigid shear studs.

2.4 Composite joints and beam-to-column connections

Cyclic behavior of a number of composite connections have been studied recently in the United States. The connection of steel girder to encased composite column has been studied by Deirlein (1992) and others since the 1980s. These studies have resulted in the development of design guidelines and models of behavior.

Studies of semi-rigid composite connections (Leon and Ammerman 1990) have indicated that these connections demonstrate good ductility. Also, the strength and stiffness of the connection is increased appreciably by using even a small amount of top rebars.

Studies of the connection of the steel beam to a circular or box composite column (Azizinamini and Shekar 1994) showed the viability of this type of connection. A tentative design procedure for this type of connection has been developed and proposed. In the procedure, by knowing the external forces and by using relatively simple equations, the state of internal stresses in the embedded steel and the concrete diagonal strut can be established.

The connection of a steel coupling beam to a reinforced concrete shear wall has been studied by Shahrooz et al. (1992). The cyclic tests of connection specimens have indicated relatively stable hysteresis loops slightly pinched due to cracking of concrete boundary elements.

At present, test data and information on the actual behavior of these and other even less studied composite connections are not extensive. Recognizing this fact, the commentary of the NEHRP Provisions (NEHRP 1994) contemplates that the design of composite connections will be done 'using basic mechanics, equilibrium, existing standards for steel and concrete construction, and research findings.' This approach has been used in the past prior to the development of the NEHRP Provisions (Griffis 1992).

2.5 Composite bracing elements

Experimental studies by Liu and Goel (1987) have shown that, by using composite concrete filled steel tube bracing, the cyclic behavior of not only the bracing but also the braced frame improves significantly. The presence of concrete adds to stiffness and strength and, more importantly, helps restrain steel tube walls and delay local buckling. By using concrete inside the steel tube bracing member the b/t of the tube wall could be reduced by up to 50%.

3 New NEHRP provisions on seismic design of composite structures

In 1985 the Building Seismic Safety Council of the Federal Emergency Management Agency of the United States formed a task group to develop the first U.S. seismic design code for steel-concrete composite structures. The intensive work of the committee and others involved in the effort since then resulted, in 1993, in the development of the first U.S. seismic design provisions for composite construction (NEHRP 1994). The efforts of the Task Group focused on two areas: (a) development of the force reduction factor, R, and drift amplification factor, Cd, for a variety of common composite systems to convert elastic forces and drifts to more realistic values, and (b) development of seismic design requirements for the composite systems and their components.

3.1 Reduction factors for composite systems

The most challenging part of the effort was to establish seismic force reduction factors, R, and drift amplification factors, Cd. There is very limited information on the actual seismic behavior of composite systems. Laboratory tests and analyses indicate that the seismic performance of a composite system or component is usually better than a similar system of either steel or reinforced concrete. Therefore, in establishing reduction factors, not only the available research data was considered but also the engineering judgment and understanding of the behavior of these systems by Task Group members and others involved.

The basic composite structural framing systems recognized by the NEHRP provisions and their corresponding force reduction factors and drift amplification factors are:

a. Composite special moment frames, R=8.0 and Cd=5.5
b. Composite ordinary moment frames, R=4.5 and Cd=4.0
c. Composite partially restrained frames, R=6 and Cd=5.5
d. Composite eccentrically braced frames, R=8.0 and Cd=4.0
e. Composite special concentrically braced frames, R=8.0 and Cd=4.5
f. Composite concentrically braced frames, R=6.0 and Cd=5.0
g. RC shear wall composite with steel elements, R=8.0 and Cd=6.5
h. Steel plate reinforced composite shear wall, R=8.0 and Cd=6.5

3.2 Composite structure design requirements

Chapter 7 of the NEHRP (1994) document begins with the design requirements for composite systems. Each system listed in Section 3.1 above is defined. Then, specific seismic deign requirements for that system are provided. For example, composite special moment frames are defined as frames consisting of either steel, reinforced or composite columns and either steel or composite beams. The strength of concrete in the columns is limited to 69 MPa (10 ksi) unless test results justify higher values.

Because of the importance of the connections and the complexity of the state of stress in composite connections, a considerable portion of the NEHRP document is concerned with the seismic design of composite connections. The results of the research projects summarized earlier in this paper, and other studies documenting cyclic behavior of composite structures and components, had significant influence in the development and formulation of the design requirements included in NEHRP (1994).

4 Conclusions

a. In recent years, a number of research projects have been conducted in the United States to understand the actual cyclic behavior of composite systems, columns, bracing members, connections and shear stud connectors. In general, the cyclic behavior of composite elements was superior to their comparable steel or concrete alternatives. Cyclic studies of composite connections have resulted in development of seismic design provisions while the studies of an innovative connector, the "semi-rigid" shear stud (Astaneh-Asl 1993), have shown its high ductility and its potential for controlling and reducing seismic forces in composite systems.

b. In 1994, the first U.S. code on seismic design of composite structures (NEHRP 1994) was released. The document addresses the seismic force reduction factor, R, and the drift amplification factor, Cd, for composite systems as well as requirements for the seismic design of the composite elements of these systems.

5 Acknowledgments

The research projects discussed here were sponsored mostly by the National Science Foundation and partially by the American Institute of Steel Construction.

The concept of "semi-rigid" shear studs was originally developed by A. Astaneh-Asl and testing of the concept was supported by the CUREe/Kajima project of the California Universities for Research in Earthquake engineering (CUREe) and the Kajima Corporation.

The NEHRP/BSSC provisions on seismic design of composite structures referred to in this paper were developed by Task Group 11 of the Building Seismic Safety Council of the Federal Emergency Management Agency. The current members of the Task Group are: G. G. Deirlein (Chairman), A. Astaneh-Asl, L. Griffis, R. Leon, J.O. Malley, I. Viest, and N. Youssef. The assistance and input of Task Group Chairman G. G. Deirlein in preparing Section 3 of this paper is sincerely appreciated. The opinions expressed in this paper are those of the author and do not necessarily represent the views of the sponsors or individuals whose name appear in this paper.

6 References

Astaneh-Asl, A. (1993) The innovative concept of semi-rigid composite beam. **Proc., Structures Congress XI**, April 1994, ASCE, Irvine, CA.

Azizinamini, A., and Shekar, Y. (1994) Rigid beam-to-column connection detail for circular composite columns. **Proc., Structures Congress XII**, April 1994, ASCE, Atlanta, GA.

Chen, C.C., and Astaneh-Asl, A. (1993) Behavior and design of axially loaded high-strength composite columns. **Report No. UCB/CE-Steel-93/15**, Dept. of Civil Engrg., Univ. of California, Berkeley, CA.

Deirlein, G.G. (1992) Research and practice in the U.S. on hybrid subassemblages and connections. **Proc., U.S-Japan Workshop on Hybrid Construction.**, Sept. 1992, Berkeley, CA.

Hanson, R.D. (1994) U.S./Japan large scale cooperative research program for steel structures. **Proc., Structures Congress XII**, April 1994, ASCE, Atlanta, GA.

Griffis, L. (1992). Composite frame construction, in **Constructional Steel Design, An International Guide**. Elsevier Applied Science, New York, NY, pp 523-553.

Leon, R.T., and Ammerman, D.J. (1990) Behavior of composite connections under seismic loading," **Proc., Fourth U.S. Natl. Conf. on Earthquake Engrg.**, Palm Springs, CA.

Liu, Z., and Goel, S. C. (1987) Seismic behavior of hollow and concrete-filled square tubular bracing members. **Rep. No. UMCE 87-11**, Dept. of Civil Engrg., Univ. of Michigan, Ann Arbor, MI.

Lu, L.W., Ricles, J.M., and Kasai, K. (1994) Research on seismic behavior of steel and composite structures at Lehigh. **Proc., Int. Workshop on Behavior of Steel Structures in Seismic Areas**, June 1994, Timisoara, Romania.

McMullin, K.M., Astaneh-Asl, A., Fenves, G. L., and Fukuzawa, E. (1993) Innovative semi-rigid steel frames for control of the seismic response of buildings. **Rep. No. UCB/CE-Steel-93/02**, Dept. of Civil Engrg., Univ. of California, Berkeley, CA.

NEHRP (1994), **Recommended Provisions For The Development Of Seismic Regulations For New Buildings**. Federal Emergency Management Agency, Building Seismic Safety Council, Washington, D.C.

Shahrooz, B.M., Remmetter, M.E., and Qin, F. (1992) Seismic response of composite coupled walls. **Proc., Engineering Foundation Conf. on Composite Construction in Steel and Concrete II**, ASCE, Potosi, MO

Viest, I.V. (1992) Composite construction, recent past, present and future" **Proc., Engineering Foundation Conference on Composite Construction in Steel and Concrete II**, ASCE, Potosi, MO, USA.

40 PROPOSALS FOR IMPROVING THE ROMANIAN SEISMIC CODE: PROVISIONS CONCERNING STEEL STRUCTURES

C. DALBAN, P. IOAN and ST. SPANU
Civil Engineering Institute, Bucharest, Romania

ABSTRACT

The severe earthquake of Romania (4 March 1977 , M = 7.2) produced heavy damages; many peoples were killed or injured. It was necessary to perform an important revision to the regulatory basis of seismic resistant design. This revision concerned all cathegories of provisions, like design principles, zonning , methods of design. Among them the chapter dealing with steel structures had an important developement.
Keywords: Seismic design code, Multistorey steel Structures, Connections.

1 INTRODUCTION

The first editions of the new Seismic Code published after 1977 were P 100 - 78 and P 100 - 81. But the most important version was published in 1992. In the provisions of this version, the conclusions of the seismic events from August 1986 (M=7.0) and May 1990 (M=7.0) were taken into account.

Our close contact beginning with 1985 , with Technical Commitee 13 of E.C.C.S. allowed to have a better overall view on the problems of seismic countries similar to those of Romania.

On this way the provisions of the last version of Romanian Seismic Code P 100 - 92 / 9 / are in good agreement with those of seismic codes from U.S.A., Japan , Italy and other seismic countries (/ 1 / , / 2 / , / 3 / , / 4 / , / 5 / , / 6 / , / 7 / , / 8 /).

The provisions of European Recommendations for Steel Structures in Seismic Zones / 6 / , / 7 / and Eurocode 8 - 88 / 8 / were also taken into consideration.

A presentation of the Romanian Seismic Code P 100 - 92 was performed in / 10 /.

The last version of the Romanian Seismic Code (P 100 - 92) includes design provisions for all types of structures used in Civil Engineering and particulary for steel structures as well.

Behaviour of Steel Structures in Seismic Areas. Edited by F. M. Mazzolani and V. Gioncu.
Published in 1995 by E & FN Spon, 2–6 Boundary Row, London SE1 8HN. ISBN: 0 419 19890 3.

Proposals for improving the Romanian seismic code 459

The chapter of P 100 - 92 devoted to general design rules for Steel Structures (Chapt. 8) is accompanied by two Annexes (E 1 , E 2) concerning : (1) the multistorey buildings and (2) the single-storey plant halls.

The present work deals with some special problems concerning the multistorey buildings and especially those using steel braced frames, aiming to ensure their resistance. That corresponds to an extension of the Annex E1 (multistorey buildings) that will be performed in the very near future .

In the following different types of multistorey structures designed in Romania are presented. They are accompanied by typical details to be applied to these structures ; some connections belong to the structures designed in Romania.

2 DESIGN OF MULTISTOREY STRUCTURES

2.1 Moment resisting frames

Some exemplifying details of frames and beam to - column connections are presented in fig. 1 .

They represents :

a) Connections with full penetration welded joints of flanges and fillet welds for web joints (fig. 1 a) .

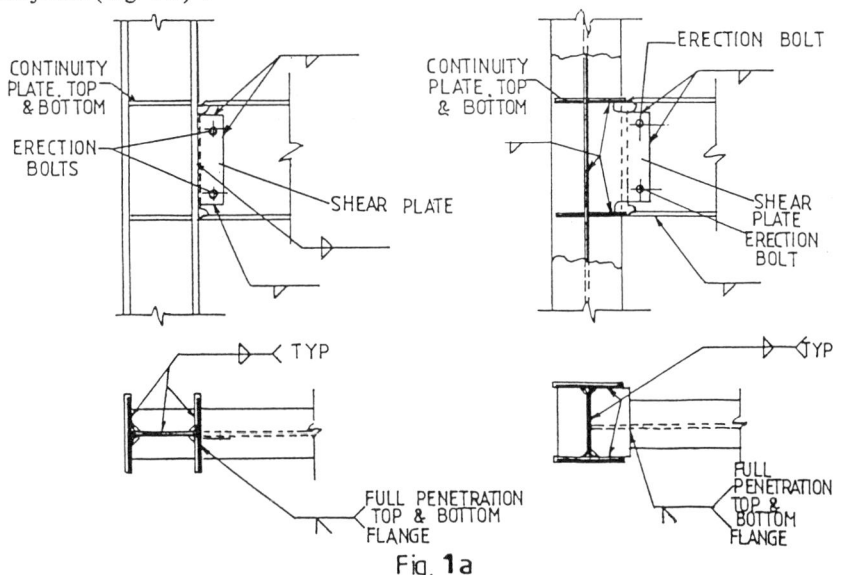

Fig. 1a

b) Connections with the same procedure for flanges but with slip critical high strength bolted joints for the web (fig. 1 b) .

c) Connections with high - strength bolted endplate (fig . 1 c) .

Fig. 1b

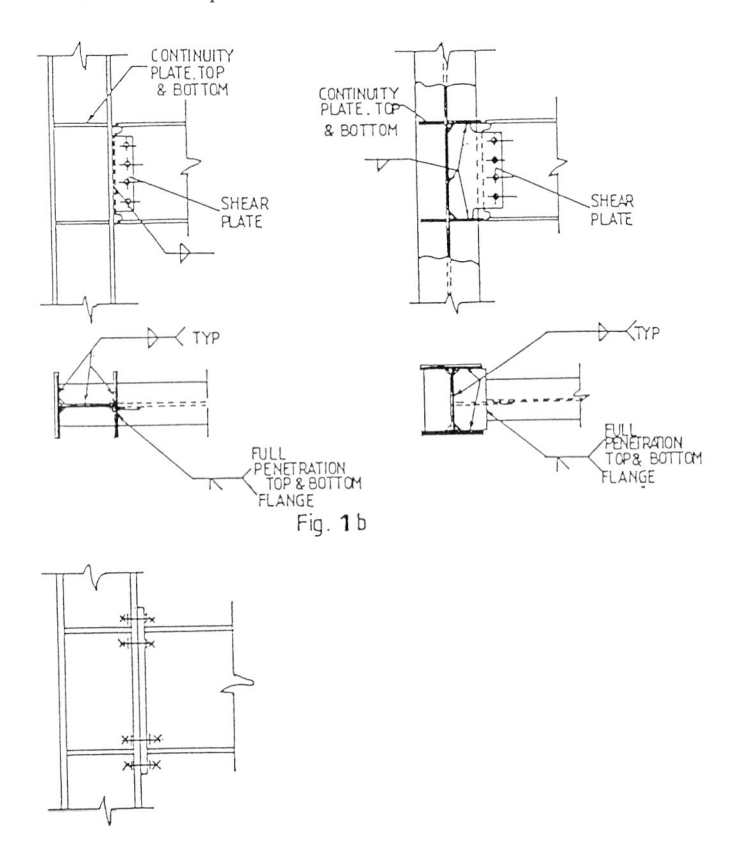

Fig. 1c

d) Connections with welded or high - strength bolted coverplates (fig.2 Detail A)

In order to avoid the formation of plastic hinges in the columns, the design value for bending moments in the columns at the level of beams are given by the sum of bending moments $M_{c,s}$, due to horizontal seismic actions multiplied by the amplification factor plus bending moments $M_{c,o}$ due to the other design loads :

$$M_d = M_{c,o} + \alpha M_{c,s}$$

where the amplification factor α is given by :

$$\alpha = \frac{\Sigma M_{r,b} - \Sigma M_{c,o}}{M_{c,s}} \quad \text{where :}$$

- $\Sigma\ M_{r,b}$ the sum of the resisting moments of the beam connected to the column.

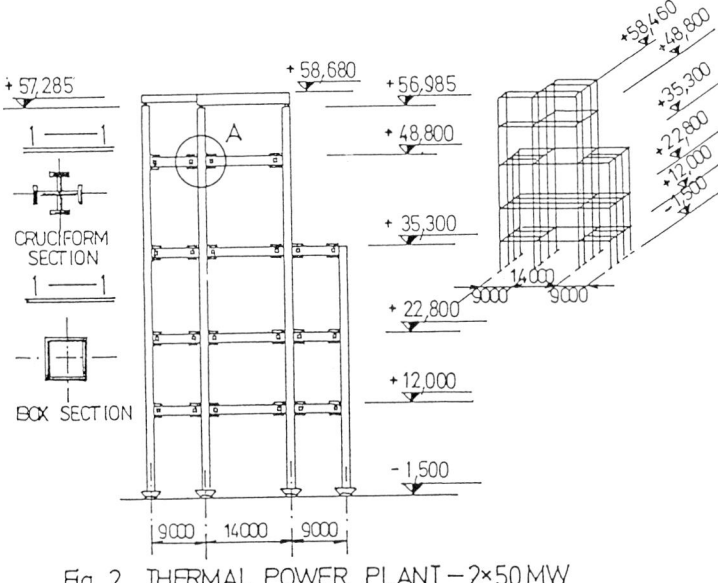

Fig. 2 THERMAL POWER PLANT — 2×50 MW
ROMANIA (BOILER-HOUSE)

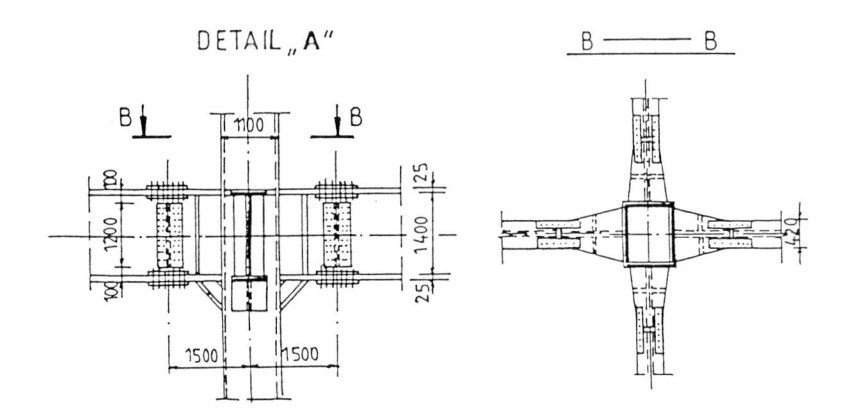

- $\Sigma\, M_{c,o}$ the sum of the bending moments in the column due to non seismic loads.
- $\Sigma\, M_{c,s}$ the sum of the bending moments in the column due to seismic loads.

At the top floor of multistorey frames $\alpha = 1$ is assumed ; at the other floors level $\alpha \geq 1.0$. In order to ensure similar stiffness of the columns in both principal directions, it is compulsory the columns should have cruciform section or box - section. Some examples of the practice existing in Romania are given in fig. 2.

2.2 Concentrically braced frames

The most typical structural configurations are presented in fig. 3.a .

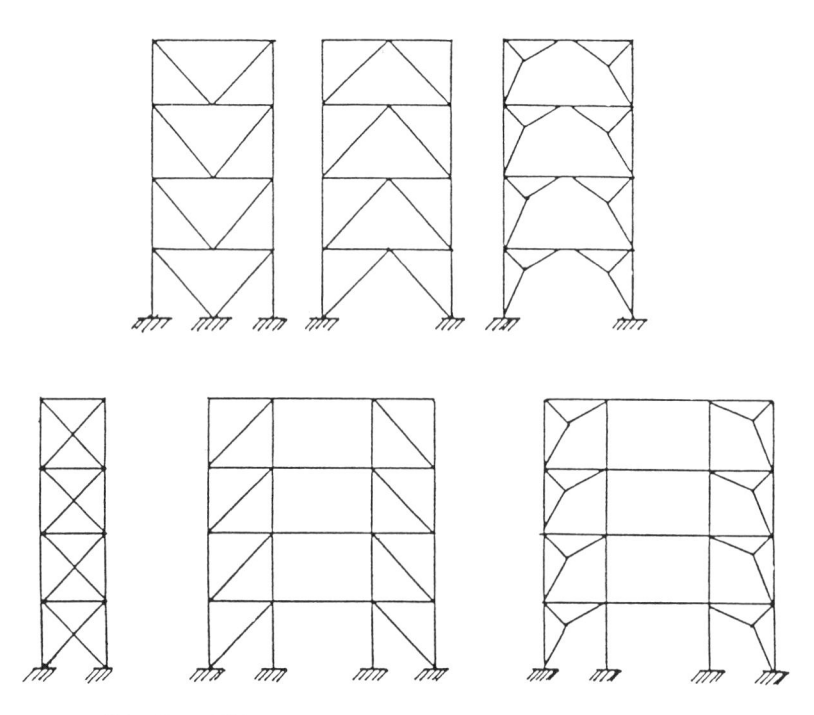

Fig. 3a – TYPOLOGIES OF CONCENTRICALLY BRACED FRAMES

Some details of concentrically braced frames and their bracing connections are presented in fig. 3 , 4 and 5. The details were mainly selected from the structures designed in Romania . Some special requirements for the gusset plates are given in the following:

a) For braces that can buckle in the plane of the gusset plate , the gusset and other parts of the connection shall have a design strength equal to or greater than the nominal in-plane bending strength of the brace.

b) For braces which can buckle out-of-plane of the gusset plate, the brace shall terminate on the gusset a minimum of two times the gusset thickness from a line about which the gusset plate can bend unrestrained by the column or beam joint. The gusset plate shall be designed to carry the compressive design strength of the brace member, without local buckling of the gusset plate (fig . 3 b) .

These requirements are resulting from full-scale structural experiments performed in U.S.A.

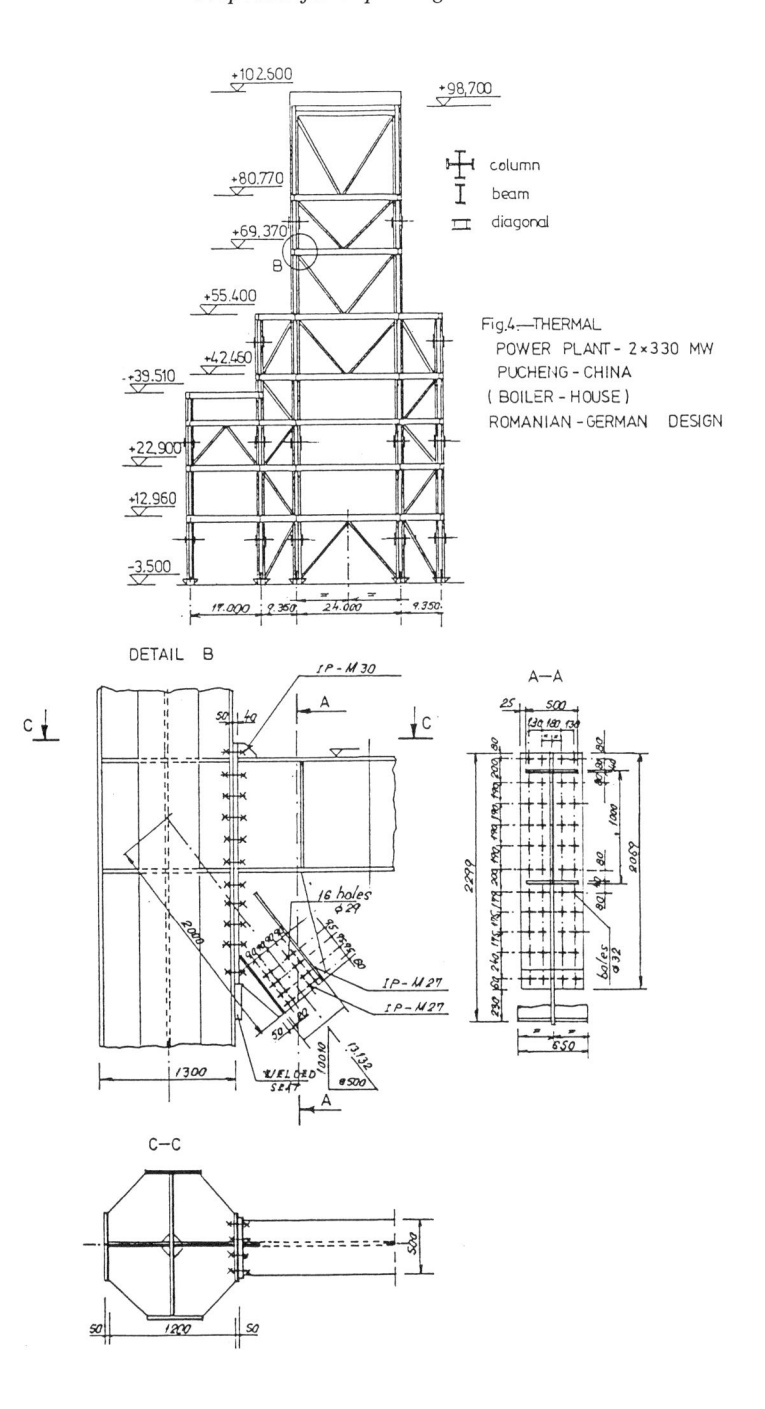

Fig.4.—THERMAL POWER PLANT - 2×330 MW PUCHENG - CHINA (BOILER - HOUSE) ROMANIAN - GERMAN DESIGN

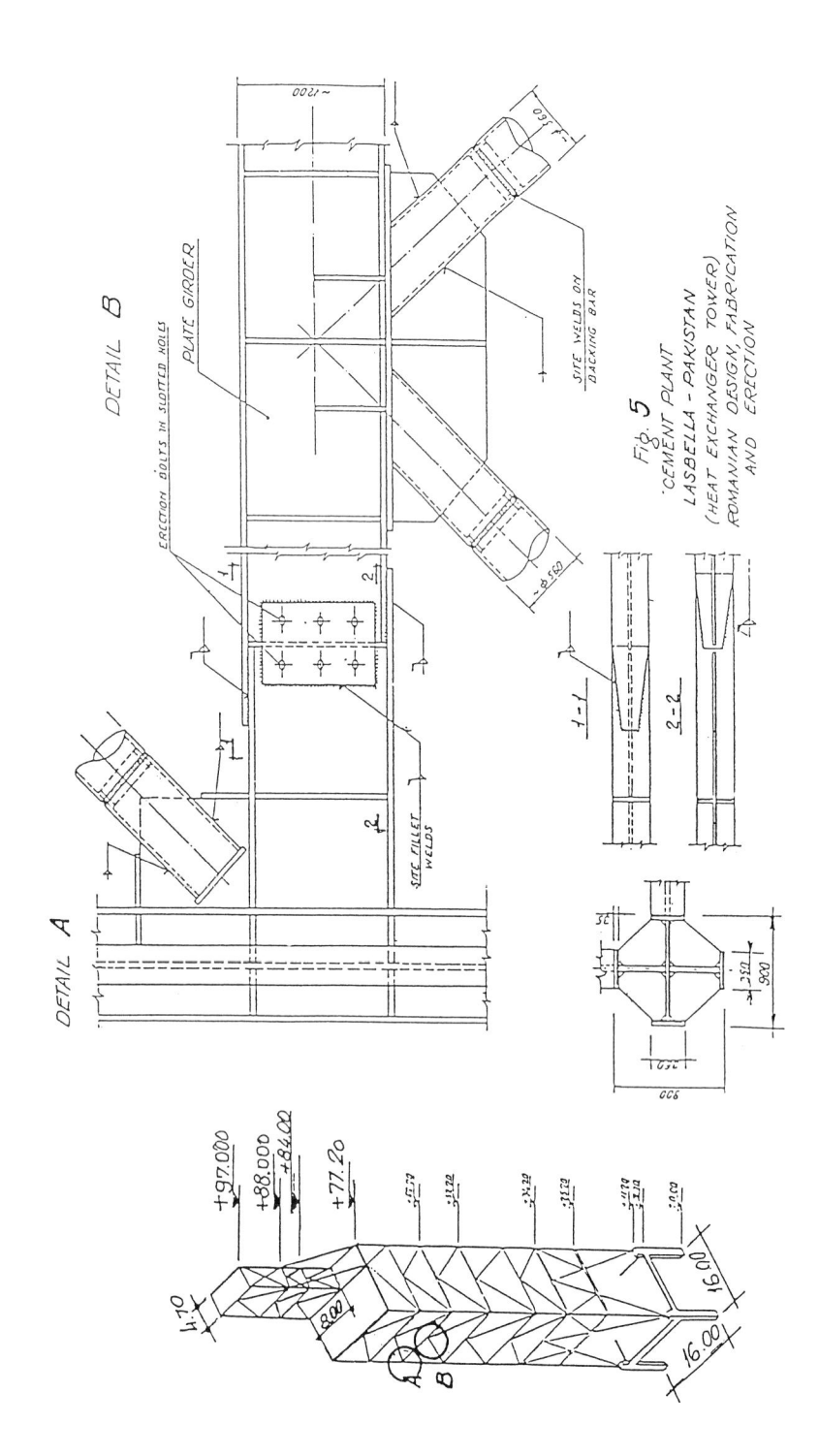

Fig. 5
"CEMENT PLANT
LASBELLA - PAKISTAN
(HEAT EXCHANGER TOWER)
ROMANIAN DESIGN, FABRICATION
AND ERECTION

Some other design procedures for the concentrically braced frames, are given in the original document in preparation, which contains all the supplementary improving provisions.

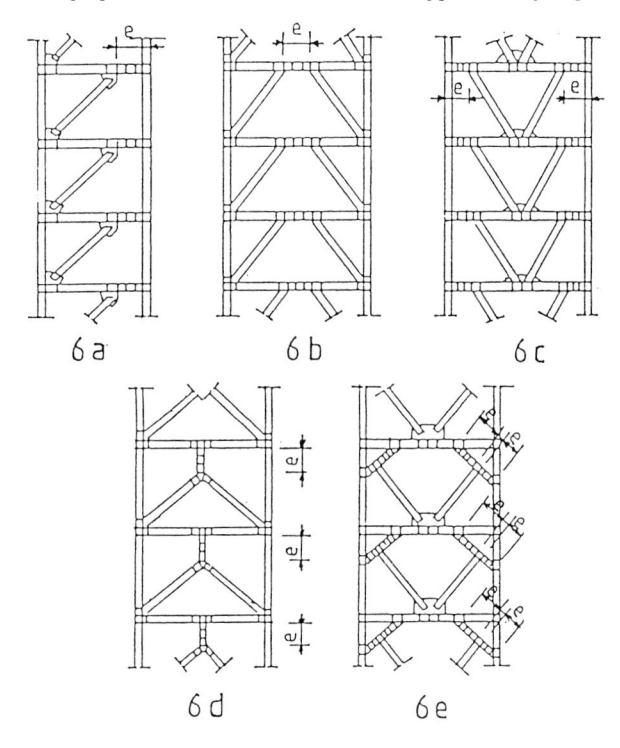

Fig. 6 TYPOLOGIES OF ECCENTRICALLY BRACED FRAMES

2.3 Eccentrically braced frames

In the existing Code P 100-92 are included only general non detailed mentions, concerning the eccentrically braced frames. Supplementary data are presented in the following :

Eccentrically braced frames shall be designed so that under severe earthquake loading, yielding will occur primarily in the links. The diagonal braces, the columns and the beam-segments outside of the links shall be designed to remain elastic under the maximum forces that can be generated by the fully yielded and strain hardened links. In eccentric braced frames, plastic hinges shall not develope in columns, at floor beam level.

Some types of eccentrically braced frames are shown in fig. 6 a, b, c, d, e.

The technical literature gives some specifical data regarding the design of the link.

Note : In the Seismic Code P 100-92 , the romanian translation of " link " is equivalent to " dissipative element ".

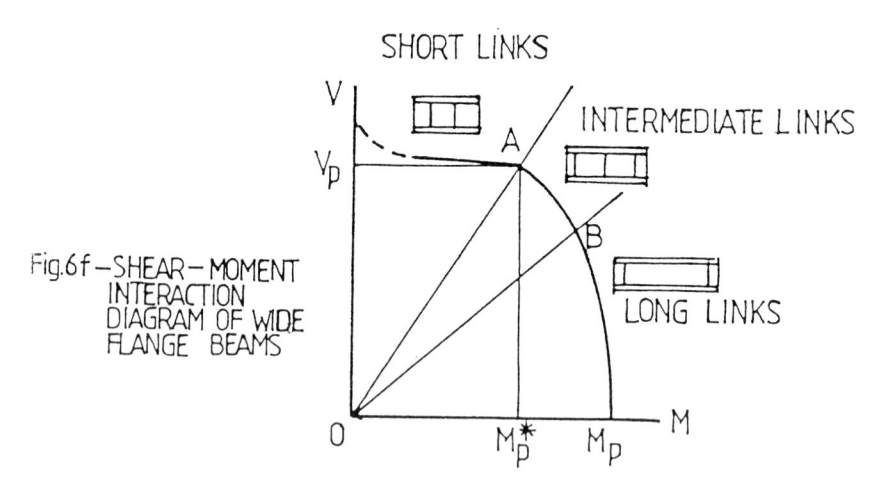

Fig.6f – SHEAR–MOMENT INTERACTION DIAGRAM OF WIDE FLANGE BEAMS

In a practical way, using the rolled wide flange beam, according to the provisions of the westeuropean and U.S.A. codes, the link elements are presented in the following classification (fig. 6. f) :
- short links (or shear links)
- intermediate links
- long links (or bending links)

These provisions may be applied to the similar welded plate girders.

Constructive provisions :
- short links

$$e \leq 1.6 \frac{M_p}{V_p}$$

- intermediate links

$$1.6 \frac{M_p}{V_p} < e \leq 3 \frac{M_p}{V_p}$$

- long links
$$e > 3 \; \frac{M_p}{V_p}$$

Where : e - the length of the link
M_p, V_p : plastic bending moment and plastic shear strength of the link

The plastic strength of links can be well in excess of V_p and M_p due to a combination of effects such as strain - hardening, influence of slabs or unintentional increase of yield stress.

Short links : $M_u = 0.75 \times e \times V_p$
$V_u = 1.5 \times V_p$
Long links : $M_u = 1.5 \times M_p$
$V_u = 2 \times M_p / e$

The design shear strength for link is given by :

$$V < \begin{bmatrix} 2 \times M_p / e \\ V_p \end{bmatrix}$$

If the link strength is controlled by shear, the flexural and axial strength within the link shall be calculated using the beam flanges only.

If the required axial strength P_u in link is less than 0.15 P_y ($P_y = A \times F_y$), the effect of the axial force on the link design shear strength may be neglected.

To avoid the inelastic shear buckling of the web (in short links) or the appearance of the local buckling of the flanges or the lateral - torsional buckling of beam , stiffeners are to be provided.

For short links the minimum stiffener spacing should be: (fig. 6. g)

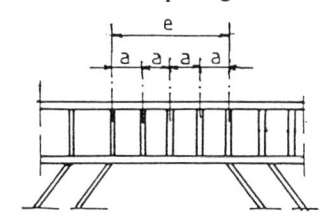

Fig6g — STIFFNER SPACING FOR SHORT LINKS

$a = 29 \; t_w - d/5$ for a plastic rotation $\gamma = 0.09$ rad
$a = 38 \; t_w - d/5$ for a plastic rotation $\gamma = 0.06$ rad
$a = 56 \; t_w - d/5$ for a plastic rotation $\gamma = 0.03$ rad

For long links the stiffener spacing should be lesser of:

$$c \leq \begin{cases} 1.5\, b_f & (\text{fig. 6 h}) \\ 0.5\, l \end{cases}$$

where : l = length of brace connection panel

The size of the stiffeners should be :

$b_s \leq \tfrac{1}{2}(b_f - 2\, t_w)$
$b_s \leq 15 \times t_s$
$t_s \geq 0.75\, t_w$
$t_s \geq 10$ mm

Web stiffener location :
Full depth web stiffeners on both sides of the beam at the brace end of the link should be located.
The link web shall be single thickness without doubler plate reinforcement and without openings.
Requirements for link beam-column connection :
 - the beam flanges shall have full penetration welds to the column;
 - for short links - web connection shall be welded;
 - for the link connected to the column web, the beam flanges shall have full penetration welds to the connection plates and the web connection shall be welded;
 - the rotation between the link beam and the column shall not exceed 0,015 rad.
Brace strength shall be 1.5 times the axial force corresponding to the plastic strength of the link beam. The same requirement is applyied to beam-segments outside of the link.
A link beam is not required in roof beams for structures over five stories.
The beam to column connection which is designed as pins in the plane of the beam web (the link beam is not adjacent to the column) must have capacity to resist a torsional moment of

$$0.02\, F_y \times b_f \times t_f \times d$$

The beam flanges shall be provided by lateral braces at intervals not exceeding $200 \times b_f / x \sqrt{F_y}$; $[F_y] = $ N/mm.
Intermediate bracing horizontal shall be designed to resist 2.0 percent of the beam flangeforce.
The distance between column and the first brace shall not exceed 0.25 L (L = unbraced beam leangth). A special attention is to be paid to the lateral buckling of the lower compressed free flange of the beam , especially at the brace connection (fig . 6 i) .

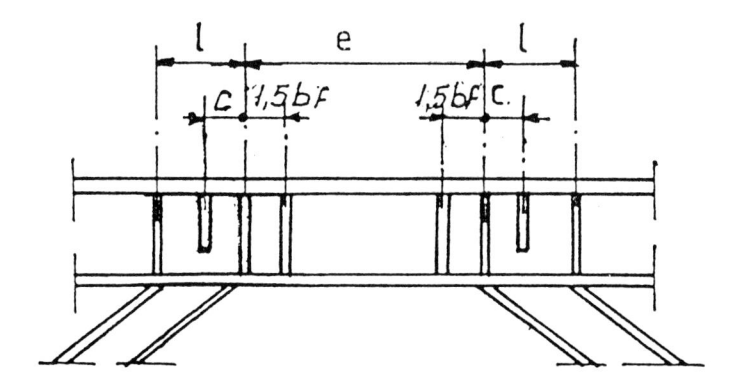

Fig. 6h — STIFFNER SPACING FOR LONG LINKS

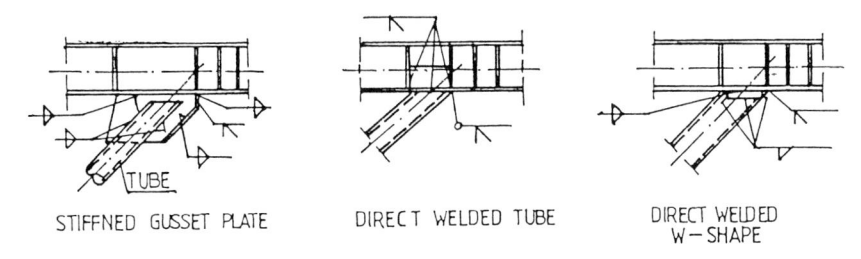

Fig. 6i — BRACE-TO-LINK CONNECTION DETAILS

STIFFNED GUSSET PLATE   DIRECT WELDED TUBE   DIRECT WELDED W–SHAPE

At beam to column connections, especially in the zone of the inavoidable plastic hinge in the column, longitudinal braces are to be provided between the columns both at the level of upper and lower flange of the beam.

In order to include in the Romanian Code P 100 - 92, supplementary provisions concerning the eccentrically braced frames, a lot of numerical testing works were performed by the authors. The results were compared with similar concentrically braced frames.

Except the classical structures with regular configuration, the numerical tests extended to some actual irregular configurated structures with irregular loading (fig. 7).

The calculation procedure, using adequate computer programmes, was established in order to ensure a quick developements of the design work.

A predimensioning procedure allows the preliminary selection of shapes for links and for other steel members which behave elastically during the plastic behaviour of the link.

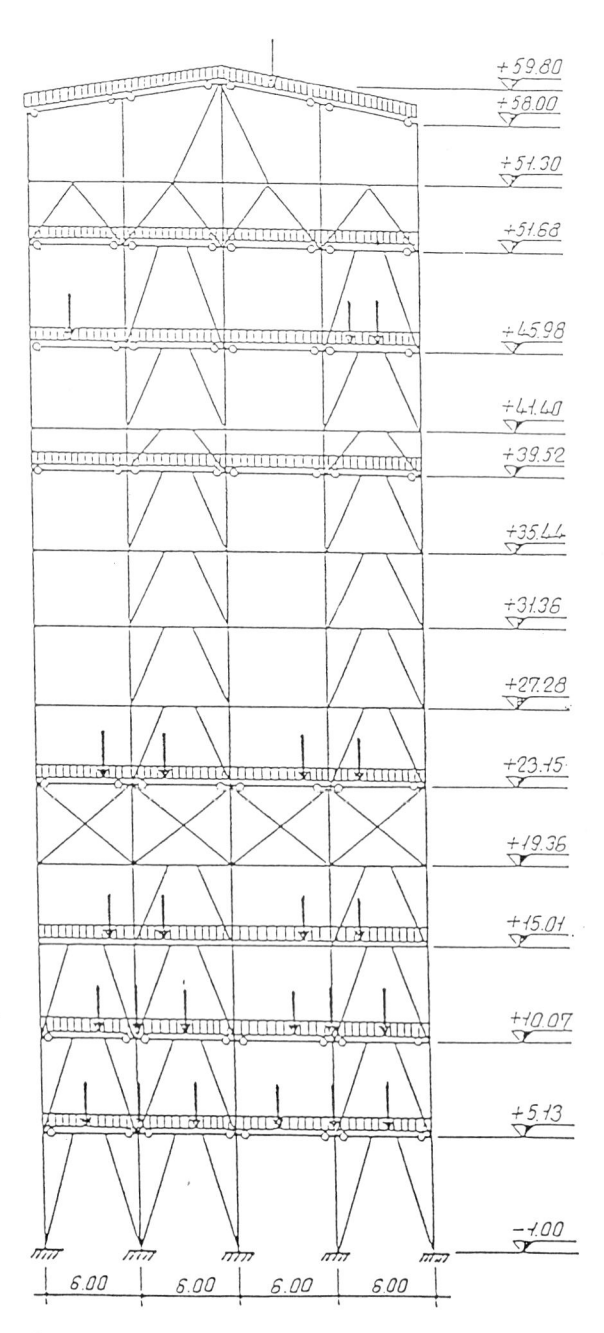

Fig. 7 – MULTISTOREY STEEL STRUCTURE METALURGICAL PLANT – ROMANIA

REFERENCES

1. U.B.C. - Uniform Building Code 1991 ; International Conference of Building Officials, Withier California
2. Seismology Commitee Structural Engineers Association of California 1990 ; Recommended Lateral Force Requirement and Commentary.
3. American Institute of Steel Construction 1990 ; Seismic Provisions for Structural Steel Buildings. Load and Resistance Factor Design.
4. A.I.J. -1986 ; Japan Seismic Code
5. C.N.R. - 1984 Italia ; Norme Tecniche per le Construzioni in Zone Sismiche.
6. E.C.C.S. - 1988; European Recommendations for Steel Structures in Seismic Zones , Working Group 1.3 ; Seismic Design.
7. E.C.C.S. - C.E.C.M. T.C. 13 ; Seismic Design (Mazzolani F.M.,Piluso V.) 1993 ; E.C.C.S. Manual on Design of Steel Structures in Seismic Zones.
8. Commision of the European Communities 1988 ; Eurocode 8 ; European Code for Seismic Regions
9. Code for aseismic design of residential buildings, agrozootechnical and industrial structures (P100-92). Romania , English version.
10. Proceedings of the Tenth World Conference on Earthquake Engineerings 1992 - Madrid,Spain; Crainic L. , Dalban C. , Postelnicu T. , Sandi H. , Teretean T. , " A New Approach of the Romanian Seismic Code for Buildings" p. 5655-5660.

41 TECHNICAL NORMS ON STEEL PLATE GIRDERS DAMAGED BY EARTHQUAKES

M. DINCULESCU
Building Research Institute, INCERC, Bucharest, Romania
D. GEORGESCU
Civil Engineering Institute, Bucharest, Romania

Abstract
This paper defines the norms, field of application, gives principles, constructional solutions and generic details concerned with repairing-strengthening of the steel plate girders strongly damaged by earthquakes.
Design methods are considered, too.
Key words: Steel Plate Girders, Repairing/Stregthening Principles & Methods, Design.

1 General scope and field of application

1.1 These norms refer to repairing and strengthening works for steel plate girders used to structures of buildings and industrial and agricultural constructions.
1.2 These norms do not apply to crane girders, to the girders pertaining to engineering structures (such as bunkers, silos,etc.), to the constructions that are part of machines/equipments, bridges, or to the nuclear or water engineering constructions.
1.3 The provisions of these norms are applicable to the existing constructions (in service or conservation), under normal temperature conditions.
1.4 The provisions of the present norms are applicable to steel plate girders subjected to static or prevailingly static loads.
1.5 These norms focus on the remedying of defects and local and /or global deteriorations such as: out-of-plane deformations, buckling, deformations accompanied or not by the rotation of the girder cross-section. The defects and deteriorations related to the material discontinuity (cracks, ruptures) are not the object of the present norms.
1.6 The norms include:

General principles
Constructional solutions and generic details.

1.7 The responsibility for applying the present norms provisions to situations, cases and conditions other than those specified at item 1.1...1.6 will be borne by the design engineer.

2 Repairing - strengthening principles

2.1 The repairing-strengthening works shall be carried out at the position of the girder (that is where the girder is located, without dismantling it from the structure).
2.2 For carrying out the repairing - strengthening works, the girder shall be subjected to unloading as much as possible (e.g. by removing the industrial dust, snow, etc. off the roof).
2.3 All the remedial works shall be based on technological documents, so that the repairing-strengthening works carried out at the position of the girder, would not cause additional unfavourable stress states into the girder.
2.4 If, when carrying out the repairing-strengthening works, the maximum bending stress caused by permanent and imposed loads do not exceed the 0.3 R value in the flanges of the girder (R = design resistance), then, these repairing-strengthening works do not require the girder propping, on condition that the provisions further on specified are observed.
2.5 The local defects can be repaired by warm straightening, with or without strengthening the cross-section in the remedied zone.
 The local deformations shall be repaired by warm straightening and by stregthening of the cross-section if this shows a diminution in its carrying capacity.
2.6 The defects and deteriorations covering large areas of the girder shall be remedied by constructional measures that would limit the extension of these defects/deteriorations under loads ("freezing" of the defects and deteriorations).
2.7 All remedial works that require welding shall be carried out with as thin as possible welds - the minimum ones resulted from analysis - executed at time intervals to avoid the material overheating. Measures to avoid overheating (beyond the necessary one) should also be taken in case of local remedies performed by warm straightening
2.8 The design engineer in charge with the repairing-stregthening solution has the duty to examine the heating effects on the physical-mechanical characteristics of the steel in the heated zones.

3 Remedying of defects

3.1 The local defects such as deformations of the girder flanges or webs caused by transportation or assembly works, can be remedied by heating the material and straightening either through manual jacks or hammering.

These operations require several successive heating-straightening phases so that the girder overheating should be avoided.

The design engineer in charge with the remedying solution shall observe the requirements specified at item 2.8.

3.2 Depending on the position (location) and severity of the defect, remedying could be limited only to straightening or will be also accompanied by a local stregthening of the girder cross-section.

4 Remedying of defects of deviation from the girder geometry, spread over considerable zones

4.1 The large, plastic, out-of-plane deformations of the girder web plate, resulting from buckling or from an inadequate execution, are recommended to be remedied by providing longitudinal stiffeners (stiffening ribs) on the entire length of the girder, supplemented or not, as the case stands, by additional transverse stiffeners (ribs).

Rigid longitudinal stiffening ribs, such as angles, shall be used only; they shall be adjusted following the distorted configuration of the girder web in order to fit tightly to it.

The following are recommended:

Location of the longitudinal stiffening ribs in the girder compression zone at a distance of about 0.3 of the girder height, distance measured from the compression flange of the girder (Fig 1).
The longitudinal ribs, provided on the entire length of the girder, shall be double sided and symmetric about the girder web plate.
Welding of the longitudinal ribs to the transverse stiffeners already existing on the initial girder (before subjection to remedial works) should be accomplished with adequate welds, capable to transmit the design resistance to the ribs cross-section, so that the longitudinal ribs also constitute a second flange in the compression zone of the girder cross-section.
The longitudinal ribs shall be welded to the girder web plate by intermittent fillet welds having the minimum dimensions required by analysis for avoiding the overheating of the girder web plate.
In case the carrying capacity of the girder must not be increased, the cross-section area of the longitudinal stiffening ribs is recommended to be about 1/3 of the cross-section area of the girder web plate.

$$A_{long.ribs} \approx 1/3 \, A_{web\,plate}$$

4.2 When the stiffness requirement results in longitudinal

ribs having a considerable cross-section area, it is rational to provide an additional flange (e.g. a plate) at the flange in tension as well (Fig 1). In this way, a significant change in the position of the girder initial neutral axis will be avoided.

Also, the newly introduced elements (longitudinal ribs in the compression zone - additional flange in the tension zone) contribute to a great extent to the girder bending capacity so that, on the whole, function of the severity of the initial defects, the repaired-strengthened girder could reach the carrying capacity of an initial girder, lacking any defects, and even an increased bending capacity.

To ensure an adequate welding (from up downwards), it is recommended that the width of the additional flange should be greater than that of the tension flange of the initial girder.

The thickness of the newly introduced additional flange should be at least equal to 1/25 of the existing flange width (Fig 1).

It is recommended that the additional flange located in the tension zone should observe the following technology:

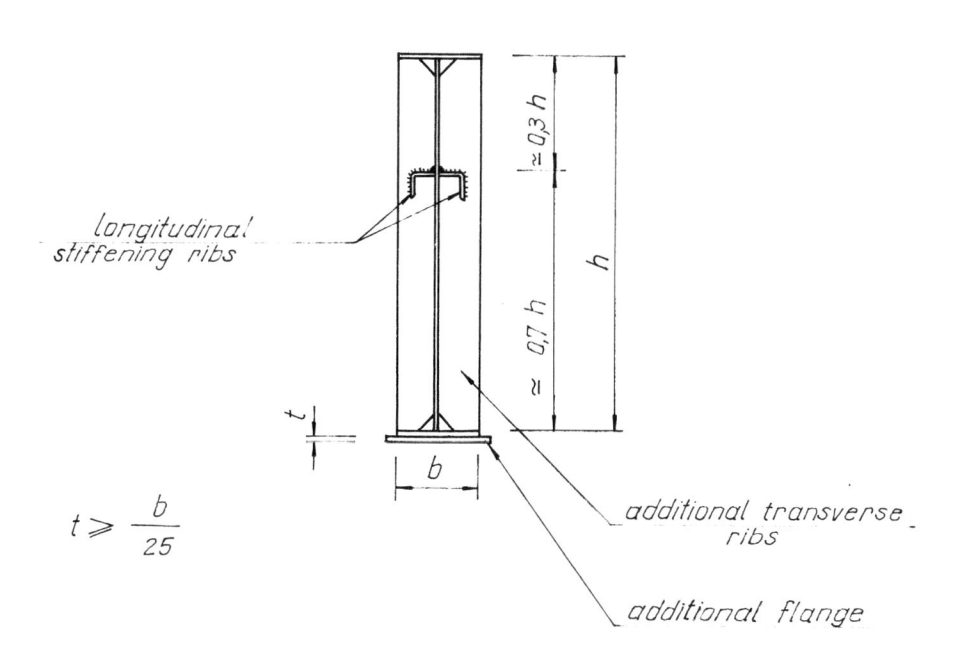

FIG. 1

The additional flange shall be attached (to the existing tension flange) at its both ends, each through 2 symmetric fillet welds, capable to transmit the design tension resistance of the cross-section of the newly introduced flange. The welds at the ends should be run first, at the beginning of the welding operation.

The newly introduced additional flange shall be further connected (to the existing tension flange) by symmetrically run intermittent fillet welds, from the ends towards the mid-span, and at time intervals carefully chosen to avoid heating throughout the flange at the same time.

4.3 Function of the severity and extent of the initial girder defects, the longitudinal ribs may be supplemented - in the zones where the web plate is significantly buckled- with additional transverse stiffeners (ribs) (Fig 1).

The additional transverse ribs shall be adjusted following the distorted configuration of the girder web plate in order to fit it tightly.

The experimental research on which the present norms are based, pointed out that in a very severely deteriorated zone of the web plate where additional transverse ribs were introduced as well, the effect of these ribs was very favourable indeed.

4.4 The global geometric deviations of the girder along its longitudinal axis - such as "wound" top flange accompanied or not by a global rotation of the girder cross-section - are recommended to be restrained (held in position) by providing additional ties both to the top flange (such as horizontal bars of post-type in transverse bracing perpendicular to the girder plane) and to the bottom flange such as braces in vertical plane connected to purlins.

The additional ties used for restraining the top flange deformations shall be carried up to the transverse bracings and connected to these ones by a geometrically undeformable system.

The design engineer of the repairing-strengthening solution has the obligation to examine the effect induced by the newly introduced ties on the existing members onto which these ties are connected.

5 Design methods

5.1 For the steel plate girders with severe defects and deteriorations -which have been subjected to repairing-strengthening works- the analysis of the carrying capacity shall be limited to the elastic range. The severe defects and deteriorations do not allow the use of the girder cross-section plastic reserves.

5.2 The girder carrying capacity analysis may be performed on the cross-section considered as free of defects and deteriorations.

5.3 In case the strengthening is accomplished without

propping the girder, and -during strengthening- in the flanges of the initial (unstregthened) girder the stresses are lower, at most equal to 0.3 R (R=design resistance), a simplified analysis is allowed in which the girder is considered to perform from the beginning with the cross-section resulting from strengthening needs.

5.4 In case the strengthening is carried out without propping the girder, but the stresses in the unstregthened girder flanges exceed 0.3 R, the following analysis is recommended:

a) the stresses induced in the girder flanges by the loads existing on the girder to be strengthened (strengthening perfomed under loads)should be determined taking into consideration the initial cross-section of the unstregthened girder;

b) the additional stresses in the initial unstregthened girder flanges, produced by the loads that will act upon the girder in addition to the loads acting during stregthening, should be determined; these additional stresses should be determined considering the geometric characteristics of the girder cross-section resulting from the stregthening needs;

c) for the more stressed of the two flanges, the stresses resulting at items a) and b) shall be summed up; the condition that the maximum sum of these two stresses do not exceed the design resistance increased by 10% (1.1 R) shall be imposed.

5.5 All the other usual verifications for girder shall be carried out according to the provisions of the norms in force concerning the elastic range.

42 ROMANIAN SEISMIC PROVISIONS FOR ONE-STOREY INDUSTRIAL BUILDINGS

D. GEORGESCU
Civil Engineering Institute, Bucharest, Romania
M. BUGHEANU
IPCT, Bucharest, Romania

Abstract
The 'Seismic Provisions for One-Storey Industrial Buildings' [1] (SPIB), in preparation, represents an application of the general provisions included in the recent Seismic Romanian Code P100-92 [2] to the particular case of one-storey industrial buildings and reflects the Romanian experience in constructing industrial buildings. Some main provisions of SPIB regarding the analysis of one-storey buildings located in seismic regions are presented in the paper.
Keywords: Industrial Buildings, Seismic Design, Steel Structure, Modal Analysis.

1 Introduction

The main provisions included in seismic codes refer especially to a certain type of constructions, largely represented by multi-storey buildings. The model of structural analysis is mainly based on the assumptions:

The area in plane of the building is quite small.
Whole foundation system behaves like a rigid one.
The masses are applied at the level of each storey of the building.
The same structural system is used both for transverse and longitudinal direction.
At each storey the floor behaves as a rigid diaphragm.

In comparison with the multi-storey buildings, the one-storey buildings present some particularities:

A large area in plane.
Individual foundations.
In addition to the masses located at the roof level, important masses due to cranes act on the stepped columns of the structure.
The structural system on the longitudinal direction differs from that along transverse direction.
The roof bracing system behaves elastically.

As a result, the assumptions above are no longer satisfied and the relationships given in P100-92 [2], based on these assumptions must be put in accordance with the actual behaviour on the one-storey industrial buildings.

In the following, the main relationships proposed in SPIB for the analysis transverse frame, longitudinal frame and roof bracing are presented.

2 Transverse frame

In SPIB the calculations of the actual transverse frame in Figure 1,a is based on the modal analysis of the condensed structure (Fig. 1,b) with two degrees of freedom. The condensed structure results by summing the masses and the moments of inertia (Fig. 1,b).

For the structure in Figure 1,b the modal equation is:

$$[[D] [M] - 1/\omega^2 [I]] = 0 \tag{1}$$

and can be written as:

$$\begin{vmatrix} \delta_{11} & \delta_{12} \\ \delta_{21} & \delta_{22} \end{vmatrix} \begin{vmatrix} m_1 & 0 \\ 0 & m_2 \end{vmatrix} - \frac{1}{\omega^2} \begin{vmatrix} 1 & 0 \\ 0 & 1 \end{vmatrix} = 0 \tag{2}$$

In Equation (2):

δ_{11}, δ_{12}, δ_{21}, δ_{22} are unit displacements, depending on the parameters $n=I_2/I_1$ and $\lambda=h_1/h$.
m_1, m_2 are masses at level 1 and level 2, respectively, where $m_1=G_1/g$, $m_2=G_2/g=\gamma m_1$.

Based on Equation (2), SPIB establishes relationships to determine:

The equivalent coefficient ε in Equation (3), in calculating the seismic horizontal shear force:

$$S = \alpha \beta k_s \psi \varepsilon \Sigma G_i \tag{3}$$

where α, β, k_s, ψ are to be introduced with their values according to P100-92 [2];

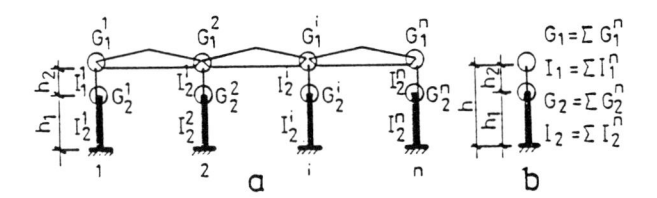

Fig.1. Transverse Frame Model Analysis

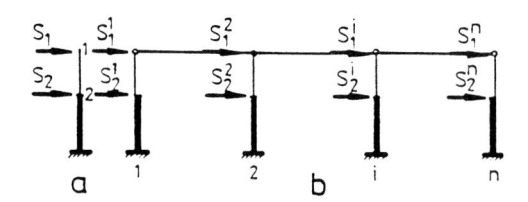

Fig.2. Shear Force S Distribution

The coefficients d_1, d_2 in calculating the horizontal seismic forces S1, S2 (Fig. 2,a) at the level 1 and 2 respectively:

$$S_1 = d_1 S$$
$$S_2 = d_2 S \qquad (4)$$

The coefficient a to calculate the fundamental period of vibration T by means of the following relation:

$$T = a \, (G_1 \, \delta_{11})^{1/2} \qquad (5)$$

Figure 2,b shows the distribution of the forces S_1, S_2 (Fig. 2,a) according to:

$$S_1^i = S_1 \frac{G_1^i}{\Sigma \, G_1^i}$$

$$S_2^i = S_2 \frac{G_2^i}{\Sigma \, G_2^i} \qquad (6)$$

With regard to the procedure of calculation proposed in SPIB it is to be observed:

(1) All the coefficients ε, d_1, d_2 and a depend on the parameters n, λ, γ and in order to simplify the design, for the most usual values of the parameters, the values of the coefficients above are given in tables, like in Table 1.

Table 1. $\gamma=1$

λ	n	ε	d_1	d_2	a
0.4	1	0.8772	0,692	0,308	0,219
0.4	2	0.8587	0,703	0,297	0,217
0.4	3	0.8469	0,713	0,287	0,215
0.4	5	0.8520	0,730	0,270	0,213
0.4	10	0.7813	0,765	0,235	0,209

(2) The procedure proposed by SPIB possesses a good degree of accuracy. The values of the horizontal forces S_1^i, S_2^i and of the fundamental period T calculated on the basis of the actual structure (Fig. 1,a) differ with a value of no more than 5% of the values S_1^i, S_2^i and T determined on the basis of condensed structure (Fig. 1,b).
(3) The horizontal forces S_1^i, S_2^i calculated according to the proposed procedure in SPIB are ε times smaller than those resulting by using P100-92. The coefficient ε varies from 0.60 to 0.95, depending on the case. More increases the parameter γ, more decreases the coefficient ε.

3 Longitudinal frames

With regard to the structural solution in Figure 3, largely used in common practice, it is to observed that, due to the great stiffness of the upper part, the actual structure loaded during an earthquake by two horizontal forces, S_1, S_2 (Fig. 3,a) can be replaced by the structure in Figure 3,b loaded by a single force $S = S_1 + S_2$, applied at the level 2 of the structure.

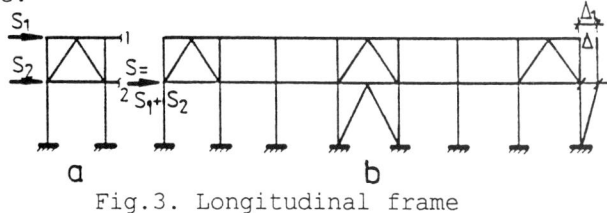

Fig.3. Longitudinal frame

Starting from this consideration, SPIB adopted the model of analysis in Figure 4.

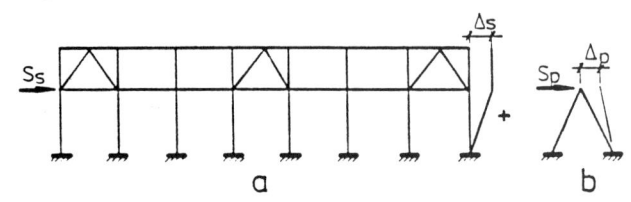

Fig.4. Longitudinal frame model analysis

In calculating the S-Δ diagram, the following steps are to be performed:

Determine the diagram S_p-Δ_p (Fig. 5) of the bracing substructure (Fig. 4,b).
Determine the diagram S_s-Δ_s (Fig. 6) of the columns substructure (Fig. 4,a).
By summing the two diagrams above, the diagram S-Δ of the actual structure (Fig. 3) is obtained (that is graphically solution of the equation of condition $\Delta_s = \Delta_p = \Delta$).

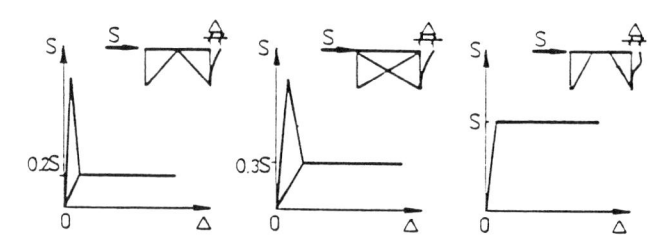

Fig.5. Diagram S_p-Δ_p

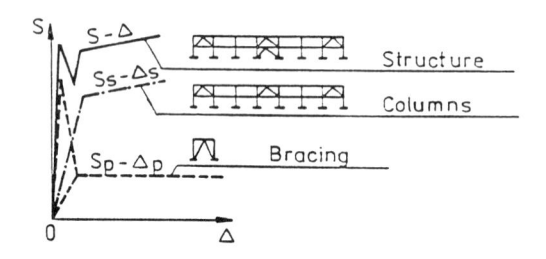

Fig.6. Diagram S-Δ

With regard to the model of analysis adopted in SPIB, it is to be observed:

(1) The model allows to determine the S-Δ diagram, both in elastic and plastic range.
(2) For the elastic and post-critical behaviour of the concentrically V and X bracing, or post-elastic behaviour of the eccentrically braced frames (Fig. 5) the simplified S_p-Δ_p diagram, adopted in SPIB, takes into account the recent theoretical and experimental results.
(3) Figure 6 shows the S-Δ diagram for a concentrically braced frame, the most common one. The diagram puts in evidence the fact that, if in the elastic range, the participation of the bracing is dominant, in the plastic range, on the contrary, the security of the structure largely depends on the behaviour of the columns. Taking into account the fact that under a severe earthquake motion, after the buckling of the braces, the stiffness and the strength capacity of the whole structure could be dramatically affected, the above mentioned remark is usually neglected in the common practice.
(4) Considering the fact that the bracing types behave differently, SPIB includes some particular provisions for each type of bracing.

4 Roof bracing system

Figure 7 shows the behaviour of a rigid diaphragm. This model, generally accepted by all the codes, is based on the assumptions:

Each transverse frame resists a part of the horizontal global seismic force S directly proportional to its stiffness (Fig. 7).
The torsion moment M is to be resisted by the whole structure, accepting a linear behaviour (Fig. 7).

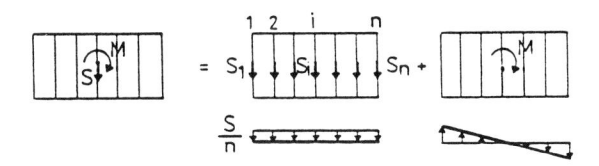

Fig.7 Rigid diaphragm

In the SPIB, this model, which can no longer be accepted for a roof bracing system which behaves elastically, is replaced by the model in Figure 8, based on the assumptions:

Each transverse frame resists the horizontal force acting in its plane (Fig. 8).
The torsion moment M is replaced by a couple of forces H, resulting from the condition $H L = M$.
Each force H is to be resisted by a number of five transverse frames in non-linear behaviour, depending on the ratio α between the stiffness of the bracings and the stiffness of the transverse frames.

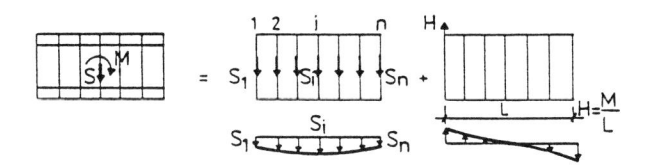

Fig.8. Elastic bracing

With respect to the model of analysis accepted by SPIB, it is to observe:

(1) The horizontal forces acting on the transverse frames as a result of the torsion moment M could be 1.2 up to 1.6 times the forces calculated by using the rigid diaphragm model, depending on the coefficient α, in the mean-

ing smaller the ratio α, greater the forces on the border transverse frames.
(2) Taking into consideration the fact that the whole behaviour of the structure could be largely influenced by the roof bracing system, SPIB includes some special provisions in order to make closer the behaviour of the elastic bracing to that of a rigid diaphragm.

5 Conclusion

Taking into account some particularities of the one-storey buildings, SPIB puts in accordance the general provisions given in P100-92 with the actual behaviour of the one-storey industrial buildings.

6 References

I.P.C. Seismic Provisions for One-storey Industrial Buildings, in preparation.
P100-92 Normativ pentru proiectarea antiseismica a constructiilor de locuinte, social-culturale, agro-zootehnice si industriale.

43 ON THE EVALUATION OF THE GLOBAL DUCTILITY OF STEEL STRUCTURES: PROPOSALS OF CODIFICATION

V. GIONCU
Building Research Institute, INCERC, Timisoara, Romania
F. M. MAZZOLANI
Istituto di Tecnica delle Costruzioni, University of Naples, Naples, Italy

Abstract
The evaluation of global ductility by means of the use of the q-factor is now-a-day a comun practice in the new generation seismic codes. But, there are still many incongruicies in the codification, which must be eliminated in the next revisions. Some proposals for improving the existing provisions are give in this paper.
Keywords: Global Ductility, Q-Factor, Collapse Mechanism

1. Introductive remarks on the existing codification

The goal of any codification is to provide design provisions which ensure public safety, i.e. to safeguard against major failure and loss of live, with limitation of damage for maintenence of the function and permission for easy repair. Based on this premise, the design methods must provide criteria for behaviour control of structures during the seismic action. The design philosophy is based on the concept that during severe earthquakes the energy input is dissipated through inelastic deformations. Due to this dissipation, the design forces can be reduced by means of the behaviour factor, namely q-factor. The exact evaluation of this factor requires several dynamic analyses for different ground motions and this is a very cumbersome and expensive way. The possibility to evaluate the q-factor by means of some simple methods adequate to be insert in a code represents an urgent need. But we have to recognize that this requirement is very difficult to be achieved, because the energy dissipation depends both on the local and global ductility. For both ductilities, the buckling phenomena in plastic field of deformations play a fundamental role.

The local ductility is tied to the shape of the members (Gioncu and Mazzolani,1994), while the global ductility depends on the structure configuration; so, a very important interaction between the two ductility types is generated during the seismic action.
Every earthquake prone Country provided in the last century its own provisions and therefore a great number of seismic codes has been developed in different ways.

But the today situation in the field of codification for seismic resistant steel structures is characterized by the polarization around of the three principal codes: EUROCODES (EC3) and (EC8) for European Countries, Limit State Design of Steel Structures (LSDSS) for Japan and Asian Countries and Uniform Building Code (UBC) for USA and American Countries. Thus, it is sufficient to analyse these three main codes for the individualization of lacks, incongruities and contradictions in the determination of the ductility factors, the so-called q-factors (Mazzolani et al.,1990a).

The codified values of this factor result from the tables, as a function of the structural types only. The following types are usually considered: moment resisting frames, concentric braced frames (diagonal bracing, V-bracing, K-bracings), eccentric braced frames and dual structures (Mazzolani,1991a).
This division just based on the typology is absolutely inadequate, because there are many other factors which have great influence on the behaviour of structures, such as:
a) **The type of accelerogram** acting at the base of structure. In fact using different accelograms (as El Centro, Tolmezzo, Bucharest, artificial ones, etc.) the values of the behaviour factor are very different, what means that each Country must assess its own q factors.
b) **The natural vibration period.** It is well known that the structures with low vibration period have a reduced behaviour factor in comparison with the structures with high vibration period.
c) **The plastic redistribution capacity** of the structures (in EC8 this factor is introduced by means of the ratio α_u/α_y).
d) **The type of collapse mechanism.** The global mechanism (plastic hinges only in the beams except at the column base) corresponds to a value of behaviour factor higher than in case of the story mechanism or mixed mechanism. The Japanise code allows the formation of story mechanism (in some conditions of column slenderness and axial force level), giving a smaller values for q than the EC8 code, which recommends global mechanisms only.
e) **The type of connections.** In some case the use of semi-rigid connections can be in favour of a better behaviour of structures under the seismic action.

f) **The local ductility**, which is one of the principal factor in the determination of q-factor. The today situation is characterized by the lack of a reliable relationship between local and global ductility. The correlations proposed in the codes, by using some cross-section behaviour classes related to the values of the behavioural factor q, are very useful for the design purposes, but are not based on a clear scientific evidence.
All these observations show that there are still many imperfections also in the modern code provisions in the evaluation of the global ductility.

2. Global ductility evaluation

Steel structures are more and more extensively used in regions of high seismic risk because of their excellent performance in term of strength and ductility.
  In comparison with the concrete structures, the steel structures are considered to be more releable to ensure the seismic safety. It is very true, but only if some basic requirements are satisfied. While for concrete structures, the reduction of rigidity due to cracking and brittle failure of concrete is the main factor for the loss of ductility, for steel structures the main problems arise because of the buckling of members and the great lateral flexibility which can introduce very important P-Δ effects. Thus, the evaluation of the global ductility of steel structures is a fundamental step in research, design and codification. In order to assess rational simplified procedures, first of all we have to recognize the major aspects and the ones which can be neglected. The q-factor for reducing the seismic forces determined for elastic structures represents a very drastic simplification which demands many very deep investigations.
For determining a correct q-factor some very important problems must be clearified:
a) **The failure criteria** must be completely different from the ones under static load conditions. In fact, in this case, under an increasing static load, the ultimate limit state is reached when a collapse mechanism is produced or some member buckling occurs. When the response to severe ground motion is considered, the prediction of the ultimate conditions is much more intricate. The decreasing in stiffness of the structure is the result of yielding in plastic hinge and the seismic forces can increase or decrease, depending on the acceleration spectrum. Thus, the variation of the forces on the structure must be replaced in the analysis with the variation of the intensity of the acceleration corresponding to some recorded earthquake acceleration.
  A different alternative can use the lateral displacements or the rotation of plastic hinges. Because in some cases

(short free periods) the use the lateral displacements give no results, it is preferable to use the rotation of plastic hinge. In this case the definition of the ultimate limit state, as the collapse is produced by the formation of a mechanism or by the loss of stability, is not significant from the dynamical point of view, the structures being still able to carry the loads, even if a mechanism is produced. The result is only a reduction of ductility. Thus, the failure criteria should refer to the limitation of displacements or plastic rotations, with respect of given deformation.

b) The **evaluation of the q-factor** must be done by an appropriate method which is selected among the existing ones. They can be grouped in three categories (Mazzolani et al.,1990a): $1°$. based of the theory of ductility factor; $2°$. extension of the results of the dynamic inelastic analysis of SDOF systems to the complex structures; $3°$. based on the energy approach. Among these methods, the most reliable seem to be the ones belonging to the first group, but they require the use of a time-history program for determining the dynamic behaviour of the structure in the field of geometrical and material nonlinearity. The determination of the q-factor is obtained as the ratio between the maximum acceleration a_u which the structure can sustain and the acceleration a_p which corresponds to the attainment of first yielding

$$q = \frac{a_u}{a_p}$$

After the determination of the acceleration a_p, using a serie of dynamic inelastic analysis in which the ground acceleration is increased step by step, a couple of values of a/a_p and Δ/Δ_p is determined (Fig.1): the bisectrix of the axes a/a_p and Δ/Δ_p represents the indefinitely elastic response. If the inelastic response is less than the indefinitely elastic one, the design based on an elastic spectrum will be on the safe side. In the opposite case the design will be on the unsafe side. The maximum value which can be assigned to q-factor is given by the intersection between the curve $a/a_p - \Delta/\Delta_p$ coming from the dynamic inelastic analysis and the bisectrix (Ballio et al.,1984). Two main problems arise in this approach. The method does not require the knowledge of rotation capacity, so the local ductility is not verified. If the member has a sufficient ductility, the method gives correct values. But when the capacity of rotation is exceeded, a degradation of cross-section behaviour is produced. This degradation leads to a changing of the $a/a_p - \Delta/\Delta_p$ curve and other values of q-factor are obtained. Thus, this method is not proper if the rotation capacity is neglected.

The second problem refers to the case of structures having

small natural period, as already mentioned, for which the $a/a_p - \Delta/\Delta_p$ curve does not intersect the elastic bisectrix (Fig.1).

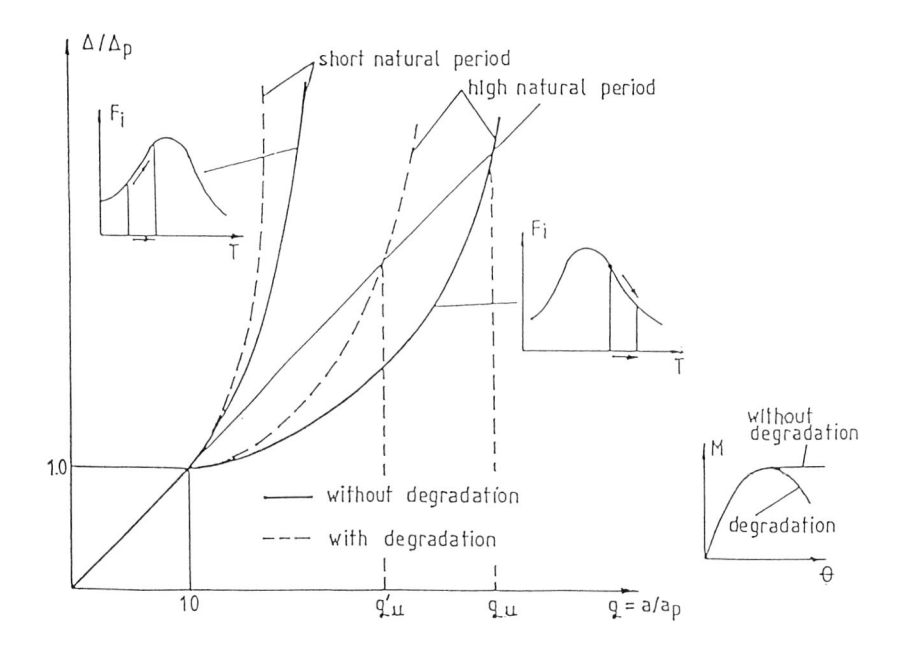

Fig.1. Determination of q-factor based on lateral displacements.

This is due to the fact that as far as the structural rigidity decreases, also the seismic forces decrease for high natural period, while for small natural period these forces increase. Thus this method falls in some situation. In these cases it is easier to use the rotation of plastic hinge θ_p and to represent the couple of values $a/a_p - \theta_p$ (Fig.2). Two behaviours are possible whether hardening is considered in the M-θ curve or not. In the first case the rotation is limited by $\theta_{p.max}$, while in the second by $\theta_{p.u}$. These limitations allow to determine the maximum value for q.

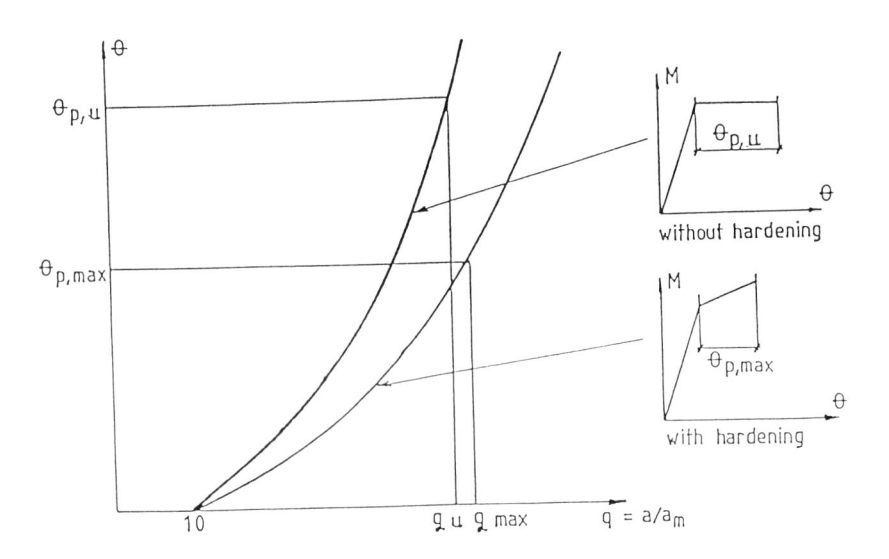

Fig.2. Determination of q-factor based on ultimate plastic rotation.

c) **The collapse mechanism type** is a very important aspect in plastic behaviour, the ductility being directly influenced by the mechanism of collapse. Two different situations are possible: the beams are weaker than the columns, known as weak beam-strong column system (WBSC), or the beams are stronger than the columns, so called strong beam-weak column system (SBWC).

In the first case, the mechanism is of global type, while in the second case, the storey mechanism is produced; in intermediate situations a mixed mechanism results (Fig.3). The studies show that the global mechanisms allow a good global ductility, while the story mechanisms have a reduced global ductility. This is a very discussed problem, because after some opinion (Akiyama and Kato,1980) it is not possible to obtain WBSC systems in economical way. While EC8 recommends the WBSC systems (Mazzolani,1991b), Japanise and American codes allow for both systems.

Using the method to plot the curves θ-a/a_p (Fig.4), two types of curves can be obtained (Gioncu et al.,1994). If the global mechanism (plastic hinge in beams) is obtained, the curve is a straight line
$$\theta = A(q-1) \quad (q>1)$$

When the plastic hinge arises in the columns the relationship is:

$$\theta = B(q-1)^2 \quad (q>1)$$

The problem is to determine the coefficients A and B for different types of structures, what seems to be possible by using numerical tests. Knowing the limit of plastic rotation, a rational q factor can be determined.

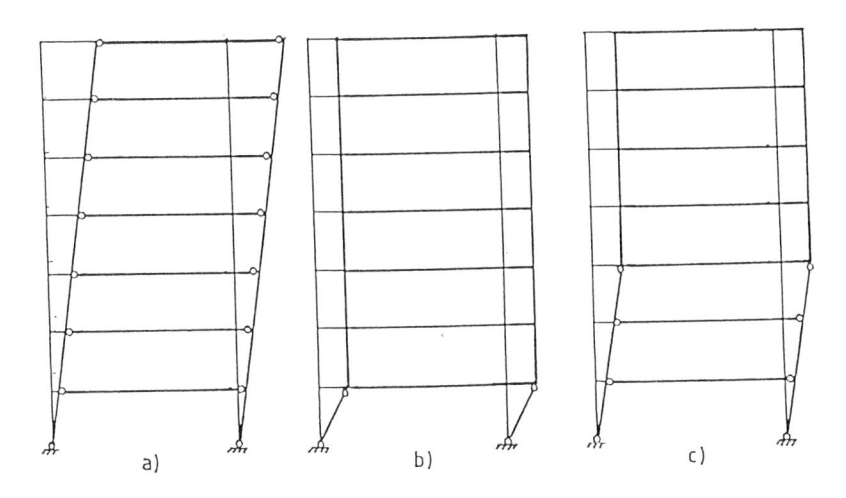

Fig.3. Collapse mechanism types.

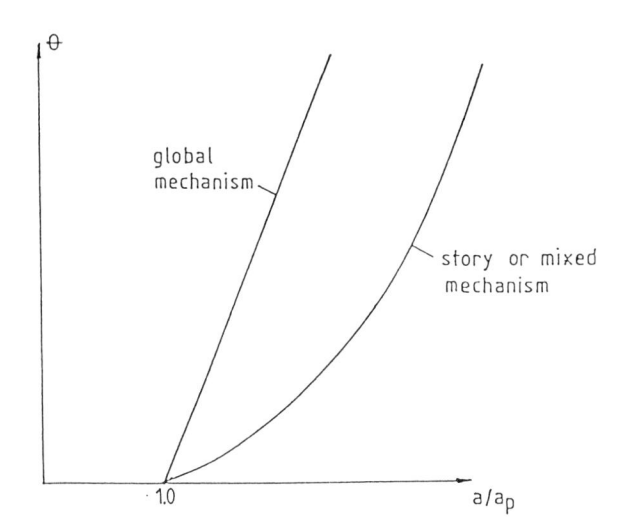

Fig.4. θ-a/a_p curve types.

d) **The design procedure** to choice the dimensions of members must guarantee the acheivement of a given type of collapse mechanism. According to the criterium of "strength hierarchy", plastic hinges have to form in the beams not in the columns. The codified design methods based on this criterium introduce an amplification coefficient for column bending moments. Nevertheless, from the practical results, they are still inadequate to guarantee that the structure fails in global mode and the global failure mode is not a sufficient condition to obtain a q-factor greater than the one assumed in preliminary design (Landolfo and Mazzolani,1990).
A releable design method based on the theory of limit design has been recently proposed (Mazzolani and Piluso,1993) for MR-frames and is going now to be extended to eccentric bracings.

e) The problem of **the definition of "regular structure"** is still open and the influence of "irregularities" on the q-factor is very unclear in the present codifications (Mazzolani,1992). In particular, for typical configurations of steel frames characterized by "set-backs" the communly adopted classification in regular and irregular frames seems to be unsatisfactory because in many cases the geometrical irregularity does not produce a lowering in the structural behaviour (Mazzolani et al.,1990b).
A correct procedure must require the use of more rational parameters, which take into account mass, stiffness and strength distributions and all other causes producing a nonuniform damage distribution under seismic loads (Mazzolani and Piluso,1994).

3. Proposed codification

The above remarks and proposals are based on the research results of the last fifteen years of research in the field of seismic resistant steel structure.

The ECCS Recommendations and, therefore, the Steel part of EC8 (Mazzolani,1992), both published in 1988 need now to be up-to-dated. The ENV version of EC8 is entering now in the period of public enquiry and this seems to be the best opportunity to operate an improvement of the existing design methods for steel and composite structures.
The revision of the text dealing with global ductility can be based on the main following steps:
a) The interaction between local and global ductility can be introduced by means of θ-q curves, in function of the characteristics of the structures (Gioncu and Mazzolani,1994).
Two situations can be taken into account (Fig.5):

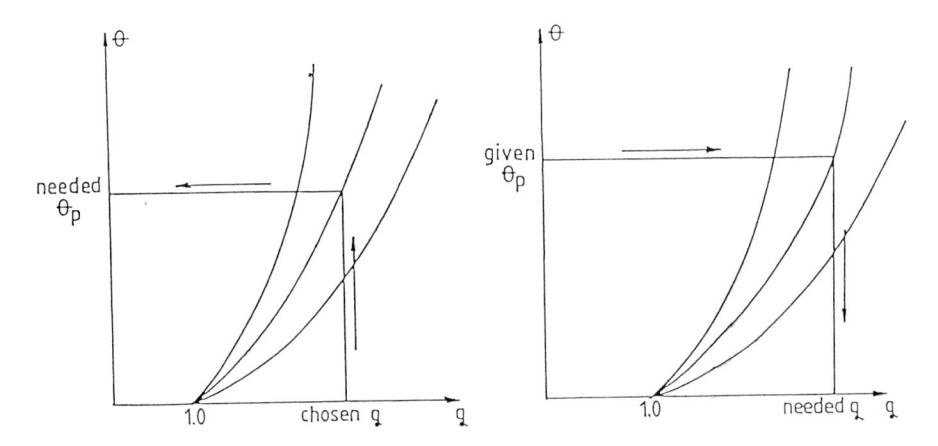

Fig.5. Interaction between local and global ductility.

- the q-factor is chosen using the nominal values given in the codes for different typologies. In this case with the help of a proper θ-q curve, the needed value of θ_p can be determined (Fig.5a) and therefore the corresponding member class;
- if for a given cross-sections the value θ_p is known according to the ductility requirements, with the help of θ-q curve the needed value for q is determined (Fig.5b).

This method requires the evaluation of the relationship between the plastic rotation limit and the q-factor for different typologies.

b) A distinction has to be made between the design q-factor (q_o) and the actual q-factor (q). This need arises because in many cases you design a structure by assuming a given value q_o of q-factor and after sizing of members you evaluate the actual one, which is usually different from the initial q_o, being

$q > q_o$ or $q < q_o$

Only the first case of course can be accepted. In the opposite case a new sizing of members is necessary. This check should require to make inelastic dynamic non linear analyses, which are very cumbersome and not easy for a practical engineer, but mostly usable in the scientific research.

In order to avoid such complex procedure, a simplified method for the evaluation of the q-factor must be proposed in the code. This can be selected among the existing approximated methods, by joining simplicity to reliability and leading to safe results.

It is appropriate to observe that this problem does not exist for concrete structures, where steel reinforcements are introduced "a posteriori" leading to the coincidence $q = q_o$.

c) The design method based on the use of the amplification factors can be revised at the ligth of the more recent results, which are not completely satisfactory. A new method based on the cynematic theorem of limit design can be proposed in principle and implemented on p.c.
d) The problem of set-backs in steel frames must be assessed on the base of new irregularity parameters, which does not penalize the majority of cases in which the reduction of meshes with the height does not produce a lowering of the q-factor.

4. References

Akiyama, H. and Kato, B. (1980) Energy concentration of multi-story buildings, in **Proceedings of 7th WCEE**, Istambul, p.553-560.

Ballio, G. Perotti, F. Rampazzo, L. and Setti, P. (1984) Determinazione del coefficiente di struttura per costruzioni metalliche soggette a carichi assiali, in **Proceedings of 2nd Congress on Seismic Engineering in Italy**, Rapallo.

Gioncu, V. and Mazzolani, F.M. (1994) Alternative methods for assessing local ductility, in **Proceedings of STESSA '94**, Timisoara.

Gioncu, V. Mateescu, G. and Iuhas, A. (1994) Contribution to the study of plastic rotational capacity of I steel sections, in **Proceedings of STESSA '94**, Timisoara.

Guerra, C.A. Mazzolani, F.M. and Piluso, V. (1990a) Evaluation of the q-factor in steel framed structures, **Ingegneria Sismica** n.2, 42-63.

Guerra, C.A. Mazzolani, F.M. and Piluso, V. (1990b) On the seismic behaviour of irregular steel frames, in **Proceedings of 9th ECEE**, Moskow.

Landolfo, R. and Mazzolani, F.M. (1990) The consequence of the design criteria on the seismic behaviour of steel frames, in **Proceedings of 9th ECEE**, Moskow.

Mazzolani, F.M. (1991a) Seismic behaviour of steel structures, in **Proceedings of 6th Conference on Steel Structures**, Timisoara.

Mazzolani, F. (1991b) The european recommendations for steel structures in seismic areas: principles and design, in **Proceedings of Annual Technical Session of SSRC**, Chicago.

Mazzolani, F.M. and Piluso, V. (1993) Failure mode and ductility control of seismic resistant MR-frames, in **Proceedings of 14th Congress C.T.A.**, Viareggio.

Mazzolani, F.M. (1992) Background document of EUROCODE 8, chapter 3: "Steel Structures", in **Proceedings of 1st State of the Art Workshop COST C1**, Strasbourg.

Mazzolani, F.M. and Piluso, V. (1994) **Theory and Design of Seismic Resistant Steel Frames**, Chapman & Hall, (in preparation).

44 SAFETY LEVELS IN SEISMIC DESIGN

F. M. MAZZOLANI
Istituto di Tecnica delle Costruzioni, University of Naples, Naples, Italy
D. GEORGESCU
Civil Engineering Institute Bucharest, Romania
A. ASTANEH-ASL
Department of Civil Engineering, University of California at Berkeley, USA

Abstract
A model of three safety levels in seismic design is proposed in this paper, in accordance with the concepts of the new generation seismic codes, such as ECCS, EC8 and ATC. An example allows some comments on the proposal.
Keywords: Security, Seismic Design, Limit States.

1. Historical background

Different periods can be identified in the path which was been followed by the structural design in high seismicity areas. A first phase started after the severe earthquakes of the second half of the 19th century and of the beginning of the 20th century, when the first conscious attempts were made in order to understand the forces arising from seismic action and to provide rational design rules. In spite of these efforts, due to the limited knowledge of the physical phenomenon, procedures for analysing structures subjected to seismic actions were not developed, but provisions were basically devoted to define height and mass limitations of buildings and to give the minimum distance between them.
  A second phase, of about five decades starting from the years between the two World Wars, has been characterized by the first, quite rough, definition of the seismic forces as inertial forces equivalent to a reduced percentage of the dead and live load.

These design forces were very small with respect to the ones arising during a severe earthquake. This limitation derived directly from the impossibility, at that time, to perform inelastic analyses; however, the greatest approximation was due to the fact that the design seismic forces were defined, for each case, without taking into account the structural typology and the nature of the foundation and supporting soil.

Moreover, the structural analyses were limited to considering each storey separately from the whole skeleton. This was due to the difficulties to analyse, by hand, highly redundant structural schemes. Furthermore, approximated laws of distribution along the building height were used for the seismic forces, due the lack of knowledge of structural dynamics.

In the past two decades, a critical review of the previous seismic design approach has been developed. In these years, due to the spreading of personal computers and the implementation of a great number of programs for structural engineering, structural analysis of highly redundant schemes, both in static and dynamic range, has become feasible for design purposes.

In this period the generic horizontal force was given as proportional to the storey weigth W_i by means of this type of equation

$$F_i = K \cdot W_i \tag{1}$$

where the proportionality constant

$$K = C \cdot R \cdot \epsilon \cdot \beta \cdot \gamma \cdot I \tag{2}$$

takes fortaitary into account:
- the sismicity of the area (C)
- the dynamic response (R)
- the foundation type (ϵ)
- the structural typology (β)
- the mass distribution (γ)
- the importance of the building (I)

Furthermore, the most important achievement of this new phase, which is continuously going on, is represented by the clarification of mechanisms allowing the reduction of the design seismic forces. It has been pointed out that the magnitude of these forces has to be assumed as a function of the deformation characteristics of the structure and of its energy dissipation capacity.

In addition, the possibility of dissipating the eartquake input energy by means of plastic excursions has been recognized and utilized, which have to be compatible with the plastic deformation capacity of the structure. As a consequence, the concepts of local and global ductility have been introduced and their importance in design of seismic resistant structures has been emphasized. The corresponding design approach, which is nowadays universally accepted, is characterized by the requirements of both strength and ductility, which represent, combined together, the "seismic toughness" of a structure. This means that the reliability of the magnitude of the seismic forces assumed in design practice has to be justified by verifying the available structural ductility and the energy dissipation capacity.

The difficulties of this approach are represented by the efforts which have to be made in order to ensure appropriate ductility levels. For this reason, the main object of the modern seismic codes is the widespreading of the design procedures oriented to the failure mode and structural ductility control.

This new generation of seismic codes is characterized by the introduction of the so-called q-factor which corresponds to an indirect evaluation of the global ductility of the structure. According to this methodology, the horizontal forces F_i, acting on the structure at any level is:

$$F_i = \frac{AR(T)}{q} \Sigma W_j \frac{z_i W_i}{\Sigma z_j W_j} \qquad (3)$$

in which A is the standard value of the ground acceleration normalized to the gravity acceleration g, R(T) is the design spectrum. A linear distribution of the forces F_i at any levels is usually assumed (Fig.1).

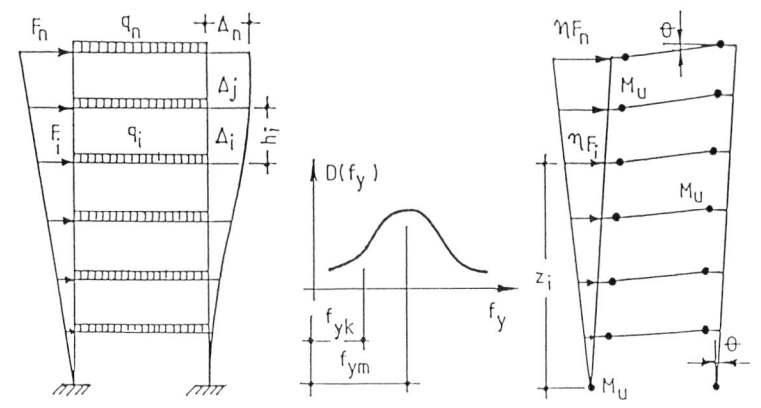

Figs.1,2,3. Distribution of forces, strength variation and collapse mechanism.

In comparison with a very rigid, non-dissipative structure (q=1), a ductile structure has to be sized for a base shear force q times smaller. The behaviour factor q depends on the type of the structure. So, a building with a concentrically braced frame (q=2) is to be sized for horizontal forces 2,5 times greater than another building, in the very close vicinity, but with a moment resisting frame structure (q=5). This important difference puts in evidence the different dissipative character of the typology.

In order to ensure a satisfactory behaviour of buildings with minor damage, in addition to the strength check, a drift condition (Fig.1) shall be satisfied:

$$\Delta_r = \Delta_j - \Delta_i < \alpha h \qquad (4)$$

The drift condition is to be checked under severe earthquake motions. As a result, in calculating the displacements Δ_i they must be amplified by the behaviour factor q.

A more refined group of codes introduces two security levels, entitled "limit states". The japonese code (JBLS,1981) considers a "moderate limit state", related to earthquake motions which would occur several times during the life of the building and an "ultimate limit state", corresponding to a severe earthquake which occurs less than once during the use of building. The structure is to be sized to actual horizontal forces F_i, given by an equation similar to Eq.3, where $A=A_m$ is the standard value of the ground acceleration associated with the moderate earthquake. The drift condition is to be checked also in the moderate state, introducing in Eq.4 the displacements produce by the above forces F_i. The behavioural factor q is used only in examining the ultimate state.

The EUROCODE 8, in the last draft 1993, introduces a "serviceability limit state" and an "ultimate limit state". The structural elements are to be sized for design force distribution given by Eq.3, by taking into account the ultimate state and the ductility of the structures, since the drift condition is related to serviciability limit state.
The American code UBC (1991) introduced also two limit states: serviceability and ultimate states.

The model proposed by the authors in present paper, considers three given limit states, so called: serviceability, damageability and collapse, which are in accordance with the philosophy design included in ECCS, EUROCODE 8 and ATC. The model is rooted in the "Capacity Design" concept.

2. Limit state requirements

2.1 Serviceability limit state
The serviceability limit state corresponds to quite moderate earthquakes with a return period of some 5 - 10 years. The structural and non-structural components should resist moderate earthquake with almost no damage and the disconfort to the inhabitans should be minimal. These requirements are fulfilled by keeping the structure in the elastic range, in order to protect the structural components and by controlling the drift, in order to protect the non structural components.
In designing the structure the following steps should be performed:
(1) Determine the efforts N,M,T, produced by the horizontal forces given by Eq.3.
(2) Choose the potential plastic zones, generally located near the beam-column joints and at the bases of columns.
(3) Size the sections where plastic hinges are expected in the beam ends under the condition:

$$M_e < M < M_{pl} \qquad (5)$$

where, according to EUROCODE 3:

$$M_e = W_{el} f_d \qquad (6)$$

$$M_{pl} = W_{pl} f_{y,k} \qquad (7)$$

where $f_d = f_{y,k}/\gamma_m$, being $f_{y,k}$ the characteristic value of the yield stress
(4) Determine M_u, the probable ultimate plastic bending moment:

$$M_u = W_{pl} f_{y,m} \qquad (8)$$

where $f_{y,m}$ is the mean value of the yield stress (Fig.2).
(5) Size all the structural components outside the plastic hinges (columns, beams, etc.) for the efforts produced by the fully yielded and strain hardened plastic hinges. This requirement is based on the "Capacity Design" philosophy in order to keep all these components essentially in the elastic range under severe earthquake motions.
The values of horizontal forces ηF_i associated to the fully yielded plastic hinges can be appreciated by using the equation of the equilibrium in the ultimate state:

$$L_i = L_e \qquad (9)$$

where L_e, L_i are the external and internal mechanical works respectively.
For the structure of Fig.1, by assuming that the failure mechanism is of global type (Fig.3), the Eq.9 turns into:

$$\Sigma M_u \theta = \eta \Sigma F_i z_i \theta \tag{10}$$

From Eq.10 results:

$$\eta = \frac{\Sigma M_u}{\Sigma F_i z_i} \tag{11}$$

This equation balances the behaviour of the structure under severe earthquake motions:
- if M_u is underestimated, the multipier η decreases and the structure could be proportioned for horizontal forces ηF_i smaller than the actual ones and, as a result, additionally undesirables plastic hinges (e.g. in the columns) may occur;
- if M_u is overestimated, the multiplier η increases and the structure could be uneconomically designed.

As a result, the M_u value must be under control and that is the reason for Eq.5.

(6) Check the drift condition (Eq.4) by means of the lateral displacements Δ_i and Δ_j produced by the forces F_i given by Eq.3 and accepting in accordance with EUROCODE 8:
   (i) for buildings having non-structural elements attached to the structure:

$$\Delta_r < 0,002h \tag{12}$$

   (ii) for buildings having non-structural elements fixed in a way as not interfere with structural deformations:

$$\Delta_r < 0,004h \tag{13}$$

In Eqs.12 and 13, h is the storey height.

2.2 Damageability limit state

The damageability limit state corresponds to quite large earthquake motions with a return period of about 50 years. Some damage to structural and non-structural components are allowed but not significant ones, meaning that a non expensive repair of the building would be done in a quite short period of time.

In order to fulfil these requirements, the following conditions are to be satisfied:
(1) Using an inelastic dynamic analysis, check that under moderate earthquake the locations of the plastic hinges are limited to the initially chosen ones.
(2) Using an inelastic dynamic analysis, check the drift, accepting as allowable values 2 times the values given by Eqs. 12 and 13.

2.3 Collapse limit state

The collapse limit state is associated to very severe earthquake motions with a return period of about 500 years (EUROCODE 8 indicates 475 years). The collapse of the structure is to be avoided and the life safety shall be maintained. It must be underlined that, due to the unpredictable nature of the phenomena, the seismic design can not fully ensure that there will be no enjury or loss of life (ATC,1978). After the earthquake, either the repair or the demolition of the building shall be decided.

In order the avoid the collapse of the building the following conditions have to be satisfied:
(1) Using an inelastic dynamic analysis, check that local collapse mechanisms, especially storey mechanism, are not expected (Landolfo and Mazzolani,1990; Chen et al.,1992).
(2) Limit the plastic rotation in the plastic hinges to 0,02 radians, which, currently, is considered the amount of required rotation capacity under distructive earthquakes.

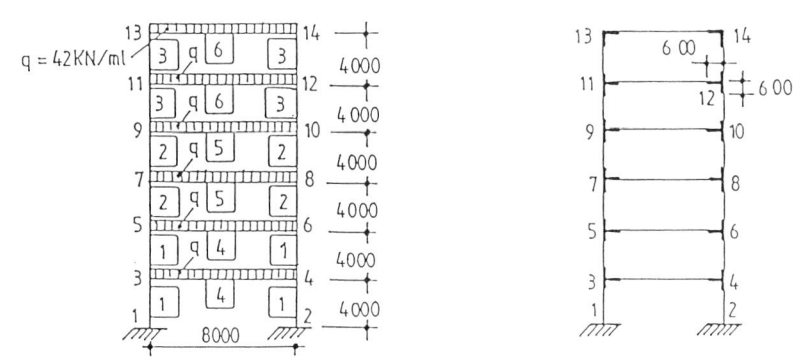

Figs.4 and 5. Design data and structural scheme.

3. Application of the model

3.1 General

The structure of Fig.4 corresponds to a building located in Bucarest. Check the structure for the accelerogram A (Vrancea N-S, 4 Mars 1977) given in Fig.6, by assuming:
- for serviceability limit state: accelerogram "0,4 A"
- for damageability limit state :   accelerogram "A"
- for collapse limit state   :   accelerogram "1,6 A"

The main data of the calculation are the following:
(1) The frame elevation, accepting rigid zones in the areas of beam-column joints, as presented in Fig.5.
(2) The steel grade is Fe 360, with $f_{y,k}$=235 N/mm^2 and $f_{y,m}$=275 N/mm^2.
(3) As design behaviour factor $q = 5,0$ is assumed.
(4) The characteristics of the structural members are given in Table 1.

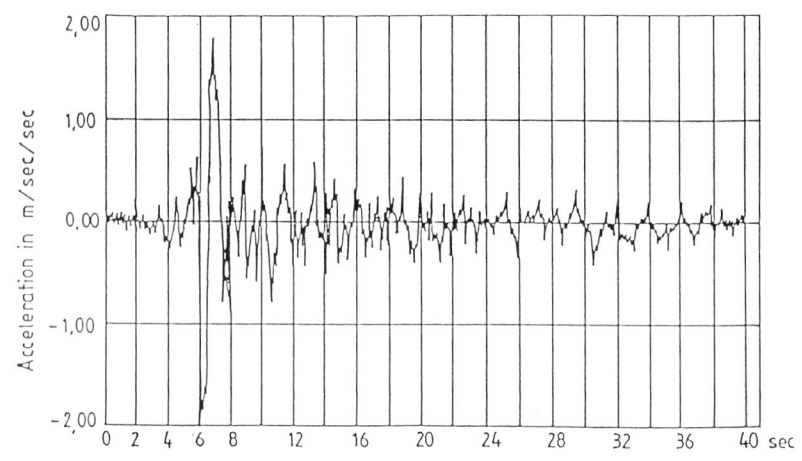

Fig.6. Vrancea N-S, 4 Mars 1977, accelerogram.

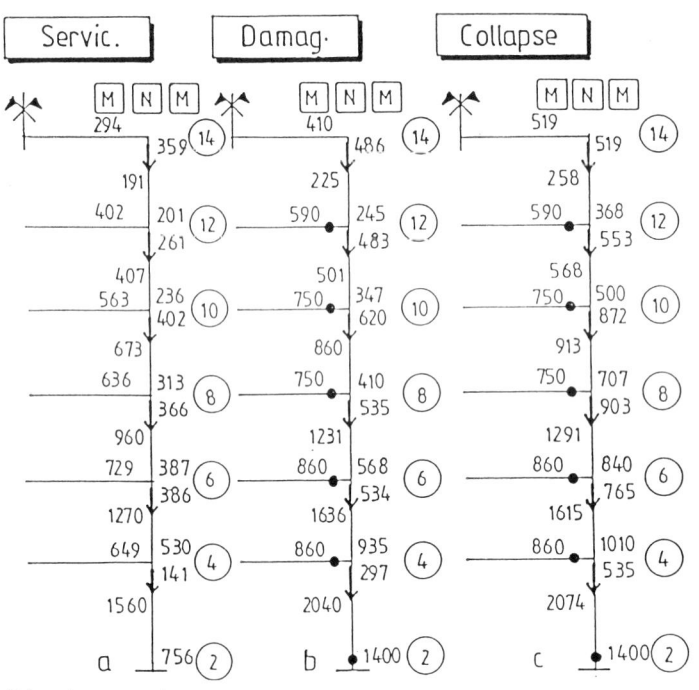

Fig.7. Maximum M(KNm) and N(KN) values.

(5) The basic elastic dynamic analysis was performed by using SAP-90 program and the results are given in Table 2.
(6) The inelastic dynamic analysis was performed by using DRAIN-2D in a Processor Program elaborated by Dan Cretu (1994).

The main results of the inelastic dynamic analysis are presented in:

Fig.7 - Maximum M and N values (in KNm and KN, respectively)
Fig.8 - Maximum lateral displacements Δ_i and drifts Δ_r (in cm)
Fig.9 - Plastic hinges history

TABLE 1: Properties of frame members

Member	Shape	A (cm^2)	I (cm^4)	M_e (KNm)	M_{pl} (KNm)	M_u (KNm)
1	I HEAA800	218	208900	-	-	1400
2	I HEAA700	191	142700	-	-	1090
3	I HEAA600	164	91870	-	-	830
4	I IPE600a	137	82920	549	738	860
5	I IPE550	134	67120	521	655	750
6	I IPE500	116	48200	412	516	590

TABLE 2: Elastic dynamic properties

Mode	Period (sec)	Participation mass (percent)	
		X-DIR	X-SUM
1	0,963	76,270	76,270
2	0,309	13,166	89,436
3	0,157	5,139	94,755
4	0,094	2,609	97,364
5	0,063	1,575	98,939
6	0,045	1,061	100,000

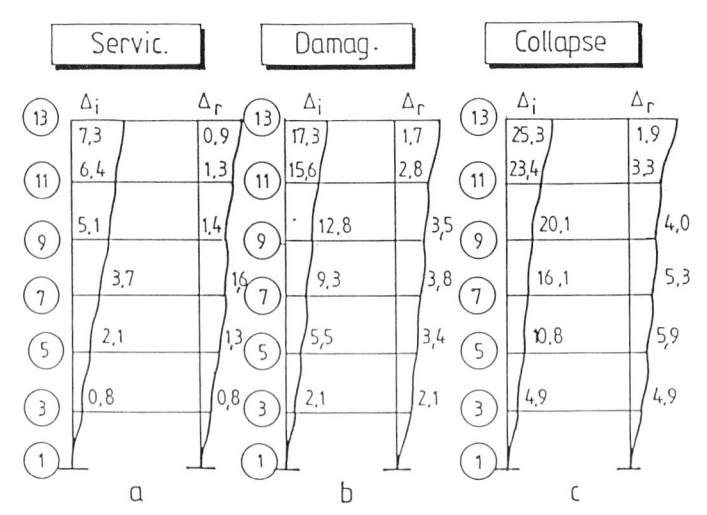

Fig.8. Maximum lateral displacements
Δ_i and drifts Δ_r (cm).

3.2 Serviceability limit state
(1) The beam sections, where plastic hinges are expected, were sized in accordance with bending moments M of Fig.7,a. The values of elastic M_e, plastic M_{pl} and ultimate M_u bending moments are given in Table 1, taking into account the N effect in columns (members 1,2,3).
The condition of Eq.5 is strictly satisfied. With respect to member (5), for instance, results:
Joint 8:   521<636<655
Joint 10:   521<563<655
(2) The maximum drift 1,6 cm in Fig.8,a satisfies the allowable drift (Eq.13):

$$\Delta_r < 0,004h = 1,6 cm$$

3.3 Damageability limit state
(1) The condition to keep the columns, except the bases, in elastic range is satisfied (see Fig.7,b and Fig.9,a)
(2) The drift condition (see Fig.8,b)

$$\Delta_r < 2 x 1,6 = 3,2 cm$$

is slightly exceeded reaching 3,8 cm at the third storey.
(3) The sequence of the formation of the plastic hinges (Fig.9,a) and the fact that the column bases are involved in yielding, very far from the beginning of the motions and for a short period of time, can be accepted as a good plastic behaviour of the structure.

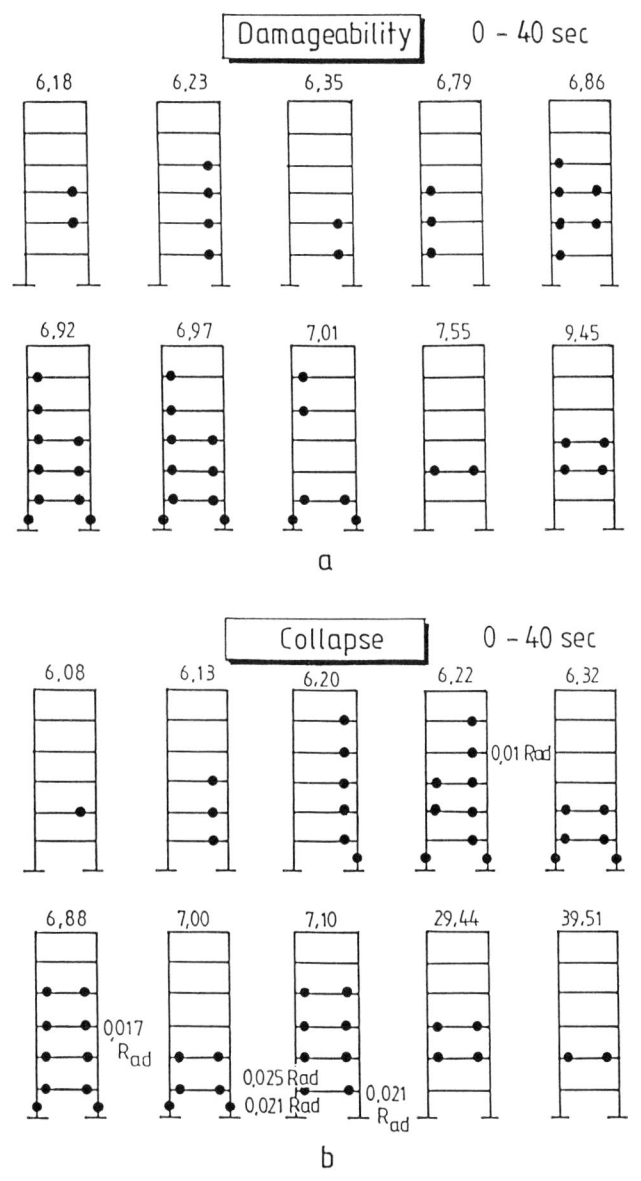

Fig.9. Plastic hinges history.

3.4 Collapse limit state

(1) No plastic hinges in the columns, except at the bases and no local mechanisms are to be expected (Fig.7,c and 9,b).
(2) The condition to limit the plastic rotation in the plastic hinges to 0,02 rad is quite satisfied (Fig.9,b).
(3) The maximum displacement at the top of the building 25,3 cm (Fig.8,c), representing some 1/100 of the height of the building, is acceptable.
(4) Taking into account all the results above and the good plastic response (Fig.9,b), it can be concluded that the general requirements to avoid the collapse of the structure are satisfied.

4. Conclusions

The proposed model of three security levels in seismic design could be a new step for the modern generation of seismic codes and at the same time a useful approach for the designers in sizing a steel structure in seismic areas.

5. References

ATC (1978) Tentative Provisions for the Development of Seismic Regulations for Buildings. Applied Technology Council.
Chen, C.C. Bonowitz, D. and Astaneh, A. (1992) Studies of a 49-Story instrumented steel structure shaken during the Loma Prieta earthquake. **Doc.UCB/EERC-92/o1**, Berkley.
Cretu, D. and Muscalu L. (1994) Plot Pre and Post Processor DR386-P.C: Version of DRAIN-2D (in preparation).
ECCS (1988) European Recommandations for Steel Structures in Seismic Zones.
EUROCODE 3 (1993) Design of Steel Structures.
EUROCODE 8 (1993) Earthquake Resistant Design of Structures.
JBLS (1981) Earthquake Resistant Regulations for Building Structures.
Landolfo, R. and Mazzolani F.M. (1990) The consequence of the design criteria on the seismic behavior of steel frames, in **Proceedings of 9th ECEE**, Moscow.
UBC (1991) Uniform Building Code.

45 SOME CONSIDERATIONS ON THE NEEDS OF FUTURE DEVELOPMENT OF EARTHQUAKE RESISTANT DESIGN PRACTICE AND CODES

H. SANDI
Building Research Institute, INCERC, Bucharest, Romania

Abstract
The paper is devoted to a critical appraisal of the state of the art of some provisions of earthquake resistant design codes and of practice. The goal is to contribute to a more consistent approach to the methodology of engineering analyses connected with safety and reliability verifications. Topics concerning the specification of seismic conditions, the design criteria, the analysis of structural performance and the more consistent implementation of the capacity design philosophy are dealt with. The conclusions are aimed at the development of codes and of methods used in advanced engineering practice.
Keywords: Seismic Conditions, Design Criteria, Seismic Risk, Capacity Design.

1. Introduction

Earthquake resistant design codes underwent during last decades a considerable evolution. It may be stated that the main goals pursued in their development were, on one hand, the adoption of more effective engineering solutions and, on the other hand, providing of means for a more accurate safety control, with the aim of gradually adopting an explicit approach to the limitation of seismic risk.

This paper is intended to present some remarks and suggestions concerning the development of the analytical side of codes with the objective to gradually overcome some inconsistencies of the present. The sequence adopted in the presentation was intended to be as logical as possible. This sequence, as any other sequence, may hide, at a first glance, the numerous cross-connections and feed-backs that are characteristic to the topics dealt with.

2. Specification of seismic conditions

2.1. General
Specification of seismic conditions relies among other on two start points:
 a) the representation of ground motion during on event and
 b) the representation of the sequence of seismic events;

Behaviour of Steel Structures in Seismic Areas. Edited by F. M. Mazzolani and V. Gioncu.
Published in 1995 by E & FN Spon, 2-6 Boundary Row, London SE1 8HN. ISBN: 0 419 19890 3.

as relevant for the engineering requirements. Given the goal of specification of seismic conditions, any of the two steps of representation must have a predictive character. The current state of knowledge, that explicitly recognizes the randomness of seismic phenomena, leads to the need of using probabilistic tools in order to predict expected seismic phenomena.
Following developments of this section are concerned with the representations referred to, up to code formats, as well as with the problem of consideration of the influence of local conditions.

2.2. Representations of ground motion during one event
The alternative representations of ground motion during one event, as referred to in some codes at present, may be classified as follows:
R.1: stochastic representations (power spectrum densities etc.);
R.2: time-history representations;
R.3: design spectrum representations.
The sequence adopted is logical [9], since specifications corresponding to any representation (except R.1) can be derived in a direct way from a previous one, while deriving specifications from subsequent over is much more debatable. So, the representations R.1 are considered to be the fundamental ones.
The R.1 approach makes it possible to account for significant features of ground motion, like its 3D character at a given point and its non-synchronous character at different points (a full model in this connection was proposed in [8]). The R.2 approach makes it possible too to account for such features, but design accelerograms (eventually multi-component ones) can be generated in a consistent way only on the basis of R.1 models. Finally ,the R.3 representations, which are inherently suited only for scalar (one- component, as a rule translational) motions cannot provide
information concerning the features referred to. The direct conclusion is that a full specification of design input must rely on some R.1 type representation and that **codes should provide guidelines on how to adopt reasonable representations of this kind, which are obviously necessary in cases when 3D analyses of structures are to be performed.**

2.3 Recurrence problems
The specification of design intensities, or of design peak ground accelerations etc., is made usually in terms of return periods of some given thresholds of design parameters. Two main comments should be kept in view in this connection:
a)the way of deriving return periods for intensities etc.;
b)the problems raised by the multi-parameter nature of a satisfactory characterization of ground motion (intensity, dominant frequency, duration etc.).
By now it is widely accepted that a consistent approach to the specification of recurrence characteristics will rely on Cornell´s method [2], that is based, essentially, on a convolution between recurrence laws of magnitudes and attenuation laws. In practice, it happens that this method is used in a wrong way, considering an apparently deterministic attenuation law, which accounts in fact for the average attenuation observed. The literature (e.g. [4]) puts to evidence the high attenuation randomness, that was observed, on the basis of

instrumental data, also in case of the recent strong Romanian earthquakes of 1986 and 1990 [5],[21]. The explicit consideration of randomness leads to a more general approach to the convolution referred to [12]. The results of considering the attenuation randomness are dramatic, leading to considerable increases in the specified severity of earthquake induced motions. It should become compulsory, in the development of seismic zonation or of specification of seismic conditions for selected sites to explicitly consider the attenuation randomness.

The concept of return period is consistent with the implicit assumption of a poissonian type recurrence of seismic events, which is not in agreement with the physics of earthquake generation. The non-poissonian nature of recurrence may be nevertheless disregarded in cases when service, or exposure durations of structures to be protected are sufficiently long. Another aspect raises more acute problems. The return period concept is considered as a rule in connection with the implicit assumption of a unique (scalar) measure of ground motion severity. The consideration of possible motions with different dominant frequencies, with different durations etc. makes it difficult to further use just return periods, for the goals referred to. A consistent reply and improvement can be given only in connection with a third-level probabilistic approach (referred to as P.3). This aspect is discussed further on, in section 3.

2.4. Consideration of local conditions

The influence of local geological conditions on the features of ground motion is widely recognized and most codes provide some specifications in this connection, usually by categorizing soil profiles. It must be mentioned here that the packages of upper layers referred to in codes (which have a total depth in the range of some tens of meters) may be definitely too thin. The experience of Romania in 1977 [1] put it to evidence that for the INCERC record a depth of at least of 125 m. was relevant and so it must be in most cases when motions are expected to occur. Another aspect to be considered is the fact that, in a few cases (e.g. Erzincan, 1992), the focal mechanism was apparently characterized by long dominant periods, to which the upper geological package is practically transparent. A critical reconsideration of the manner of specifying criteria for considering local geological conditions is thus necessary.

2.5. Zonation format

Codes in force assign generally one single value for the intensity, or peak ground acceleration of a zone specified. In an increasing number of cases the return period attributed is also specified. Given the needs of design in relation to the design criteria (that are dealt with in Section 3), more information on the seismic conditions is necessary. A possible approach (using the concept of return period, in spite of the reservations expressed previously)is to develop in a code zonation maps for several return periods (e.g.: 10, 100, 1000 years) and to prescribe corresponding rules of interpolation. Another approach is to specify, for any zone of the map, the intensities, or accelerations etc. corresponding to various return periods.

2.6. Addenda to code specifications

Keeping in view previous comments, it turns out that response spectra, which are (and will remain in the previsible future) the most frequently used representation of seismic conditions, cannot provide sufficient input information for structural analysis, especially in cases when a consistent 3D input is required. Two suggestions can be kept in view in order to surpass this shortcoming. On one hand, it is desirable to prescribe (as in the Annex C of the Romanian code [19]) explicit disturbing rotations of the ground-structure interface, in cases when the interface behaves as a rigid body. On the other hand, it is desirable to prescribe also space correlation or ˙ coherence characteristics, aimed to characterize the non-synchronousness of ground motion along a same direction, at different ground-structure contact points. A proposal presented in [8] could be considered in this connection, in order to develop full information on the input in case of any kind of ground-structure interface.

3. Design criteria

3.1. Limit states and importance classes

Advanced codes consider several limit states and, in some cases, specify also different values of design accelerations to be used for verification against exceeding limit states referred to. There is nevertheless a lack of an integrated approach, which should consider in a consistent way various limit states (that may considerably differ for different functions of structures), the concern for importance (in more detail, for the functional importance, for potential effects of exceedance of limit states), for the specific equipment with which a structure is endowed, for the role of a structure in a definite larger system (e.g. a lifeline). A consistent approach should be that of specifying in a consistent manner return periods of design accelerations etc. with a differentiation with respect to all factors referred to.

Another important aspect, of qualitative nature, is the need of a flexible approach to the calibration of design criteria, keeping in view, as previously mentioned, the possible role of a structure
in a larger, integrated, system, and/or the protection needs for equipment, for some specific architectural components etc.

3.2. Admissibility of post-elastic deformation

A problem of apparently qualitative nature that is obviously linked to design criteria is that of the admission of post-elastic deformation. This may be converted (as in EC-8 for r.c. structures) also to a problem of quantitative nature, in case one specifies the admissible ductility demands. Options in this field must be linked to two fundamental aspects: the application of the capacity design philosophy and the consideration of the possibilities of exceedance of the specified design loading. This latter aspect is further on dealt with, in connection with the possibility of use, at least for some limited goals, of a more consistent probabilistic approach then the first level one (referred to as P.1).

3.3. Higher order probabilistic approaches

The format of safety verification rules corresponds currently in all codes to the first level approach. This involves some considerable shortcomings from the view point of the real requirements of the engineering profession. The use any fixed level of action severity implies a lack of concern for the sensitivity of structural performance with respect to a possible exceedance (which always exists) of the levels prescribed. There is no possibility of direct quantification of the risk of exceedance of various limit states. The approach is limited practically to the consideration of mechanical phenomena and does not permit a direct linkage to the analysis of risk of losses of various natures, which is of central importance for the calibration of earthquake protection criteria. The first level approach is more or less compatible with the design spectrum representation (R.3), but its use in connection with other representations (R.1),(R.2) involves a lack of consistency, primarily due to the multi-parameter nature of representations R.1 and R.2, which makes it difficult to state e.g. whether an accelerogram is more severe than another. The use of the second level probabilistic approach (in FOSM format) cannot compensate for all these shortcomings, so the solution to be envisaged is to use, at least in thinking about code development, the P.3 philosophy.

A full use of the P.3 approach makes it possible to finally perform explicit estimates of risk or, conversely, of safety and reliability, putting to evidence, among other, the dependence of
the outcome on the duration of exposure, or of service. Such estimates are derived on the basis of appropriate convolutions between the characteristics of seismic hazard (or of recurrence of events with various severity levels) and the characteristics of seismic vulnerability. This leads to the conclusion that the basic objective of engineering analyses concerning structural performance should be to determine vulnerability characteristics of structural systems, to serve as an input for the convolutions referred to. **This has direct implications on the manner in which structural analysis is to be carried out** and some further comments are presented on this subject in Section 4.

One cannot imagine, for the near future, that the P.1 approach can be widely replaced in practice by a P.3 approach. The P.3 approach must be kept in view nevertheless by code writers, since this would considerably contribute to the development of more consistent codes. In the longer term the P.3 approach should gradually pervade also practical activities, as this happens in other engineering branches.

4. Analysis of structural performance

4.1. Modelling of structures

A common factor to be considered in this connection in all cases is the need of representing in some appropriate way the interaction between ground and structure. Neglecting this important aspect may lead to coarse errors in the estimate of natural periods or of distribution of forces between various bearing components. In most of the cases interaction can be taken into account in a satisfactory way without greatly adding to the sophistication of analyses. Starting from the substructure approach it is possible to define for simple portions of the ground-structure interface stiffness characteristics (and, eventually, inertia and equivalent viscous stiffness ones).

Modelling of structures in case of linear behaviour does not raise, as a rule, special problems. On the contrary, in case of non-linear behaviour considerably additional problems are raised.

Since the analysis of non-linear performance is of recognized importance, structural modelling must deserve special attention. It must be kept in view, in this connection, that the models developed must account as a rule for 3D behaviour since, even in case of dynamically symmetrical structures, it is not correct to use superposition of results concerning oscillations in different subspaces (e.g.: oscillations in different orthogonal vertical planes representing planes of dynamic symmetry). On the other hand, non-linear analysis requires the use of realistic constitutive laws, which should correspond, in most cases, to multi-parameter states of loading and deformation. It is known that, unfortunately, the literature does not offer sufficiently rich information for the various situations that can occur in practice. **Nevertheless codes must specify requirements aimed to lead to a satisfactory structural modelling especially in cases when non-linear oscillations are to be analyzed.**

4.2. Modelling of seismic action for time-history analyses

Three representations of actions, referred to as R.1, R.2, R.3, were dealt with in Section 2. Following comments are related to the representation R.2, which is the one appropriate for non-linear dynamic analyses. Codes hesitate in general between requiring the use of natural (recorded) or artificial (generated) accelerograms. There is a tendency in practice to give natural accelerograms a credit they do not deserve, given the fact that they will not be repeated in future. The instrumental data obtained in Romania put to evidence the considerable differences that occur between accelerograms recorded at a same place during different earthquakes originating more or less in the same source zone [5]. In more general terms, it should be the task of education provided to structural and earthquake engineers, to emphasize the considerable differences between observed actions and design actions.

To discuss in more specific terms the use of design accelerograms, it is necessary that they be compatible with the goals of structural analyses. In case one adopts the P.3 probabilistic approach, as discussed in Section 3, it turns out that design accelerograms should be specified in a way compatible with the needs of engineering vulnerability analysis. Design accelerograms should satisfy in this case two basic requirements:

a) to cover (in discrete terms) the whole domain of possible values of parameters that characterize in a macroscopic way the (multi-component) accelerograms (e.g.: peak ground acceleration or effective peak acceleration, peak ground velocity, relevant duration etc.);

b) to be specified in a way that provides, for each system of values referred to at item (a), several sample accelerograms, in order to perform analyses according to a Monte-Carlo philosophy and to perform, at the end, a statistical analysis of results, such as required for deriving vulnerability characteristics of structures dealt with.

4.3. Modelling of action for static biographical analyses

Static biographical analyses are not considered in codes at a level in agreement with their importance. Indeed, such analyses can be extremely useful from at least two viewpoints:

a) in order to provide data for condensation of structural models used in time-history analyses, which is of high practical interest, given the huge amount of time involved by time-history analyses, especially in cases when all requirements of sections 4.1 and 4.2 are kept in view;

b) in order to provide data required by capacity design.

Static biographical analyses organized for any of the purposes referred to should be performed for all alternative loading schemes judged to be relevant for a structure dealt with, applying, of course, a multiplying factor to the loads defined.

4.4. Analyses for non-structural components.

Some non-structural components, which should deserve special attention in earthquake resistant design, are more or less disregarded in many codes. The most important components of this category are as a rule non-building, equipment components, playing special roles. There exists a wide range of possible variants of performing analyses of non-structural components, depending on representations adopted, on the domain of performance (linear or non-linear), of the dynamic features of components, on the way in which they are supported or connected etc. It is necessary to provide in codes at least a general view of the problem, with consideration of the possible representations, models, techniques etc., as well as of a design strategy defining the level of protection (e.g.: should a kind of equipment keep its function up to the eventual collapse of the main structure ? is it necessary to consider in non-linear analysis the overstrength of the main bearing structure ?).

5. Developments connected with the capacity design philosophy

5.1 Capacity design versus quantitative verifications.

Capacity design is rightly recognized as a powerful design tool, able to provide a good earthquake performance of structures. Yet, there exists a tendency to forget that, even in case capacity design was correctly implemented, a structure cannot withstand unlimited ductility demands. The ability of structures to withstand ductility demands should be checked in a probabilistic way, such as to make sure that the risk to exceed the existing reserves is limited at an acceptable level. Several case studies show that there exists often a kind of instability of ductility demand, i.e. while for relatively moderate ductility demand the amplification of accelerograms leads to a moderate increase of the demand itself, for high ductility demand an eventual limited accelerogram amplification may lead to a type of instability i.e. to very fast increase of ductility demand. Another aspect, illustrated e.g. in [6], refers to the statistical aspects of scatter of ductility demand. The scatter of response observed (which is constant for a given set of sample accelerograms, in terms of coefficients of variation, as far as the accelerograms depend on a single multiplicative parameter and the behaviour is linear) tends initially to decrease for moderate average ductility demand, but tends afterwards to increase in an unstable way in case the average ductility demand is exceeding some moderate level. This increase of scatter means, of course, a loss of control with obvious implications for structural safety. To conclude, it must be stated that **capacity design should be accompanied, in order to limit risks, by some kind of suitable analyses aimed to control the risk of exceedance of available ductility.**

5.2. Capacity design versus 3D analysis

Capacity design is considered, in most cases, in relation to the design of frames. In this case it is usual to consider the bending moments around a beam-column joint and to make sure that the sum of ultimate moments in columns incident in a joint exceeds by a convenient margin the sum of ultimate moments in beams incident in the same joint. This condition may be satisfactory in case of plane frames loaded exclusively in their plane, but in case of spatial frames it is necessary to reconsider the situation, since in that case beams oriented in all directions, incident in a joint, join their ultimate moments to exceed the ultimate moment of columns incident in a joint referred to. Taking into account this case, which is not the single one, one may state that capacity design cannot be thoroughly implemented without the consideration of the 3D behaviour of structures.

5.3. Capacity design versus verification of foundations.

Code requirements specify quite frequently the need to avoid overstresses in ground due to the redistribution of pressures during earthquakes. Such requirements could be formulated in agreement with the capacity design language, asking that plastic deformation should occur in structures rather than in ground. This requirement may prove to be non-realistic and this may happen especially in case of taller shear wall buildings, for which there may exist considerable overstrength, able to impose post-elastic deformation to happen in the ground. This possibility cannot be considered automatically as an alarming situation. Some estimates based on instrumental data make it possible to conclude that, during the 1977 earthquake, many taller (about ten storeys tall) shear wall buildings should have rotated on the ground under conditions of transient, alternating, overpressure at the edges of foundations. After the earthquake uniform settlements of a few centimeters were measured at numerous such buildings and one may state that these settlements (produced in time) led to a recovery of practically full contact between foundations and ground. Comparative data on natural periods, obtained from microtremors prior to and after the earthquake have shown only slight increases of natural periods as far as the damage degree was slight. Consequently, it may be stated that the requirement of completely avoiding overpressure on the ground should be relaxed and replaced by the requirement that loss of overall stability of the active zone of the ground is to be avoided with a high degree of reliability. More concrete specifications on this subject should require, of course, in-depth studies on non-linear ground-structure interaction.

5.4. Capacity design versus calibration of behaviour factor

Behaviour factors, aimed to reduce seismic design loading as compared with what would correspond to linear design spectra are present in all codes, yet one may remark a difference in the philosophy of their calibration. While some codes prescribe a calibration based on direct experience and expert judgment, that may seen often unconservative if compared with an analytical approach, other codes, which are based on Housner's classical approach appear to be unduly conservative. Of course, any unconservative calibration is to be avoided in this field. Nevertheless, there is a source of even more dangerous underestimates in cases when rather courageous values are prescribed for the behaviour factor without consideration of the concrete mechanisms of post-elastic behaviour. As an example (and this is by far not the single code doing so) the Romanian code P.100-92 [19] prescribes low values (0.15 to 0.3) for the factor ψ (which corresponds to 1/q of EC-8) for usual reinforced concrete structures, without any warning, in case of prefabricated structures, that failure of joints prior to the post-elastic deformation of bent members is bound to lead to brittle failure of the structure as a whole. As a conclusion, **prescribing the calibration of behaviour factors must rely first on qualitative criteria aimed to confirm the appropriate implementation of the capacity design strategy and only thereafter on the examination of the ductility resources which show how far from a unit value of the factor referred to it is possible to safely go.**

6. Final remarks

A complementary aspect, not dealt with in previous sections, that deserves attention, is the fact that there must exist an appropriate relationship between codes for earthquake resistant design, as EC-8, and codes of general design principles, as EC-1. A survey of the current state of the art shows that the relationship between EC-1 and EC-8 should become more consistent, in the following sense: EC-1 should, of course, not enter details that are specific to earthquake protection, but the general principles expressed in EC-1 should not be in conflict with the earthquake protection needs. To give a few counterexamples on EC-1: the blunt classification of actions into "direct" and "indirect" is in clear disagreement with the need of considering interaction, e.g. between ground and structures; the system and format of characteristic and design values etc. is not well suited for dealing with such entities as design accelerograms; there is an obvious gap between the semi-probabilistic format and the capacity design philosophy and needs.

To conclude, the needs of consistent engineering analyses raised by the challenge of earthquake protection, that made earthquake engineering to play a leading role, being by now at the forefront of structural engineering, as well as of some other engineering branches, should be thoroughly considered and answered by developing appropriate tools.

References

1. Balan, St., Cristescu, V. and Cornea I. (editors): The Romania Earthquake of 4 March 1977. Editura Academiei, Bucharest, 1982.
2. Cornell, C.A.:Engineering Seismic Risk Analysis. Bull. SSA, 1968.
3. Crainic, L., Dalban, C., Postelnicu, T., Sandi, H. and Teretean, T.: A New Approach of the Romanian Seismic Code for Buildings.

Proc. 10-th WCEE, Madrid, 1992.
4. Lomnitz, C. and Rosenblueth, E. (editors): Seismic Risk and Engineering Decisions. Elsevier, Amsterdam - Oxford - New York - Tokyo, 1976.
5. Radu, C., Radulescu, D. and Sandi, H. :Some Data and Considerations on Recent Romanian Earthquakes. AFPS, Cahier Technique No.3,1990.
6. Sandi, H.: Contributions to the Theory of Structural Design. Thesis. Institute of Civil Engineering, Bucharest, 1966.
7. Sandi, H.: Conventional Seimic Forces Corresponding to Non-Synchronous Ground Motion. Proc.3-rd ECEE,Sofia,1970.
8. Sandi, H.: Stochastic Models of Ground Motion. Proc. 7-th ECEE, Athens,1982.
9. Sandi, H,: Conceptual Framework for a Full Specification of Local Conditions. Proc. 9-th WCEE, Tokyo - Kyoto, 1988.
10. Sandi, H. : Prerequsites for the Use of Artificial Accelerograms. Proc. 14-th RSEE, Ossiach, 1988, edited by ÖGE, Vienna,1989.
11. Sandi, H.: Alternative Stochastic Approaches to Seismic Response Analysis. Proc. SMiRT 10, Los Angeles, 1989.
12. Sandi, H.: Use of Instrumental Data for Evaluation of Ground Motion and for Specification of Seismic Conditions. Some Data on Recent Romenian Experience. Symp. on Earthquake Disaster Prevention, CENAPRED, Mexico City, 1992.
13. Sandi, H. : Implications for Design of the Spatial Nature of Seismic Action. Proc. 2-rd Turkish National Symp. on Earthquake Engineering, Istanbul, 1993.
14. Sandi, H.: Development of Romanian Earthquake Resistant Regulations.Some Suggestions for Work at an International Level. Proc. Symp. D-A-CH on Seismic Actions on Structures with Various Risk Potential; European Regulations. Weimar, 1993.
15. Commission of the European Communities. Industrial Processes. Building and Civil Engineering: Eurocode No.8. Structures in Seismic Regions. Design. Part 1. General and Building. May 1988.
16. International Conference of Building Officials: Uniform Building Code. 1991.
17. Applied Technology Council: Tentative Provisions for the Development of Seismic Regulations for Buildings. ATC 3-06, 1978, revised 1986.
18. IAEE: Earthquake Resistant Regulations. A World List. 1992.
19. MLPAT: Code for the Earthquake Resistant Design of Buildings and Industrial Structures, P.100-92, (in Romanian, Englishj translation in English also). Bucharest, 1992.
20. CEN: Basis of Design and Actions on Structures. European Standard. Ref. No. ENV 1991-1: 1993.
21. INCERC: Studies on the Development of a Realistic Seismic Zonation of the Territory of Romania (in Romanian). Res. Contract No.135/1990, 1993.
22. INCERC: Comparative Analysis of the Conceptual Basis and of the Calibration of Design Parameters in Earthquake Resistant Design in the Codes of Romania, Western Europe (EC-8), USA, Japan and New Zealand (in Romanian). Res. Contract No. 850/1993. 1993.

PART SIX
SEMI-RIGIDITY

46 SEMI-RIGIDITY: GENERAL REPORT

D. GEORGESCU
Civil Engineering Institute, Bucharest, Romania

Abstract
The first part of this general report presents a state-of-the art in the field of steel semi-rigid structures.
 The second part gives brief informations about contributions to Session 6 of the International Workshop and Seminar on Behaviour of Steel Structures in Seismic Areas - STESSA '94.

1 State of the art

A building is a complex in space system consisting of structural elements, beams and columns, and non-structural elements, cladding, floors and partitions, assembled together by means of different connection types.
 The actual response of such a complex system to various static and dynamic loading is complicated to be predicted, as depending on geometrical, material and loading effects and may be largely affected by a number of structural and non-structural interactions, the characteristics of which are not very well known yet. In addition, the structure itself is a complex three-dimensional system, whose response to different actions is deeply influenced by the behaviour of its beam-to-column joints.
 As a result, in order to assess an analysis model of a building some reasonably simple approaches are to be accepted, depending on the level of knowledge of the physical phenomena, as well as on the level to perform structural analysis.
 Looking back at the time when theory of structures started to be sketched and all calculations were conducted by hand it is easy to understand the necessity to produce very simple approaches. In this condition the following assumptions were accepted:

 The building was identified with its structure.
 The ideal model of 'pinned joint' and 'rigid joint' was adopted.
 All the floors were considered as perfect rigid diaphragms.
 The structural analysis was limited to the elastic range.

By introducing the 'rigid floor' concept the actual three-dimensional structure was replaced by a system of a lot of orthogonal two-dimensional frames and so the complex problem of an in space structure was reduced to the analysis of a frame sized to resist the vertical and horizontal forces acting in its plane. Consequently, for a long period of time, the most theoretical and experimental works were dedicated to understanding of the elastic frame behaviour in the plane of loading.

Nowadays the inelastic analysis of 3-D systems incorporating out-of-plane joint response may be performed and simplified procedures for second order analysis were developed. However, the 'rigid floor' concept is still strongly regarded, not only as a requirement to allow a simplified analysis, but as a basic concept in designing buildings in seismic areas. In order to ensure a rational behaviour under severe earthquake motions and to reduce the unfavourable torsional effects all the seismic codes are largely based on this concept.

As far as the beam-to-column connection is concerned it is worth to underline that the concept of 'semi-rigid' connection was known even some fifty years ago. Johnstone et al (1942) noticed that every 'rigid' connection does possess a certain degree of flexibility and every 'pinned' connection does possess a certain capacity of resisting bending moments.

In Figure 1 the angle θ results from:

$$\theta = \theta_b - \theta_c \qquad (1)$$

where θ_b and θ_c are the rotations of the beam and of the column respectively. In terms of θ different joint types are characterized by $\theta=0$ for rigid joints, $\theta_b>\theta>0$ for semi-rigid joints and $\theta=\theta_b$ for pinned joints.

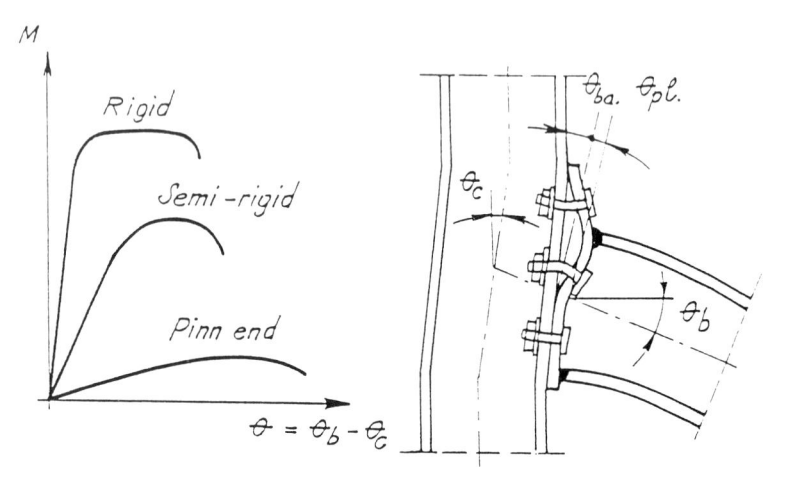

Fig.1. Classification of connections

For sake of simplicity the steel frame analysis generally refers to the limit cases of frames with rigid joints and frames with rigid pinned joints, but semi-rigid joints permitted by the codes were used in the current design for years ago.

The British code BS 449 (1959) permits three types of construction: simple design, semi-rigid design and fully rigid design.

Simple design assumes that the beams and girders are connected to columns for shear only and that their ends are free to rotate under gravity loading.

Semi-rigid design permits account being taken of the rigidity of the connections.

Providing that some conditions are satisfied, a reduction of 10% in the maximum bending moment in beam, assuming this is to be simply supported, is permitted. This simple alternative calculation is still allowed by the recent British code BS 5950 (1985).

Fully rigid design is based on the assumption that beam-to-column members are connected with sufficient rigidity to transfer the design moments with no rotation of the members relative to each other.

The American code AISC Specification (1969) permits three basic types of construction: Type 1, Type 2 and Type 3.

Type 1 construction (rigid frame) assumes that the beam-to-column connections possess sufficient rigidity to keep 'virtually unchanged' the original angles between the intersecting members.

Type 2 construction (simple framing) assumes that the ends of beams and girders are connected for share only and are free to rotate under gravity load.

Type 3 construction (semi-rigid framing) assumes an intermediate behaviour of connections between the Type 1 and Type 2.

In 1971 a very important test program was initiated at Lehigh University to study the behaviour of moment resisting beam-to-column connections. As the benefits of semi-rigid design were not clearly known, the tests were dedicated to rigid connections. This test program was very important in assessing design recommendations for rigid connections and, at the same time, in opening the way to the semi-rigid design.

The results of this test program were largely reported by Huang and Chen (1973), Huang et al (1973a), Chen et al (1974), Rentschler and Chen (1975), Parfitt and Chen (1976), Standing et al (1976a) and are summarized in the following:

(1) In the theoretical analysis the joint was modelled on the basis of a cantilever (Fig. 2) and the total deflection Δ at the end of the beam was expressed as the sum:

$$\Delta = \Delta_1 + \Delta_2 + \Delta_3 \tag{2}$$

where Δ_1 and Δ_2 are deflections of the beam due to bending deformation and shear deformation respectively, and Δ_3 is the deflection induced by the connection panel zone deformation. The panel zone deformation was calculated on the basis of a refined model (Fig. 3).

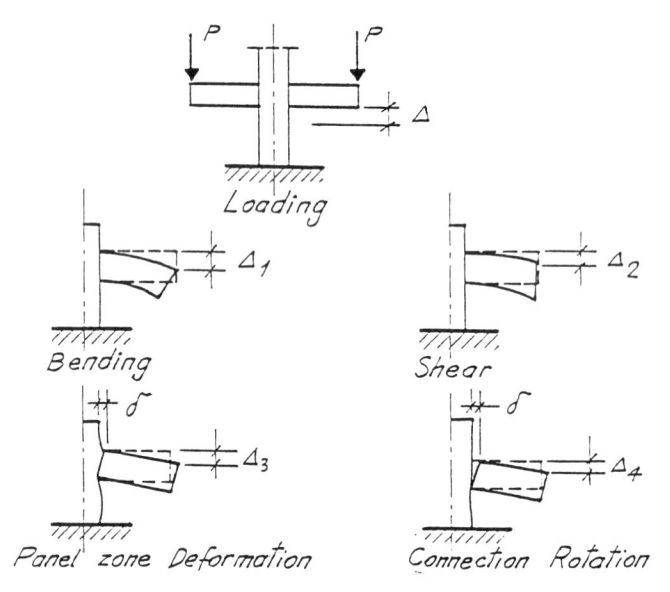

Fig.2. Connection deflection components

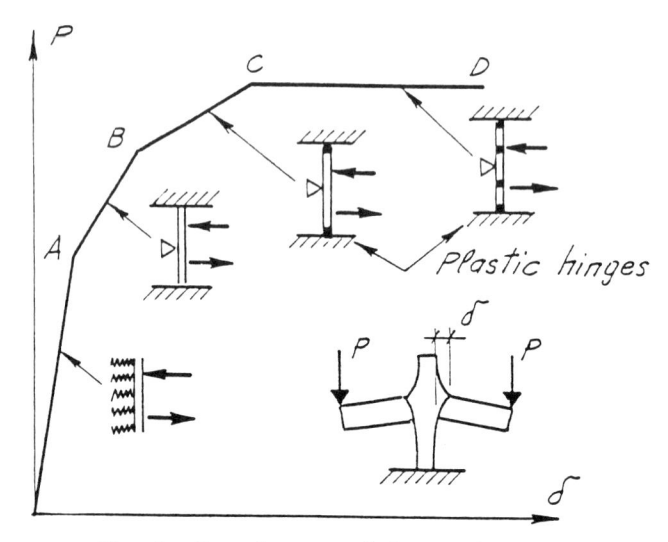

Fig.3. Panel zone deformation

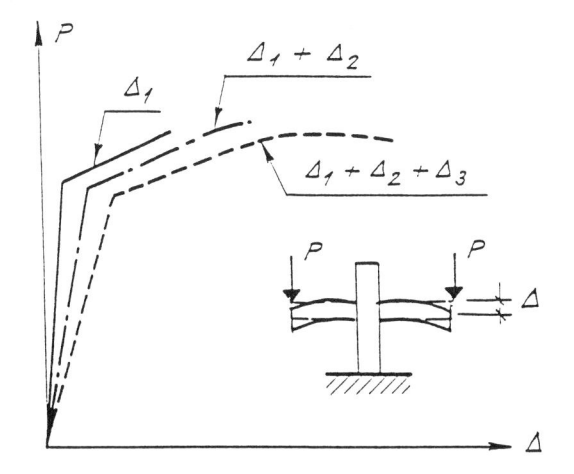

Fig.4. Predicted deflection components

Figure 4 shows the influence of the three computed deflection components on the total deflection Δ. A dominant effect of the panel zone deflection Δ_3 in the ultimate plastic range CD was observed. On the basis of this remark and on the results of other theoretical and experimental works, (Fielding and Chen, 1973b, Kato, 1974) in the recent codes special provisions concerning the shear stiffening of the panel zone were introduced. At the same time, the experimental results clearly showed that in fact all connections behave like semi-rigid ones and differ each other by their degree of rigidity.

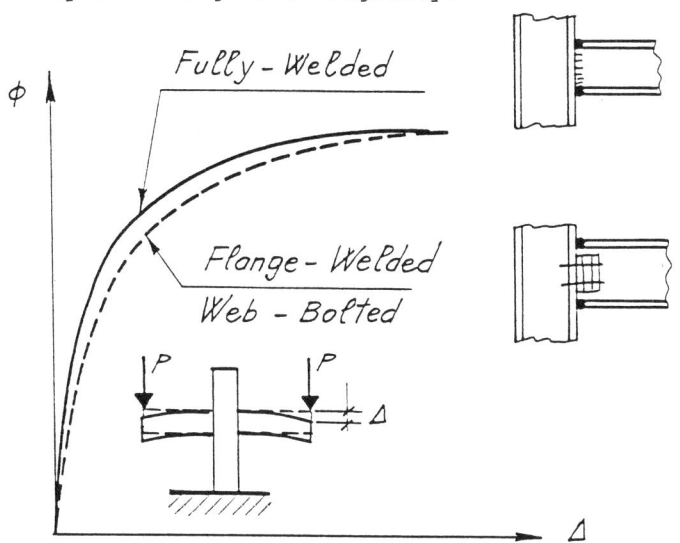

Fig.5. Load-deflection diagrams

(2) Figure 5 shows the experimental results obtained in the test program on a fully-welded web-bolted connection and on a flange-welded web-bolted connection and put in evidence that the replacement of the welded web connection by a bolted one does not affect the M-θ behaviour. On the basis of this experimental results and, as the advantages of bolted connections in decreasing fabrication and erection costs were very well known, the interest in bolted connections increased and research on these types of connections was developed. Replacing the welded-flange by bolted-flange, the connections became more flexible (Figures 6,a and 6,b), and so the way to semirigid connections was opened.

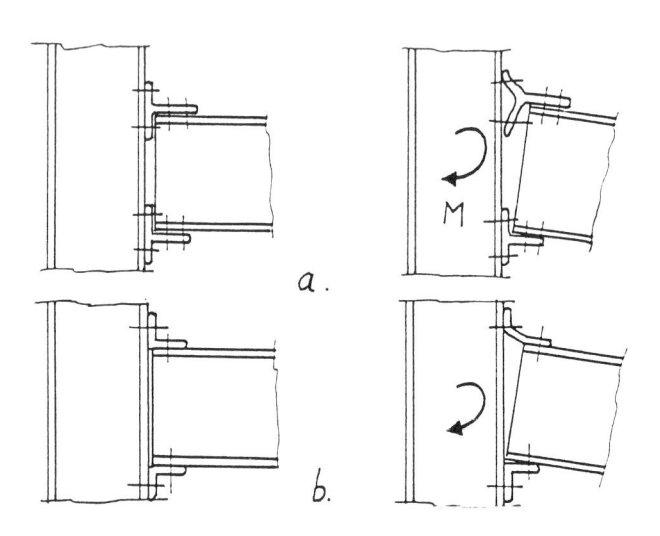

Fig.6. Flexible connections

In the last years, the cost of labour has increased more rapidly than the cost of the material and, consequently, in order to produce more economic structures, designers were determined to balance material and labour costs. As a result, the interest in developing semi-rigid connections increased. Besides this economical reason, the development of the semi-rigid connections theory was encouraged by the development in computer technology, allowing the analysis of complex systems.

In the past years, extensive theoretical and experimental works have been carried on to investigate the main features of the semi-rigid connection behaviour from both of theoretical and of constructional point of view, in order to offer to the designers all the informations needed to take full advantage from the semi-rigid structures.

With regard to the present situation in the field of semi-rigid construction it is to be observed:

(a) Nowadays it is quite generally accepted that any connection has to be considered as a semi-rigid one and consequently, the tendency to replace the classical classification (rigid, semi-rigid and pinned connections) by a single class of semi-rigid connections is to be observed. The recent American code AISC LRFD (1986) reflects this tendency by introducing only two types of construction, i.e. type FR (fully restraint) and type PR (partially restraint), including in the type PR both semi-rigid and pinned connections.

(b) Taking into consideration the recent theoretical and experimental results and the experience in semi-rigid construction, the recent codes permit the use of the semi-rigid connections.

The British standard BS 5950(1985) allows the use of the semi-rigid connections on the base of experimental evidence.

The American code AISC LRFD (1986) permits the use of semi-rigid connections on the evidence of predictable proportion of full end restraint.

The European code Eurocode 3 (1992) permits the use of the semi-rigid connections subjected to static loading. The recent American AISC Seismic Provision (1992) allows the use of the semi-rigid connections in ordinary moment frames located in seismic zones.

A first proposal for a recommended design procedure regarding semi-rigid frames in seismic areas has been developed by Astaneh and Nader (1992e).

(c) A large number of test programs has been conducted for the purpose of defining moment-joint diagram M-Φ under monotonic loading (Lewitt et al, 1966, Azizinami, 1987, Ballio et al, 1987a, Kiski and Chen, 1990, Jaspart, 1991). The experimental diagram M-Φ puts in evidence three phases i.e. elastic, plastic and hardening phases (Fig. 7).

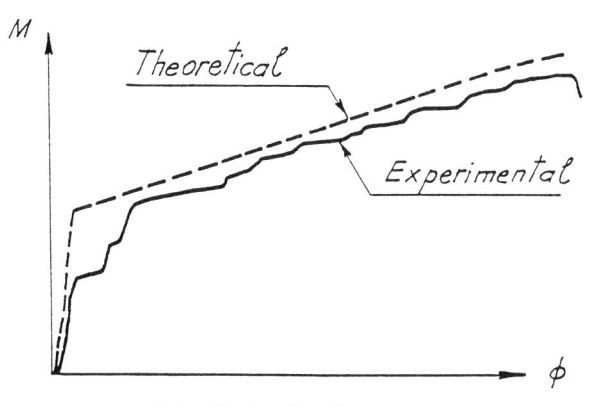

Fig.7. M-Φ diagram

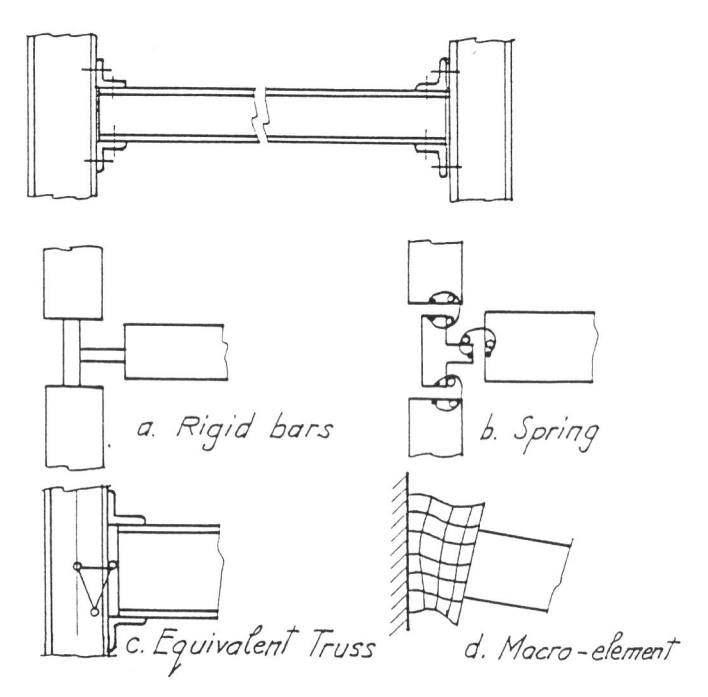

Fig.8. Joint models

On the basis of different joint models (Fig. 8) and using the experimental data, various theoretical relationships were proposed in order to predict in full security (Fig. 7) the actual behaviour of the semi-rigid joints in all phases. On the basis of theoretical and experimental works, Eurocode 3 (1992c) admits the use of semi-rigid connections subject to static loading and gives in the Annex J the M-Φ relationship.
In an important contribution to Session 6, De Stefano, Bernuzzi, De Luca and Zandonini present a very interesting comparison between Eurocode 3 and other formulations and some improvements to Eurocode 3 provisions are proposed.
(d) The behaviour of the semi-rigid connections under cyclic reversal loading are examined in some research works (Ballio et al, 1987a, Astaneh, 1991a, Bernuzzi, 1992a and 1992b). The tests performed have proved that the behaviour of the semi-rigid connections under cyclic loading is largely affected by the bolt slippage.
Two contributions are dedicated to this aspect:
The work by Bernuzzi, De Stefano, De Luca and Zandonini presents the results of a very interesting test program concerning the behaviour of a top-and-seat angle connections under cyclic reversal loading. A new concept, 'slip capacity design', is proposed in order to avoid the unfavourable effects of the slippage of bolts.

Semi-rigidity: general report 527

The work by Grecea and Mateescu refers to the experimental results concerning the behaviour of semi-rigid connections under alternate static loads.

(e) In analysing a semi-rigid moment resisting frame, the moment-rotation M-Φ behaviour must be included, i.e. the deflection Δ_4 due to the connection rotation in addition to the deflection Δ_1, Δ_2 and Δ_3 in Figure 2 is to be considered.

Analyses may be conducted at any required level of refinement, depending on the choosed joint model (Fig. 8).

The spring model is largely used, as in this model the incorporation of the joint behaviour does not increase very much the size of the problem. This model is used by Lewitt in a very important contribution. An original method for advanced analysis of semi-rigid frames, called 'notional plastic hinge' is proposed. The spring model is also used in the contribution of Dubina, Grecea and Zaharia.

An original procedure, based on the thermodynamics of irreversible processes, is proposed by Flejon.

A simplified procedure based on the model of a substructure extracted from the general structure is proposed by Faella, Piluso and Rizzano.

(f) The connection rotation largely affects the rigidity of the structure, the sensivity at the second order effects and the dynamic response.

As far as rigidity (drift condition) and second order effects are concerned, it seems that, presently, is quite generally accepted to limit the use of the semi-rigid frames to 10-15 stories.

With respect to dynamic behaviour of semi-rigid frames in seismic zones there are a lot of problems including a general structural concept to be discussed. First of all it is to underline that rigid joint is not only a type connection, but a part of the 'capacity design' concept. This structural concept assumes plastic hinges locations only in beams and at the base of columns, and requires the formation of the beam hinges not at beam-column contact face, but in the imediate vicinity of the joint. As a consequence, the recent American Code AISC Seismic Provisions (1992) requires beam-to-column connections with full penetrated welded flanges, i.e. a fully restraint connection.

The tendency to replace rigid joints by semi-rigid ones, which allows large rotations into the connection itself, requires, besides the full informations concerning the behaviour of such a connection type, under severe erthquake motions, a new general structural concept. In the writter opinion, such a concept is on the way to be constructed.

Recently, a first proposal for a recommended design procedure has been developed (Astaneh and Nadar, 1992e).

The provisions included in this recommandations are mainly based on the investigation of the behaviour of an one sto-

rey one bay steel structure with semi-rigid connections subjucted to base excitations (Astaneh and Nader, 1991a) as well as on the results of the analysis by means of numerical simulations of a 4-storey, 7-storey and 10-storey semi-rigid frames (Astaneh and Shen, 1990a).
On the basis of this theoretical and experimental works, a main recommandation regarding the concept of the semi-rigid connections has been proposed in the meaning that a semi-rigid joint must yield only in the plate elements of the connection, while all the connecting bolts have to remain in the elastic range.
In the general reporter opinion, in addition to this recommandation, a limitation of the connection rotation is to be imposed.
In the contributions to this session, important recommandations refering to the connection behaviour were proposed by Bernuzzi, De Stefano, De Luca and Zandonini by introducing the concept of 'slip capacity design', and by Faella, Piluso, Rizzano regarding the use of full strength connections in seismic areas.
Taking into account both the effects of the recent January 1994 Northridge earthquake and the poor experience in using semi-rigid connections in seismic areas, a very interesting inovative semi-rigid frame solution is proposed at this Workshop in the paper presented by Astaneh. The system consists of 'column trees' and floor girders connected to each other by ductile semi-rigid connections located away from the face of the column. The proposed system represents an attempt to combine the experience in using rigid connection frames with the benefits of the semi-rigid connections.
(g) In the last years, an increasing interest in quantifying the economical benefits of semi-rigid connections is to be observed. Recently, Colgon and Bjorhovde (1992c) have investigated the advantages of the semi-rigid connections in comparison with fully rigid connections, finding a saving in cost of 20% in a moment resisting frame and 10% in a concentrically braced frame respectively.
(h) In a very interesting contribution to this session, Pinto outlines the research program pf the COST-C1 seismic working group, dedicated to assess the seismic behaviour of civil engineering structures designed and/or constructed with semi-rigid connections. The performed and ongoing works and group numbers are presented.
(i) During past years, numberous theoretical and experimental works were dedicated to the use of semi-rigid connections in seismic areas, a subjected of extensive discussion. The papers presented to Session 6 of the Workshop represent an important contribution in improving the state of knowledge related both to the main parameters affecting the cyclic response of the semi-rigid connections and to the behaviour of semi-rigid frames.

2 Survey of conference papers

The papers allocated to Session 6 are:

(1) **Amado, C. Benussi, F. and Noe, S.** Behaviour of unbraced semi-rigid composite frames under seismic actions.
In this paper the seismic response of semi-rigid composite frames is analysed on the basis of a numerical analysis performed on three bays frame. Four different kinds of beam-to-column connections has been considered for the internal joints. Three of them are semi-rigid composite and the fourth is a full strength steel connection. The external joints have been always considered as pinned. The response to seismic actions has been numerically analysed under the four registered ground motions, assuming a mass proportional damping equal to 3% of the critical value. By subdividing the behaviour factor q in the overstrength factor q_Ω and ductility factor q_μ a very important conclusion for the design purposes results. The semi-rigid composite frames resist seismic actions especially due to their high overstrength factor ($q_\Omega \sim 2q_\mu$) since, on the contrary, for semi-rigid frames the ductility factor q_μ is dominant ($q_\mu \sim 2q_\Omega$).

(2) **Astaneh, A.** Seismic behaviour and design of steel semi-rigid structures.
The paper presents the major results of the research works performed by the author since 1986, as well as a summary of seismic code procedures and their application in design. An inovative semi-rigid frame solution is proposed. This solution is based on the damage phenomena observed after the January 1994 Northridge earthquake and on the large experience of the author in designing semi-rigid frames in seismic areas.

(3) **Bernuzzi, C. De Stefano, M. De Luca, A. and Zandonini, R.** Moment-rotation behaviour of top-and-seat angle connections.
In this paper the experimental behaviour of top-and-seat angle connections under cyclic reversal loading is analysed. The response of a lot of top-and-seat angle connections is examined with regard to their behaviour and to the possibility of introducing a model to predict the cyclic response. Preliminary to the cyclic analysis, one test under monotonic loading was performed, in order to define a conventional yielding displacement. The tests were performed controlling the displacements and accepting four reversal loading history. Observing that the cyclic response is largely influenced by slippage of bolts, a design philosophy called 'slip capacity design' is proposed, in order to avoid the bolt slip before the yielding of the angle legs. Finally, a numeric model of calculation able to describe the actual cyclic behaviour is described.

(4) **Liew, J.Y.R. and Chen, W.F.** Advanced analyses for

semi-rigid frame design.
An original procedure for advanced analyses of semi-rigid frames, called 'notional load plastic hinge method', is discussed in this paper. This method takes into account the effects of the initial geometric and structural imperfections by means of an equivalent initial sway imperfection Φ. In calculating moment resisting frames, MRF, the initial sway imperfections are replaced by equivalent horizontal notional forces $\Phi\Sigma P_i$, representing a fraction of the total gravity loads ΣP_i acting on each storey of the building. The notional forces $\Phi\Sigma P_i$ are to be applied at all level, in addition to the design lateral load. The proposed method, representing a second order plastic analysis for frame design, is utilized in the paper to design frames with semi-rigid connections including the non-linear moment rotation behaviour of the connections. A relationship to describe the moment-rotation characteristic, calibrated against experimental data, is proposed.
(5) **Stefano, M. Bernuzzi, C. Luca, A. and Zandonini, R.** Semi-rigid top-and-seat cleated connections: A comparison between Eurocode 3 approach and other formulations.
In this paper a very important and extensive analysis with regard to some prediction methods for semi-rigid connection behaviour is performed. A comparison between the EC3 approach and other formulations leads to very important conclusions with reference to the behaviour of top-and-seat angle connection, a joint type largely used in current design. The design approach of the Annex J to EC3 is presented into detail. A particular attention is paid to slippage, bearing and strain hardening, in order to realise a more realistic prediction of the moment-rotation curve.
(6) **Dubina, C. Grecea, D. and Zaharia, R.** Numerical simulation concerning the response of steel frames with semi-rigid joints under static and seismic loads.
The results of some numerical simulations with regard to the response of steel frames with semi-rigid joints under static and dynamic loads are presented in this paper. Four different types of calibrated frames are investigated. The influence of the initial stiffness on the bilinear M-Φ curve on the structure static behaviour and the influence of the semi-rigid joints on the seismic analysis are discussed. Some interesting conclusions result from the performed analysis concerning the sensibility at second order effects, the influence of the initial stiffness of the connections on the period of vibration, the behaviour factor q and the yield mechanism.
(7) **Faella, C. Piluso, V. and Rizzano, C.** Connection influence on the seismic behaviour of steel frames.
In this paper the influence of the beam-to-column connections on the seismic behaviour is analysed, considering both full strength and partial strength semirigid connections. The analysis is performed on a simplified model of

a substructure, extracted from the general structure and consisting of a column between two floors and two adjacent beams, by assuming that beams are subjected to double curvature bending with zero moment in midspan. The results obtained by means of the simplified model are compared with the ones provided by the numerical simulation of the dynamic inelastic response to severe ground motions. A very important conclusion results with regard to the location of the discontinuity characterizing the passage from the partial strength to the full strength. In the writer opinion on the basis of the actual backround the use of the full strength connections in seismic area appears to be preferred.

(8) **Faella, C. Piluso, V. and Rizzano, C.** Connection in fluence on the elastic and inelastic behaviour of steel frames.
The influence of the beam-to-column connections on the elastic and inelastic behaviour of steel frames is investigated in this paper. A simplified model described in the previous paper was utilised in the analysis performed, both on strength and partial strength connections. The influence of the connection type on the period of vibration, the frame sensitivity to second order effects and the actual ductility are examined. On the basis of the performed analysis some important conclusions result, especially concerning a significant reduction of ductility in case of partial strength connections.

(9) **Flejou, J.L.** Seismic effects on steel frames with rigid joints: Modelling, Vibrations, Damping.
A semi-rigid connection model for steel construction, based on thermodynamical approach is proposed in this paper. The model permits a good understanding of the connection phenomena concerning non-linear behaviour, different stiffness under loading and unloading, and ultimate state. The model was implemented in a layered finite element program for frame static and dynamic analysis. The numerical analysis fits well the test results.

(10) **Grecea, D. and Mateescu G.F.** Behaviour of semi-rigid connections under alternate static loads.
This paper presents the experimental results of a research program developed on two types of semi-rigid welded and bolted joints under monotonous and alternate static loads. Some conclusions resulting from the tests are presented in the paper.

(11) **Pinto, A.V.** COST-C1-Seismic working group activities.
The research program of the COST-C1 seismic working group is presented in this paper. COST-C1 is an European research project, intended to study the behaviour of semi-rigid connections. A number of thirteen European countries are now participating in this project. The seismic working group research program and the working group members are presented in this paper.

3. References

AISC (1969) **Specifications for the Design, Fabrication and Erection of Structural Steel for Buildings**, American Institute of Steel Construction, INC, Chicago, Illinois

AISC (1992) **Seismic Provisions for Structural Steel Buildings**, American Institute of Steel Construction, Chicago, Illinois

AISC LRFD (1986) **Load and Resistance Factor Design Specification for Structural Steel Buildings**, American Institute of Steel Construction, Chicago, Illinois

Astaneh, A. and Nader, N. (1991a) Cyclic behaviour of frames with semi-rigid connection, **Proceedings** of the Second International Workshop Connections in Steel Structures II: Behaviour, Strength and Design, Pittsburg, Pennsylvania, April.

Astaneh, A. and Nader, N. (1992e) Proposed Code Provision for Seismic Design of Steel Semirigid Frames, submitted to **AISC Engineering Journal** for Review and Publication.

Astaneh A. and Shen, J. (1990a) Seismic Response Evaluation of an Instrumented Six Storey Steel Building. Earthquake Engineering Research Center, **Report UBC/EERC-90/20**, University of California at Berkeley, August.

Azizinamini, A. Bradburn, I.H. and Radziminski, I.B. (1987) Initial stiffness of semi-rigid steel beam-to-column connections. **J. Construct. Steel Res.**, 8, 71-90.

Ballio, G. Calado, L. De Martino, A. Faella, C. and Mazzolani, F.M. (1987a) Cyclic Behaviour of Steel Beam-to-column Joints Experimental Research. **Costruzioni metalliche**, No. 2

Bernuzzi, C. (1992a) Cyclic Response of Semi-rigid Steel Joints. **Proceedings**, First State of the Art Workshop Semi-rigid Behaviour of Civil Engineering Structural Connections, COST-C1, Strasbourg, France.

Bernuzzi, C. Zandonini, R. and Zanon P. (1992b) Semi-rigid Steel Connections under Cyclic Loads. **Proceedings**, First World Conference on Constructional Steel Design, Acapulco, Mexico.

Bernuzzi, C. Zandonini, R. and Zanon P. (1991) Rotational Behaviour of End Plate Connections, **Costruzioni metalliche**, No. 2.

Bijloard, F.S.K. Nethercot, D.A. Stark, I.W.B. and Tschemmernegg, F. (1989) Structural Properties of Semi-rigid Joints in Steel Frame, IABSE, **Survey** S-42/89, May.

BS 449 (1959) **Specification for the Use of Structural Steel in Building**, British Standard Institution.

BS 5950 (1985) **Structural Use of Steelwork in Building.**, Part 1, Code of Practice for Design in Simple and Continious construction: hot rolled sections, British Standard Institution.

Chen, W.F. Huang, T.S. and Beedle, L.S. (1974) Recent Results on Connection Research at Lehigh. **Proceedings** of the Regional Conference on Tall Buildings, Bangkok,

Thailand.
Colson, A. et Bjorhovde, R. (1992b) Interet economique des assemblages semi-rigides, **Construction metallique**, Nr. 2.
Eurocode No 3 (1992c) **Design of Steel Structures**, European Committee for Standardization.
Fielding, D.J. and Chen, W.F. (1973b) Steel Frame Analysis and Connection Shear Deformation, **Journal of the Structural Division**, ASCE, Vol. 99, No ST1, Proc. Paper 9481, 1-18.
Huang, J.S. and Chen, W.F. (1973) Steel Beam-to-column Moment Connections, **ASCE National Structural Engineering Meeting**, No. 1020, San Francisco.
Huang, J.S. Chen, W.F. and Beedle, L.S. (1973a) Behaviour and Design of Steel Beam-to-column Moment Connections. Bulletin 188, **Welding Research Council**, New York.
Jaspart, J.P. (1991) Study of Semi-rigidity of Beam-to-column Joints and of Its Influence on the Strength and Stability of Steel Frames (in French), Ph. D. Thesis, University of Liege, Dept. of Applied Sciences, Belgium.
Johnstone, B. and Mount, E.H. (1942) Analysis of Building Frames with Semi-rigid Connections, **ASCE Transactions**, Vol. 107, 993-1019.
Kato, B. (1974a) A Design Criterion of Beam-to-column Joint Panels, New Zealand National Society for Earthquake Engineering **Bulletin**, Vol. 7, No. 1.
Kishi, N. and Chen, W.F. (1990) Moment-rotation Relations of Semi-rigid Connections with Angles. **J. Struct. Engrg.**, ASCE 116(7), 1813-1834
Lewitt, C.W. Chesson, E. and Munse, W.H. (1966) Restraint Characteristics of Flexible Riveted and Bolted Beam-to-column Connections. **Dept. Civ. Engrg.**, University of Illinois, Urbana.
Liew, J.Y.R. White, D.W. and Chen, W.F. (1993) Limit States Design of Semi-rigid Frames Using Advanced Analysis. Part 1: Connection Modelling and Classification, Part 2: Analysis and Design, **J. Constructional Steel Research**, Elsevier Science Publishers, London.
Liew, J.Y.R. White, D.W. and Chen, W.F. (1993a) Second Order Refined Plastic Hinge Analysis for Frame Design. Part 1 and Part 2, **J. Structural Engineering**, ASCE, 119.
Mazzolani, F.M. and Piluso, V. (1993b) P-Δ Effect in Seismic Resistant Steel Structures. **Structural Stability Research Council**, Anual Technical Session and Meeting, Milwakee, Wisconsin, April.
Nethercot, D.A. and Zandonini, R. (1988) Methods of Prediction of Joint Behaviour: Beam-to-column Connections. Stability and Strength Series, Vol. 8, Connections, Ed. Narayana, R, **Elsevier Applied Science Publ.**, London.
Parfitt, J. and Chen, W.F. (1976) Tests of Welded Steel Beam-to-column Moment Connections, **Journal of the Structural Division, ASCE**, Vol 102, No. ST1, Proc Paper 11854, 189-200.
Rentscler, G.P. and Chen, W.F. (1975) Test program of Mo-

ment-Resistant Steel Beam-to-column Web Connections, Fritz Engineering Laboratory **Report** No. 405, 4, Lehigh University Bethlehem.

Standing, K.F. Rentschler, G.P. and Chen, W.F. (1976a), Tests of Bolted Beam-to-column Flange Moment Connections, **Welding Research Council Bulletin**, No. 218.

47 BEHAVIOUR OF UNBRACED SEMI-RIGID COMPOSITE FRAMES UNDER SEISMIC ACTIONS

C. AMADIO, F. BENUSSI and S. NOE
Department of Civil Engineering, University of Trieste, Italy

Abstract

The static behaviour of semi-rigid composite frames is sufficiently known to allow their use in non seismic zones. Some recent studies have shown that this type of frames which offer interesting structural advantages, can be utilised also in seismic zones. A direct relation exists between the characteristics of the cyclic response of the semi-rigid joints and the seismic resistant capacity of the frames. Even if their overall behaviour under horizontal forces may be considered as brittle the high hardening range of the joints leads to high values of the behaviour factors. The key parameter that determines the dissipative capacity of the frames is the ductility of the joints, generally related to the structural details of the steel part of the connection.

Keywords: Semi-rigid composite frame, seismic behaviour.

1 Introduction

Many theoretical and experimental studies have indicated that steel-concrete composite non-sway frames with semi-rigid joints represent a reliable and suitable structure for non-seismic zones. If compared with frames with rigid or pinned joints they offer significant improvements of the cost-effectiveness, as the structural continuity can be increased and calibrated at very limited cost simply by placing reinforcing bars in the concrete slab, even if the steel connection is very simple and flexible. Nevertheless, this type of frame is not yet commonly used also owing to the lack of simple prediction methods for its mechanical behaviour. Recently some important attempts to develop a consistent theoretical approach have been made, Amadio et al. (1993a), Arbed (1991). If the moment-rotation curve of the joint under hogging moment is known, the design of these frames under vertical loads is very simple. Usually the available ductility, as highlighted by static tests, is greater than the required and only the stability of the lower flange of the beam needs to be checked with simple methods, Amadio et al. (1993a). The beam-to-column interaction problem, Puhali et al. (1990), may be neglected if the concrete slab is in contact with the column and the loads are not strongly asymmetric, Amadio et al. (1989). The advantages that semi-rigid composite joints offer in braced construction could be exploited also for unbraced frames and for seismic zones, as they already are for steel structures, Astaneh and Nader (1993), but some difficulties arise in the appraisal of the mechanical behaviour of the joints. Their cyclic behaviour, with the real hysterethic loops, the available ductility and the effect of fatigue for low number of cycles must be determined. These characteristics, strictly related to the interaction between the concrete slab and the steel connection, are not easily deducible from a low number of tests or estimated by means of simple theoretical models. Presently only few experimental data are available for symmetric, Benussi et al. (1991) and asymmetric ,Leon (1990), load conditions. The tests have pointed out that, even for simple steel connections, the cyclic behaviour of the joints can be suitable and reliable for use under seismic actions. A numerical analysis identifying and checking the key parameters that affect this type of frames has been recently performed, Amadio et al. (1993b,1994), preliminary to a new especially programmed theoretical and experimental campaign. Based on the experimental data of some semi-rigid composite joints, a more accurate study of the seismic response of a 3 bay-4 floor frame is presented in this paper.

2. Numerical analysis on semi-rigid composite frames

2.1 Characteristics of the frames and of the numerical models

The three bays frame of figure 1 has been analyzed using four different kinds of beam-to-column connections for the internal joints. Three of them (CT1C, CT2C, CT3C in figure 2) are semi-rigid composite, the forth is a full strength steel connection (FSSC), without any contribution of the slab. The external joints have been always considered as pinned, in the hypothesis of slab not extended out of the external columns. Each of the four storey frames has been considered as one of the 4.8 m spaced frames of an office block with double corridors. The values of dead and live loads have been fixed in 4.0 kN/m^2 and 2.0 kN/m^2 respectively. The cladding weight is 7.5 kN/m. The semi-rigid joints CT1C,CT2C and CT3C have been designed

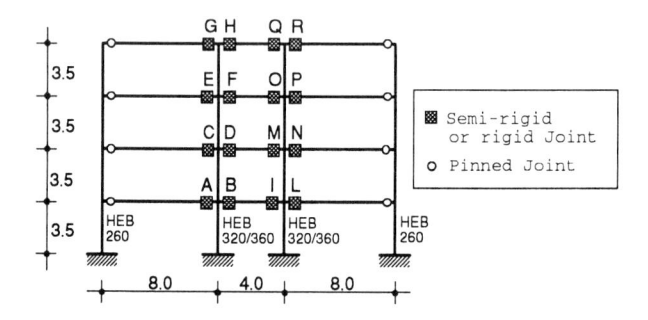

Fig.1. Semirigid composite frame

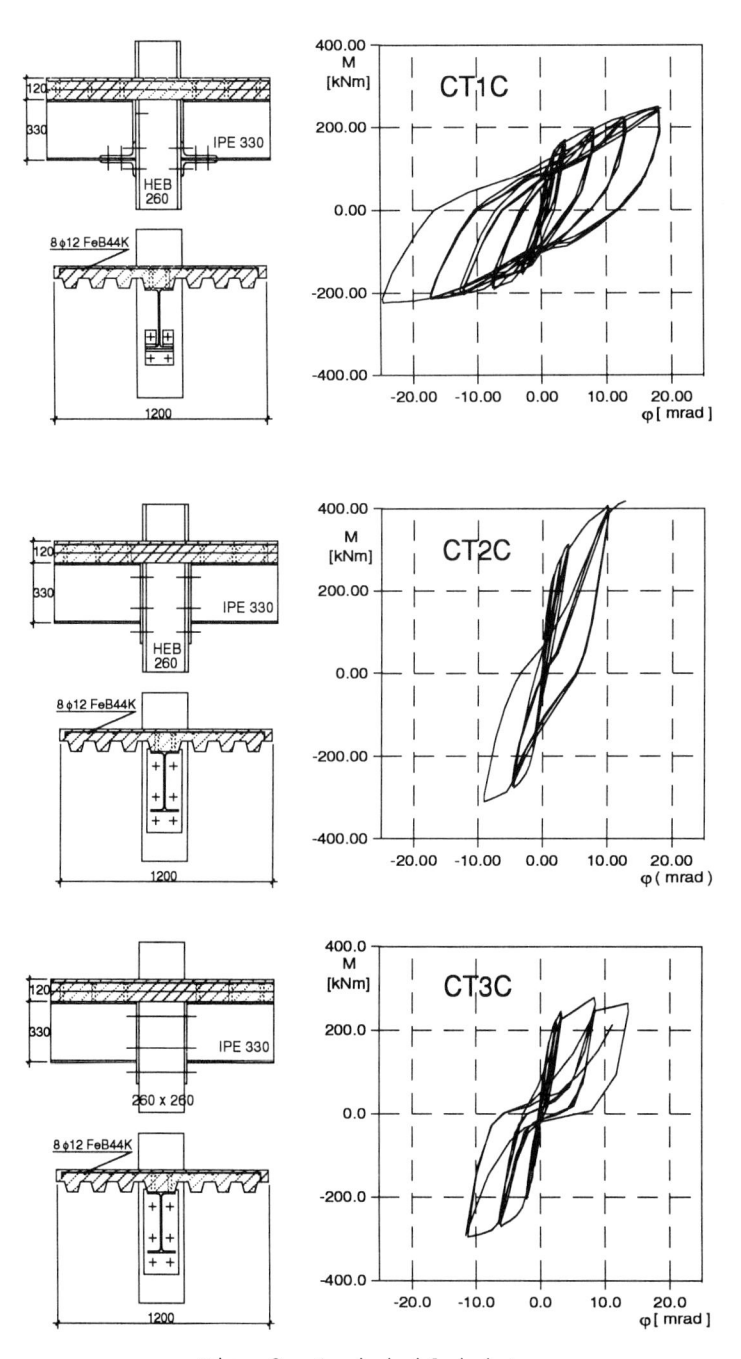

Fig. 2. Semirigid joints

to be used in braced frames assuming that the semi-rigid frames
will be loaded only by 30% of the horizontal seismic loads.
Therefore they have been designed to achieve a sufficient
strength under negative bending moments, with a positive moment
in the region of to the negative. No particular ductility
requirement has been considered.
Different specimens of the joints have been tested under
hogging monotonic, Bernuzzi et al. (1991), and cyclic bending
moments, Benussi et al. (1991), all with symmetrical loading.
For the cyclic tests the ECCS short loading procedure have been
followed. Some of the main results of the cyclic test are
summarized in table 1. The joint CT1C has a cleated connection
between the beam lower flange and the column. The collapse in
the cyclic test corresponds to the breakage of the lowest cleat
under sagging moment. The maximum positive rotation reached is
about 5 times the elastic value, while for negative rotation
the ratio is higher than 8. The over-strength (ratio between
collapse and yielding moment) is equal to 1.57 and 1.44 for
positive and negative moments respectively. The steel
connection of joint CT2C is obtained with an extended end
plate. The slab is equal to that of joint CT1C and CT3C. The
stronger steel connection increases the stiffness and strength
of the joint. The collapse under positive bending moment was
due to the fracture of the lower part of the end plate. The
bending collapse moment is higher than in CT1C but the values
of ductility are lower (3.3 for positive moment and higher than
6 for negative moments). Even for joint CT2C the over-strength
is remarkable (1.33 and 1.65 for positive and negative
moments). The joint CT3C differs from CT2C only because the
column is made up with a concrete filled square hollow steel
profile. The extended end plates are connected to the column by
means of screwed bars. The collapse for sagging rotations was
due to the fracture of the lowest bars. The loosening of the
bolted connection, due to the elongation of the screwed bars
before collapse, causes the evident pinching in the M-φ curve.
The maximum allowable rotation under positive bending moment
for the semirigid joints of the numerical models is the
experimental one (table 1). Under negative bending moment the
maximum value reached during the tests by one of teh joints
(CT1C) was adopted for all of them for these reasons: i) the
collapse of the three joints always occurred under positive
bending moment while the collapse under negative bending moment
was far to be reached; ii) the nominal ultimate hogging bending
moment of the three joints are very similar.

Table 1. Experimental response of the semirigid joints

Joint	φ_y^+	M_y^+	φ_u^+	M_u^+	φ_y^-	M_y^-	φ_u^-	M_u^-
	(mrad)	(kNm)	(mrad)	(kNm)	(mrad)	(kNm)	(mrad)	(kNm)
CT1C	3.4	158.9	18.5	250.8	3.0	155.7	24.7	224.5
CT2C	3.9	313.6	12.8	417.5	1.5	188.1	9.1	310.1
CT3C	3.2	245.3	13.7	278.3	2.1	204.0	11.6	296.1

The experimental values of the elastic limits (table 1) have been used in the analyses. They well correspond to the first plastic deformation in the steel connection (cleats, extended end plate or screwed bars) and the first yielding of the slab rebars for sagging and hogging moment respectively. It must be stressed that the experimental moment-rotation relationships used in the numerical analyses were obtained under symmetric loads, while seismic actions leads to asymmetric load conditions. This approximation can be however justified by the fact that the cyclic experimental response of the joints in semi-rigid composite frames under lateral load is qualitatively very similar to that here adopted, Leon (1990).The rigid joint FSSC has been employed as reference for the characteristics of the semi-rigid joints. The value of the plastic moment for the FSSC connection is 311 kNm, according to the experimental yielding stress for the IPE 330 steel beam. The behaviour of the joint, assumed as rigid-perfectly plastic, was determined neglecting the contribution of the concrete slab both in tension and compression, as indicated by one of the criteria defined by EC8 for dissipative zones in composite frames. The ultimate rotation of the plastic hinge is 25 mrad for positive and negative rotation according with experimental results obtained for IPE 300 steel beams under cyclic loads, Castiglioni and Di Palma (1989); the value is about half the experimental or the theoretical value, Kato (1990),for monotonic loads.

The composite full interaction beams of the frames have the same characteristics of the experimental specimens. In the numerical analysis, the beams have been considered elastic with stiffness equal to the 40% of the stiffness after slab cracking added to the 60% of the stiffness before cracking.

The frame CT1C has been designed for high sismicity zones according to EC8 code, with a behaviour factor of 5. The internal columns are HEB 320 profiles. In the other frames the slab and steel beam are the same of CT1C frame, whereas the internal columns are HEB 360 profiles in order to avoid the formation of plastic hinges in the columns.

Both static and dynamic numerical analyses have been performed by means of the finite element code Abaqus, including geometrical non-linearities. The non-linear moment-rotation relation in the semirigid joints have been simulated by a phenomenological model based on the experimental results, Noè and Spanghero (1993).

2.2 Response of the frames to monotonic and cyclic loads

A preliminary appraisal of the strength and ductility properties of the frames was obtained by analysing the response under a set of statically applied horizontal forces (vertically distributed according to the first modal shape) and the same amount of constant vertical loads considered for the dynamic analyses (100% of dead loads plus 30% of live loads). Table 2 summarizes the main results.

Table 2: Static responses of the frames

Frame	V_y/W	V_u/W	δ_y/h	δ_u/h	V_u/V_y	δ_u/δ_y	$\bar{q}$
CT1C	0.060	0.229	0.0021	0.0173	3.816	8.38	6.6
CT2C	0.108	0.350	0.0027	0.0160	3.241	5.90	4.9
CT3C	0.120	0.289	0.0034	0.0134	2.402	3.96	3.4
FSSC	0.143	0.321	0.0031	0.0215	2.247	6.99	4.9

In figure 3a, where the ratio between the base shear V and the weight of the frame W is plotted vs the ratio of the top floor drift δ and the total height of the frame h, the higher initial stiffness and ductility of the FSSC frame is evident. Frames CT2C and CT3C show higher strength at collapse.

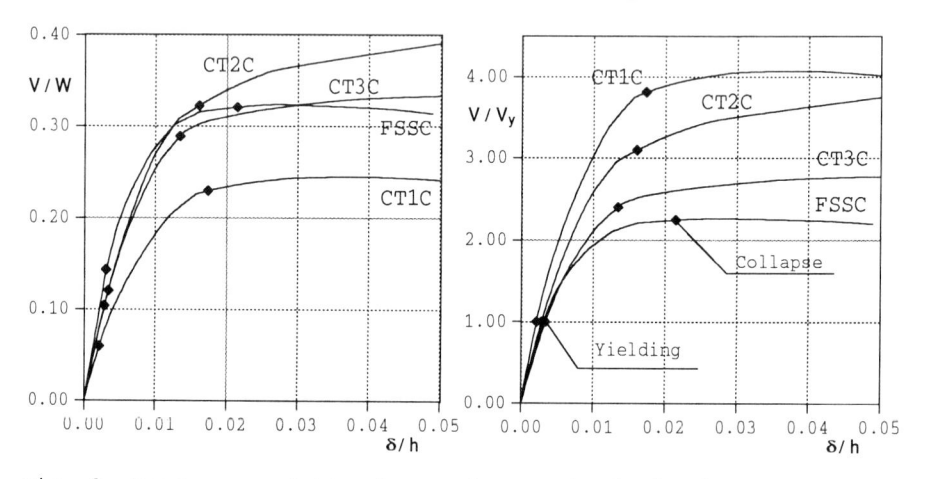

Fig. 3. Responses of the frames to monotonic loads

The semirigid frames attain the maximum rotation in one of the joints well before a collapse mechanism is created, while the frame FSSC behaves more ordinarily reaching the maximum rotation just before the formation of a collapse mechanism with a lower value of over-strength. Figure 3b, where the same curves are reported using in ordinate the ratio V/V_y instead of V/W, shows clearly the higher collapse to yielding shear ratio and ductility of the CT1C frame. In frame CT1C the joints are close to the yielding under the vertical load, thus the high over-strength is mainly due to the low value of V_y rather than to the hardening in the joints.

Figure 4 shows the responses of the frames for the same set of horizontal loads applied cyclically. The cycles are approximately between the limits $+/-V_y, +/-3V_y$, up to $+/-5V_y$ and $+/-7V_y$, until the frame ductility allows it. The frame CT1C shows dissipative capacity much more better than the other two semirigid frames and almost as good as the FSSC response.

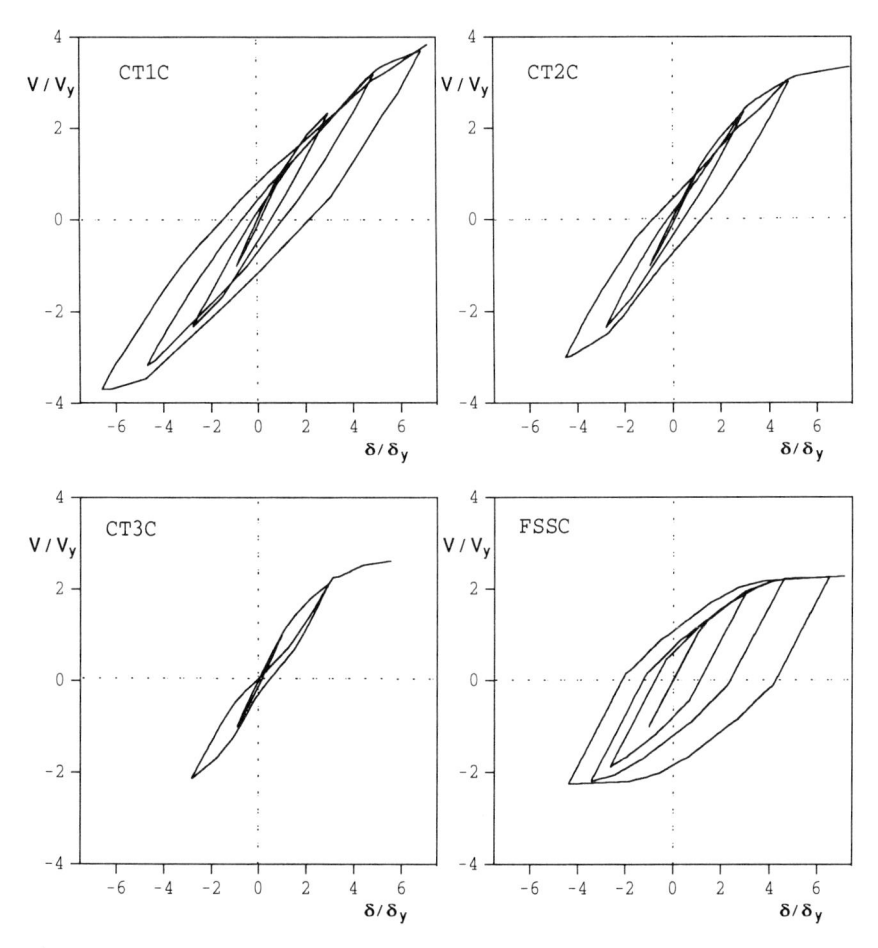

Fig.4 Frame responses to cyclic statically applied loads

2.3. Response of the frames to seismic loads

The response to seismic actions of the four frames has been numerically analysed under the four registered ground motions of table 3, while the first two natural periods of the frames are reported in table 4.

Table 3. Earthquakes

Seismic event	Comp.	PGA(g)
El Centro 1979	140°	0.267
El Centro 1940	NS	0.349
Tolmezzo Friuli 1976	WE	0.314
Miyagi Ken-Oki 1978	N00E	0.263

Table 4. Natural periods(s)

Frame	mode 1	mode 2
CT1C	1.05	0.32
CT2C	0.94	0.28
CT3C	0.97	0.29
FSSC	0.87	0.27

In all the numerical analyses a mass proportional damping equal to 3% of the critical value one has been assumed. The main results are summarized in tables from 5 to 8 where a_y and a_u are the p.g.a. values at first yielding and collapse.

Table 5: Frame CT1C

Seismic Event	$\frac{a_y}{g}$	$\frac{a_u}{g}$	$\frac{\delta_y}{h}$	$\frac{\delta_u}{h}$	$\frac{V_y}{W}$	$\frac{V_u}{W}$	q	q_Ω	q_μ
ElCentro 40	0.060	0.644	0.0021	0.0165	0.071	0.301	10.7	4.18	2.55
ElCentro 79	0.067	0.555	0.0018	0.0171	0.076	0.321	8.3	4.19	1.98
Tolmezzo	0.077	0.766	0.0019	0.0128	0.068	0.315	9.9	4.63	2.14
Miyagi	0.031	0.290	0.0021	0.0171	0.064	0.269	9.4	4.18	2.25

Table 6: Frame CT2C

Seismic Event	$\frac{a_y}{g}$	$\frac{a_u}{g}$	$\frac{\delta_y}{h}$	$\frac{\delta_u}{h}$	$\frac{V_y}{W}$	$\frac{V_u}{W}$	q	q_Ω	q_μ
ElCentro 40	0.074	0.690	0.0026	0.0143	0.112	0.401	9.3	3.59	2.59
ElCentro 79	0.093	0.528	0.0026	0.0147	0.118	0.356	5.7	3.00	1.88
Tolmezzo	0.122	0.767	0.0026	0.0148	0.113	0.423	6.3	3.72	1.70
Miyagi	0.028	0.284	0.0028	0.0154	0.109	0.358	10.3	3.29	3.12

Table 7: Frame CT3C

Seismic Event	$\frac{a_y}{g}$	$\frac{a_u}{g}$	$\frac{\delta_y}{h}$	$\frac{\delta_u}{h}$	$\frac{V_y}{W}$	$\frac{V_u}{W}$	q	q_Ω	q_μ
ElCentro 40	0.091	0.439	0.0034	0.0130	0.126	0.340	4.81	2.69	1.79
ElCentro 79	0.127	0.442	0.0032	0.0129	0.141	0.374	3.49	2.64	1.32
Tolmezzo	0.139	0.604	0.0031	0.0118	0.125	0.396	4.33	3.17	1.37
Miyagi	0.035	0.196	0.0034	0.0132	0.129	0.313	5.65	2.43	2.33

Table 8: Frame FSSC

Seismic Event	$\frac{a_y}{g}$	$\frac{a_u}{g}$	$\frac{\delta_y}{h}$	$\frac{\delta_u}{h}$	$\frac{V_y}{W}$	$\frac{V_u}{W}$	q	q_Ω	q_μ
ElCentro 40	0.071	1.170	0.0029	0.0206	0.159	0.444	16.43	2.80	5.87
ElCentro 79	0.093	0.882	0.0029	0.0194	0.143	0.448	9.39	3.12	3.01
Tolmezzo	0.123	1.060	0.0032	0.0195	0.140	0.451	8.64	3.23	2.67
Miyagi	0.051	0.493	0.0031	0.0219	0.169	0.409	9.57	2.42	3.95

As foreseeable from the static responses, the frame FSSC has better seismic properties owing to its higher ductile capacity. The collapse p.g.a. values are between 1.4 to 1.8 times the correspondent values of the frame CT1C that has the best behaviour among the semi-rigid frames. Nevertheless, to this difference in seismic resistance does not correspond a similar difference in the values of the behaviour factors. For all the four frames these values are rather high. For the semi-rigid frames the reason is the high over-strength. The elastic dynamic response, the envelope of the dynamic non linear response for different p.g.a. values up to collapse and the static response already shown in figure 2, are reported in figure 5 for the frames CT1C and FSSC, as proposed by Astaneh and Nader (1993). Subdividing the behaviour factor in the overstrength factor q_Ω and ductility factor q_μ, it is evident the greater importance of the former for the semi-rigid frames.

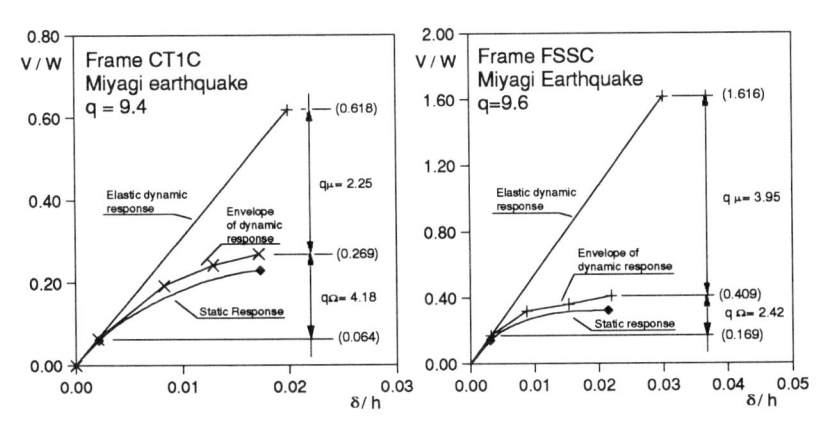

Fig. 5 Behaviour factors for CT1C and FSSC frames

The value of q_Ω and q_μ are reported for all the frames and accelerograms in the tables 5-8. The over-strength of the semi-rigid frames is strictly related to the quantity and distribution of the vertical loads concurrent with the seismic action , Amadio et al.(1994). The closer the joints are to yielding under the vertical loads before the earthquake the lower is the value of base shear at yielding under seismic loads and consequently the higher is the behaviour factor and the overstrenght the frame shows. This obvious consequence has limited effects on rigid steel frames due to the reduced overstrength values of plastic hinge sections. If the plasticity is concentrated in semi-rigid joints with high collapse to yielding moment ratio the attainment of a stress level close to the nominal elastic limit under service loads may not reduce the proper safety coefficient to collapse. In these case the assumption of an usual low value for the behaviour factor could be excessively conservative. To achieve a correct appraisal of the behaviour factor $\bar{q}$ a sufficiently accurate and simple method uses an equivalent static analysis by of an extension of the energetic method of Como e Lanni (1983). The $\bar{q}$ value is evaluated by the relation:

$$\bar{q} = \sqrt{\frac{W_u}{W_y}} \quad (1)$$

where W_y is the stored elastic energy at the elastic limit and W_u the total energy stored and dissipated at collapse. The comparison between the results of table 2 with those of tables 5 to 7 shows the applicability of the method for semi-rigid frames. The dynamic analyses indicate another advantage of semi-rigid composite or steel frame. The lower stiffness and resistance of CT1C frame does not imply a remarkable reduction of the carrying capacity to seismic actions as the increase of the natural period corresponds to the reduction of the seismic forces. For the analyzed frames it is sometimes necessary to consider the effect of the higher modes in the evaluation of the base shear. Figure 6 shows the dynamic response of CT2C frame under Tolmezzo and Miyagi Ken Oki collapse ground motions. In the case of Tolmezzo earthquake, characterised by low values of the spectral acceleration for long period, the influence of the second mode when the maximum base shear is reached is evident. For Miyagi Ken Oki earthquake, with high values of spectral accelerations for high periods, the first mode is the more important. This is shown also in table 5-8 where the maximum dynamic shear at the base for Tolmezzo earthquake is greater than the collapse base shear base of table 2. For this frame, a parabolic distribution of static forces with an increment of values near the base is therefore opportune.

3. Conclusions

The numerical analysis performed has indicated that there is a direct relation between the characteristics of the cyclic response of the semirigid joints and the seismic resistant capacity of the frames. The frames are characterized by a high over-strength with brittle failure and their ductility is lower than in frames where the steel connection is complete and the slab is considered interrupted. This is due to low ductility of the tested joints that have premature collapse in some particular structural details.
The joints must be designed with reference to the ductility and not only to the strength, avoiding failure in the bolts and in the welds or local instabilities in the flanges, thus approaching the level of ductility of rigid steel connections.
The frames, even if they collapse in a brittle way, have a high behaviour factor, due to the high hardening and to the low elastic limit of the joints and consequently of the frame.
If evaluated with reference to the elastic limit reached in the most stressed joint, the q-factor is strongly related to the amount of vertical loads, since the further rotation capacities are mainly in the high hardening range chacterizing the joints. A sufficiently accurate information can be obtained by analyzing the frame under a system of static horizontal forces acting according to the first mode. A good agreement with the results of the dynamic analysis is obtained for the displacements at the elastic limit and at collapse if this is correlated to the rotation limit capacity. The appraisal of the behaviour factor with an energetic method is sufficiently accurate and on the conservative side.
A proper practical use of semi-rigid unbraced frames will be however possible only when further studies will be performed on the partial beam-to-column interaction for asymmetrical (seismic) loads and on the development of reliable and simple

prediction method, thus avoiding the need of experimental data for every different joint.

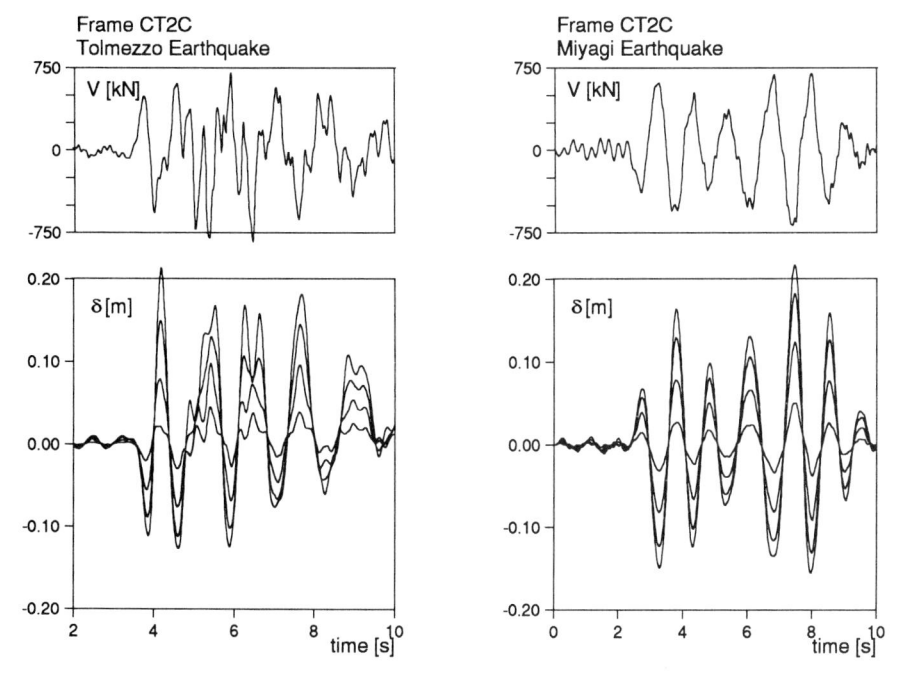

Fig. 6 Drift and base shear time histories for CT2C frame

4 References

Amadio C., Puhali R., Zandonini R.,(1989), L'effetto della continuità parziale nei telai composti. **Proc. XII C.T.A. Congress**, Capri, Italy.

Amadio C., Benussi F., Noè S.,(1993) Modellazione di giunti semirigidi per la progettazione di telai composti, **Proc. 1st Italian Workshop on Composite Structures**, Trento, Italy

Amadio C., Benussi F., Noè S.,Spanghero F.(1993) Il comportamento di telai composti con giunti semirigidi soggetti ad azioni sismiche. **Proc.XIV C.T.A. Congress**. Viareggio, Italy.

Amadio C., Benussi F., Noè S.(1994) Seismic behaviour of semi-rigid composite frames. **Proc. Ercad Conference**, Berlin, Germany

Arbed Reserches (1991), Mathematical model for the prediction of the moment-rotation curves of composite joints with cleated connections - program Springs, **C.E.C. Agreement N.7210-SA/507 - Semi Rigid Action in Steel Frame Structures.** Annex 3

Astaneh A., Nader M.N. (1993) Proposed code provisions for seismic design of steel semi-rigid frames. Presented at **XIV C.T.A. Congress**. Viareggio, Italy

Benussi F., Bernuzzi C., Noè S. & Zandonini R. (1991) Giunti semirigidi in telai composti soggetti ad azioni cicliche. **Proc. XIII C.T.A. Congress**. Abano, Italy.

Bernuzzi C., Noè S., Zandonini R.,(1991) Semi-rigid composite joints, experimental studies. **Int. Workshop, Connections in Steel Structures**. Pittsburg

Castiglioni C.A., Di Palma N., (1989) Experimental behaviour of steel members under cyclic bending. **Costruzioni Metalliche** n.2/3

Commission of the European Communities (1992) **Eurocode No.4: Design of composite steel and concrete structures - Part 1: General rules and rules for buildings.**

Commission of the European Communities (1988) **Eurocode No.8: Structures in seismic regions - Design - Part 1: General and buildings.**

Como M. & Lanni G. (1983) Aseismic Toughness of Structures. **Meccanica** 18: 107-114.

ECCS - European Convention for Constructional Steelwork. Technical Committee 1 - Technical Working Group 1.3. (1986) **Recommended Testing Procedure for Assessing the Behaviour of Structural Steel Elements under Cyclic Loads**. First Edition.

Hibbitt, Karlsson & Sorensen Inc. (1991) **Abaqus version 4.9** - User Manual.

Kato B.(1990) Deformation capacity of Steel Structures. **Journal of Constructional Steel Research** 17, 33-94

Leon R.T. (1990) Semi-rigid composite construction. **Journal of Constructional Steel Research** 15, 99-120.

Leon , R.T. & Zandonini R. 1993. Composite connections. In P.J. Dowling, J.E. Harding & R. Bjorhovde (eds), **Constructional Steel Design - An International Guide**: 501-522. London: Elsevier Applied Sciences.

Noè S., Spanghero F. (1993) Un modello del comportamento ciclico di giunti semirigidi composti. **Proc. XIV C.T.A. Congress**. Viareggio, Italy.

Puhali R., Smotlak I. & Zandonini R. (1990) Semi-Rigid composite action: experimental analysis and a suitable model. **Journal of Constructional Steel Research** 15 1&2: 121-151.

48 SEISMIC BEHAVIOR AND DESIGN OF STEEL SEMI-RIGID STRUCTURES

A. ASTANEH-ASL
Department of Civil Engineering, University of California at Berkeley, USA

Abstract
The paper first summarizes research projects completed at the University of California at Berkeley since 1986 on the seismic behavior of steel semi-rigid building frames and their semi-rigid connections. The studies indicated that semi-rigid frames have potential for being used in highly seismic areas to reduce seismic forces. Then, the paper presents proposed code-formatted seismic design procedures developed for semi-rigid steel building frames, followed by a summary of projects on the development of a new steel semi-rigid framing system. Finally, the paper describes the development and application of a number of semi-rigid systems currently used to retrofit major existing steel bridges.
Keywords: Steel Structures, Seismic Design, Partially Restrained Frames, Semi-Rigid Connections and Frames.

1 Introduction

Most riveted and bolted steel building structures and many steel bridges built prior to the development of welding technology are semi-rigid to some extent. In addition, many connections of modern welded-bolted steel structures are semi-rigid to some extent. The concept of semi-rigidity has been used in the design of many steel building structures to carry gravity loads and to resist wind loads. However, the intentional use of semi-rigidity to control and reduce seismic forces in building and bridges is fairly recent. In the following sections, several research projects are summarized which were conducted at the University of California at Berkeley on the development and application of seismic design procedures for steel structures of buildings and bridges.

2 Semi-rigid steel building structures

2.1 Seismic behavior of rigid and semi-rigid frames
In steel rigid as well as semi-rigid frames, resistance to seismic effects is provided primarily by bending action. The main difference of rigid and semi-rigid steel frames is in the bending strength and rotational stiffness of the beam-to-column connections relative to the

connected beams. In rigid frames, the connections are designed to be stronger and stiffer than the beam and are expected to remain essentially elastic during earthquakes. However, in semi-rigid frames, the connections are intentionally designed to have less bending capacity and stiffness than the connected beams. As a result, in the event of an earthquake, the semi-rigid connections are the primary inelastic element of the system. Figure 1 shows typical steel semi-rigid connections.

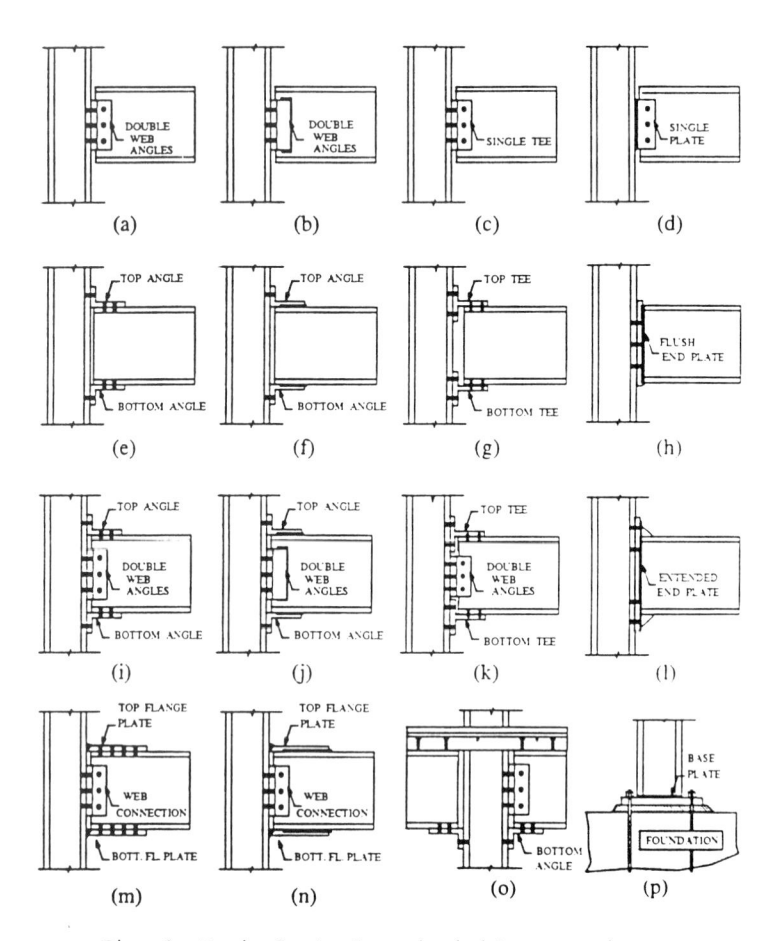

Fig. 1. Typical steel semi-rigid connections.

Typical moment-rotation hysteresis responses of steel rigid and semi-rigid connections are shown in Figure 2. Rigid connections show bi-linear moment-rotation response with two distinct regimes of behavior: (1) the initial elastic behavior and (2) the post-yielding behavior. In steel semi-rigid connections, the moment-rotation behavior in general has four distinct regions: (1) the initial elastic region, (2) the first stage of softening due to yielding of steel elements or friction slippage of the bolts, (3) the secondary stiffening mostly due to kinematic hardening, and (4) the final yielding and strain hardening of the material of the connection. In rigid connections with a given moment capacity, it is very difficult to control the initial rigidity which is generally very high. However, in today's common semi-rigid connections, almost all parameters of behavior such as initial stiffness, secondary stiffness, initial yield or slip moment, and final moment capacity can be controlled by choosing appropriate connection geometry and material properties (Astaneh-Asl & Shen 1993).

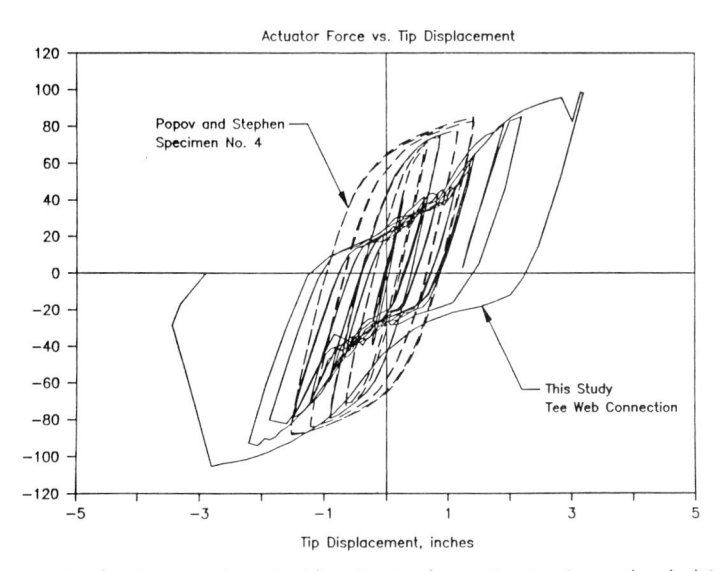

Fig. 2. Typical moment-rotation behavior of steel semi-rigid and rigid connections.

The inelastic local behavior of semi-rigid connections in a semi-rigid frame affects the global behavior. Figure 3 shows the typical seismic responses of steel rigid and semi-rigid frames recorded during shaking table tests (Astaneh-Asl et al. 1991). Other studies conducted later, some of which are listed in the references, have shown similar behavior. A comparison of the seismic behavior of rigid and semi-rigid steel moment frames reveals that the seismic forces generated in semi-rigid frames are generally less or on the same order as forces in comparable rigid frames. The lateral displacements of semi-rigid frames are usually slightly more than for rigid frames.

The decrease of forces and some increases in displacement in semi-rigid frames are attributed to elongation of period, increase in damping, decrease of stiffness at early stages of behavior, and the 'isolation effects' due to gap opening and closing in semi-rigid frames. If a semi-rigid steel structure has connections with sufficient ductility, the studies done so far and experience of past earthquakes indicate that the behavior of bolted semi-rigid steel frames is superior to the behavior of welded rigid frames.

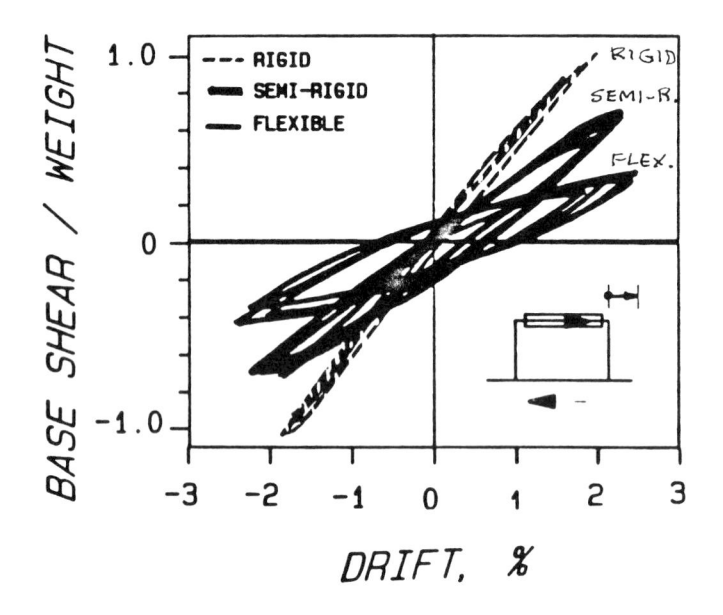

Fig. 3. Base-shear lateral-drift response of rigid and semi-rigid steel frames.

2.2 Seismic design of steel building semi-rigid frames

Currently most seismic design codes permit the use of semi-rigid steel building frames. However, the codes have very limited guidelines and provisions on how these structures should actually be designed. In U.S. codes such as the Uniform Building Code (1994), steel semi-rigid frames are categorized as 'Ordinary Moment Frames' with a reduction factor of R equal to 4, which is half of the reduction factor of 8 for Special Ductile Moment Resisting Frames. In addition, the AISC Seismic Provisions (AISC 1992) requires that for Ordinary Moment Frames the connections have been demonstrated by cyclic tests to have adequate rotation capacity for a drift calculated using a factored earthquake load of 0.4RwE.

As a result of application of current provisions in the U.S. Codes, semi-rigid frames are required to be designed for twice as much force as comparable Special Ductile Moment Frames, yet are required to provide much larger ductility. There is little research or actual performance data to support these restrictive requirements.

It appears that because of the high ductility of bolted semi-rigid frames, the design forces in these frames should be even less than in special moment frames (see Figure 2). Because of the lack of rational provisions in the code, use in the United States of steel semi-rigid frames for seismic resistance has been very limited. As a result, almost all moment frames built in recent years in highly seismic areas such as California have been Special Ductile Moment Resisting Frames. Weld fractures caused by the January 1994 Northridge earthquake have been discovered in more than 100 welded "Special Ductile Moment

Resisting" frames.

It is interesting to note that no damage has been reported to the bolted steel semi-rigid moment frames in the aftermath of past earthquakes. The damage reports on the great 1906 San Francisco earthquake show a number of steel riveted semi-rigid frames that were standing after the quake amid the rubble of downtown San Francisco and no collapse of any steel frames is mentioned. One can hope that the costly damage that occurred during the Northridge earthquake will act as a wake-up call to the earthquake engineering community to re-examine the data on the seismic behavior of steel moment frames and utilize the many advantages of bolted semi-rigid structures.

In the aftermath of the Northridge earthquake and its damage, the wisdom of formulating code provisions so that structural engineers are almost forced to use only Special Moment Frames with one unique detail is seriously questioned. Because of the very low reduction factor assigned to other systems, to achieve economical design the structural engineers are almost forced by the code to use only welded special ductile moment resisting frames with the exclusion of other potentially well-behaved and economical systems such as steel semi-rigid frames.

In 1992, a tentative proposed seismic design procedure for semi-rigid steel building frames was developed (Astaneh-Asl, Nader and Herriott 1991). The procedure which is an "equivalent static load" procedure has a format similar to the current code procedures for seismic design of steel moment frames. Table 1 provides a summary of the seismic design procedures for steel moment frames according to UBC and the proposed seismic design procedures for steel semi-rigid frames.

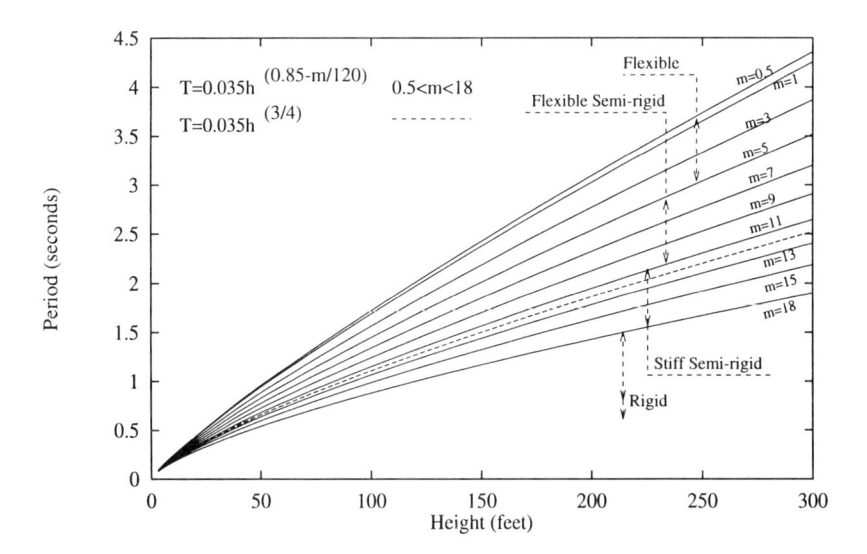

Fig. 4. Variation of period versus semi-rigidity of frame in semi-rigid frames.

The main difference between the procedures for rigid and semi-rigid frames is in the introduction of the connection stiffness term m which is the ratio of initial rotational stiffness of the connection to EI/L of the connected beam. If m is greater than 18, the connection is fully rigid. If m is less than 0.5, the connection is almost pin. For m between 0.5 and 18, the connection is semi-rigid. However, our studies (Nader and Astaneh-Asl 1992) have indicated that until further research on stability of semi-rigid frames is conducted, the use of the procedure is limited to m values greater than 5. The m value that is to be used in the procedure is the average of m values for the connections of the frame under consideration.

The effect of connection semi-rigidity on the period is shown in Figure 4. As connection stiffness decreases, the period of vibration of structure increases.

Similar to current code procedures, in the proposed method for semi-rigid frames, an equation is provided to distribute the base shear over the height of the structure. To account for the effect of higher modes, a portion of the lateral force equal to Ft is applied at the roof level. In addition, the distribution of lateral force of the height is not linear. The amount of nonlinearity depends on m as well as the ratio of effective rigidity of semi-rigid girder to rigidity of columns as shown in Table 1.

2.3 New developments

One of the projects currently underway at the Department of Civil Engineering of the University of California at Berkeley (Astaneh-Asl 1988) is to develop an innovative steel semi-rigid frame. The system shown in Figure 5 consists of 'column trees' and floor girders connected to each other by ductile semi-rigid connections at a point away from the face of the column. Column tree rigid frame systems have been used in the past. In column tree rigid frames the splice is designed to be stronger than the girder, and after erection the system acts as any other rigid moment frame. However, in the proposed semi-rigid column tree system, the splice is intentionally made semi-rigid and less stiff than the connected girder. As a result, the semi-rigid connections at the location of girder splices act as the primary point of inelasticity and hinge formation during the seismic event. By controlling the stiffness and strength of the semi-rigid connection acting as a fuse, one can control and reduce the global seismic response of the building.

In order to establish the seismic behavior of the proposed semi-rigid column tree system, a series of inelastic dynamic analyses of a four-story and a 24-story structure was conducted (McMullin et al. 1993). As part of the analyses, a parametric study was conducted to establish the effects of various parameters on the global response and demand of components of the system. Figure 6 shows the seismic drift response of the 24-story column tree moment frame with rigid and semi-rigid splices with m values of 1, 3, and 5. The parameter, m, is the ratio of rotational stiffness of connection to EI/L of the girder. As the figure indicates, the drift response of the column tree system with semi-rigid splices was **less** than the drift for the rigid frame. The seismic forces in the system were also less than the forces generated in a comparable rigid moment frame system.

Behaviour and design of semi-rigid structures

Table 1. Tentative proposed procedure for steel semi-rigid frames

$$V = \frac{ZIC}{R} W$$

$$C = \frac{1.25 S}{T^{2/3}}$$

$$R = 8$$

$$T = 0.035 \, h_n^{(0.85-m/120)} \quad \text{(When } 5 < m < 18; \text{ semi-rigid frame)}$$

$$T = 0.035 \, h_n^{3/4} \quad \text{(When } m \geq 18; \text{ rigid frame)}$$

$$m = \frac{(K_e)_{Conn.}}{(EI/L)_{Beam}} \quad \text{(m is frame semi-rigidity and should not be less than 5)}$$

$$V = F_t + \sum_{i=1}^{n} F_i$$

$$F_x = \frac{(V - F_t) \, w_x \, h_x^\delta}{\sum_{i=1}^{n} w_i \, h_i^\delta}$$

$$\delta = \begin{cases} 1.0 & \text{(When } T \leq 0.5 \text{ sec.)} \\ 0.5T + 0.75 & \text{(When } 0.5 < T \leq 2.5 \text{ sec.)} \\ 2.0 & \text{(When } T > 2.5 \text{ sec.)} \end{cases}$$

$$F_t = \begin{cases} 0 & \text{(When } T \leq 0.7 \text{ sec.)} \\ 0.07TV & \text{(When } T > 0.7 \text{ sec.)} \end{cases}$$

F_t need not exceed 0.25V.

(Note: For definition of terms see Uniform Building Code (1994).

COLUMN TREE STEEL STRUCTURE

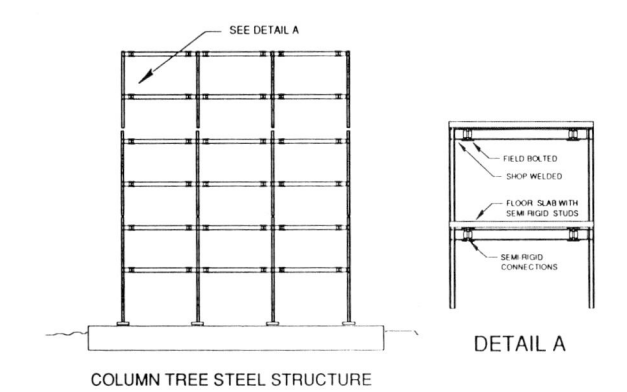

DETAIL A
- FIELD BOLTED
- SHOP WELDED
- FLOOR SLAB WITH SEMI RIGID STUDS
- SEMI RIGID CONNECTIONS

Fig. 5. A view of the proposed semi-rigid column tree framing system (Astaneh-Asl 1991).

The fact that the drifts in a semi-rigid frame can sometimes be less than the drifts in a comparable rigid frame appears to be surprising to some structural engineers who have worked with code procedures using static equivalent forces in analysis and design. However, one should realize that seismic forces are not static and external forces. The seismic forces generated in a building depend on ground motion, stiffness, mass and damping. The mass of semi-rigid and rigid frames are similar. However, semi-rigid connections add to damping and, because of gap openings and closings in these connections, there is a certain 'energy isolation' effect that acts like a base isolator. In addition, by decreasing stiffness, the period of vibration of semi-rigid frames elongates. All of these effects can result in reduction of seismic forces and drifts.

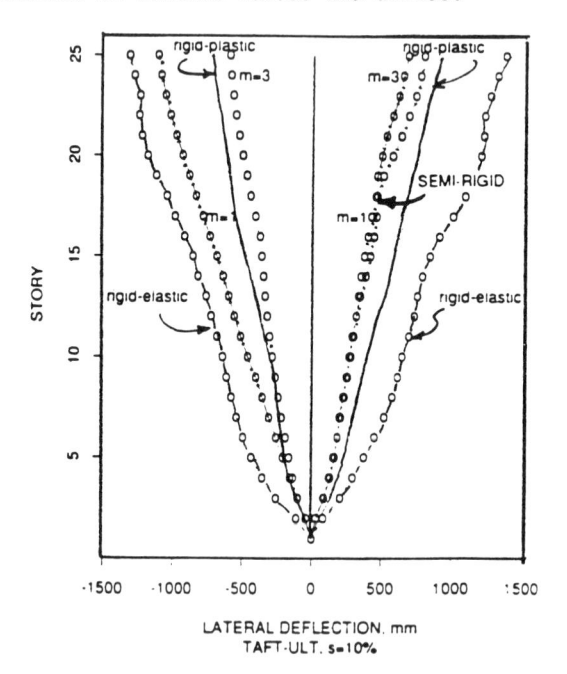

Fig. 6. Lateral displacement of 24 story rigid and semi-rigid frames with various degrees of semi-rigidity.

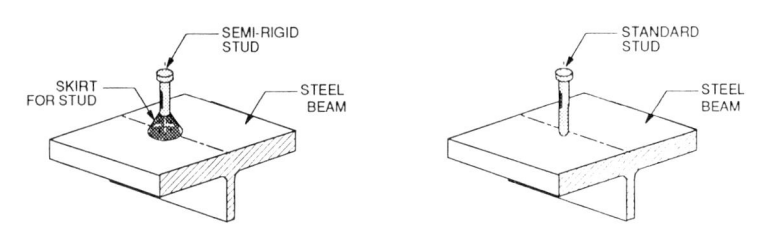

Fig. 7. The semi-rigid stud proposed by A. Astaneh-Asl and a standard stud.

Another new development has been the testing of the 'semi-rigid shear studs' proposed by A. Astaneh-Asl (Astaneh-Asl 1993). The proposed semi-rigid shear studs are similar to standard studs but at the base of the stud a cone shaped 'skirt' is placed to prevent concrete from engulfing the total length of the shear stud (Figure 7). As a result, under cyclic loading during an earthquake, the base of the stud can yield in a combination of shear and bending. The benefits of using a composite system with shear studs are: more ductile studs, almost no damage to concrete, protection of stud welds from premature fracture, increased damping of the system, and better control of the stiffness and strength. More information on the semi-rigid shear studs is provided in Astaneh-Asl (1994).

3 Semi-rigid steel bridges

The San Francisco-Oakland Bay Bridge, one of the major bridges in the United States, is currently undergoing seismic retrofit using the concept of semi-rigidity (Astaneh-Asl 1993). More information on the bridge is provided in a companion paper in this volume by A. Astaneh-Asl (1994). Figure 8 shows a segment of the Bay Bridge as planned to be retrofitted to become a semi-rigid structure. During service earthquakes, the tower will act as a vertical cantilever with its base fixed. However, when seismic forces are large enough, the semi-rigid connections at the base of the tower, shown in Figure 8, will yield and permit the tower to slightly rock. The rocking of the semi-rigid tower, E9, will result in the desirable semi-rigid behavior of the structure and considerable reduction in seismic response. As a result of application of the concept of semi-rigidity, the seismic forces in the structure shown in Figure 8 were reduced from about 60% to about 40% of the weight of the structure.

Similar concepts of semi-rigidity have been proposed (Astaneh-Asl and Shen 1993) for the other parts of the 8-km-long eastern portion of the San Francisco Bay Bridge. The comprehensive study of a variety of seismic retrofit options such as strengthening, adding damping, and base isolation indicated that the most efficient solution is converting the structure to a semi-rigid system. This was done by placing steel semi-rigid connections made of bolted steel angles and plates at strategic locations of the bridge and by allowing critical towers to rock in the event of maximum credible earthquakes.

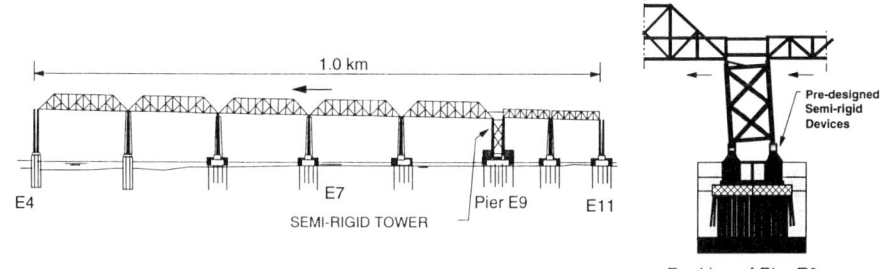

Fig. 8. A view of the E4-E11 segment of the Bay Bridge with proposed semi-rigid retrofit concept.

4 Acknowledgments

The research leading to the information presented here on buildings was funded by the National Science Foundation, University of California, California Universities for Research in Earthquake engineering (CUREe), and the Kajima Corporation. The bridge research projects of the author are primarily funded by the California Department of Transportation. Many individuals, particularly those whose names appear in the references have contributed to the studies summarized in the paper. Their valuable contributions are acknowledged and sincerely appreciated.

5 References

American Institute of Steel Construction (AISC). (1992) **Seismic Provisions for Structural Steel Buildings**. AISC, Chicago, IL.

American Institute of Steel Construction (AISC). (1993) **Manual of Steel Construction, Volume I**. AISC, Chicago, IL.

Astaneh-Asl, A. (1988) Use of steel semi-rigid connections to improve seismic response of precast concrete structures, **Research Briefs**, Precast Seismic Structural Systems Workshop, University of Calif., San Diego, Nov., 1988.

Astaneh-Asl, A. (1993) The innovative concept of semi-rigid composite beam. **Proc., Structures Congress XI**, April 1993, ASCE, Irvine, CA.

Astaneh-Asl, A. (1994) Recent seismic research on steel bridges. (1994) **Proc., STESSA '94, Behav. of Steel Struct. in Seismic Areas**, 26 June-1 July 1994, Timsoara, Romania, Eur. Convention for Constructional Steelwork (ECCS).

Astaneh-Asl, A., Nader, M., and Harriott, J.D. (1991) Behavior and design considerations in semi-rigid frames. **Proc., AISC National Steel Construction Conference**, 5-7 June 1991, AISC, Washington D.C.

Astaneh-Asl, A., and Shen, J.-H. (1993) Seismic evaluation and retrofit concepts. Vol. 10 in Seismic condition assessment of the East Bay Crossing of the San Francisco-Oakland Bay Bridge. **Rep. No. UCB/CE-Steel-93/12**, Dept. of Civil Engrg., Univ. of California, Berkeley, CA.

McMullin, K.M., Astaneh-Asl, A., Fenves, G.L., and Fukuzawa, E. (1993) Innovative semi-rigid steel frames for control of the seismic response of buildings. **Rep. No. UCB/CE-Steel-93/02**, Dept. of Civil Engrg., Univ. of Calif., Berkeley, CA.

Uniform Building Code. (1994) International Conference of Building Officials, Whittier, CA.

49 MOMENT-ROTATION BEHAVIOUR OF TOP-AND-SEAT ANGLE CONNECTIONS

C. BERNUZZI
University of Trento, Trento, Italy
M. DE STEFANO
Istituto di Tecnica delle Costruzioni, University of Naples, Italy
E. D'AMORE and A. DE LUCA
University of Reggio Calabria, Reggio Calabria, Italy
R. ZANDONINI
University of Trento, Trento, Italy

Abstract
In this paper the experimental behaviour of top-and-seat angle connections under cyclic reversal loading is examined in order to improve the state of knowledge related both to the main parameters affecting the cyclic response and to the dissipation capabilities of the energy associated to cyclic loads. In particular, the effect of bolt slippage, which increases to a large extent the joint flexibility, is quantitatively evaluated. On the basis of standard codes for steel design some practical indications are also given to reduce such effect. Finally, a numerical model, based on a pure mechanical approach, is used for simulating the joint behaviour.
Keywords: Semirigid Connections, Cyclic Response, Bolt Slippage, Mechanical Model

1 Introduction

The behaviour of semirigid connections represents a subject of extensive discussion within the fields of research in constructional steel design. It is now agreed that any connection has to be considered as semirigid and therefore the appropriate values of stiffness, ultimate moment capacity and ultimate rotation have to be assumed within the analysis of a semicontinuous frame. The studies are now being more and more specialized since each form of connection (header plates, flush-end-plate, extended-end-plate, top-and-seat angle, web angle, etc.) deserves specific studies devoted both to the experimental behaviour and to the analytical simulation.
During past years, numerous tests have been conducted to study the monotonic flexural response of semirigid connections (Nethercot 1985; Goverdan 1983), whereas only few cyclic tests have been performed in order to achieve better understanding of their behaviour under earthquake type loadings. Among the latter test programs, it is worth mentioning a series of cyclic tests conducted by Ballio et al. (1987) on flange plates connections, flange and web cleated connections, extended-end-plate connections and welded connections. This study evidenced the influence of column web stiffeners on the connection behaviour and the large bolt slippage phenomena, which affect to a large extent the shape of the experimental loops. Azizinamini et al. (1989) investigated the response of top-and-seat connections with double web angle. Different types of loading sequences were applied: low-to-high amplitude, high-to-low amplitude and constant amplitude loading histories were considered. Furthermore, geometry of angles and bolts was varied to identify how joint properties and behaviour were affected. Astaneh et al. (1989) examined hysteresis moment-rotation behaviour of double angle connections subjected to earthquake type loadings, focusing on performances in terms of ductility and capacity for energy dissipation.

Behaviour of Steel Structures in Seismic Areas. Edited by F. M. Mazzolani and V. Gioncu.
Published in 1995 by E & FN Spon, 2–6 Boundary Row, London SE1 8HN. ISBN: 0 419 19890 3.

Bernuzzi et al. (1991) have tested top-and-seat angle (TSC) and flush-end-plate (FPC) connections under cyclic loadings, evidencing the influence of the loading history (Bernuzzi 1992) and of the connection details (Bernuzzi et al. 1992).

In this paper the response of TSC connections (Figure 1) is analyzed with respect to their experimental behaviour and to the possibility of implementing a mechanical model capable to simulate the cyclic response. Test behaviour has been examined in order to improve the state of knowledge related both to the main physical phenomena affecting the cyclic response and to the dissipation capabilities of the energy associated to cyclic loads. In particular, a special attention has been paid to evaluate effects of bolt slippage, that appears to increase to a large extent the joint flexibility. On the basis of the experimental response, some guide-lines are provided for reducing slip through a 'slip capacity design'.

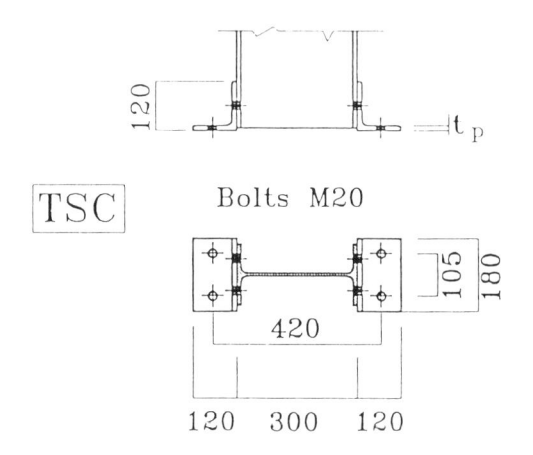

Fig. 1. Characteristics of the TSC connections.

Finally, a numerical model, based on a pure mechanical approach, is used to simulate the connections cyclic moment-rotation response. The model, developed by De Stefano et al. (1994) for web angle connections, is extended to top-and-seat angle connections and appears to be capable to provide numerical predictions in acceptable agreement with the experimental data.

2 Tests

2.1 Specimen and testing equipment

The investigation on the response of steel connections was mainly devoted to appraise the rotational response and to quantify the influence of the connection components (i.e., bolts, plate and angles). The specimen consisted of a beam stub attached by means of the connection to a counterbeam having negligible deformation (Figure 2) and the horizontal force was applied at the free end of the beam stub. The testing apparatus was designed in order to simulate the conditions of beam-to-column joint with negligible column deformability, focusing the attention on the sole connection behaviour. A system composed by a hydraulic jacks permits to apply the reversal loads.

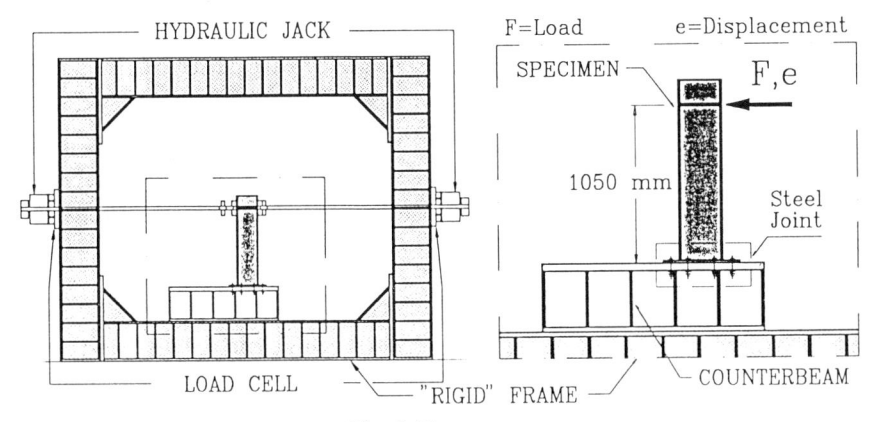

Fig. 2. Test set-up.

The yielding and tensile stresses of the materials of the specimens (beam, bolts, angles or plate) were slightly greater than the values specified by the Italian code. Beams were made of steel grade Fe 360 and the values of elastic and plastic moment of the beam were 148 kNm and 173 kNm respectively. The steel used for the angles was Fe 360, with yield stress $f_y=0.315$ kN/mm^2 and ultimate stress $f_u=0.460$ kN/mm^2. All bolts were grade 8.8 bolts, preloaded to 80% of the nominal yield strength according to the CNR Standards (1988).

2.2 Measuring arrangement
The measuring set-up was designed so to enable evaluation of both the global response of the nodal zone and the local behaviour of the connection and its components.

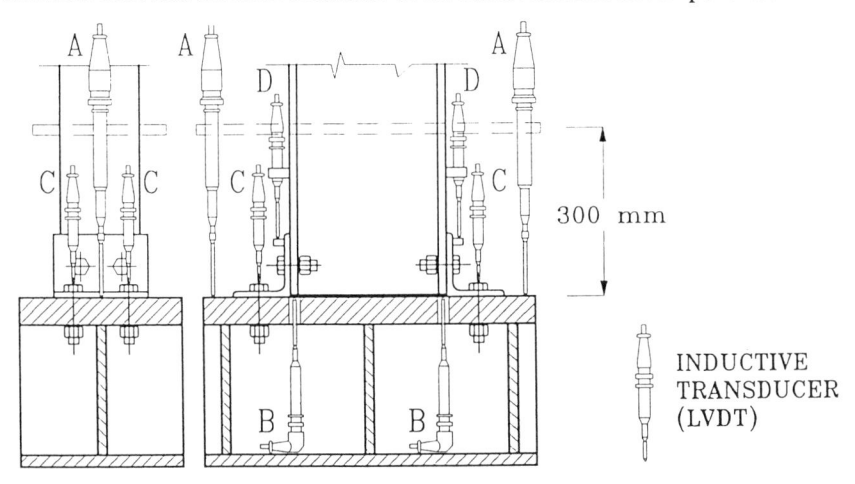

Fig. 3. Measuring arrangement.

Each specimen was equipped with linear voltage displacement transducers (LVDT) connected to a computer-assisted-data-logging system, whereby the specimen response could be monitored in real time during the test by controlling a number of significant parameters. Referring to Figure 3, transducers B were used to evaluate the rotation of the connection, ϕ, transducers C evaluated the contribution to rotation due to the bolt elongation, ϕ_b. Transducers A allowed the evaluation of the rotation ϕ_{tot} at a cross-section close to the joint zone. In the case of connections TSC, transducers D measured the slip between the angles and the beam flanges.

2.3 Testing procedures

Preliminary to the cyclic analysis, one test under monotonic loading (TSC-M) was conducted; on the basis of the monotonic load-displacement curve (F-e) a conventional yielding displacement, e_y, was so defined as the intersection between the line of the initial stiffness (K) and the line of the stiffness tangent to the inelastic branch of the curve (K_T), according to complete testing procedure of the ECCS Recommendations [6], as can be seen from Figure 4.

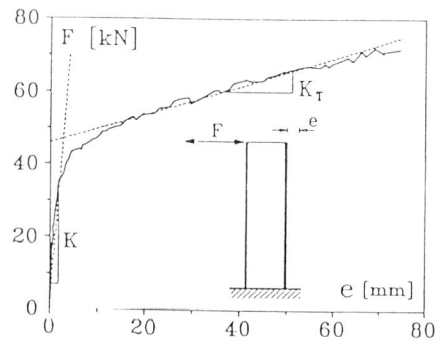

Fig. 4. Stiffness parameters of the load-displacement curve.

The tests were carried out controlling the displacement at the end of the cantilever beam. As reported in Figure 5, four reversal loading history, performed for integer multiples of e_y, have been considered. There are differences relative to the number of cycles at the same level of displacement and for the increments of the cycle amplitude.

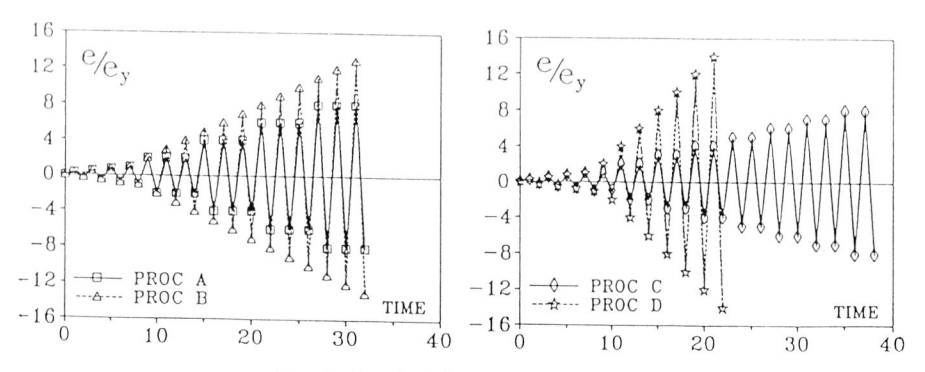

Fig. 5. Applied displacement history.

In the first part of all cyclic tests, the increment of displacement were sufficiently small to ensure that at least four levels of displacement had been reached before the conventional yielding displacement e_y. The main characteristics of the loading histories can be summarized in the following:
- proc. A: three cycles for each even multiplier of e_y (test TSC-A);
- proc. B: one cycle for each integer multiplier of e_y (test TSC-B);
- proc. C: two cycle for each integer multiplier of e_y (test TSC-C);
- proc. D: one cycle for each even multiplier of e_y (test TSC-D).

The procedure A is proposed by the ECCS Recommendation for assessing the structural behaviour under cyclic loads.

2.4 Test results

The key behavioural factors as well as determination of the parameters characterizing joint responses can be evaluated on the basis of the experimental data. In the case of monotonic tests on the top and seat connection (TSC-M), after a first slip for a load F equal to 24 kN, corresponding to a moment at the joint of 25.2 kNm, non negligible slips between the angles and the beam flanges were observed for a load equal to 45 kN. It was reached a very high rotational capacity (> 90 $mrad$) and collapse was due to the fracture of one bolt connecting the angle to the counterbeam.

Table 1. Parameters of joint behaviour

Test	$S_{j,init}$ (kNm/mrad)	M_{max} (kNm)	ϕ_{max} (mrad)
TSC-M	25.2	83.8	95.3
TSC-A	24.2	70.6	74.2
TSC-B	23.7	72.0	69.0
TSC-C	21.0	67.1	67.1
TSC-D	29.1	70.6	68.7

In the following tests on the same connection (TSC-A, TSC-B, TSC-C and TSC-D), the inelastic phenomena due to the slippage, were observed for a load approximately equal to 24 kN. As it will be shown in the next section, they affected to a large extent the cyclic responses, characterized in fact by a noticeable pinching increasing with the level of imposed displacement. The collapse was due, also in these cases, to fracture of the bolt in tension. Table 1 summarizes results that have been shown elsewhere (Bernuzzi 1992), namely the initial stiffness $S_{j,init}$, the maximum moment M_{max} and the corresponding rotation ϕ_{max}.

No significant effect of the type of cyclic loading history has been detected. As already shown by Bernuzzi (1992), the monotonic envelopes of all four cyclic moment-rotation responses are quite similar. However, for a given rotation, a small decrease in the sustained moment is seen if more than one constant amplitude cycle is applied (Procedures A and C). Nonetheless, as the imposed rotation further increases, the moment reaches in the first cycle the same values as those obtained with procedures B and D. The experimental loops moment M vs rotation ϕ of all tests are presented in Figure 6, confirming the above observations.

3 Effect of bolt slippage

As introduced earlier, the cyclic response of the tested connections is largely influenced by slippage of bolts connecting angle legs with top and bottom beam flanges. The major effect of such phenomenon is clearly explained by the fact that the values of the moment M reached during tests are much larger that the maximum value M_{slip} that could be applied without bolt slippage. An approximate evaluation of M_{slip} can be done in the following manner. The used fasteners were 20 mm diameter high strength (grade 8.8) bolts, having nominal ultimate stress f_u of 0.8 kN/mm^2 and nominal yield stress of 0.64 kN/mm^2.

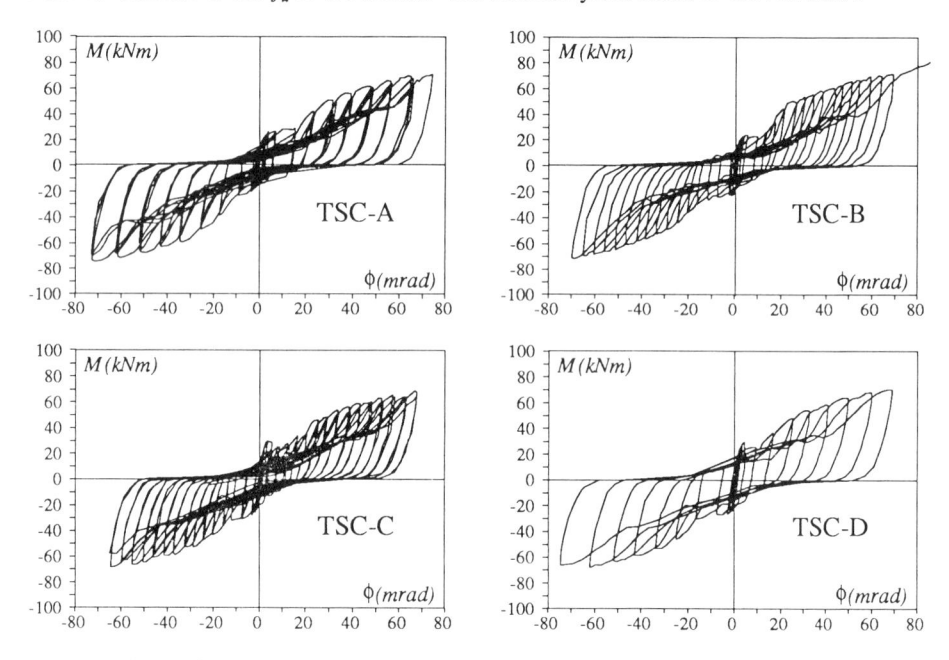

Fig. 6. Cyclic moment-total rotation response of the tested connections

According to the Italian CNR Standards, the applied pretensioning force F_p is 110 kN. Therefore, considering a slip (friction) coefficient $\mu=0.30$, which is considered by the Italian Regulations, the maximum force that can be sustained by each bolt is equal to:

$$F_{slip} = \mu \cdot F_p = 33 kN$$

Therefore, since each angle is fastened to the beam flange through two bolts, the initial slip moment M_{slip} is equal to:

$$M_{slip} = 2 \cdot F_{slip} \cdot d_b = 19.8 kNm$$

where d_b is the beam depth. The value of M_{slip} is clearly smaller than the applied moments, which reach 80 kNm. As a matter of fact, it can be seen that the design of the fasteners according to the AISC Allowable Stress Design rules leads to adoption of bolts having quite similar properties compared to those suggested by the Italian Regulations. In particular, if reference is made to high strength grade 8.8 bolts, considered by the Italian Standards, and to high strength bolts ASTM A325 bolts, considered by the U.S. Standards, which present ultimate stress equal to 0.828 kN/mm^2, the bolt diameter and the pretensioning force required by both specifications to carry a certain shear force are very similar.

As demonstrated in Figure 7, the computed value of M_{slip} is a conservative estimate of that corresponding to initial bolt slippage during the monotonic test TSC-M. In fact, the experimental monotonic moment-rotation curve shows a plateau, which is clearly due to bolt slippage, for an applied moment larger than the theoretical M_{slip}. It should be also noticed that if the slip coefficient μ was taken equal to 0.45, which is the maximum value given by the Italian Standards, an upper bound (M_{slip}=29.7 kNm) of the actual initial slip moment would have been obtained.

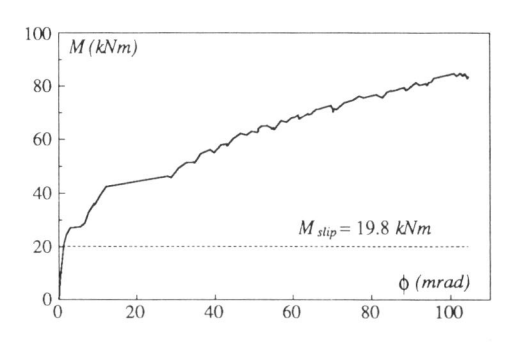

Fig. 7. Moment-rotation response for test TSC-M.

As evidenced before, the measuring arrangements enabled to evaluate slip between angles and beam flanges. Therefore, it has been possible to exclude the portion of the rotation due to slip from the total rotation, thus defining a net rotation ϕ_{net}, due to deformation of the joint elements only. Plots of the moment M versus net rotation ϕ_{net} are reported in Figure 8, for all cyclic tests. From comparison with Figure 6, it can be observed that the maximum applied net rotation is about one third of the total one. Going further into detail, the smallest values of the net rotation are those for test TSC-A, whereas larger values of ϕ_{net} are found for test TSC-D, where only one constant amplitude cycle was applied for the same imposed total rotations ϕ as test TSC-A. This means that, for a certain total rotation, the rotation due to slip increases as the connection is subjected to more than one constant amplitude deformation cycle. Similar results are found also for the other two cyclic tests (TSC-B and TSC-C).

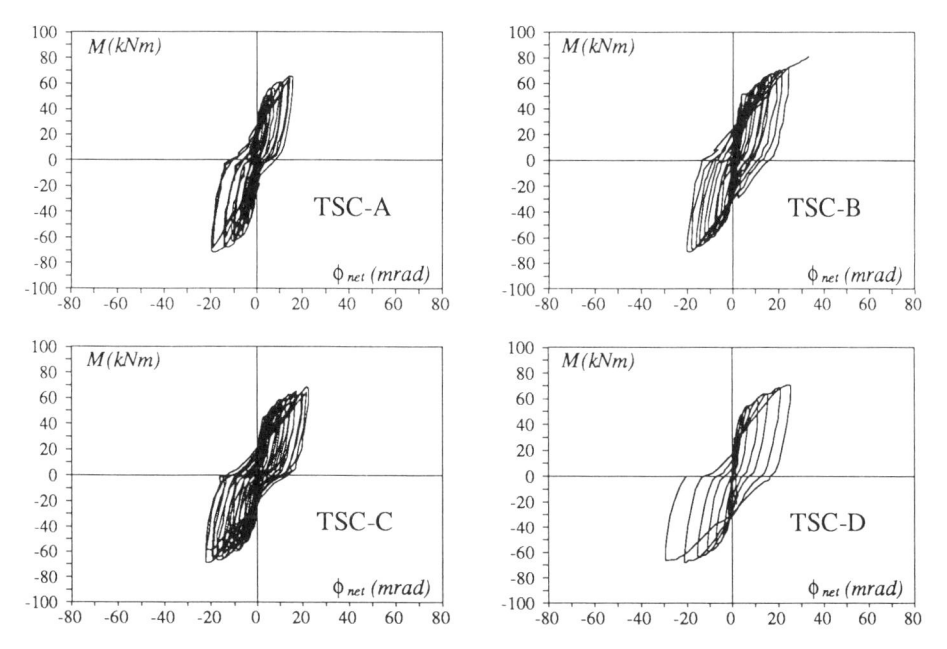

Fig. 8. Cyclic moment-net rotation response of the tested connections

It can be concluded that a large portion of the joint flexibility is due to bolt slip: this phenomenon becomes dominant if the joint is subjected to large cyclic loading, thus notably influencing the overall structural response.

4 Slip behaviour: comparison with other tests and design aids

The above described results demonstrate that the moment-rotation behaviour of the tested top-and-seat angle connections is largely dependent on bolt slippage. However, other results have evidenced quite a smaller influence of such phenomenon. As an example, tests conducted by Azizinamini et al. (1989) on beam-to-column bolted connections are considered herein. The connection elements comprised top and seat angles bolted to the flanges of the beam and of the column, with double web angles bolted to the beam web and to the column flange. The moment-rotation hysteresis loops reported in Figure 9 refer to a joint connecting a W14 x 38 beam with a W12 x 96 column. The top and seat angles were L6 x 4 x 3/8 and the web angles were 2L4 x 3-1/2 x 1/4 (geometrical dimensions, given in US units, are expressed in inches). The fasteners were 7/8 *in* diameter ASTM A325 bolts, having F_{slip} equal to 45.39 kN; namely, four bolts were used to fix the top and seat angles to the beam flanges. Neglecting the contribution of the web angles to M_{slip}, a lower bound value of the slip moment is given by:

$$M_{slip} = 4 \cdot F_{slip} \cdot d_b = 65.02 \quad kNm = 575.28 \quad kip-in$$

It can be seen that this value, which underestimates the actual M_{slip}, is quite close to the largest values of applied moment M. This explains why cyclic bending behaviour of such a connection is essentially unaffected by bolt slip.

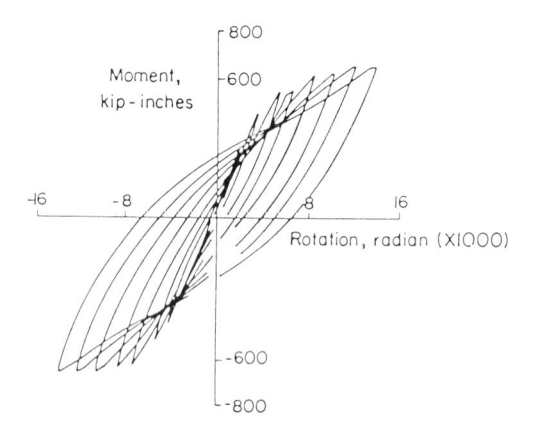

Fig. 9. Cyclic moment-rotation response of top-and-seat angle connections with web angles (after Azizinamini et al. 1989).

The tests conducted by Bernuzzi et al (1992) have proved that bolt slippage can be a major source of deformability in bolted joints designed according to the Italian Regulations, whereas the tests conducted by Azizinamini et al. (1989) evidenced that such effect is dramatically reduced if design results in proper values of M_{slip}. The need for development of design specifications to avoid bolt slippage is also justified by the significant reduction in the capacity for energy dissipation due to slip. For this purpose, the design philosophy should consist of sizing bolts diameter and/or the pretensioning force such that yielding of the angle legs precedes the bolt slip, i.e. the sum of F_{slip}'s of the bolts subjected to shear is larger than the force corresponding to angle yielding F_y. This approach could be called "*slip capacity design*". A value of F_y can be predicted by considering that, during cyclic loading, two plastic hinges lines usually form in the angles, as shown, among the others, by Astaneh et al. (1989). One plastic hinge line forms along the edge of the bolt line and a second plastic hinge line forms adjacent to the fillets of the angles. Let M_p denote the plastic moment of the angle leg cross section, then the value of F_y can be expressed as:

$$F_y = \frac{2M_p}{L}$$

where L is the between the distance between the plastic lines.
For the top-and-seat angle connections tested by Bernuzzi et al. (1992) the computation of F_y leads to a value of 154 kN. If the aforementioned 'slip capacity design' criterion was used, 30 mm diameter bolts class 8.8, pretensioned with a force F_p equal to 256 kN (μ=0.30) would be needed, instead of 20 mm diameter bolts.

5 Numerical prediction by a mechanical model

A numerical model based on a complete mechanical approach, i.e. without the introduction of any empirical or curve-fitting parameter, was developed by De Stefano et al. (1994) for simulating the cyclic rotational response of web angle connections. The model generates the moment-rotation curves by coupling the load-deformation axial response of double angle segments. As shown in Figure 10, the model can be directly extended to prediction of rotational response of top-and-seat angle connections, which is obtained by coupling the cyclic axial response of the top and of the seat angles. The two rigid bars represent the column (counterbeam) and the beam, which are considered undeformable relative to the two angles, which are represented by two nonlinear springs.

As discussed elsewhere (De Stefano et al. 1994), the numerical simulation accounts both for the bending behaviour of the angle leg fastened to the column flange and for the contact phenomena which alternatively arise at the top and the bottom angles. No allowance is made for slip phenomena, which come out to significantly affect the moment-rotation response of the connections considered herein. Therefore, the simulation has been conducted with reference to the moment M - net rotation ϕ_{net} loops presented in Figure 8. The parameters geometrical and mechanical parameters needed as input for the model have been determined according to definitions given in (De Stefano et al. 1994).

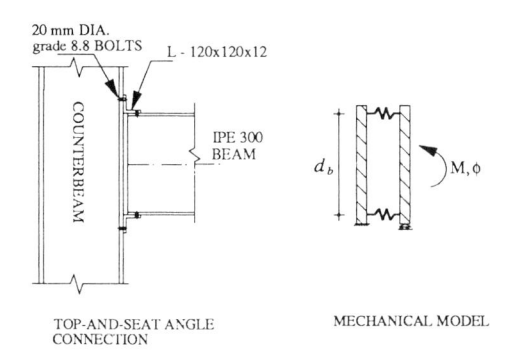

Fig. 10. Mechanical model for top-and-seat angle connections

Figure 11 shows a comparison between experimental and numerical loops for TSC-D. It can be seen that the numerical model provides a good prediction of the joint initial stiffness and strength since it captures the monotonic envelope of the experimental loops well. However, the shape of the numerical loops is different from that of the experimental loops,

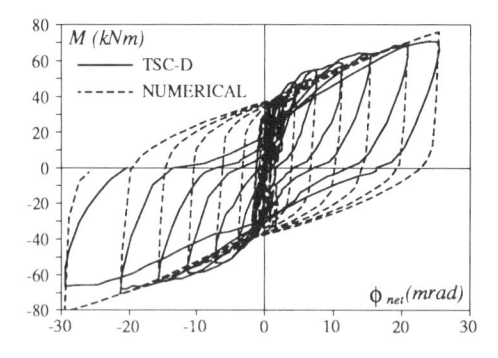

Fig. 11. Comparison between of test and numerical hysteresis loops for test TSC-D.

thus leading to an overestimate of the energy dissipation. Furthermore, the stiffening effect, due to the contact occurring alternatively at the top and bottom angles, is almost not represented by the model. In fact, when the angle in compression touches the counterbeam, the model neutral axis suddenly shifts close to that angle. Therefore, the increase in rotational stiffness provided by the model is given only by the increment of the lever arm of the force sustained by the angle yielded in tension. Such effect appears not to be sufficient to cover the actual one.

6 Conclusions

In this paper, the cyclic moment-rotation response of top-and-seat angle connections, tested at University of Trento, has been examined in order to evidence the main behavioural aspects. In particular, since the measuring arrangement enabled the evaluation of the behaviour of the joint components, in addition to the global rotational response, the effects of bolt slippage have been investigated quantitatively. By comparing the joint total rotations with the net rotations, the latter ones obtained by excluding the portion due to bolt slippage, it has been seen that the such phenomenon affects to a large extent the overall moment-rotation response if the bolt diameter and the pretensioning force are not specifically designed. Finally, a numerical model, based on a pure mechanical approach, has been used to simulate the test response, showing its capability to capture the envelope of the actual cyclic behaviour without any curve-fitting parameter.

7 References

American Institute of Steel Construction (AISC) (1989) Manual of Steel Construction - Allowable Stress Design - Ninth Edition.

Astaneh, A., Nader, M. and Malik, L. (1989) Cyclic behaviour of double angle connections. **Journal of Structural Engineering**, ASCE, Vol. 115, No. 5.

Azizinamini, A. and Radziminski, J.B. (1989) Static and cyclic performance of semirigid steel beam-to-column connections. **Journal of Structural Engineering**, ASCE, Vol. 115, No. 12.

Ballio, G., Calado, L., De Martino, A., Faella, C., Mazzolani, F.M. (1987) Cyclic behaviour of steel beam-to-column joints experimental research. **Costruzioni Metalliche**, No. 2.

Bernuzzi, C. (1992) Cyclic response of semi-rigid steel joints. **Proceedings**, First State of the Art Workshop - Semi-Rigid Behaviour of Civil Engineering Structural Connections, COST-C1, Strasbourg, France.

Bernuzzi, C., Zandonini, R. and Zanon, P. (1991) Rotational behaviour of end plate connections. **Costruzioni Metalliche**, No. 2.

Bernuzzi, C., Zandonini, R. and Zanon, P. (1992) Semirigid steel connections under cyclic loads. **Proceedings**, First World Conference on Constructional Steel Design, Acapulco, Mexico.

CNR-UNI 10011 (1988) Steel constructions: Recommendations for design, execution and maintenance (in Italian)

De Stefano, M., De Luca, A. and Astaneh, A. (1994) Modeling of cyclic moment-rotation response of double-angle connections. **Journal of Structural Engineering**, ASCE, Vol. 120, No. 1.

European Convention for Constructional Steelwork (1986) Recommended testing procedures for assessing the behaviour of structural elements under cyclic loads. Technical Committee 1, TWG 1.3 - Seismic Design, Publ. 45

Goverdan, A.V. (1983) A collection of experimental moment-rotation curves and evaluation of prediction equations for semi-rigid connections. M.S. Thesis, Vanderblit University, Nashville, Tennessee.

Nethercot, D.A. (1985) Steel beam-to-column connections - a review of test data and its applicability to the evaluation of joint behaviour in the performance of steel frames. CIRIA Project Study.

50 SEMIRIGID TOP-AND-SEAT CLEATED CONNECTIONS: A COMPARISON BETWEEN EUROCODE 3 APPROACH AND OTHER FORMULATIONS

M. DE STEFANO
Istituto di Tecnica delle Costruzioni, University of Naples, Italy
C. BERNUZZI
University of Trento, Trento, Italy
E. D'AMORE and A. DE LUCA
University of Reggio Calabria, Reggio Calabria, Italy
R. ZANDONINI
University of Trento, Trento, Italy

Abstract
This paper is devoted to assessment of adequacy of some methods of prediction of semirigid joints moment-rotation behaviour. In order to carry out such comparison, which includes the formulation of the Eurocode 3, reference is made to cleated connections, which constitutes a joint typology of great interest for design purposes. The comparison has led to better understanding of the influence of some behavioural factors, such as bearing and strain hardening, the knowledge of which allows to obtain a more realistic prediction of the joint flexural behaviour both in the elastic and in the inelastic range of behaviour.
Keywords: Semirigid Connections, Methods of Prediction, Strain Hardening

1 Introduction

The concept of semirigid connection response is well known since the beginning of this century, and has been used in design since the '30s also for tall building framing. Studies were carried out of joint behaviour and the main problems related to elastic frame analysis incorporating joint flexibility were solved (Zandonini, 1992). Theoretical knowledge did not actually have an immediate impact on practice. Rather, simplified models were adopted in many countries, enabling the designers to use the traditional ideal models of "rigid" and "pinned" joint (e.g., the so-called wind connection method). Codes were not providing any specific provision for semi-rigid frame design until very recently.
A comprehensive approach to design of steel structures, comprising extensive guide-lines, has been included in the Eurocode 3 (1992), which can be considered the pioneer of the last generation of steel codes. This code, in fact, makes allowance for partial strength and/or semirigidity of the connections when performing analysis of the so-called semi-continuous frames. A complete Annex is devoted to the definition of the properties of semirigid connections according to the philosophy of components. Such philosophy is based on identification of all components that can affect the joint response. In particular, the present version of the Annex J (1994) to the EC3 introduces the components that refer to fully welded connections, flange cleated connections and end-plate connections.
The Annex J rules allow to construct a design moment-rotation relationship for the joint by combining the properties of the components, that are considered relevant for the joint typology, in order to define both the connection strength and the connection stiffness. In this manner, the philosophy of components allow a wide applicability of the Annex J. The drawback of this approach is represented by the complexity deriving from the application of all the formulas provided in the code.
This paper is aimed at appraising some methods of prediction of semirigid joints moment-rotation behaviour. In particular, the well known models by Richard and Abbott

(1975) and Kishi and Chen (1990), as well as the EC3 formulation contained in the Annex J, have been considered. In order to carry out such comparison, reference is made to cleated connections, which constitutes a joint typology of great interest for design purposes, even though in the past larger research efforts have been devoted to different types of connections, such as end-plate connections. The comparison has led to better understanding of the influence of some behavioural factors, the knowledge of which allows to obtain a more realistic prediction of the joint flexural behaviour both in the elastic and in the inelastic range of behaviour.

2 Experimental research

This section contains a brief review of experimental research in the field of monotonic behaviour of semirigid connections. This is essentially aimed at identifying the main features of the moment M - rotation ϕ curves that are of interest for the purpose of developing analytical methods of prediction. In particular, the significant phases of behaviour that are common to most test curves are evidenced, even if it is to be pointed out that the relative importance of each phase varies from case to case.

In past years, a large number of test programs has been conducted in order to provide a better understanding of the behaviour both in USA and in the European countries. Among the former ones, reference is made to tests performed by Azizinamini et al. (1987) on top-and-seat angle connections with web angles and to those by Lewitt et al (1966), reported by Kishi and Chen (1990), on double web angle connections. For what concerns the tests conducted in Europe, curves obtained by Jaspart (1991) and by Bernuzzi et al. (1992) for top-and-seat angle connections are considered. Samples of all the above test programs are reported in Figure 1. It can be seen that significant slip phenomena affected the moment-rotation response of top-and-seat angles tested by Jaspart and Bernuzzi et al., whereas joints tested by Azizinamini et al. and by Lewitt were able to develop continually increasing moments through the full range of imposed rotations. Effect of bolt slippage is examined in a companion paper (Bernuzzi et al. 1994) with reference to tests made by Bernuzzi et al. (1992). Except for this aspect, all curves denote similar behaviour, characterized by three phases: 1) elastic phase, with rotational stiffness almost constant and approximately equal to its initial value $S_{j,init}$; 2) yielding phase, within which the moment-rotation curves exhibit the largest nonlinearity; 3) hardening phase, within which the response can be considered linear, as in the first phase, with a rotational stiffness $S_{j,h}$ far smaller than $S_{j,init}$.

As far as the hardening phase, De Stefano and Astaneh (1991) and Jaspart (1991) have investigated the physical phenomena that are involved in this state, leading to the conclusion that for monotonic behaviour the ratio $S_{j,h}/S_{j,init}$ is related to the ratio of the steel hardening modulus E_h to the Young's modulus E. Namely, in case of top-and-seat angle connections, Jaspart (1991) suggested the following relationship:

$$\frac{S_{j,h}}{S_{j,init}} = \frac{E_h}{E} \qquad (1)$$

A precise evaluation of the ratio E_h/E is rather complex, since there is no well-established definition for E_h. Furthermore, E_h is dependent on several factors. Massonet and Save (1980), based on a wide collection of experimental data, showed that E_h/E is affected by the strength of the steel. However, for low and medium strength (mild) steels, that are the most widely used, values of E_h/E were always close to 1/50. Kato et al. (1990) conducted a statistical analysis on mechanical properties of steels usually used in Japan. The mean values of E_h/E reported in their study was close to 1/60. It has to be said, though, that if one wants to assume a reasonable bilinear approximation for the entire stress-strain curve, this value grows up to approximately 1/30. In any case, it has to be said that there is no

general agreement about values of E_h/E to be considered for different types of steels.
At the connection level, experimental information on the ratio $S_{j,h}/S_{j,init}$ can be inferred from the results obtained by Attiogbe and Morris (1991), when fitting moment-rotation curves of five types of semirigid connections with analytical expressions depending on both stiffness parameters. Referring to top-and-seat angle connections, the ratio $S_{j,h}/S_{j,init}$ was ranging between 1/15, for low stiffness connections, and 1/43 for high stiffness.

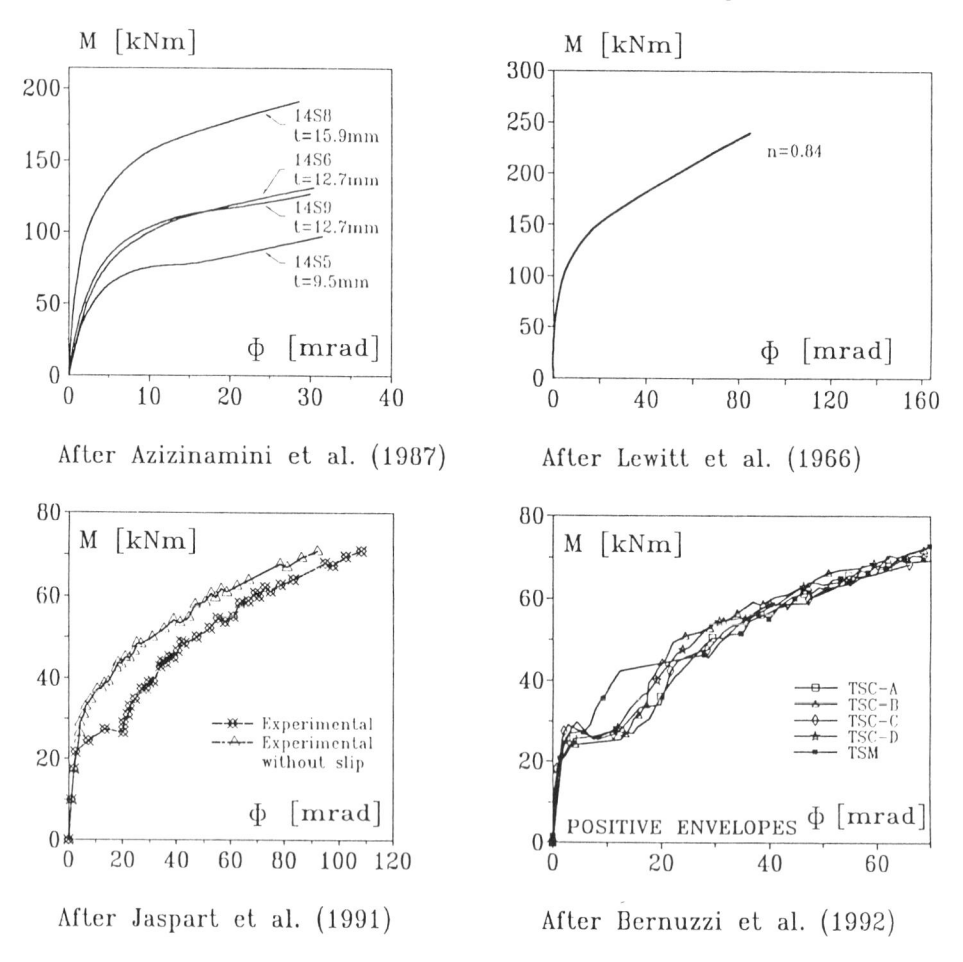

Fig. 1. Experimental moment-rotation curves of semirigid cleated connections

3 An appraisal of different models

Several types of prediction models have been developed in past years to model semirigid behaviour of steel beam-to-column connections. The available models can be subdivided into mathematical expressions and analytical models. Among the former models, it is worth mentioning the polynomial expression by Sommer (1969) and by Frye and Morris (1975), whose coefficients are to be determined through curve-fitting procedure. Also the analytical models consist of standardized models, but the parameters which they are dependent upon have a physical meaning. Among these methods, the Richard and Abbott (1975) power model, expressed as follows:

$$M = S_{j,h}\phi + \frac{(S_{j,init} - S_{j,h})\phi}{[1 + ((S_{j,init} - S_{j,h})\phi/M_o)^n]^{1/n}} \quad (2)$$

where M_o and ϕ_o are reference moment and rotation respectively, and n is a shape parameter, has been widely used. It should be noticed that all parameters of Equation (2) are to be defined by curve-fitting. Such model was adopted by Attiogbe and Morris (1991) to fit rotational response of several types of connections, including top-and-seat angle connections, as mentioned in the previous section.

A simplified version of the Richard and Abbott model was considered by Kishi and Chen (1990), which basically is obtained by Equation (2) by setting $S_h=0$ and taking M_o as the ultimate moment capacity of the connections. A mechanical approach was used to define the initial stiffness and the ultimate moment, considering different types of cleated connections, while the shape parameter n was determined by curve fitting. It has to be said that the assumption by Kishi and Chen leads to flatten the moment-rotation curve as the moment approaches M_o. This somehow apparently contradicts the experimental evidence, since hardening affected always moment-rotation curves; however, it will be shown later that the method used for estimating M_o leads to an overestimate of such parameter so that the hardening branch of experimental curves is always included into the steeper branch of the analytical curves. In this sense, the parameter M_o is to be considered as a curve-fitting one.

The formulation introduced into the Annex J to the EC 3 in order to define the joint moment-rotation characteristic relationship belongs to the analytical models since the moment-rotation response is given by:

$$\phi = \frac{(1.5M/M_o)^\xi}{S_{j,init}} \cdot M \quad (3)$$

where M_o is the moment resistance of the connection and ξ depends on the joint typology: for top-and-seat angle connections $\xi=3.1$. Equation (2) holds for $M \leq M_o$. The following limitations apply to Equation (2):

$$\left(1.5\frac{M}{M_o}\right)^\xi \geq 1$$

thus implying that the slope of the moment-rotation curve is constant and equal to $S_{j,init}$ if $M \leq 2/3 M_o$.

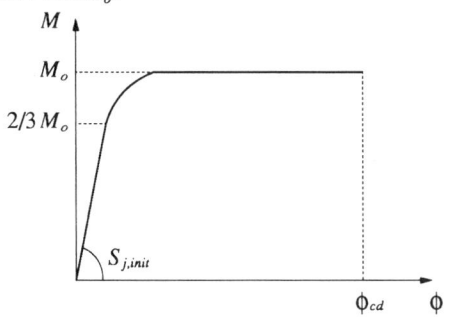

Fig. 2. Joint characteristic curve of Annex J

After reaching the moment resistance M_o, the moment characteristic curve presents a plateau, without strain hardening, up to the ultimate rotation ϕ_{cd}, as shown in Figure 2. The Annex J analytical formulation (3) is dependent on three parameters: initial stiffness $S_{j,init}$, moment capacity M_o and a shape factor ξ. The main difference among the EC3 approach and the other formulations relies on the definition of $S_{j,init}$. In the EC3, in fact, a component method is adopted for the definition of this property.

Within the component method, all the contributions of the components existing in the connections are singled out and, for each of them, a stiffness coefficient is evaluated and then all the coefficients are coupled as in a series system. More in detail up to ten components are embodied in the EC3 and therefore several computations, which are demonstrated in the Appendix A, are needed. It emerges from the above discussion that all analytical models, including those not considered herein for brevity, do not include any formulation or parameter for introducing: 1) the effect of bolt slippage as dependent upon load level and 2) a failure criterion, i.e. an ultimate rotation capacity ϕ_{cd}. For the rotation capacity, the Annex J, though introducing such parameter, does not provide specific indications for its computation.

4 Comparison among different formulations

In order to carry out a meaningful comparison among the EC3 approach and other formulations, extensive test data, also coming from different laboratories, should be considered. It has to be said, though, that the analysis herein is restricted to top-and-seat angle connections for which the available experimental data, as shown in the previous section, are consistently characterized by the same typical features: elastic phase, yielding phase, hardening phase. For this reason and having the purpose of taking simple the comparison, the test data by Bernuzzi et al. (1992) relative to a top-and-seat angle connection has been selected for the comparison.

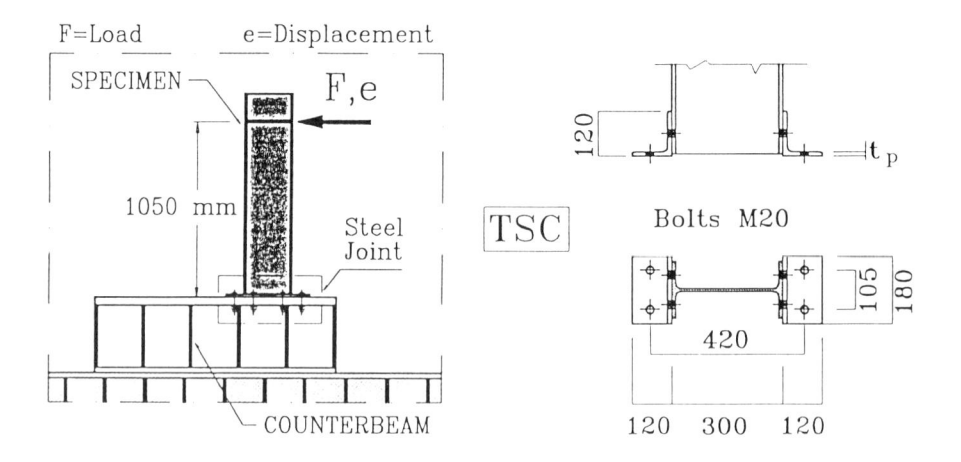

Fig. 3. Test set-up and geometry of the top-and-seat angle connection.

It has to be underlined that the adopted monotonic moment-rotation curve has been shown elsewhere (Bernuzzi 1992) to be representative of the monotonic envelope of four tests.
Therefore, the test curve under examination can be considered as the characteristic monotonic moment-rotation curve for this type of connection, being free of any test specificity.
The test set-up and the geometrical dimensions of the joint, completely described along with the mechanical properties in a companion paper (Bernuzzi et al., 1994), are given in Figure 3.

4.1 Application of the Richard-Abbott model

As already pointed out, the Richard-Abbott model depends on parameters defined by curve-fitting. The initial stiffness $S_{j,init}$ and the hardening stiffness $S_{j,h}$ have a clear physical meaning: for the connection under examination values of 25 $kNm/mrad$ and 0.5 $kNm/mrad$ respectively were taken. Attiogbe and Morris (1991) suggested methods for determining values of the reference moment M_o and of the shape parameter n. According to the procedure called 'method of selected points', the function has been forced to pass through the origin and two experimental data points, thus obtaining for M_o a value of 36.80 kNm and for n a value of 0.7799.

Figure 4 shows a comparison between the experimental moment-rotation curve and the Richard-Abbott function, showing an excellent agreement within all three phases of behaviour. It has to be underlined, though, that the Richard-Abbott analytical model is completely based on a curve-fitting approach, thus not evidencing any physical component of behaviour.

4.2 Application of the Kishi-Chen model

The Kishi-Chen model is dependent on three parameters: initial stiffness $S_{j,init}$, moment capacity M_o and the shape parameter n.

For what concerns the initial stiffness, Kishi and Chen (1990) proposed a procedure based on a mechanical approach. It is assumed that the center of rotation of the connection coincides with the center of the seat angle leg adjacent to the beam flange and that the joint flexibility is due to flexural and shear deformation of the top angle leg adjacent to the column flange only. Namely, the top angle leg is idealized as a propped cantilever, whose cross section bending stiffness is denoted as EI_t (equal to 5.3395 kNm^2 for the tested connection). Then, the rotational stiffness $S_{j,init}$ is expressed as:

$$S_{j,init} = \frac{3EI_t}{1+0.78t^2/g_1^2} \cdot \frac{d_1^2}{g_1^3} \qquad (4)$$

In the previous Equation, according to definitions given by Kishi and Chen (1990), d_1 represents the distance between the centers of legs of the top and seat angles (equal to 312 mm for the tested connection), whereas g_1 is given by:

$$g_1 = g_t - w/2 - t/2 \qquad (5)$$

where g_t is the gauge distance from the top angle's heel to the center of bolts hole, w is the nut's width and t is the angle thickness. For the tested connection g_t, w and t are equal to 60 mm, 37 mm and 12 mm. By inserting the above values into Equation (4), the value of $S_{j,init}$ obtained through the simple formulation by Kishi and Chen is 30.756 $kNm/mrad$. This value slightly overestimates the experimental value of 25 $kNm/mrad$, since only one source of deformability is accounted for.

For what concerns the moment capacity M_o, it is considered a collapse mechanism consisting of two yield lines located along the top angle leg at a distance equal to:

$$g_2 = g_1 - k_t \qquad (6)$$

where k_t indicates the distance from the top angle heel to the toe of the fillet. The distance g_2 is equal to 10.5 mm for the connection tested by Bernuzzi et al.

The complete description of the used procedure, which includes also effects of moment-shear interaction through the Drucker yield criterion, is reported by Kishi and Chen (1990). It is worth underlining that the steel yield stress is introduced into the formulas. Therefore, the obtained M_o should be considered as a yield moment, instead of an ultimate capacity. It comes out, indeed, that M_o is equal to 88.221 kNm which is clearly greater than the load level corresponding to onset of significant inelasticity. However, the above value of M_o is close to the failure moment, sustained by the connection at the end of the hardening phase. Therefore, by selecting a proper shape factor n, it is possible to include most of the experimental moment-rotation curve in the steeper part of the analytical function, thus overcoming the drawback due to the explicit absence of an hardening branch. Figure 4 contains the Kishi-Chen curve obtained by setting $n=0.60$.

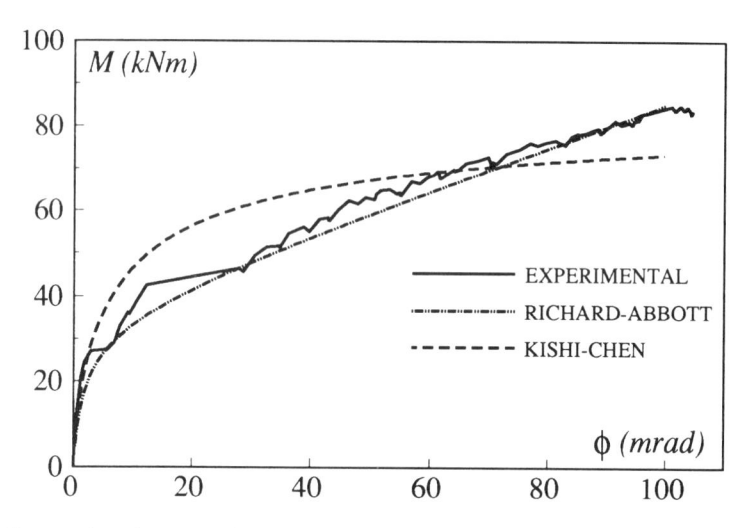

Fig. 4. Comparison between the experimental and the analytical moment-rotation curves

4.3 Application of the Eurocode 3 design model
The Annex J to Eurocode 3 gives rules to model semirigid joints by defining through Equation (3) a characteristic moment-rotation curve as a function of the following joint properties: initial stiffness $S_{j,init}$, design moment resistance M_o.
Namely, the design moment-rotation relationship of a joint is built up by combining the properties of the relevant joint components. The following components are considered:
- column web in shear
- column web in compression
- column web in tension
- column flange in bending
- end plate in bending
- bolted flange cleat in bending
- plate in tension or compression
- plate in bearing (beam flange or end-plate or cleat)
- bolts in tension
- bolts in shear
- beam flange and web in compression
- beam web in tension

For each component of behaviour, the properties in terms of strength, stiffness and deformation capacity are determined and, subsequently, are combined in order to derive the overall joint properties.
For what concerns the initial stiffness semi-rigid connections, the components are assumed to form a series system; therefore, the joint rotational stiffness $S_{j,init}$ is obtained by a combination of the flexibilities of the components as follows:

$$S_{j,init} = \frac{Eh_t^2}{\sum_i 1/k_i} \tag{7}$$

where E is the Young's modulus, h_t is the lever arm and k_i is the stiffness coefficient of the i-th component.

For top and seat angle connections, the following components are considered relevant: column web in shear without stiffener (k_1), column web in tension (k_2), column web in compression (k_3), column flange in bending (k_4), bolts in tension (k_5), flange cleat in bending (k_7), flange cleat in bearing (k_8), bolts in shear (k_9), beam flange in bearing (k_{10}). Appendix A contains all computations needed to evaluate $S_{j,init}$ and M_o. It comes out that the direct application of the Annex J rules leads to $S_{j,init}=0.96$ $kNm/mrad$ and to $M_o=41.99$ kNm. Introducing these values into Equation (3), the characteristic curve has been developed: it can be seen from Figure 5 that the moment-rotation curve predicted by the Annex J is quite far from the experimental one. This is primarily due to the fact that the predicted value of $S_{j,init}$ is much lower than the actual one, whereas the value of M_o appears close to the load level corresponding to onset of yielding. Results reported in Appendix A demonstrate that the bearing components introduce very large deformabilities, since values of the corresponding stiffness coefficients, k_8 and k_{10}, are far smaller than those related to the other components. However, it is to be underlined that those deformabilities do not arise in the elastic range of behaviour since they essentially depend on the bearing capacity of the flange cleat and of the beam flange.

If the bearing components were neglected, i.e. values of the stiffness coefficients k_8 and k_{10} are taken as infinity, the curve, predicted by modifying in this fashion the Annex J formulation, would provide an acceptable approximation of the actual behaviour for $M \leq M_o$. In fact, omitting the bearing components results in a value of $S_{j,init}$ equal to 26.319 $kNm/mrad$, which is quite close to the experimental one. However, the Annex J formulation must be further modified in order to introduce strain hardening. To accomplish this scope, a linear variation in the moment M with the rotation ϕ can be assumed with slope equal to $S_{j,h}$. Based on the indications provided in the review of the experimental research, it has been assumed:

$$S_{j,h} = 1/50 S_{j,init} = 0.5 kNm/mrad$$

Figure 5 shows the moment-rotation curve obtained by modifying the Annex J rules as above described.

From the above discussion, it emerges that the formulation proposed by the Annex J for defining the joint properties and the joint moment-rotation characteristic curve can be improved by implementing the following two modifications:
1) when defining the initial *elastic* rotational stiffness the terms accounting for the bearing, which arises in the inelastic range of behaviour, should be neglected; alternatively, the simple formulation by Kishi and Chen appears to be suitable;
2) the moment-rotation characteristic curve should present a strain hardening branch, after reaching the moment resistance M_o, which essentially corresponds to onset of significant yielding.

5 Conclusions

In this paper, an analysis of applicability of some prediction methods for semirigid connection behaviour has been carried out with reference to top-and-seat angle connections. The comparison with the experimental moment-rotation curves has evidenced that some behavioural factors, such as bolt slippage and damage, still need to be introduced explicitly in all examined formulations in order to achieve a more realistic prediction of the joint moment-rotation curves. The design approach of the Annex J to EC3, which is to be considered the most advanced one among the current steel codes, has been presented and appraised into detail. It has been found that the prediction of the initial stiffness can be improved by not including the coefficients which account for the bearing, since this factor usually does not appear in the initial stage of behaviour. Furthermore, a more realistic prediction of the moment-rotation curve can be obtained by characterizing the post-yielding behaviour with a strain hardening branch. In any case, the behaviour of this form of connections under cyclic reversal loadings is complex due to the influence of several physical phenomena, as evidenced before (i.e., damage, slippage, bearing, strain hardening). Therefore, further studies, both experimental and theoretical, are planned by the authors in order to improve both the state of knowledge and the existing methods of prediction.

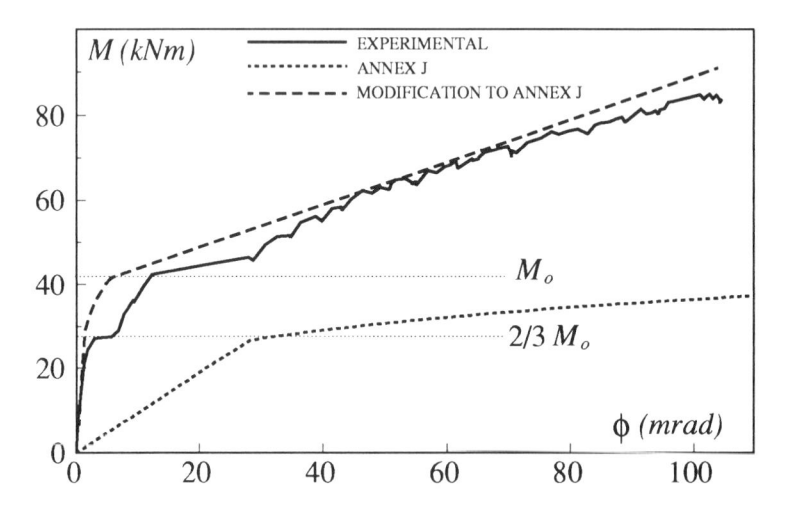

Fig. 5. Comparison between the experimental and the Annex J moment-rotation curves

References

Attiogbe, E. and Morris, G. (1991) Moment-rotation functions for steel connections. **J. Struct. Engrg.**, ASCE, 117(6), 1703-1718.
Azizinamini, A., Bradburn, J.H. and Radziminski, J.B. (1987) Initial stiffness of semi-rigid steel beam-to-column connections. **J. Construct. Steel Res.**, 8, 71-90.
Bernuzzi, C. (1992) Cyclic response of semi-rigid steel joints. **Proceedings**, First State of the Art Workshop - Semi-Rigid Behaviour of Civil Engineering Structural Connections, COST-C1, Strasbourg, France.
Bernuzzi, C., Zandonini, R. and Zanon, P. (1992) Semirigid steel connections under cyclic loads. **Proceedings**, First World Conference on Constructional Steel Design, Acapulco, Mexico.
Bernuzzi, C., De Stefano, M., D'Amore, E., De Luca, A. and Zandonini, R. (1994) Moment-rotation behaviour of top-and-seat angle connections, in **Behaviour of Steel Structures in Seismic Areas** (ed. F.M. Mazzolani), E & FN Spon, London.
CEN (1992) - Eurocode 3 - Design of steel structures.
CEN (1994) - Eurocode 3 , Part 1.1 - Joints in Building Frames (Revised Annex J). 4th draft prepared for PT 9 - NTC's meeting in Delft.

De Stefano, M. and Astaneh, A. (1991) Axial force-displacement behaviour of steel double angles. **J. Construct. Steel Res.**, 20, 161-181.

Frye, M.J. and Morris, G.A. (1975) Analysis of flexibly connected steel frames. **Can. J. Civ. Eng.**, 2(3), 280-291.

Jaspart, J.P. (1991) Study of semirigidity of beam-to-column joints and of its influence on the strength and stability of steel frames (in French), Ph.D. Thesis, University of Liege, Dept. of Applied Sciences, Belgium.

Kato, B., Aoki, H. and Yamanouchi, H. (1990) Standardized mathematical expression for stress-strain relations of structural steel under monotonic and uniaxial tension loading. **Mater. Struct.**, RILEM, 23, 47-58.

Kishi, N. and Chen, W.F. (1990) Moment-rotation relations of semirigid connections with angles. **J. Struct. Engrg.**, ASCE, 116(7), 1813-1834.

Lewitt, C.W., Chesson, E., Jr. and Munse, W.H. (1966) Restraint characteristics of flexible riveted and bolted beam-to-column connections. Dept. Civ. Engrg., Univ. of Illinois, Urbana.

Massonet, C. and Save, M. (1980) **Calcolo plastico a rottura delle costruzioni** (Italian edition of the book Calcul Plastique des Constructions), edited by CLUP, Milan, Italy.

Richard, R.M. and Abbott, B.J. (1975) Versatile elastic-plastic stress-strain formula. **J. Engrg. Mech. Div.**, ASCE, 101(4), 511-515.

Sommer, W.H. (1969) Behaviour of welded header plate connections, M.A.S. Thesis, University of Toronto, Toronto, Canada.

Zandonini, R. (1992) Analysis and design of steel frames with semi-rigid joints. ECCS Publication No.67

Acknowledgments

The financial support of C.N.R. (Italian National Council of Research) is gratefully acknowledged.

Appendix A: evaluation of the joint parameters through the Annex J

In the following, the computations needed for evaluating the joint parameters $S_{j,init}$ and M_o through Annex J are reported, with reference to the release published in February 1994.

In order to identify the parameters that are involved in the computation of the stiffness coefficients for the tested connections, it is to be underlined that the parameters referred by the Annex J to the column in this application are those of the counterbeam of the experimental set-up. In the Annex J it is stated that, if the column web is stiffened, the relevant coefficients should be taken as infinity. Since the counterbeam web was stiffened, the above indication has been applied, thus setting k_1, k_2 and k_3 equal to infinity.

According to the Annex J, the stiffness coefficient k_4 is expressed as:

$$k_4 = \frac{0.85 \cdot l_{eff} \cdot t_{fc}^3}{m^3} = \frac{0.85 \cdot 204.1 \cdot 50^3}{32.5^3} = 631.7159$$

where parameters l_{eff}, t_{fc} and m are to be defined with reference to the counterbeam flange (see Annex J for details). This term is very large since the counterbeam is very stiff, thus giving a negligible contribution in the Equation (6).

The coefficient k_5 is given by:

$$k_5 = 1.6 \cdot \frac{A_s}{L_b} = 1.6 \cdot \frac{245}{68} = 5.7647$$

where A_s is the net area of bolts and L_b is the elongation length of bolts.

The coefficient k_7 is given by:

$$k_7 = \frac{0.85 \cdot l_{eff} \cdot t_a^3}{m^3} = \frac{0.85 \cdot 90 \cdot 12^3}{35^3} = 3.0832$$

where parameters l_{eff}, t_a and m are to be defined with reference to the flange cleat (see Annex J for details). This term introduces the flexural behaviour of the top angle adjacent to the counterbeam, which is the only source of deformability considered by Kishi and Chen when defining $S_{j,init}$ through Equation (4).

The coefficient k_8 is provided by:

$$k_8 = \frac{2.4 \cdot n_b \cdot K_b \cdot K_t \cdot f_{u,a} \cdot d}{E} = \frac{2.4 \cdot 2 \cdot 0.375 \cdot 1.125 \cdot 460 \cdot 20}{206000} = 0.0904$$

where parameters n_b, K_b, K_t, $f_{u,a}$ and d have been introduced according to definitions of the Annex J.

The coefficient k_9 is given by:

$$k_9 = \frac{8 \cdot n_b \cdot f_{u,b} \cdot d^2}{E \cdot d_{M16}} = \frac{8 \cdot 2 \cdot 1000 \cdot 20^2}{206000 \cdot 16} = 1.9417$$

The coefficient k_{10} is defined as:

$$k_{10} = \frac{2.4 \cdot n_b \cdot K_b \cdot K_t \cdot f_{u,fb} \cdot d}{E} = \frac{2.4 \cdot 2 \cdot 0.375 \cdot 1.0031 \cdot 360 \cdot 20}{206000} = 0.0631$$

The above stiffness coefficients are combined through Equation (7), where the lever arm, for top and seat angle connections, is taken as the distance from the bolt row in tension and the centre of compression, which is located at the interface between the beam bottom flange and the seat angle. Therefore, parameter h_t has been set equal to 360 mm and the initial rotational stiffness is equal to:

$$S_{j,init} = \frac{206000 \cdot 360^2}{\sum_i 1/k_i} = 0.9563 \quad kNm/mrad$$

The value of $S_{j,init}$, estimated through the formulation of the Annex J, is much lower than the experimental value, which is about 25 kNm/mrad.

If the stiffness coefficients k_8 and k_{10} were set equal to infinity, i.e. if the deformabilities corresponding to bearing of flange cleat and beam flange were neglected, the initial stiffness would be equal to:

$$S_{j,init} = 26.319 \quad kNm/mrad$$

thus providing a value which is quite close to the one obtained by the monotonic test.
For what concerns the moment resistance M_o of top and seat angle connections, the following formula is provided by the Annex J:

$$M_o = F_o \cdot h_t$$

where F_o is the strength of the weakest component, i.e. the flange in tension. For the evaluation of F_o the Annex J idealizes the bolted cleats by means of equivalent T-stubs. The three possible modes of failure of the flanges of these equivalent T-stubs are similar to those which are likely to occur in the cleats:

complete yielding of the flange:

$$F_o = \frac{4 \cdot M_{pll,o}}{m} = \frac{4 \cdot 0.25 \cdot l_{eff} \cdot t_f^2 \cdot f_y}{m} = \frac{4 \cdot 0.25 \cdot 90 \cdot 12^2 \cdot 315}{35} = 116640N$$

bolt failure with yielding of the flange:

$$F_o = \frac{2 \cdot M_{pl2,o} + n \Sigma B_{t,o}}{m + n} = \frac{2 \cdot 0.25 \cdot 90 \cdot 12^2 \cdot 315 + 43.75 \cdot 2 \cdot 196000}{35 + 43.75} = 243698N$$

bolt failure:

$$F_o = \Sigma B_{t,o} = 2 \cdot 196000 = 392000N$$

Therefore, the controlling failure mode is the first one, giving a moment resistance equal to:

$$M_o = \frac{116640 \cdot 0.36}{1000} = 41.99 kNm$$

This value of the strength appears to be close to the one which corresponds to onset of significant inelasticity in the experimental moment-rotation behaviour.

51 NUMERICAL SIMULATION CONCERNING THE RESPONSE OF STEEL FRAMES WITH SEMI-RIGID JOINTS UNDER STATIC AND SEISMIC LOADS

D. DUBINA, D. GRECEA and R. ZAHARIA
Technical University, Timisoara, Romania

Abstract

This paper represents the results of some numerical simulations concerning the response of steel frames with semi-rigid joints under static and seismic loads. It was followed the influence of M-ϕ model curve on the stability and seismic behaviour of steel structures. Also, it was studied the influence of semi-rigid joints on the seismic behaviour in comparison with the steel structures with rigid connections. It was tried to make a connection between the critical yield factor from the stability analysis and the q-factor from the seismic analysis.
<u>Keywords:</u> Stability, Seismic Behaviour, Yielding, Hinge, Mechanism, Eigenperiod, q-factor.

1 Introduction

Steel structures with semi-rigid joints have a greater possibility of deformation, in comparison with the rigid ones, which means that they are more sensitive at the second order effects, including the influence of imperfections. Also, the semi-rigid joints are influencing the dynamic response of the steel structures, modifying the structural coefficient q and the plastic hinges biography.

Promotion of steel structures with semi-rigid joints in Romania has to satisfy the seismic criteria of strength and deformability. Following this objective there were in Timisoara many research proposals in the domain, looking for: joint solutions, M-ϕ type curves including experimental researches and numerical simulations on structures.

The purpose of this paper is the influence of M-ϕ model curve on the stability and seismic behaviour of steel structures. The reason is the necessity of adopting a M-ϕ bilinear model, required by the practic analysis and design of steel frames with semi-rigid joints and according with the material behaviour model Prandtl.

2 Analysed structures

The analysed and tested structures were the four calibration frames recomandated by the ECCS (1992) and used also in many papers (Jaspart,1991). The types of profiles used for the beams are: IPE 200 and IPE 300 and for the columns: HE 160B, HE 160A and HE 200B with the dimensions and loading system presented in Fig. 1.

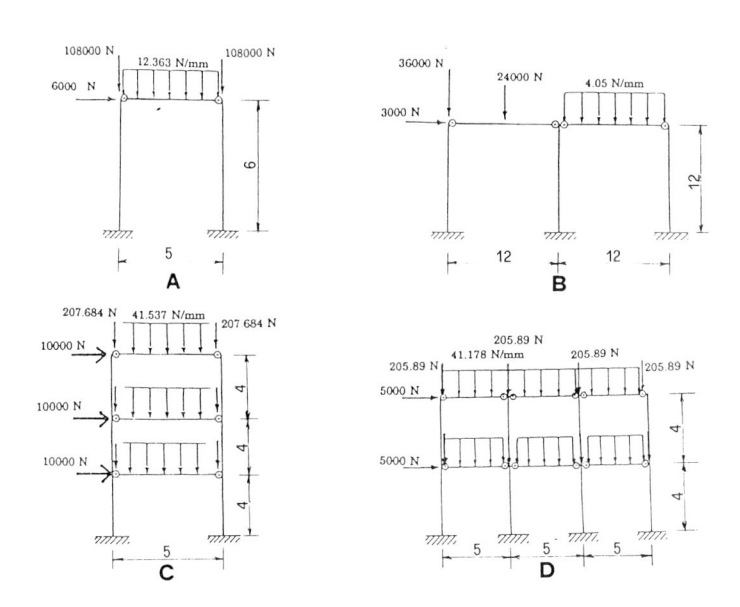

Fig1. Analysed structures

For the numerical analysis there were used real and bilinear M-ϕ curves for the semi-rigid joints. There are two different types of semi-rigid connections for each structure, the characteristics (M_u, K_i) were determined after the theoretical models from the literature: EC 3 (1992) and Chen, Kishi (1987) and are presented in Table 1.

Table 1. Analysed structures: Characteristics

Type of structure Characteristics	A	B	C	D
Ultimate moment	47.1	72	103	54.5
M_u (kNm)	31.4	70	100	37
Initial stiffness	12962	26101	32861	16974
K_i (kNm/rad)	8641.2	22100	17420	9459

3 Influence of the initial stiffness K_0 in the bilinear M-ϕ simulation of the structure static behaviour

Using the programme PEP-micro of CTICM (1992) it was tried to determine the influence of the initial stiffness K_0 in the M-ϕ simulation of the structure static behaviour. With the programmes ROBOT (1991) and respectively, PEP-micro it was made a second order elastic and plastic analysis of the structures, obtaining the elastic and plastic yielding factors λ_{el} and λ_p. Also it was looked for λ_{p1} the load multiplier for the first plastic hinge moment.

This multiplier λ_{el} is equivalent with α_{cr} from EC 3 because it can be said that:

$$\alpha_{cr} = V_{cr}/V = \lambda_{cr}V/V = \lambda_{cr} = \lambda_{el} \qquad (1)$$

In the bilinear model of the M-ϕ curve, the rising branch has different values of the rigidity as K_{sec}, $K_{i/2}$, K_i, and K_{IPE}, where K_{IPE} means the rigidity of the beam. For the elastic analysis, these rigidities were introduced as elastic springs.

In Table 2 there are given the values of the elastic critical coefficient $\lambda_{el}=\lambda_{cr}$ for these different bilinear models of the joints.

Table 2. Analysed structures: Loading critical factor

Structure	Loading critical factor (λ_{el})			
	K_{sec}	$K_{i/2}$	K_i	K_{IPE}
A	6.504	6.884	7.337	7.699
B	8.677	8.841	9.011	9.161
C	10.660	12.210	14.430	17.080
D	5.710	6.056	6.433	6.804

Looking on the results it can be seen that the values of the loading critical factor λ_{el} are quite influenced by the model of the bilinear curve. Also, it can be observed that only structure C can be considered as a rigid structure because $\lambda_{el} = \lambda_{cr} > 10$.

For a better analysis it was represented the elastic critical factor as a function of rotational stiffness. The critical factor was computed for different values of stiffness like $0.1K_i$, $0.2K_i$, $0.3K_i$, K_{sec}, $K_{i/2}$, K_i, $2K_i$ and K_{IPE}. Analysing the represented curves for the four types of structures in Fig. 2, it can be concluded that the factor influenced zone by the rotational stiffness is till the value of K_i, so that the critical factor is influenced only by the models with the rising branch equal with K_{sec} and $K_{i/2}$.

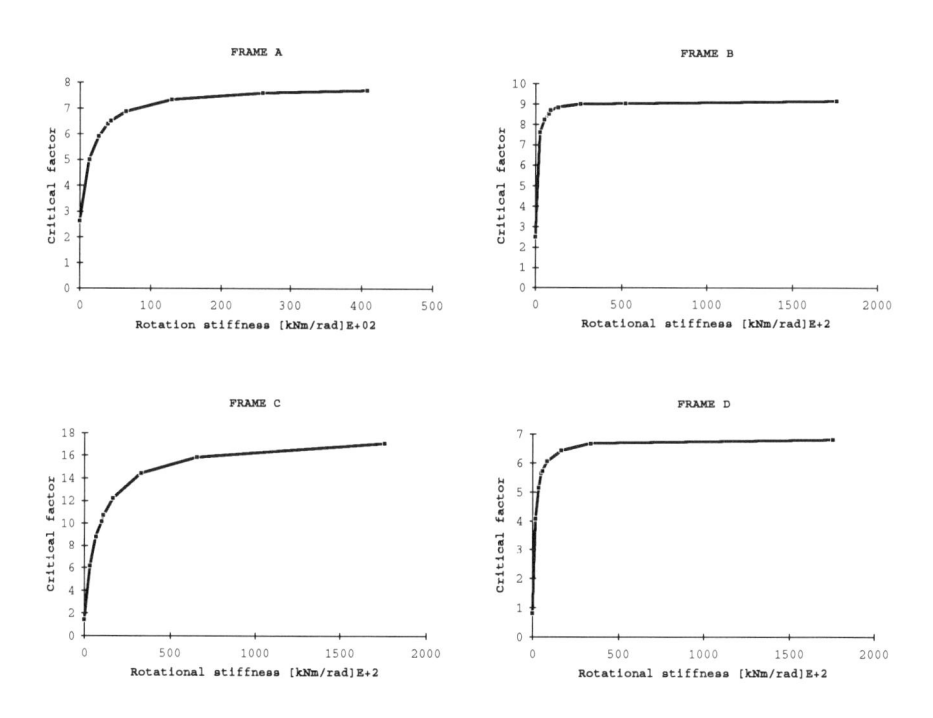

Fig.2. Elastic critical factor

Doing the same curves, but on the same diagram like in Fig.3 and looking once again in Table 2, it can be said that for frame C, which is a rigid structure, the critical factor λ_{el} is more influenced by the different values of the rotational stiffness.

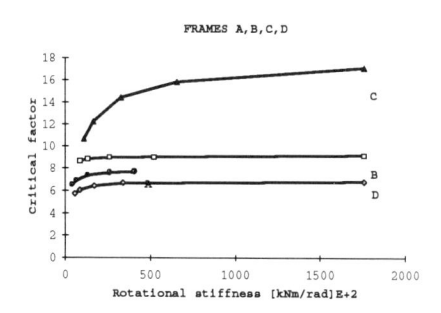

Fig. 3 Analysed structures. Elastic critical factor

The multipliers λ_p, plastic yielding factor and λ_{p1}, load factor for the first plastic hinge moment were computed with the programme PEP-micro with two types of semi-rigid joints for each structure and with different values of initial rotational stiffness K_0 for the bilinear model of M-ϕ curve. The values are given in Tables 3 and 4.

Table 3. Second order elasto-plastic analysis

Structure	M_u (kNm)	Elastoplastic yielding factor (λ_p)			
		K_{sec}	$K_{i/2}$	K_i	K_{IPE}
A	47.00	2.037	2.036	2.039	2.042
	31.40	1.855	1.855	1.859	1.863
B	72.00	2.758	2.757	2.756	2.756
	70.00	2.735	2.734	2.734	2.733
C	10.30	1.545	1.542	1.553	1.553
	10.00	1.554	1.531	1.530	1.537
D	54.50	1.464	1.461	1.452	1.403
	37.00	1.350	1.346	1.341	1.307

Table 4. Second order elasto-plastic analysis. Moment of first plastic hinge

Structure	M_u (kNm)	First plastic hinge in structure (λ_{p1})			
		K_{sec}	$K_{i/2}$	K_i	K_{IPE}
A	47.00	1.924	1.833	1.736	1.665
	31.40	1.418	1.344	1.233	1.141
B	72.00	1.778	1.585	1.385	1.209
	70.00	1.765	1.611	1.383	1.176
C	10.30	1.254	1.135	1.017	0.913
	10.00	1.404	1.306	1.090	0.887
D	54.50	1.259	1.034	0.778	0.538
	37.00	1.121	0.947	0.671	0.366

Analysing the values from Tables 3 and 4 it can be observed that λ_p is weakly influenced by different values of the M-ϕ curve rising branch from K_{sec} to K_{IPE}, because it depends especially of the ultimate moment value M_u.

The only one which is really influenced by these values from K_{sec} to K_{IPE} is λ_{p1}, the first plastic hinge coefficient, which has quite different values for each type of M-ϕ curve rising branch. So the different bilinear M-ϕ curves influence the moment of the first plastic hinge in the structure, which associated with practically the same moment of mechanism development means different intervals from the first plastic hinge to the mechanism for the same structure and the same joint characteristics.

4 Influence of the semi-rigid joints in the seismic analysis

For the dynamic analysis it was studied the influence of the initial stiffness K_0 of the bilinear M-ϕ simulation on the structure first three eigenperiods. This analysis was made with the programme ROBOT(1991) and the results can be seen in Table 5.

Table 5. Dynamic analysis. First three eigenperiods

Structure		Eigenperiod (s)			
		K_{sec}	$K_i/2$	K_i	K_{IPE}
A	T1	0.175198	0.171358	0.167100	0.163918
	T2	0.032476	0.032473	0.032470	0.032467
	T3	0.018573	0.017887	0.017068	0.016418
B	T1	0.815971	0.812056	0.804035	0.797005
	T2	0.127903	0.127901	0.127898	0.127895
	T3	0.099208	0.095677	0.091184	0.086338
C	T1	0.291325	0.271369	0.247826	0.224860
	T2	0.081410	0.077652	0.073036	0.068350
	T3	0.040670	0.040024	0.039181	0.038271
D	T1	0.243130	0.232369	0.220405	0.208205
	T2	0.078673	0.077190	0.075482	0.073662
	T3	0.022613	0.022608	0.022605	0.022600

Looking on the results it can be seen that only the first eigenperiod is major influenced by the different values of the initial rotational stiffness K_0, the second and the third eigenperiods remaining quite constants. Comparing the values of the first eigenvalue T_1 of each structure, it can be seen in Fig.4 that the greatest is for frame B.

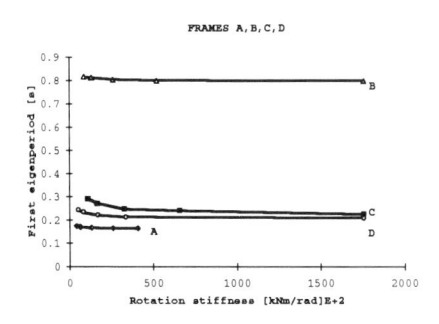

Fig.4. Analysed structures. First eigenperiod.

The seismic analysis was made on the four structures with the programme DRAIN-2D of Berkeley (1973). DRAIN-2D was developed to determine the dynamic inelastic response of the structures. This programme realizes a time-history inelastic analysis in which it is shown the mechanism of the plastic hinges appearance on elements, at each step of the accelerogram.

There were dynamically analysed the four structures already presented before, with the joint beam-column as semi-rigid and rigid. The semi-rigid joint behaviour was modelled by the semi-rigid element of the programme with a bilinear M-ϕ curve characterised by the initial rotational stiffness K_0 and the ultimate moment M_u. Being a dynamical analysis after the attainment of M_u it was considered a degradation of 5%. As accelerogram, it was used the earthquake of Bucharest from the 4th of March 1977, component N-S.

For each structure it was analysed the yield plastic mechanism and the structural coefficient q. To determine the q factor it was used the elasto-plastic dynamic analysis method, recomandated by EC 8, known also as Ballio-Setti method:

$$q = \lambda_{max} / \lambda_e \qquad (2)$$

where: λ_e - multiplier of the accelerogram for the first plastic hinge;

λ_{max} - multiplier of the accelerogram for which the elasto-plastic curve of the displacement joints the elastic one.

There were represented in Fig.5, the elasto-plastic and elastic curves d/de as a function of $\lambda/\lambda e$ for the four frames. This way the q factor was obtained directly on the abscissa.

In Table 6 are given the values for the accelerogram multipliers, the maximum displacements and the structural factor q.

Table 6. Dynamic analysis

Structure		Accelerogram multiplier		Displacement (mm)		Structural coefficient
		λ_e	λ_m	d_e	d_m	q
A	rigid	0.13	0.26	82	139	3.40
	semi-rigid	0.15	0.40	86	143	4.65
B	rigid	0.89	1.30	387	460	3.10
	semi-rigid	0.60	1.40	235	509	4.20
C	rigid	0.12	0.66	89	447	2.40
	semi-rigid	0.04	0.50	68	381	30.80
D	rigid	0.06	0.20	44	122	3.50
	semi-rigid	0.03	0.27	25	133	9.00

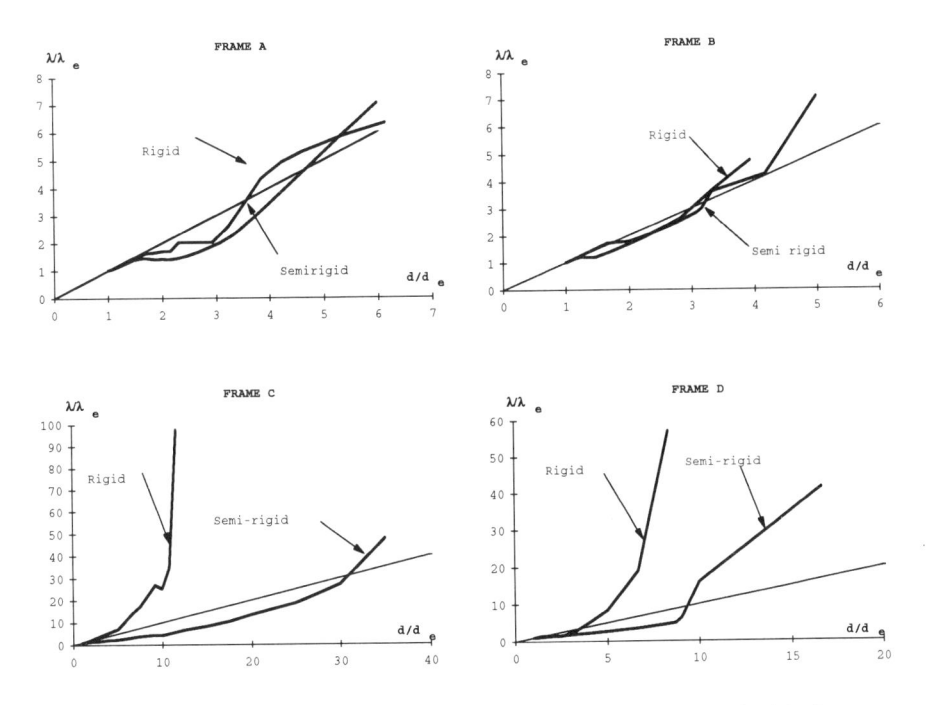

Fig.5 Inelastic behaviour for rigid and semi-rigid frames.

From the dynamic analysis it can be taken some interesting conclusions.

The frames behaviour is characterised by the plastic hinges appearance in the structures, that is the yield mechanism. At the rigid structures can appear a partial floor mechanism with plastic hinges in columns and beams, but at the structures with semi-rigid joints the yield mechanism is always global with plastic hinges at the ends of the beams and finally at the columns base.

Looking in Table 6 it can be seen that the values of the q factor for structures with semi-rigid joints are greater, which means a smaller seismic force. The structural coefficients q of semi-rigid structures and rigid structures are influenced by different factors as:
- the joint flexibility, which can be characterised by the ratio between the connection last moment M_u and the plastic moment of the beam M_{pl};
- the eigenperiods are greater in the case of structures with semi-rigid joints and also the P-Δ effects from the same reason, of greater displacements.

The horizontal displacements of the structures with semi-rigid joints are greater than the horizontal displacements of the rigid structures, but not with differences greater than 15-35%.

The theoretical seismic load is smaller at the semi-rigid structures also by the response factor which is smaller because the semi-rigid structures are more flexible, have greater displacements and so, smaller eigenperiods.

Another advantage may be realised by the fact that the ultimate moments of semi-rigid connections are smaller than those of the connected beams and columns, which makes them passiv controlling devices that protect the columns and the beams from being overstressed (Sedlacek et al.1987).

5 Conclusions

From the static analysis it can be concluded that:

Initial rotational stiffness $K_0=K_i$, $K_i/2$, K_{sec} for the bilinear model of the semi-rigid joint M-ϕ curve influences the structures behaviour, taken into account the specifications of EC 3 about the sensibility at second order effects. The rigid structures with the elastic yielding factor $\lambda_{cr}=\alpha_{cr}=\Sigma V/\Sigma V \geq 10$ are more sensible at the diminution of the semi-rigid joints stiffness, so that they demand a finer modelation of the M-ϕ curve.

First plastic hinge coefficient λ_{p1} is really influenced by the different values of K_0, from K_{sec} to K_{IPE} in comparison with plastic yielding factor λ_p which is not. That means that the yielding plastic mechanism remains the same, being modified only the moment of the first plastic hinge appearance, which means different intervals from the first plastic hinge to the mechanism, for the same structure and the same joint characteristics.

From the dynamic analysis, it can be retained the following main conclusions:

Different initial stiffness of the semi-rigid joints K_0 influence the values of the structural coefficient q, because as it was already said for the same structure with the same joints, there were obtained different intervals from the first plastic hinge to the yielding mechanism, which means different values of the structural coefficient, the q factor.

Only the first eigenperiod is major influenced by the different values of the initial stiffness K_0, the other ones remaining practically constant.

Displacements at structures with semi-rigid joints are greater in comparison with structure with rigid joints, (about 15-35%), so they have greater eigenperiods, which means smaller response factors and smaller calculus seismic loads.

The q factor is greater at structures with semi-rigid joints, than the structures with rigid joints, which means also a smaller calculus seismic load.

Yield mechanism is always a global mechanism in structure with semi-rigid joints, in comparison with a structure with rigid joints, where it can appear partial floor mechanism.

Ultimate moments of semi-rigid connections are smaller than those of the connected beams and columns, which makes them elements that protect the columns and the beams from being overstressed.

After the static and dynamic analysis we consider that only this four structures, also included as calibration frames in ECCS Recommendations (1992), are not enough to take general conclusions about the connection between the static and dynamic analysis, especially between λ_{cr} and q factor.

6 References

Building Research Institute (INCERC) Timisoara (1993) Calculul static, dinamic si de stabilitate al structurilor in cadre etajate cu noduri semi-rigide pentru cladiri civile si industriale. Contract nr.1523/1992, Faza 4/1993.

EUROCODE 3. Design of Steel Structures. (1992) Comission of the European Communities, February 1992.

European Convention for Constructional Steelwork-ECCS, T.C.8/T.W.G.8.1/8.2 (1992) Analysis and Design of Steel Frames with Semi-Rigid Joints,First Edition,Re.N.67/1992

Galèa, Y. Bureau, A. (1992) Logiciel PEP-micro. Analyse de structures à barres en plasticité. Manuel d'utilisation. Version 2-3/92.

Guerra, C.A. Mazzolani, F.M. Piluso, V. (1990) Evaluation of the q factor in steel framed structures. **Ingegneria Sismica**, anno VII, no 2, 1990.

Jaspart, J.P. (1991) Etude de la semi-rigidité des noeuds poutre-colonne et son influence sur la résistance et la stabilité des ossatures en acier. Thése de doctorat. Université de Liège,Faculté de Sciences Appliquées,1991.

Kishi, N. Chen, W.F. Matsuoka, K.G. Nomachi, S.G. (1987) Moment Rotation Relation of Top-and Seat-Angle with Double Web-Angle Connections. **Connections in Steel Structures**, Cachan, 1987.

Powell, G.H. (1973) DRAIN-2D User's Guide. University of California, 1973.

ROBO-BAT (1991) ROBOT STRUCTURES 5.63 Manuel d'utilisation et de reference. Ed. Avril 1991.

Sedlacek, G. Koo, M.S. Ballio, G. (1987) The Response of Steel Structures with Semi-Rigid Connections to Seismic Actions. **Connections in Steel Structures**, Cachan, 1987.

52 CONNECTION INFLUENCE ON THE SEISMIC BEHAVIOUR OF STEEL FRAMES

C. FAELLA, V. PILUSO and G. RIZZANO
Department of Civil Engineering, University of Salerno, Italy

Abstract
In this paper, the influence of the beam-to-column connections on the seismic behaviour of steel frames is analysed by means of a simplified model able to include all the key parameters governing the inelastic response. Both full strength and partial strength semirigid connections are considered. Finally, the results obtained by means of the simplified model are compared with the ones provided by the numerical simulation of the dynamic inelastic response of a group of frames with a specific connection typology subjected to severe ground motions.
Keywords: Connections, Seismic Design, q-factor

1 Introduction

In the european seismic code, Eurocode 8 (Commission of the European Communities, 1993), it is generally requested that connections in dissipative zones have to guarantee sufficient overstrength to allow the yielding of the ends of connected members. It is deemed that the above design condition is satisfied in case of welded connections with butt welds or full penetration welds. On the contrary, in case of fillet weld connections and in case of bolted connections, the design resistance of the connection has to be at least 1.20 times the plastic resistance of the connected member. This means that the use of full-strength connections is suggested and dissipative zones have to be located at the member end rather than in the connections.

The use of partial strength connections, i.e. the contribution of the connections in dissipating the earthquake input energy, is not forbidden. Notwithstanding, it is strongly limited because, in such a case, the experimental control of the effectiveness of such connections in dissipating energy is requested.

In both cases, either full-strength or partial-strength semirigid connections, there isn't any recommendation regarding the value of the behaviour factor to be used in design.

Despite numerous experimental tests regarding the behaviour of beam-to-column connections under monotonic and cyclic loads have been carried out by many researchers, the seismic behaviour of semirigid frames has not been exhaustively investigated.

Recently, Astaneh and Nader (1991, 1992a) have investigated the behaviour of a one storey one bay steel structure with rigid, semirigid and flexible connections subjected to a variety of base excitations using the shaking table of the Earthquake Engineering Research Center of the University of California at Berkeley. The same Authors have investigated, by means of numerical simulations, the seismic response of 4-storey, 7-storey and 10-storey semirigid frames (Astaneh and Nader, 1992b). Moreover, in (Astaneh and Shen, 1990) the seismic inelastic behaviour of an instrumented 6-storey semirigid steel building has been analysed.

Behaviour of Steel Structures in Seismic Areas. Edited by F. M. Mazzolani and V. Gioncu.
Published in 1995 by E & FN Spon, 2–6 Boundary Row, London SE1 8HN. ISBN: 0 419 19890 3.

On the base of the above analyses, a first proposal for a recommended design procedure has been developed in (Astaneh and Nader, 1992a and 1992b). In particular, as in semirigid frames with partial strength connections large plastic deformations have to be experienced by the connecting elements, the connection design has to be done forcing the yielding to occur in a desiderable location within the connection. It is suggested (Astaneh and Nader, 1992b) to design the connections in such a way that yielding occurs only in the plate elements, such as angles, splices and end-plates, while the connectors, such as bolts and welds, have to remain in elastic range. Moreover, design equations for calculating the base shear and the corresponding distribution over the height have been proposed, including also equations for estimating the period of vibration and the design value of the behaviour factor (Astaneh and Nader, 1992b). Nevertheless, the problem does not seem exhaustively studied and its clarification deserves further investigations due to the great number of parameters governing the seismic behaviour of moment resisting frames. In addition, this number increases as soon as the wide range of different connection typologies is concerned.

For these reasons, aiming at a clarification of all parameters affecting the seismic inelastic behaviour of full/partial strength semirigid frames, a simplified model has been introduced in (Faella, Piluso and Rizzano, 1993a), with reference to full strength-semirigid connections. In this paper, starting from the knowledge of the connection influence on the elastic and inelastic behaviour of the model which has been investigated in a companion paper (Faella, Piluso and Rizzano, 1994), the seismic response is analysed with reference also to the case of partial strength-semirigid connections. The approach presents also the advantage to allow the use of the great amount of studies concerning the seismic response of the SDOF systems and, therefore, includes also the effects of the random variability of the ground motion.

2 The analysed substructure

The study of the seismic response of steel frames, including the connection behaviour, can be carried out by means of the substructure represented in Fig.1 (Faella, Piluso and Rizzano, 1993a and 1994). It represents a subassemblages which has been extracted by an actual frame by assuming that beams are subjected to double curvature bending with zero moment in the midspan section and by considering each beam as contemporary belonging to two storeys, so that their mechanical properties has been halved. The flexural stiffness of the columns of the original frame, from which the substructure has been derived, is equal to EI_c/h; the flexural stiffness of the beams is equal to EI_b/L. The moment versus rotation curve of the connection is modelled with a bilinear relation which is completely defined by means of only two parameters: the elastic rotational stiffness K_φ and the ultimate moment M_{uc}. It is evident that, with reference to the elastic range, the comparison between the semirigid substructure and the rigid one is governed by the following nondimensional parameters:

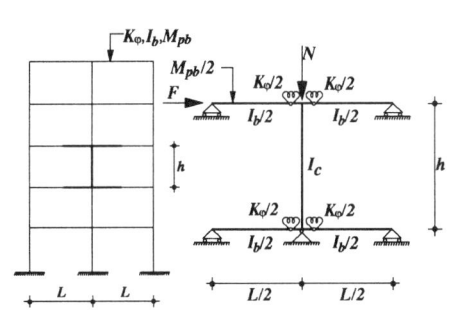

Fig.1
The analysed substructure

$$T1 = \frac{E I_b / L}{E I_c / h} \qquad K = \frac{K_\varphi L}{E I_b} \tag{1}$$

being $T1$ the beam-to-column stiffness ratio and K the nondimensional rotational stiffness of the connection.

In addition, with reference to the inelastic behaviour, the nondimensional ultimate flexural resistance of the connection:

$$\overline{m} = \frac{M_{uc}}{M_{pb}} \tag{2}$$

provides the distinction between the two fundamental cases of full strength connections for which $\overline{m} > 1$ and partial strength connections for which $\overline{m} < 1$.

In (Faella, Piluso and Rizzano, 1994), it has been evidenced that the ratio between the period of vibration of the model with semirigid joints and the one of the model with rigid joints is given by:

$$\frac{T_k}{T_\infty} = \left(\frac{K(1+T1)+6}{K(1+T1)} \right)^{1/2} \tag{3}$$

In addition, also the frame sensitivity to second order effects is affected by the connection deformability. It can be is expressed through the stability coefficient γ, which represents the slope of the softening branch of the behavioural curve relating the multiplier of the horizontal forces α to the nondimensional top sway displacement δ/δ_1 (being δ_1 the top displacement under the horizontal forces corresponding to $\alpha = 1$),

According to (Faella, Piluso and Rizzano, 1994), the ratio between the stability coefficient of the semirigid model γ_k and the one of the rigid model γ_∞ is given by:

$$\frac{\gamma_k}{\gamma_\infty} = \frac{K(1+T1)+6}{K(1+T1)} \tag{4}$$

As soon as the inelastic behaviour is concerned, the distinction between full strength and partial strength connections becomes determinant, because in the first case dissipative zones are located at the beam ends while, in the second one, they are in the connecting elements. In case of full strength connections, a decrease of the available ductility arises while the ultimate resistance of the frame is slightly decreased due to second order effects (Faella, Piluso and Rizzano, 1993b and 1994).

The available ductility of the substructure with full strength-semirigid connections can be expressed as (Faella, Piluso and Rizzano, 1994):

$$\mu_k^{(FS)} = 1 + \frac{3}{2} R_b \frac{K}{K + 6 + K\,T1} \quad \text{for } K \to \infty: \quad \mu_\infty^{(FS)} = 1 + \frac{3}{2} \frac{R_b}{1 + T1} \tag{5}$$

being R_b the beam rotation capacity.

On the contrary, in case of partial strength connection, a decrease of the lateral load resistance of the frame occurs, while connection influence on the available ductility is strictly related to the ratio between the connection plastic rotation capacity and the beam plastic rotation capacity.

This is due to the fact that, in this case, yielding is located in the connecting elements rather than at the beam ends.

The ductility of the substructure with partial strength semirigid-connections is given by (Faella, Piluso and Rizzano, 1994):

$$\mu_k^{(PS)} = 1 + \left(\frac{5\,d_b\,K}{L} \frac{5.4 - 3\overline{m}}{2\overline{m}} - 1 \right) \frac{6}{K + 6 + K\,T1} \tag{6}$$

3 Seismic behaviour

3.1 Evaluation of the substructure q-factor as a SDOF system

The analyses presented in the previous sections have pointed out the advantages of investigating the seismic behaviour of semirigid steel frames, with full strength or partial strength joints, by means of a simplified model. In fact, the use of the simplified model shown in Fig.1 has led to closed form solutions which clearly evidence all the effects due to the beam-to-column connections. In particular, it has been underlined that the joint deformability produce a beneficial increase of the period of vibration, but also a dangerous increase of the frame sensitivity to second order effects. In addition, in case of full strength joints, the decrease of the frame lateral stiffness is also responsible of the reduction of the available ductility. Regarding to the available ductility, a different behaviour has been highlighted in case of partial strength connections. In this case, the connection deformation capacity increases as far as the stiffness of the connection decreases, but this phenomenon can effectively lead to an advantageous increase of the global ductility only in case of beams having a very small plastic deformation capacity. In other words, in case of beams made of ductile sections (first class sections) is better to adopt full strength connections so that the yielding is forced to occur in the beam ends rather than in the connection and the beam rotation capacity is rationally exploited. On the contrary, the use of partial strength connections can prevent the complete exploitation of the plastic reserves of the frame leading to a critical reduction of the global ductility, due to severe local conditions.

Another important advantage, related to the use of a simplified model, is represented by the possibility to predict its seismic behaviour through the great amount of informations regarding the seismic response of the SDOF (Simple Degree Of Freedom) systems. In fact, it is easy to recognize that the lateral load versus top displacement curve of the substructure given in Fig.1 is equivalent to the one of an elastic perfectly plastic SDOF system with geometrical nonlinearity, i.e. second order effects included (Faella, Piluso and Rizzano, 1994).

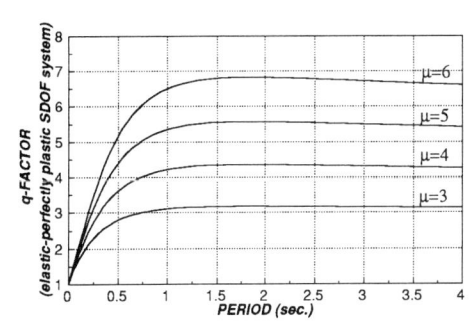

Fig.2
Relationship $q_o(\mu, T)$

Therefore, the q-factor of the simplified model can be evaluated as:

$$q(\mu, T, \gamma) = \frac{q_o(\mu, T, \gamma=0)}{\varphi(\mu, \gamma)} \tag{7}$$

being $q_o(\mu, T)$ the q-factor of an elastic perfectly plastic SDOF system having μ and T as available ductility and period of vibration respectively, while $\varphi(\mu, \gamma)$ is a coefficient which takes into account the influence of second order effects (Cosenza et al., 1988; Cosenza, Faella and Piluso, 1989; Faella, Mazzarella and Piluso, 1993).

The value of the q-factor, in absence of second order effects, can be evaluated by means of the relation proposed by Krawinkler and Nassar (1992):

$$q_o = [c(\mu - 1) + 1]^{\frac{1}{c}} \quad \text{with} \quad c = \frac{T}{1+T} + \frac{0.42}{T} \tag{8}$$

This relation is represented in Fig.2.

Moreover, the influence of second order effects can be taken into account through the following relation (Cosenza, Faella and Piluso, 1989; Mazzolani and Piluso, 1993):

$$\varphi = \frac{1 + \psi_1 (\mu - 1)^{\psi_2} \gamma}{1 - \gamma} \tag{9}$$

where the mean values of the coefficient φ can be obtained through the parameters

$\psi_1 = 0.62$ and $\psi_2 = 1.45$.

As a consequence, the ratio between the q-factor of the substructure including the effects of the beam-to-column joints and the one of the ideal model with full strength rigid joints is given by:

$$\frac{q_k^{(FS)}}{q_\infty^{(FS)}} = \frac{q_{0k}^{(FS)}}{q_{0\infty}^{(FS)}} \frac{\varphi_\infty^{(FS)}}{\varphi_k^{(FS)}} \tag{10}$$

in case of full strength semirigid joints, and by:

$$\frac{q_k^{(PS)}}{q_\infty^{(FS)}} = \frac{q_{0k}^{(PS)}}{q_{0\infty}^{(FS)}} \frac{\varphi_\infty^{(FS)}}{\varphi_k^{(PS)}} \tag{11}$$

in case of partial strength semirigid joints.

In equations (10) and (11), $q_\infty^{(FS)}$ is the q-factor of the substructure with full strength rigid joints and, therefore, it is evaluated through the value $q_{0\infty}^{(FS)}$ of q_0 provided by equation (8) with T_∞ and $\mu_\infty^{(FS)}$ and through the value $\varphi_\infty^{(FS)}$ of φ provided by equation (9) with γ_∞ and $\mu_\infty^{(FS)}$.

In equation (10), $q_k^{(FS)}$ is the q-factor of the substructure with full strength semirigid joints and, therefore, it is evaluated through the value $q_{0k}^{(FS)}$ of q_0 provided by equation (8) with T_k and $\mu_k^{(FS)}$ and through the value $\varphi_k^{(FS)}$ of φ provided by equation (9) with γ_k and $\mu_k^{(FS)}$.

Finally, in equation (11), $q_k^{(PS)}$ is the q-factor of the substructure with partial strength semirigid joints. It has to be evaluated like $q_k^{(FS)}$, but using $\mu_k^{(PS)}$ instead of $\mu_k^{(FS)}$.

It can be observed that $\mu_\infty^{(FS)}$ and $\mu_k^{(FS)}$ can be expressed through equations (5) as a function of K, $T1$ and R_b. In addition, by means of equation (4), γ_k can be expressed as a function of K, $T1$ and γ_∞.

Furthermore, being the value q_0 of the q-factor in absence of second order effects affected also by the period of vibration, as shown by equation (8), the influence of the connection deformability on the period value should be taken into account through the ratio T_k / T_∞ expressed by equation (3). Notwithstanding, in the usual period range of steel structures ($T \geq 0.8\ sec$), this dependence on T can be neglected as it is evidenced in Fig.2. With this simplification, the normalized value of the q-factor is given as a function of four parameters:

$$\frac{q_k^{(FS)}}{q_\infty^{(FS)}} = f(K, T1, \gamma_\infty, R_b) \tag{12}$$

In case of partial strength connections, a significant increase of the number of parameters involved in the seismic behaviour of the model arises. In fact, taking into account that the ductility $\mu_k^{(PS)}$ can be expressed through equation (6), it is easy to recognize that the normalized value of the q-factor is affected by six parameters:

$$\frac{q_k^{(PS)}}{q_\infty^{(FS)}} = f'(K, T1, \gamma_\infty, R_b, \overline{m}, L/d_b) \tag{13}$$

Fortunately, the number of parameters can be reduced when it is taken into account that K, $\overline{m}$ and L/d_b can be related through the following relation derived from experimental data:

$$\overline{m} = C_1 \left(\frac{L}{d_b K}\right)^{-C_2} \tag{14}$$

where the coefficients C_1 and C_2 are dependent on the connection typology (Faella, Piluso and Rizzano, 1994).

In addition, this relation allows to define for each connection typology a range of K in which $\overline{m} \leq 1$, i.e. partial strength connections are obtained, and another range in which, being $\overline{m} \geq 1$, full strength connections are attained. As a consequence, in the first range the relation (13) has to be considered, while in the second one the equation (12) is valid.

The resulting behaviour of the substructure is represented in Fig.3 for some values of the parameters governing the connection influence on the seismic response of the substructure.

It can be noted that, in case of beams having an high rotation capacity ($R_b = 6$), a

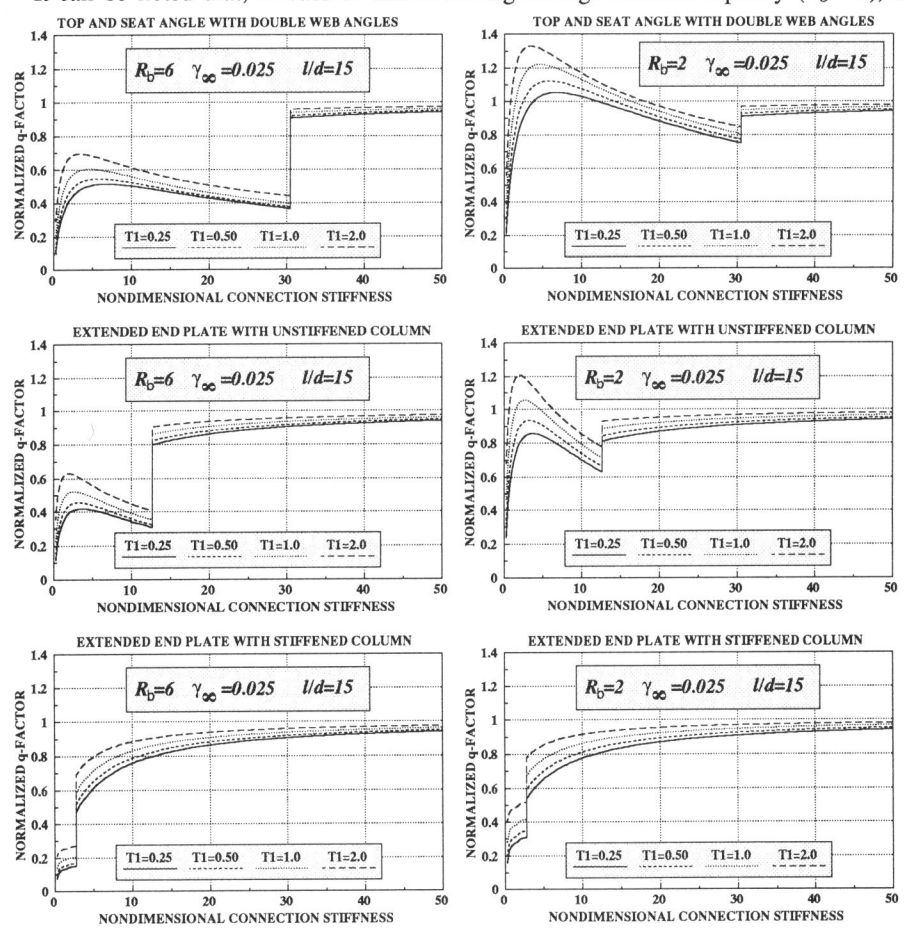

Fig.3
Connection influence on the normalized q-factor

significant reduction of the q-factor is attained, independently of the connection typology, in the whole range of the connection stiffness. In particular, this reduction is critical, in the partial strength zone, for connection stiffness values near to the one governing the passage from the full strength to the partial strength condition. This is due to the loss of ductility which arises when yielding occurs in the connections, rather than in the beam ends. In addition, in the partial strength range, the structure ability to dissipate the earthquake input energy is partially recovered decreasing the connection stiffness. This behaviour has to be ascribed to the fact that the rotation capacity of partial strength connections increases as far as their stiffness decreases. On the contrary, in case of full strength connections the reduction of the q-factor is less significant, becoming important only for small value of the connection stiffness.

In case of beams having a small rotation capacity a similar behaviour is recognized, but the discontinuity due to the passage from the full-strength to the partial strength condition is strongly reduced. As a consequence, the increase of ductility in the partial strength zone can lead to particular situations in which the value of q-factor is greater than the reference one $q_\infty^{(FS)}$, due to the connection influence. Notwithstanding, for very small values of the connection rotational stiffness, the reduction of the q-factor is always critical.

3.2 Seismic performance of the subassemblage

In the previous Section, the seismic response of the subassemblage of Fig.1 has been investigated from the design point of view. Therefore, the attention has been focused on the value of the q-factor. In fact, the normalized value of the q-factor, expressed by equation (13) in the partial strength range and by equation (12) in the full strength range, allows to establish the required amplification or reduction of the design horizontal forces with respect to the reference case represented by the ideal frame with rigid full strength connections.

Another point of view is constituted by the comparison based on the seismic performance of the subassemblage, i.e. its ability to withstand severe earthquakes. This comparison has to be developed taking into account that the ability of a structure to resist to destructive earthquakes depends not only on its ductility and energy dissipation capacity, expressed by the q-factor, but also on its strength.

A brittle structure can be able to resist to a severe ground motion relying on its strength only, because it is a non dissipative structure (q=1). Therefore, its ultimate strength has to be at least equal to the following value:

$$F_y = M\ A\ R(T) \qquad (15)$$

being M the mass of the system, A the peak ground acceleration and $R(T)$ the ordinate of the normalized elastic design response spectrum corresponding to its period T.

Considering that the normalized elastic design response spectrum is assigned by the seismic codes on the base of the site soil conditions, it can be recognized that such a structure is able to sustain an earthquake having a peak ground acceleration given by:

$$A = \frac{F_y}{M\ R(T)} \qquad (16)$$

On the contrary, a dissipative structure can resist to destructive earthquakes relying both on its strength and on its ductility and energy dissipation capacity ($q > 1$). Therefore, a dissipative structure is able to withstand a peak ground acceleration given by:

$$A = \frac{q\ F_y}{M\ R(T)} \qquad (17)$$

being F_y the first yielding resistance of the structure.

Therefore, it is clear that, in the case of the subassemblage with full strength-semirigid connections, the seismic performance can be compared with the reference case of full strength-rigid connections by means of the following ratio:

$$\frac{A_k^{(FS)}}{A_\infty^{(FS)}} = \frac{F_{y_k}^{(FS)}}{F_{y_\infty}^{(FS)}} \frac{q_k^{(FS)}}{q_\infty^{(FS)}} \frac{R(T_\infty)}{R(T_k)} \qquad (18)$$

where $F_{y_k}^{(FS)} / F_{y_\infty}^{(FS)} = 1$.

Analogously, in the case of partial strength-semirigid connections, the comparison with the reference case of full strength-rigid connections leads to the introduction of the ratio:

$$\frac{A_k^{(PS)}}{A_\infty^{(FS)}} = \frac{F_{y_k}^{(PS)}}{F_{y_\infty}^{(FS)}} \frac{q_k^{(PS)}}{q_\infty^{(FS)}} \frac{R(T_\infty)}{R(T_k)} \qquad (19)$$

where, in this case and with reference to the simplified model, it results

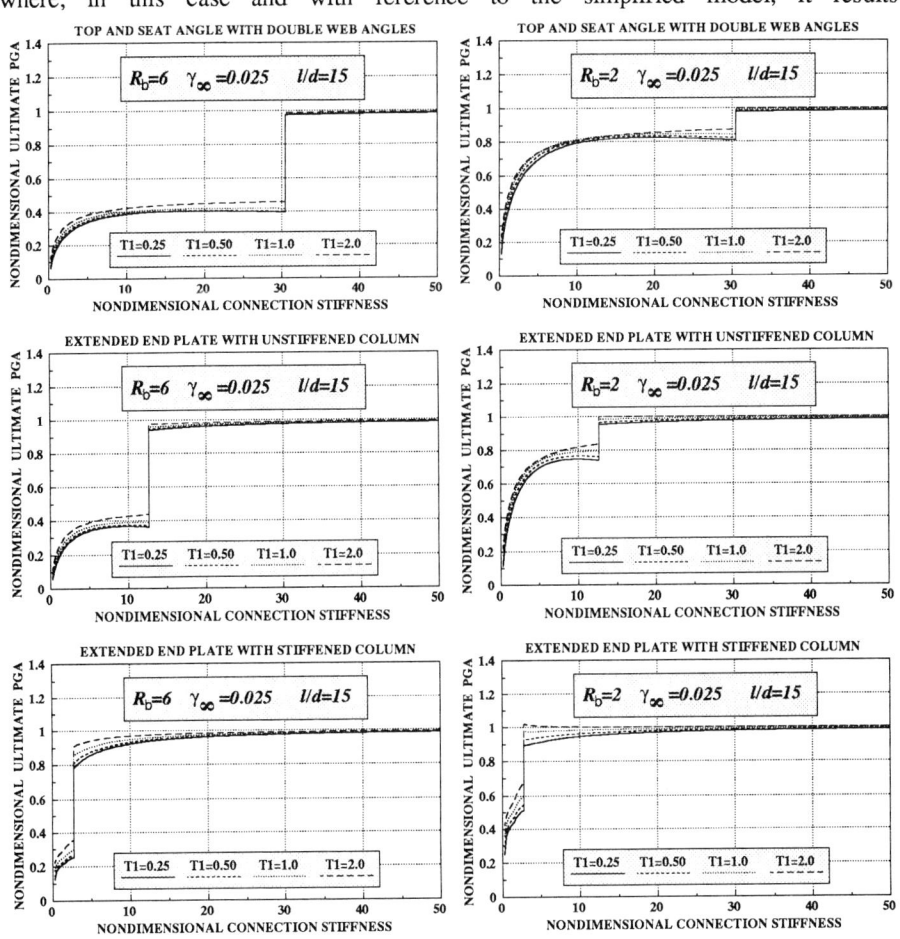

Fig.4
Connection influence on the nondimensional ultimate PGA

$F_{y_k}^{(PS)} / F_{y_\infty}^{(FS)} = \overline{m}$.

As steel structures are usually characterized by a period of vibration located in the softening branch of the elastic design response spectrum, it results:

$$\frac{R(T_\infty)}{R(T_k)} > 1 \qquad (20)$$

Therefore, from the seismic performance point of view, equations (18) and (19) evidence the beneficial effect due to the influence of the period of vibration.

The resulting behaviour is governed by three parameters as it is pointed out by equations (18) and (19), where it can be recognized a strength ratio, a q-factor ratio and a period ratio corresponding respectively to the first, second and third term of the right-hand side (being, as in Eurocode 8, in the softening branch of the elastic design response spectrum $R(T_\infty) / R(T_k) = T_k / T_\infty$).

Also in this case, equations (18) and (19) have to be applied according to the behaviour of the considered connection typology. Therefore, equations (19) has to be used in the range of partial strength connections and equation (18) in the range of full strength connections. The resulting behaviour is represented in Fig.4 for some values of the parameters governing the seismic behaviour of the subassemblage.

These figures show that generally, with reference to the comparison between the seismic performance of the model with full/partial strength semirigid connections and the one of the same model with full strength rigid connections, a decrease of the peak-ground acceleration leading to collapse arises. This reduction is significant.

In order to investigate the ability of the simplified model to clarify the connection influence on the seismic behaviour of actual MR-frames, a first comparison between the results coming from the application of the model and the ones obtained by means of numerical simulations has been carried out with reference to the two bay-four storey frame shown in Fig.5. As an example, with reference to extended end plate connection with unstiffened column, this comparison is given in Fig.6. In this figure, the points represent the results obtained by evaluating, for each value of the connection stiffness, the peak ground acceleration leading to collapse. This value has been computed through dynamic inelastic analyses which have been repeated for increasing values of the peak ground acceleration up to failure. The collapse condition has been defined as the occurrence of the ultimate plastic rotation at the beam end, in case of full strength connections, or in the connection itself, in case of partial strength connections. The analyses have been performed by using an accelerogram, which has been generated in order to match the elastic design response spectrum provided in seismic codes. This comparison evidences a sufficient agreement between the curves obtained from the analysis of the simplified model and the numerical simulations of the dynamic inelastic response of the considered frame.

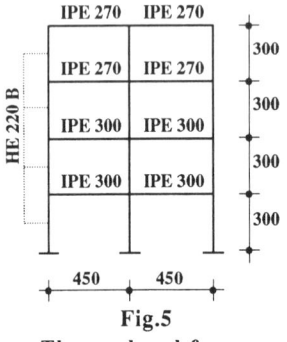

Fig.5
The analysed frame

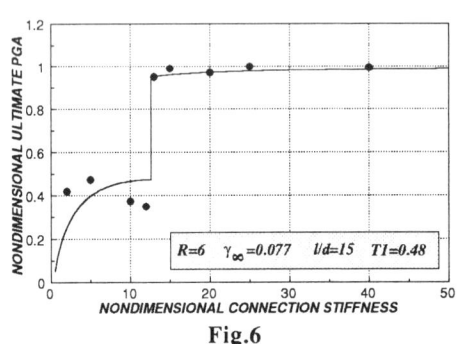

Fig.6
Comparison between the model and the analysed frame

4 Conclusions

The above results seems to be encouraging in view of the possibility to investigate the complex influence of the connection behaviour on the seismic response of actual frame. However, it has to be underlined that the most important role in the seismic behaviour of full/partial strength semirigid frames is played by the ratio between the beam plastic rotation capacity and the connection one. In addition, as the location of the discontinuity characterizing the passage from the partial strength to the full strength condition is governed by the connection strength versus stiffness relation, also this aspects deserves further investigations.
Furthermore, in the writer opinion, due to the fundamental role exhibited by the beam to connection ductility ratio, it seems that the use of partial strength connections in seismic zones requires further investigations. On the base of the actual background, the use of full strength connections, as suggested by the modern codes, appears to be preferred.

5 References

Astaneh, A. and Nader, N. (1991) Cyclic behaviour of frames with semi-rigid connection, Proceedings of the Second International Workshop «**Connections in Steel Structures II: Behavior, Strength and Design**», Pittsburg, Pennsylvania, April.

Astaneh, A. and Nader, N. (1992a) Shaking Table Tests of Steel Semi-Rigid Frames and Seismic Design Procedures. **1st COST C1 Workshop**, Strasbourg, October.

Astaneh, A. and Nader, N. (1992b) Proposed Code Provision for Sesimic Design of Steel Semirigid Frames. Submitted to **AISC Engineering Journal** for review and publication.

Astaneh, A. and Shen, J. (1990) Seismic Response Evaluation of an instrumented six story steel building. Earthquake Engineering Research Center, **Report UBC/EERC-90/20, University of California at Berkeley**, August, 1990.

Commission of the European Communities (1993) Eurocode 8: Earthquake Resistant Design of Structures, October (Draft).

Cosenza, E. Faella, C. and Piluso, V. (1989) Effetto del Degrado Geometrico sul Coefficiente di Struttura. IV Convegno Nazionale, **L'Ingegneria Sismica in Italia**, Milano, Ottobre.

Faella, C. Mazzarella, O. and Piluso, V. (1993) L'Influenza della non-linearità geometrica sul danneggiamento strutturale sotto azioni sismiche. VI Convegno Nazionale, **L'Ingegneria Sismica in Italia**, Perugia, 13-15 Ottobre.

Faella, C. Piluso, V. and Rizzano, G. (1993a) Sul Comportamento Sismico Inelastico dei Telai in Acciaio a Nodi Semirigidi. VI Convegno Nazionale, **L'Ingegneria Sismica in Italia**, Perugia, 13-15 Ottobre.

Faella, C. Piluso, V. and Rizzano, G. (1993b) L'Influenza del Comportamento Nodale sulla Risposta Inelastica dei Telai Sismo-Resistenti. **Giornate Italiane della Costruzione in Acciaio**, C.T.A., Viareggio, 24-27 Ottobre.

Faella, C. Piluso, V. and Rizzano, G. (1994) Connection Influence on the Elastic and Inelastic Behaviour of Steel Frames. **International Workshop and Seminar on Behaviour of Steel Structures in Seismic Areas**, STESSA 94, Timisoara, Romania, 26 June-1 July.

Krawinkler, H. and Nassar, A.A. (1992) Seismic Design based on Ductility and Cumulative Damage Demands and Capacities, in «**Nonlinear Seismic Analysis and Design of Reinforced Concrete Buildings**» edited by P. Fajfar and H. Krawinkler, Elsevier, London.

Mazzolani, F.M. and Piluso, V. (1993) P-Δ Effect in Seismic Resistant Steel Structures. **Structural Stability Research Council**, Annual Technical Session & Meeting, Milwaukee, Wisconsin, April.

Acknowledgements

This work has been partial supported with Research Grant MURST 40% 1992.

53 CONNECTION INFLUENCE ON THE ELASTIC AND INELASTIC BEHAVIOUR OF STEEL FRAMES

C. FAELLA, V. PILUSO and G. RIZZANO
Department of Civil Engineering, University of Salerno, Italy

Abstract

In this paper the influence of the beam-to-column connections on the elastic and inelastic behaviour of steel frames is investigated by means of a simplified model. Both the case of full strength connections and the case of partial strength connections are considered and the differences in the resulting behaviour of the model are evidenced. Finally, the results obtained from the analysis of the simplified model are compared with the ones, derived by means of numerical simulations, of some actual frames.
Keywords: Connections, Global Ductility, Local Ductility, Second order effects

1 Introduction

The response of steel frames subjected to seismic action is strongly affected by the moment-rotation behaviour of their beam-to-column joints, i.e. the one resulting from the interaction between the connection and the panel zone.

Two main parameters are involved in the connection classification (Eurocode 3, 1992): the flexural strength and the rotational stiffness.

With reference to the flexural strength, three connection classes can be identified: full strength connections, partial strength connections and nominally pinned connections.

In addition, with reference to the rotational stiffness, the following classification can be made (Eurocode 3, 1992): rigid or fully restrained connections, semirigid or partially restrained connections and nominally pinned connections.

Excluding the case of nominally pinned connections, which do not correspond to the case of moment resisting frames, four fundamental cases can be recognized:
a) full strength-rigid connections; **b)** full strength-semirigid connections; **c)** partial strength-rigid connections; **d)** partial strength-semirigid connections.

Such a distinction is necessary, because the seismic behaviour of moment resisting frames is considerably affected both by the strength and by the stiffness of the beam-to-column connections. Notwithstanding, the term semirigid frame is often used undifferently both for frames with full strength-semirigid connections (case b) and for frames with partial strength connections (cases c and d). On the contrary, the term rigid frame is often adopted to denote frames with full strength-rigid connections, but the distinction between cases a and c should be made, representing two limit cases.

It is evident that the case of full strength-rigid connections (case a) represents the reference case, i.e. the case in which the beam-to-column connection exhibits the ideal behaviour and does not represent an imperfection within the frame.

In particular, from the point of view of the location of the dissipative zones, completely different behaviours are developed in case of full strength connections and in case of partial strength connections. Therefore, in order to simulate the behaviour of the beam-to-column joints, both the connection and the cross-section of the connected

Behaviour of Steel Structures in Seismic Areas. Edited by F. M. Mazzolani and V. Gioncu.
Published in 1995 by E & FN Spon, 2-6 Boundary Row, London SE1 8HN. ISBN: 0 419 19890 3.

member have to be taken into account.

Three different types of behaviour can be recognized depending on the ultimate moment of the connection and on the plastic moment of the connected beam (Fig.1) (Cosenza, De Luca and Faella, 1984): a) $M_{uc} > M_{pb}$ (Fig.1a); b) $M_{uc} \approx M_{pb}$ (Fig.1b); c) $M_{uc} < M_{pb}$ (Fig.1c); being M_{uc} and M_{pb} respectively the ultimate moment of the connection and the plastic moment of the connected beam.

In the first case, the shape of the moment versus rotation global (beam plus connection) curve can be accurately represented with a bilinear elastic-perfectly plastic model. A possible plastic hinge will develop at the beam end, while the connection remains in elastic range. Plastic rotations involve only the beam end.

In the third case, the accurate modelling of the shape of the moment versus rotation global curve requires a nonlinear representation. Moreover, the dissipative zone is located within the connection, while the beam remains in elastic range. In this case, the connection has to be able to experience large plastic rotations.

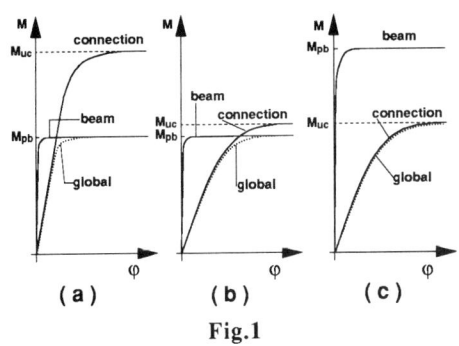

Finally, in the intermediate case, representing a transition condition, a nonlinear modelling is still required. Furthermore, yielding occurs both in the connection elements and at the end of the beam.

Aiming at a clarification of all parameters affecting the inelastic behaviour of full/partial strength semirigid frames, a simplified model has been introduced in a previous work (Faella, Piluso and Rizzano, 1993a), with reference to full strength-semirigid connections, and is herein extended to the case of partial strength-semirigid connections. This approach, which will be described in the following sections, allows to derive closed form solutions able to highlight the role of all connection parameters involved in the behaviour of actual frames, which affect the period of vibration, the frame sensitivity to second order effects and the available ductility. The consequences of these effects on the seismic response of steel frames are investigated in a companion paper (Faella, Piluso and Rizzano, 1994).

Fig.1
Beam plus connection behaviour

2 The analysed substructure

The study of the seismic response of steel frames, including the connection behaviour, can be carried out by means of the simplified model represented in Fig.2 (Faella, Piluso and Rizzano, 1993a). It represents a subassemblage, which has been extracted by an actual frame by assuming that beams are subjected to double curvature bending with zero moment in the midspan section and by considering each beam as contemporary belonging to two storeys, so that their mechanical properties has been halved. The flexural stiffness of the columns of the original frame, from which the substructure has been derived, is equal to EI_c/h; the flexural stiffness of the beams is equal to EI_b/L. The moment versus rotation curve of the connection is modelled with a bilinear relation which is completely defined by means of two parameters only: the elastic rotational stiffness K_φ and the ultimate moment M_{uc}. It is evident that, with reference to the elastic range, the comparison between the semirigid substructure and the rigid one is governed by the following nondimensional parameters:

$$T1 = \frac{EI_b/L}{EI_c/h} \qquad K = \frac{K_\varphi L}{EI_b} \qquad (1)$$

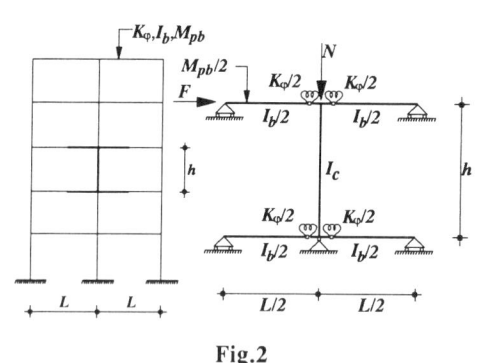

Fig.2
The analysed substructure

being $T1$ the beam-to-column stiffness ratio and K the nondimensional rotational stiffness of the connection.

In addition, with reference to the inelastic behaviour, the nondimensional ultimate flexural resistance of the connection:

$$\overline{m} = \frac{M_{uc}}{M_{pb}} \qquad (2)$$

provides the distinction between the two fundamental cases of full strength connections, for which $\overline{m} > 1$, and partial strength connections, for which $\overline{m} < 1$.

In Section 3 only the connection influence on the elastic behaviour of the model will be investigated, as it is common both to full strength and to partial strength connections. The analysis of the parameters influencing the inelastic behaviour of the model will be performed in the following Section 4.

3 Connection influence on elastic parameters

3.1 Period of vibration

In order to investigate the influence of the connection rotational stiffness on the period of vibration, it is sufficient to evaluate the influence of K on the lateral stiffness of the substructure. In fact, it is easy to recognize that the ratio between the period of vibration of the semirigid model T_k and the one of the rigid model T_∞ is given by:

$$\frac{T_k}{T_\infty} = \left(\frac{K_{S\infty}}{K_{Sk}}\right)^{1/2} \qquad (3)$$

being K_{Sk} and $K_{S\infty}$ respectively the lateral stiffness of the semirigid model and the one of the rigid model.

The study of the substructure subjected to an horizontal force F applied at the top level provides the following relation:

$$\begin{bmatrix} \dfrac{6EI_b}{L}\dfrac{K}{K+6} + \dfrac{6EI_c}{h} & -\dfrac{6EI_c}{h^2} \\ -\dfrac{6EI_c}{h^2} & \dfrac{6EI_c}{h^3} \end{bmatrix} \begin{Bmatrix} \varphi \\ \delta \end{Bmatrix} = \begin{Bmatrix} 0 \\ \dfrac{F}{2} \end{Bmatrix} \qquad (4)$$

being φ the rotation of the joints and δ the top sway displacement.

The lateral stiffness of the substructure is derived as F/δ leading to the following relation:

$$K_{Sk} = \frac{12EI_c}{h^3}\frac{K\;T1}{K+6+K\;T1} \qquad \text{for } K \to \infty: \qquad K_{S\infty} = \frac{12EI_c}{h^3}\frac{T1}{1+T1} \qquad (5)$$

Therefore, equation (3) provides:

$$\frac{T_k}{T_\infty} = \left(\frac{K(1+T1)+6}{K(1+T1)} \right)^{1/2} \quad (6)$$

This relation is represented in Fig.3. It evidences that the connection deformability produces an increase of the period of vibration. This is, generally, an advantageous effect, because it corresponds to the shifting in a period range of the design response spectrum where the spectral acceleration is less severe.

3.2 Stability coefficient

The frame sensitivity to second order effects is expressed through the stability coefficient γ. It represents the slope of the softening branch of the behavioural curve relating the multiplier of the horizontal forces α to the nondimensional top sway displacement δ/δ_1 (being δ_1 the top displacement under the horizontal forces corresponding to $\alpha = 1$). In case of simple degree of freedom systems, such as the simplified model herein analysed, the stability coefficient can be expressed as:

$$\gamma = \frac{N}{K_s h} \quad (7)$$

being N the vertical load and K_s the lateral stiffness.

As a consequence, the ratio between the stability coefficient of the semirigid model γ_k and the one of the rigid model γ_∞ is given by:

$$\frac{\gamma_k}{\gamma_\infty} = \frac{K_{s_\infty}}{K_{s_k}} = \frac{K(1+T1)+6}{K(1+T1)} \quad (8)$$

This relation is represented in Fig.4. It points out that the connection deformability leads to an increase of the frame sensitivity to second order effects. This phenomenon is undesiderable, because it gives rise to an amplification of the ductility demand under severe ground motion (Cosenza, Faella and Piluso, 1989; Mazzolani and Piluso, 1993; Faella, Mazzarella and Piluso, 1993).

4 Connection influence on inelastic parameters

4.1 General

It has been evidenced that the connection deformability has two effects on the elastic

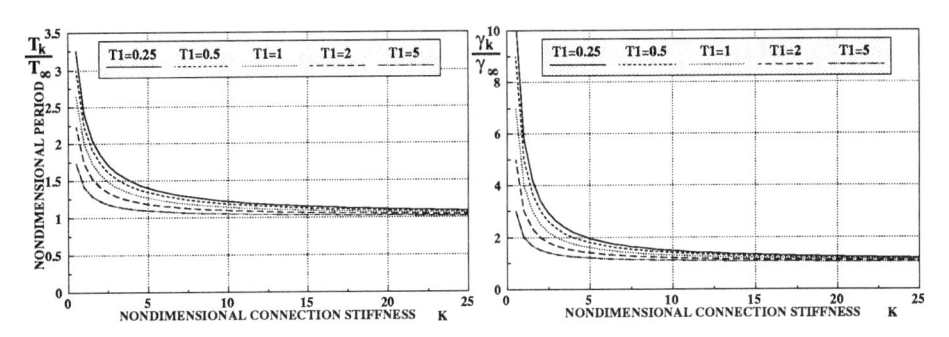

Fig.3 Influence on the period of vibration

Fig.4 Influence on the stability coefficient

behaviour of the substructure, i.e. on the frame elastic behaviour. In fact, on one hand, it determines an increase of the period of vibration and, on the other hand, it leads to the increase of the frame sensitivity to second order effects. From the seismic point of view, the first effect is desiderable as it locates the structure in a more favourable zone of the design response spectrum. On the contrary, the increase of the frame sensitivity to second order effects is undesiderable, due to the amplification of the negative effects of the geometric non-linearity under seismic loads. The role of these two important effects has been widely investigated in a previous work (Faella, Piluso and Rizzano, 1993a).

As soon as the inelastic behaviour is concerned, the distinction between full strength and partial strength connections becomes determinant. In fact, it governs the location of the dissipative zones, which are concentrated at the beam ends in the first case and in the connecting elements in the second one. According to the type of connection, the ultimate strength and/or the available ductility are affected. In case of full strength semirigid connections, a decrease of the available ductility arises, due to the increase of the lateral deformability of the structure, while the ultimate resistance of the frame is slightly decreased due to second order effects (Faella, Piluso and Rizzano, 1993b). On the contrary, in case of partial strength connections, a decrease of the lateral load resistance of the frame occurs, while the connection influence on the available ductility is strictly related to the ratio between the connection plastic rotation capacity and the beam plastic rotation capacity.

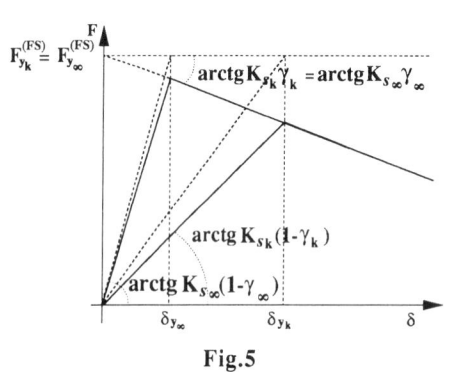

Fig.5
F-δ relation for full strength connections

4.2 Frames with full strength connections

Fig.5 shows the lateral load versus top displacement curve both for the substructure with full strength-rigid connections and for the same substructure with full strength-semirigid connections. First of all, it can be observed that the first order yield resistance is the same in the two cases:

$$F_{y_k}^{(FS)} = F_{y_\infty}^{(FS)} \qquad (9)$$

Due to second order effects, the yield resistance decreases assuming the following values:

$$F_{y_k}^{(FS)''} = F_{y_k}^{(FS)} (1 - \gamma_k) \qquad F_{y_\infty}^{(FS)''} = F_{y_\infty}^{(FS)} (1 - \gamma_\infty) \qquad (10)$$

Therefore, being $\gamma_k > \gamma_\infty$, the actual yield resistance of the simplified model is decreased, due to the amplification of second order effects produced by the connection deformability. The slope of the softening branch of the lateral load versus top displacement curve is unaffected, as it is evidenced also by equation (8) which shows that $\gamma_k K_{s_k} = \gamma_\infty K_{s_\infty}$.

The available ductility of the substructure can be expressed as:

$$\mu_k^{(FS)} = \frac{\delta_y + \delta_p}{\delta_y} = 1 + \frac{\varphi_{pb} h}{\delta_y} \qquad (11)$$

being δ_p the plastic ultimate displacement and φ_{pb} the plastic rotation that the beams are able to withstand.

The maximum plastic rotation that a member is able to develop is usually derived by means of a three point bending test (Fig.6). As a consequence, it can be expressed as:

$$\varphi_{pb} = R_b \, \varphi_y = R_b \, \frac{M_{pb} \, L}{4 \, E \, I_b} \tag{12}$$

being R_b the plastic rotation capacity of the beams.

In addition, equation (4) provides the following relation between the member end rotation and the top displacement:

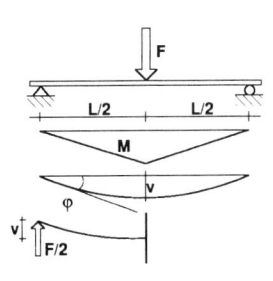

Fig.6 Three point bending test

$$\varphi = \frac{K + 6}{K + 6 + K \, T1} \, \frac{\delta}{h} \tag{13}$$

The bending moment at the end of the beam is given by:

$$M = \frac{3 \, E \, I_b}{L} \, \frac{K}{K + 6} \, \varphi \tag{14}$$

By imposing the yielding condition, $M = M_{pb}/2$, the following relation is obtained:

$$\frac{h}{\delta_y} = \frac{6 \, E \, I_b}{M_{pb} \, L} \, \frac{K}{K + 6 + K \, T1} \tag{15}$$

Therefore, by substituting equations (15) and (12) in equation (11), the available ductility of the simplified model with full strength-semirigid connections can be expressed as:

$$\mu_k^{(FS)} = 1 + \frac{3}{2} R_b \frac{K}{K + 6 + K \, T1} \quad \text{for } K \to \infty: \quad \mu_\infty^{(FS)} = 1 + \frac{3}{2} \frac{R_b}{1 + T1} \tag{16}$$

This relation represents the ductility of the model with full strength-rigid joints.

The above equations point out that the connection deformability leads to a reduction of the available ductility, as it has been already evidenced through numerical simulations in a previous work (Faella, Piluso and Rizzano, 1993b).

4.3 Frames with partial strength connections

In case of frames with partial strength-semirigid connections, a significant decrease of the yield resistance of the substructure arises, while the connection influence on the available ductility is strictly related not only to the connection deformability, but also to the ratio between the connection plastic rotation capacity and the beam plastic rotation capacity. This is due to the fact that, in this case, yielding is located in the connecting elements rather than at the beam ends.

Fig.7 shows the lateral load versus top displacement curve of the substructure with partial strength-semirigid connections and its comparison with the same model having full strength-rigid connections.

The reduction of the first order yield resistance is given by:

$$\frac{F_{yk}^{(PS)}}{F_{y\infty}^{(FS)}} = \overline{m} \tag{17}$$

The actual yield resistance furtherly decreases due to second order effects. In fact, for the model with partial strength-semirigid connections, it is given by:

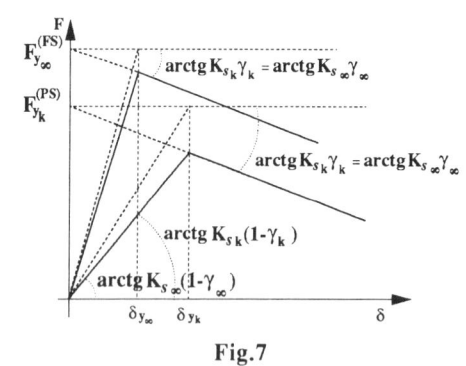

Fig.7
F-δ relation for partial strength connections

$$F_{y_k}^{(PS)''} = F_{y_k}^{(PS)} (1 - \gamma_k) \quad (18)$$

while, for the model with full strength-rigid connections, it is still given by the second one of equations (10).

The available ductility of the model with partial strength connections is obtained taking into account that, in this case, yielding occurs in the connection. Therefore, it can be expressed by:

$$\mu_k^{(PS)} = 1 + \frac{\varphi_{pc} h}{\delta_y} \quad (19)$$

being φ_{pc} the maximum plastic rotation that the connection is able to withstand.

Moreover, in this case, the yielding condition is still given by equation (15) provided that M_{pb} is substituted by M_{yc}, being M_{yc} the yielding resistance of the connection.

By substituting this yielding condition and the second one of equations (1) into equation (19) and taking into account that $\varphi_{yc} = M_{yc} / K_\varphi$ is the yield rotation of the connection, the following relation is obtained:

$$\mu_k^{(PS)} = 1 + \left(\frac{\varphi_{uc}}{\varphi_{yc}} - 1 \right) \frac{6}{K + 6 + K\, T1} \quad (20)$$

being $\varphi_{uc} = \varphi_{pc} + \varphi_{yc}$ the ultimate rotation of the connection.

According to Bjorhovde, Colson and Brozzetti (1990), on the base of experimental data, the ultimate rotation of the connections can be expressed as:

$$\varphi_{uc} = \frac{5.4 - 3\overline{m}}{2} \frac{5 M_{pb}\, d_b}{E I_b} \quad (21)$$

being d_b the depth of the beam.

In addition, the yield rotation of the connection is given by:

$$\varphi_{yc} = \frac{\overline{m}\, M_{pb}}{E I_b} \frac{L}{K} \quad (22)$$

By substituting equations (21) and (22) into equation (20), the available ductility of the substructure with partial strength-semirigid connections can be expressed through the following relation:

$$\mu_k^{(PS)} = 1 + \left(\frac{5 d_b K}{L} \frac{5.4 - 3\overline{m}}{2\overline{m}} - 1 \right) \frac{6}{K + 6 + K\, T1} \quad (23)$$

This equation points out that the available ductility of the substructure with partial strength semirigid joints depends on the beam depth-to-span ratio d_b/L, the beam-to-column stiffness ratio $T1$, the nondimensional stiffness of the connection K and the corresponding nondimensional ultimate moment $\overline{m}$.

4.4 Actual ductility of the substructure

The connection stiffness can be also expressed as (Bjorhovde, Colson and Brozzetti, 1990):

$$K_\varphi = \frac{EI_b}{L_e} = \frac{EI_b}{\eta\, d_b} \qquad (24)$$

being $L_e = \eta\, d_b$ the equivalent beam length of the connection, i.e. the length of a beam having a flexural stiffness equal to the rotational stiffness of the connection.

By combining the second one of equations (1) and equation (24), the following relation is obtained:

$$\eta = \frac{L}{d_b\, K} \qquad (25)$$

The analysis of the actual ductility of the substructure can be simplified by taking into account that the stiffness of the connection and its ultimate moment are related. In fact, from the analysis of the experimental data collected in the SERICON data bank (Weynand,1992), for extended end plate connections, and by Azizinamini and Radziminski (1988) for top and seat angle with double web angles connections, the following relation can be derived with a satisfactory correlation coefficient:

$$\overline{m} = C_1\, \eta^{-C_2} \qquad (26)$$

The curves $\overline{m} - \eta$ with the values of C_1 and C_2 are represented, with the corresponding experimental data, in Fig.8.

Due to the great variety of structural details which can be adopted for the beam-to-column joints, the relation between the joint stiffness and its ultimate moment deserves further investigations and supplementary experimental tests. Notwithstanding, the above equation can be useful for a better comprehension of the inelastic behaviour of the model and, therefore, of semirigid frames.

In particular, taking into account equation (25), the use of equations (26) and (23) allows, for the corresponding joint typologies, to express the ductility of the model through a relation of the following type:

$$\mu_k^{(PS)} = \mu_k^{(PS)}\left(K, T1, \frac{L}{d_b}\right) \qquad (27)$$

However, the actual ductility of the substructure has to be derived by considering the existence of a value of the nondimensional stiffness of the connection leading to the full strength condition. This value, which is dependent on the beam depth-to-span ratio d_b/L and on the connection typology, allows to evidence two ranges of the connection stiffness. The first one corresponds to $M_{uc} < M_{pb}$, i.e. to partial strength connections, so that the ductility of the substructure is provided by $\mu_k^{(PS)}$ (Eq.23) and yielding occurs in the connections rather than in the beam ends. The second one

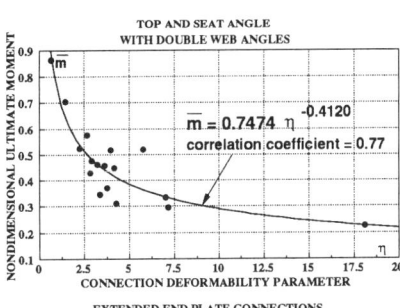

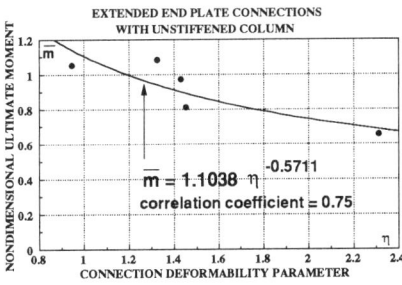

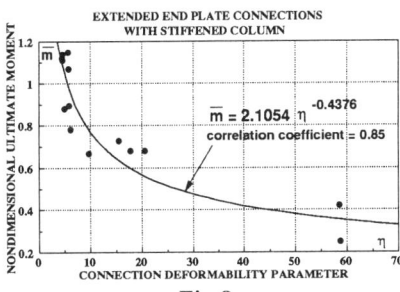

Fig.8
$\overline{m} - \eta$ **curves**

Top and seat angle with double web angles: $\overline{m} = 0.7474\, \eta^{-0.4120}$, correlation coefficient = 0.77

Extended end plate connections with unstiffened column: $\overline{m} = 1.1038\, \eta^{-0.5711}$, correlation coefficient = 0.75

Extended end plate connections with stiffened column: $\overline{m} = 2.1054\, \eta^{-0.4376}$, correlation coefficient = 0.85

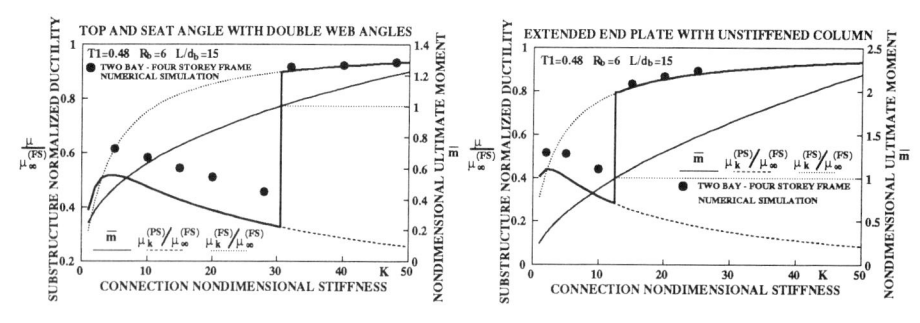

Fig.9
Normalized ductility

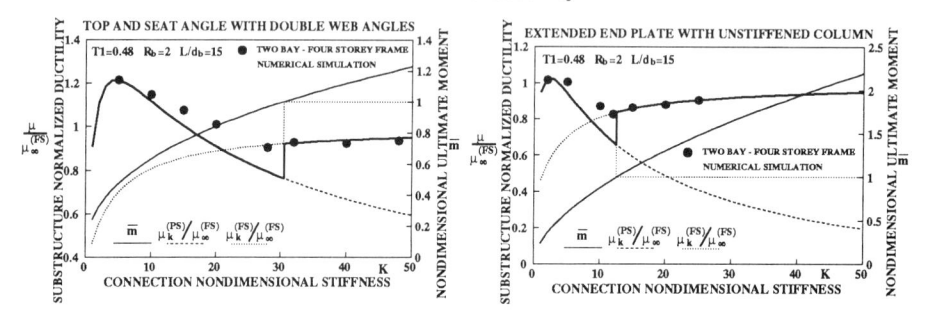

Fig.10
Normalized ductility

leads to the condition $M_{uc} > M_{pb}$, i.e. to full strength connections, so that the ductility of the substructure is provided by $\mu_k^{(FS)}$ (Eq.16), because yielding arises in the beam ends rather than in the connections.

The above behaviour is represented in Figs.9 and 10, for $R_b=6$ and $R_b=2$, where reference is made to extended end plate connections with unstiffened column and to top and seat angle with double web angle connections. It can be recognized that for $R_b=6$, i.e. for ductile sections (first class sections), the connection influence leads always to a worsening of the inelastic behaviour, leading to a reduction of the available ductility with respect to the case of full strength-rigid connections. In addition, this worsening seems to be untolerable in case of partial strength connections.

On the contrary, in case of beam sections having a small rotation capacity ($R_b=2$) different behaviours are exhibited in the two ranges of K corresponding to full strength and partial strength connections.

In the range of full strength connections, the connection deformability always leads to a reduction of the available ductility. On the contrary, in the partial strength range, the connection effects depends on the ratio between its plastic deformation capacity, which increases as far as the connection stiffness decreases, and the beam rotation capacity. Therefore, depending on the connection typology, a range of the connection stiffness can arise in which an improvement of the available ductility is attained.

In the same figures, the results of some numerical simulations of the inelastic response of the frame shown in Fig.11 are also given. It can be noted that these results are in good agreement with the theoretical curves obtained from the analysis of the simplified model. Therefore, it can be stated that the simplified model of Fig.2 is able to interpret the complex influence of the beam-to-column connections on the elastic and inelastic response of steel frames.

5 Conclusions

This paper has pointed out that the beam-to-column connections influence the period of vibration, the frame sensitivity to second order effects and its available ductility. In particular, as soon as the available ductility is concerned, the strength of the connection plays a fundamental role. In fact, in case of partial strength connections a significant reduction of ductility arises. This reduction increases as far as the ratio between the beam plastic rotation capacity and the connection plastic deformation capacity increases. In addition, it has been evidenced that the knowledge of the conditions governing the passage from the partial strength to the full strength behaviour is of primary importance. Therefore, in order to establish simplified relationships among all parameters characterizing the joint behaviour, additional experimental investigations are helpful.

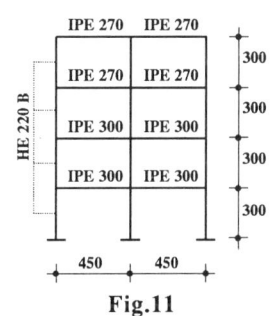

Fig.11
The analysed frame

6 References

Azizinamini, A. and Radziminski, J.B. (1988) Prediction of Moment-Rotation Behaviour of Semi-Rigid Beam-to-Column Connections. **Connections in Steel Structures: Behaviour, Strength and Design** Edited by R. Bjorhovde, J. Brozzetti and A. Colson, Elsevier Applied Science, London and New York

Bjorhovde, R. Colson, A. and Brozzetti, J. (1990) Classification System for Beam-to-Column Connections. **Journal of Structural Engineering**, ASCE, Vol.116, No.11, November.

Commission of the European Communities (1992) Eurocode 3: Design of Steel Structures

Cosenza, E. De Luca, A. and Faella, C. (1984) Nonlinear behaviour of framed structures with semirigid joints. **Costruzioni Metalliche**, N.4.

Cosenza, E. Faella, C. and Piluso, V. (1989) Effetto del Degrado Geometrico sul Coefficiente di Struttura. IV Convegno Nazionale, **L'Ingegneria Sismica in Italia**, Milano, Ottobre.

Faella, C. Mazzarella, O. and Piluso, V. (1993) L'Influenza della non-linearità geometrica sul danneggiamento strutturale sotto azioni sismiche. VI Convegno Nazionale, **L'Ingegneria Sismica in Italia**, Perugia, 13-15 Ottobre.

Faella, C. Piluso, V. and Rizzano, G. (1993a) Sul Comportamento Sismico Inelastico dei Telai in Acciaio a Nodi Semirigidi. VI Convegno Nazionale, **L'Ingegneria Sismica in Italia**, Perugia, 13-15 Ottobre.

Faella, C. Piluso, V. and Rizzano, G. (1993b) L'Influenza del Comportamento Nodale sulla Risposta Inelastica dei Telai Sismo-Resistenti. **Giornate Italiane della Costruzione in Acciaio**, C.T.A., Viareggio, 24-27 Ottobre.

Faella, C. Piluso, V. and Rizzano, G. (1994) Connection Influence on the Seismic Behaviour of Steel Frames. **International Workshop and Seminar on Behaviour of Steel Structures in Seismic Areas**, STESSA 94, Timisoara, Romania, 26 June-1 July.

Mazzolani, F.M. and Piluso, V. (1993) P-Δ Effect in Seismic Resistant Steel Structures. Structural Stability Research Council, 1993 Annual Technical Session & Meeting, Milwaukee, Wisconsin, April.

Weinand, K. (1992) SERICON - Databank on joints in building frames Proceedings of the **1st COST C1 Workshop**, Strasbourg, 28-30 October.

Acknowledgements

This work has been partially supported with Research Grant CNR.93.02287.CT07 and MURST 40% 1992.

54 BEHAVIOUR OF SEMI-RIGID CONNECTIONS UNDER ALTERNATE STATIC LOADS

D. GRECEA and G. F. MATEESCU
Building Research Institute, INCERC, Timisoara, Romania

Abstract
This paper represents the experimental research programme developed on two types of semi-rigid joints welded and bolted, under monotonous and alternate static loads. Paper contains the description of the experimental investigations and the characteristic results. Also, it is presented a comparison between experimental results and the theoretical model, with some conclusions.
Keywords: Semi-Rigid Connections, Alternate Static Loads, Behaviour, Experimental Investigation, Results, Comparison, Conclusions.

1 Introduction

This paper presents the experimental research programme developed at the Building Research Institute Timisoara, initiated for the improvement of the design and execution actual conception of the steel structures beam-to-column connections.
   The aim of the research programme was to analyse the behaviour of some semi-rigid connections till yielding under monotonous and alternate static loads.
   Experimental models were made as described in paragraph 2 and represent two types of semi-rigid joints:
   Joint A - welded full strength connection with seat
            angle and top and web plates.
   Joint B - bolted partial strength connection with top,
            seat and double web angle.
The main parameters which had determined the choise of these experimental models were the followings:
   Beam and column cross-section type are double T
   symmetric sections maded of welded plates.
   Beam-to-column joint relative stiffness was determined
   by calculus, using the model proposed by Chen and Kishi
   (1987) for the top, seat and double web angle
   connection.

Behaviour of Steel Structures in Seismic Areas. Edited by F. M. Mazzolani and V. Gioncu.
Published in 1995 by E & FN Spon, 2–6 Boundary Row, London SE1 8HN. ISBN: 0 419 19890 3.

Load type was monotonous and alternate produced by a concentrated load applied at the end of the cantilever. A general description of the experimental programme on models is presented in Table 1.

Table 1. General description of the experimental research

Nr.	Model	Connection Type	Column (mm)	Beam (mm)	Loading Type
1	A1	welded with seat angle	flange	flange	monotonous
2	A2	and top and web plates	18x240	10x140	alternate
3	B1	bolted with top, seat	web	web	monotonous
4	B2	and double web angle	10x204	6x250	alternate

2 Types of tested beam-to-column connections

Beam-to-column experimental elements are plane structural subassemblages made up of two components: a portion of beam in cantilever of 1.5m connected to a portion of column of 1.2m. These models were designed to allow for realistic types of loading, while having sizes which prevent external load introduction effects from altering the response of the joint itself. Column and beam lengths were chosen after the dimensions of the experimental stand. The models can be considered like some exterior T joints, so the beam is connected in the middle of the column. The bending moment of the joint is produced by a concentrated load applied to the free end of the cantilever. Experimental elements were made up of welded plates of OL 37 steel.

Joints A and B are schematically represented in Fig.1 and are made up of the following components shown in Table 2.

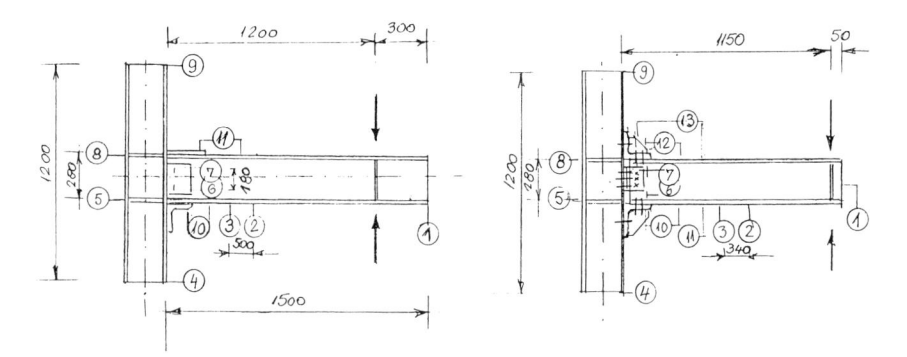

Fig.1. Testing elements (A,B) and displacement devices location

From each model were made up two identical elements.

Table 2. Components of joints A and B

Joint	Components (mm)							
	Column		Beam		Connection			
	Flange	Web	Flange	Web	Upper flange	Lower flange	Web	Connector
A	18x240	10x 204	10x140	6x 250	Plate 20x120 -220	Angle L150x 150x18	Plate 18x150 -230	Corner welding
B	18x240	10x 204	10x140	6x 250	L150x 150x14	L150x 150x14	2L120x 120x12	Bolts M 20

3 Testing arrangements

Experimental programme was realised using the ECCS Recomandations for testing unitary procedure for assessing the behaviour of structural steel elements under cyclic loads. ECCS Recomandations indicate the loads application steps and also parameters which must be determined after the experimental tests.

Experimental programme was focused on the cyclic behaviour of some full scale plane structural subassemblages as marginal T beam-to-column joints with an increasing static load (monotonous and alternate) at the end of the cantilever, to realise in connection a bending stress.

Experimental results analysis followed to put in evidence the characteristics:
- characteristic diagrams load-displacement (P-Δ) and moment-rotation (M-θ);
- behaviour stages and yielding mechanism;
- load reserve capacity through the difference of the experimental stress and the theoretical one;
- influence of the component parameters on the general joint behaviour;
- influence of some factors of the models execution technology.

The test specimens were inserted in a horizontal steel testing closed frame with 3.00x1.50m interior dimensions (Fig.2). The steel testing rig was so formed to place and test the models in a horizontal way. This frame was aimed to take the reactions produced by the loads applied to the test specimens.

The ends of the columns were supported by devices like rigs which were able to support vertical and horizontal reactions, while providing no moment restraint. Guide-plates were located all along both the beam and the column to prevent any spacial behaviour and out-of-plane displacement of the subassemblage, whose in-plane behaviour was investigated only.

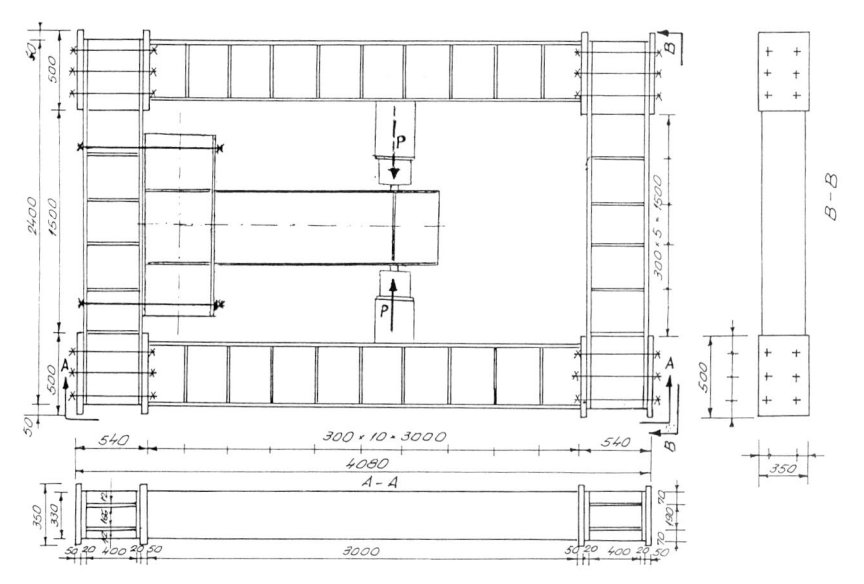

Fig.2. Steel testing frame and loading system.

Load was applied by hydraulic jacks fitted with load cells at the end of the cantilever (Fig.3). At the monotonous tests A1 and B1, the load was increased up to the joint collapse and at cyclic tests according with ECCS Recomendations.

4 Instrumentation and measurements

The tests were instrumented so that the measurements allow for determining the amplitude of all the components of the joint deformability at any loading level. It was useful to perform measurements of horizontal and vertical displacements as well as of rotations in appropriate sections. For this purpose were used mechanical devices for mesuring displacements with wire and with rod. Their disposement is shown in Fig.1.

As it was required to identify and measure separately all the components of the joint deformability, it was very important that the measurements to be scheduled accordingly. So, it was necessary to perform supplementary measurements which allowed to compute a specified rotation by at least two different manners and it was warranted to get results even one device was not functioned or something wrong was likely to occur during the test. The direct measure is always preferable and the most thrusting but the searched results have also to be deduced by indirect measurements, if it is necessary.

Following measurements were considered necessary to be made for steel T models (Fig.1):
- vertical and horizontal displacements at the end of the beam and the column;
- rotations of the beam and of the column;
- rotation associated to the load introduction deformability of column web;
- rotation due to the slip of the cleat-beam flange junction.

Maximum displacements of the beam portion resulted from the devices(1) measurements placed at the free end of the beam (Fig.1).

In view to compute the rotation of the beam were used the displacements measured by the devices(1), or (2)and(3), or (6) and(7), and for column, (4)and(9) placed at the ends of the columns.

Rotation of the column resulted also from displacements measured by the devices (5)and(8) placed on the column flange and located symmetrically to the beam axis (Fig.1).

Devices (2)and(3) were located at the lower flange of the beam as near as possible to the connection cross-section. Devices (6)and(7) were pointed to some small stitches welded transversally on the web and were located symmetrically to the beam axis.

The measurements of the rotation resulting from the slip at the interface of the flange cleat and the beam flange were made by measuring the relative displacement between two points located as close as possible, respectively on the cleat and on the flange, on both upper and lower flanges.(Devices (10), (11) and respectively (10), (11), (12), (13) in Fig.1)

5 Experimental characteristic curves

The characteristic deformations necessary for the curves moment -rotation were obtained by computing the experimental measurements.

The following characteristic curves associated to the different components of the joint deformability have been recorded for steel joints.

Main components:
- joint rotation

$$\phi = \theta + \gamma \quad \text{or} \quad \phi = F((1),(4)\text{and}(9)) \tag{1}$$

where: θ - connection rotation

γ - rotation due to shear of column web panel
- connection rotation

$$\theta = \theta_b - \theta_c \tag{2}$$

where: θ_b - beam rotation, θ_b by (1) or (2),(3) or (6),(7)

θ_c - column rotation, θ_c by (4),(9) or (5),(8).

- rotation due to shear of column web panel
$$\gamma = \phi - \theta \tag{3}$$

Components of connection deformability:
- slip of cleat flange
$$\theta s = ((12)-(10))/h_b \tag{4}$$
where: h_b - height of beam

6 Experimental characteristic results

6.1 Experimental behaviour

Test of joint **A1** was a monotonous static loading at the free end of the cantilever. From 0 to 90kN the load steps were by 10kN and after that 5kN. The maximum load was 190kN and displacement 56.6mm. The first local buckling in the compressed beam flange were observed at 110kN. Yielding was produced by local buckling of the compressed beam flange.

Joint **A2** was tested with a cyclic static alternate load. At this test it was tried to work after the ECCS Recommendations with displacements (multipliers of elastic displacement e_y) and the load was determined by the load cells. The maximum loads were 203kN (cycle I of $4e_y^+$) and -170kN (cycle III of $2e_y^-$). For the last three positive cycles of $4e_y^+$ the joint riched the loads: 203kN, 200kN and 190kN and for the last negative cycle of $4e_y^-$, -157kN. Maximum attained displacement was about 72.2mm. Yielding was produced by buckling of the beam flange and web in the compressed zone.

Joint **B1** was loaded statically monotonous. The load increased with a step of 10kN till 70kN and with 5kN till the maximum of 95kN and a displacement of 69.9mm. The flange angles stiffness modified the theoretical yielding mechanism, moving it from the angle to the bolts of the column flange from the tensioned zone.

Joint **B2** was loaded statically alternate in the same way like at joint A2. The loading reached only the positive semi-cycle I of $(4e_y^+)$ with a load of 46 kN, but the maximum load was reached in cycle II $(2e_y^+)$ with a value of 93kN. Yielding was produced by the bolts of the column flange tension zone. This quit different mechanism, in comparison with the theoretical (Kishi et al.,1987) and experimental (INCERC,1990) ones was modified by the flange angles stiffness.

6.2 Results interpretation

Curves load-displacement (P-Δ) for tested joints A1 and B1 (Fig.3) were used to determine the elastic displacement e_y and the corresponding load P_y (Table 3), which served to decide the cycles limits for the static alternate loading after ECCS Recommendations.

Table 3. Elastic displacements and loads

Joint	e_v (mm)	P_v (kN)
A1	21	130
B1	20	72

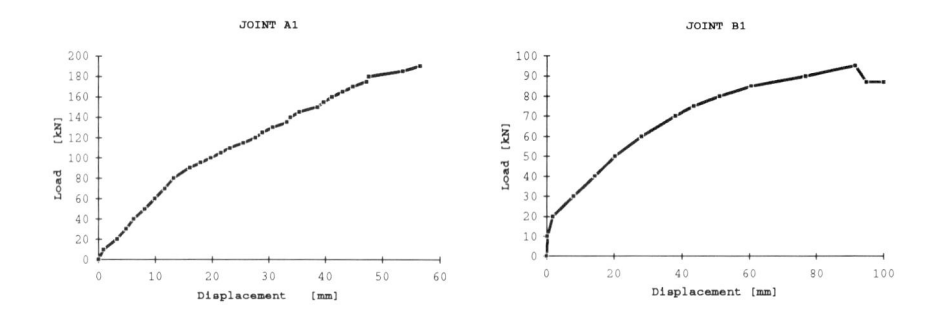

Fig.3. Load-displacement curves (P-Δ).

Analysing the results it was concluded that joint B is more flexible with lower forces and greater displacements. At joint B, after the bolt yielding from tension zone the connection is still able to assume loads by the beam web zone connection, at a lower load level. Curves moment-rotation (M-θ) were represented using the experimental parameters recorded during the loading programme and were determined using the relations (1)-(4). In Fig.4 are represented for joint A1 the curves: moment-joint rotation (M-ϕ), moment-connection rotation (M-θ) and moment-web column shear deformation rotation (M-γ).

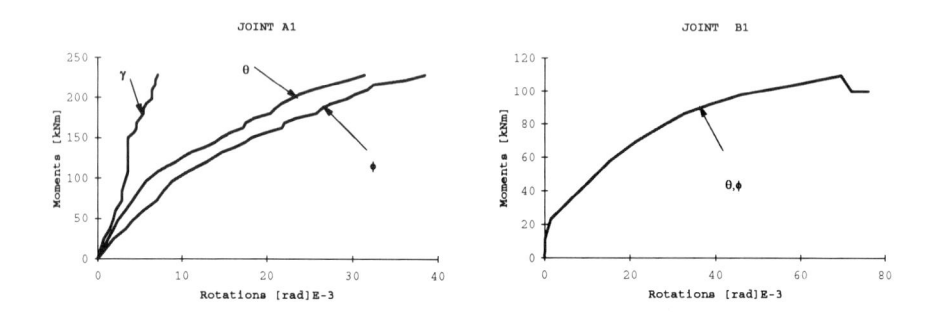

Fig.4. Moment-rotation curves for monotonous loading.

The initial stiffness determined on this curve is around 22×10^3 kNm/rad. For the joint B1 it was represented only the curve M-θ, because from the experimental results the values of the column web shear deformations were very small, so could be neglected and joint rotation φ is practically identical with connection rotation θ. The main component of the connection rotation is given by the beam rotation $θ_b$, because the column rotation is very small. Initial stiffness determined on the experimental curve is about 20×10^3 kNm/rad.

Figure 5 represents the M-θ curves for the cyclic static alternate loading of the joints. The obtained M-θ curves for A2 and B2 joints have a tipical form for the welded and bolted joints known also from the literature.

In comparison with the welded M-θ curve, the bolted M-θ diagram is more extended on horizontal direction, because of the greater deformations. At B2 M-θ curve, it is remarked a flattening in the central zone at high cycles loading produced by great displacements at low increase of the load.

6.3 Comparison between experimental and theoretical results

Theoretical results were computed using the theoretical models well known in literature like EC3 (1991) and Chen, Kishi model for bolted connection with top, seat and double web angles.

Experimental results were determined from the loading tests programme on the four connection specimens and from the obtained curves.

In Table 4 are presented the obtained experimental and theoretical results.

Table 4. Experimental and theoretical results

Connection type	Initial stiffness K_i (kNm/rad)		Maximum load P_{max} (kN)		Ultimate moment M_u (kNm)	
	exper.	theor.	exper.	theor.	exper.	theor.
A1	28.8×10^3	22×10^3	190	120	247	156.5
A2			203		263	
B1	25.4×10^3	20×10^3	95	83.7	109.3	96.3
B2			93		106.9	

It can be observed that the values of the initial stiffness are quit apropriate with greater values for the experimental ones. Also, the values of maximum load and ultimate moment for bolted joints B1 and B2 are quite similar, with greater values for the experimental ones.

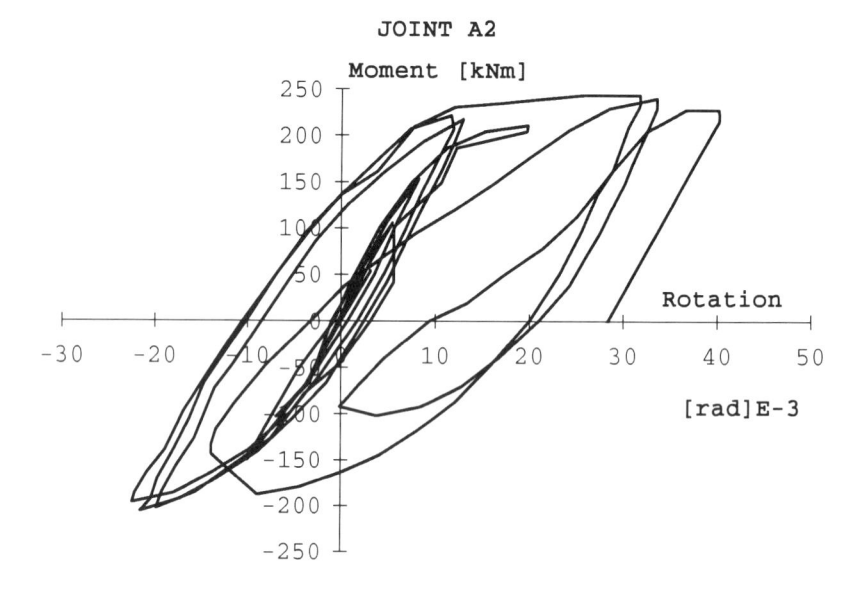

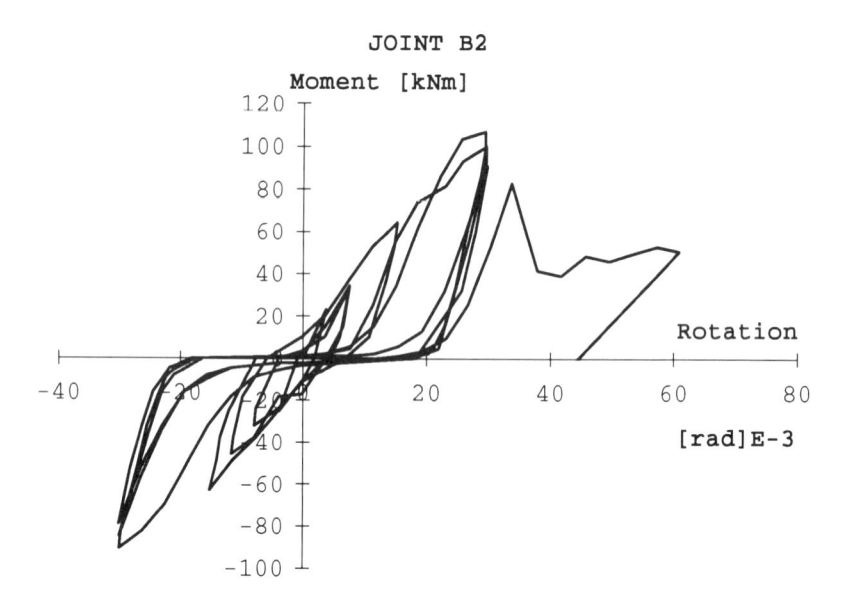

Fig.5. Moment-rotation curves (M-θ) for alternate loading.

For welded connections A1 and A2, which are full strength connections, the difference between the two coresponding values is quite important, the experimental ones being the greatest.

For a better interpretation, it can be computed the reserve of the loading capacity, apreciated between the experimental and theoretical values of the maximum load or ultimate moment as presented in Table 5.

Table 5. Reserve of the loading capacity

Joint type	Reserve of the loading capacity (%)
A1	36.80
A2	40.90
B1	12.00
B2	10.00

7 Conclusions

Analysis of joints behaviour, cyclic curves and yielding mechanism determined the following conclusions:

$M-\theta$ curves have specific forms for welded and bolted connections. The bolted connections $M-\theta$ curve is more extended on the horizontal, in comparison with the welded one. This thing is more significant especially at the high cycles, where in the central zone there is no important increase of the load for significant displacements.

At the bolted joints it can be remarked greater displacements for lower loads in comparison with the welded ones.

The angle flange stiffness modified the yielding mechanism, transforming the theoretical flange angle mechanism into a bolt yielding mechanism.

For the loading capacity reserve it can be remarked a greater one for the full strength welded connection, than for the partial strength bolted connection.

After the bolts yielding from the tension zone, it can be seen the loading capacity of the connection through the web shear zone connection, but at a lower level of the load.

This research developed in Timisoara in the last 3-4 years could not intend to cover the hole domain of the semi-rigid connections and especially in seismic areas there are necessary other more researchs. Although, the experience accumulated in the last years determines us to recomandate in seismic areas only the use of semi-rigid joints with full strength connections at sway frames according to the EC3 and EC8.

8 References

C.E.C. Agreement Nr.7210-SA/507 (1991) Semi-rigid action in steel frame structures, Draft of the Final Report, November 1991.

E.C.C.S.-Technical Committee 1-Structural Safety and Loadings. Technical Working Group 1.3-Seismic Design. Recommended Testing Procedure for Assessing the Behaviour of Structural Steel Elements under Cyclic Loads.

EUROCODE 3 (1992) Design of Steel Structures, Commission of the European Communities, February 1992.

Building Research Institute (INCERC) Timisoara (1990) Procedee moderne de realizare a imbinarilor de montaj. Imbinari grinda-stilp, Contract nr. 1506/1990.

Building Research Institute (INCERC) Timisoara (1992) Calculul static, dinamic si de stabilitate al structurilor metalice in cadre etajate cu noduri semi-rigide pentru constructii civile si industriale, Contract nr.1523/1992.

Jaspart, J.P.(1991) Etude de la semi-rigidité des noeuds poutre-colonne et son influence sur la résistance et la stabilité des ossatures en aciers, Ph.D.Thesis, M.S.M. Departement, Université de Liège, 1991.

Kishi, N. Chen,W.F. Matsuoka, K.G. Nomachi, S.G.(1987) Moment-rotation relation of top- and seat-angle with double web-angle connections, **Connections in Steel Structures**, Cachan, 1987.

Mazzolani, F.M. (1987) Mathematical model for semi-rigid joints under cyclic loads, **Connections in Steel Structures**, Cachan, 1987.

55 COST-C1 – SEISMIC WORKING GROUP ACTIVITIES

A. V. PINTO
European Laboratory for Structural Assessment, AMU, STI,
Joint Research Centre, Commission of the European Communities,
Ispra, Italy

Abstract
The present paper outlines the research programme of the COST-C1 seismic working group focusing mainly on the aims and working methods. In addition, a general description of the ongoing and programmed work by the working group members will be done. The general objective of the working group is to assess the seismic behaviour of civil engineering structures designed and/or constructed with semi-rigid connections. Improvement of construction technologies and manufacturing leading to a more competitive position of the European Industry in earthquake prone zones are envisaged. Relevant research achievements will be used to improve design codes, specifically Eurocode N.8 (EC8) which is the European design code for seismic regions.
Keywords: COST-C1, Seismic Working Group, Semi-rigid Connections, Earthquake Behaviour.

1 Introduction

COST-C1 is a European research project intended to study the behaviour of semi-rigid civil engineering structural connections. It is mainly a framework for R&D cooperation in the field, allowing for both the coordination of national research projects and/or the participation of third countries in Community programmes, taking the form of pre-competitive or basic research or activities of public utility. In addition to the European Union (EU) Member States, six of the EFTA countries are now participating in the project (Austria, Finland, Iceland, Norway, Sweden, Switzerland) as well as Czech Republic, Slovakia, Hungary, Poland, Turkey, Slovenia and Croatia.
  Prof. A. Colson from the 'Ecole Nationale Superieure des

Arts et Industries de Strasbourg', France and Prof. Tschemmernegg From the University of Innsbruck, Austria are, respectively, the Chairman and Vice-chairman of COST-C1 and the Management Committee (chairman, vice-chairman, WG chairmen and national representatives) meets twice a year.

The first COST-C1 Workshop took place in Strasbourg, Colson (1992), leading to the setting up of six working groups (WGs): Steel and Composite, Concrete, Timber, Numerical Methods, Seismic and Data-base. The 'organization' is schematically presented in Table 1. In addition to the three 'material' working groups, three other WGs, dealing with common subjects, have been set up.

This paper presents the proposed research programme of the Seismic working group focusing mainly on the aims and working methods. In addition, a general description of the ongoing and programmed work by the working group members is included.

Table 1: COST-C1 working groups and corresponding chairmen (from COST-C1 leaflet)

Working Groups	Chairmen
Concrete	K.S. Elliot (UK)
Timber	P. Haller (CH)
Numerical Methods	K. Virdi (UK)
Seismic	A. Pinto (CEC)
Data-base	K. Weynand (D)

2 Seismic working group research programme

2.1 Objectives

The general objectives of the working group are to study the cyclic behaviour of semi-rigid connections and to assess the seismic performance of civil engineering structures having semi-rigid connections. Relevant research achievements will be used to improve design codes, specifically Eurocode N.8 (EC8), CEC (1988), which is the European design code for seismic regions.

The secondary objectives are the improvement of construc-

```
┌─────────────────────────────────────────────────────────────┐
│ • Study of the Cyclic        - Cyclic Testing,              │
│   Behaviour of Semi-                                        │
│   rigid Connections  ──────▶                                │
│                              - Development and/or Cali-     │
│                                bration of Suitable Analyt-  │
│                                ical Models.                 │
│                                            │                │
│ ─────────────────────────────────────────▼───────────────   │
│                                                             │
│ • Evaluation of Seis-        - Numerical Seismic Analy-     │
│   mic Performance of           sis,                         │
│   Structures with    ──────▶                                │
│   Semi-rigid Connec-         - Large/Full Scale testing:    │
│   tions                        PSD, Shaking Table, etc.     │
│                                            │                │
│ ─────────────────────────────────────────▼───────────────   │
│                                                             │
│     • Identification of Design Code Improvements            │
│                                                             │
│                              and                            │
│                                                             │
│     • Development of Appropriate Construction and           │
│                Manufacturing Technologies                   │
│                                                             │
└─────────────────────────────────────────────────────────────┘
```

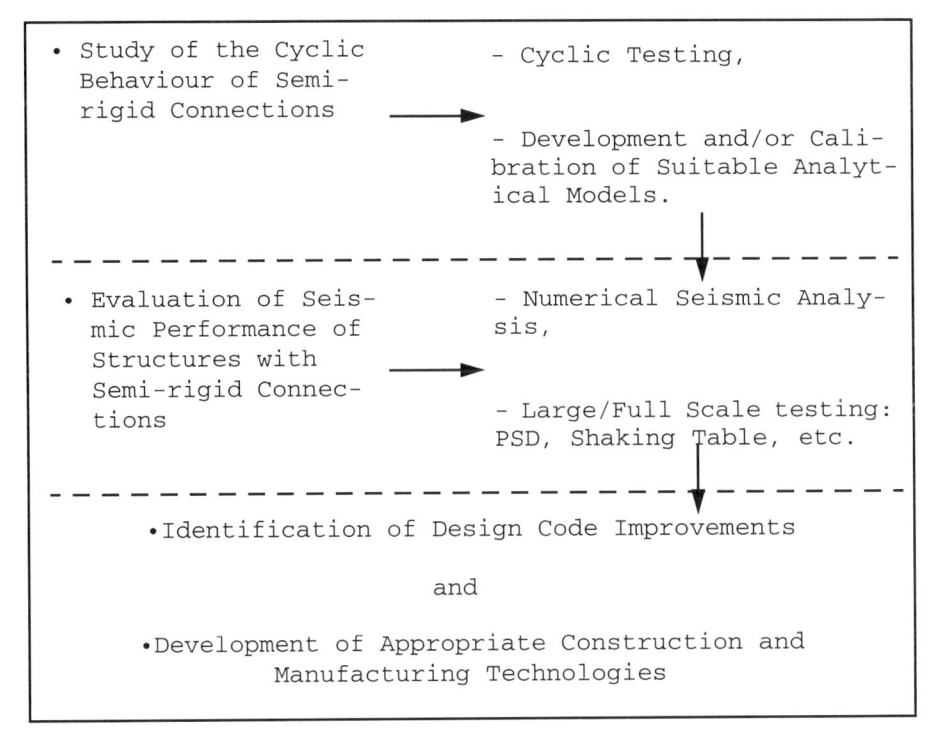

Figure 1. Tasks of the COST-C1 Seismic Working

tion technologies and manufacturing leading to a more competitive position of the European industry in earthquake prone zones.

2.2 Current state of knowledge

Design of seismic resistant structures is currently performed assuming that connections are rigid. However it is often impractical (feasibility and economical aspects) to provide for that. Moreover, modern construction technologies make use of 'pre-fabricated' components that are assembled in place with connections exhibiting a semi-rigid behaviour, often not well known.

This applies in particular to pre-cast concrete (P/C) structures made from reinforced or prestressed concrete elements and to steel (ST), timber (TB) and composite (S/C) structures. Concerning the (P/C) it is foreseen to have an Annex to EC8 which covers this subject. However, the necessary basic information is not available yet.

In other countries special attention is now being devoted to the semi-rigid behaviour of structural connections and the development of suitable design codes is pursued (e.g. Japan is currently preparing a first draft of a design code for pre-cast concrete, Shinsuke and Yoshikazu (1992), and the USA are updating the design code for steel structures with semi-rigid connections, Astaneh-Asl (1992)).

2.3 Description of the research project and specific requirements

The project includes experimental and analytical studies of the seismic behaviour of structures with semi-rigid connections and three main structural materials will be covered namely: pre-cast concrete, steel & composite and timber. In addition, development and assessment of strengthening and repair techniques for R/C structures will be performed.

The seismic performance of such structures will be evaluated on the basis of the experimental results and exploitation of developed models.

Guidelines for inclusion in design codes will be sought.

This research is to be carried at three distinct levels, involving numerical and experimental work, namely (see figure 1):

a) study of behaviour of single connections to be carried out in the different research institutes with funding from national governments and/or industry,

b) study of the behaviour of complete structures to be performed by the research institutes and the European Laboratory for Structural Assessment (ELSA) - JRC - Ispra (large scale tests). The large scale tests already planned and described later will be funded by current running programmes of the JRC. However if new tests are identified funds for their realization must be obtained from industry or from the national governments,

c) identification of relevant results for design codes for construction and for manufacturing industry. Funds should be available from national governments, EC and industry.

2.4 Existing and/or planned work

Development and calibration of models representing local (connection) behaviour will be performed on the basis of existing experimental results. Tests on models (structures and sub-assemblages) are also foreseen. In particular, the results of the tests already planned in ELSA will be used to assess the structural seismic performance. These tests are included in the programme of the European Association of Structural Mechanics Laboratories (EASML) Donea et al.(1992) and are the following:

- Pseudo-dynamic (PSD) testing of a three storey steel frame with R/C slabs and semi-rigid connections (1992/1993),
- PSD testing of a 4-storey R/C frame (1993/1994)
- PSD testing of a 3-storey composite (steel & concrete) frame (1994/1995)

The following work is programmed for the different National Research Institutes:

- PSD testing of twelve (12) one-storey timber frames - University of Florence- Italy (1993)
- Testing and analysis of steel semi-rigid frame connections (beam/column and 'infil panel'/structure) - University of Salerno and University of Napoli - Italy (1993/94)
- Experimental and numerical work on strengthening and repair of R/C connections - University of Bogazici, Istanbul - Turkey (1993/94)
- Modelling of the cyclic/dynamic behaviour of semi-rigid connections and structures - LMS-Nantes - France, DCE - University of Ljubljiana - Slovenia, LMT Cachan- ENS de Cachan/CNRS/UP6 - France & LBS - INSA de Lyon - France
- Modelling and analysis of semi-rigid connections in bridges - DMECI - University of Madrid - Spain

2.5 Working Group Members

STEEL & COMPOSITE
- Prof. F. M. MAZZOLANI - Istituto di Tecnica delle Costruzioni, Università di Napoli - Italy
- Dr. V. PILUSO - Istituto di Ingegneria Civil - Università di Salerno - Italy
- Prof. D. LE HOUEDEC - LMS - Ecole Centrale de Nantes France
-Prof. R. ZANDONINI - Dipart. Meccanica Srutturale e Progettazione Automatica TRENTO - Italy

PRE-CAST   CONCRETE
- Dr. F. COMAIR - French Pre-cast Concrete Industry Study & Research (CERIB), EPERNON - France -

TIMBER
- Dr. P. HALLER - IBOIS-Construction en Bois, EPFL Lausanne - Switzerland
- Prof. A. CECCOTTI - Dipartimento di Ingegneria Civile - Università degli Studi di Firenze - Italy
- Dr. E. FOURNELY - Laboratoire Génie Civil - Université Blaise Pascal, Clermont-Ferrand - France

BRIDGES
- Prof. E. ALARCON - Dep. de Mecanica Estructural y Construcciones Industriales - Universidad Politecnica de Madrid - Spain

REPAIR & STRENGTHENING
Prof. G. ASKAR, Mr. A. KOYLUOGLU and Mr. Y. YUVA - Civil Engineering. Department - Bogazici University, Istanbul - Turkey

MODELLING & ANALYSIS
-Prof. P. FAJFAR - Dep. of Civil Engineering - Inst. of Structural and Earthquake Engineering - University of Ljubljiana, Slovenia
-Prof. J. MAZARS, Dr. C. La BORDERIE & Dr. J.L. FLEJOU - LMT Cachan, ENS de Cachan/CNRS/ Université Paris 6, - France
-Prof. J.M REYNOUARD, Dr. O. MERABET & F. FLEURY - DGCU - LBS - INSA Lyon - France

2.6 Duration of the project

Duration of the proposed research project is estimated to be four (4) years (1993 -1996).
   The work described in the research programme is to be completed to the end of 1996. The experimental and modelling work is to be performed during the first three years. Exploitation of the experimental results and numerical modelling, aiming at the improvement of design codes (simplified design models and procedures, calibration of design parameters, etc.) is to be performed during 1995 and 1996.

3 Support from the Commission of the European Communities

Funding for COST projects comes from national governments and industry. However, the Commission provides a basis to coordinate and set up the present cooperative research project. This support facilitates the platform meetings and conferences essential to the work (e.g. travel expenses). In addition, it helps in the coordination between research institutes and interested industries and in the writing of the reports. Moreover, funding for short term scientific missions in the COST framework will be available.

4 Performed and ongoing work

Two seismic working group meetings have already been organized. The first one aiming mainly at the establishment of research collaborations and the second devoted to the presentation of the first preliminary results. Definition and optimization of testing procedures, analytical modelling of semi-rigid behaviour have been discussed.

In particular, the results of the PSD tests of the 3-storey steel moment resisting frame performed in ELSA, (Kakaliagos 1994), (Kakaliagos et al. 1994), and the final design of the beam-to-column connections (evaluation of strengthening and repair techniques) proposed by the University of Bogazici in collaboration with ELSA have been presented.

Analytical approaches (conventional and energy based) to the evaluation of seismic performance of structures with semi-rigid connections were suggested.

Despite the potential of the working group and the 'novelty' of the subject some difficulties related to the number of different aspects under research have been recognized.

The first WG report is to be prepared by the end of May, 1994.

5 Conclusion

The research programme and the ongoing work of the COST-C1 (Civil Engineering Structural Connections) Seismic Working Group were shortly described.

Participation, in the activities of the working group, of further institutes and industry dealing with semi-rigid behaviour of structural connections is strongly encouraged.

6 References

Astaneh-Asl, A. (1992) Shaking Table Tests of Steel Semi-rigid Frames and Seismic Design Procedures, in **Proceedings of the First COST-C1 Workshop**. ENSAI de Strasbourg, 28-30 October, Strasbourg, France.

CEC (1988) **Eurocode No 8 - Structures in Seismic Regions - Design, Part 1 - General and Building**. Report EUR 12266 EN, Commission of the European Communities, Luxembourg.

Colson, A. (1992) (editor) **Proceedings of the First COST-C1**

Workshop. ENSAI de Strasbourg, 28-30 October, Strasbourg, France.

Donea, J., Jones P. M. and Pinto, A.V. (1992) **European Association of Structural Mechanics Laboratories (EASML).** Technical Note No. I.92.146, STI, JRC, Commission of the European Communities, Ispra (VA), Italy.

Kakaliagos, A. (1994) **Pseudo-Dynamic Testing of a Full Scale Three-Storey One-Bay Steel Moment-Resisting Frame Experimental and Analytical Results.** Report EUR 15605 EN, ELSA, STI, JRC, Commission of the European Communities, Ispra (VA), Italy.

Kakaliagos, A., Verzeletti, G., Magonette, G. and Pinto, A.V. (1994) PSD Testing of a Full Scale 3-Storey Steel Frame in **Behaviour of Steel Structures in Seismic Regions** (Edited by F M Mazzolani). E&FN Spon, London.

Shinsuke, N. and Yoshikazu, K. (1992) US-Japan Cooperative Research Program (Phase 4): Research project activity of pre-cast seismic structural system (PRESSS) from Japan side, in **Proceedings of the 10th World Conference on Earthquake Engineering**, Balkema, Rotterdam.

PART SEVEN
BASE ISOLATION AND ENERGY DISSIPATION

56 STATE OF THE ART REPORT ON BASE ISOLATION AND ENERGY DISSIPATION

E. MELE
Istituto di Tecnica delle Costruzioni, University of Naples, Italy
A. DE LUCA
University of Reggio Calabria, Reggio Calabria, Italy

Abstract
In the very last years the seismic design approach, based on the structural response control, has gained more and more attention by the research community and by the professionals. An increasing number of applications of structural control in the design of buildings has been recently registered in Japan, USA, New Zealand and Europe. In this report the basic principles and the most recent applications of the different control systems are reviewed. A special attention, concerning design and research items, is devoted to base isolation system, which is the most widespread control technique to date. In the last section of this report the papers presented in the session on base isolation and energy dissipation are reviewed.
Keywords: Structural Response Control, Energy Dissipation, Active Control, Base Isolation, Seismic Design.

1 Introduction

A significant progress has been recently made in the development and application of innovative systems for seismic protection. The common feature of the different proposed approaches is the modification of the dynamic interaction between structure and earthquake ground motion in order to minimize damage and vibrations in all the components of the construction.
  In such a way both the structural and the non structural parts do not experience significant damage and, at the same time, the floor accelerations, responsible of damage to contents of the building, are remarkably reduced. As a result of this enhanced behaviour the structural safety is greatly increased, the functionality of the building can be preserved, and a high level of comfort (mainly derived from reduction of floor accelerations) can be obtained even in the case of severe seismic events.
  In other words the innovative design strategies are "not just a disaster prevention technology" (Kobori, 1994) because they lead to constructions characterized by a superior global performance.
  Both in the scientific research context and in the professional world the new seismic design approaches have rapidly gained increasing interest, as it is proved by the number of International Conferences and Workshops organized world-wide on this subject. In addition to the last 10th WCEE held in Madrid on July 1992, where approximately 10% of the papers dealt with topics in the area of

Behaviour of Steel Structures in Seismic Areas. Edited by F. M. Mazzolani and V. Gioncu.
Published in 1995 by E & FN Spon, 2-6 Boundary Row, London SE1 8HN. ISBN: 0 419 19890 3.

structural control (Buckle, 1993), in the last two years (1992-1994) several international and national meetings and conferences specifically dealing with base isolation, energy dissipation and control have been held (Tokyo, 1992; San Francisco, 1993; Capri, 1993; Los Angeles, 1994). At the same time, in the very recent years, the number of designed and realized applications of unconventional seismic-resistant systems has greatly grown in Japan, USA, New Zealand and Europe (Kelly, 1993a and 1993c; Mazzolani, 1993). Furthermore significant efforts have been spent by national and international codification committees in order to provide design guide-lines and application rules (Blakeley et Al., 1979; SEAONC, 1986; SEAOC, 1990; AFPS, 1990; UBC, 1991; SSN, 1993; UBC, 1994).

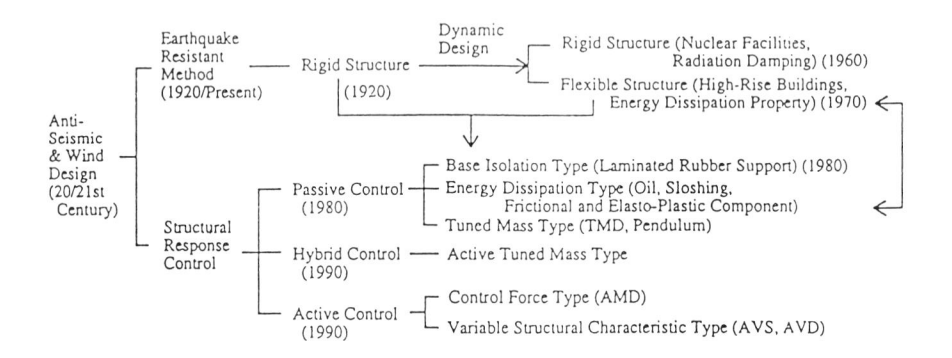

```
                          ┌ Earthquake                               Dynamic     ┌ Rigid Structure (Nuclear Facilities,
                          │ Resistant ─── Rigid Structure ─Design──→ │           Radiation Damping) (1960)
                          │ Method            (1920)                 └ Flexible Structure (High-Rise Buildings,
                          │ (1920/Present)                             Energy Dissipation Property) (1970) ←
Anti-                     │                      ↓
Seismic                   │                                          ┌ Base Isolation Type (Laminated Rubber Support) (1980)
& Wind ──────────────┤                    Passive Control ─┤ Energy Dissipation Type (Oil, Sloshing,
Design                    │                      (1980)              │   Frictional and Elasto-Plastic Component) ←
(20/21st                  │ Structural                               └ Tuned Mass Type (TMD, Pendulum)
Century)                  │ Response ──┤
                          │ Control       Hybrid Control ─── Active Tuned Mass Type
                          │               (1990)
                          │
                          │               Active Control ─┬ Control Force Type (AMD)
                          │               (1990)          └ Variable Structural Characteristic Type (AVS, AVD)
```

Tab.1 Evolution process of seismic design (Kobori, 1994)

The new strategies of seismic design can be regarded as the latest step of the evolution process of the earthquake resistant design (Tab. 1 (Kobori, 1994)). Starting from the beginning of the century, specific design methods were proposed for designing earthquake resistant structures. The intuition of Gustave Eiffel, to model the earthquake forces by means of an equivalent wind, was somehow implemented into the first design methods based on equivalent static forces procedures. The earthquake-proof buildings were conceived as rigid structures, being at the time the concept of "solidity and massiveness" equated to the concept of "safety and reliability" (Soong, 1990).

Only at the end of the 60's the dynamic response analysis became a mature technique and a powerful tool in modelling and understanding the actual seismic behaviour of constructions. Design of more flexible structures capable of attracting less severe seismic forces was started at that time. In addition, the enhancement in the comprehension of the structural behaviour beyond the elastic limit suggested to reduce the seismic design action accounting for the ductility capacity, i.e. the deformation and energy dissipation capacity, of the structure itself, thus allowing the design of more economical buildings.

In more recent years the traditional approach of seismic design has been critically reviewed due to the inherent drawbacks: among them the concept of allowing significant damage through ductile behaviour under a severe earthquake is clearly criticizable. In order to minimize the damage, the possibilities offered by advancements in technological and industrial research, as well as in the scientific context, have allowed for the definition of a new approach to seismic design, mainly based on the idea of controlling the response of the structure by reducing the dynamic interaction between the external source (earthquake) and the structure itself.

2 Structural Response Control

The control of the structural response can be based on two different approaches, namely either the modification of the dynamic characteristics or the modification of the energy absorption capacity of the structure. In the first case the structural period is shifted away from the predominant periods of the seismic input, thus avoiding the risk of resonance occurrence. In the second case the capacity of the structure to absorb energy is enlarged through appropriate devices which preserve from damage the structure.

Both these approaches can be implemented either in passive or in active systems. In passive systems the modified properties of the structure (period and/or damping capacity) do not vary depending on the seismic ground motion, whereas in the active systems the structural characteristics are modified just as a function of the specific seismic input.

The actively controlled structures are therefore characterized by the "adaptiveness" to the dynamics of an ever-changing environment (Soong, 1990).

	ACTIVE AND HYBRID Systems	PASSIVE Systems
Damping Effect	Active Mass Driver (AMD) Hybrid Mass Damper (HMD)	Oil Damper Friction Damper Lead Damper Steel-Hysteretic Damper Visco-elastic Damper Tuned Mass Damper (TMD) Tuned Liquid Damper (TLD)
Non-reasonance Effect	Active Variable Stiffness (AVS)	
Isolation Effect		Base Isolation

Tab.2 Classification of structural response control systems (Kobori 1994)

A comprehensive classification of the structural response control systems is provided in Tab.2, taken from (Kobori, 1994). The classification is made both on the basis of the "degree of adaptiveness" of the system (passive, active or hybrid system), and on the basis of the approach adopted in implementing the response control (non-reasonance, isolation, damping). As a consequence of the above delineated multitude of approaches to the seismic design, in the very next future, the structural designer will be able to select the most appropriate structural system among both the conventional and the innovative ones, depending on the required structural performance and the economical background (Akiyama, 1993). In this perspective the level of accepted damage in a construction can be considered as an input data of the design process.

In the following, recent developments of the different control systems are reported and the main applications to structures are reviewed. A special attention is devoted to base isolation system (BIS) which is the most widespread control technique to date.

3 Passive Control Systems Based on Damping Effect

In the classification provided in table 2 of the previous section the passive systems have been classified on the basis of the main underlying strategy: damping or isolation. Section 5 will be entirely devoted to the base isolation system, while this section deals with the passive systems based on the principle of improving the damping capacity of the structure. Among the latters, a further distinction has to be made between the systems which preferentially dissipate the earthquake induced energy in specific devices, especially designed and introduced in the structure for this purpose, and the ones which, utilizing an additional vibrating mass (solid or liquid), tuned to match the principal frequency of the building, produce a counterforce which offsets movements of the building.

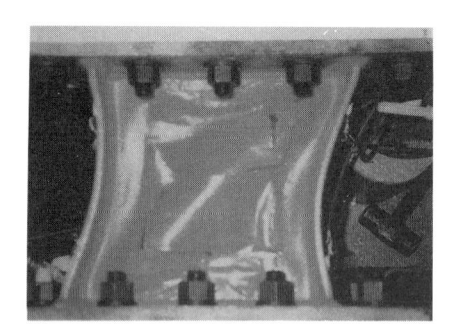

(a) **(b)**
Fig.1 (a) Deflection of steel damper and (b) its location in the bracing System (Obayashi Co., from Fujita, 1991)

A variety of different mechanical devices have been developed and utilized in applications of this type of passive systems; they can be roughly grouped into three main classes, depending on the energy dissipation mechanism: friction, hysteretic (lead, steel) and viscous / viscoelastic (oil, rubber, acrylic materials) dampers. Different dampers utilising vibrating masses (TMD) or liquid tanks (TLD) have also been developed and applied in several buildings.

An example of the steel hysteresis type damper is the one developed by the Obayashi Company and introduced in the bracing system of the Sumitomo Irifune Office Building, a 14-story steel structure building built in Tokyo in 1989 (fig.1). Under shearing deflection the steel panel reaches the yielding condition and thus provides the energy absorption through the high hysteretic properties of steel especially when appropriately designed for this purpose.

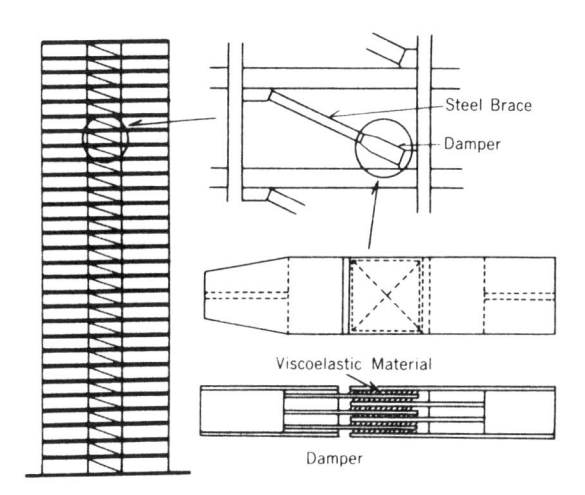

Fig.2 Viscoelastic damper (Hazama Co., from Fujita, 1991)

Passive energy dissipation can be obtained through the use of viscoelastic materials. A damper of this type, developed by the Hazama Corporation, is made of a sandwich of steel plates and acrylic viscoelastic material (fig.2). It is placed at the end of the steel bracings and when the building is shaked, viscoelastic material experiences shear deformations, absorbing and transforming into heat the vibration energy.

The development of damping devices allows to use irregularity of buildings as a means for reducing seismic induced vibrations. It is in fact possible, by taking advantage of the different vibration modes of two adjacent buildings, to reduce the vibrations of both structures if an appropriate damper is placed between them. This concept has been put into practice in an actual building, the KI-Building, shown in fig.3 (a), which is composed of two bodies, a five-story building A and a nine-story building B, connected each other through three corridors. A specifically developed damper has been located at the connections between the different bodies. The proposed damper is of elasto-plastic type and utilizes the energy dissipation capacity through hysteretic mechanism of a steel center-hollowed bell-shaped device (fig.3(b)); several static and dynamic cyclic tests have proved the good deformation capacity, the high energy dissipation ability and the stable hysteretic characteristics of the bell-damper. The effectiveness of the system has been assessed on the basis of the results of numerical analysis, which showed about 30% of reduction of the building maximum response.

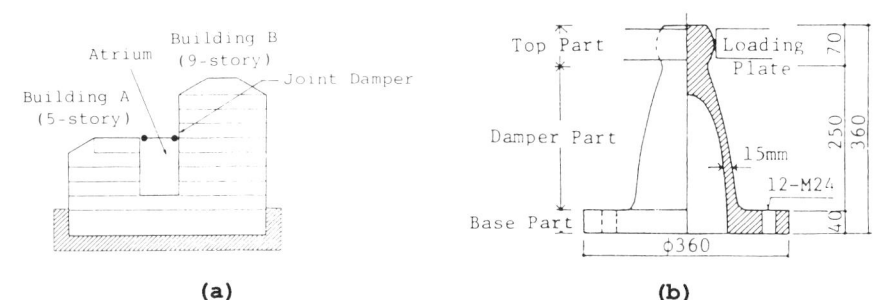

Fig.3 (a) Cross section of KI-building and (b) utilised bell-damper (Kajima Co., from Fujita, 1991)

Still based on the concept of differently vibrating systems is the implementation of the tuned mass damper. TMD is an apparatus usually installed at the top of the building which is tuned to the natural frequency of the building and, shaking sympathetically with the building, reduces the structural vibrations. Applications of TMD usually employ large tanks of sloshing water or steel mass rollers. The Bridgestone Corporation, a leader Japanese company among the manufacture industries of rubber devices, has recently proposed the use of multi-stage rubber bearings for TMD applications (fig.4). These particular devices present some advantages with respect to the other ones usually utilised, allowing for a compact and maintenance-free apparatus.

Fig.4 TMD using multi-stage rubber bearing (Bridgestone Co., from Fujita, 1991)

The above mentioned devices represent only some examples among the several ones involving the passive energy dissipation approach which have been widely applied both in retrofit of existing buildings and in new designed structures. A more complete review of the most significant proposed devices and actual application can be found in (Hanson, 1993; Ciampi, 1993).

Base isolation and energy dissipation 637

4 Active Control Systems

The working principle of the active control is the one of an "intelligent" structure, which detects vibrations induced by external events by means of sensors installed in the building and transmits the collected data to a computer system, which analyses the data and governs the activation of the control system in order to counterwork vibrations. As previously outlined, the practical systems proposed to realize this idea can involve the damping or the non-reasonance strategy. In the former case the damping of the structure is increased by activating mass dampers through computer-controlled actuators, whereas in the latter case the vibration response is reduced by adjusting the fundamental period, i.e. the stiffness, of the building, thus avoiding the possibility of reasonance with the predominant frequency of the seismic input.

Based on these alternative principles some systems have been developed, fabricated and installed in full-scale structures. The applications concern essentially steel structures located in Japan and employ different control systems, namely the active mass dampers (AMD), active bracing systems (ABS), active variable stiffness systems (AVS) and semi-active, or hybrid mass dampers (HMD).

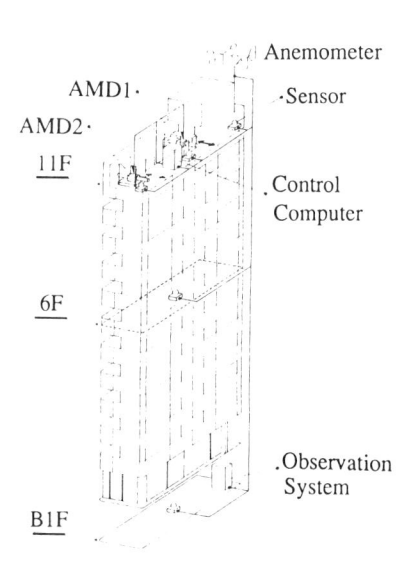

(a) (b)
Fig.5 (a) View of building and (b) Composition of the AMD system (Kajima Co., from Fujita, 1991)

The Kyobashi Seiwa Building, built in Tokyo in 1989 by the Kajima Corporation (fig.5) adopts an Active Mass Driver (AMD). The building, an eleven-story steel framed structure, is very slender and sensitive to earthquake and strong wind induced vibrations. The installed AMD system has therefore the objective of bringing the structural response within the serviceability limit conditions.

The system is composed of two auxiliary masses placed on the roof, which are activated by servo-hydraulic actuators and provide the control forces. A forced vibration test was carried out to assess the dynamic behaviour of the structure and time history analyses were performed for verifying the working conditions of the system even at ultimate limit state.

An Active Variable Stiffness (AVS) system has been proposed and developed by the Kajima Corporation. In this case the computer, which analyzes the seismic external action, selects the optimum value of stiffness and supplies it to the building through variable stiffness devices which are able to change the connecting conditions between braces and beams at each floor. As the stiffness continuously changes in order to achieve the non-reasonance condition, the response of the structure is always reduced, resulting in an optimization of seismic performance. The system was installed in a prototype actual building, the Experiment Control Building of the Kajima corporation, completed in 1990 in Tokyo. In fig.6 (b) is provided the installation of the variable stiffness device in the building, which is a three-story steel frame (fig. 6(a)). A forced vibration experiment has been carried out on the building and has confirmed the effectiveness of the AVS; furthermore earthquake simulation analyses have proved the system capacity even under severe earthquake conditions.

Fig.6 (a) View of building and (b) Installation of the AVS device (Kajima Co., from Fujita, 1991)

As a result of a more consolidated experience in the field of passive tuned mass dampers, some proposals of active-passive system (the so-called hybrid systems) have been formulated. These typology tend to combine the advantages of both passive and active systems and to minimize the drawbacks. Therefore, even though they roughly preserve the same high effectiveness and reliability in damping vibrations than the active systems, the dimensions and the required driving power of the dampers is remarkably reduced.

The hybrid system developed by the Ishikawajima-Harima Heavy Industries Co. (fig.7) combines a pendulum passive mass damper with active control drivers, requiring a relatively small electric power. Shaking table experiments of the device mounted on a model tower structure have been carried out to test the working conditions of the system.

Fig.7 Hybrid mass damper (Ishikawajima-Harima Heavy Industries Co., from Fujita, 1991)

A similar system, the "rolling pendulum method", has been developed by the Kumagami Gumi Co.; also in this case the idea is to obtain a system which works with smaller added masses, and therefore with less external energy supply.

In the context of mass dampers it has to be reminded the pendulum-type tuned active damper developed by the Mitsubishi Heavy Industries and installed on the seventy-story MM21 Land Mark Tower built in 1993 in Yokohama. The structure, with 296 meters of height, is the highest in Japan, and uses a tuned active damper in order to reduce the effects of wind. The system can be effective also in absorbing vibrations induced by moderate earthquake, but cannot nullify the effects of a major seismic event.

The significant progress registered in active control technology in the very last years will make possible the realization of the most ambitious project ever conceived in civil constructions. It is a super-skyscraper 1000 meters high - the first "vertical kilometre" - to be built in Japan, which will have over 300 floors and will be able to contain apartments, offices, and every kind of facilities for the autonomous living of 100,000 people. The underlying idea is a radically different concept of city, spreading in vertical sense, which can be assumed as the landmark not only for the Japanese high-technology, but also for the humankind challenge towards new frontiers.

The realization of this structure will involve new constructional materials, like carbon-fibers and ceramic materials, as well as advanced active control systems. All the major Japanese construction companies (Kajima, Takenaka, Obayashi, Shimizu, Taisei and Kohinoike) have already presented projects and solutions, but this feasibility phase of study will probably go on over 1998, while the construction is planned to be completed around the 2010 - 2015.

This project which will significantly contribute to the development of new technical solutions, confirms the validity of active systems for structural response control. It should be underlined, though, that active control systems could never be a current seismic protective system, being essentially devoted to high technological value constructions. Passive systems instead are best fitting for a wide application, also to ordinary buildings. In particular the base isolation system, already applied in a large number of buildings, can be considered as a mature technique deserving to be evaluated as a competitive alternative to classical solutions for earthquake resistant structures. The competitiveness derives from wide possibilities of application, economics, no real maintenance costs, high seismic performance of the building.

5 Base Isolation Systems

5.1 Introduction

The concept of BIS is well known: it relies on decoupling the structure from the damaging component of the seismic ground motion, which is practically obtained by supporting the structure on a highly flexible layer (elastomeric bearings). In this manner the fundamental period of the construction is shifted in the high-periods region, where the energy content of the earthquake is usually lower. Due to the reduction of the input energy into the building, an improved dynamic behaviour with respect to conventional constructions is obtained during earthquake, with significant reduction of interstory drifts, of floor accelerations and of stresses in structural members (fig.8).

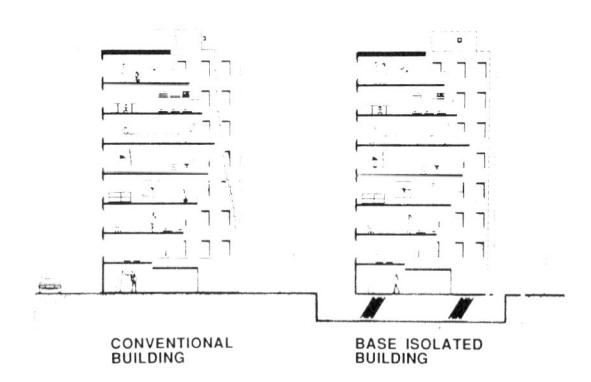

CONVENTIONAL BUILDING   BASE ISOLATED BUILDING

Fig.8 Working principle of BIS (Shimizu Co., from Fujita, 1991)

The most widely utilised seismic isolation system consists of laminated rubber bearings, made of rubber layer with steel shims interleft. These devices are characterized by a high flexibility and large deformation ability in the horizontal direction, thus producing the isolation effect, while show high vertical stiffness (about 500÷1000 times the horizontal stiffness) and a stable, large load bearing capacity (fig.9).

Base isolation is the most widespread control technique and several applications have been already realised all over the world. The numerous applications have marked the evolution of this system deriving from the strong research effort enforced by companies and researchers. These efforts are still underway for improving the performance of the devices and for optimizing the superstructure plus isolation system behaviour also through appropriate design of the complex.

Fig.9 Rubber bearing during horizontal testing (Bridgestone Co., from Fujita, 1991)

In the following the major items of the research activity are briefly outlined; in particular section 5.2 is devoted to the problem of the definition of the shear coefficient, which is strictly related to the design choice of allowing a certain amount of inelastic deformation in the superstructure. Section 5.3 deals with the modelling of isolation and global structural behaviour, while the cost/benefit analysis, which is a key-point in the feasibility study of a base isolated structure, is discussed in section 5.4; finally some design and implementation aspects, as well as the position of current codes, are presented in section 5.5, and some concluding remarks with reference to the existing applications are provided in section 5.6.

5.2 Definition of Shear Coefficient

A wide research activity has been developed on the definition of the shear coefficient to be adopted for the design of base isolated structures. According to UBC 91, the total base shear to be applied to the upper structure is given by:

$$V = C_s W$$

where C_s is a function of the isolated period and accounts for the zone seismicity, near fault effect, soil profile, damping and for reduction factor R_W.

The definition of the appropriate values of the reduction factor of seismic forces represents a problem not yet completely solved for conventional fixed base (f.b.) structures. In the case of base isolated (b.i.) structures this problem is completely open since different, sometimes opposite, positions are registered in the codes and in the relevant bibliography.

The two different positions can be synthesized in the following opposite design philosophies:
a) b.i. structures should provide higher performance with negligible damage in the superstructure under the Design-Basis Earthquake (DBE)

and/or the Maximum Credible Earthquake (MCE);
b) there is no reason for adopting a different philosophy in the design of b.i. structures, with respect to what is currently done in the case of f.b. structures.

Position a) leads to advising reduction factors equal to one, while position b) yields to suggesting reduction factor values similar to the ones adopted in conventional structures. It has to be said though that the previous considerations might represent an oversimplification of the problem since there are significant differences between the performance and the behaviour of b.i. structures when compared to the f.b. ones. The discussion presented hereafter is aimed at evidencing some of these differences through a review of the contributions provided by different researchers.

Among the Authors in favour of position a) we recall Chaloub & Kelly (1990), Yamaguchi et Al. (1993) and Occhiuzzi et Al. (1994). In particular Chaloub & Kelly (1990) suggest a conservative design for the upper part of b.i. structures (R_w factor equal to one) "before a full understanding of the effect of yield in the superstructure on the interaction between the isolation system and the superstructure is reached". Yamaguchi et Al. (1993) confirms that the present position of Japanese design code "does not make use of reduction factor for lateral design force of the superstructure". The Authors explain this design philosophy by reminding that "this is because seismic isolation system is based on the concept that *energy input is designed to be concentrated on an isolation interface*. The idea of expecting the energy absorption capacity by superstructure's plastic deformation does not meet this basic concept." A similar position, at least with reference to the DBE, is shared by Occhiuzzi et Al. (1994), who conclude that "it is inappropriate at the DBE level to use any significant response reduction factor R." The Authors demonstrate their conclusions through an extensive parametric analysis of several MDOF systems simulated by means of a simplified procedure based on equivalent linearization and modal analysis.

A completely different position is the one reported in (Lin & Shenton, 1992; Shenton & Lin, 1993), where, on the basis of numerical analyses carried out on b.i. steel and r.c. structures, the Authors point out that b.i. structures show a better or comparable behaviour than analogous f.b. structures even when designed with the f.b. values of the reduction factor (i.e., for steel moment resisting frame R_w=12). It should be emphasized that these conclusions are based on a limited number of structural typologies analyzed. Also the selection of input ground motions deserves a deeper investigation before extrapolating the results obtained.

The effects of allowing inelastic deformations in the superstructure of a b.i. system cannot be directly referred to the observed behaviour of analogous f.b. structures. This question has been raised by several Authors (Di Pasquale et Al. (1989); Vestroni et Al. (1991); De Luca et Al. (1994a and 1994b); Pinto & Vanzi (1992); Calderoni et Al. (1993); Palazzo & Petti (1993); Dolce & Quinto (1994)) who have compared the inelastic behaviour of f.b. and b.i. structures, the latters being analyzed at increasing level of Peak Ground Acceleration (PGA), or, alternatively, designed at increasing values of reduction factor.

In particular Di Pasquale et Al. (1989) and Vestroni et Al. (1991) have shown that by adopting large values of the reduction factor the response of b.i. structures is remarkably worse than the one of analogous conventional structures because of a sharp increase of the ductility demand. In these papers a 2-DOF model is adopted for analysing different b.i. and f.b. structures under artificial acceleration signals scaled at increasing values of PGA.

A sharp increase in ductility demand at large values of R_w is also observed by De Luca et Al. (1994a and 1994b). In these papers real MDOF systems are analysed by means of inelastic time-step analysis and the influence of the inherent stiffness of the structure (different number of story), of the design criteria (low or high ductility), as well as of the soil type (stiff and soft) is pointed out. A distinction is also made between the *actual* reduction factor (defined for each structure under each specific seismic input) and the design reduction factor, in order to account for the different spectral acceleration values of the considered historical ground motions at the isolation period.

Another confirmation of the significant differences between the behaviour of f.b. and b.i. structures comes from Pinto & Vanzi (1992), who point out that the influence of the strength capacity on the ductility demand in b.i. structures is greater and less predictable than for f.b. structures and the current deterministic approach to the design, based on the adoption of reduction factor, may lead the structure to unexpected and unsafe conditions. The Authors recommend the use of base isolation in cases where the ductility demand can be avoided.

Similar results are also observed in Calderoni et Al. (1993), Palazzo & Petti (1993), Dolce & Quinto (1994). In these last papers an attempt is also made for suggesting a value of reduction factor to be adopted in the design of b.i. structures. The Authors reach similar results even though through different procedures. In particular a value around 1.5 is commonly agreed. It has to be said though that further investigations are needed before assuming a general validity for these results.

In the numerical analyses aimed at comparing the performance of b.i. and f.b. structures and at deriving the values of reduction factor, one of the main aspects which deserves more attention, as underlined in De Luca et Al. (1994b), is represented by the actual safety level of the b.i. structure, defined as the ratio of the specific seismic strength demand (i.e. the spectral acceleration value at the isolation period) to the strength capacity of the structure. Concerning this point, some numerical analysis carried out by adopting seismic inputs characterized by a sharp decrease of spectral values at increasing values of the period, might be strongly misleading.

In the light of all the contributions here shortly reviewed it can be clearly stated that yielding in the structural members above isolation may lead to the drop in the fixed-base frequency of the upper structure, thus reducing the isolation efficiency and leading to high ductility demand, which the superstructure, even though properly designed (high ductility design), cannot supply.

In conclusion the scenario here presented reflects the controversy on the definition of the reduction factors, which is strictly connected to the allowance for nonlinear behaviour in base isolated structures.

5.3 Modelling of Isolation System and Superstructure

Different models have been proposed in the existing literature for the analysis and design of base isolation systems. A simplified approach to the simulation of the actual behaviour is the one which utilises SDOF models, i.e. global relationships in terms of force-displacement.

The simplest SDOF model which can be adopted for the simulation of rubber devices is the equivalent viscous linear elastic, which characterizes the global behaviour of the system by means of two parameters, i.e. the stiffness and the equivalent viscous damping coefficient. Being the actual response of rubber materials significantly nonlinear, the stiffness is usually defined as the secant value; furthermore, in the high damping rubbers the energy dissipation is essentially due to a hysteretic mechanism, therefore the equivalent viscous damping coefficient is defined through energy equivalence criterion. This model is often used because, preserving the advantages of linearity, it is simple to be implemented; the main limitation of this model is the assumption of constant stiffness, which, on the contrary, strongly depends on deformation. However the use of the model leads to an acceptable degree of approximation of the peak response (De Luca et Al., 1993c; Kelly, 1993b).

The main limitations of the linear viscous model are overcome by the hysteretic models. Among them widely used (Dolce & Quinto, 1994; Di Pasquale, 1989; Vestroni, 1991; Lin & Shenton, 1992; Shenton & Lin 1993; Occhiuzzi et Al., 1994) is the bilinear model which is defined by means of three parameters, the pre-yield and post-yield stiffnesses and the yield point (force or deformation). The bilinear model, still rather simple to be implemented, allows to obtain different loops varying the parameters (Skinner et Al., 1993). An increasing degree of approximation can be obtained passing to three-linear, and hence to multilinear models.

More refined are the differential hysteretic models, which, coupling a high computational efficiency and a capacity of reproducing a variety of behaviour, are well-fitted for the simulation of complex constitutive relationships. One of them is the Bouc-Wen model which has been frequently adopted in the simulation of high-damping rubber bearings (Pinto & Vanzi, 1992; De Luca et Al., 1993c; 3D-BASIS, 1989). The model requires the selection of seven parameters, among which five depend on the rubber mixture and device geometry, while the remaining two have to be calibrated on experimental measured secant stiffness and equivalent viscous damping. A further refinement of the model has been proposed by De Luca et Al. (1993c and 1994d) where the experimental parameters are updated depending on the level of deformation, thus allowing for the precise reproduction of hysteresis loops, even when high strain-hardening occurs at very large strains.

Also the nonlinear hysteretic model of Serino et Al. (1992) is able to simulate the decay of stiffness at increasing deformation. In particular the model adopts an exponentially decrease of rubber tangential modulus with increasing shear strain and depends on three parameters, which respectively are defined as the initial (maximum) and the asymptotic (minimum) value of the shear stiffness, and the rate of stiffness decrease.

All these SDOF models, capable of simulating the force-deformation relationship of an isolation device, can be implemented into a MDOF model representative of the global isolated structural complex. In fact a real isolated building can be modelled at different levels, by adopting respectively a SDOF, a 2DOF or MDOF model. In particular MDOF models are essential when the actual inelastic behaviour of the superstructure together with the values of interstory drifts and floor accelerations is of major concern. A very accurate structural model should be fully nonlinear, i.e. should account for the inherent nonlinear behaviour of the isolation and of the upper part. Depending on the specific context of analysis, an acceptable degree of approximation can be also reached only partially accounting for the nonlinearities.

If the purpose of analysis is targeted at evaluating the behaviour of the isolation device, different and more complex type of modelling should be implemented. In particular in order to assess the local stress and strain distribution in the rubber layers and steel shims of the device, a FEM numerical analysis should be undertaken. Few complete studies of this type have been carried out to date. In fact the FEM analysis involving elastomeric materials is a very arduous problems due to the highly non linear elastic constitutive law, the large displacement and strain which the model is subjected to, and the nearly-incompressible behaviour. Specific studies on this subject can be found in (Seki et Al., 1987; Takayama & Tada, 1992; Martelli et Al., 1992; Sheperd & Billings, 1992; Forni et Al., 1993; Imbimbo et Al., 1994a and 1994b).

5.4 Cost Effectiveness and Optimization

Even though significant improvements of the seismic structural behaviour can be achieved through BIS, "it is by no means a panacea" (Mayes et Al., 1992). Therefore advantages and limitations of BIS have to be clearly stated both in technical and economical terms, in order to assess the range of optimum applicability.

For this purpose the comparison to analogous fixed-base structure has to be considered as a judgement factor. However the comparative analysis between conventional and isolated can be made following different approaches.

From a purely technical point of view the fixed and isolated solutions can be compared in terms of several performance parameters accounting for the level of damage in the structural elements as well as in the non-structural components and to contents. Proper measures of structural damage are mainly connected to the ductility demand, while the response parameters responsible of damage to non structural components are the interstory drifts and floor accelerations. Therefore, a comparison of the structural behaviours in terms of precise performance parameters for different environmental conditions (soil type, seismic zone, etc.) and superstructure typologies, can provide the optimum applicability of BIS.

Shifting the comparison in the economical context, the approach is no more unique, and the definition of the terms of the comparison involves the assessment of the b.i. structures performance requirements. In fact the isolated structure project can be based either on the "equal performance" or on a "superior performance" (with respect to fixed base) criterion. The definition of such criteria is strictly related to the code requirements and essentially reflects in the value of the reduction factor R_w to be adopted in the design. The philosophy of the available design code specifications to date matches the "superior performance" criterion.

In the first case ("equal performance") the economical comparison between fixed and isolated structural solution can be simply carried out by comparing the construction costs. This approach has only been applied to nuclear facilities, where the high structural performance are required regardless to construction method, and to the project for the Los Angeles County Fire Department (Mayes et Al., 1992), where a global saving of 6% was estimated for the BIS solution.

In this case the differences between costs of fixed and isolated structures is based on well defined items, involving only constructional costs. For the base isolated, in addition to the superstructure cost, it has to be accounted for some further sources of cost, due essentially to: isolator units, additional excavation, special foundation structure, special flexible joints for structures and equipments at isolation interface. On the other hand, reductions of costs are possible, mainly due to the reduced seismic forces which allow for savings both in the structural system and in the electrical and mechanical equipments.

However as previously stated, the philosophy of current codes usually leads to base-isolated structures characterized by a seismic performance better than conventional construction. The major difficulty in a cost comparison between alternative solutions is just the economic quantification of this enhanced behaviour. In fact, while initial construction costs are well defined, some specific criteria have to be outlined for other earthquake cost impacts, which are essentially the following (Mayes et Al., 1992): earthquake insurance premium, physical damage reparation costs to restore the building's pre-earthquake value, disruption due to building and contents damage, loss of market share or clients. With regard to the earthquake insurance premium, special savings are possible if any earthquake mitigation measures is adopted. The consequence of business disruption can be economically evaluated depending of the specific activity and base isolation is able to improve the chances of full business survival after a major earthquake.

The estimate of the damage repairing actual costs is the most difficult point, and can be based on three different approaches. The first method (ATC-13, 1985) involves a "damage matrix" developed for different structural systems on the basis of data collected from past earthquake. The second procedure (Ferritto, 1984) is based on "damage ratios", which are related to the main structural response parameters responsible of damage, i.e. interstory drift and floor acceleration, and on repair multipliers, which are defined for each component of the construction. The third method (Parducci, 1988 and 1993) is based on a "damage correlation" defined on 2DOF systems by evaluating the damage as a function of the global ductility demand.

Application cases of the three procedures are reported in the existing literature (Mayes, 1992). With reference to the same structure, the "damage matrix" and the "damage ratio" procedure provide very close estimates, which usually indicate for the isolated solution a reduction of cost due to earthquake damage by factors of 4 to 7.

Base isolation and energy dissipation **647**

In conclusion it has to be pointed out that in addition to the possible decrease of damage costs, the BIS is also aimed at improving the economical value of the building.

5.5 Design Problems and Code Specifications

In this section only some of the design problems of BIS are dealt with, since it is beyond the purpose of the paper to address all the design aspects. Very exhaustive reviews on this subject can be instead found in Skinner et Al. (1993) and in Kelly (1993b).

Two main aspects governing the strength design of the single bearing are the vertical bearing capacity and the horizontal deformation capacity. Adequate limiting criteria for both these capacities have been stated in order to restrain the maximum stress and strains occurring in the rubber layers and in the steel shims.

In order to eliminate the excessive strains at the bottom and top opposite corners observed in devices under vertical and horizontal load conditions, the first elastomeric bearings utilised in U.S. and New Zealand applications were of the dowel type, i.e. neither firmly fixed to the foundation nor to the elevation structure (fig.10 (a)). However this typology presented the problem of rocking motion during earthquakes. More recent devices, bolted both to foundation and to superstructure, are adequately manufactured and designed to reduce the occurrence of local high strains (fig.10 (b)). A special attention in the manufacturing technology is also paid to the quality of the adhesion between steel shims and rubber, which represents a critical point in the reliability assessment of the device.

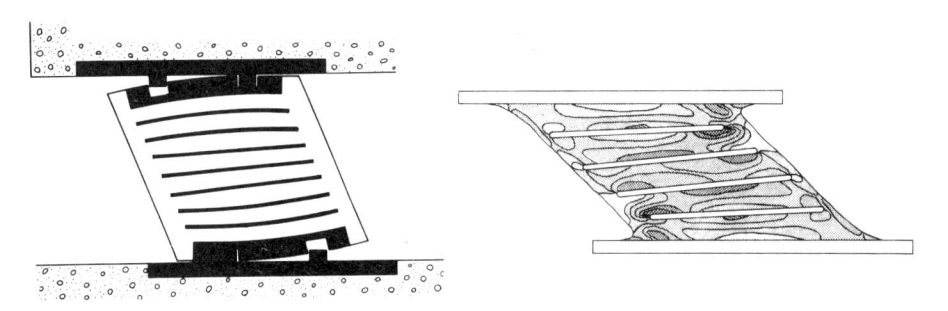

(a) **(b)**
Fig.10 (a) Dowel type and (b) bolted type rubber bearing (Bridgestone Co., from Fujita, 1991)

For what concerns the damping capacity, it has to be reminded that a damping function is essential in order to reduce the large displacement at the base level. This amount of damping can be obtained either by means of a separate mechanism, i.e. providing some additional devices able to absorb energy, or by incorporating the dissipative function in the rubber itself, by artificially enlarging the natural damping capacity of the material.

The former possibility involves the coupling of rubber bearings with oil and/or steel dampers, which dissipate energy due respectively to the viscous resistance of oil and to the yielding of steel; this solution has been widely used in the earlier applications, particularly in Japan (fig.11 (a)). Alternatively lead-rubber bearings, which embody a lead plug into the rubber device, have been developed by New Zealand researcher (Skinner, 1993), realized and widely installed in base isolated structures in many countries (fig.11 (b)).

(a)   (b)
Fig.11 (a) Rubber bearing plus rod steel damper (Kumagai Gumi Co., from Fujita, 1991) and (b) lead rubber bearing (Oiles Co., from Fujita, 1991)

The second case, i.e. the possibility of providing enough damping capacity to the rubber itself, allows for remarkable advantages from the stand-point of the design and performance of the building. Therefore high-damping rubber bearings have been developed and widely used in applications.

The typologies of isolators and dampers have different characteristics, thus they differently influence the behaviour of the superstructure. Therefore the selection of the isolation bearing and the possible coupling with a specific type of damper has to be appropriately made according to the configuration, morphology and destination of the building, as shown in fig.12.

Code specifications and design guidelines for base isolated structures have been developed in different countries since 1979 (Blakeley et Al., 1979) The first document in the USA was published by the SEAONC in 1986 (SEAONC, 1986); in 1990 the "Blue Book" of the SEAOC included in appendix the "Tentative General Requirements for the Design and Construction of Seismic-Isolated Structures", which have also been adopted by the UBC 1991. In Japan two types of documents have been published in 1989 with reference to base isolation design, i.e. the "Recommendations for the Design of Base Isolated Buildings" published by the Architectural Institute of Japan, and "The Safety Evaluation Guidelines of Base Isolated Buildings", provided by the Building Center of Japan. In Europe a significant activity is aimed to the development of design guidelines specifically devoted to isolated structures (AFPS, 1990; SSN, 1993).

It has to be underlined, however, that recommendations and design guidelines must be highly flexible in order to reflect the continuing progress and to embody the acquired results of the research activities. Very significant in this perspective is the evolution observed from the 1986 document of SEAONC to the 1990 publication of SEAOC. The latter in fact expresses a radical departure from the previous as well as from the conventional seismic resistant design, while is closer to the Japan recommendations. The main point which denote the new approach is related to the explicit reference to a two seismic input level design, which corresponds to different levels of structural performance requirements.

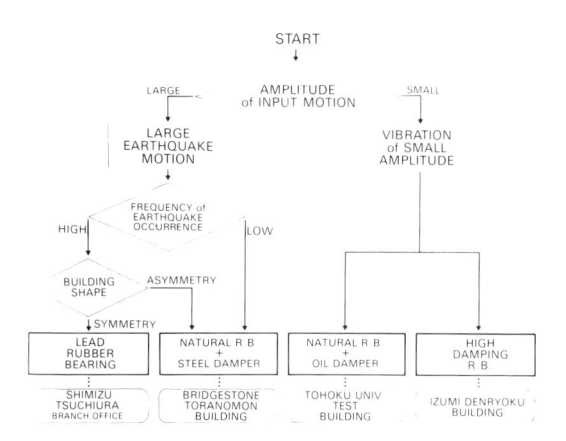

Fig.12 Flow chart for the selection of isolation device (Bridgestone Co., from Fujita, 1991)

5.6 Applications

The base isolation system has today reached a mature technique stage and the applications in different countries are several. The idea of base isolation is not new and a variety of proposed or realized ancient applications can be found in literature (Buckle, 1990). However the first modern base-isolated building was the William Clayton Building, built in Wellington, New Zealand, in 1981. The first base isolated building in the USA, the Foothill Communities Law and Justice Center, was built in 1982-1985 in Rancho Cucamonga, California. Further applications were designed and realized only some years later (1989), but after this quite slow start of the widespread utilisation of BIS, in the second half of 80s a sharp increase of proposed and built isolated structures has been registered. The leading country in the technology of isolation systems and in the number of applications (more than 70) is Japan, even though a large number of new isolated buildings (13) have been realised in the USA, in New Zealand and in Europe. An aspect which has to be mentioned is the proposed, and in some cases realized, application to retrofit of existing structures, even in the case of historic building. In USA 14 projects, some in phase of construction, have been proposed, while in Italy the possibility of applying the BIS to ancient monumental buildings has been suggested by different authors (Mazzolani, 1993).

Some isolated structures in USA have experienced the first severe field test, when in January 1994 were subjected to the Northridge earthquake. Three of them, in particular, experienced strong ground shaking (PGA in excess of 0.20g). A set of very interesting records has been obtained and some indication on structural performance have been gathered in (Clark, 1994). In particular the University of Southern California Teaching Hospital experienced a peak foundation acceleration of 0.37 g, which was attenuated to values included between 0.10 and 0.21 g in the upper structure, thus confirming the expected capability of isolation system to reduce seismic input energy transferred to the superstructure. A remarkably better performance of b.i. structures compared to neighbouring f.b. constructions, which showed significant damage and partial collapse, is the outcome of the field observations. In addition, the records and the collected data provide a valuable basis for the future research activities on BIS.

6 Review of papers presented in the session on base isolation and energy dissipation

Eight papers have been presented in the session "Base Isolation And Energy Dissipation". Four of them deal with specific problem of design and of performance assessment of base isolated structures. Three papers address passive energy dissipation systems, while one paper describes an advanced system of passive control. In the following a brief review of the main aspects dealt with in each paper is given.

The paper "**Design of base isolation devices for steel structures**" presented by G. Serino and M. Imbimbo deals with the procedures to be adopted in the preliminary design process of the isolation devices. The Authors refer to the most widely used typology of devices, the High Damping Laminated Rubber Bearings (HDLRB); for what concerns the displacement spectrum to be considered in the design, reference is made to the EC8 provisions. After having outlined the two available design approaches, i.e. the so-called "stress approach", aimed to limit the maximum compression stress in the single device, and the so-called "strain approach", which is based on the criterion of fixing the maximum compression strain in the rubber, the Authors provide a design chart for each soil profile type defined by EC8 (fig.13). The charts comprehend all the design steps to be followed in the two alternative procedures and make also allowance for a compact design, which implicitly satisfies the global failure (i.e. roll-out and buckling) checks of the device.

Still in the context of BIS applications to steel structures is the paper presented by De Luca, Faella and Mele entitled "**Serviceability and ultimate performance of base isolated steel frames**". More specifically in the paper the seismic behaviour of base isolated and fixed base moment resisting frames, designed in accordance to UBC 91 provisions, is compared both at serviceability and ultimate conditions, in terms of interstory drift, top sway, input energy, hysteretically and viscously dissipated energy.
The parametric analysis carried out by the Authors at each limit state is extended to different soil-type ground motions and to different number of stories of the steel frames. For what concerns the serviceability conditions the performance improvement obtained by means of BIS is so significant as no damage even to non-structural elements are expected, particularly in the case of stiff soil and for lower rise structures. Also at ultimate limit state the response

of base isolated frames is remarkably reduced with respect to the analogous fixed base ones (fig.14). In addition it is clearly showed that, even in the case in which the input energy in the structure is not reduced through BIS, due to the particularly high spectral value in the long period field, the dissipation is essentially obtained through viscous work of the isolation devices, thus preserving from damage the superstructure. The Authors therefore emphasise that BIS can be considered an effective seismic protective system even for very flexible structures, like steel moment resisting frames.

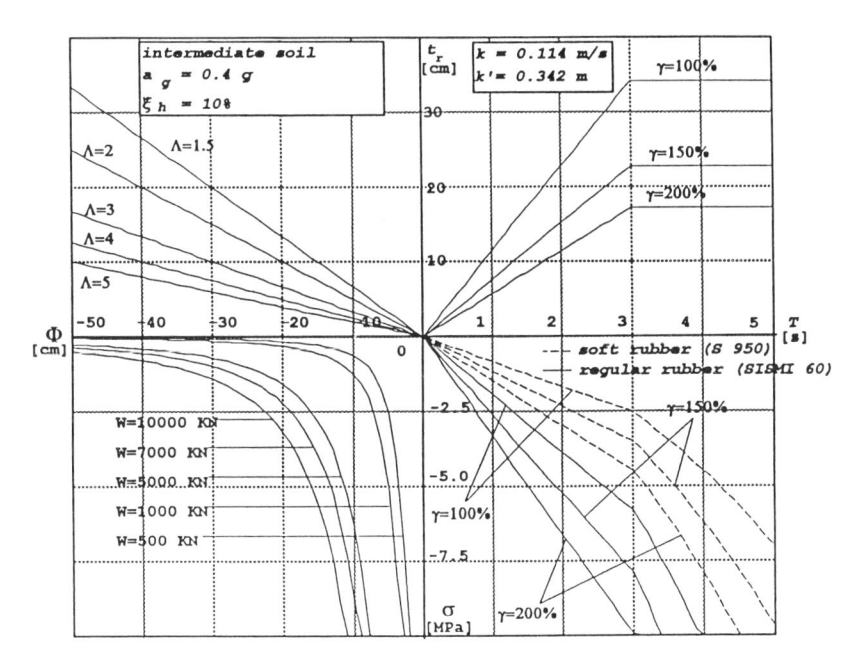

Fig.13 Design Chart for a_g= 0.4 g and intermediate soil type

Another paper dealing with BIS of structures is **"Seismic base isolation for framed structures"**, where the Authors, Lenza, Pagano and Rossi, propose a low-technological content isolation system for multistory frames. The system, which can be roughly considered as a refinement of the soft-story concept, consists of two elements: an elasto-fragile device and an elastic recoil system, both to be placed at the first column-hinged floor (the "soft-story"). The former element provides high lateral stiffness for ordinary wind loads, while, breaking at strong earthquake occurrence, allows for a significant lengthening of the structural fundamental period (up to 4.72 seconds); the latter element instead, provides the necessary balancing of the P-δ effect, which is the main drawback of a practical application of the soft-story concept. Several dynamic analyses are carried out by the Authors, both on a simplified (SDOF) and a complete (MDOF) model, in order to evaluate the influence of the

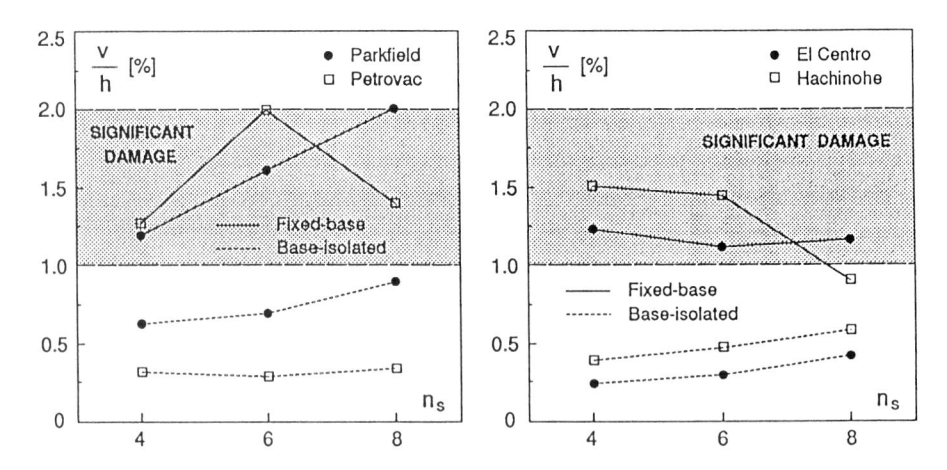

Fig.14 Interstory drift for severe earthquakes

different parameters on the global response and to define the values to be adopted in the design. It has to be pointed out that the base displacements corresponding to very high values of the period, like the ones adopted by the Authors, are expected to be very large and probably technically unacceptable. Some perplexities also arise on installation and feasibility of the system as well as on the lack of experimental verification.

The next paper presented in this session is **"Passive control methods for diminution of seismic response of secondary systems"**, by Csak and Kegyes. The Authors starting from a critical overview of the conventional design approach to seismic resistant structure design, which allows to point out the main limitations and drawbacks inherent to this approach, analyse some of the alternative methods today available thanking the recent researches and developments world-wide carried out. Some of the earlier applications of BIS implemented in Japan are then reported and discussed.

The paper **"Story dissipative system for steel structures"** by Palmaru, Mihai and Rotaru, is the first of the ones dealing with passive energy dissipation and can be inserted in a highly developed research context. In fact the first studies on steel ADAS (Added Damping And Stiffness) systems have been carried out at the beginning of the 70s, particularly in New Zealand and in USA; since that time significant research efforts have been spent, specifically at the EERC of the University of Berkeley, in order to optimize the design and applications to seismic protection of buildings. In this paper the results of an experimental research carried out on X-shaped steel dampers are presented. Two damper models, composed respectively of eight and four X-shaped steel sheets, and two prototype r.c. structures have been selected. Different combinations of dampers and structures, even the case of no-damper, have been tested under alternating cycles of imposed displacement, with amplitude increased up to failure. After comparison between experimental data and analytical predictions, some general design recommendations are then provided by the Authors, following the observed behaviour of the models. In particular the maximum number of steel sheets for each damper, as well as the maximum number of dampers to be used in different multistory buildings, is suggested in order obtain an effective improvement of seismic performance both in terms of stiffness and damping of the system. Finally the possibility of replacing the dampers to restore the seismic capacity of the structure after seismic events is pointed out by the Authors. It is not clear why the Authors of the presented paper, even though dealing with a highly developed research subjects, do not refer at all to the wide existing literature.

In the paper **"Design methodologies for energy dissipation devices to improve seismic performance of steel buildings"** by Serino the basic aspects of the designing metallic and friction energy dissipation devices to be adopted as seismic protection system for steel braced frames are reviewed. The Author provides some general indications on the optimum applicability field of energy dissipation devices, depending on the dynamic characteristics of the structure and on the acceleration spectrum shape of the specific ground motion (fig.15). In addition the Author analyses the design procedures suggested by different Authors both in USA and in Italy in order to minimize structural response and damage. Among the considered methodologies some common features are identified which lead to the definition of the values to be adopted for the most significant design parameters, i.e. the ratio of brace plus device stiffness to the structural story stiffness and the ratio between the device plus brace limit force and the frame yield force. Some additional recommendations are provided by the Author about the distribution of brace stiffness and device limit force suitable to avoid damage concentration at single story.

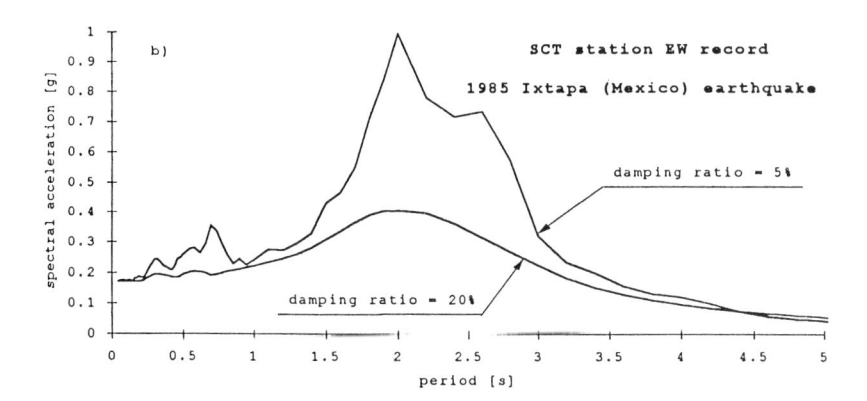

Fig.15 Acceleration spectrum showing period range where supplemental damping is beneficial

Still dealing with energy dissipation devices is the paper **"The use of steel plates to improve the energy-dissipation characteristics of a precast frame system"**, by Proença and Azevedo. In this paper an innovative precast frame system, consisting of one-piece column elements, with protruding beam stubs and central beam elements, is presented (fig.16). The peculiarity of the system is the presence of a so-called "continuity plate" which is factory-welded to the longitudinal bars of the beam stubs and in-situ welded to the central beams. This type of joint guarantees the force continuity in the beams and is expected to provide high dissipation capacity to the structure. In order to verify the behaviour of this system varying the location and the mechanical characteristics of the joint, a wide experimental investigation has been carried out by the Authors on different full-scale prototypes, as well as on single steel specimens, which the continuity plates were made of. The obtained test results, differently processed, have pointed out a large dissipation capacity of the joint and the significant influence of the joint behaviour on the overall structural performance. In addition the Authors develop an analytical model of the joint to be adopted in the simulation analysis of the global structures under earthquake time-histories, with the final purpose of defining the q-factor of the proposed precast structure.

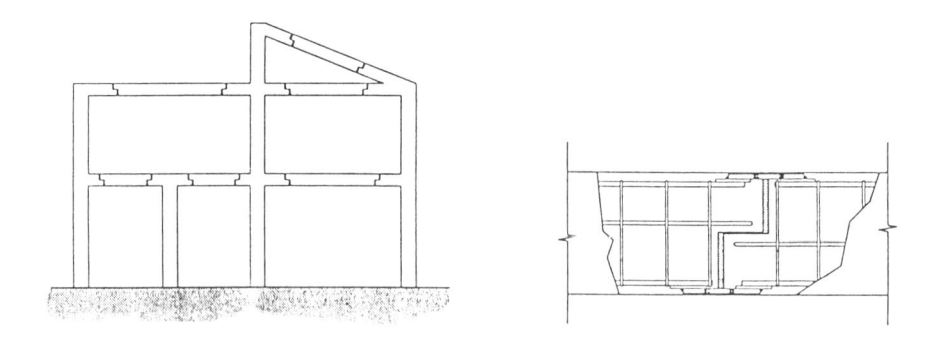

Fig.16 Possible joint locations and joint detailing

A very interesting paper is **"Development of vibration control system using U-shaped water tank"**, presented by Shimizu and Teramura. The Authors describe an advanced system of passive control which couples a Tuned Liquid Column Damper (TLCD) with a Period Adjuster (PA) (fig.17). The typical weak point of the TLCD, i.e. the need of employing a number of small tanks in order to tune the liquid mass period to the building natural period, is overcome in the TLCD-PA, which utilises in addition to the classical U-shaped tank for LCD a small U-shaped tank. The two tanks are connected through air-ducts and the liquid movements in the small tank activate a mechanical spring, which stiffness controls the period and the behaviour of the fluid in the LCD. Due to this PA equipment it is possible to apply a single large tank to buildings characterized by different dynamic behaviour. After experimental results obtained by means of shaking table tests, the Authors provide an example of TLCD-PA application to an actual building, the Cosima Hotel in Tokyo. The steel structure of the building, an EBF 106 m high, would be very sensitive to wind and earthquake induced vibrations due to its slenderness. The results of free vibration experiments carried out on the building show a significant improvement in damping and reduction of the transmission factor. Very effective reduction of the displacements are obtained under normally occurring slight motions, while under severe earthquake ground motions the calculated response of the structure shows a not very sensitive reduction of the peak acceleration, but an excellent reduction of aftershocks.

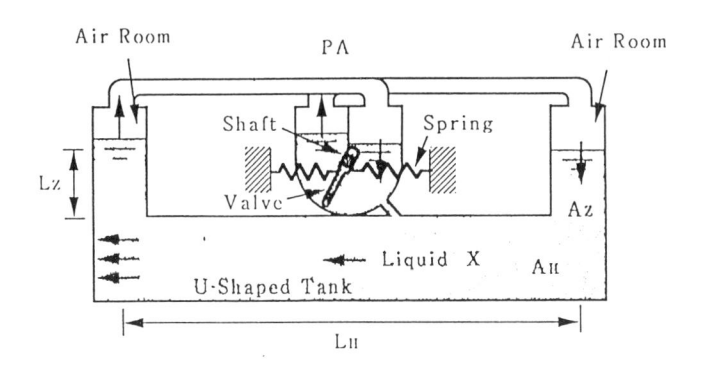

Fig.17 Sectional view of LCD-PA

Acknowledgment
This research has been partially supported by CNR. The contribution of major Japan companies providing most of the material and pictures referenced and utilised in this paper is gratefully acknowledged.

References

AFPS (1990), Recommendations AFPS90 - Chapitre 22: Appiuis parasismiques, **Association Française du Gènie Parasismiques, Presses des Ponts et Chausses**, Paris, France.
Akiyama H. (1993), State-of-the-art in th world on seismic isolation of buildings based on the results of the International Workshop held in Tokyo in April 1992. Most recent developments for buildings and bridges in Japan. **Proc. of Int. Post-SMiRT Conf. Seminar on Isolation, Energy Dissipation and Control of Vibration of Structures**. Capri (Napoli), Italy.
Bertero V.V. (1988), State of the Art Report - Ductility based structural design. **Proc. of Ninth World Conference on Earthquake Engineering**, Tokyo-Kyoto, Japan.
Blakeley R.W.G., Charleson A.W., Hitchcock H.C., Megget L.M., Priestley M.J.N., Sharpe R.D., Skinner R.I. (1979), Recommendations for the design and construction of base isolated structures, **Bulletin of the New Zealand National Society for Earthquake Engineering**, Vol.1, No.2.
Buckle I.G., Mayes R.L. (1990) Seismic Isolation: history, application, and performance - A world view, **Earthquake Spectra**, Vol.6, No.2.

Buckle I.G. (1993), Future directions in seismic isolation, passive energy dissipation and active control, **Proc. of ATC-17-1**, Seminar on seismic isolation, passive energy dissipation and active control, Vol.1, San Francisco, California.

Calderoni B., De Crescenzo A., Ghersi A. & Serino G. 1993. La definizione del livello di progetto per le strutture isolate alla base, **Proc. 6° Convegno Nazionale "L'ingegneria Sismica in Italia"**, Perugia, Italy.

Chaloub M.S. and Kelly J.M. (1990), Comparison of SEAONC base isolation tentative code to shake table tests, **J. of Structural Engineering**, ASCE, Vol.116, No.4.

Ciampi V. (1992), Development of passive energy dissipation techniques for buildings, **Proc. of Int. Post-SMiRT Conf. Seminar on Isolation, Energy Dissipation and Control of Vibration of Structures**. Capri (Napoli), Italy.

Clark P.W., Higashino M., Kelly J.M. (1994), Response of seismically isolated buildings in the January 17, 1994 Northridge earthquake, **Mostra-Convegno EDILEXPO 94 - Ambiente 2000, I Sistemi Di Protezione Sismica Non Convenzioanli: Stato dell'Arte e Risultanze sul Loro Comportamento alla Luce degli Ultimi Eventi Sismici**, Ancona, Italy.

De Luca A. and G. Faella (1993a), Response of fixed base and base isolated r.c. frames. **Proc. of Int. Post-SMiRT Conf. Seminar on Isolation, Energy Dissipation and Control of Vibration of Structures**. Capri (Napoli), Italy.

De Luca A. and G. Faella (1993b), Comportamento dinamico di strutture multipiano isolate alla base. **6° Convegno Nazionale "L'Ingegneria Sismica in Italia"**. Perugia, Italy.

De Luca A., Faella G., Pellegrino C., Ramasco R., Siano F. (1993c) Un modello per la simulazione di dispositivi elastomerici di isolamento sismico, **6° Convegno Nazionale "L'Ingegneria Sismica in Italia"**. Perugia, Italy.

De Luca A., Faella G. & Mele E. (1994a), Effects of design level on dynamic behaviour of multistory base isolated structures, **Second Int. Conf. on Earthquake Resistant Construction and Design**, Berlin, Germany.

De Luca A., Faella G. & Mele E. (1994b), Serviceability and ultimate performance of base isolated steel frames, **Proc. of Int. Workshop and Seminar on Behaviour of steel structures in seismic areas**, Timisoara, Romania.

De Luca A., Faella G. & Mele E. (1994c), Design level and damage in base isolated steel structures, **Proc. of Tenth European Conference on Earthquake Engineering**, Vienna, Austria.

De Luca A., Faella G., Pellegrino C., Ramasco R., Siano F. (1994d), A refined model for simulating the behaviour of base isolation devices, **Proc. of V USNCEE**, Chicago.

Di Pasquale G., Vestroni F., Vulcano A. (1989) Risposta inelastica di una struttura isolata alla base, **Proc. of Int. Meeting on Base Isolation and Passive Energy Dissipation**, Assisi, Italy.

Dolce M., Quinto G. (1994), Non linear response of base-isolated buildings, **Proc. of Tenth European Conference on Earthquake Engineering**, Vienna, Austria.

Ferritto J.M. (1984) Economics of seismic design of buildings, **J. of Struct. Eng.**, ASCE, Vol.110, No.2.

Forni M., Martelli A., Bonacina G., Mazzieri C., Serino G. (1993), Most recent developments on seismic isolation of nuclear facilities in Italy, **Proc. of Int. Post-SMiRT Conf. Seminar on Isolation, Energy Dissipation and Control of Vibration of Structures**. Capri (Napoli), Italy.

Fujita T. (1991) Editor of Special Scientific Events Subcommittee of the Executive Committee for the 11th Int. Conf. on Structural Mechanics in reactor Technology (SMiRT 11) (1991), Seismic isolation and response control for nuclear and non-nuclear structures. Special issue for the exhibition of the 11th Int. Conf. on Structural Mechanics in Reactor Technology, Tokyo, Japan.

Hanson R.D., Aiken I.A., Nims D.K., Richter P.J., Bachman R.E. (1993), State-of-the-art and state-of-the-practice in seismic energy dissipation, **Proc. of ATC-17-1**, Seminar on seismic isolation, passive energy dissipation and active control, Vol.1, San Francisco, California.

Imbimbo M., Mele E., De Luca A. (1994a), Finite element modelling of rubber bearings under large strains, **Proc. of ABAQUS Users' Conference '94**, Newport RI, USA.

Imbimbo M., Mele E., De Luca A. (1994b), Analisi tensioanle agli elementi finiti di un dispositivo elastomerico soggetto a carico assiale, **Proc. of 23rd AIAS National Conference**, Rende, Italy.

Kelly J.M, Buckle I.G., Koh C.G. (1987) Mechanical characteristics of base isolation bearings for a bridge deck model test, **Report No.UCB/EERC-86/11**, EERC, University of California at Berkeley, USA.

Kelly J.M. (1993a), State-of-the-Art and State-of-the-Practice in Base Isolation, **Proc. of ATC-17-1**, Seminar on seismic isolation, passive energy dissipation and active control, Vol.1, San Francisco, California.

Kelly J.M. (1993b), **Earthquake resistant design with rubber**. Springer-Verlag, London.

Kelly J.M. (1993c), Most recent developments on isolation of civil buildings, **Proc. of Int. Post-SMiRT Conf. Seminar on Isolation, Energy Dissipation and Control of Vibration of Structures**. Capri (Napoli), Italy.

Kelly T.E., Mayes R.L., Weissberg S. (1989), Earthquake damage estimates of conventional and isolated structures, **Proc. of Int. Meeting on Base Isolation and Passive Energy Dissipation**, Assisi, Italy.

Kobori T. (1994), Current development in active seismic response control of building structures, **Proc. of 2nd Int. Conf. on Earthquake Resistant Construction and Design**, Berlin, Germany.

Lin A.N. and Shenton H.W. (1992), Seismic performance of fixed-base and base-isolated steel frames, **J. of Structural Eng.**, ASCE, Vol.118, No.5.

Martelli A., Parducci A., Forni M. (1993), State-of-the-Art on development and application of seismic isolation and other innovative seismic design techniques in Italy, **Proc. of ATC-17-1**, Seminar on seismic isolation, passive energy dissipation and active control, Vol.1, San Francisco, California.

Mayes R.L., Jones L.R., Kelly T.E. (1992), The economics of seismic isolation in buildings, **Proc. of Int. Workshop on Recent Developments in Base-Isolation Techniques for Buildings**, Tokyo, Japan.

Mazzolani F.M., Serino G. (1993), Most recent developments and applications of seismic isolation of civil buildings in Italy, **Proc. of Int. Post-SMiRT Conf. Seminar on Isolation, Energy Dissipation and Control of Vibration of Structures**. Capri (Napoli), Italy.

Ministero dei LL. PP., Servizio Sismico Nazionale (1993) **Linee guida per il progetto di edifici con isolamento sismico.**

Mizukoshi K., Yasaka A., Izuka M., Takabayashi K. (1992), Failure test of laminated rubber bearings with various shapes, **Proc. of Tenth World Conference on Earthquake Engineering**, Madrid, Spain.

Nagarajaiah S., Reinhorn A.M., Constantinou M.C. (1989), Non linear dynamic analysis of three-dimensional base isolated structures (3D-BASIS), **Technical Report No.NCEER-89-0019**, NCEER, State University of New York at Buffalo, USA.

Occhiuzzi A., Veneziano D. & Van Dyck J. (1994), Seismic design of base isolated structures, **Proc. of Second Int. Conf. on Earthquake Resistant Construction and Design**, Berlin, Germany.

Palazzo B., Petti L. (1993), Fattori di riduzione per strutture isolate alla base, **6° Convegno Nazionale "L'Ingegneria Sismica in Italia"**. Perugia, Italy.

Parducci A. (1993), Aspetti economici dell'isolamento sismico, **6° Convegno Nazionale "L'Ingegneria Sismica in Italia"**. Perugia, Italy.

Pinto P.E., Vanzi I. (1992) Base-ioslation: reliability for different criteria, **Proc. of Tenth World Conference on Earthquake Engineering**, Madrid, Spain.

SEAONC (1986), **Tentative Lateral Force Requirements**. Structural Engineering Association of Northern California. San Francisco, USA.

SEAOC (1990), **Tentative General Requirements for the Design and Construction of Seismic-Isolated Structures**. Base Isolation Subcommittee of the Seismology Committee of the Structural Engineers Association of California. San Francisco, USA.

Seki W., Fukahori Y., Iseda Y., Matsunaga T. (1987), A large deformation finite element analysis for multilayer elastomeric bearings, **Rubber Chemistry and Technology**, Vol.60.

Serino G., Bonacina G., Spadoni B. (1992) Implications of shaking table tests in the analysis and design of base isolated structures, **Proc. of Tenth World Conference on Earthquake Engineering**, Madrid, Spain.

Shenton H.W. & Lin A.N. (1993), Relative performance of fixed-base and base-isolated concrete frames, **J. of Struct. Eng.**, ASCE, Vol.119, No.10.

Sheperd R., Billings L.J. (1992), Mechanics of elastomeric seismic isolation bearings, **Proc. of Tenth World Conference on Earthquake Engineering**, Madrid, Spain.

Skinner R.I., Robinson W.H., McVerry (1993), **An Introduction to Seismic Isolation**, John Wiley & Sons.

Soong T.T. (1990), **Active structural control: theory and practice**, longman Group, UK.

Takayama M., Tada H. (1992), Development of new isolators, **Proc. of Int. Workshop on Recent Developments in Base-Isolation Techniques for Buildings**, Tokyo, Japan.

UBC (1991), **Uniform Building Code**, International Conference of Building Officials, Whittier, California.

UBC (1994), **Uniform Building Code**, International Conference of Building Officials, Whittier, California.

Vestroni F., A. Vulcano and G. Di Pasquale (1991), Earthquake response analysis of a nonlinear model of a base-isolated structure. **Proc. of Int. Meeting on Earthquake Protection of Buildings**. Ancona, Italy.

Yamaguchi S., Akiyama H., Wada A., Nakazawa T. (1993), Seismic isolation system design procedure in Japan and its comparison to the U.S., **Proc. of ATC-17-1,** Seminar on seismic isolation, passive energy dissipation and active control, Vol.1, San Francisco, California.

57 SERVICEABILITY AND ULTIMATE PERFORMANCE OF BASE-ISOLATED STEEL FRAMES

A. DE LUCA
University of Reggio Calabria, Reggio Calabria, Italy
G. FAELLA and E. MELE
Istituto di Tecnica delle Costruzioni, University of Naples, Italy

Abstract
The performance at the serviceability and the ultimate limit states of fixed-base and base isolated unbraced steel structures is compared. The inelastic behaviour of low-to-medium rise steel frames, designed according to UBC 91 requirements, is evaluated in order to assess the improvement of performance in base isolated structures. Moderate and severe earthquakes are considered in the analyses for checking the serviceability and the ultimate conditions while the response of the frames, computed in terms of drift ratios and energy dissipation, evidences the benefits obtained through the adoption of base isolation systems.
Keywords: Base Isolation System, Steel Frames, Serviceability and Ultimate States, Seismic performance.

1 Introduction

The main idea under base isolation design relies on the reduction both of the total energy input into the structure and of the hysteretic energy demand in the superstructure. The first purpose is achieved through a shift of the fundamental period in a region where the spectra are characterized by smaller energy content while the second one is obtained by dissipating a large amount of the input energy through the viscous work of the devices.

The adoption of base isolation to steel moment resisting frames might be questionable since these structures are already characterized by fundamental periods which are far larger than the ones of corresponding reinforced concrete frames.

The applications in the professional practice (Kelly 1993) and in the research field (Chaloub and Kelly 1990, Griffith et Al. 1990) have been so far devoted to braced frames where the fixed base fundamental period is still in the high energy region of the spectrum. More recently Lin and Shenton (1992), with specific reference to 4-story frames, have underlined how a benefit in terms of weight of the structure can be gained in steel structures by the adoption of base isolation systems both in the case of braced and unbraced frames.

Main purpose of this paper is to investigate on the possibilities of utilisation of base isolation in steel structures and in particular on the benefit which might be gained in terms of performance. The behaviour of the frames is evaluated, in accordance to (Uang and Bertero 1991), by referring to the serviceability and ultimate limit conditions. For this purpose the approach adopted in the japanese seismic code (Earthquake 1988) is assumed for defining a moderate earthquake and a severe earthquake. The response of the frames is evaluated in engineering terms by considering the values of story drift and the input energy, separated into viscous damped, representative of energy absorption in the isolation devices, and hysteretic energy, representative of damage in the superstructure.

2 Seismic design checks: serviceability and ultimate

The actual code format which can be found in most of the existing regulations (UBC, SEAONC, NEHRP, EC8) implicitly requires that the structures should not undergo major damage under a moderate earthquake (ME) and should not collapse under a severe earthquake (SE). The design is based on internal forces resulting from a linear elastic analysis using forces reduced by a factor which takes into account the actual non-linear response of the structure and, in particular, the energy dissipation capacity through ductile behaviour. Therefore, codes simplify the structural analysis requiring only one elastic analysis under reduced forces. In this way, somehow, "one single shot" allows to control serviceability and ultimate conditions.

It has to be said though that the simplicity of this "one shot" approach does not guarantee clarity: it is not clear, in fact, which are the strength and drift requirements both under moderate and severe earthquakes. The philosophy instead is clear: the serviceability limit state requires to prevent, under earthquakes with large probability of occurrence (ME) during the life of the structure, damage whose repair costs are high in comparison with the costs of the structure itself. For major levels of earthquake ground motion (SE) having an intensity equal to the strongest forecast at the site (ultimate limit state), the structure has to ensure an adequate degree of safety against collapse by undergoing structural as well as non-structural damage.

Base isolation can represent an effective means for improving the performance both under ME and SE. It is therefore important, for better evidencing the benefits gained through the use of base isolation systems (BIS), to distinguish the serviceability and the ultimate limit conditions. This allows to assess under each of them the performance of the structure to be compared to the one of analogous fixed-base structure.

Aim of this paper is to underline, through a parametric analysis which considers low-to-medium rise steel frames and different earthquakes, if the benefits in terms of performance are still possible in steel moment resisting frames (SMRF). The analysis is not aimed to quantify a potential decrease in structural weight but to assess the improvement of performance deriving from BIS. For this purpose, the natural records considered in the analyses are scaled to a PGA equal to 0.08 g for the ME and to a PGA value equal to 0.40 g for the SE; the latter value is consistent with the UBC provisions for high sismicity zones, while the former follows the japanese seismic code indications for scaling the ultimate to service conditions.

3 Parametric analysis and frame design

As anticipated in the previous section, the analysis is aimed to investigate on the improvement of performance deriving from the application of BIS in a wide range of SMRF. For this purpose 3-bays 4, 6 and 8 story frames have been considered.

The design of the frames, which has been carried out in accordance with UBC provisions (1991), is more in detail provided in the following section. The isolation period for all the frames has been set equal to 3.0 seconds, and the stiffness of the isolation system has been consequently evaluated.

For what concerns the input ground motions, four different historical records, whose main parameters are reported in Table 1, have been utilised in the analyses. The selection of the seismic input has been made in order to account for different soil conditions: the Petrovac and the Parkfield signals can be assumed as representative of a stiff soil condition, while the El Centro and the Hachinohe records are characterized by a spectrum closer to a soft soil condition. Each signal, as said above, has been used two times scaling the PGA value in order to obtain the ME and the SE ground motions.

Table 1. Characteristics of the input ground motions.

Earthquake	Date	Site	Component	Duration [sec]	PGA [g]
Montenegro	15.4.79	Petrovac	NS	19.62	0.438
Parkfield	27.6.66	Parkfield	N65E	43.78	0.495
Imperial Valley	18.5.40	El Centro	S00E	53.80	0.348
Tokachi-Oki	16.5.68	Hachinohe	NS	36.00	0.229

In the design phase an elastic analysis of the fixed base frames has been executed, both with the equivalent lateral force procedure and with modal superposition technique, for defining performance of the frames in accordance with UBC provisions. Subsequently, inelastic dynamic analyses under ME and SE have been conducted both for the fixed base and the base isolated frames in order to carry out the comparison in terms of performance.

The program DRAIN-2DX (Prakash et Al. 1992), which makes allowance for P-Δ effects and axial-bending interaction, has been used for the analyses; isolation devices have been simulated by means of an equivalent viscous damping.

3.1 Design of the frames

Several choices have to be made when designing a steel frame. The first one pertains to the lateral resisting structural system: in this paper it has been assumed that all of the frames are designed with moment-resisting connections, thus being SMRF, in order to provide the necessary resistance against seismic loads. For what concerns the conventional seismic forces, reference has been made to UBC 91 which has been also considered for implementing the requirements on column to beam rotational strength ratios aimed at developing strong-columns-weak-beam frames; other code requirements referring to stress and drift limitations have been accounted for in the design.

Another fundamental step in the seismic design is the choice of an appropriate value of the force reduction factor in order to account for overstrength and ductility which affect the inelastic structural behaviour. The value suggested by UBC has been adopted for the fixed-base frames ($R_w=12$). It is worth pointing out that in the case of base-isolated frames no complete agreement about the values to be adopted for such coefficient has been yet reached by the scientific community, even though smaller values than the ones used for the conventional structures are generally recommended ($R_{wI}=R_w/4$). Hence, a proper design of BIS should consider the base isolated period spectral value reduced by an appropriate factor.

The structures have been designed as fixed-base and isolation has been applied to these structures. In this manner the comparison can be carried out for the same values of structural weight. Further improvements which will consider a refinement of the design of BIS structures might be performed in the future also after a more complete agreement is reached about the definition of the reduction factors in BIS.

The analysed frames have beam spans equal to 5.0 m and story heights equal to 3.5 m. A dead load equal to 2.25 t/m and a live load equal to 1.0 t/m have been assumed in design. Furthermore, joint vertical perimeter loads equal to 3.0 t are applied to consider the presence of exterior claddings. The cross sections of beams and columns, changing every two floors, are given in Table 2.

Table 2. Cross sections of beams and columns.

	Stories	Soil profile type S1			Soil profile type S3		
		Beams	External Columns	Internal Columns	Beams	External Columns	Internal Columns
4 STORY FRAMES	3-4	IPE 330	HE 200B	HE 260B	IPE 360	HE 220B	HE 300B
	1-2	IPE 330	HE 220B	HE 300B	IPE 360	HE 240B	HE 340B
6 STORY FRAMES	5-6	IPE 330	HE 200B	HE 260B	IPE 360	HE 220B	HE 300B
	3-4	IPE 330	HE 220B	HE 300B	IPE 360	HE 240B	HE 340B
	1-2	IPE 360	HE 240B	HE 340B	IPE 400	HE 260B	HE 400B
8 STORY FRAMES	7-8	IPE 330	HE 200B	HE 260B	IPE 360	HE 220B	HE 300B
	5-6	IPE 330	HE 220B	HE 300B	IPE 360	HE 240B	HE 340B
	3-4	IPE 360	HE 240B	HE 340B	IPE 400	HE 260B	HE 400B
	1-2	IPE 360	HE 260B	HE 400B	IPE 400	HE 280B	HE 450B

Within the approach which has been adopted in this paper, which consists in firstly designing the fixed-base frames and then applying BIS, the design of fixed-base frames represents a key-point. Therefore in the following section the elastic performance of some frames chosen for the parametric analysis is reported.

3.2 Elastic response parameters

The results of the elastic analyses provided hereafter allow to identify the main structural properties of the frames, thus giving a tool for defining the expected performance. Three major properties are essential within the framework of our purpose. These properties, which are all related each other, are the fundamental period, the flexibility and the sensitivity to second-order effects.

The fundamental periods of the frames are reported in Table 3 which shows that they are quite high, ranging from 0.82 to 1.67 seconds, thus providing considerably different results than the

Table 3. Fundamental periods of the analysed frames.

	4 Stories	6 Stories	8 Stories
Soil profile type S1	T = 0.98	T = 1.34	T = 1.67
Soil profile type S3	T = 0.82	T = 1.12	T = 1.41

ones obtainable through the approximate formula suggested in the UBC. The isolation ratio ranges from 1.80 to 3.66.

In Figure 1 the global drift ratio v_{top}/H (v_{top} is the top displacement and H is the global frame height) is plotted for the design related to stiff soil (S1) and soft soil (S3) versus the number of stories n_s. Solid lines represent the results obtained from modal analysis while dashed lines provide the values computed by static analysis through equivalent lateral force procedure. Solid and empty squares represent frame design for S1 soil condition, solid and empty circles represent S3 soil condition.

The same type of representation, but in terms of interstory drift ratio v/h (h is the story height), is provided in Figure 2. In this case, all the values have been obtained by applying the equivalent lateral force procedure since no indication is given in the code on the procedure for defining these values through the modal analysis.

The values of α_c (multiplier of vertical loads corresponding to the attainment of elastic critical load), which represent an indicator of sensitivity to second order effects, are also provided in this figure. It is widely accepted that a value of α_c equal to 10 should guarantee that second order effects are negligible, while a smaller value indicates that bending effects due to vertical loads somehow should be included in the analysis.

It has to be underlined that UBC 91 advises a limit value for the interstory drift equal to 0.3% which is evidenced in the figure by a dashed line; a second dashed line is plotted in the figure to show the region where α_c is larger than 10: all the frames fall in the range, displayed in the figure by a shaded area, bounded by the above limitations on the interstory drift ratio and on α_c

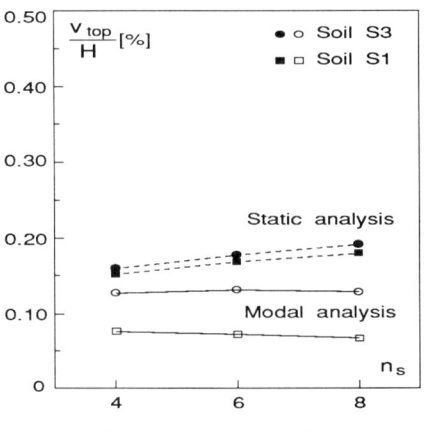

Fig. 1 Global drifts

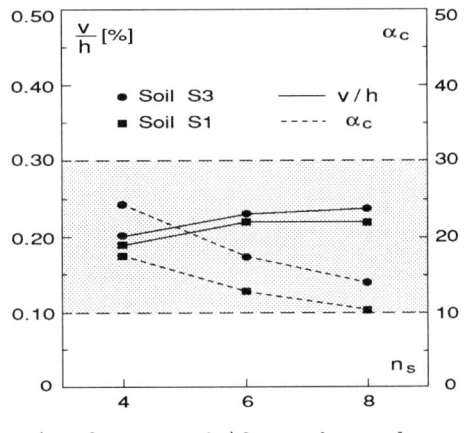

Fig. 2 Story drifts and α_c values.

parameter and therefore it is possible to conclude that the frames design is satisfying.

4 Results from inelastic analyses

4.1 Serviceability performance (Moderate Earthquake ground motion)

Non-structural damage is usually controlled by limiting the deformations in the structure. The main parameter commonly adopted for checking the serviceability performance is the interstory drift but another significant parameter is the top lateral displacement which is representative of an average response of the entire frame in terms of story drift along the total height.

The results of the serviceability check under ME are reported in Figures 3 and 4. In the first one the interstory drift ratios of the structures under the appropriately scaled earthquake signals are provided. In particular, in both Figures 3a and 3b, representative of stiff and soft soil condition respectively, the response of fixed-base frames (solid lines) is compared to the one of base-isolated frames (dotted lines). Each point has been computed through a time history inelastic analysis of the structure subjected to a record obtained scaling the peak ground acceleration to a value equal to 0.08 g. The appropriate stiffness values of devices have been adopted in the analysis taking into account that at service conditions the shear deformations are smaller than the design shear strain (usually assumed equal to 100% at ultimate conditions).

In a classification provided in (Hasselman et Al. 1980), on the basis of available data, a value of interstory drift equal to 0.1% is indicated as a threshold below which no non-structural damage is registered in common buildings. This value bounds the shaded part of the figures, appointed as no-damage region. The results given in Figure 3, therefore, allow to conclude that the performance of base isolated steel frames is generally significantly improved in the service condition. Larger benefits are gained in the case of 4 and 6 story frames, especially under stiff soil conditions.

The same results, in terms of global drift ratios, are represented

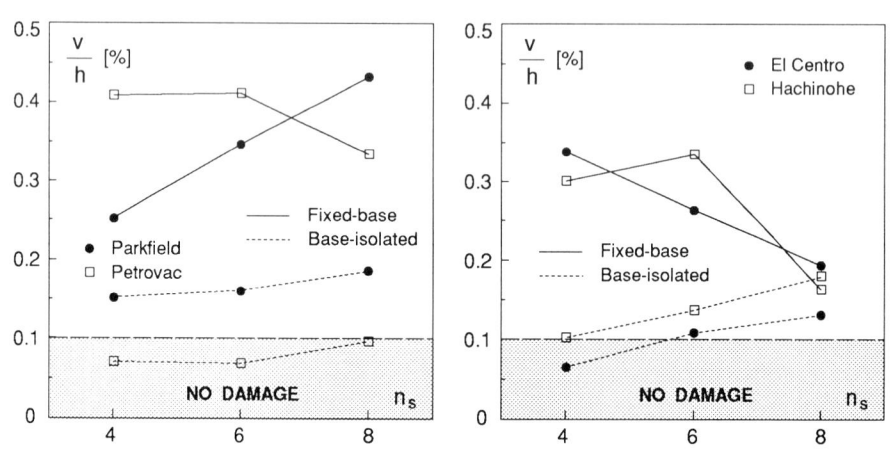

Fig. 3 Interstory drifts for moderate earthquakes

in Figure 4. In this case, it is more evident that no significant improvement is obtained for the 8 story frames under soft soil earthquake ground motions (Figure 4b).

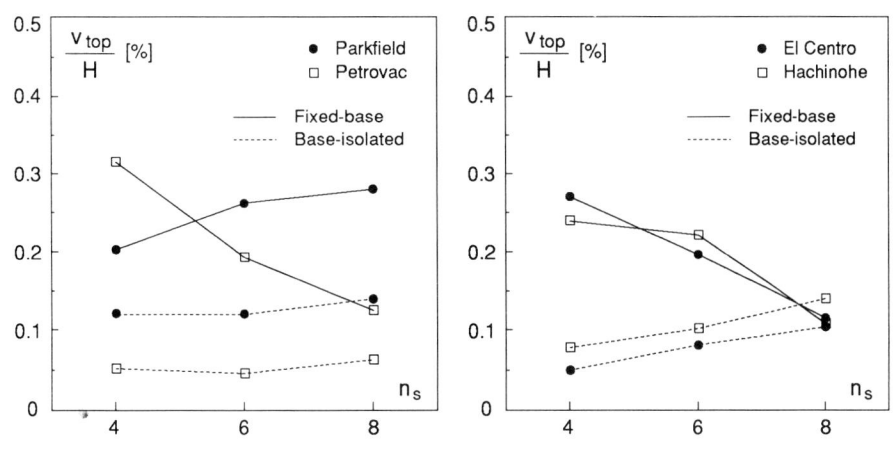

Fig. 4 Global drift ratios for moderate earthquakes.

4.2 Ultimate performance (Severe Earthquake ground motion)

In this section the performance of the frames at the ultimate limit condition is evaluated. The parameters adopted for characterizing the response of the frames are the interstory drift, which is related to the inelastic rotation demand in beams and columns, and the input energy. In fact, it is well accepted that the hysteretic part of the input energy represents damage in the structure (Banon et Al. 1981, Lin and Mahin 1985).

The response in terms of interstory drift v/h is provided in Figure 5 which reports the results of the analyses obtained using SE. The representation is of the same type of the one previously

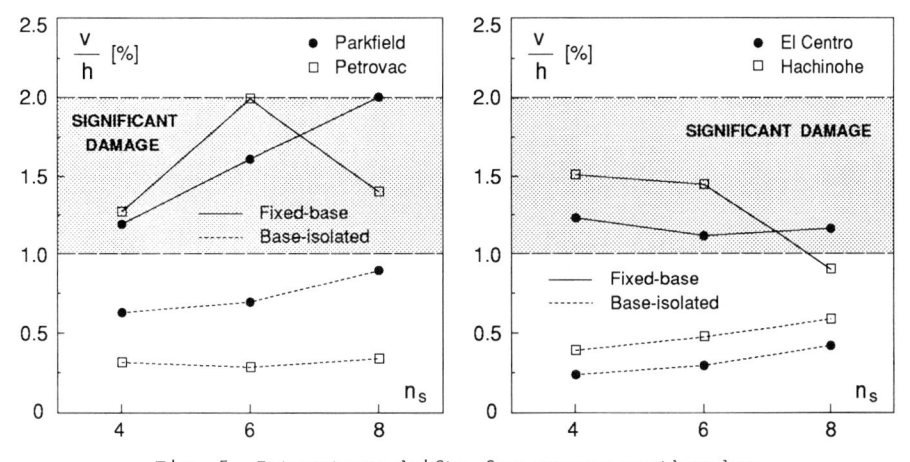

Fig. 5 Interstory drifts for severe earthquakes.

adopted in Figure 3 for the serviceability limit state. Again a comparison is carried out among fixed-base and base-isolated frames. The computed values of v/h confirm that, also with respect ultimate condition, the response is remarkably improved through the use of BIS. In Figure 5 the shaded area is bounded by two limits respectively corresponding to values of interstory drift equal to 1.0% and 2.0%. This field, appointed as significant damage area, corresponds to important or complete damage in non-structural elements (Hasselman et Al. 1980). Furthermore, it corresponds to noteworthy damage in the structural elements and can be related to values of inelastic rotations in the joints of difficult attainment in the case of welded connections which are the ones used in SMRF.

In conclusion, from Figure 5 it can be derived that the fixed-base frames will undergo significant damage or collapse while the behaviour of the base-isolated frames is more suitable. Base isolated structures are in most of the cases characterized by a value of interstory drift smaller than 0.5% which respects the requirement provided in the ANSI/ASCE 7-88 (1990) with reference to service conditions.

In order to assess the improvement in terms of input energy and of hysteretic energy demand due to the adoption of BIS, the different contributions to the total energy have been computed for all the frames analysed under SE. With reference to a 6-story frame subjected to the Parkfield record, the energy time histories are plotted in Figure 6: in particular, the viscous damped, the hysteretic and the kinetic portions of the input energy are provided in Figure 6a for the fixed-base case while the analogous results for the base isolated frame are represented in Figure 6b. The comparison shows how, in this specific case, the adoption of the BIS reduces both the input energy and the hysteretic energy demand.

The results for all the examined frames are given in Figure 7 where the input energy for fixed-base frames is compared to the one of base-isolated frames for several input ground motions and for different number of stories. The figure shows that BIS allows to reduce the input energy in frames subjected to stiff soil ground

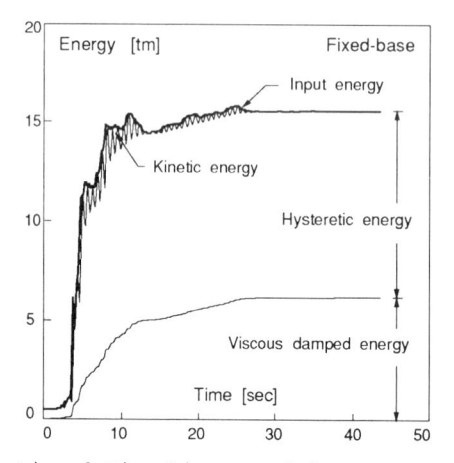

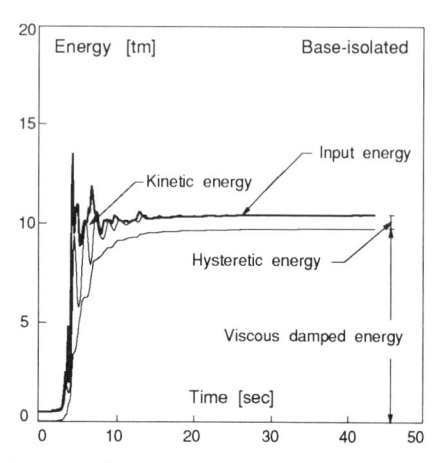

Fig. 6 Time history of input energy in the fixed-base and base-isolated 6-story frame subjected to Parkfield record scaled to 0.40 g.

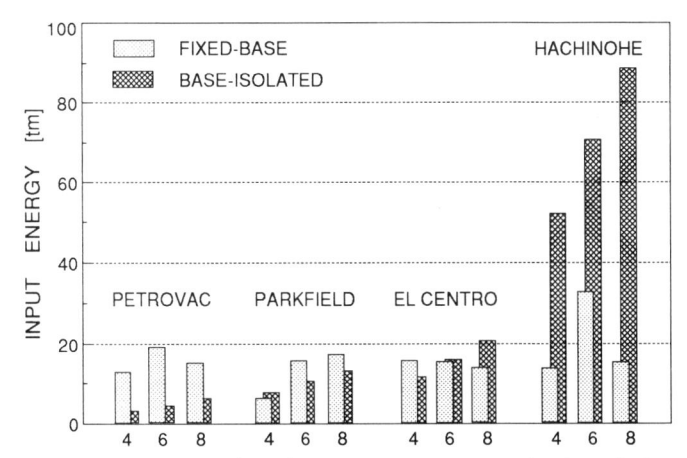

Fig. 7 Input energy in fixed-base and base-isolated frames.

motions while both the soft soil records (El Centro and Hachinohe) lead to an increase of the input energy with exception of the 4-story frame subjected to El Centro earthquake. In particular, the input energy computed in the base-isolated frames using the Hachinohe signal is consistently larger than the one computed in the fixed-base frames under all the other records, but such comparison is not enough representative of the inelastic behaviour of the superstructure.

A more significant assessment is the one which provides the amount of the hysteretic energy which is representative of damage in the superstructure. Then, in Figure 8 only the hysteretic portion of the input energy is plotted and the figure shows that a significant reduction is constantly achieved in the BIS.

The amount of viscous damped energy, non-dimensionalized to the input energy, is instead given in Figure 9 which shows how in the

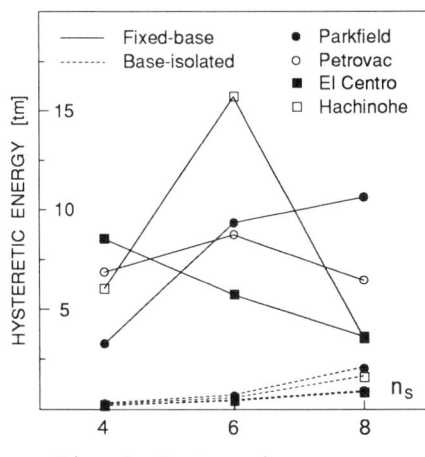

Fig. 8 Hysteretic energy.

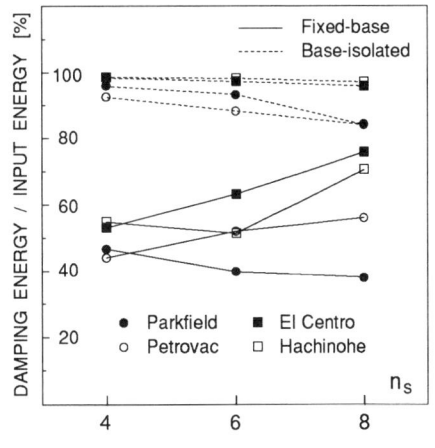

Fig. 9 Viscous damped energy.

base isolated structures about 90÷95% of the input energy is dissipated through the viscous work of the devices and, therefore, damage is restricted mostly to the isolation system thus preserving the superstructure.

Figures 8 and 9, hence, allow to draw the conclusion that even in the cases in which base isolation leads to an increase of input energy, damage in the superstructure is decreased due to the presence of isolation devices.

5 Conclusions

The analyses carried out in this paper have proved how base isolation represents an efficient mean to improve the performance of steel moment resisting frames. Such improvement is achievable for low-to-medium rise structures (4 to 8 story frames) under earthquake ground motions representative of different soil conditions.

The performance of fixed-base and base-isolated structures at the serviceability and the ultimate conditions have been compared and the benefits which the base isolation allows to gain have been assessed. Further numerical analyses should be undergone for confirming this encouraging results.

Acknowledgment
This research has been partially supported by Ministry of Education and Research (M.U.R.S.T. 40%).

6 References

ANSI/ASCE 7-88 (1990), American Society of Civil Engineers, **Minimum design loads for buildings and other structures**, ASCE Standard, New York

Banon H., Biggs J.M. and Irvine H.M. (1981), Seismic damage in reinforced concrete frames, **J. of Structural Engineering**, ASCE, Vol.107, No.ST9.

Chaloub M.S. and Kelly J.M. (1990), Comparison of SEAONC base isolation tentative code to shake table tests, **J. of Structural Engineering**, ASCE, Vol.116, No.4.

Earthquake Resistant Regulations - A world list (1988), Int. Assoc. for Earthquake Engineering, Tsukuba, Japan.

Griffith M.C., Aiken I.D. and Kelly J.M. (1990), Displacement control and uplift restraint for base-isolated structures, **J. of Structural Engineering**, ASCE, Vol.116, No.4.

Hasselman T.K., Eguchi R.T. and Wiggins J.H. (1980), Assessment of damageability for existing buildings in a natural hazards environment, **J.H.Wiggins Company, Technical Report N. 80-1332-1**.

Kelly J.K. (1993), State-of-the-Art and State-of-the-Practice in Base Isolation, **Proc. of ATC-17-1**, Seminar on seismic isolation, passive energy dissipation and active control, Vol.1, San Francisco, California.

Lin J. and Mahin S.A. (1985), Effect of inelastic behaviour on the analysis and design of earthquake resistant structures, **Earthquake Engineering Research Center, Report N. UCB/EERC-85/08**, Dept. of Civil Engineering, University of California, Berkeley, California.

Lin A.N. and Shenton H.W. (1992), Seismic performance of fixed-base and base-isolated steel frames, **J. of Structural Eng.**, ASCE, Vol.118, No.5.

Prakash V., Powell G.H. and Filippou F.C. (1992), DRAIN-2DX: Base Program User Guide, **Report N. UCB/SEMM-92-29**, Dept. of Civil Engineering, University of California, Berkeley, California.

Uang C. and Bertero V.V. (1991), UBC Seismic serviceability regulations: critical review, **J. of Structural Engineering**, ASCE, Vol.117, No.7.

UBC (1991), **Uniform Building Code**, International Conference of Building Officials, Whittier, California.

58 DESIGN OF BASE ISOLATION DEVICES FOR STEEL STRUCTURES

M. IMBIMBO and G. SERINO
Istituto di Tecnica delle Costruzioni, University of Naples, Italy

Abstract
Different design procedures most commonly adopted in the preliminary design of base isolation devices are reviewed and discussed. Based on this survey, a unified view of the design process is derived, which represents a completion and extension of a proposal previously developed (Augenti and Serino, 1991). Comprehensive design charts which allow a clearer understanding of the different possible procedures and their fast and reliable application are then presented. In developing the design charts, constitutive relations which are valid for two rubber compounds most commonly adopted in Italy in manufacturing isolation bearings are used.
Keywords: Seismic Isolation, Laminated Rubber Bearings, Design Charts

1 Introduction

Base isolation represents an effective way to reduce seismic stresses in buildings. In the case of steel structures, it has the further advantage that isolation devices may be considered as special prefabricated elements that need to be mounted on site, like other steel structural members. A solution often adopted is to attach the isolation bearings, with their mating end plates, to the superstructure column in the steel fabrication plant, where bolt torques and weldments could be made under controlled shop conditions, and then erect by crane on site the assembled base plate, isolator and column elements onto preset anchor bolts, held in place with anchor nuts. Such a process can be performed fast and easily in an ordinary building yard. In other words, for steel framed structures, the isolation devices may be considered as an extension of the column base plate with the assembly erected no differently than in traditional fixed-base construction.

In this paper, reference is made exclusively to isolation systems based on High Damping steel-Laminated Rubber Bearings (HDLRB), which are today the most widely used isolation devices in Italy due to both their excellent performance characteristics and the significant R&D carried out in Italy on this type of system (Mazzolani and Serino, 1993). The design procedures considered herein are aimed toward a preliminary sizing of the devices and selection of rubber parameters, as well as to allow initial decisions to be made regarding superstructure vertical and lateral load resisting system. These preliminary design choices will then form the basis of a second design stage, which will lead to final determination of isolation devices and structural members through a more accurate evaluation of seismic response of the isolated structure. Since the elastomeric bearing is usually an intrinsically non-linear system, the preliminary design, in which as it will be shown later an equivalent linear model of the isolation system is used, represents only an approximate solution. More refined analyses performed adopting a finite element model of the isolated system, which incorporates non-linear behaviour of the isolation devices determined from accurate experimental results, is usually required to derive final design quantities. This further optimisation of the aseismic design is not considered in this paper as it is carried out in a different way for each specific design case.

2 Available procedures for preliminary design

The preliminary design of an elastomeric bearing can be performed by means of two different procedures, namely the stress and the strain approach, which are herein outlined and discussed.

The design process is based on the performance required for the isolated system and the characteristics of the utilised elastomeric material. The quantities supposed to be the known *input data* of the design process are thus the following:

M = mass of the structure above the isolation system

W = vertical load to be carried by a single bearing: in a building this quantity is equal to the maximum design load attainable at the column base immediately above the isolator, whose size is usually chosen so as to minimise the number of different bearings to design; in what follows we suppose for simplicity that one single bearing geometry is necessary for the building so that $W = Mg/n$, where n is the number of isolation bearings (the procedure can be easily generalised in the case of multiple bearing geometry)

T_h = horizontal period of the isolated system; since preliminary design is generally performed by modelling the isolated structure as an equivalent linear SDOF system, the horizontal period defines the required horizontal stiffness K_h through the relation:

$$K_h = \left(\frac{2\pi}{T_h}\right)^2 M = \left(\frac{2\pi}{T_h}\right)^2 n\frac{W}{g} \qquad (1)$$

T_v = vertical period of the isolated system; similarly as above, the required vertical stiffness K_v is given by:

$$K_v = \left(\frac{2\pi}{T_v}\right)^2 M \qquad (2)$$

$\bar{\gamma}$ = maximum permissible shear strain of the elastomer

$\bar{\sigma}$ = allowable compression stress of the elastomer, or $G_s(\bar{\gamma})$ = its secant shear modulus, when the stress approach is adopted

ε_u = ultimate elongation strain in the elastomer, when the strain approach is adopted

$S_d(T_h,\xi_h)$ = maximum relative horizontal displacement in the isolation devices, as a function of horizontal period T_h and equivalent viscous damping ratio of the isolation system ξ_h, obtained from design response spectrum or code formula

In what follows $S_d(T_h,\xi_h)$ is derived from the elastic response spectra given by the latest version of Eurocode 8 (EC8, 1993), although this code does not include provisions for isolated buildings and suggests to refer to specific guidelines set by competent National Authorities to design such structures. The reason of our adoption of EC8 is that, in absence of significant soil-structure interaction effects which practically never occur in the case of isolated structures, design input motion cannot be influenced by the fact that the structure is isolated. Furthermore, EC8 elastic response spectra have been adopted in a recent Italian proposal of design guidelines for restraint devices in seismic zones (UNI, 1993) and will be included in the European document CEN TC-167 SC1 on aseismic devices, presently in its drafting process. According to EC8, in the long period range characteristic of isolated structures, the relative design displacement in the isolation devices may be assumed equal to:

$$T_C \leq T_h \leq T_D: \qquad S_d = \left(\frac{1}{4\pi^2} a_g S\eta\beta_o T_C\right) T_h = kT_h \qquad (3)$$

$$T_h \geq T_D: \qquad S_d = \frac{1}{4\pi^2} a_g S\eta\beta_o T_C T_D = k' \qquad (4)$$

where the quantities a_g, S, η, β_o, T_C and T_D are given in EC8. More specifically, a_g is the design peak ground acceleration, S the soil parameter (equal to 1.0 for rock, stiff and intermediate soil and 0.9 for soft soil), η is the damping correction factor given as a function of the equivalent viscous damping ratio ξ_h (in percent) by:

$$\eta = \sqrt{7/(2+\xi_h)} \geq 0.7 \qquad (5)$$

and β_o the maximum amplification factor with respect to a_g in the acceleration elastic response spectrum (equal to 2.5 for all soil sites). The periods T_C and T_D correspond to the lower limits of the constant velocity and constant displacement branches of the design spectrum: T_C is 0.4 s for rock or stiff soil, 0.6 s for intermediate soil and 0.8 s for soft soil, while T_D is 3.0 s for all soil sites. It is worthy to mention that the presence of a constant displacement branch in the elastic response spectrum is an innovation of the latest version of EC8. In previous versions of the code, the constant velocity branch had no upper limit and this was the cause of very conservative results obtained when designing isolated structures whose isolated period was larger than 3.0 s (De Luca and Serino, 1992).

The design process allows to select the geometrical characteristics and the type of the elastomer to be adopted for the bearing in order to provide the required support capacity and the horizontal and vertical stiffness. Hence, the following *output data* are derived:

t_r = total rubber thickness of the bearing

$G_s(\bar{\gamma})$ = secant shear modulus of the elastomer, or σ = maximum compression stress under vertical loads, when the stress approach is adopted

σ = maximum compression stress under vertical loads, when the strain approach is adopted

Φ = diameter of the bearing

t = thickness of single rubber layer

The two design procedures are now explained. They are identical in evaluating total rubber and single layer thicknesses, but different in deriving bearing diameter. In the stress approach, the allowable compression stress of the elastomer or the secant shear modulus of the selected rubber compound are chosen as input quantities of the design process, while maximum strain in the elastomer is verified *a posteriori*; on the contrary, a limiting shear strain criterion is adopted in the strain approach, and the compression stress is verified *a posteriori*.

2.1 Stress approach

The starting point of the design process is the determination of total rubber thickness, obtained through the following relation:

$$t_r = \frac{\alpha_t S_d}{\bar{\gamma}} \qquad (6)$$

where α_t is an appropriate relative displacement magnification factor to include the effects of accidental torsion and potential eccentricities between the centre of mass and the centre of stiffness of the isolation system. In most cases $\alpha_t \leq 1.1$, and $\alpha_t = 1.0$ is assumed for simplicity in what follows. Taking this into account and adopting the S_d values given by (3) and (4), equation (6) becomes:

$$T_C \leq T_h \leq T_D: \qquad t_r = \frac{kT_h}{\bar{\gamma}} \qquad (7)$$

$$T_h \geq T_D: \qquad t_r = \frac{k'}{\bar{\gamma}} \qquad (8)$$

The design process according to the stress approach can proceed following two different options:

a) the allowable compression stress of the elastomer is known, the required secant shear modulus and bearing diameter are derived

b) the secant shear modulus is known, the compression stress in the elastomer and consequently bearing diameter are derived

Suppose the elastomeric bearings to have a value of the shape factor S (equal to compressive area divided by lateral area of single rubber layer free to bulge - for a circular pad $S = \Phi/4t$) large enough to neglect the effects of bending deformation and a vertical compressive load much lower than the critical value so that its influence on transverse stiffness is negligible (both conditions are verified in most cases). Each isolation device can then be considered as a shear type beam, and the horizontal stiffness of the isolated system is given by:

$$K_h = n \frac{G_s(\bar{\gamma})A}{t_r} \tag{9}$$

where $A = \pi \Phi^2/4$ is the area of each circular bearing. Substituting (9) in (1) and dividing both terms by nA we obtain:

$$\frac{G_s(\bar{\gamma})}{t_r} = \left(\frac{2\pi}{T_h}\right)^2 \frac{\sigma}{g} \tag{10}$$

being $\sigma = W/A$. From equation (10), imposing $\sigma = \bar{\sigma}$ the required secant shear modulus $G_s(\bar{\gamma})$ is derived (option a) or, alternatively, knowing $G_s(\bar{\gamma})$ the compression stress σ is computed (option b). Then $A = W/\sigma$ is derived and thus the bearing diameter Φ.

Substitution of (7) and (8) into (10) allows to explicitly express the dependence of compression stress from horizontal period:

$$T_C \leq T_h \leq T_D: \qquad \sigma = \frac{g}{4\pi^2 k} G_s(\bar{\gamma})\bar{\gamma} \cdot T_h \tag{11}$$

$$T_h \geq T_D: \qquad \sigma = \frac{g}{4\pi^2 k'} G_s(\bar{\gamma})\bar{\gamma} \cdot T_h^2 \tag{12}$$

Once bearing diameter is known, the thickness of single rubber layer is derived from the bearing shape factor, determined from the required vertical period of the isolation system. Denoting with E_c the effective compression modulus of the elastomer constrained by steel shims, the vertical stiffness of the isolated system is given by:

$$K_v = n \frac{E_c A}{t_r} \tag{13}$$

For circular bearings with moderate shape factor $E_c = 6S^2G$ and the average shear strain $\gamma_{c,ave}$ produced by a compression strain ε_c is $\gamma_{c,ave} = \sqrt{6S\varepsilon_c}$ (Chaloub and Kelly, 1990). In the case of a non linear rubber compounds, like those adopted in HDLRBs, it is just logical to assume $E_c = 6S^2 G_s(\gamma_{c,ave}) \cong 6S^2 G_s(\bar{\gamma})$, as both $\gamma_{c,ave}$ and $\bar{\gamma}$ are always far above 20%, i.e. in a range where the secant shear modulus is not very sensitive to variations of shear strain. Using this last relation for E_c, from (1), (2), (9) and (13) we get:

$$\frac{K_v}{K_h} = \left(\frac{T_h}{T_v}\right)^2 = \frac{E_c}{G_s(\bar{\gamma})} = 6S^2 \tag{14}$$

which allows to derive the required shape factor S, and thus the thickness of single rubber layer, once the target T_h/T_v ratio is known.

It is interesting to notice that definition of single rubber layer thickness is related to the required performance of the isolation system in the vertical direction only, whereas definition of the other design unknowns is obtained from the required horizontal response.

2.2 Strain approach

As mentioned above, in the strain approach determination of total rubber and single layer thickness is done as in the stress approach, using equations (7), (8) and (14). For what concerns the evaluation of bearing diameter, in the strain approach the secant shear modulus $G_s(\bar{\gamma})$ is considered known and Φ is determined from the area A obtained through the following equation:

$$A = \frac{W}{E_c \varepsilon_{c,max}} \qquad (15)$$

where $\varepsilon_{c,max}$ represents the maximum allowable compression strain derived from a limiting shear strain criterion used in some bridge bearings design codes (AASHTO, 1991). The limiting strain criterion can be concisely expressed by:

$$\gamma_{c,max} + \gamma \leq f\varepsilon_u \qquad (16)$$

where $\gamma_{c,max}$ represents the maximum shear strain caused by the compression strain ε_c, given by $\gamma_{c,max} = 6S\varepsilon_c$ according to (Chaloub and Kelly, 1991), γ is the shear strain under horizontal loads and ε_u is the elongation at break of the elastomer (usually $\varepsilon_u = 500\%$). In the AASHTO code: $f = 0.33$ under vertical loads only; $f = 0.5$ under vertical and horizontal loads resulting from creep, post-tensioning, shrinkage, thermal effects and imposed rotation; $f = 0.75$ under vertical and horizontal seismic loads and imposed rotation. Thus:

$$\varepsilon_{c,max} = \frac{f\varepsilon_u - \gamma}{6S} \qquad (17)$$

Since the stress approach does not take account of the compression strain due to shear loading, it generally provides more conservative results compared to those obtained through the limiting shear strain criterion.

3 Buckling and roll-out checks

After that bearing geometry and type of elastomer have been determined, two limit states still need to be analysed: roll-out and buckling.

3.1 Roll-out failure

In the case of doweled bearings, since they are connected to the foundation and the superstructure through shear connections only, they are unable to transmit tensile loads. Equilibrium under a transverse load is ensured by a change in the line of action of the resultant vertical forces acting at the top and bottom of the bearing. The limit of this migration of resultants is reached when the relative displacement becomes (Buckle and Liu, 1993):

$$\Delta_{ro} = \frac{W\Phi}{K_h H + W} \qquad (18)$$

where H is the total bearing height. Roll-out failure is associated to this limit displacement, which represents the beginning of the decreasing branch in the transverse force-displacement curve. Since axial and transverse forces acting on bearings generally vary in an isolation system, roll-out instability in a single bearing does not necessarily mean that the system as a whole is unstable: the presence of bearings far from the critical state can restrain the unstable bearings from failure.

In the case of bolted bearings, it is possible to apply a reactive moment at the top and bottom ends of a bearing through tension and compression in the bolts, and roll-out instability cannot occur. In early designs, dowels were preferred to bolts as bearing end restraints because it was felt that rubber should not be subjected to tension. Today, bolted bearings are much more commonly adopted, though tensile stress must be limited to avoid the rubber cavitation phenomenon and rupture of the bond between rubber layers and steel shims.

3.2 Buckling failure

Elastomeric bearings exhibit a buckling phenomenon similar to that of an ordinary column, but dominated by the low value of the shear stiffness. The problem can be very effectively analysed through the energy method, and the expression of the buckling load is given by (Raithel and Serino, 1993):

$$W_{cr} = \frac{P_s}{2} \left(\sqrt{1 + 4 P_E / P_s} - 1 \right) \tag{19}$$

with:

$$P_s = G_{to} A_s \qquad P_E = \pi^2 \frac{E_c (I / 3)}{H^2} \tag{20}$$

where G_{to} is the initial tangent shear modulus, A_s and I are the shear area and moment of inertia of the bearing cross section. The buckling load is generally much larger than the design vertical load, particularly in the case of squat bearings, but its value must be always checked to avoid this type of instability.

3.3 Compact bearing design

Sometimes it may be convenient to design isolation bearings squat enough to automatically guarantee safety against buckling or roll-out. In this way overall modes of failure are prevented, and only local failures like rubber tearing and bond rupture can occur.

Current rubber compound can sustain shear strains up to 400-500% without rubber tearing and debonding. In this case, computation for roll-out and buckling failures show that these type of instabilities cannot occur if bearing diameter is 4-5 times larger than total rubber thickness, i.e. if:

$$\Phi = \Lambda t_r \tag{21}$$

with Λ being at least 4 and preferably around 5. The above relation can be introduced as a further equation in the design process: in this case, to equate the number of equations with that of the design unknowns, the horizontal period becomes an output data instead of an input data of the design process.

4 Design charts

The previously outlined design procedures can be easily translated into a graphical format which provides an effective automatic tool for the preliminary sizing of an isolation bearing. Figures 1 to 3 refer to the three soil sites considered by EC8 (rock or stiff soil, intermediate and soft soil, respectively), a design peak ground acceleration a_g = 0.4 g and an equivalent viscous damping ratio ξ_h = 10%.

In the right upper part of each chart the relations (7) and (8) between the total rubber thickness t_r and the horizontal period T_h are visualised, being the maximum permissible shear strain $\bar{\gamma}$ a known parameter selected by the designer. In the right bottom part, the dependence through equations (11) and (12) of the compression stress σ from the horizontal period T_h is shown. The $G_s(\bar{\gamma})$ values adopted to trace the curves refer to the two carbon-black loaded natural rubber vulcanizates most commonly adopted in Italy (ALGA SISMI 60 and ALGA S 950), and have been computed using a non-linear constitutive law for the rubber which provides excellent comparison with the experimental results (Serino et al., 1992).

The compression stress σ is related to the vertical load W and the diameter Φ through the relation:

$$\sigma = \frac{4W}{\pi\Phi^2} \tag{22}$$

shown in the left bottom part of each chart. Finally, in the left upper part of the chart the relation (21) between bearing diameter Φ and total rubber thickness t_r is represented.

The described charts can be used in the preliminary design of a bearing, adopting the **stress** approach, in the following way:
- once the horizontal period T_h has been selected, the set of curves in the right upper part allows to derive total rubber thickness t_r, for a given maximum permissible shear strain $\bar{\gamma}$
- an allowable compression stress $\bar{\sigma}$, normally around 6.9 MPa, is selected, so that it is possible to choose the required $G_s(\bar{\gamma})$ and thus the elastomer type from the right bottom curves (option a)
- a type of compound is chosen, then the curves in the right bottom part provide the compression stress σ (option b)
- the curves in the left bottom part allow to compute the bearing diameter Φ form the selected or derived compression stress σ and the vertical load to be carried W

The same charts can also be used in a compact bearing design. In this case Λ, $\bar{\gamma}$, elastomer type and W are known, and thus all the curves to be used in the four quadrants. The design solution in terms of t_r, T_h, σ and Φ is obtained by graphically finding the closed rectangular path which connects these four curves.

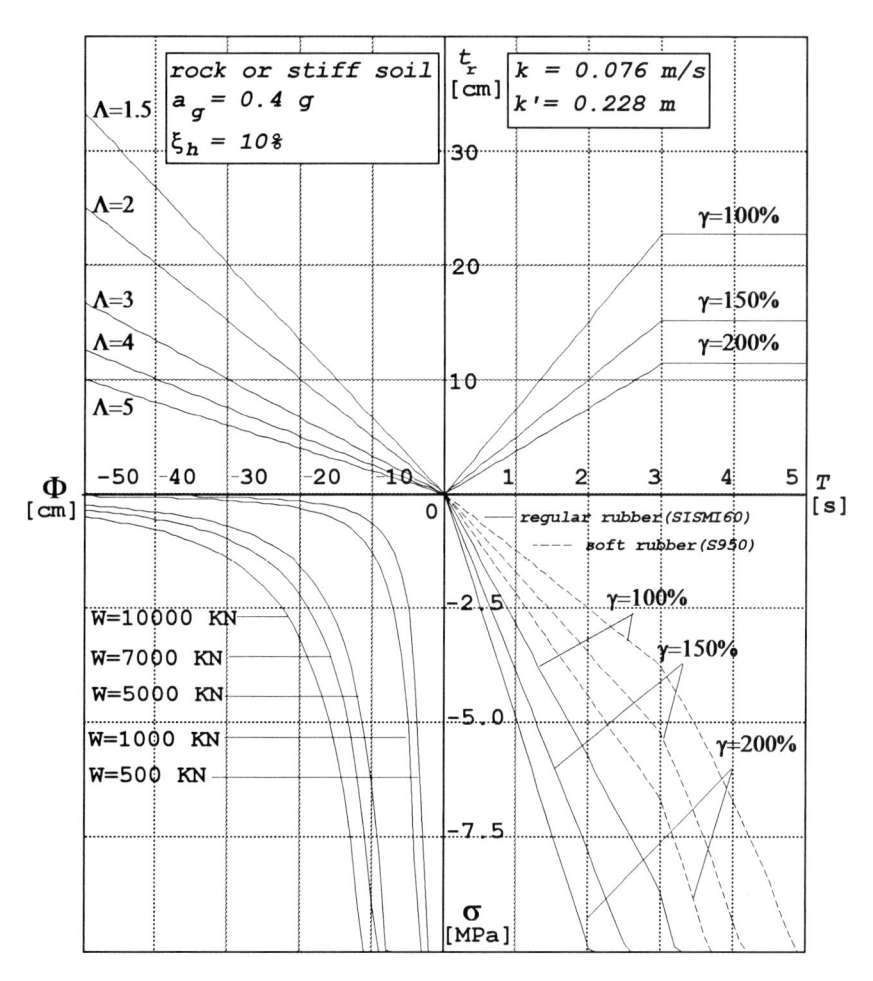

Fig. 1. Design chart for a_g=0.4 g and rock or stiff soil (ξ_h=10%).

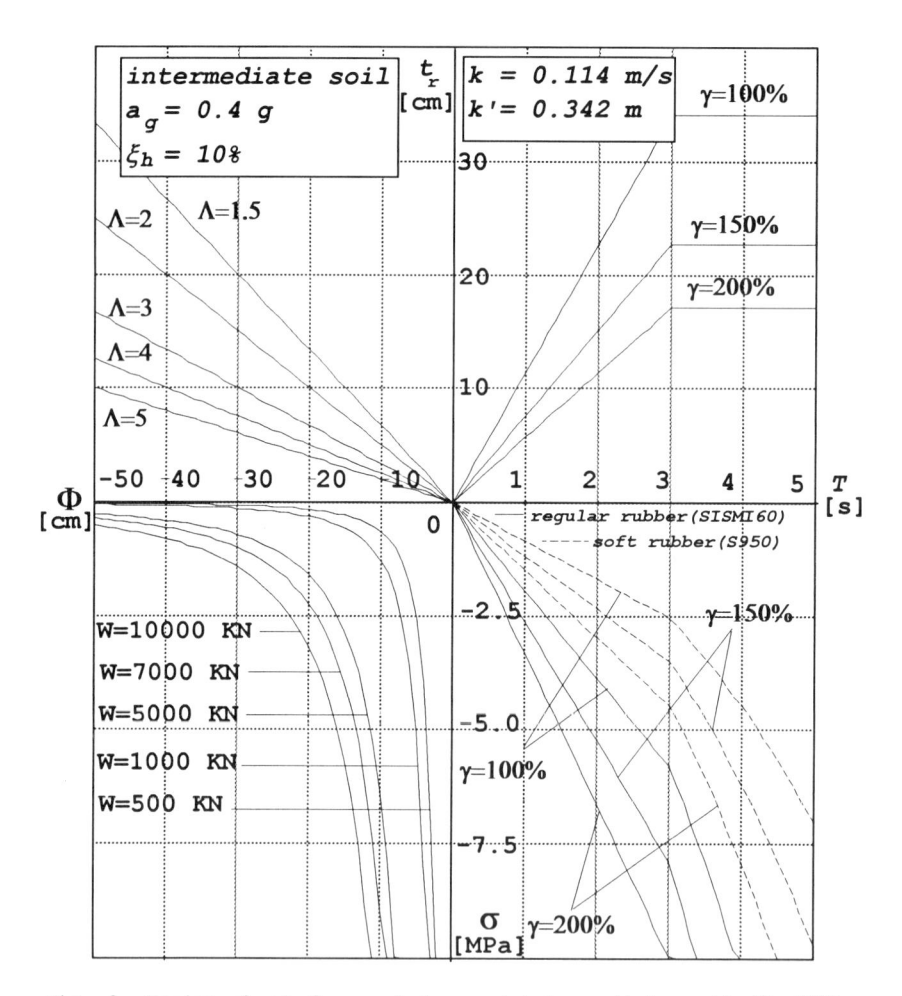

Fig. 2. Design chart for a_g=0.4 g and intermediate soil (ξ_h=10%).

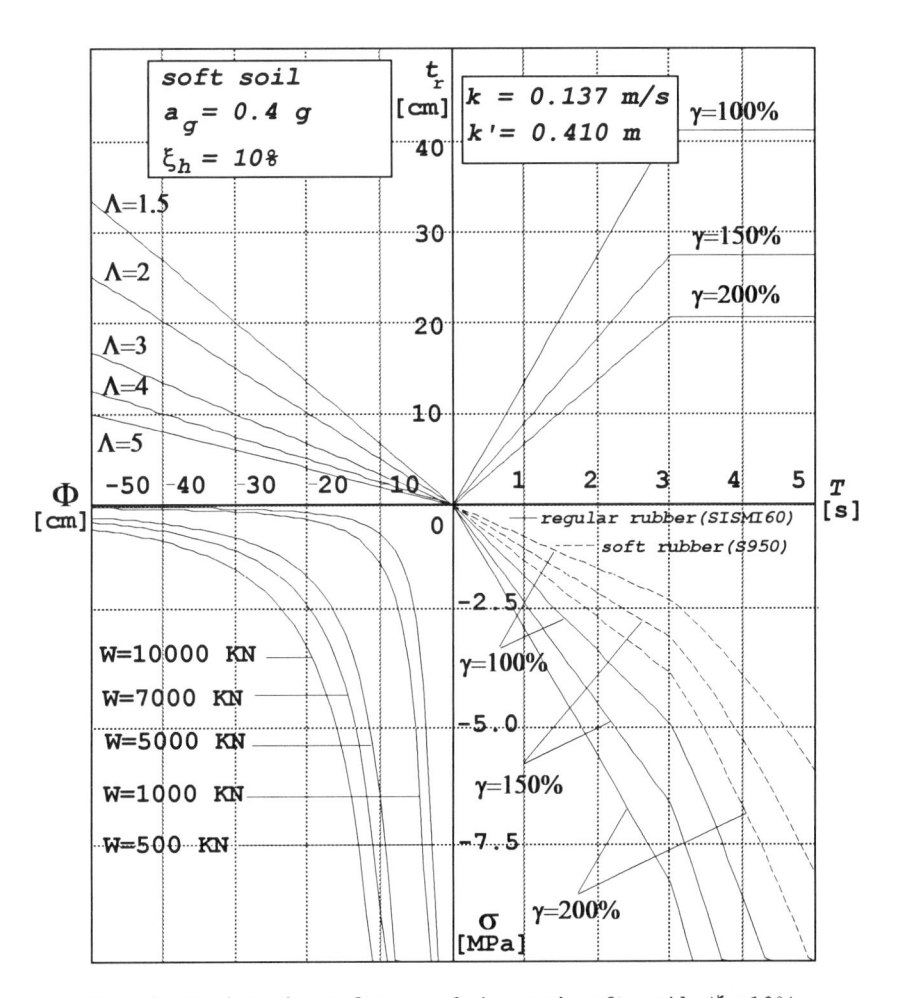

Fig. 3. Design chart for a_g =0.4 g and soft soil (ξ_h=10%).

5 References

Augenti, N. and Serino, G. (1991) Proposal of a design methodology for base-isolated structures, in **Proc. International Meeting on Earthquake Protection of Buildings**, Ancona, Italy, pp. 203/C-214/C.

AASHTO (1991) **Guide Specifications for Seismic Isolation Design**, American Association of State Highway and Transportation Officials, Washington, D.C.

Buckle, I.G. and Liu, H.(1993) Stability of Elastomeric Seismic Isolation Systems, in **Proc. of a Seminar on Seismic Isolation, Passive Energy Dissipation and Active Control**, San Francisco, California, Vol. 1, pp. 293-305.

Chalhoub, M.S. and Kelly, J.M. (1990) Effect of bulk compressibility on the stiffness of cylindrical base isolation bearing, **International Journal of Solids and Structures**, Vol. 117, No. 7.

De Luca, A. and Serino, G. (1992) Evaluation of displacement design spectra for base isolated systems, in **Proc. 10th World Conference on Earthquake Engineering**, Madrid, Spain, Vol. 10, pp. 5829-5834.

EC8 (1993) **Eurocode 8: Earthquake Resistant Design of Structures**, Commission of the European Communities, CEN/TC250/SC8, Bruxelles.

Mazzolani, F.M. and Serino, G. (1993) Most recent developments and applications of seismic isolation of civil buildings in Italy, in **Proc. International Post-SMiRT Conference Seminar on Isolation, Energy Dissipation and Control of Vibrations of Structures**, Capri, Italy, paper #5.

Raithel, A. and Serino, G. (1993) Stability and post-critical behaviour of laminated elastomeric isolators (in Italian), in **Proc. 6th Italian National Conference on Earthquake Engineering.**, Perugia, Vol. 1, pp. 329-338.

Serino, G. Bonacina, G. and Spadoni, B. (1992) Implications of shaking table tests in the analysis and design of base isolated structures, in **Proc. 10th World Conference on Earthquake Engineering**, Madrid, Spain, Vol. 4, pp. 2405-2410.

UNI (1993) Restraint devices for constructions in seismic zones (in Italian), guidelines proposal U73.08.007.0, Ente Nazionale Italiano di Unificazione.

59 DEVELOPMENT OF VIBRATION CONTROL SYSTEM USING U-SHAPED WATER TANK

K. SHIMIZU and A. TERAMURA
Obayashi Corporation, Tokyo, Japan

ABSTRACT
This report summarizes a serie of studies for the realization of a bi-directional vibration control system (Tuned Liquid column damper with period adjustment equipment:LCD-PA) which can be used to provide reductions in the movements of a steel high rise building loaded by wind or earthquakes of medium strength. At first, the effectiveness of this system was proved through theoretical and experimental study using a small simple model in a laboratory; secondly, this vibration control system was applied for a steel high rise building of 112 m and gained good results.

1 Introduction

It is a well known fact that stability or dynamic response problems in vibration system are in most cases directly related to resonance of the system. Tuned mass dampers(TMD) are effective in suppressing the vibrational resonance motions of highrise, flexible steel framed structures which are subjected to wind or earthquake excitation.
  Previous TMD devices consisted of steel weights, metal springs and oil or viscous dampers. Recently, a tuned liquid damper (TLD) employing a tank filled with water, utilizing liquid motion and based on a free water wave theory, has been developed. Furthermore, a tuned liquid column damper (LCD) of which the tank is similar to a U-shaped pipe, utilizing fluid current between horizontal and vertical portions has been developed[1][2]. While initial studies examined the behaviour in one direction, later experiments were performed using a square based bi-directional LCD[3][4], as shown in Fig.1. The bi-directional LCD has advantages when compared to a TLD as adjustability of the frequency and damping by selection of the cross-sectional area between horizontal portions and vertical reservoirs. These TLD's and LCD's have a number of advantages, such as the use of safe material water, low cost, almost zero trigger level, and so on. In addition, the size of a tank is small because of the need to tune its fre-

Fig.1 A square based LCD.

Fig.2 Sectional view of LCD-PA.

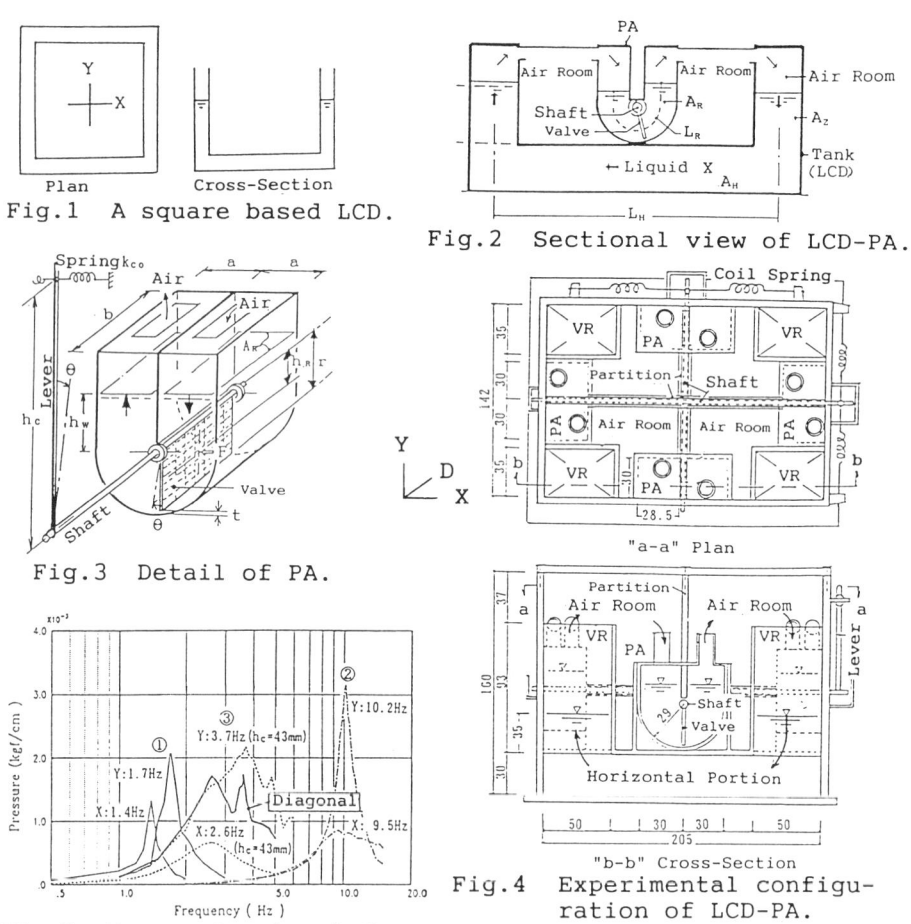

Fig.3 Detail of PA.

Fig.4 Experimental configuration of LCD-PA.

Fig.5 Measured natural frequencies in X,Y and diagonal directions.

quency to the natural frequency of structure as well as to control the damping of liquid motion. It is necessary to study the frequency adjustment system under a varing size of the tank (and its mass) in order to apply the system to buildings with different characteristics.

This paper, based on these studies, is concerned with a bi-directional liquid column damper (LCD) with period adjustable equipment (PA). The PA's function is to control the behavior of liquid motion in the LCD.

2 Configuration of the LCD-PA

2.1 Concept of the LCD-PA
Figure 2 shows an original sectional view on the LCD-PA

which has a rectanguler U-shaped tank (LCD), a pair of air rooms, and period adjustable equipment (PA). When the tank is moved in horizontal direction, fluid travels in both horizontal and vertical directions. Accordingly, at one side air is compressed, while at the other side the air pressure reduced. The sinusoidal pressure changes induce a fluid movement in the subsidiary U-shaped tank, resulting in the movement of the valve and shaft and movements in the springs. In detail, the PA consists of a U-shaped tank, a valve supported by a shaft and a lever supported by a mechanical spring as shown in Fig.3. Consequently, the behavior of fluid in the LCD tank is controled by the stiffness of the spring of the PA.

2.2 Configuration and tests of the LCD-PA

The configuration of LCD-PA, which was investigated, is presented in Figure 4. This device is a rectangular based bidirectional LCD with four PA's.The tank has a horizontal portion where liquid is free to move in any horizontal direction, four vertical reservoirs (VR) at each corner above the horizontal portion and four air rooms and separated by partitions. The PA is arranged between two vertical reservoirs(VR).

Tests were carried out to investigate the natural frequency of fluid motion. A harmonic base displacement in the horizontal direction with a constant amplitude was imposed to the tank by shaking table. The excitation was varied in the frequency range of 0.5Hz - 15Hz. The dynamic pressure of air and water were measured by using pressure gages. The testsinvolved the investigation of varing natural frequency of the fluid in horizontal X and Y directions. The cases in the experiment were ① opened tank without using the PA, ② airtight closed tank without using the PA, ③ tank with operating the PA. The test results revealed three distinct types of natural frequencies as shown in Figure 5. Measured frequencies of case① were 1.4 Hz and 1.7 Hz in respectively horizontal X and Y direction, those of case② were respectively 9.5 Hz and 10.1 Hz, and those of case③ were respectively 2.6 Hz and 3.7 Hz. Futhermore, the natural frequencies in diagonal direction were 2.3 Hz and 3.7 Hz which are the same as the frequencies in X and Y directions in case ③.

Hence, the period adjustment equipment (PA) has a major influence upon the natural vibration frequency of the LCD.

3 Experimental investigation of LCD-PA-Structure model

The efficiency of the LCD-PA was experimentally investigated by using the LCD-PA-structure interaction model as shown in Fig.6. The structure is a four story steel structure. The natural frequency for first vibration mode is 2.76 Hz, the total weight is 13.48 kg and the critical da-

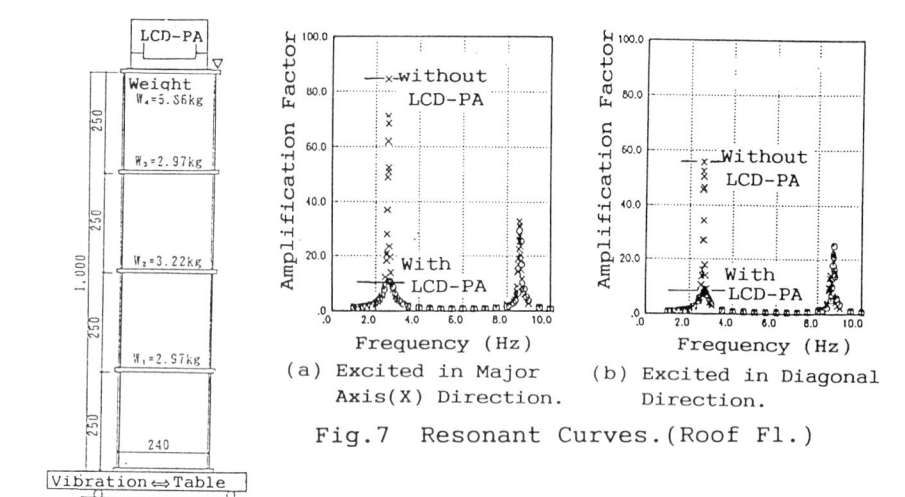

Fig.6 Experimental model.

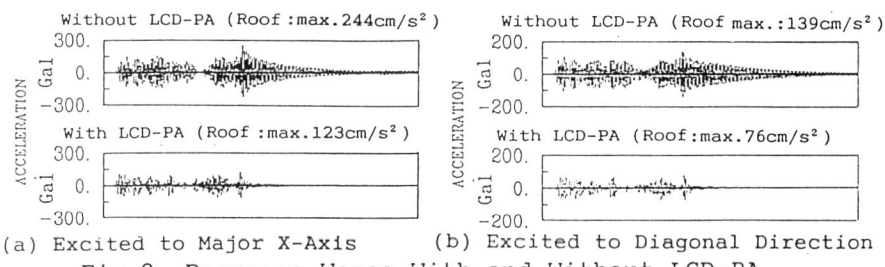

Fig.7 Resonant Curves.(Roof Fl.)
(a) Excited in Major Axis(X) Direction.
(b) Excited in Diagonal Direction.

Fig.8 Response Waves With and Without LCD-PA.
(a) Excited to Major X-Axis
(b) Excited to Diagonal Direction

mping ratio is 0.6%. External force on the structure is given by vibration table. The LCD-PA tank is mounted on the top floor of the structure and has an effective liquid weight of 0.66 kg and a frequency of 2.60 Hz. The acceleration of both the top floor and the vibration table were measured. The resonant response curve for the structure with and without the LCD-PA is shown in Fig.7. Magnified peak value of resonance is about 9.0 in case of with the LCD-PA, and 84.0 without. Fig.8 shows the acceleration response waves on the roof floor with and without the LCD-PA when for the vibration table the El-Centro 1940 EW records are used (max.50 Gal). The efficiency of vibration-suppression is improved by using the LCD-PA.

4 Equations of motion of LCD-PA-Structure model

For dynamic systems with a larger number of degrees of freedom, such as flexible structures, assuming a modal response, the N different equations of motion become uncoupled from one another and each one describes dynamic the behavior of a single, separated "spring-mass-damper" vibrational system. The interaction vibration model of structure and LCD-PA is shown in Fig.9. The following equations of motion (1) and (2) are derived from the model.

$$M_1 \ddot{x}_1 + K_1 x_1 + C_1 \dot{x}_1 + K_2(x_1 - x_2) + C_2(\dot{x}_1 - \dot{x}_2) = -M_1 \ddot{y} \quad (1)$$

$$\{m_1 + (A_H/A_Z)^2 m_2\} \ddot{x}_2 + K_2(x_2 - x_1) + C_2(\dot{x}_2 - \dot{x}_1) = -m_1 \ddot{y} \quad (2)$$

where, M_1, K_1, C_1, and x_1 are the generalized mass, stifness, damping, and displacement related to the eigenmode of the structure and y denotes the amplitude of the input acceleration. Following m_1, m_2, K_2 C_2, and x_2 are related to the LCD-PA, in which the fluid mass $m_1 = \rho A_H L_H$, $m_2 = \rho A_Z L_Z$ neglecting fluid in the PA. The stiffness K_2 is shown in equation (3).

$$K_2 = 2\rho g A_H^2 / A_Z + (2np_0/Q)A_H^2 - (2np_0/Q)^2 A_H^2 A_R^2 / \{2\rho g A_R + (2np_0/Q) A_R^2 + K_c\} \quad \cdots (3)$$

where, ρ =unit mass of fluid, g= the gravity acceleration, n= the specific heat of air, p_0 = an atmospheric pressure and Q= air volume. The A_H and L_H are cross-sectional area and length of the horizontal portion, A_Z and L_Z are those of the vertical reservoirs, and A_R and L_R are those of the PA. K_c denotes the stiffness derived from the relationship between fluid displacement of a vertical reservoir (VR) and air pressure force in the PA which inclds the coil spring stiffness of k_{co}.

$$k_c = 2ah_c^2 k_{co}/(h_R r^2) \quad \cdots (4)$$

where a, h_c, h_R, and r are described in Figure 3.

Tuned optimum values for frequency and critical damping of the LCD-PA are found by simulations using equation (1) and (2) and also using the conventional tuned mass theory. In comparing the calculated response to the measured one, though omitted the evidence here, it becomes apparent that only little difference exist.

5 Realized example: Cosima Hotel, Taito-ku, Tokyo

5.1 Outline of the structure and the vibration control system LCD-PA

The concerning building is a 26 story hotel with a height of 106.2 m of which the main structure is a steel structure (the underground part is a steel-concrete structure). This high rise building has a high height to width ratio and accordingly resembles a tower (Fig. 10).

The foundation of the building is firmly connected with ground by pretensioned grout anchors using high strength steel pretension wires. In addition, a super structure is adopted as the frame of the building in order to resist earthquake and wind loads. A vibration control system, a LCD-PA, is installed on top of the building. The

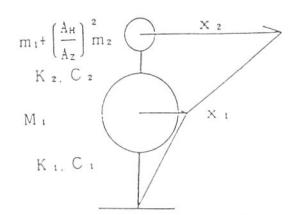

Fig. 9   Interaction Vibration Model.

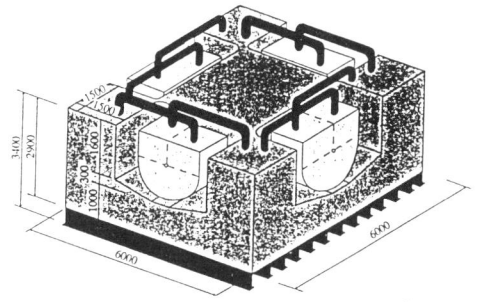

Fig. 11   The employed vibration control system, a LCD-PA.

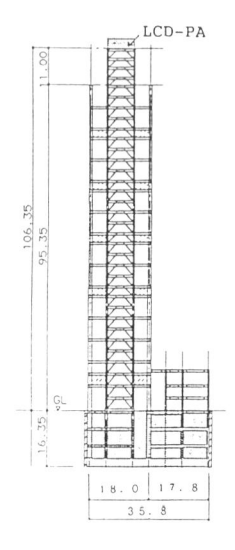

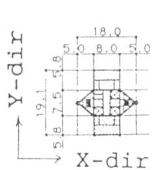

Fig. 10   Cross section, typical floor plan and side view Cosima Hotel.

Number of stories: above ground: 26 stories;
                         underground: 3 stories;
                         penthouse: 2 stories
Maximum building   106.2 m (eaves height: 95.2 m,
height: maximum height: 112 m)
Structure: above ground: steel structure;
                         combination of eccentric
                         K-braced and rigid frames
                         underground: steel concrete structure;
                         combination of reinforced concrete
                         earthquake resistant walls and rigid
                         frames
Standard floor   multi floor part: 17.5 m x 8 m
plan: (width including staircases: 13 m)
Floor area: 9798 m^2
Building area: 777 m^2
Weight of above
ground part: 4635.7 ton

LCD-PA is shown in Fig. 11 and has a total and effective liquid weight of respectively 51 and 36 ton.

5.2 Structural movements characteristics

In order to understand the moving characteristics of this structure, a forced vibration experiment was conducted. The results of the experiment were used to investigate the following characteristics.

5.2.1 Natural frequency

On the time-displacement measurements of the experiment a frequency analysis was conducted to obtain the natural frequencies of the building. In the direction parallel to the length of the building (X-direction) the first natural frequency is 0.50 Hz, while the second natural frequency is 1.69 Hz. In the direction parallel to the width of the building (Y-direction) the first and second natural frequency are respectively 0.48 Hz and 1.69 Hz. The period of the displacement is only slightly dependent on the value of the displacement and similar periods are found under normally occurring slight motion and under maximum displacement (3.5 mm). There is no influence of existence of water in the (vibration control) tank as similar periods are found for a full and empty tank.

5.2.2 Damping ratio

With the measurements obtained from the vibration experiment, the dampingt ratio for the first natural frequency in the case of free vibration was investigated. If there is no vibration control system the displacement reduction factor is 0.0068 for the X-direction and 0.0055 for the Y-direction. If the vibration control system is employed displacement reduction factor increases by about a 10 fold to give values of respectively 0.0550 and 0.600 for X and Y-direction (Fig. 12).

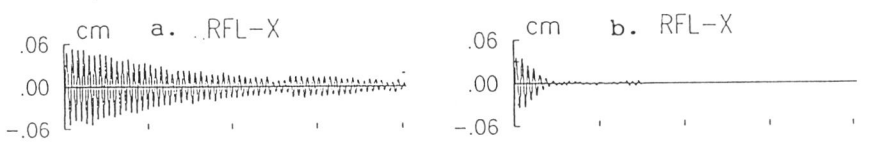

Fig. 12   Time-displacement measurements (X-direction, 26th floor) of free vibration experiment:
a.   Without vibration control,
b.   With vibration control.

5.3 Vibration control efficiency

5.3.1 Transmission function

From the structural movements characteristics the transmission function for the acceleration between the 1st and 26th floor is calculated (transmission function = v_{26}/v_1) and the results are shown in Fig. 13. If there is no vibration control system the transmission factor is about 105-120 for the first natural frequency. Employment of the

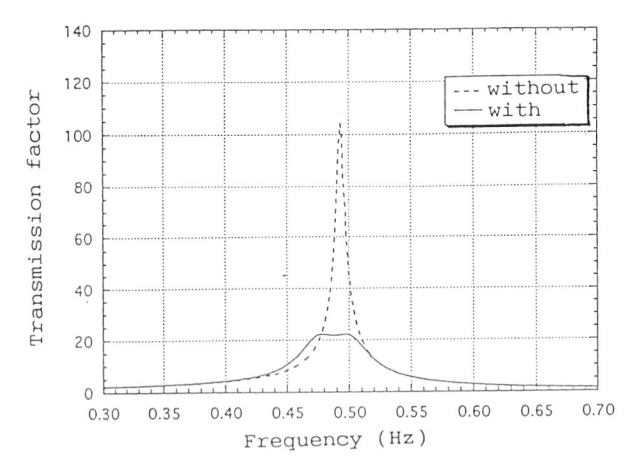

Fig. 13 Transmission factor ($=v_{26}/v_1$) as function of frequency under ground motion for cases with and without vibration control.

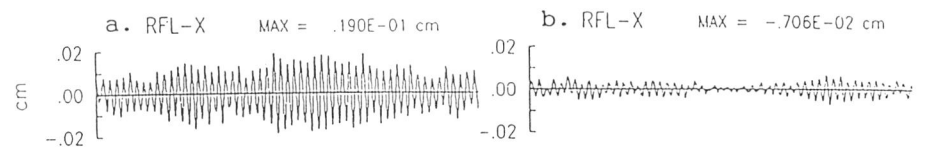

Fig. 14 Time-displacement measurements (X-direction, 26th floor) under normally occurring slight motion:
a. Without vibration control,
b. With vibration control.

vibration control system reduces the transmission factor to about 25, about 1/4 to 1/5 of the original value for the situation with no vibration control system.

5.3.2 Microtremor
Using displacement measurements under microtremor in the cases with and without vibration control, the effect of vibration control under a small earthquakes is investigated. The time-displacement measurements are shown in Fig. 14 for the cases with and without vibration control. Looking at the result, a reduction in the displacement is observed confirming the vibration reduction influence of the vibration control equipment.

5.4 Efficiency prediction of vibration control under strong wind load
Using the various vibration control factors of this building obtained from the experiment, an efficiency prediction of vibration control under strong wind load is made. In Fig. 15 the relation between the return period and the average velocity is shown while in Fig. 16 the relation between the return period and the acceleration response of the 26th floor of this building is given (1 gal = 1 cm/s^2).

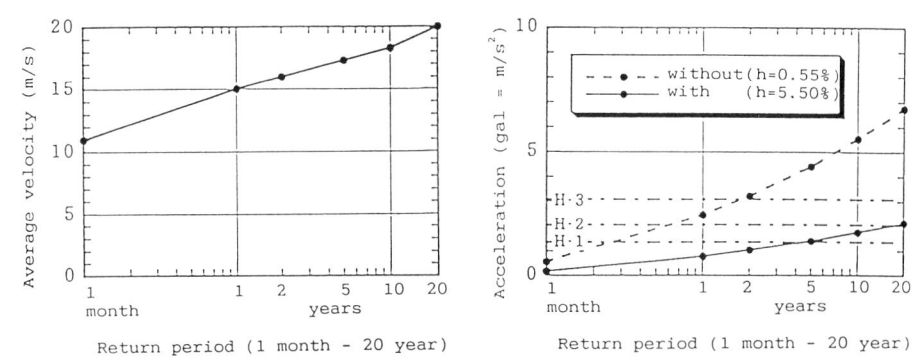

Fig. 15 Average velocity (X-direction, 26th floor) as function of the return period.

Fig. 16 Acceleration (X-direction, 26th floor) as function of the return period for cases with and without vibration control.

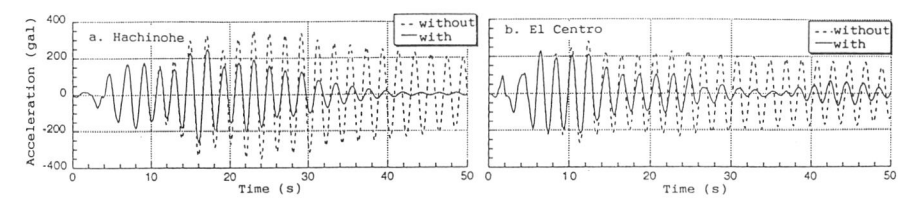

Fig. 17 Time-displacement measurements (X-direction, 26th floor) under earthquake load for cases with and without vibration control.

From the calculation results it is confirmed that the vibration reduction influence of the vibration control system is satisfactorily.

5.5 Efficiency prediction of vibration control under earthquake load

Using the various vibration control factors of this building obtained from the experiment, and efficiency prediction of the vibration control under earthquake load is made. The two earthquake wave records which are used in the analysis are normalized to the same maximum velocity of 25 kine (1 kine = 1 cm/s).

The following two earthquake wave records are used:

- The North-South (NS) component of the Tokachi-oki earthquake records (1968, recorded in Hachinohe, maximum acceleration after normalization: 165.1 gal).
- The North-South (NS) component of the Imperial Valley earthquake records (1940, recorded in El Centro, maximum acceleration after normalization: 255.4 gal).

Figure 17 shows the acceleration response in X-direction of the 26th floor of this building under the input of both the earthquake records in cases with and without vibration control. The analysis results show a small reduction for maximum acceleration and confirms the vibration reduction influence of the vibration reduction system in case of aftershocks. Especially to be noticed is that even in the case of an earthquake with a rather long period (like the Hachinohe wave) introducing a lot of resonance in this particular structure the favorable characteristics as reduction of the maximum acceleration and reduction of the aftershocks can be expected.

6 Conclusions

From this series of studies the following conclusions can be drawn:
- The employment of a tuned liquid damper (TLD) reduces displacement and accelerations caused by wind or earthquakes of medium strength.
- The tuned liquid damper (TLD) is improved by adding a period adjustment system (PA), so that the same vibration control system can be used for buildings with different vibration characteristics.
- The presented tuned liquid damper with period adjustment system (TLD-PA) uses only one large tank in comparison with conventional liquid damper systems where several smaller tanks are employed.
- To achieve sufficient reduction of the displacements and accelerations under earthquake load with a maximum velocity of 25 cm/s the height of the vertical reservoir should be 1600 mm, making water level movements of ±800 mm possible.

7 Acknowledgments

The writers express appreciation to I.Hujita and M.Yoshimura of Mitsubishi Heavy Industries,LTD for their joint study.

8 References

1) Sakai,F.,Takaeda,S.,and Tamaki,T., Tuned Liquid Column Damper, Proc.of International Conf.on Highrise Building, Nanjing,(1989),p.926.
2) Watkins,R.D.and Hitchcock,P.A., Tests on Various Liquid Column Vibration Absorber, Proc.of International Conf. on Motion and Vibration Control,YOKOHAMA,1992.
3) Watkins,R.D.and Hitchcock,P.A.,Model Tests on a Two-way Liquid Column Vibration Absorber, Reseach Report Dept. of Civ.and Min.Eng.,University of Sydney,1992, p.656.
4) P.A.Hitchcock,Kenny C.S.Kwok.,Vibration Control of Structures Using Liquid Column Vibration Absorber, Asia-Pacific Vibration Conf.'93.KITAKYUSHU,Nov.1993, p.799

60 SEISMIC BASE ISOLATION FOR FRAMED STRUCTURES

P. LENZA, M. PAGANO and P. P. ROSSI
Istituto di Tecnica delle Costruzioni, University of Naples, Italy

Abstract
This paper describes a base isolation system for normal framed buildings that is characterized by a prevailingly elastic behaviour and a low technological level. In the light of dynamic analyses, the authors emphasize the mechanic and dynamic characteristics of this system that must be adequately designed in order to produce a favourable seismic response, compatibly with the present thecnological and market possibilities. Attention is focused on the need to reduce the relative displacements of the building increasing the low dissipation capacity of the structure, on the system economic aspects and finally on the applicability of the isolation system also to old buildings.
Keywords: Multistoreys Buildings, Base Isolation, Seismic Response

1 The isolation system

In a previous paper [1], the authors described a proposal of base isolation for framed buildings having a structure that can be considered as a spatial frame with stiff floors constructed in steel, reinforced concrete or in composite concrete-steel material. Moreover, for the main isolation elements the technology of metallic materials was used. The design strategy is characterized by devices with low thecnological contents that is also consistent with the capacities of normal size enterprises and by a cost that, at present, appears to be competitive with other solutions [1].
According to what proposed in [2], the intervention includes (fig.1-2):
a) a normal foundations system (direct or indirect) with first floor columns hinged at both ends (labile first floor);

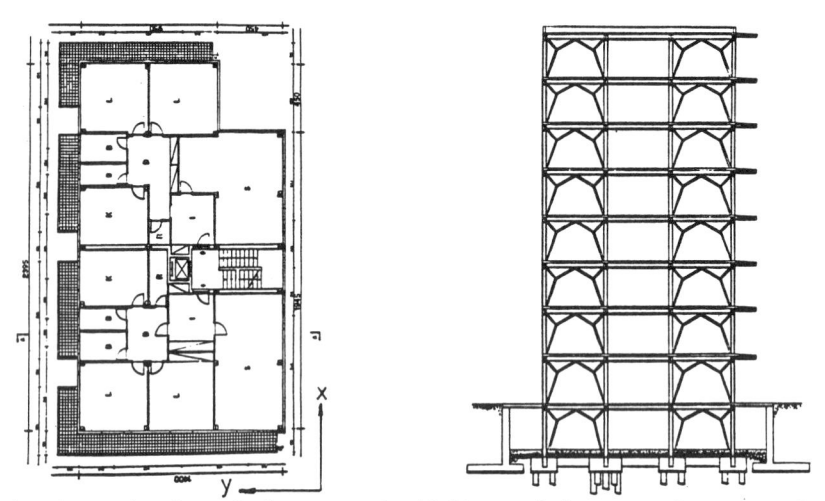

Fig.1 Typical multistorey building with an underground level, designed for low horizontal forces and verified in seismic area provided with the proposed base isolation system.

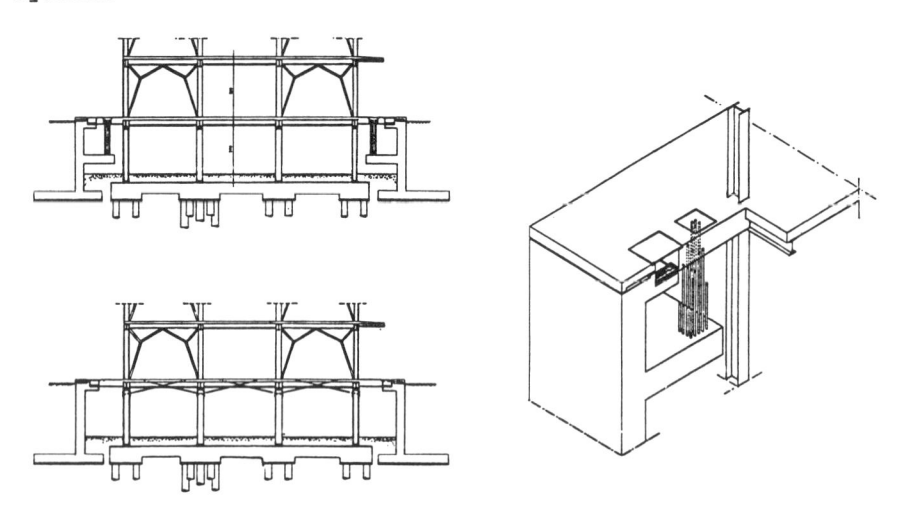

Fig.2 The arrangement of the first storey oriented to an useful seismic isolation of the building.
The columns are hinged at a normal foundation system. The elastic recoil system is obtained either with bundles of metallic bars or with steel ropes. Brick elasto-fragile elements stiffen the building in normal service.

b) an elastic recoil system applied to the first floor so as to assure, to an acceptable extent, the balancing of the P-δ effect [3]. The overall stiffness of such a system has to assume the following value:
$$K' > P / h$$
where P is the weight of the building and h is the height of the hinged columns.
This system must remain in the elastic field also with respect to values of displacements greater than the maximum ones that can be expected at the level of the first floor, so as to avoid its degradation and the decrease of its reliability vis-à-vis future earthquakes;
c) an elasto-fragile system that stiffens the first floor having such a size that it can disengage itself at a shear value slightly greater than the maximum shear produced by the wind.
If K'' denotes the lateral stiffness given to the frame by the system c) in correspondance with the first floor, at this level the effective initial lateral stiffness of the system will be equal to K'- P/h +K''; at this stage, because of the high value of K'', the balancement of the wind forces occurs without large displacements that could disturb the building's users.
In case of earthquakes, the breaking of the elasto-fragile system suddenly reduces the lateral stiffness to K'- P/h which determines a more favourable seismic response.
After the earthquake, the stiffening mentioned at point c) must be restored keeping in mind that, until it doesn't, the building will be particurarly deformable under the wind force.
In deciding the value of K', we shall be also guided by such an evaluation, considering that a low value of K' - P/h improves the seismic response, but reduces the degree of safety to the instability, due to vertical loads, and makes the users of the structure "perceive" wind displacements in the period following the earthquake (until the stiffenings are not restored).
The maximum displacement in the elastic phase of the recoil system will have to be always greater than the maximum displacement of the first floor with respect to the foundations base both during the earthquake and during the following stage characterized by the wind forces acting on the frame that has been transiently deprived of the stiffenings.
It seems that the order of magnitude of the first period of buildings so isolated can exeed 4 seconds as observed in numerical tests with particularly beneficial effects in terms of seismic response.
Nor does it seems that one should fear the well known phenomenon which due to plasticizations in the superior structure, makes the period of the space frame close to the period of the isolation system which might cause considerable phenomena of strain amplification in case of violent earthquakes. The main problem is to limit the

relative displacements between the building block and the foundations because the magnitude of these displacements involves techological problems.

2 Analysis models of the structure

The building, endowed with a semi-labile floor, with a high deformability of the first floor compared to the other, behaves in a way similar to a SDOF system. Therefore it seemed convenient to decide to use this simplified model to analyse the structure. This model should be formed by a simple linear elastic oscillator (SDOF) with a sudden decrease in stiffness vis-à-vis a given shear (corresponding to the breaking of the elasto-fragile systems).
A more accurate analysis, more complex from the computational viewpoint, was performed which considered the building as a spatial frame with stiff floors and took account of the destablizing effect of vertical loads [3]. The dynamic structural analysis was made in the elastic field by using Newmark's alghorhytm [4].
Historical and artificial accelerograms were taken into consideration for such an analysis As written former, Tolmezzo and Bagnoli Irpino's accelerograms were selected in order to use accelerometric signals completely different as to frequency contents and duration. Concerning the latter, an accelerogram was used that returns the response spectrum proposed by CNR-GNDT for buildings located on compressible soils adjusted to the PGA of Tolmezzo earthquake [5].

Earthquake	Station	Component	Duration (sec)	PGA (cm/scc^2)
Friuli 6-5-76	Tolmezzo	EW	36.38	260.97
Campano-Lucano	Bagnoli Irp.	EW	79.19	183.57
23-11-80	Bagnoli Irp.	UP	79.19	101.57
Artificial			20.00	260.97

3 Numerical tests

At this stage of the study, having set as a priority goal the correct evaluation of the magnitude of the displacements, the impact of some basic parameters on such a magnitude has been measured. Indeed, the numerical tests carried out [2] had shown the importance in this respect, of viscous or hysteretic dissipation and suggested that the vertical component of the seismic motion might negatively affect the results.
Our study has dealt with a nine-storey building (29.2 m), rectangular in shape (24x14 m) (fig.1), isolated by means of the proposed system. A previous study [2], by a stiff

floor spatial model, has already shown that the main
problem in analysing an isolated building as described
above, consists in the rather high values of the
displacements of the first floor. Moreover, this study has
demonstrated that, in order to reduce the above
displacements to acceptable levels, vis-à-vis the elastic
recoil system characteristics, it is convenient to ensure a
viscous damping coefficient equal to 3% with respect to
the critical value of the isolated system, so it is
necessary to assure, at least, a mild dissipation capacity
(fig.3). First, a dynamic analysis has been made of the
SDOF model with a reduction in stiffness in order to
evaluate the response in terms of maximum shear and
displacement for variable values of effective stiffness
and, therefore, of the fundamental period (fig.4).
As regards a structure with a quite high first period of 3-
4 seconds, like the one that for technico-economic
considerations we want to built, the results reveal a
favourable behaviour consistent with the possibilities of
the elastic recoil system vis-à-vis values of the
equivalent viscous damping coefficient greater than or
equal to 3%.
Coefficient $\beta=2*\nu*\omega$ that appears in Newmark's algorhytm and
that takes into account the system dissipation capacities
has been derived on the basis of the first period of
vibration of the isolated structure.
The possibility of introducing two coefficients β has then
been taken into account in order to better schematize the
system capacities of dissipating energy referring both to
the initial stage (with intact stiffening brick supports)
and to the following one (with disrupted supports). This
study has permitted to control how such an analysis can
produce results which are not very different from those
mentioned above and how it is therefore possible to proceed
without making particular errors with the SDOF model by
neglecting the greater dissipation capacities of the system
in the short initial stage.
A further numerical test allowed us to evaluate the
influence of the instability produced by vertical ground
accelerations on the structural response.
The building, schematized as a simple oscillator, has been
subjected to the simultaneous force of the vertical and
horizontal (EW) components of Bagnoli Irpino accelerogram
in order to estimate a possible influence of the changes in
the vertical forces on the effective stiffness of the
system and, therefore, on the structure's response. Also in
this case, the analysis made have revealed the scarce
influence of the factor on the oscillator response.

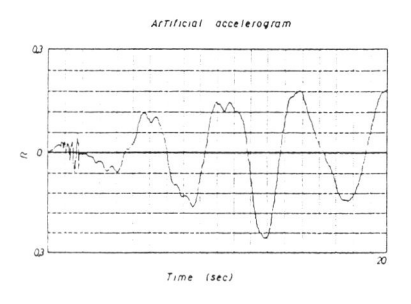

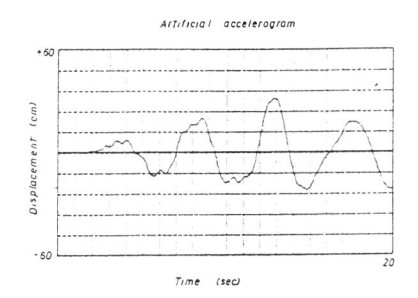

Fig.3 Dynamic analysis results of the nine-storey building for earthquake acting in "y" direction. The first period of the structure changes from 1.3 sec (not isolated) to 4.72 sec (isolated, after brick disengagement)

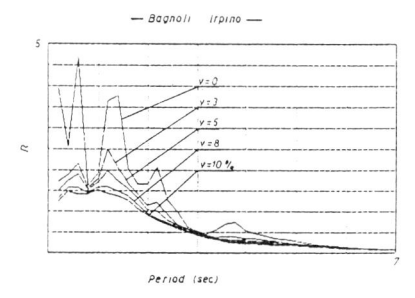

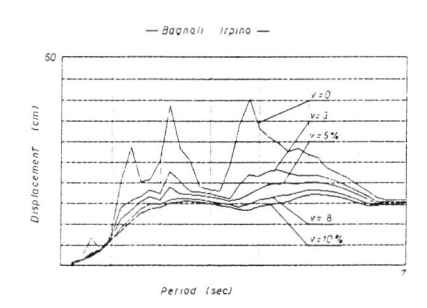

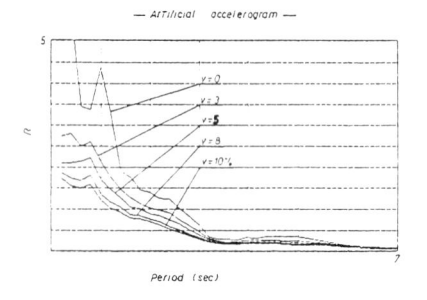

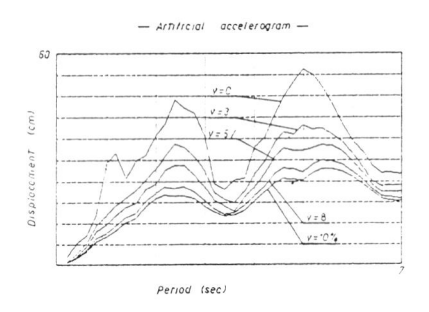

Fig.4 Response spectrum for historical (Bagnoli Irpino) and artificial (CNR-GNDT) accelerograms in terms of relative displacements and response coefficient
R= Shear max / (Mass*PGA)
for some values of damping viscous coefficient.

Since the presence of a given dissipation capacity in the
system appeared to be desirable if not necessary, an
analysis has been developed where this is no longer
schematized through a viscous type of force, placed in
particular in the hinges, but it is hysteretic and exerted
by the external recoil system by means of some mild steel
bars having an elasto-plastic behaviour. In the oscillator
model a spring must be included which has a perfectly
plastic-elastic behaviour to be parallely combined with the
previously introduced elastic spring and having a force-
displacement law set by the values of the elastic stiffness
and of the displacement corresponding to its
plasticization.

Within the periods that concern us, this analysis has
supplied encouraging results, also for very high values of
the elastic stiffnesses to the total stiffness ratios,
which is an evidence of the fact that a slightly plastic
behaviour in the recoil system is already sufficient to
bring the displacements back to the range of acceptable
values vis-à-vis the characteristics of the system under
consideration (fig.5).

The dynamic analysis of the spatial model carried out by
assuming, for this structure, a fundamental period of 4.5
seconds and an equivalent viscous damping coefficient equal
to 3% supplies results that reflect the results obtained
with the simpler SDOF model.

Table 1 reports both the geometric characteristics of the
building analyzed and the mechanical parameters of the
isolation system proposed for the building under
examination.

TABLE I - MECHANIC PARAMETERS -

P	2.77 E6		kg
h	360		cm
K'	13.000		kg cm-1
K''	4.75 E5	//x	kg cm-1
	8.07 E5	//y	kg cm-1
Q wind	50.000	//x	kg
	85.000	//y	kg

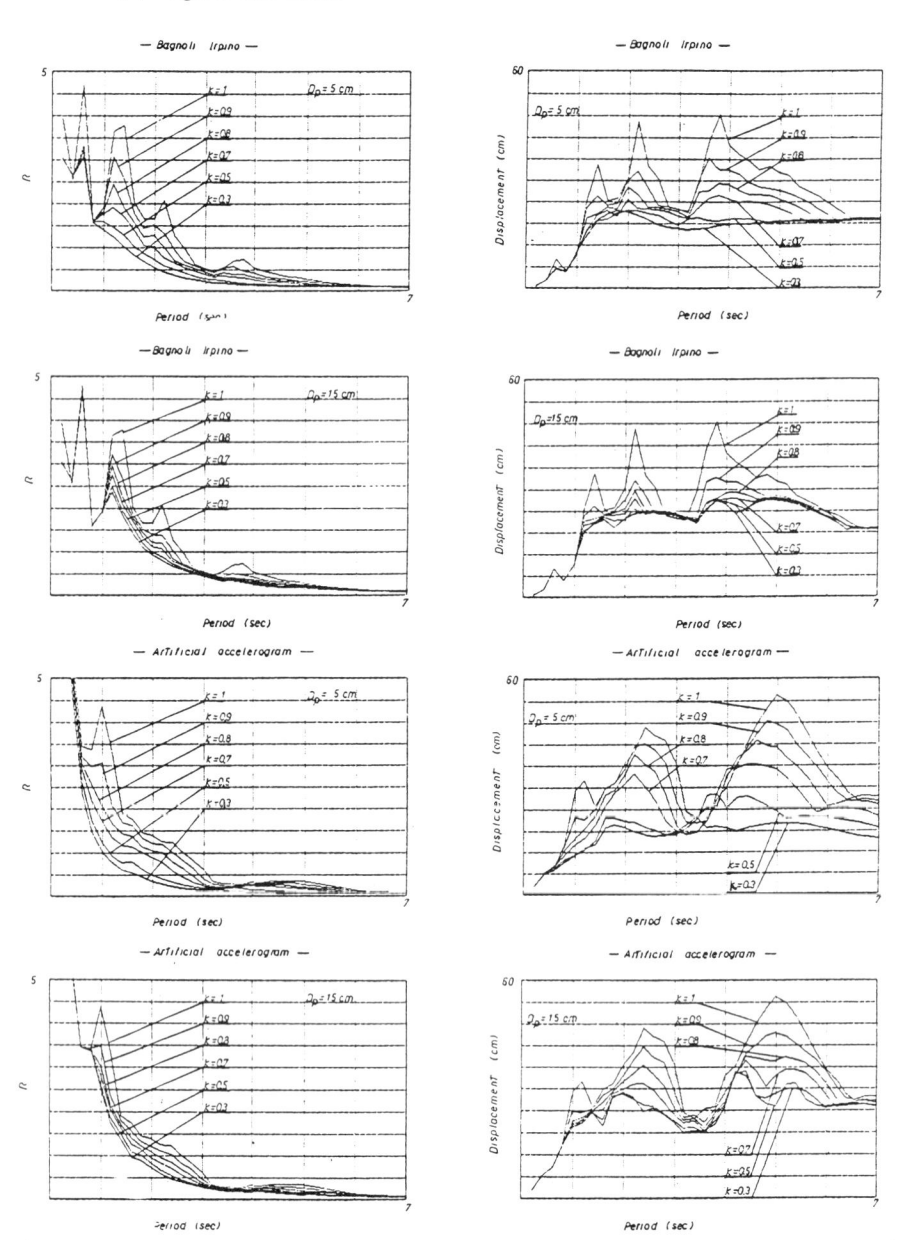

Fig.5 Diagrams show, in terms of relative displacements and ground acceleration amplification, the response of the oscillator provided with two springs, whose behaviour is elastic in one, perfectly plastic-elastic in the other. We consider variable both the plasticization displacement and the elastic system stiffness to the overall one ratios. The period on the "x" axis is referred to the initial stiffness

4 The design of the isolation system

4.1 Technological and economic aspects

Remarkable difficulties have been met in constructing an isolation system having adequate mechanical characteristics by using the available materials and technologies.
No particular hindrance has been met in designing the elasto-fragile system. Some elements consisting of common bricks have been located in lateral "pockets" along the edge of the building (for example in correspondance of the frames), that contrasts with the external retaining wall of the building (fig.2). Where the latter is not included (buildings without underground floors), some brick struts can be constructed and placed between the top of the columns of the first storey and the horizontal beam having such a size that they can disengage themselves when a given shear occurs.
The design of the elastic recoil system was a more difficult task. Indeed, it must assure a given linear stiffness equal to K' while remaining in the elastic field also in the case of several centimeter (up to 30-35 cm) displacements.
Some possible technical solutions were found:
a) construction of bundles of metallic bars of variable height forming a whole an elastic cantilever with variable inertia, by using bars in mild steel or with high resistence (fig.2).
b) construction of steel ropes used as stays considerably inclined on their vertical in correspondance with the hinged columns of the first storey. In this case normal precompressed strands can be employed connected on the top of the columns and on the above beam, with the usual ancorage systems. It should be noted that for the function they have to perform, no significant pre-tension is required, hence a normal strecher will suffice (fig.2)
It must be added that in case an external recoil system having an elasto-plastic behaviour is required, this can be constructed by using mild steel beams of such a size that they can reach the plasticization for a given displacement (5-15 cm).
Fig 6 shows the proposed design of a hinge with a viscous behaviour based on the physical principle of viscous dissipation produced by internal material attrition. The latter, in addition to transmitting vertical loads, is obliged to extrude within the system since it has to undergo constant volume strains, while the steel bar, centrally placed, transmits the shear forces.
The rigorous evaluation of the economic impact of the proposed design strategy, also compared to other isolation systems, is complex and difficult in addition so being absolutely outside the author's field of competence. At any rate, the solution taken into consideration appears at first sight undoubtetly interesting from the economic standpoint too, also because of the single foundation

system and of the use of relatively small quantities of not very sophisticated material. However, account must be also taken of some additional costs due to the hinges and, in the case of the external elasto-plastic recoil system, to the restoretion of plasticized bars.

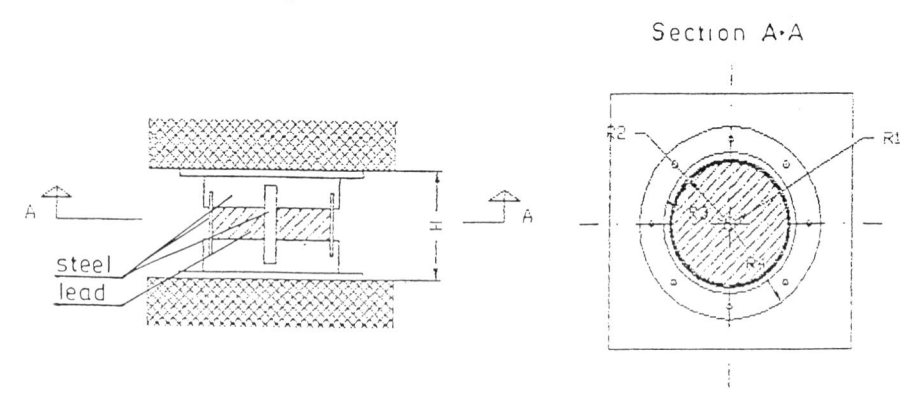

Fig.6 Dissipating hinge scheme, based on the principle of viscous dissipation produced by internal lead attrition

4.2 Thermal effects, installation compatibility, system's vulnerability.

The first floor can undergo a few millimiter changes in size due to thermal effects; it is advisable that the brick have a free space of the same size.

The pipes of the building's installations, due to the considerable relative motion existing between the rising body and the foundations base (of about 30 cm) will have to be connected to the road network by means of flexible pipe sections.

The brick systems are certainly vulnerable (sabotage or other actions); however, they do not jeopardize the stability of the building. This is also true in case of the desired breaking of the above elements. Among the elastic recoil systems, that if broken would impair the structure stability, only the system endowed with stays (which is certainly economically more convenient) seems to be relatively apter to be sabotaged.

At any rate, the efficiency and integrity of the system over time have necessarily to be controlled.

5 Isolation of the existing framed structures

In principle, the isolation system proposed seems suitable also for adjusting the existing framed structures both in reinforced concrete and in metallic structure. Since there is no need to change the foundations system and being the recoil system construction difficulties negligible, the

only real problem arises when introducing the hinges at the
ends of the first floor columns.

6 Possible developments

The isolation system proposed relies on a strategy of
seismic protection proposed by other authors too [6] [7].
Numerical evidence has been positive also in relation to
the same building isolated by means of elastomeric supports
[8] [9] [10]. The most important verification seems to be
that of the limitation of the recoil system displacements
which are considerably affected by the correct evaluation
of the dissipation capacity [2].
At the present stage of the research, the system
illustrated seems to offer concrete possibilities of
technical application even though the various solutions
must only be considered as indicative because they can be
certainly improved also in the ligth of drawbacks that only
tests on the field will be able to identify.
Since one of the main limitations of the use of elastomeric
devices seems to consist in the influence of the vertical
load on the lateral stiffness that might cause critical
phenomena, the isolation system proposed could also use the
modern elastomeric supports performing the function of
elastic recoil system at the first labile floor, also using
the dissipation capacities. Such systems could contrast the
lateral retaining wall, as already discussed [6] or to be
inserted in some braces at the first floor.

References

/1/ P.Lenza,M.Pagano : "Isolamento di base per edifici
   intelaiati";Proposed to " Ingegneria Sismica"; Maggio 93
/2/ P.Lenza,M.Pagano,P.P.Rossi : "Sull'isolamento di base
   degli edifici" ; 6° National Congress "L'ingegneria
   sismica in Italia",Perugia October 1994
/3/ M. Pagano: "teoria degli edifici: edifici in cemento
   armato"; Naples 1970
/4/ P.Lenza: "Analisi dinamica di telai spaziali sottoposti
   ad eccitazione sismica del suolo", from
   A.Ghersi,P.Lenza: "Teoria degli Edifici: Telai spaziali
   per edifici regolari a piani rigidi", Naples 1988
/5/ CNR GNDT: "Norme tecniche per le costruzioni in zone
   sismiche", Dicember 1984.
/6/ A. Carotti,F.De Miranda : "Contributo all'applicazione
   del "base isolation" nel progetto di strutture in
   c.a. di edifici multipiano in zona sismica";
   "Ingegneria sismica", n.3 1986
/7/ A.Parducci: "Isolamento sismico e sistemi dissipativi-
   Relazione Generale, Proc. International Meeting on

Base Isolation and Passive Energy Dissipation",
Assisi, June 8-9, 1989.
/8/ Imbimbo M.,Serino G.:"Modelli analitico-sperimentali per gli isolatori elastomerici ad elevato smorzamento",
XX A.I.A.S. Congress , Palermo, September 1991.
/9/ F.M.Mazzolani,A.Mandara :"Nuove strategie di protezione sismica per edifici monumentali; il caso della Collegiata di San Giovanni Battista in Carife"
/10/ ALGA S.p.a. : "Strategia antisismica" ;
Elastomeric devices; ALGASISM

61 DESIGN METHODOLOGIES FOR ENERGY DISSIPATION DEVICES TO IMPROVE SEISMIC PERFORMANCE OF STEEL BUILDINGS

G. SERINO
Istituto di Tecnica delle Costruzioni, University of Naples, Italy

Abstract

In the last few years, many passive energy dissipation systems have been proposed and effectively adopted for seismic response control. In the case of steel frame structures, which are able to accommodate large interstorey drifts without significant damage, bracing systems equipped with energy dissipation devices are very well suited. In this paper, the most promising procedures for the design of buildings equipped with friction or metallic yielding devices are reviewed and compared, pointing out the advantages and drawbacks of each method. It is shown that, although some very interesting results have already been obtained, research efforts are still needed before a completely satisfactory comprehensive design methodology will be available.

Keywords: Passive Energy Dissipation, Damped Braced Moment Resisting Frames, Seismic Design Procedures

1 Suitability of supplemental damping to improve seismic performance

Supplemental damping devices have been used for more than 20 years to reduce wind response in tall buildings, particularly in the United States and Japan. In the last few years, passive energy dissipation systems have also been proposed and effectively adopted for seismic response control, both in new designs as well as in retrofit projects, so that supplemental damping represents today a valid new and innovative option available to designers to reduce earthquake demand in a structure.

There are a variety of structures for which supplemental damping can be very beneficially used, but there are also some for which is not practical. All supplemental damping devices presently available require some form of relative deformation in the structure to activate the damper. Therefore, in the case of steel frame structures, which are able to accommodate large interstorey drifts without significant damage, bracing systems equipped with energy dissipation devices characterised by a stable and non degrading behaviour for many repeated cycles, are very well suited.

Another point which is worth to be mentioned regards the expected reduction in structural response caused by supplemental damping. Increasing the energy dissipation capacity of the structural system is beneficial only if the fundamental periods of the construction

Behaviour of Steel Structures in Seismic Areas. Edited by F. M. Mazzolani and V. Gioncu.
Published in 1995 by E & FN Spon, 2-6 Boundary Row, London SE1 8HN. ISBN: 0 419 19890 3.

incorporating the damping devices fall within the range where dynamic amplification occurs for the expected earthquake motion at the site. For an acceleration time history recorded on stiff soil, like the Tolmezzo record of the 1976 Friuli (Italy) earthquake, additional damping is beneficial in the case of short intermediate period (0.2 ÷ 1.0 s) structures, see Figure 1a. On the contrary Figure 1b shows that for the same structures supplemental damping provides very little benefit in the case of an acceleration time history like the one recorded at the SCT station in Mexico City during the 1985 Ixtapa (Mexico) earthquake, but it would significantly reduce dynamic response for a construction having a fundamental period around 2.0 s. The acceleration response spectra relative to the Bagnoli Irpino record of the 1980 Irpinia (Italy) earthquake, see Figure 1c, show dynamic amplification over a wider period range, so that in this case the number of candidate structures for supplemental damping applications is larger.

It is well known that, for typical buildings having 5% inherent equivalent viscous damping, providing 10% to 20% supplemental damping is very effective in reducing earthquake response. Providing more than 20% damping, response is not reduced significantly and is difficult to justify from a cost-benefit perspective. The effectiveness of additional energy dissipation decreases as the magnitude of the added equivalent viscous damping increases: the cost effective limit on the energy dissipation systems depends upon the structural system and the device selected. For structures having only a small amount of inherent damping (1% to 2%), the addition of only 10% damping will reduce the earthquake response significantly.

2 Overview of today currently adopted energy dissipation devices

Today currently adopted energy dissipation devices for seismic protection are based either on friction mechanism, or on the hysteretic behaviour of metals in the plastic range or on viscous damping. As mentioned above, in order to be effective, supplemental damping devices must be inserted at points of the structure where significant relative displacements are expected to occur during an earthquake. For this reason, steel braces in moment resisting frames generally represent an excellent choice for damping device location, particularly in the case of steel braced moment resisting frames, and most of the studies and completed applications of supplemental damping for seismic protection refer to this structural scheme.

Regarding friction devices, a variety of systems differing in their mechanical complexity and in the materials used for the sliding surfaces have been proposed and effectively adopted in earthquake engineering applications. Most of them (De Luca and Serino, 1989) generate rectangular hysteresis loops characteristic of Coulomb friction and may exhibit permanent offsets after an earthquake. Recently, a unique type of friction device having self-centring capabilities and a friction force proportional to displacement has been developed (Nims et al., 1993). Friction devices have very good performance characteristics and their behaviour is not significantly affected by load amplitude, frequency or number of applied load cycles.

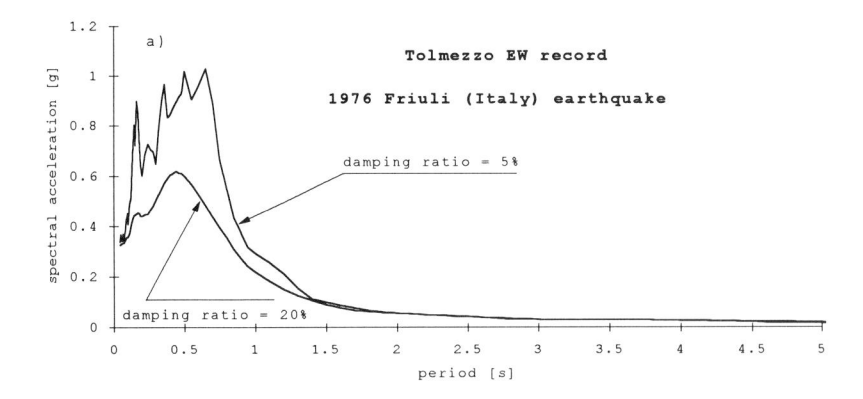

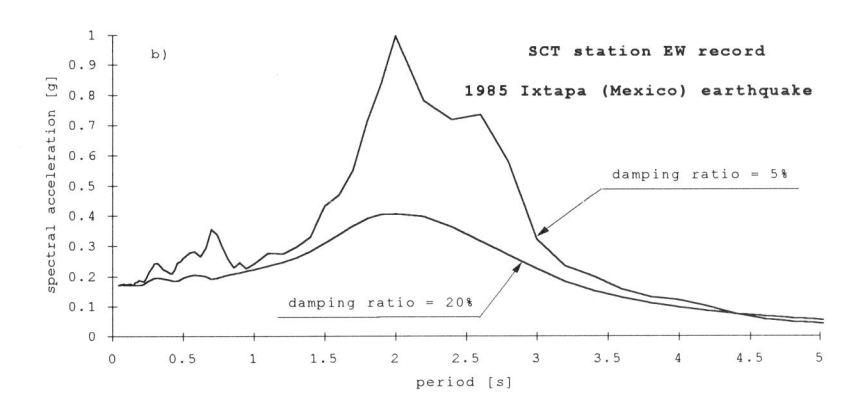

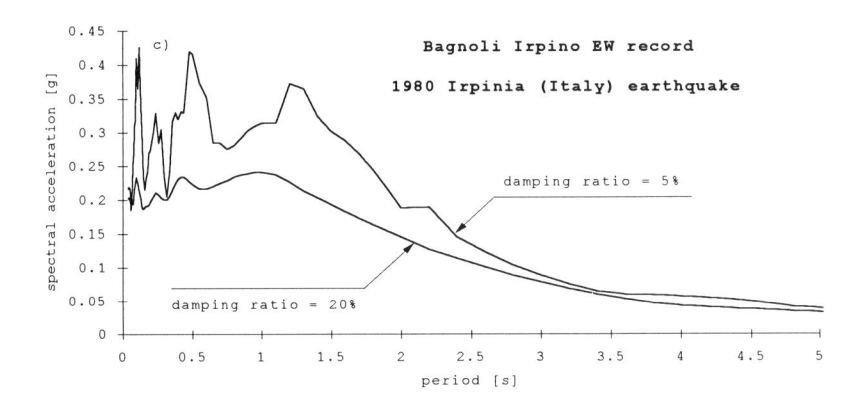

Fig. 1. Acceleration response spectra of different records showing period ranges where supplemental damping is beneficial.

An issues of great importance with friction devices is their long-term reliability and maintenance: sliding surfaces are required to slip at a specific load during an earthquake, even after decades of nonuse. Another point which may sometime cause some problems is the introduction of higher frequencies as the devices undergo stick-slip behaviour.

Metallic systems take advantage of the energy dissipation mechanism of metals when deformed beyond yielding. A wide variety of different types of devices have been developed which utilise flexural, shear or extensional deformation modes into the plastic range. A particularly desirable feature of these systems is their stable behaviour, long-term reliability and generally good resistance to environmental and temperature effects.

Metallic devices using mild steel are usually composed of triangular or X-shaped steel plates so that yielding is spread almost uniformly throughout the material, resulting in a device able to sustain repeated inelastic deformations in a stable manner, without concentrations of yielding and premature failure (De Luca and Serino, 1989). A number of different steel cross-bracing dissipators have been developed and tested in Italy (Ciampi, 1993). In Naples, a 29-story steel suspended building, having floors hanging from two concrete cores, utilises tapered steel devices as dissipators between the core and the suspended floors (Ciampi, 1993). Other metallic devices use lead extrusion as energy dissipation mechanism. Their hysteretic behaviour is very similar to that of many friction devices, with essentially rectangular hysteresis loops. Lead extrusion dampers show stable and repeatable cycles, are insensitive to environmental conditions and show insignificant ageing effects.

Recently, some new types of metallic devices based on Shape Memory Alloys (SMA) have been proposed and started to be studied. SMAs have the ability to dissipate energy without sustaining any permanent deformation, because they undergo a reversible phase transformation as they deform (contrary to the intergranular dislocations, typical of steel deforming in the plastic range). Nitinol (nickel-titanium) tension devices as part of a cross-bracing system have been tested at the Earthquake Engineering Research Center of the University of California at Berkeley (Aiken et al., 1993), while copper-zinc-aluminium SMAs devices with torsion, bending and axial deformation modes have been investigated at the National Center for Earthquake Engineering Research in Buffalo, New York (Witting and Cozzarelli, 1992). SMAs devices must be designed such that deformations do not occur beyond their elastic limit, as this results in permanent yield in the material. Many SMAs also exhibit excellent fatigue resistance. Moreover, Nitinol has outstanding corrosion resistance, superior even to that of stainless steel and other corrosion-resistant alloys.

Viscoelastic dampers have been used in structural applications for vibration and wind control for more than 20 years, but the extension to the seismic domain has occurred more recently. Wind vibration control applications have typically involved providing the building with only about 2% of critical damping, while the level of damping required for a feasible seismic energy dissipation system is significantly higher, of the order of 10 to 20%. Viscoelastic shear dampers using special polymers for seismic applications have been developed in the United

States by 3M, and by Shimizu and Bridgestone Corporations in Japan. Fluid viscous dampers possessing linear viscous behaviour, which have been used for many years in the military and aerospace fields, are also starting to be used for structural applications (Constantinou et al., 1993). These dampers can be very compact in size in comparison to force capacity and stroke. A particularly desirable feature of these dampers is that the damping force is out-of-phase with the displacement: this has the advantage that, when they are inserted in the diagonal braces of a frame structure, the peak damper force does not occur simultaneously with the peak induced column moments.

It is worth to point out some other important differences in the mechanical behaviour between the friction/yielding devices and the viscoelastic/viscous ones. First of all, friction and metallic damping devices show a strong nonlinear behaviour at deformation levels exceeding their slip or yielding limit, while this does not occur in the case of viscoelastic and viscous devices. This implies that the expected level of earthquake response must be known when friction/yielding systems are adopted, as the structure secant stiffness and equivalent viscous damping values are amplitude dependent. Another major difference regards the maximum force that each device develops during an earthquake. In the viscoelastic or viscous devices, the maximum seismic force is function of the maximum displacement and velocity, while in the friction or yielding systems maximum earthquake force equals the design slip or yield force plus strain hardening, and can be thus more easily controlled.

3 Available procedures for the design of the energy dissipation devices

Since the adoption of supplemental damping devices for seismic protection still represents a novelty in earthquake engineering, a comprehensive design methodology for the selection and sizing of the energy dissipation devices is not yet available. Anyway, some authors in the USA and Italy have studied the problem and also developed some interesting design procedures, but these proposals usually refer to a specific type of energy dissipation system. In what follows, the most promising procedures for the design of the supplemental damping devices in steel braces of moment resisting frames are reviewed and compared. Reference is made to the case of friction and metallic yielding devices only, which require more accurate analyses due to the above mentioned intrinsic nonlinear behaviour. Design procedures for viscoelastic and viscous devices, whose behaviour can be more easily modelled through simpler linear models, will not be considered in this paper.

3.1 Design procedures developed by USA authors

Filiatrault and Cherry (1990) developed a design procedure for a particular type of damped bracing, originally proposed by two Canadian researchers (Pall and Marsh, 1982). The system basically consists of a special mechanism containing slotted friction brake lining pads introduced at the intersection of the frame cross-braces. The device is designed not to slip under normal service loads and moderate earthquakes. During severe seismic excitations, the device slips at a predetermined load, before any inelastic deformation of the main members has occurred. A peculiarity of the system is the presence of a special mechanism, composed of four link elements in the shape of a rectangle around the slotted slip joints at the braces intersection, which eliminates a problem occurring when the braces are slender and thus effective in tension only, caused by the fact that the friction joint slips in tension but does not slip back in the compression (buckled) regime.

The design procedure is based on the definition of an "Optimum Slip Load Distribution", representing the slip force pattern in the braces able to minimise structural response under seismic action. From a series of nonlinear time history analyses, Filiatrault and Cherry (1990) determined, for a large number of friction braced frames and earthquake time histories, the slip load distribution which minimises a relative performance index (RPI) given by:

$$RPI = \frac{1}{2}\left[\frac{SEA}{SEA_{(o)}} + \frac{U_{max}}{U_{max(o)}}\right] \quad (1)$$

where SEA and U_{max} are the total strain energy area (area under the entire strain energy time history) and the maximum total strain energy for a friction damped structure, respectively, while $SEA_{(o)}$ and $U_{max(o)}$ are the same quantities for the identical structure, but without bracing. The authors determined the most important parameters controlling the slip load and proposed the design slip load spectrum given in Figure 2 to obtain the adimensionalised total optimum slip shear V_o/ma_g from the T_g/T_u ratio, where a_g and T_g are the peak ground acceleration and predominant period of the design ground motion, T_u is the fundamental period of the unbraced structure and m the total mass of the structure. The slip spectrum is completely described by specifying the ordinates α and β, which correspond to $T_g/T_u = 1$ and $T_g/T_u = 15$ respectively, given by:

$$\alpha = (-1.24NS - 0.31)\frac{T_b}{T_u} + 1.04NS + 0.43 \quad (2)$$

$$\beta = (-1.07NS - 0.10)\frac{T_b}{T_u} + 1.01NS + 0.45 \quad (3)$$

where NS is the total number of storeys and T_b the fundamental period of the fully braced structure. The optimum slip shear for each storey is then simply obtained by dividing the total optimum slip shear V_o by the number of storeys NS. The slip shear is thus the same for all the storeys (constant slip shear distribution). The procedure is valid for a wide range of T_b/T_u ratios ($0.20 \leq T_b/T_u \leq 0.80$) and the authors considered only the case of braces having the same cross sectional area at all storeys (constant bracing stiffness distribution). The fact that only uniform distribution of slip shear and bracing stiffness were considered certainly represents a limit of the design procedure.

Nonlinear time history analyses were also performed by Xia and Hanson (1992), though in their work the authors referred to a metallic yield system, the steel-plate added damping and stiffness (ADAS). They identified in the device yield displacement, device strain-hardening ratio, ratio of bracing member stiffness to device stiffness and ratio of device + brace stiffness to structural story stiffness without the device in place as the most important parameters which influence the response. On the basis of the limited analyses performed (three structural systems and three earthquake records only), a complete design procedure for buildings incorporating ADAS devices was not provided, but the following important suggestions regarding the selection of the above mentioned parameters were given: 1) the device yield displacement should be in the range of 0.0014 ÷ 0.0020 times the story height to avoid device ductility ratios beyond 10; 2) the device strain-hardening ratio has little influence on the response but its effect should be taken into account in the design of the braces and other structural members supporting the devices; 3) a brace stiffness to device stiffness value of 2 was recommended, provided that the bracing has enough strength to yield the ADAS devices; 4) considered that the structural response decreases with increase in the ratio SR between the device + brace stiffness to structural story stiffness without the device and that its effectiveness is relatively constant at larger SR values and varies depending on the characteristics of the earthquake, an SR ratio of about 2 was indicated as appropriate in most case.

Similar results were also obtained by Tsai et al. (1993), who studied the nonlinear response of SDOF ADAS frame systems subjected to four historical earthquake excitations scaled to an effective peak ground acceleration of 400 cm/s^2. They found that the most important parameters affecting the response are the above indicated stiffness ratio SR, the strength ratio U between total restoring force developed in the system when the frame yields and the one reached when the device yields (obviously $U \geq 1$ as the device starts yielding before the frame), and the frame lateral displacement when the device yields. The authors also found that the response of the system was little sensitive to both the device and bare frame strain hardening ratios and indicated a strength ratio U of about 2 as optimal, and SR values less than 4 and 2 as appropriate for systems having short and medium to long vibration periods, respectively.

A SDOF consisting of a dual system composed of a linear frame provided with an elastic perfectly plastic element of the friction or yielding type acting in parallel was considered by Scholl (1993). Using a simple derivation first proposed by Thiel et al. (1986), Scholl (1993) derived an expression which gives the equivalent viscous damping ratio for the dual system as a function of the stiffness ratio SR and of the ratio FR between the device slip or yield force and the corresponding force in the frame only. This expression is graphically represented in Figure 3 and can be used as a guide for establishing an appropriate FR value following the selection of a practical SR value.

3.2 Design procedures developed by Italian authors
With the aim of developing a design procedure for the energy dissipation devices, the nonlinear dynamic response of r.c. frames with friction braces has been extensively studied in Italy by Vulcano (1991) and Vulcano and Guzzo-Foliaro (1993). The authors adopted an approach similar to that used by Filiatrault and Cherry (1990), whose main

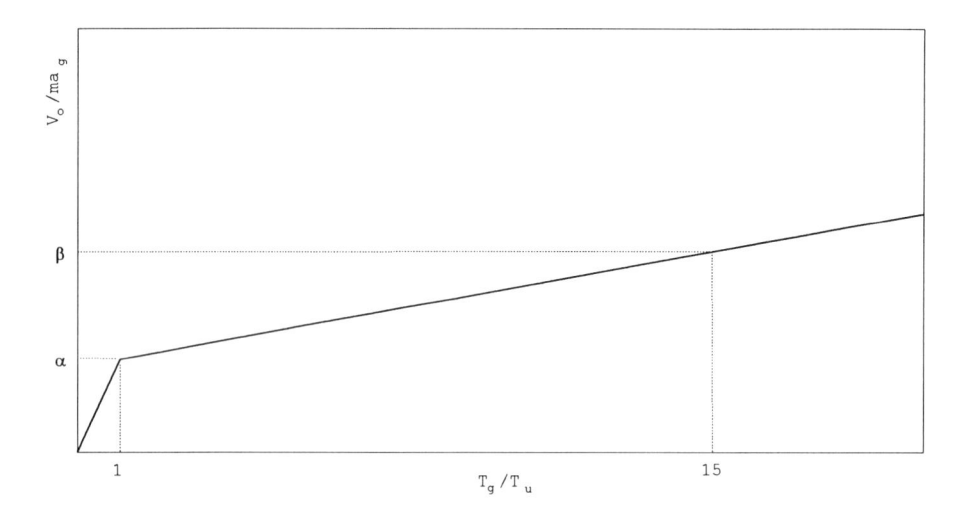

Fig. 2. Design slip load spectrum proposed by Filiatrault and Cherry (1990).

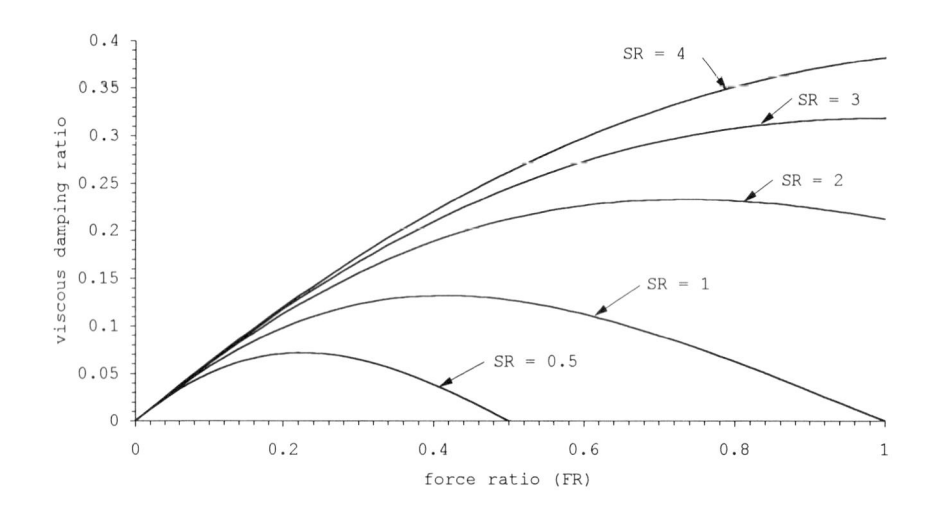

Fig. 3. Equivalent viscous damping for a damped braced frame as a function of stiffness ratio SR and limit force ratio FR (Scholl, 1993).

results have been reported above, but included the nonlinear behaviour of the frame in the analyses and considered also nonuniform slip shear and bracing stiffness distributions. In the paper by Vulcano (1991), two nonuniform slip shear and bracing stiffness distributions were considered: a) bracing stiffness proportional to the frame only stiffness at each story and slip shear proportional to the axial force in the tensile brace under the static design seismic loads before the friction devices start slipping; b) bracing stiffness and slip shear linearly decreasing along the height of the building. The analyses carried out clearly indicated that both nonuniform distributions are able to better spread the energy dissipated in the friction devices at the different storeys compared to the constant distribution proposed by Filiatrault and Cherry (1990), which on the contrary tend to concentrate the dissipated energy in the lower storeys. Further insight into the problem was obtained in the paper by Vulcano and Guzzo-Foliaro (1993), where a nonuniform distribution of the type a) above was adopted but computing story stiffness and axial force in the tensile braces considering the dynamic inertial loads associated to the first vibration mode of the braced frame instead of the static design loads. Is was also shown that very similar results in terms of optimal slip load values are obtained considering four different performance indexes quantifying the response reduction of the braced frame with respect to the same frame when unbraced. The authors also demonstrated that a simplified SDOF with a trilinear force-displacement constitutive law and mechanical properties derived from those of a braced frame is able to catch the essence of the seismic response and can thus be profitably used in the design process to select the optimum slip load and brace stiffness values.

A complete procedure for the selection of brace stiffness and limit force for a SDOF similar to that considered by Vulcano, representing a simple braced frame equipped with a friction or metallic yielding supplemental damping device, has been developed by Ciampi et al. (1993). Two different design problems have been considered: the retrofit of a conventional frame structure with the objective of reducing damage under a strong earthquake to a degree compatible with the one considered acceptable by current Italian code, and special protection of a strategic building with the objective of having no damage in the structure even in the case of a severe earthquake. Denoting β the ratio between the displacement which causes yielding in the brace and the corresponding yielding displacement of the frame, it is shown that the optimal value of β is around 0.5 in the case of conventional frame retrofit and is approximately 0.15 in the case of special protection. The above values of β are the lowest compatible with a damage level in the frame corresponding to a required ductility ratio $\mu_f = 4 \div 6$ and $\mu_f = 1 \div 2$ for the two cases, respectively, and allow the brace to remain elastic under an earthquake level corresponding to 1/4 of the design level (moderate event). In the design procedure a strength ratio between the brace and the frame yield forces $\rho = 1$ is implicitly recommended. The proposed design procedure can be easily generalised and applied in the design of real MDOF damped braced frames, but this requires the definition of proper stiffness and limit force distributions for the braces at the different storeys, with the objective of favouring uniform engagement of the bracings in the energy dissipation process and avoiding concentration of frame damage at specific locations. The authors propose to adopt linear distributions of limit force along the height of the building and observe that limit force distribution of uniform resistance, i.e. which conform with the required seismic shear distribution, tend to produce concentration of damage at specific storeys, whichever the associated distribution of stiffness both in the unbraced and in the braced frames, and that the same occurs for constant strength distributions.

4 Comparison among the different procedures and conclusions

The previously presented overview of today available procedures for the design of friction or metallic yielding energy dissipation braces for seismic protection of moment resisting frame structures has clearly shown that the core of the problem is the selection of appropriate values for brace stiffnesses and limit (slip or yielding) force in the devices which are able to minimise structural response and damage. In order to understand the essence of the phenomenon and to find out the optimal brace stiffness and device limit force, many authors (Tsai et al., 1993; Scholl, 1993; Vulcano and Guzzo-Foliaro, 1993; Ciampi et al., 1993) have studied the dynamic seismic response of a simple SDOF with a trilinear force-displacement constitutive law. Although the analysis approaches, system parameters, response indexes quantifying the beneficial effects of supplemental damping and earthquake records considered by the different authors vary considerably, very similar results have been found in terms of optimal brace stiffness and device limit force: 1) an optimal value of the ratio SR between the device + brace stiffness and structural story stiffness is around 2 for conventional seismic protection problems, i.e. when limited damage in the frame is accepted under a strong earthquake, while larger SR values (around 4 or more) are needed in special protection problems, i.e. when no damage is allowed even under a destructive event; 2) the optimal value of the ratio ρ between the device + brace limit force and the frame yield forces is around 1. Similar results were also obtained by Xia and Hanson (1992) from nonlinear time history analyses of braced frames with supplemental damping devices.

The optimal values of the SR and ρ ratios obtained from the analysis on the SDOF system can be profitably used in the design of MDOF damped braced frames, but the actual distribution of brace stiffnesses and limit force in the devices play a major role on the issue of how damage distributes more or less uniformly within the frame. The constant bracing stiffness and limit force distributions indicated in the design procedure by Filiatrault and Cherry (1990) produce concentration of damage at the lower levels of the structure, while much better results can be obtained in terms of diffusion of damage adopting brace stiffnesses proportional to the bare frame storey stiffnesses and limit forces proportional to the static or 1st mode dinamic design storey shears (Vulcano, 1991) or linear distributions of brace stiffness and limit force along the height of the building (Ciampi et al., 1993).

5 References

Aiken, I.D., Nims, D.K., Whittaker, A.S. and Kelly, J.M. (1993) Testing of passive energy dissipation systems, **Earthquake Spectra**, Vol. 9, No. 3, pp. 335-370.

Ciampi, V. (1993) Development of passive energy dissipation techniques for buildings, in **Proc. Int. Post-SMiRT Conf. Sem. on Isol., Energy Dissip. and Control of Vibrations of Struct.**, Capri, Italy, pp. 495-510.

Ciampi, V., De Angelis, M. and Paolacci, F. (1993) On the seismic design of dissipative bracing for seismic protection of structures (in Italian), in **Proc. 6th Italian Nat. Conference on Earthquake Engineeering**, Perugia, Vol. 1, pp. 141-150.

Constantinou, M.C., Symans, M.D., Tsopelas, P. and Taylor, D.P. (1993) Fluid viscous dampers in applications of seismic energy dissipation and seismic isolation, in **Proc. of Sem. on Seismic Isol., Pass. Energy Dissip. and Active Control**, San Francisco, California, Vol. 2, pp. 581-591.

De Luca, A. and Serino, G. (1989) Innovative structural systems for earthquake protection (in Italian), **Ingegneria Sismica**, Vol. 6, No. 1, pp. 3-22.

Filiatrault, A. and Cherry, S., (1990) Seismic design spectra for friction-damped structures, **Journal of Structural Engineering**, Proc. ASCE, Vol. 116, No. 5, pp. 1334-1355.

Nims, D.K., Inaudi, J.A., Richter, P.J. and Kelly, J.M. (1993) Application of the energy dissipating restraint to buildings, in **Proc. of Sem. on Seismic Isol., Pass. Energy Dissip. and Active Control**, San Francisco, California, Vol. 2, pp. 627-638.

Pall, A.S. and Marsh, C., (1982) Response of friction damped braced frames, **Journal of the Structural Division**, Proc. ASCE, Vol. 108, No. 6, pp. 1313-1323.

Scholl, R.E. (1993) Design criteria for yielding and friction energy dissipators, in **Proc. of Sem. on Seismic Isol., Pass. Energy Dissip. and Active Control**, San Francisco, California, Vol. 2, pp. 485-495.

Thiel, C.C., Elsesser, E., Jones, L., Kelly, T., Bertero, V., Filippou, F. and McCann, R. (1986) A seismic energy absorbing cladding system: a feasibility study, in **Proc. of a Sem. and Works. on Base Isol. and Passive Energy Dissipation**, Redwood City, California, pp. 251-260.

Tsai, K.C., Chen, H.W., Hong, C.P. and Su, Y.F. (1993) Design of steel triangular plate energy absorbers for seismic resistant construction, **Earthquake Spectra**, Vol. 9, No. 3, pp. 505-528.

Vulcano, A. (1991) Control of seismic response of frame structures by means of dissipative braces (in Italian), in **Proc. 5th Italian Nat. Conference on Earthquake Engineeering**, Palermo, Vol. 1, pp. 359-368.

Vulcano, A. and Guzzo-Foliaro, G. (1993) Design criteria for damped braced frame structures (in Italian), in **Proc. 6th Italian Nat. Conference on Earthquake Engineeering**, Perugia, Vol. 1, pp. 131-140.

Witting, P.R. and Cozzarelli, F.A. (1992) Shape memory structural dampers: material properties, design and seismic testing, **Report No. NCEER-92-0013**, State University of New York at Buffalo, May.

Xia, C. and Hanson, R.D., (1991) Influence of ADAS element parameters on building seismic response, **Journal of Structural Engineering**, Proc. ASCE, Vol. 118, No. 7, pp. 1903-1918.

PART EIGHT
DESIGN AND APPLICATIONS

62 STRENGTHENING AND REPAIRING OF LATTICE STRUCTURES

N. BALUT and A. MOLDOVAN
Building Research Institute INCERC, Timisoara, Romania

Abstract
The best way to strengthen a structure if the design is governed by seismic loads is to increase the q-factor. In the case of lattice structures, this can be done by modifying the bracing system in order to take advantage of the dissipative capacity of tension diagonals. This paper is an attempt to apply the above principle to certain categories of lattice structures (trusses subjected mainly to vertical loads, lattice towers, concentrically braced multistorey frames, and bracings of one-storey mill-building structures). The classic methods of strengthening and repairing, which are also discussed, are applicable mostly to non-dissipative members and connections.
Keywords: Steel Structures, Lattice Structures, Seismic Action, Behaviour Factor, Strengthening, Repairing.

1 Introduction

The strategy to be followed when strengthening or repairing a structure depends on the nature of the loads acting on it. If these are static or quasistatic loads, they can be regarded as input data of the problem (although in some cases a partial removal of the self load and/or a reduction in the live loads may represent a good solution). The designer's task is to identify the 'weak' elements and to decide on the most convenient strategy: the strengthening and/or repairing of individual members, or the modification of the structure. These two types of interventions are often combined in practice. Anyway, their purpose is in all cases to enhance the strength and/or the rigidity of the structure, in order to improve its behaviour under the action of given loads.

On the contrary, dynamic loads, like wind loads acting on a structure (which will not be discussed herein), and especially seismic loads, are integrant parts of the structural response. Such loads are not input data, because they can be influenced in a favourable or in an unfavourable sense by modifying the structure.

Behaviour of Steel Structures in Seismic Areas. Edited by F. M. Mazzolani and V. Gioncu. Published in 1995 by E & FN Spon, 2-6 Boundary Row, London SE1 8HN. ISBN: 0 419 19890 3.

The basic principles to be observed when performing an intervention on a structure the design of which is governed by seismic action are specified in Part 1.4 of Eurocode 8. The main concern of the present paper is an attempt to apply these principles to the following categories of steel lattice structures:

(a) trusses subjected mainly to vertical loads;
(b) lattice towers;
(c) concentrically braced multistorey frames;
(d) bracings of one-storey mill-building structures.

The very important case of eccentrically braced multistorey frames will not be discussed here.

By lattice structures, the best way to reduce the seismic loads is to increase the behaviour factor q. In some special cases this is achieved by using base isolation or energy dissipation devices, but generally it is preferable to modify the structure in order to increase its ductility.

It must be emphasized that any effort aiming at the increase of the q-factor is only justified if the design is governed by seismic action. Conversely, if self load, live loads, snow, wind etc. are decisive, and even if earthquake is predominant but seismic loads assuming q=1 are still within acceptable limits, one should resort to the classic strategy. This is also applicable to non-dissipative members and connections of dissipative structures, which must be given an overstrength in order to make sure that none of them will attain its limit state before the dissipative members can develop their full hysteretic capacity.

An indefinite part of the existent structures can be expected to have insufficient ductility, because this concept was generally ignored in the past. What can be done is to identify and to upgrade those which are considered to be the most important.

2 The behaviour of different bracing systems

A considerable research effort has been directed towards the investigation of the behaviour of braced structures under cycling loading. This paper, which is not intended to be a state-of-the-art report, will only mention a few aspects.

Until recently, it was believed that a bracing system satisfying the classic strength and rigidity requirements was as good as any other. In fact, the dissipative capacity differs to a great extent from one system to another.

In a lattice structure, the component members are subjected mainly to axial tension or compression. The ductility depends on the capacity of a part of these members to dissipate the earthquake input energy. First of all, those members must be made of a steel with mechanical proper-

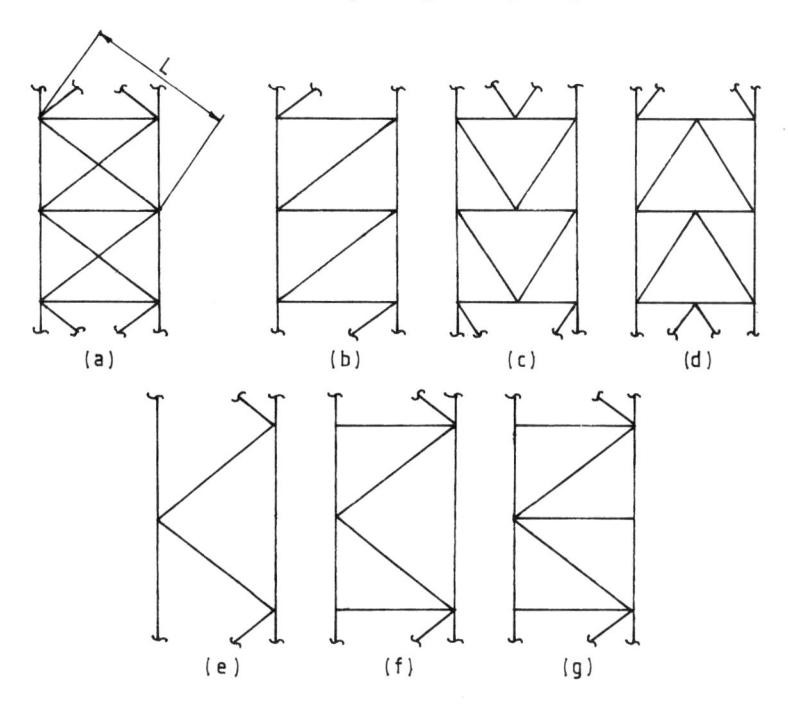

Fig.1. Bracing systems.

ties satisfying the requirements of Eurocode 3 (1993); otherwise, the structure is non-dissipative (q=1). Members subjected to stress reversals can dissipate energy mainly in tension, while hysteretic behaviour in compression is adversely influenced by overall and local buckling. Besides rigidity degradation, brittle fracture of the connections can occur, as well as low-cycle fatigue cracks and rupture of the member. In the case of built-up members, the behaviour is affected by the individual buckling of the component members.

The following discussion is based to a great extent on Mazzolani and Piluso's (1993) synthetic presentation of the behaviour of steel structures in seismic areas.

Diagonal bracing systems, like the X-bracing in Fig.1a, have the best dissipative capacity among the concentric bracings, due to yielding of the diagonals in tension.

The dissipative capacity of the diagonals in compression

is usually neglected, but their non-dimensional slenderness is limited to $\bar{\lambda}$ = 1.5 in EC8 and the Romanian code P100 (1992). The effective length for in-plane buckling can be reduced by achieving rigid connections between the diagonals and the adjacent members. The effective length for out-of-plane buckling depends on the nature of the axial force in the other diagonal (tensile, zero or compressive) and on whether the intersection joint provides continuity for out-of-plane bending of both diagonals. According to the Romanian standard 10108 (1978), in case that one of the diagonals is in tension, the effective length of the compression diagonal is 0.5 L if both diagonals have bending continuity at the central joint, and 0.7 L if one of them is interrupted and connected to a gusset plate which only provides continuity for axial forces.

The N-bracing system (Fig.1b) is worse than the X-bracing because it can dissipate substantial energy only when the diagonals are in tension. It can be equivalent to the X-bracing if two N-bracings are placed symmetrically, to have tension diagonals for both senses of the horizontal loads. Anyway, the effective lengths of the diagonals are greater than in the case of an X-bracing.

The performance of the V- (Fig.1c) and of the inverted V- (Fig.1d) bracings depends on the strength and rigidity of the horizontal member where the two diagonals meet, as have shown Georgescu et al. (1992). Unless this member is a sufficiently strong beam, the tension diagonal cannot attain the yield stress after the compression diagonal has buckled. If the strength of the horizontal member is negligible, the buckling load of the diagonals drops to about 15% of the value corresponding to monotonic loading. This means a much poorer behaviour than that of X-bracings. However, the V-bracing can behave much better than the inverted V if the structure is located in an area of moderate seismicity and if the magnitude of the tensile stresses arising from vertical loads is sufficient to avoid stress reversals under seismic action.

P100 restricts the slenderness to $\bar{\lambda}$ = 0.75. Bertero et al. (1989) suggest that the ratio $P_{cr}/P_y \geq$ 0.8. In terms of EC3, this means $\bar{\lambda} \leq$ 0.792 for buckling curve a and $\bar{\lambda} \leq$ 0.575 for curve c.

Triangular systems (Figs.1e, 1f and 1g) are non-dissipative. Those represented in Figs.1e and 1f (the latter being known as K-bracing) are particularly unfavourable since the lateral displacement of a joint having two diagonals and no horizontal member (which occurs after the buckling of the diagonal in compression) favours the buckling of the vertical member. The system in Fig.1g can become disssipative if two bracings of this kind are placed symmetrically, to have diagonals in tension for both senses of the load.

In all cases, the width/thickness ratio of the component walls of a dissipative member must be limited, as specified in EC3 and EC8.

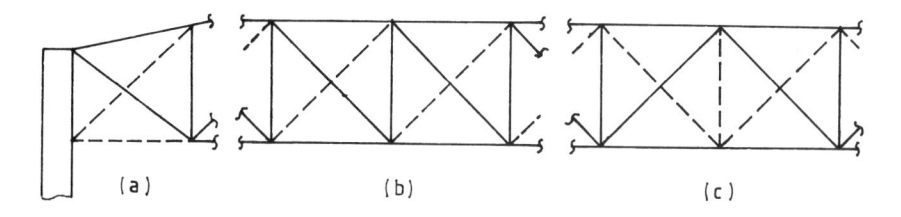

Fig.2. Modification of trusses.

3 Some suggestions for modifying lattice structures

3.1 Trusses subjected mainly to vertical loads

Snow or wind loads being normally more important than earthquake for the design of a roof truss, there seems to be little point in any effort to increase the q-factor of the truss itself. However, for mill-buildings located in high seismicity areas it can be advisable to transform the hinged connections between the trusses and the columns into moment-resistant ones, by adding new elements. In Fig.2 and all the following, the existent members are drawn in solid lines and the newly added ones in dashed lines. It is recommended to introduce new diagonals, at least in the marginal panels of the truss (in order to reduce the compressive forces arising in the bottom chord from vertical loads) and to fix the joints of the bottom chord compression members against out-of-plane buckling. The transformation of the truss-to-column connection (Fig.2a) improves the hysteretic behaviour of the frame. If the dissipative zones are restricted to the columns (as recommended in P100), then all the truss members are active; otherwise, it is usually assumed that compression diagonals are inactive under seismic loads.

Among other categories of trusses that are subjected mainly to vertical loads (self- and live loads), there can be some cases when seismic action is decisive. If increasing the q-factor is assumed to be worthwhile, this can be achieved by introducing additional members to obtain an X-braced structure where tension diagonals are the dissipative elements. One could possibly admit the idea of the bottom chord yielding in tension, but this would lead to unacceptable permanent vertical deflections.

3.2 Lattice towers

Earthquake can be more important than wind for the design of towers carrying substantial vertical loads, especially if they are placed on top (inverted pendulum structures).
Lattice towers are space structures. They must be pro-

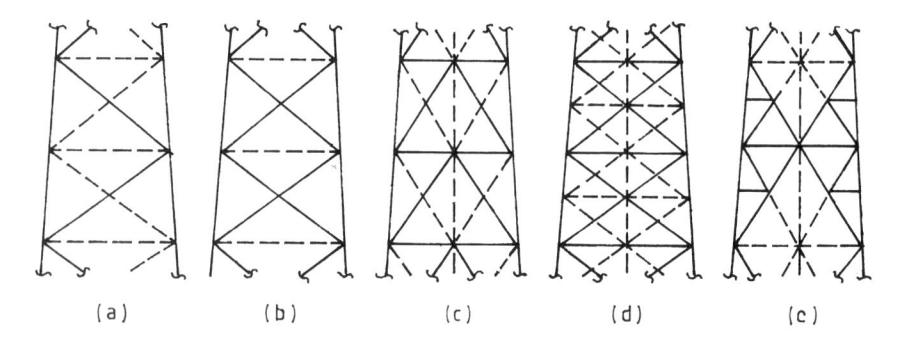

Fig.3. Modification of lattice towers.

perly braced in horizontal planes to prevent cross section distortion. Then new members must be added in the planes of the component trusses to obtain a tension diagonal system (Fig.3).

The efficiency of this modification is greater if the original structure has a low q-factor, like those represented in Figs.3a and 3b, which are transformed into X-bracings. The same solution can be applied to the systems shown in Figs.1f and 1g.

The V- or inverted V-structures can be improved either by strengthening the horizontal members, or by introducing new vertical members and diagonals (Fig.3c).

If the original system is an X-braced structure, the q-factor cannot be increased by introducing new members. The solution shown in Fig.3d is too cumbersome. Strengthening the existent members and their connections seems to be a better alternative.

In the case of Fig.3e, it may be sufficient to add only the horizontal members (to obtain an X-bracing) and perhaps to reduce the effective length of the compression diagonal for out-of-plane buckling. Then the buckling load of the compression diagonal (even degraded by cyclic loading) could possibly be sufficient to prevent the secondary lattice system from becoming a mechanism and to avoid the necessity of introducing new vertical and diagonal members.

According to P100, the slenderness of the component members of inverted pendulum lattice towers is limited to $\bar{\lambda} = 0.6$ for chord members and $\bar{\lambda} = 0.7$ for diagonals.

3.3 Concentrically braced frames

The same solutions should be applied to braced frames as to lattice towers. The difference is that in this case one could possibly accept plastic zones in the frame members (provided their rotation capacity is sufficient) and take into account a collaboration between the bracings and the frames in energy dissipation.

If the solutions based on the addition of new bracing members are not feasible owing to functional or aesthetic reasons, the existent bracings should be removed and replaced by eccentric bracings. Due attention must be paid in this case to the correct conformation of the link beams.

3.4 Bracings of one-storey mill-building structures

Vertical bracings between columns in the longitudinal direction are intended to take over the longitudinal forces caused by crane braking, earthquake and wind. Crane braking or earthquake govern the design of the bracings under the crane girders (depending on the crane hoisting capacity and working conditions on one side, and on the ground acceleration and seismic behaviour of the structure on the other side). When comparing these forces, one should make sure that the q-factor assumed in the analysis reflects the actual dissipative capacity of the structure.

Inverted V-bracings are very popular. In the case of cranes with low or medium hoisting capacity the diagonals are directly connected to the runway girder (Fig.4a), while in the case of heavy cranes it is preferred to have an independent horizontal member (Fig.4b), to avoid the transmission of vertical forces from the runway girder to the

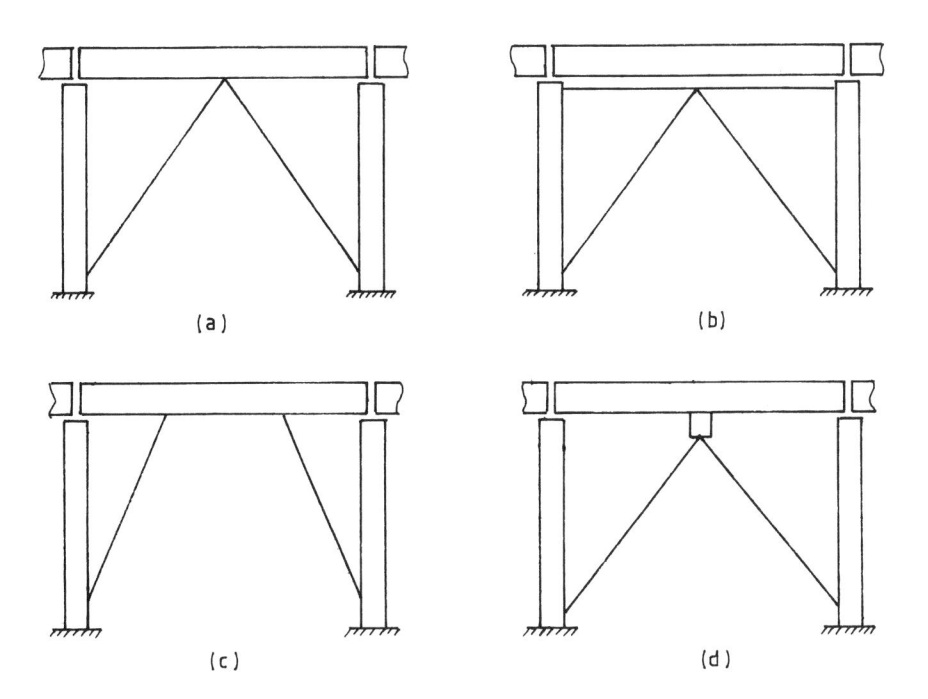

Fig.4. Bracings under the runway girders.

diagonals. The disadvantage of this system is its poor dissipative capacity. If seismic action is decisive for design, the alternatives for strengthening are:

(a) to strengthen the existent horizontal member;
(b) to connect the central joint to the runway girder;
(c) to transform the hinged runway girder-to-column connection into a moment resistant one;
(d) to replace the V-bracing by an eccentric bracing system (Fig.4c or 4d);
(e) to replace the V-bracing by an X-bracing.

The first three variants are rather complicate; especially (b) and (c) can favour the risk of fatigue failure. The eccentric bracing system shown in Fig.4c is not as good as in the case of eccentrically braced frames because the runway girder is generally too strong to act as a dissipative link beam. An eccentric system including a vertical link beam (Fig.4d) or a dismountable dissipative device is preferable. Replacing the V-bracing by an X-bracing is also a good solution, but its disadvantage is that it obstructs the braced bay. The two columns connected by an X-bracing may need strengthening in some cases, to make possible the yielding of the diagonal in tension.

The design of bracings placed over the runway girders is governed by earthquake only in some particular cases, when the roofing is very heavy (e.g. reinforced concrete slabs). Their strengthening is a simpler problem.

In the case of tall and rather short mill-buildings with heavy cranes, an adequate system of horizontal and vertical bracings should be introduced in order to take advantage of the spatial collaboration between the plane frames.

4 Strengthening and repairing of individual members

As far as possible, the centroid of the strengthened cross section should coincide with the initial one.

Members which can be subjected to tensile forces and show visible overall or local deformations should be straightened or replaced. Locally deformed zones can be slotted by flame cutting, straightened and rejoined by butt welding. It is recommended to rely only on cover plates for the transmission of forces in zones which have undergone such operations. Special attention shall be paid to repairing of cracks (if any), by using an adequate technology.

In the case of a compression member, the buckling load must be determined taking into account the influence of the measured initial deformations, as well as those arising from the welding of strengthening elements. Rebrov (1988) has suggested a simplified method for estimating the buckling load when strengthening elements are welded to a mem-

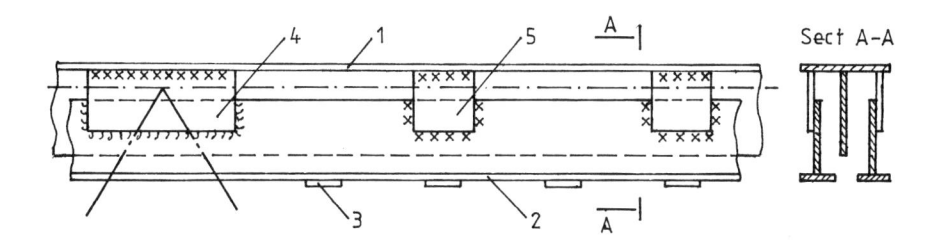

Fig.5. Repairing of a deformed truss chord.

ber which is subjected to a considerable axial load (up to 80% of the buckling load if the structure is statically determinate and 60% if it is indeterminate). The results of tests carried out at INCERC Timișoara on welded T sections (unpublished report, 1991) are in good agreement with that simplified method.

Non-dissipative members which are excessively deformed and cannot be straightened must be repaired in an unloaded state. Fig.5 shows a possible solution for repairing the top chord of a truss which is only accessible from the inside if it is wished to leave the roofing in place. The existent chord (1) as well as the strengthening elements (2) are T sections. The latter are connected by shop welded battens (3). Battens (4) are shop welded to member (1) and site welded to members (2), while battens (5) are site welded to members (1) and (2) as well, to accommodate to the geometry of the deformed chord. Welded T sections subjected to axial compression up to buckling and then upgraded by this method were tested at INCERC Timișoara (1991). The buckling loads of the repaired members (subjected to eccentric compression due to the centroid shift) were greater than those of the initial members. It must be mentioned that the initial members, although deformed, collaborate with the strengthening elements in carrying the load.

5 Strengthening and repairing of connections

The quality of the existent connections (including stitches between the component elements of built-up members) shall be thoroughly controlled and all necessary repairs be accomplished. The design and construction rules stipulated in EC3 and other codes must be observed.

Welded connections are usually strengthened by welding. For fillet welds, the simplest way is to increase their length and/or thickness. The results of monotonic loading tests have shown that their strength is not influenced by the force acting on the connection when strengthening takes place; however, an unloading (at least partial) before this intervention is advisable.

6 Conclusions

The strategy for strengthening and/or repairing a given structure depends in the first place on whether the design is governed by seismic action and whether ductility requirements are essential. In that case, the designer's effort should aim at the reduction of the seismic loads by increasing the q-factor. For lattice structures, the best way to attain a higher ductility is to modify the bracings in order to obtain a tension diagonal system.

Damaged members must be replaced if they are assumed to be dissipative elements, but it is acceptable in many cases to repair non-dissipative members and connections.

While a permanent reduction of the self load and of the live loads acting on the structure may be justified in some cases, a temporary unloading before the intervention is always advisable.

7 Acknowledgement

The authors are grateful to Professor Victor Gioncu for the useful discussion they had on the problems dealt with in this paper.

8 References

Bertero, V.V. Uang, C.M. Llopiz, C.R. and Igarashi, K. (1989) Earthquake simulator testing of concentric braced dual system. **J. Struct. Eng.**, 115, 1877-1894.

Georgescu, D. Toma, C. and Goşa, O. (1992) Post-critical behaviour of 'K' braced frames. **J. Construct. Steel Research**, 21, 115-133.

Commission of the European Communities (1993) **Eurocode 3: Design of Steel Structures.**

Commission of the European Communities (1993) **Eurocode 8: Earthquake Resistant Design of Structures.**

Mazzolani, F.M. and Piluso, V. (1993) **ECCS Manual on Design of Steel Structures in Seismic Zones.**

Ministry of Public Works, Romania (1992) **P100. Code for the Earthquake Resistant Design of Civil, Industrial and Agricultural Buildings** (in Romanian).

Rebrov, B.S. (1988) **Strengthening of Metal Structures** (in Russian). Stroiizdat, Leningrad.

Romanian Institute of Standardization (1978) **STAS 10108/0. Design of Steel Elements** (in Romanian).

63 USE OF STEEL STRUCTURES FOR CONSOLIDATING A MULTI-STOREY REINFORCED CONCRETE STRUCTURE DAMAGED BY SEISMIC ACTIONS

C. DALBAN and S. DIMA
Civil Engineering Institute, Bucharest, Romania
R. ANGELESCU, V. MUSTATA and A. LEONTE
Institute of Power Studies and Design, Bucharest, Romania

Abstract
The aim of this work is to present strengthening measures carried out at the main building of TPP (Thermal Power Plant), Brazi-Prahova, damaged during the 4 march 1977 earthquake.
Keywords : Strengthening of r.c. structures, steel members used for strengthening.

1. Introduction

The main building of TPP Brazi has been designed in 1959-1960 and it consists of the turbine house, the intermediate building and the boiler house.
  The structural elements (fig.1) are cast-in-place, reinforced concrete frames on both directions, except for the roof structure over the turbine house, which is in steelworks.
  The transverse frames have two unequal spans: 30 m between grid lines A and B (turbine house) and 9 m between the grid lines B and V (the intermediate building), and have different elevations.
  The longitudinal frames (grid lines A, B, V), are built of 6 m bays and formed out of three longitudinal segments of eight bays each. The engine room (30 m span) has a composite structure: cast-in-place reinforced concrete columns and longitudinal beams, and steel roof structure (trusses, purlins, bracings). Connections of trusses to the columns are pinned at 22.0 m elevation.
  The intermediate building (9 m span) structure consists of cast-in-place reinforced concrete transverse and longitudinal frames, and slabs placed at 31.68 m (the roof), 20.0 m, 12.7 m, 8.0 m and 4.0 m elevation.
  The slab over the basement (+0.00 m) rests on beams supported by their own columns. The basement extended under the whole building, has a height of approximately 3.5 m.
  Structural columns are supported by pad foundations having the upper face located between 3.50 m and 3.80 m bellow the 0.00 m surface.

---

1) Dr. Eng., Professor
2) Lecturer
3) Dr. Eng., Deputy Chief Eng.
4) Dr. Eng. Design Group Chief Eng.
5) Eng. Designer

Behaviour of Steel Structures in Seismic Areas. Edited by F. M. Mazzolani and V. Gioncu.
Published in 1995 by E & FN Spon, 2-6 Boundary Row, London SE1 8HN. ISBN: 0 419 19890 3.

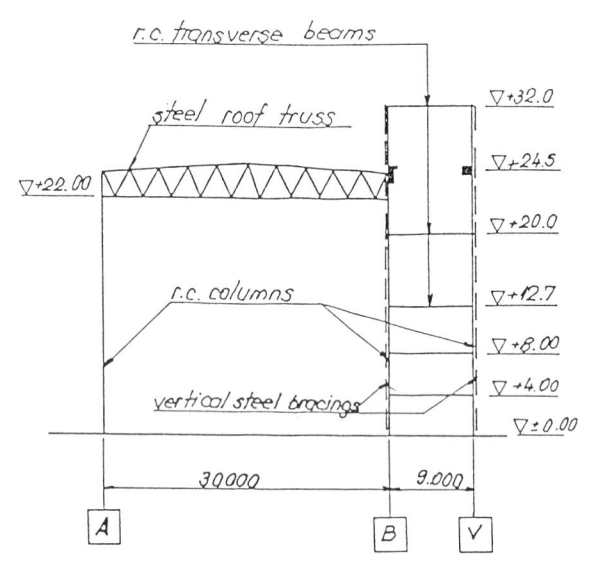

Fig. 1 — Transverse frames

2 Damages caused by the 1977 earthquake

Heavy damages have occurred in the structural elements:

- horizontal cracking of some columns at connections with transverse beams and ruin of reinforcement concrete layer covering
- inclined cracking of some transverse beams at the supports
- failure of some longitudinal beams at connections with columns.

Repairs carried out immediately after the 1977 earthquake had a local character and mainly consisted in strengthening the damaged areas of the columns by metal encasing. This did not ensure the initial stiffness of the members. The research Institute INCERC Bucharest presumed a decrease of stiffness for the transverse frame of 10 to 15 percent for columns, and of 15 to 35 percent for beams.

The following eartquakes on 31.08.1986 and 30.05.1990 caused no further damage but showed up the necessity of an overall strengthening action.

3 Safety state evaluation

Stresses calculation in the structural elements of the damaged building were carried out by using the IMAGES-3D program. The considered loads have been brought up to date for the year 1993. Relevant load combination assumed for calculations has included seismic effect.

The evaluation work has put forth the fact that substantial sub-unit safety factors ($n = 0.7.....0.1$) are a general characteristic of the building. This fact was shown by 1977 earthquake.

Strength analysis performed on transverse frames have come out close to the actual safety requirements, and only small local strengthening interventions were required (epoxy resin injection, metal encasing).
On longitudinal direction more severe deficiencies have come out:

- strong deformability (along longitudinal frames), elastic deflection at uppermost level has reached ~ 20.0 cm (~ h/160)
- inadequate connections at joints, which have been therefore damaged
- improper reinforcement (poor anchorage) on the bottom face of the beams at supports area which led to beams failure.

In order to increase safety level on longitudinal direction, an overall consolidation of concrete frames was performed using steel members. A general bracing is this obtained of "Π" form (see fig. 2).

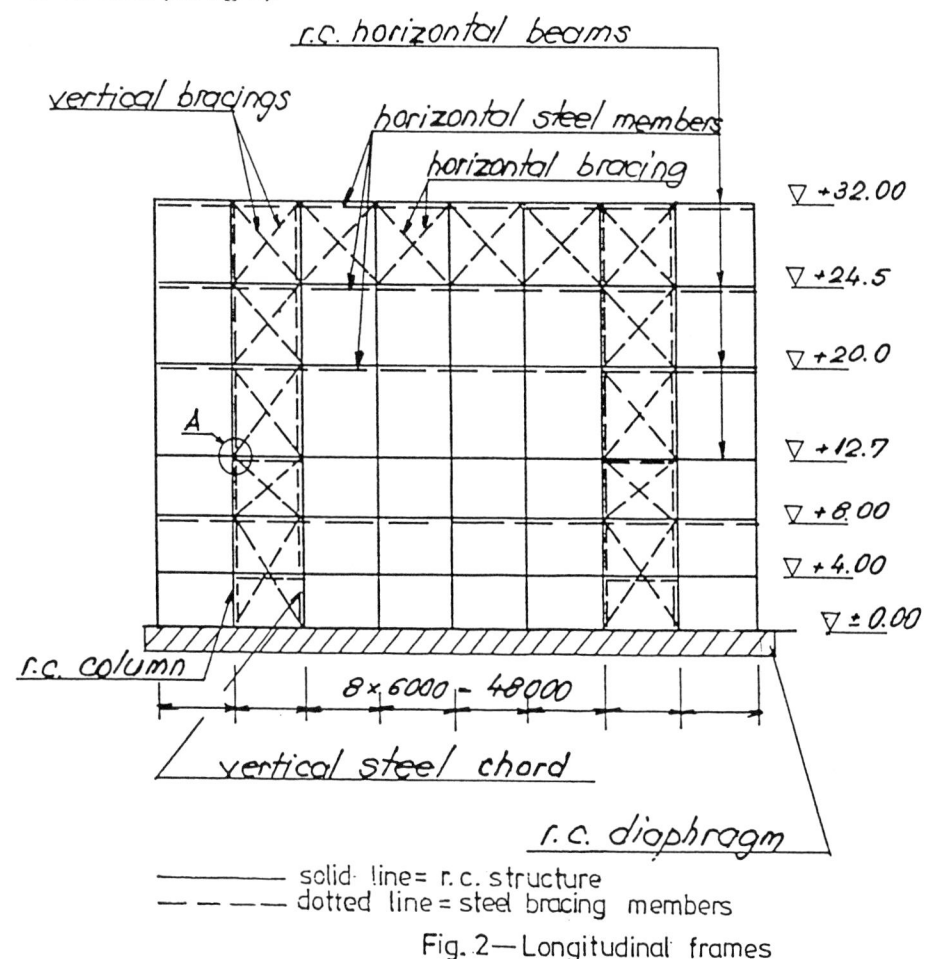

Fig. 2—Longitudinal frames

4 Strengthening works

4.1 General aspects

Local strengthening of the damaged elements and connections proved to be sufficient for the transverse frames. For the longitudinal frames a change in the structure was required.

At the same time, considering the decision of restraining structure life to a determined period of time, it has been possible to decrease the class of importance from the first class to the third class (according to P-100 Code). The importance factor for the building may be this way considered equal to 1.0 instead of 1.4, while the safety factor against earthquakes takes the minimal value 0.5 (instead of 0.7 which is the value for buildings in the first class of importance). A less severe strengthening work and lower costs were achieved.

4.2 Aspects referring to the consolidating solution adopted

Repair of the columns areas damaged by the earthquake was made using metal encasing and trimmers welded to the longitudinal reinforcement. Some of these repairs were carried out just after the 1977 earthquake, but their efficiency has only partially succeeded to bring the structure back into initial state (Young modulus of concrete being decreased).

On the purpose of increasing safety level, an intervention on concrete framework has proved necessary, both by consolidating damaged members (a sufficient measure for traverse frames) and by changings in structural system (a measure performed on longitudinal frames which are much more sensitive to earthquake loading).

Repair of heavy damaged areas of the columns produced during 1977 earthquake, is performed using metallic encasing and trimmers, together with suitable measures of ensuring collaboration between concrete and steel members (concrete covering removement and encases welding to longitudinal column reinforcements).

Quake-strength increasing of column rows on longitudinal direction by structural system modification of longitudinal frames, was the main purpose of strengthening activity. This may be achieved by introducing two vertical "**X**" bracings for each building segment along the longitudinal frames. The longitudinal column rows are connected to these vertical bracings by using horizontal longitudinal members (grid line A). Longitudinal "X" bracing were developed on the height between two horizontal connections at the upper level of longitudinal frames (grid line B and V - fig. 2). This solution does not induce stresses in the beams of existing structure. (see fig. 2a)

On some areas, because of insufficient available space, a inverted "**V**" bracing was adopted, having a central joint on a metallic horizontal member located bellow the reinforced concrete beam.

Adopted cross section for horizontal members and bracing consists of channels, lattice and battens connected.

Reinforced concrete columns of the vertical bracing are enclosed all over their height by four built-up hollow sections, each consisting of two welded channels (U 30) placed on column wide sides edges.

These hollow section chords are connected together and to the reinforced concrete columns by metal trimmers built using channel sections and supplementary fastened on concrete column by CONEXPAND $\varnothing$ 20 bolts. (fig.3). This solution is allowing for an

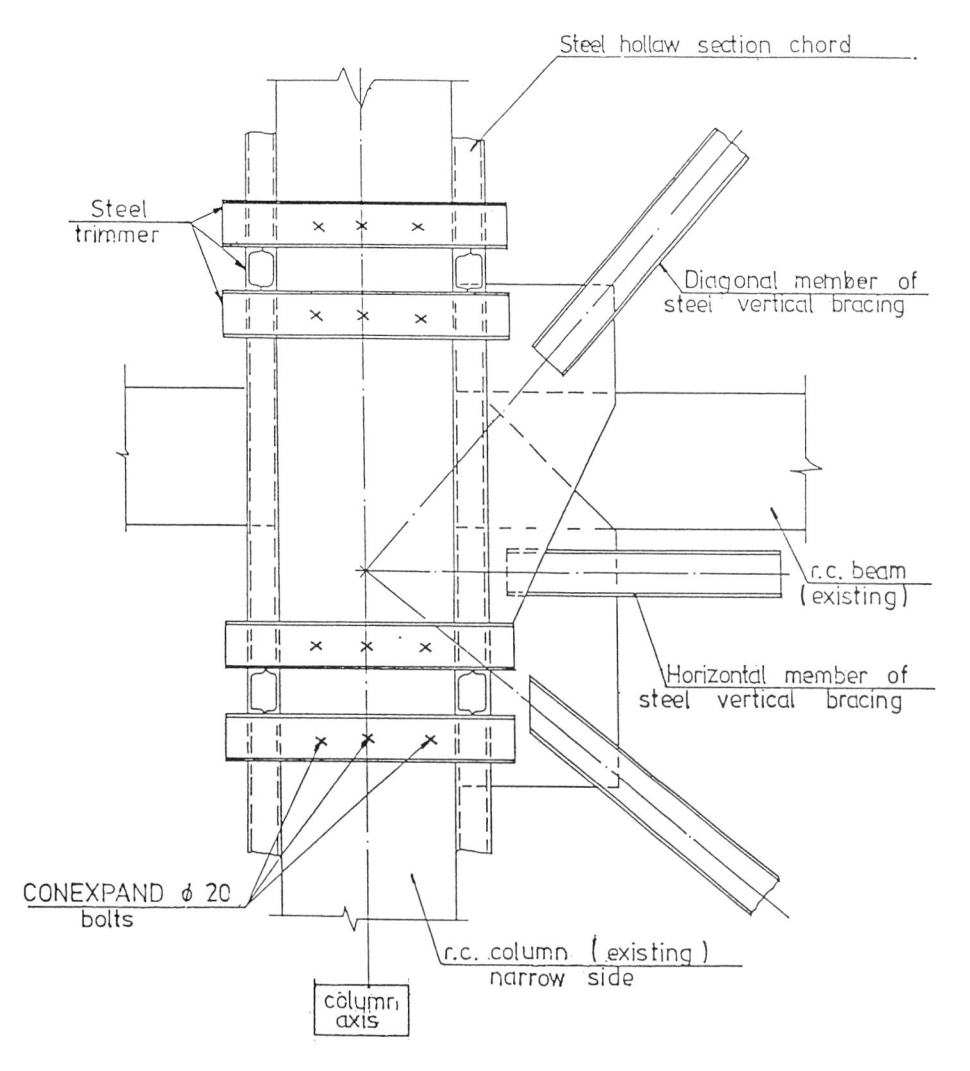

Fig. 2a—Detail A

equilibrated distribution, as axially as possible, of horizontal longitudinal loading to vertical bracing.

The connections and spliced joints of bracing elements are realised by welding.
Changing initial dominating bending loading of columns in the unbraced structure, with axial loading significant in the columns belonging to vertical bracing only, didn't lead to the necessity of modifying existing foundations. Their initial dimensioning has permitted to transmit the new solicitations to the ground.

Anchoring to the ground of the four built-up hollow sections of vertical bracing which are framing concrete columns was performed as follows:

Fig. 3—Detail

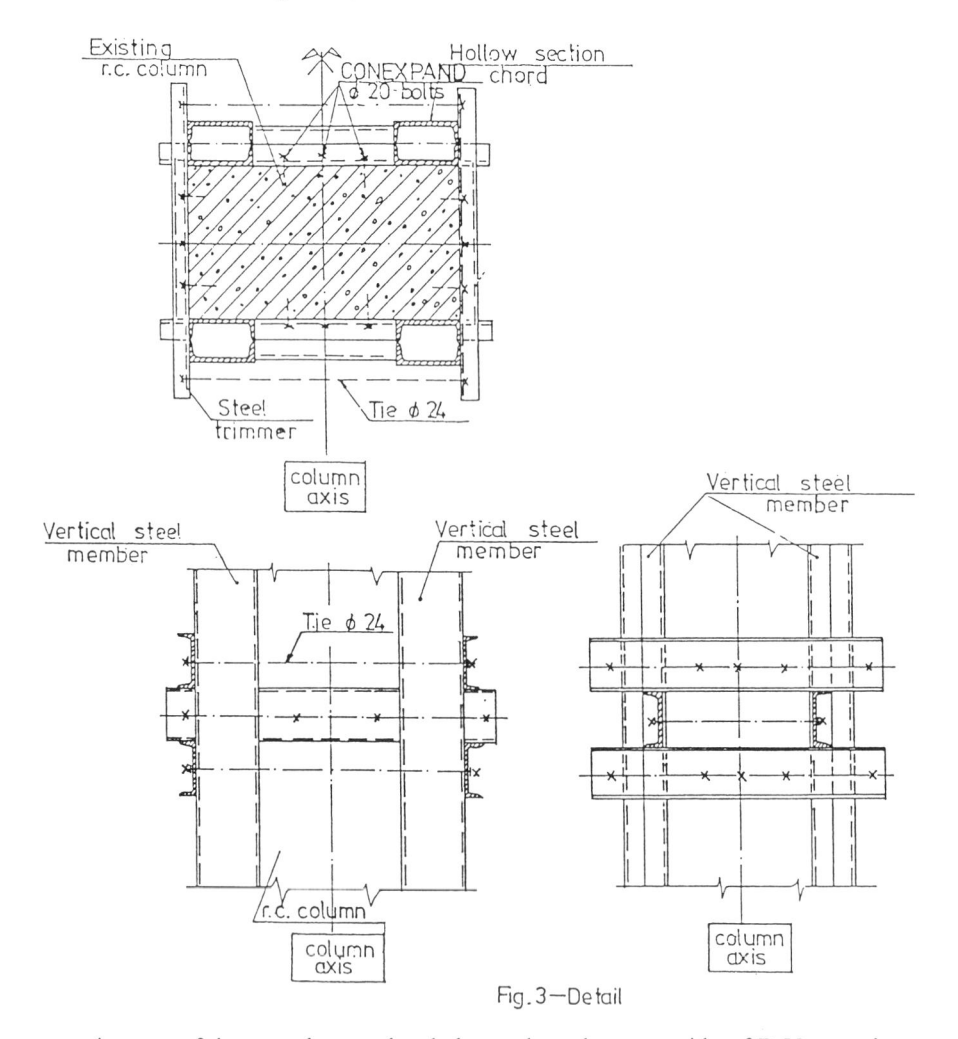

- in case of the two shorter chords located on the outer side of B-V span, by anchoring chords on two trimmers situated at upper and lower level of basement slab.
- in case of the two longer chords, situated on the inner side of B-V span, by chord prolongation to reach upper foundation level; a reinforced concrete strengthening bush of 75 cm height and 40 cm width, cast over existing foundations and anchored upon them by 36 mm diameter steel reinforcements introduced in drilled holes epoxy resin filled up, has served in anchoring the longer chords (fig. 4).

4.3 Aspects referring to structural analysis
Stress calculation was made using IMAGES-3D programme. Static and dynamic

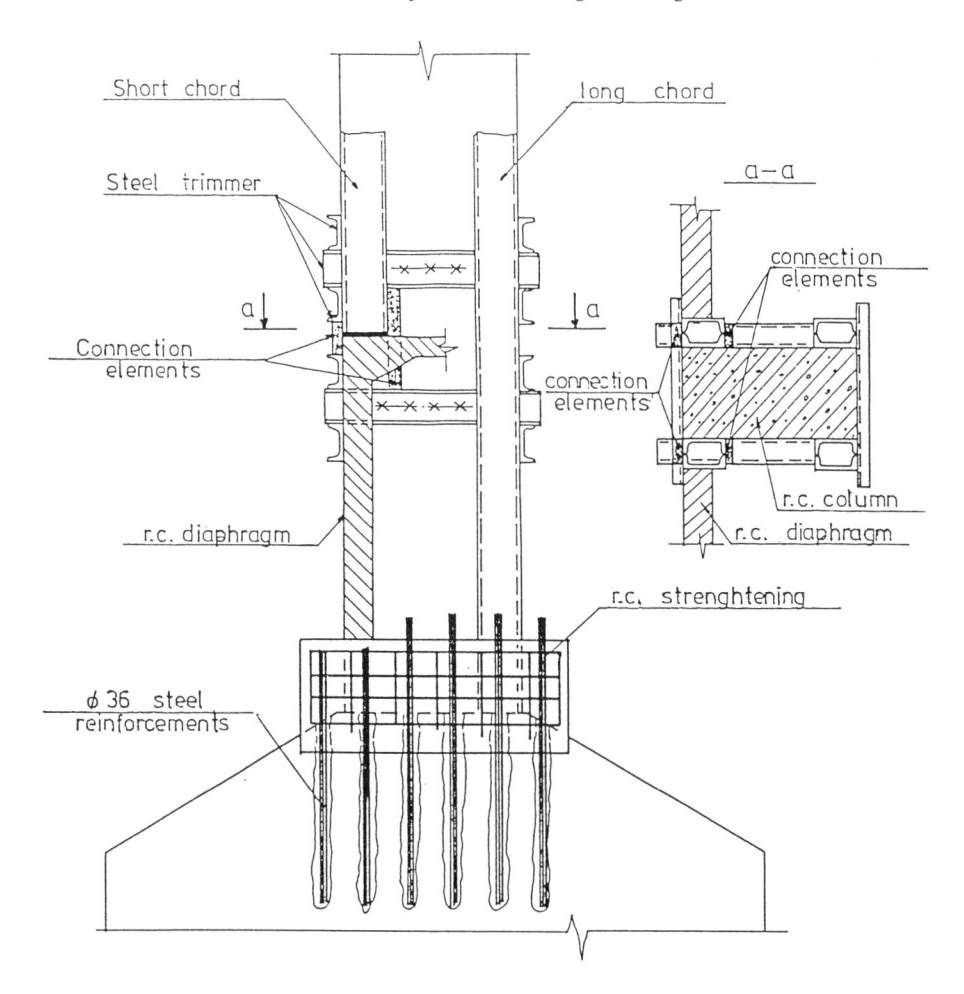

Fig. 4 — Detail

response of main building framework were determined on plan models. The seismic horizontal force at the column base level has been evaluated using elastic-plastic time history approach. Eartquake safety level (R), related to this base strength (for transverse frame) was calculated according to Romanian Seismic Code P100-92 .

Strengthening steel members were designed using the "plane longitudinal models".
For the strengthened structure, the following types of frames (structural analysis using FEM "beam elements") were considered:

1 - equivalent steel section frames (according to P83/81 Code) for the vertical bracing afferent columns; reinforced concrete columns for the rest
- reinforced concrete beams with adjacent steel beams were necessary (fig. 2)

Connections between the reinforced concrete structure and steel strengthening members have been assumed rigidly connected to reinforced concrete structure.

Fundamental vibration period of the building had decreased from 1.851 sec. (for unstrengthened structure) to 0.634 sec.

2 - frames for which the steel and the concrete members have been distinctly modelled, assuming that a perfect structural cooperation is achieved at joints. The same value as before was obtained for vibration period.

Modelling of elastic connections between concrete and steel elements (using distinct rigidities springs) which should allow for unequal displacements of these elements in connection zone has led to inconvenient situations. Realising of perfect collaboration between steel and concrete has proved necessary (and also establishing a computing model as realistic as possible for the actual connection, which should exhibit a certain elasticity degree).

3 - model *2.* was used in the design of steel strengthening elements, in which reinforced concrete frames should undertake vertical (gravity) loading while steel bracing and concrete columns not included in bracing should entirely undertake earthquake. Only longitudinal steel beams were taken into account in this model.

A 10% diminished elasticity modulus was considered for the concrete, according to INCERC estimations.

Loading on cross-sections induced in the structure by earthquake were calculated using spectral analysis method presented in Romanian Seismic Code P100-92.

Seismic force according to P100-92 Code, was computed using the following relation:

$$S = \alpha \; k_s \; \beta \; \psi \; G$$

where :

$\alpha = 1.0$ is the importance factor for buildings of class 3

$k_s = 0.25$ is taken according to the inelastic design response spectrum available for the Brazi-Ploiesti location

$\psi = \dfrac{1}{q} = 0.3$ is ductility factor established based on the assumption that the new, strengthened structure is a composite one (steel + concrete) and theconcrete elements, do not fully respect actual provisions required for a satisfactory ductility.

When establishing physical and mechanical characteristics of former reinforced concrete elements, test results and conclusions from INCERC - Bucharest (8676 / 1990 contract) were considered.

5 Conclusions

The following aspects may be emphasised:
  a) The steel strengthening solution provides:

- a small size, which is convenient considering the fact that a greatdeal of structural volume is occuped by equipment.
- welding connection are the only interventions to be carried out in situ
- considering a proper structural concept a good collaboration may be obtained between concrete and steel members

b) Changing the structural system in longitudinal direction by introducing two vertical bracing connected together at the upper side to longitudinal members (beams on grid line A, bracing on grid lines B and V) results in:

- decreasing bending moments on columns on longitudinal direction
- increasing horizontal strength along the building; the horizontal seismic forces are mainly resisted by bracings and not by r. c. column bending.
- decreasing of elastic deflection in longitudinal direction for about 8 times (2.5 cm ~ h/1280 instead of ~ 20.00 cm ~ h/160)

c) The strengthening works, which is realistic in view of the available limited space, has provided minimum level of seismic safety (R_{min} = 0.5) on a limited term, considering the decreasing of building importance class from 1 to 3.

6 References

1. Code for Aseismic Design of Residential Buildings, Agrozootechnical and Industrial Structures (P100-92) - Romania, English Version
2. Eurocode 8 - Design provisions for earthquaqe resistance of structures - Part 1 - 1 : General rules - seismic actions and general requirements foe structures. Final draft, oct. 1993
3. Eurocode 8. Part 1.4. Repair and Strengthening. Draft. oct. 1993

64 ON THE SEISMICAL BEHAVIOUR OF TWO BUILDING MODELS WITH SUSPENDED STOREYS

I. DIMOIU and A. BOTICI
Civil Engineering Faculty, Timisoara, Romania

Abstract
Building having suspended storeys on a reinforced concrete core can give a more favourable response to the earthquake events than the usual one. The behaviour of inertial masses is similar to the behaviour of a common pendulum. A very small part of the R/C core motion is transfered to the inertial a mutual masses.

Two different steel models have been constructed and tested in order to receive a dynamical response both in quantitative and qualitative manner. The both models are investigated by using the spectral analysis method and the time history procedure. Different practical realisation of beam - core connection are considered: rolled joints, hinged joints and welded joints, respectively.

The both models have been constracted of the 1: 10 scale. They have been tested on a shaking table with one degree of freedom in harmonic motion. The dynamic response is emphasized by displacement of the points noticed at the floor levels.

Keyword: Suspended Storeys, Rolled Hinged, Welded, Joints, Reversed Pendulum.

1 Bearing structure

Steel and concrete are efficiently used if steel is subjected to tension effort while concrete undergoes compression. Some highrise buildings are made up of few reinforced concrete cores and a steel space structure on the top of the cores. The suspended storeys are hung of the steel structure by means of steel bars or cables. The above mentioned steel bars undergo axial effort, so their cross areas are of reduced magnitude. No instability phenomena do occur. The steel space structure support a large concentrated loads which act at the steel truss nodes. A lot of truss members are tensioned but some of them are compressed.

Buildings with suspended storeys can record a more favourable response in active earthquake area than the usual ones. Sometimes is it also possible to obtain a dangerous response under special circumstances. That is related to the steel beam connection of the floor to the R/C core.

Behaviour of Steel Structures in Seismic Areas. Edited by F. M. Mazzolani and V. Gioncu. Published in 1995 by E & FN Spon, 2-6 Boundary Row, London SE1 8HN. ISBN: 0 419 19890 3.

A reverse pendulum is attained if the beams of the floor are built in the R/C core. Such building structure had been shown the worst seismic response due to its discontinuity in lateral rigidity at the ground floor. A reverse pendulum offers a decreased rigidity at just the level which undergoes the most shear force.

In the case of building whith hung storeys, the inertia storey mass behaviours like a common pendulum if the storey steel beam has a free end or a roller end on R/C core. A very small part of the core motion is transferred to inertia mass during the earthquake. The storey steel connection to R/C core can be realised as a damping roller support. In such a case a very great amount of induced seismic energy is dissipated and the inertia mass motion is essentially reduced.

2 Studed models

2.1 Umbrella type building model

The model has only three hung storeys, Fig.1. The R/C core is modelled by a steel tub bar formed of two equal angles L 45.45.5 and two U8 channels welded one with the other. The steel space structure at the top of the core is made of two U8 profile. The U profiles are screwed at the top of the core. The columns between two succesive floors are of 3 mm diameter cable. The current floor beam is a 5x40 mm steel plate. There is a lumped mass at the beam – tie rod connection.

Three different types of beam core connection are considered shown in Fig. 3. They are: rolled connection damped rolle connection and built in connection respectively.

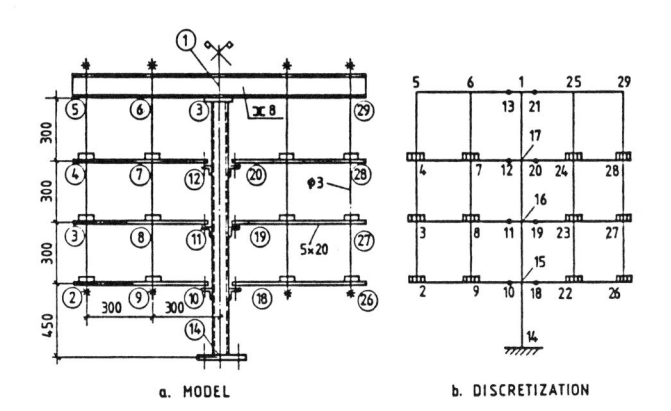

FIG. 1 UMBRELLA TYPE BUILDING

2.2. Model of building with two reinforced concrete cores

The Fig.2 is illustrating the model of a building with two R/C cores. Every core is destinated for lift cage and staircase. The model is three storied. The R/C core is modelled of a tub steel bar, formed of two equal angles L 45.45.5 and two U8 channels welded one with the other.

The steel superstructure at the top of the core consists of a steel beam formed at two U8 profile. The beam is screwed at the top at the core at every of its end. This steel beam catches the hung storeys. The column of the model between two succesive floors are of 3 mm diameter cable. The current floor beam is a 5x40 mm steel plate.

The above mentioned connections between floor beam and core are considered Fig.3. The lumped masses are fixed at the beam - tie rod connections.

3 Studing of seismic respons

3.1. Spectral analysis

The current design activity of the common building structure is based on spectral analysis. Thus the models have been investigated by using this procedure. The acceleration spectrum values Sa/g are noticed in the Table 1 corresponding to the period and 0.02 damping fraction.

Table 1. Acceleration spectrum values.

Period	sec	.00	1.50	2.50	3.00
Sa/g		.125	.125	.050	.050

In order to compute seismic response the discretized models of the Fig.1 and Fig.2 have been used. A set of short beam elements was located near the cores to carry out a detailed

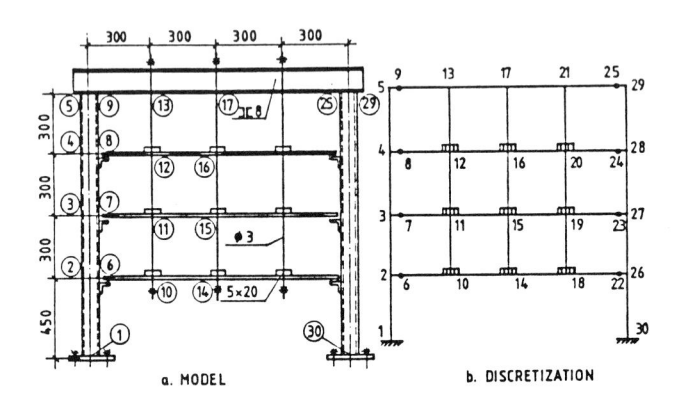

FIG. 2 FRAME TYPE BUILDING

research concerning the influence of connection rigidity on the seismic response.

Modal and spectral displacements in the 1 st oscillating mode are given in the Table 2 and Table 3 respectively. They are refered to the umbrella type building and the frame type building respectively.

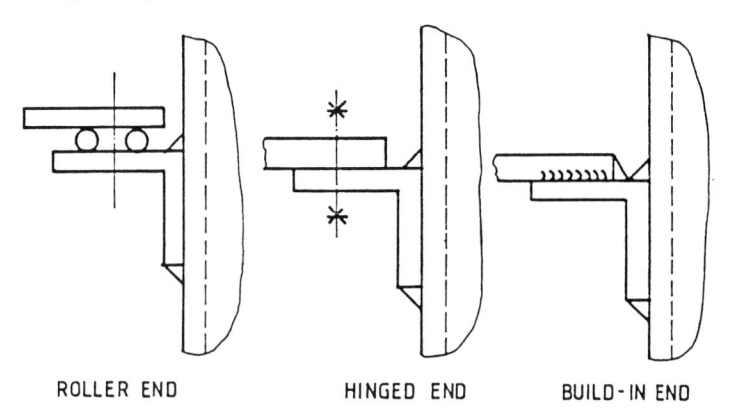

FIG.3 TYPES OF FLOOR BEAM-CORE CONNECTION

Table 2   Modal and spectral displacements

	Current Beam - Core Connection					
	Roller End		Hinge End		Rigid End	
Period	T = 1.51s.		T = 0.25s.		T = 0.25s.	
Nodes	Mod	Spec.	Mod.	Spec.	Mod	Spec.
0	1	2	3	4	5	6
2	0.514	8.51	0.278	0.15	0.278	0.15
3	0.428	7.08	0.3453	0.19	0.343	0.19
3	0.270	4.46	0.389	0.22	0.389	0.22
10	0.514	8.51	0.276	0.15	0.276	0.15
11	0.428	7.08	0.341	0.19	0.341	0.19
12	0.270	4.46	0.378	0.22	0.378	0.22
15	0.010	0.17	0.378	0.22	0.378	0.22
16	0.008	0.14	0.347	0.19	0.341	0.19
17	0.006	0.11	0.276	0.15	0.276	0.15

A few nodes have been equipped with dampers. The damping value has been changes in three stages. Thus, the magnitude of horizontal stiffness was of 500 N/m, 1000 N/m and 2000 N/m respectively. The following nodes of the umbrella

Table 3 Modal and spectral displacements

	Current Beam - Core Connection					
	Roller End		Hinge End		Built in End.	
Period	T = 1.51s.		T = 0.25s.		T = 0.25s.	
Nodes	Mod	Spec.	Mod.	Spec.	Mod	Spec.
0	1	2	3	4	5	6
2	.0029	.0396	.321	.060	.321	.060
3	.0037	.0514	.393	.074	.393	.074
4	.0046	.0632	.446	.084	.446	.084
6	.596	8.211	.321	.061	.321	.061
7	.494	6.808	.394	.075	.394	.075
8	.308	4.243	.445	.084	.445	.084

building model are equipped with dampers: 10, 11, 12, 18, 19 and 20. The Table 4 ilustrates the first three natural shapes of free oscillations.

The first row presents the natural shapes of the model having no dampers. The second row illustrates the natural shapes of the model equipped with the same kind of damper at every storey. On the third row of Table 4 there are different dampers at the storeys. The horizontal stiffness of the dampers increases downward. On the fourth row of the Table 4, there are different dampers too, but the horizontal stiffness of dampers decreases downward. One can see various natural shapes in the vertical sens or horizontal sens. They are related to the magnitude of damper stiffness and the disposal of the damper devices.

The modal and spectral displacements of the damped umbrella model are noticed in the Table 5.

The frame model has been equiped with dampers too.

The damper devices have been placed at the following nodes of the actual model: 6, 7, 8, 22, 23 and 24. The magnitude of damping has been changed in three stages. The horizontal stiffness of dampers was: 5000 N/m, 1000 N/m, and 2000 N/m respectively. The modal and spectral displacements at few nodes of the damped model are noticed in the Table 6.

3.2 Spectral analysis remarks

The seismic response of the both models considered is similar to a common pendulum when the current beam - core connection is a roller one. The period at free oscillations is higher than the period attained in the other cases. The kinetic energy of the core motion is not transffered to the suspended inertia masses. The nodal displacements and the spectral displacements are of a small values. The small inertial forces

Table 4   MODAL SHAPES OF UMBRELLA TYPE

Table 5   Modal and spectral displacements for damped model

	Horizontal Stiffness kN/m					
	$k_{xx}=500$		$k_{xx}=1000$		$k_{xx}=2000$	
Period	T = 0.27s.		T = 0.19s.		T = 0.15s.	
Nodes	Mod.	Spec.	Mod.	Spec.	Mod.	Spec.
2	0.512	0.277	0.507	0.143	0.0035	0.0004
3	0.428	0.232	0.428	0.120	0.0172	0.0020
4	0.273	0.148	0.280	0.079	0.0997	0.0118
10	0.509	0.276	0.502	0.141	0.0034	0.0004
11	0.426	0.231	0.424	0.119	0.0168	0.0020
12	0.272	0.141	0.277	0.079	0.0976	0.0116
15	0.015	0.008	0.028	0.008	0.577	0.0686
16	0.012	0.006	0.020	0.006	0.469	0.0558
17	0.009	0.005	0.106	0.005	0.362	0.0430

Table 6   Modal and spectral displacements of damped model

	Horizontal Stiffness kN/m					
	$k_{xx}=500$		$k_{xx}=1000$		$k_{xx}=2000$	
Period	T = 0.141s.		T = 0.127 s.		T = 0.109s.	
Nodes	Mod.	Spec.	Mod.	Spec.	Mod.	Spec.
2	.313	.046	.306	.036	.291	.026
3	.388	.057	.381	.045	.368	.032
4	.443	.065	.441	.053	.437	.038
6	.313	.046	.302	.036	.291	.026
7	.388	.057	.382	.045	.368	.032
8	.443	.067	.441	.053	.437	.038

are the reason of the above mentioned reduced displacements.

The smallest values of the spectral displacement attained in the case of dampers are imposed at the ends of the floor beams. The dampers must be arranged in a special way to obtain the most favourable seismic response.

The inertia mass displacements are greater than displacement in the case of a weak stiffness of the dampers (k_{xx} = 500 N/m). The displacements of suspended masses are smaller than the core - node displacements in the case of an increased stiffness of dampers (k_{xx} = 2000 N/m). In other words the damping must be imposed by means of an optimum design calculation.

Concerning the other types of beam core connection like built in, or hinged ones, the behaviour of the models is changed. The umbrella type model becomes a reverse pendulum.

The two - core type model passes to a common frame structure with a rigidity gap at the ground floor. The lack of lateral rigidity of the two models are placed at just the level where the shear seismic force is the greatest one. The induced energy reaches the core of the model and it is transferred to the inertia masses. There are not significant differences between the displacements of the inertia masses and the corresponding nodes of the cores.

4 Time history analysis

4.1. Time history parameters

An harmonic acceleration diagram has been used in the time history analysis of the models investigated. The time - acceleration pairs defining ground motion in horizontal direction are moticed in the Table 7. The time increment equals 0.01 sec. The period of ground motion is 0.13 sec. The magnification factor applied to the acceleration values is 1.23 m/s^2 The time increment for priting out the nodal displacements is of 2 step.

Table 7   Acceleration of base motion

Step	1	2	3	4	5	6	7	8	9	10	11	12
Time	.01	.02	.03	.04	.05	.06	.07	.08	.09	.10	.11	.12
Acc.	1	2	3	2	1	0	-1	-2	-3	-2	-1	0

The Willson's method has been used. In order to express the damping matrix, the mass proportion was taken = 0.02 and the tangent stiffness proportion was = 0.007. Every model has been analized for two alternatives related to the following two beam - core connection: a) roller end b) built in end.

The Table 8 is presenting the horizontal displacements of some nodes of the umbrella type model. The Table 9 is showing the horizontal displacements of some nodes of the two - core type model.

4.2. Remarks of time history analysis

The time history analysis has shown a common pendulum behaviour of hung storeys when beam - core connections are roller ends. The displacements of lumped masses are less than the displacements of the corresponding nodes of the core.

Table 8   Horizontal displacements – umbrella model

a. Roller Ends

Nodes	Time					
	.04	.06	.08	.10	.12	.14
2	1	4	7	12	14	19
3	1	4	8	15	18	23
4	1	4	11	20	22	30
10	1	4	6	11	13	18
11	1	4	8	16	17	22
12	1	4	12	19	23	30
15	1	4	8	16	27	42
16	1	2	5	10	17	27
17	0	1	2	4	8	13

b. Built in Ends

2	1	4	8	14	22	32
3	1	4	9	17	27	40
4	1	4	10	19	32	49
10	1	4	8	14	22	32
11	1	4	9	17	27	40
12	1	4	10	19	32	49
15	1	4	10	19	31	49
16	1	4	9	17	26	39
17	1	3	8	15	22	31

In the case of rigid beam – core connection the umbrella type model becomes a reverse pendulum. The two – core type model becomes a common frame structure. The displacements of the lumped masses are of a symilar magnitude with the displacements of the corresponding nodes at the core.

5   Experimental tests

The used device for experimental tests consists of a shaking table with one horizontal degree of freedom. The motion of shaking table is harmonic one of T = 0.10 sec.period.
. The oscillating amplitude can be changed from zero up to 12 mm in running time.
The testing models have been fixed on the shaking table. The freedom degree has been in the plane of the model. Every storey of the model has been equiped with KD 42 – VEB transduser. One of traducers was also located on the shaking table. The transducer signal has been introduced into a SM

Table 9 Horizontal displacements – frame model

a) Roller Ends

Node	Time					
	.04	.06	.08	.10	.12	.14
2	1	3	9	17	26	39
3	1	2	10	20	35	49
4	1	2	12	22	37	52
6	1	2	4	5	7	18
7	1	1	6	8	19	29
8	1	1	8	10	20	32

b) Built in Ends

2	1	2	13	20	25	29
3	1	2	14	19	28	34
4	1	2	13	23	30	38
6	1	2	12	19	26	32
7	1	2	13	20	29	36
8	1	3	14	24	31	39

61 – RET oscillator and further transferred to a TSS 101 – RFT paper recorder. The storey displacements have been recorded.

The above mentioned three types beam – core connetion were taken into account: roller end, hinge and and built – in end.

The Fig. 3a, is representing the motion of the shaking table on the testing excitation.

The Fig.4b 4c and 4d is illustrating the storey motion of umbrella type model in the case of beam roller end.

The records present the different amplitude of motion at different level onsidered which show the elevation ofthe motion with respect to the level of storey. The Fig. 5b, 5 and 5d, are the storey motion record of the two – core model.

The above remark thing is valuable too in this case.

The common pendulum is realised in the case of roller ends of beam core connection and the dynamic response is the most favourable.

The hinge connetion or the build in connetion concelled the effect of inertial masses suspention.

The experimental tests have been confirmed the analytical developments. When the lumped masses are suspended by the tie – rods their recorded motion is, obiously, amplificated that the stiffness of the cables are not able to transfere the motion energy of the cores of the storeys.

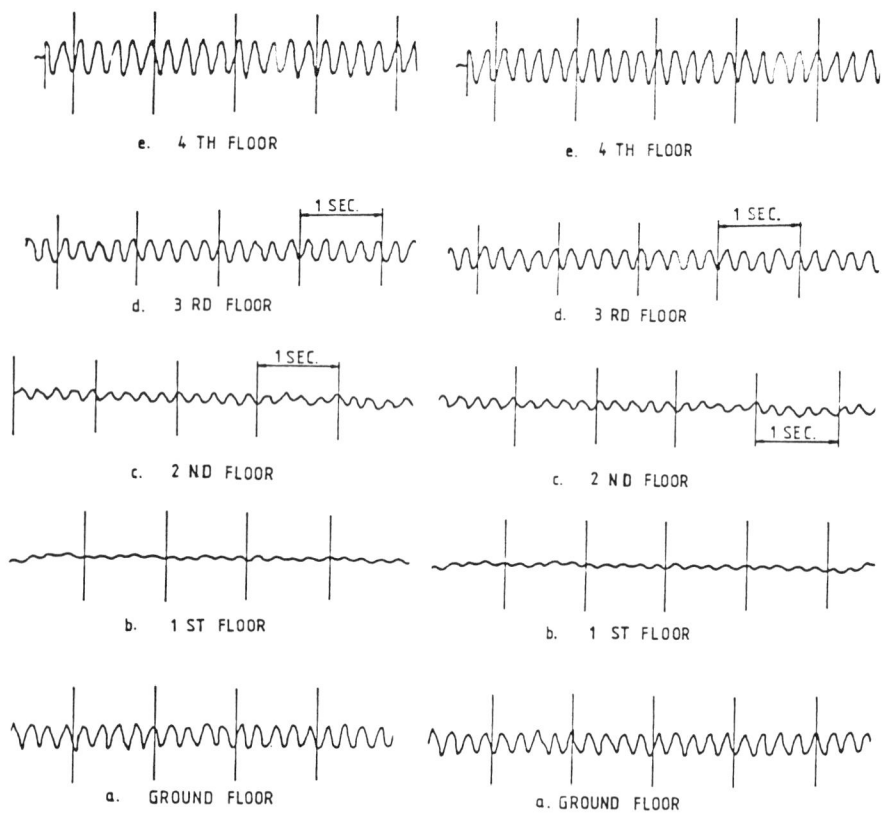

FIG. 5 HORIZONTAL FLOOR DISPLACEMENT TWO CORE BUILDING MODEL-ROLLER END CONNECTION

FIG. 4 HORIZONTAL FLOOR DISPLACEMENT UMBRELLA BUILDING MODEL-ROLLER END CONNECTION

6 References

1. Wilson L.E., Habibulah A. SAP 90 Users Manual.
2. * * * International Meeting on Base Isolation and Passive Energy. Dissipation. Ancona. 1989.
3. Dimoiu I., - Earthquake Engineering (in Romanian) Lito. I.P.Timisoara. 1992.

65 SEISMIC UPGRADING OF CHURCHES BY MEANS OF DISSIPATIVE DEVICES

F. M. MAZZOLANI and A. MANDARA
Istituto di Tecnica delle Costruzioni, University of Naples, Italy

Abstract
The seismic upgrading of a church building located in the Southern Italy, which was severely damaged by the earthquake of 1980, is described in this paper. Special dissipative devices have been adopted for the first time in a historical building in Italy order to improve its seismic response. A stiff steel gridwork has been also fitted on the top of the church for achieving the box-like behaviour under horizontal actions. The results of the dynamic analysis confirm the appropriateness of the use of such innovative techniques in the structural rehabilitation of monumental buildings, mainly when they allow for a limitation of the traditional strengthening operations on the existing structure.
Keywords: Seismic Retrofitting, Monumental Building, Dissipative Devices.

1. Introduction

The use of non conventional techniques for the seismic protection are more and more spreading in the field of new buildings also in Italy (Mazzolani and Serino, 1993).
In case of monumental buildings a new approach is now in progress, giving the bases for the assessment of general rules for new applications (Mazzolani et al., 1994). In particular, when the restoration of old church buildings is faced, a widely used practice is represented by the substitution of the existing wooden roof with new steel trusses. In such operations, the connection between the roof elements and the r.c. beam usually casted on the top of masonry walls plays a significant role in the structural behaviour of the whole building. Particularly, when pinned to the r.c. top beams, the new roof structure is able to behave as a rigid diaphragm, ensuring the required box-like behaviour under seismic actions. This can be generally obtained by adopting the well known steel trussed structures integrated by corrugated sheets and r.c. slab.

On the other hand, this "diaphragm effect", even though quite suitable from the seismic point of view, does not solve the problem of the thermal variations produced by the sun rays periodically acting on the roof. Such a loading condition, often underestimated in the current design practice, is indeed responsible of the cyclic damage in the masonry, which can result in a heavy degradation of its mechanical features (ageing). The use of sliding bearings would certainly avoid the arising of thermal stresses but, at the same time, would involve a poor seismic behaviour, due to the lack of connection between roof and vertical walls. This is rather not acceptable in church buildings, in which the roof usually represents the only element able to behave as a rigid diaphragm. Such a problem can be overcome by adopting suitable oleodynamic devices, whose main feature is that to realize provisional restraint conditions at the base of roof trusses: fully pinned ends, in the case of quickly applied loads, such as e.g. seismic quakes; sliding bearings, in the case of slow actions, namely the loads produced by thermal expansions. In addition, when the external acting

Behaviour of Steel Structures in Seismic Areas. Edited by F. M. Mazzolani and V. Gioncu.
Published in 1995 by E & FN Spon, 2-6 Boundary Row, London SE1 8HN. ISBN: 0 419 19890 3.

load results in a force on the devices higher than their nominal plastic threshold, an energy dissipation takes place, which allows for a reduction of the seismic impact on the construction.

This paper refers to such an application carried out on the S. Giovanni Battista Church in Carife (Avellino) in the framework of the restoration intervention supported by the "Soprintendenza ai B.A.A.A.S." of Salerno and Avellino (Mazzolani and Mandara, 1992). The church under consideration, placed not far from the epicentre of the Irpinia earthquake occurred on 23 November 1980, represents the first case of seismic protection of monumental buildings operated by means of dissipative oleodynamic devices. This application belongs to a general research program concerning the use of non conventional systems, including base isolation of new and old buildings (Mazzolani and Serino, 1994). In such context, the authors wish to underline the quite innovative character of the intervention at the light of the advantages obtained in the examined case, as well as in the general perspective of implementation of such a kind of devices in the seismic protection of monumental buildings.

2. The restoration design

2.1 Brief description of the church

The church, whose first erection dates back to the middle of the 18th century, has the typical latin cross plant with a single bay and many crypts at underground level (fig .1). All the original bearing structure is made of irregular sandstone masonry with very poor mechanical features and thickness going from 0.80 to 2.50m. Only the bell tower, placed on the right hand of the church, presents a more regular texture with outer skin made of brick masonry. Each longitudinal wall is stiffened by three counterforts, extended to the whole height of the church. The underground crypts, used as burial place, are covered by vaults and are interrupted by masonry walls connecting the afore-mentioned counterforts from one side of the church to the other. The horizontal structures are made with cylindrical vaults based just below the roof level; round vaults are used only on the presbytery and on the end abside. The original roof structure consists of wooden trusses, in very bad conditions of conservation (Candela et al., 1991).

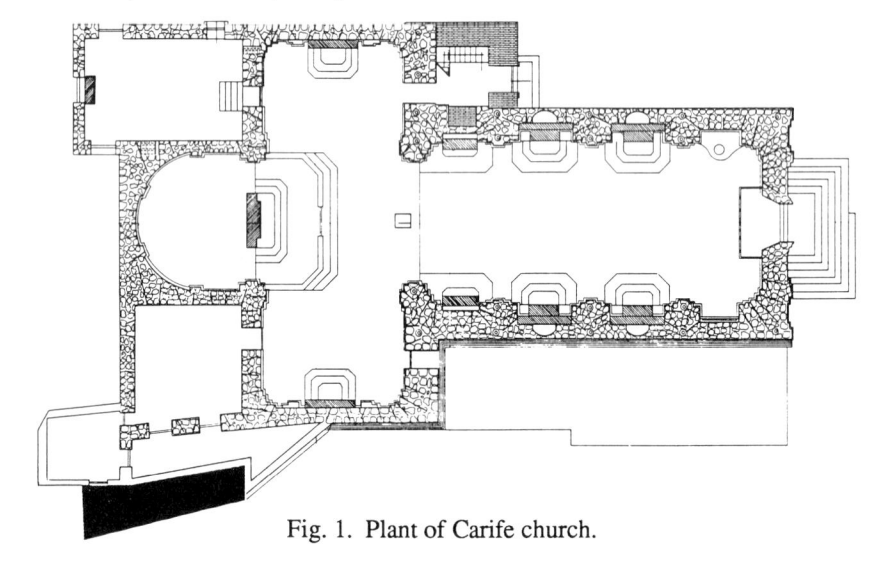

Fig. 1. Plant of Carife church.

2.2 Damage analysis

The structural damage existing in the building before intervention can be summarized in the disconnection and partial collapse of the front facade, collapse of the existing wooden roof with partial failure of the underlying cylindrical vault, great damage at the top of the bell tower, collapse of cornices, spreaded cracking of all masonry elements, strong damage of masonry crossings, etc.

All damage can be considered as a consequence of the several earthquakes occurred in Irpinia in the last two centuries and appears to be justified by the typical features of the church building, consisting in highly slender masonry walls, lack of horizontal diaphragms, thrust of arches and vaults, lack of connection in masonry crossings, ageing of structural material.

2.3 Strengthening interventions

The consolidation design has been carried out according to the following guide-lines:

increase of both shear and bending strength of the masonry, above all in the crossings, in order to improve the behaviour of the building under seismic actions;
achievement of the box-like behaviour for the whole structure, so to ensure the distribution of the seismic loads to the effective shear walls;
elimination of the co-active thermal stresses in the masonry due to the temperature variations acting on the roof;
dissipation of an amount of the seismic energy by means of suitable hysteretical devices.

The adopted solutions are listed below:

strengthening of both masonry walls and crossings by means of reinforced concrete injections;
installation of horizontal and vertical post-tensioned steel tie-beams alongside the masonry panels;
cast of r.c. beams on the top of the building walls in order to improve the connection among them;
consolidation of existing vaults by means of an upper r.c. slab;
erection of a new roof structure made of steel triangular trusses connected at the lower chord by a horizontal steel gridwork;
installation of dissipative devices at the end supports of the roof trusses;

3. The problem of thermal stresses

The effect of the thermal variations acting on the roof has been evaluated for the walls of the main bay of the church. Their structural scheme has been assumed to consist of three masonry males on each side (fig. 1), connected by thin masonry links of negligible stiffness. Thus, the evaluation of thermal stresses has been done by considering the simplified plane model shown in fig. 2. By referring to the most stressed males (the central ones), and assuming a perfectly elastic behaviour of materials, the following equilibrium condition on the top of each male may be imposed:

$$\frac{3E_m I_m \delta}{h_m^3} + \frac{48 E_c I_c \delta}{L_c^3} = \frac{2 E_s A_s}{L}\left(\alpha \frac{L}{2}\Delta T - \delta\right) \qquad (1)$$

with:

δ, the unknown displacement at the top of the masonry male;
h_m, the height of the masonry male;
L, the axis-to-axis span of the roof structure;
L_c, the length of the r.c. beam on the top of the masonry wall;
α, the average value of the steel thermal expansion coefficient, equal to 0.000012;
ΔT, the applied thermal change on the roof;
E_m, the masonry elastic modulus;
E_c, the concrete elastic modulus, assumed equal to 30.000 N/mm²;
E_s, the steel elastic modulus, equal to 210.000 N/mm²;
A_s, the whole cross sectional area of the steel section pertaining to the masonry male under consideration;
I_c, the inertia moment of the r.c. beam against its vertical centroidal axis;
I_m, the inertia moment of the masonry section equal to $bt_m^3/12$ for fully compressed section, and $b[3(t_m/2 - e)]^3/12$ for partially compressed section;
t_m, b the thickness and the width of the masonry males, respectively;
e, the eccentricity of the vertical load.

By remembering that the bending moment M at the base of male is given by $M=Fh_m=3E_mI_m\delta/h_m^2$, it may be written:

$$e = \frac{M}{N} = \frac{Fh_m}{\gamma_m h_m t_m b} = \frac{F}{\gamma_m t_m b} = \frac{3E_m I_m}{\gamma_m t_m b h_m^3}\delta \qquad (2)$$

where the axial load N has been put equal to the male selfweight $\gamma_m h_m t_m b$ (with γ_m masonry unit weight), by neglecting the little contribution of the steel roof trusses. From eq. (2) it may be observed that, with the adopted assumptions, the load eccentricity e is independent of the wall height h_m; as a consequence, also the tensioned portion in the masonry section, if present, will be constant along the male, so making possible to refer to a free-standing column with constant section equal to $b \times 3(t_m/2 - e)$.
By eliminating δ from eqs (1) and (2), one may obtain:

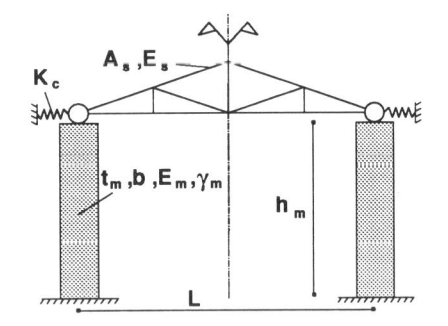

Fig. 2. The simplified model adopted for the thermal analysis.

$$\frac{e}{t_m} = \frac{\alpha L \Delta T}{2\gamma_m b t_m^2}\left(\frac{K_s}{K_m+K_c+K_s}\right)K_m \qquad (3)$$

being:

$$K_m = \frac{27E_m(t_m/2-e)^3 b}{4h_m^3}; \qquad K_c = \frac{48E_c I_c}{L_c^3}; \qquad K_s = \frac{2E_s A_s}{L};$$

the shear stiffness at the top of the free-standing masonry male, the midspan shear stiffness of the top r.c. beam and the axial stiffness of the steel roof trusses, respectively.

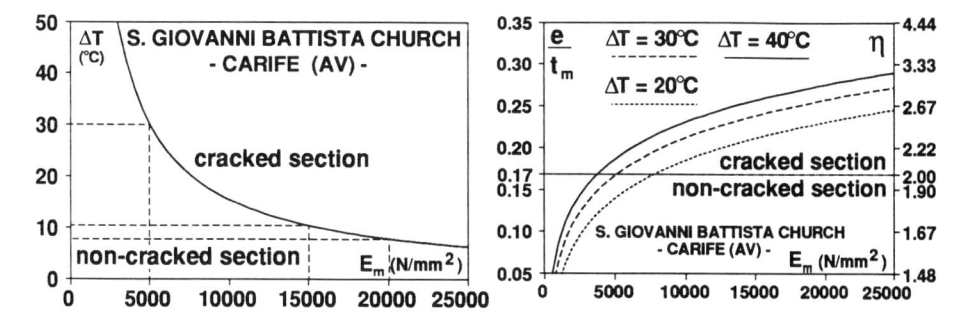

Fig. 3. Carife church: values of ΔT necessary to produce tensile stresses into the masonry as a function of E_m.

Fig. 4. Carife church: values of the stress amplification factor η as a function of E_m and ΔT.

The following geometrical and mechanical features have been measured for the Carife church: $h_m = 12$m; $t_m = 2.2$m; $b = 2.2$m; $L = 11.2$m; $A_s = 300$cm^2; $L_c = 19$m; $I_c = 28.6 \times 10^6$cm^4; $\gamma_m = 1800$kg/m^3. Fig. 3 shows that for E_m ranging from 5.000N/mm^2 to 15.000N/mm^2, tension in the masonry walls occurs for a thermal variation going from 30°C for $E_m = 5.000$N/mm^2 to 10.3°C for $E_m = 15.000$N/mm^2. For $E_m = 20.000$N/mm^2, which can be considered as an upper limit for heavily strengthened masonry, just 7.8°C of thermal change are sufficient to cause tension in the masonry walls.

It should be observed that the arising of tensile stresses in the masonry, which occurs for $e/t_m > 1/6$, also produces a significant increase of the maximum compressive stress in the section. This can be evaluated by means of the stress amplification factor η:

$$\eta = \frac{\sigma_{max}}{\sigma_{med}} = \frac{2}{3(1/2 - e/t_m)} \qquad (4)$$

By supposing $\Delta T = 40$°C, the e/t_m values calculated with eq (3) are equal to 0.1861, 0.2317, 0.2572 and 0.2745 for $E_m = 5.000, 10.000, 15.000$ e 20.000N/mm^2, respectively. For such e/t_m values η is equal to 2.12, 2.49, 2.75 and 2.96 (see fig. 4). These η values can induce a dangerous state of crushing in the masonry, above all at the base of masonry males. In addition, the degradation of mechanical features of the masonry due to the tension and compression cycles should be considered. This could involve the onset of microcracks along the tensioned masonry surfaces, with disconnection of the plaster cover and full exposition of the masonry to the atmospheric agents.

Based on eq. (3), it may be easily observed that both K_m and K_c are much lower than the trusses axial stiffness K_s. This allows to recognize that the effect of thermal variations is also significant for the males adjacent to the transverse walls of the church, in spite of the higher shear stiffness of the r.c. top beam. On the other hand, it would be safer to assume that the roof steel trusses are infinitely stiff. According to this assumption, tensile stresses arise in the masonry for ΔT equal to 28°C, 14°C, 9.4°C e 7°C for $E_m = 5.000, 10.000, 15.000$ e 20.000N/mm^2, respectively. Of course these values of ΔT are slightly lower than those obtained by keeping into account the actual stiffness of the trusses.

This hypothesis allows for the problem to be generalized by introducing the non-dimensional parameter $\chi = (Lt_m E_m)/(h_m^3 \gamma_m)$, representing the ratio between the quantities which enhance the effect of thermal variations (L, t_m e E_m) and those which tend to reduce it (h_m e γ_m).

By substituting the expressions of K_m, K_c and K_s into eq. (3), one may obtain with easy algebraic manipulation:

$$\frac{e}{t_m} = \frac{27}{8} \alpha \Delta T \frac{L t_m E_m}{h_m^3 \gamma_m} \left(\frac{1}{2} - \frac{e}{t_m} \right)^3 = \frac{27}{8} \alpha \Delta T \left(\frac{1}{2} - \frac{e}{t_m} \right)^3 \chi \quad (4)$$

The values of e/t_m, are plotted in fig. 5 as a function of χ and ΔT. In the figure the corresponding values of the coefficient η ratio are also shown. By expressing ΔT as a function of $1/\chi$ the curves plotted in fig. 5 become straight lines passing through the origin (fig. 6). Therefore, a simple criterion for assessing the necessity to apply interventions able to reduce the effect of thermal changes, can be defined by referring to $1/\chi$ values. Particularly, by considering a ΔT values equal to 30°C, it is possible to define four ranges for $1/\chi$, limited by values of the stress amplification factor η equal to 2, 2.5 and 3, respectively (see fig. 6). These ranges, represented in table 1, can be used to classify the building situations and to identify the necessity of intervention by means of special devices.

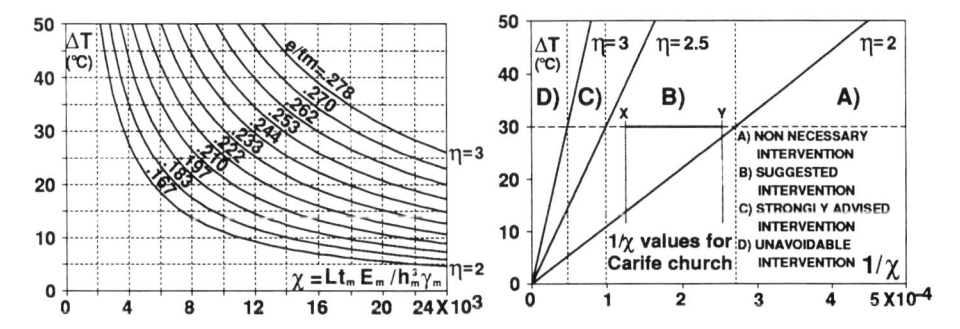

Fig. 5. Values of the stress amplification factor η as a function of χ.

Fig. 6. Feasibility ranges for the application of special devices in church buildings as a function of $1/\chi$.

Table 1. Feasibility ranges for the application of special devices in church buildings.

A) $1/\chi > 2.70 \times 10^{-4}$ buildings with very narrow bays, slender walls and masonry of poor mechanical features; tension arises for thermal variations higher than 30°C; for $\Delta T=30°C$, $\eta<2$: *intervention is not necessary.*	B) $1.00 \times 10^{-4} < 1/\chi < 2.70 \times 10^{-4}$ buildings with wider bays as compared to A), made of masonry with higher mechanical features; for $\Delta T=30°C$, $2<\eta<2.5$: *intervention is suggested.*
C) $0.50 \times 10^{-4} < 1/\chi < 1.00 \times 10^{-4}$ buildings with wide bays and made of high strength masonry; for $\Delta T=30°C$, $2.5<\eta<3$: *intervention is strongly advised.*	D) $0 < 1/\chi < 0.50 \times 10^{-4}$ buildings with high strength masonry and bay width higher than in C); for $\Delta T=30°C$, $\eta>3$: *intervention is unavoidable.*

In the Carife church, by assuming an average value of elastic modulus E_m falling between 5.000N/mm^2 and 10.000N/mm^2, the parameter $1/\chi$ ranges from 1.25×10^{-4} e 2.50×10^{-4}; as they fall in zone B), these values are sufficient to justify the use of special devices for eliminating the effect of thermal variations. The interval of $1/\chi$ values is represented in fig. 6 with a thick segment (X - Y).

4. The roof steel gridwork

The roof structure has been conceived in order to achieve the seismic box-like behaviour of the church thanks to a rigid diaphragm put on the top of the masonry walls. For this purpose a steel lattice gridwork has been built at the level of the lower chord of the roof trusses. It has been extended to all the plant of the church, according to the scheme shown in fig. 7. The main function of the gridwork is that to absorb the inertia forces of the weak walls - namely these orthogonal to the seismic action - and to transfer them to the walls which are parallel to the quake direction. This is possible thanks to the great in-plane stiffness of the gridwork, which can be considered at a first glance as an infinitely rigid diaphragm.

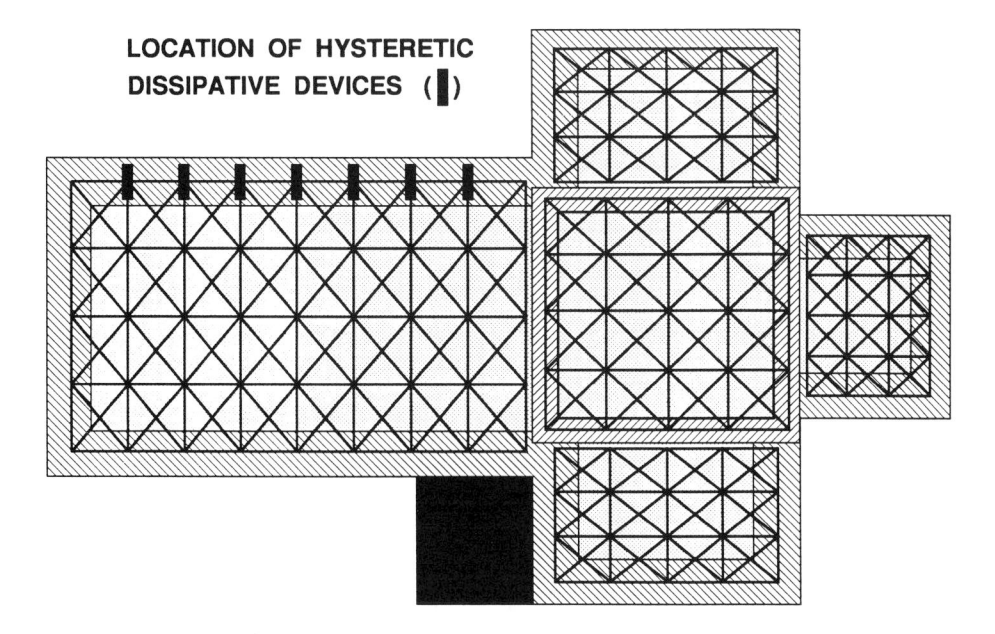

Fig. 7. Schematic view of the steel gridwork arranged on the top of Carife church.

The structural scheme corresponding to this idealization allows to consider the gridwork as a high beam supported at the top of the effective walls. In this way a simple design procedure can be followed, by assuming that, for a given earthquake direction, the weak walls are restrained at the base by a fixed support and on the top by a hinged one. The inertia forces acting on the wall can be approximately evaluated by adopting a static approach, as referred to in the Italian Seismic Code, by considering a force distribution directed according the first vibration mode. A more conservative design for the gridwork can be achieved by assuming the walls as perfectly hinged both on the base and on the top. In such a way, the inertia forces acting on the gridwork can be put equal to the half of the whole seismic weight of the wall plus the seismic weight of the roof structure. According to the Italian Code, these forces are given by $F = KW = cR\varepsilon\beta I\gamma W$; For Carife church, by considering the $2.50m$ spanned strip of masonry wall related to each bearing of the gridwork, it results $W = G + sQ = 1.400kN$, while $R = \varepsilon = \gamma = 1$, $\beta = 2$, $I = 1.2$. The coefficient β has been put equal to 2 instead of 1 - as recommended for r.c. and steel dissipative structures - in order to limit the gridwork plastic deformations to values compatible with the ductility of the underlying masonry walls. On the other hand the value 4, recommended by the code for non-dissipative structures - namely masonry buildings - seemed too high at the light of the ductility increase obtained by the consolidation of the masonry structure. The coefficient I is equal to 1.2 because the church is to be considered as a public building. Under these hypotheses it results $F = 336kN$; this corresponds to the value according to which the gridwork structure and the truss bearings should be designed if a perfectly elastic behaviour under design seismic action is required.

5. The oleodynamic devices

As already stated in the introduction, the use of oleodynamic devices instead of pinned or sliding supports allows for the structural response under both thermal and seismic loads to be fully optimized (fig. 8). The behaviour of the such devices strongly depends on the way of application of the external load: when the load is slowly applied, as in the case of thermal variations, the force transmitted through the device is very small and this behaves practically as a sliding bearing. When the load is applied quickly, such as in the case of seismic quakes, the device opposes a strength which is a function of the rate of application of the load itself: as higher such rate, as stiffer the device. The maximum force allowed for the device, which can be transmitted for a very speedy application of the external load, defines its nominal strength. For external quick loads below its nominal capacity, the device practically behaves as a pinned restraint.

When the rate of application of the external load is higher than that corresponding to the threshold strength, a perfectly plastic behaviour takes place, with an energy dissipation depending on the magnitude of bearing elongation. If a cyclic application of load is involved, the bearing exhibits hysteretic features, which turn to be very useful for the dissipation of earthquake kinetic energy.

According to the calculations referred to in the previous paragraph, in order to resist the required seismic action of $336kN$ without plastic slips, two oleodynamic devices with a nominal strength of $200kN$ and a maximum elongation of $20mm$ have been installed for each truss bearing, so to reach a nominal capacity of $400kN$. In such a case, the hysteretic behaviour of the device occurs for strongly destructive earthquakes.

The devices have been placed on one side of the main bay of the church, as shown in fig. 7; a constructional detail can be observed in the picture of fig. 9; the coupled configuration allowed for a rational constructional arrangement, both from structural and functional point of view. Furthermore, stainless steel has been used for the devices, so to eliminate the problems of inspection and manteinance.

In order to assess the mechanical features of the adopted devices, an experimental testing program has been carried out at the Politecnico of Milano, aiming to investigate the response of the device in the following service conditions: a) slowly applied displacements; b) rapidly applied displacements; c) cyclic load.

Condition a) corresponds to the displacements involved by the thermal changes. In this case the transmitted force should be rather low; nominally, it may be accepted that it should not be higher than 10% of the maximum load capacity of the device. Two load conditions have been considered: a full cycle with a displacement of ±3mm impressed in both directions with a speed of *0.01mm/s* (fig. 10a); a displacement impressed at the speed of *0.001mm/s* for a time of *200s* (fig. 10b). The strength opposed by the device, equal to *20kN* and *4kN* in the first and second case respectively, confirms its actual capacity to behave in service condition as a sliding support at the base of roof trusses.

Condition b) reproduces a quick load, as generally occurs during an earthquake; in such a case, the device should oppose the impressed displacements with its maximum load bearing capacity. In the test a full cycle with maximum displacement of ±*10mm* has been applied at the speed of *1mm/s* (fig. 10c) and *2mm/s* (fig. 10d), in order to evaluate the rate of application of external actions exploiting the full strength of the device: this occurred for a rate of *2mm/s*.

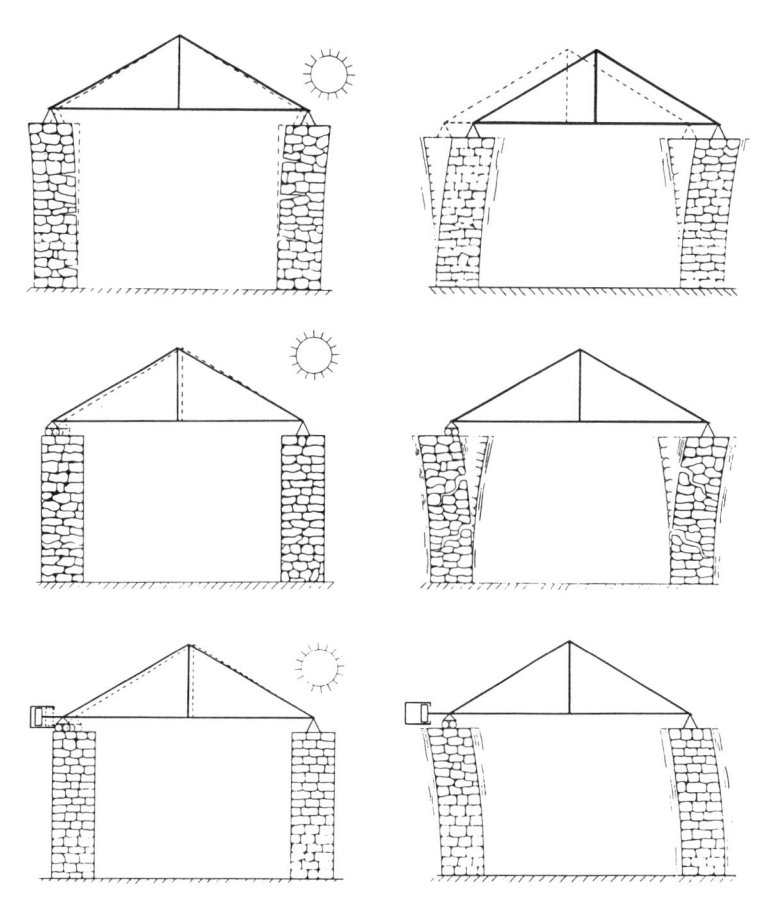

Fig. 8. Effectiveness of the special devices as compared to pinned and sliding supports.

A further test consisting of 3 cycles with width of ±*10mm* has been performed with such rate value, in order to investigate the effect of heat dissipation on the mechanical degradation of the devices. This test has shown the hysteretic features of the device, by confirming its aptitude to dissipate a part of the seismic kinetic energy. The cyclic load-to-displacement diagram has approximately a rectangular form (fig. 10e); this allows to assume a rigid-perfectly plastic behaviour of the device in practical calculations.

Fig. 9. A constructional detail of the adopted devices.

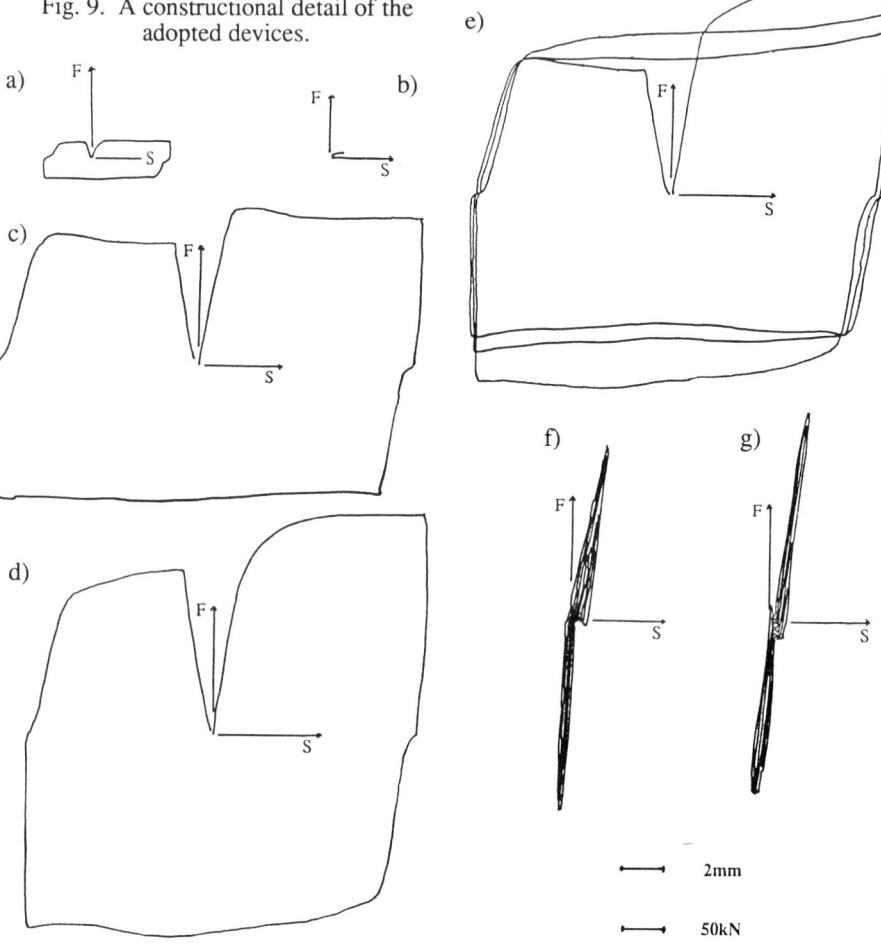

Fig. 10. Results of experimental tests on the adopted devices.

Finally, under condition c) the cyclic behaviour under service repeated loads has been investigated. Two tests, each of 10 cycles, at *0.4Hz* (fig. 10f) and *0.6Hz* (fig. 10g) have been carried out. A pulsating load of magnitude gradually increasing up to the nominal maximum capacity of the device has been applied. The obtained results show that, for load values lower than the nominal capacity, the behaviour is quite stable and free from significant reduction of stiffness. This reached the average values of *114kN/mm* and *144N/mm*, for *0.4Hz* and *0.6Hz*, respectively. Thus, it is clear that the device practically behaves as a pinned restraint up to values of the external load corresponding to its nominal capacity.

6. Dynamic analysis

An inelastic dynamic analysis of the church has been performed in order to investigate the influence of the plastic threshold of the device on the amount of dissipated energy. In fact, such threshold can be considered as a design parameter, conditioning both the obtained protection level and the structural damage.

Generally, two limit cases can be defined: the first one, in which the load capacity tends to zero and the device practically behaves as a sliding restraint; the second one, in which the plastic threshold tends to infinite and consequently the device behaves as a pinned bearing. In the first case the maximum relative displacements between the roof structure and the masonry walls take place, whereas in the second one such displacements vanish. Of course in both cases the dissipated energy will be zero, because the transmitted force or the allowed displacement are zero. Evidently, there will be an intermediate value of the device nominal capacity for which a maximum value of the dissipated energy is attained. A correct design procedure of such intervention should follow this main guide-line, by controlling that the displacement accomplished by the adopted device are not as high as to damage the underlying masonry walls.

Since a rigourous approach would have involved a quite complex analysis of the inelastic seismic behaviour of the whole spatial structure of the church, a simplified plane model has been defined (fig. 11), in order to highlight the influence of the connection between the walls and the roof trusses on the seismic response of the church.

The adopted calculation code has been the well known DRAIN-2D. It allowed for modelling the actual plastic behaviour and the cyclic softening of the masonry, by means of a moment-to-rotation relationship based on the bi-linear Takeda model. Each wall has been modelled with five beam elements connected by non-linear springs, accounting for both the actual plastic behaviour and softening of the masonry. The inertial masses of the walls have been concentrated in each node; the mass of the trusses has been concentrated at the top of the walls. The dissipative devices has been modelled as a truss element with a rigid-perfectly plastic behaviour.

Accounting for what already discussed in section 4, a ground motion transverse to the main bay of the church has been considered. Due to the great stiffness of both the steel gridwork and the transverse walls, the top of the right wall has been kept fixed and hinged to the roof trusses; on the contrary, the restraint on the top of the left wall depends on the behavioural phases of the oleodynamic devices.

For the examined case, by referring to a masonry strip of *2.50m*, the following material features have been considered for the masonry: shear modulus $G = 110N/mm^2$, elastic modulus $E = 6G = 660N/mm^2$, ultimate compressive strength $f_{yc} = 6.25N/mm^2$, ultimate shear strength $f_{vc} = 0.25N/mm^2$. Fe 360 steel with $f_y = 235N/mm^2$ has been adopted.

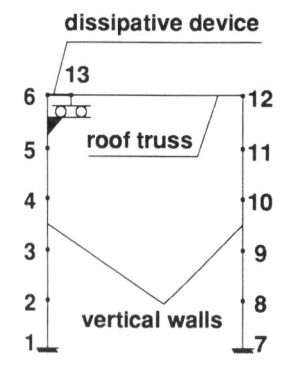

Fig. 11. The simplified scheme adopted for the dynamic analysis.

The dynamic behaviour of the model under consideration has been investigated for a plastic threshold of the device ranging from *100kN* to *500kN*. An artificial accelerogram - simulated with *SIMKQE* Code by referring to rigid soil - has been used. A maximum ground acceleration equal to *0.35g* has been considered. The validity of using the adopted devices has been confirmed by the dissipated energy (fig. 12), which exhibits a maximum placed around *400kN*, corresponding to the plastic threshold of the installed devices. The dynamic analysis also confirms the benefits produced by the devices, by showing a significant reduction of displacements at node 6 as respect to a simple sliding bearing (fig. 13).

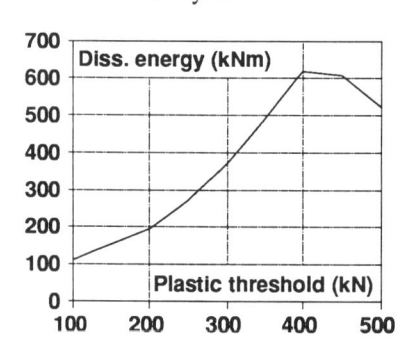

Fig. 12. Diagram of dissipated energy as a function of device threshold.

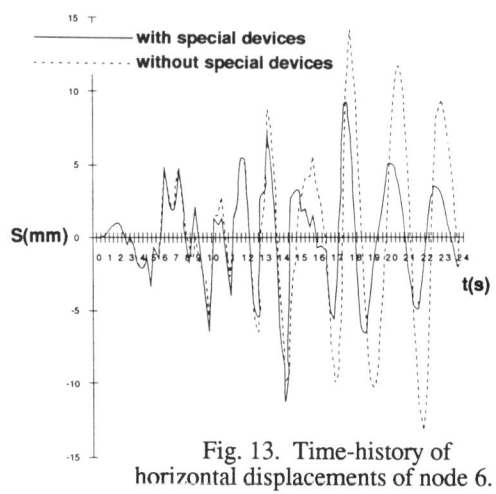

Fig. 13. Time-history of horizontal displacements of node 6.

7. References

Candela, M., Mandara, A., Mazzolani, F.M., Piluso, V. and Rozza, F. (1991), L'impiego dei dispositivi a vincolo provvisorio nel restauro di un edificio di culto, **Proc. of ANIDIS Congress L'ingegneria sismica in Italia**, Palermo.

Mazzolani, F.M. and Mandara, A. (1992) **Nuove strategie di protezione sismica per edifici monumentali: il caso della Collegiata di San Giovanni Battista in Carife.** Segno Ass.

Mazzolani, F.M. and Serino, G. (1993) Most recent developments and applications of seismic isolation of civil buildings in Italy, **Proc. of the Int. Post-SMIRT Conf. Seminar on Isolation, Energy Dissipation and Control of Vibrations of Structures**, Capri (Italy).

Mazzolani, F.M., Serino, G. and Zampino, G. (1994), Seismic protection of italian monumental buildings with innovative techniques, **Proc. of the Int. Workshop on Application and Development of Base Isolation**, Shantou (China).

Mazzolani, F.M. and Serino, G. (1994), Innovative techniques for seismic retrofit: design methodologies and recent applications, **Proc. of the 2nd Franco-Italian Symposium of Earthquake Engineering - Strengthening and Repair of Structures in Seismic Areas**, Nice (France).

66 MODELS, SIMULATIONS AND CONDENSATIONS IN THE DESIGN OF A STEEL-CONCRETE COMPOSITE STRUCTURE PLACED IN SEISMIC ZONE

V. STOIAN and I. OLARIU
Technical University, Timisoara, Romania

Abstract
Design of a structure placed in a seismic zone impose some specific aspects. The paper deals with the design process for a steel-concrete composite structure. The structural system consist of a skeletal space frame with steel braces.
The structural solution combines steel-concrete composite columns reinforced with stiff and elastic steel and steel-concrete composite beams.
Influence of the braces and initial stresses in the steel structure is studied.
Based on a condensed model the nonlinear analysis is made, for different accelerograms.
Qualitative and quantitative conclusions are presented.
Keywords: Steel-concrete, Composite elements, Structural design, Seismic code, Nonlinear analysis.

1 Structural system

The multistorey building of 9, partially 12 levels is placed in a D seismic zone, according to Romanian Seismic Code (Fig.1).

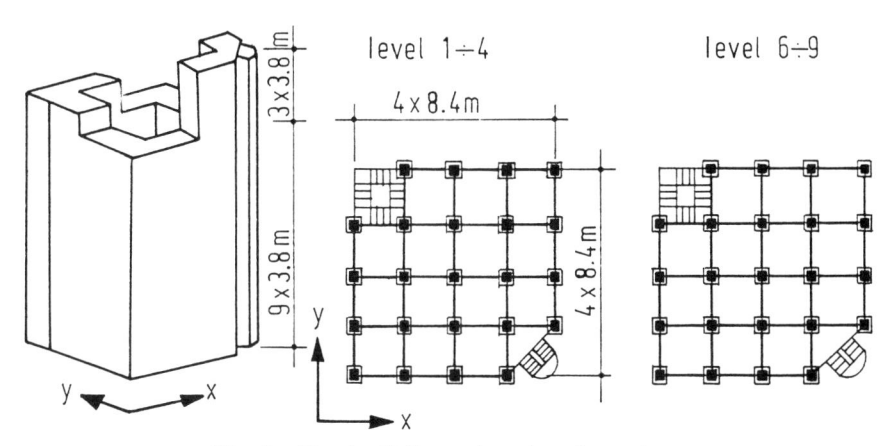

Fig.1. The building structural system

Behaviour of Steel Structures in Seismic Areas. Edited by F. M. Mazzolani and V. Gioncu.
Published in 1995 by E & FN Spon, 2-6 Boundary Row, London SE1 8HN. ISBN: 0 419 19890 3.

Structural system consist of a stiff spatial reinforced concrete box as infrastructure, in which are embedded the component elements of the structure. The columns and the beams of the spatial skeleton of the structure are designed as steel-concrete composite elements.

2 Structural stiffening

2.1 Braces positioning
The structure is stiffened with braces made from steel diagonals, placed between the bottom end of the column-beam joint and the bottom part of the cross section of the beam frame at the next level.
In order to stiffen the structure due to the effect of the horizontal loads, two variants of braces positioning were studied:
- A variant - at which the braces are placed on the exterior part of the building in two consecutive spans on each side (Fig.2a);
- B variant - at which the braces are placed in the four corners of the exterior part of the building (Fig.2b)

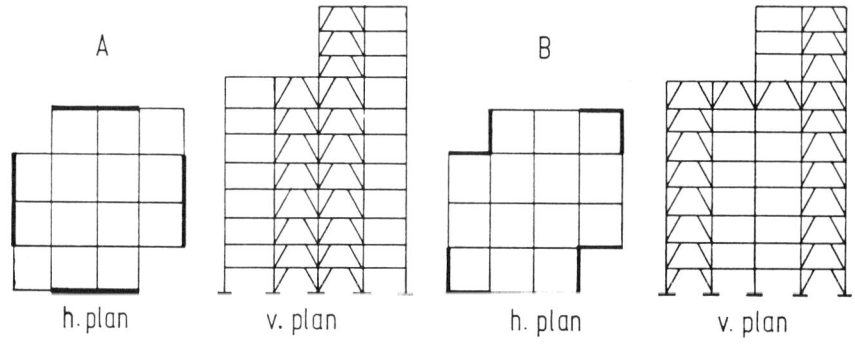

Fig.2. Braces positioning

In both two variants the braces consist of diagonal bars subjected to axial forces. Therefore, the modal response of the structure, lateral displacements and the relative level stiffness are affected by the number of braces rather than their position (Tab.1).

Table 1. Structural modal response

PERIOD (s)	1	2	3	4	5	6
VARIANT A	1.48	1.43	1.12	0.55	0.54	0.40
VARIANT B	1.53	1.47	1.37	0.55	0.53	0.48

Modal response of the structure and axial forces in the columns are highly dependent by the braces position, as it is shown in Tab.2.

Table 2. Displacements, horizontal forces and stresses

VARIANT		A			B		
MODE		1	2	3	1	2	3
d_x	(m)	-0.002	0.034	-0.007	-0.009	0.024	-0.023
d_y	(m)	0.034	0.002	0.006	0.023	0.022	0.014
teta	(rad)	0.0003	-0.0002	-0.002	0.0014	-0.0004	-0.002
F_x	(t)	-63	1076	-57	-246	734	-537
F_y	(t)	1083	84	50	571	647	317
M_{teta}	(tm)	1574	1634	5369	7232	-4760	-8614
M_{maxsup}	(tm)	-32	87.4	8.6	-69	69	-9
M_{maxinf}	(tm)	109.9	-244.3	-35.8	1940	-175.5	30.3
N	(t)	-450.6	-102.4	70.2	94.0	-355.6	105
M_{minsup}	(tm)	35.2	8.2	-5.2	34	14	-14
M_{mininf}	(tm)	-185.4	-31.4	28.2	-152.9	-83.5	36.7
T_{max}	(t)	20.9	-42.2	-7.3	33.6	-28.6	5.7
T_{min}	(t)	-39.5	-6.1	6.1	-31.1	-18.1	5.9

Dynamic analysis reveals that the structure has an geometrical and mechanical symmetry toward one of the diagonal axis. This direction becomes principal inertial direction in the second mode. Due to the little mass inertia moment at all levels, the general torsional moment is not very high, the structural response being uncoupled (Fig.3). As a conclusion, the effect of braces can be studied on a plane frame.

The square form of the building plane admits as symmetry axes both the sides and the diagonal. These axes will be found in modal analysis as principal axes, thus:
- side, if it will be accentuated the stiffness upon the side (variant A);
- diagonal, if it will be placed corner braces (variant B) and eccentrically mass positioning. In this situation the dynamic structural response is coupled.

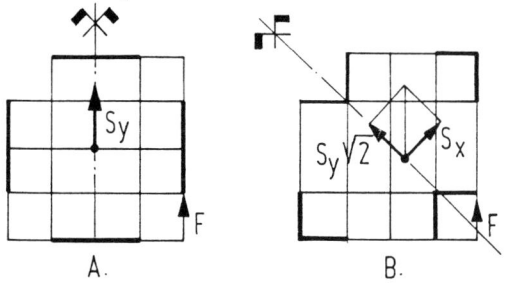

Fig.3 Principal structural axes

Based on these conclusions the local effect of the braces positioning was analyzed on an plane frame.

Table 4. Seismic forces acting on unbraced and braced frames

LEVEL	TOTAL (t)	UNBRACED (t)	BRACED (t)	%
12	15459	10129	5330	34.5
11	10699	3895	6804	63.6
10	47619	30426	17193	36.1
9	80818	46575	34243	42.3
8	88405	49995	38410	43.4
7	91356	51137	40219	44.0
6	92720	52256	40464	43.6
5	100313	60897	39416	39.3
4	109363	69581	39782	36.3
3	129824	87237	42587	32.8
2	160521	112922	47599	29.6
1	340507	254331	86176	25.3

Table 5. Relative level stiffness of unbraced and braced frames

LEVEL	MASS (t)	TOTAL (t)	UNBRACED (t)	BRACED (t)
12	344	49	30	19
11	344	47	33	14
10	903	114	76	38
9	1341	163	67	96
8	1341	152	77	75
7	1341	137	82	55
6	1341	119	60	59
5	1635	119	99	20
4	1635	92	67	25
3	1635	64	70	-6
2	1635	37	75	-38
1	1635	13	188	-175

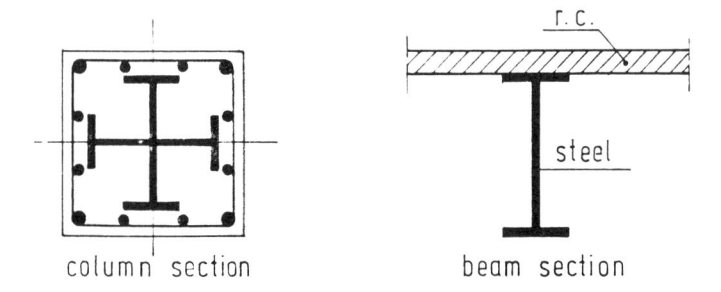

Fig.5. Typical cross section for columns and beams

As a conclusion, it is obviously that providing the braces in the central spans (variant A) has advantage in the distribution of the axial forces due to horizontal loads (Fig.4).

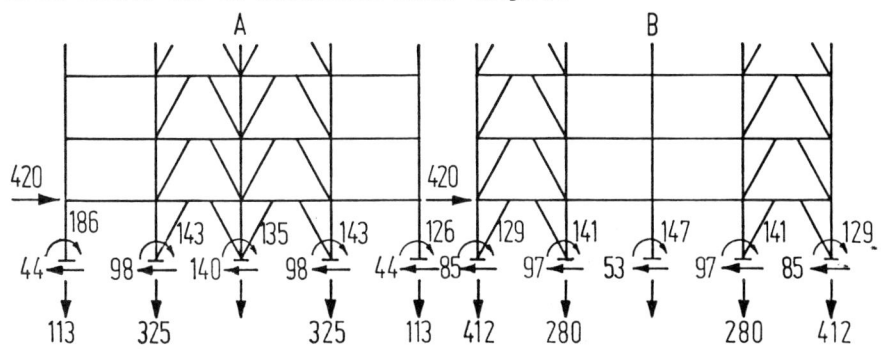

Fig.4. Axial force distribution in a plane frame

Variant B give an coupled dynamic response of the structure and determine a bigger general torsional moment.

Absolute and relative displacements reflect the monotonic character of the structure, with different values at the top levels (Tab.3).

Table 3. Extreme relative structural displacements

VARIANT	A	B
d_{rmax}/H_{level} (mm/m)	10.356	22.372
d_{rmin}/H_{level} (mm/m)	3.3	3.48

2.2 Braces efficiency

Braces efficiency can be evaluated by the seismic forces absorbed by the unbraced and braced frames (Tab.4) or by the relative level stiffness (Tab.5).

3 Structural detailing

The columns has squared section because of the beam spans which are identically on both directions. The main stress state in columns is eccentric compression with bending moments approximately equal on both directions. Load bearing capacity to torsion is lost due to the existence of the floor slab and due to the minor value of the torsional capacity of the steel beams. Rigid reinforcement is realized by two I profile manufactured by welded steel bands. The beams are also steel-concrete composite elements (Fig.5).

The floor system consist of precast concrete slabs resting on an orthogonal grid. The grid consist of steel beams, simply supported on main beams. All the joints are welded. For the steel elements of the structure were utilized the specific dimensions presented in Tab. 6, in which A is the cross sectional area and U is the perimeter of the cross section.

Table 6. A/U ratio for structural elements

ELEMENT	A/U (m)
Frame beams	100-165
Secondary grid beams	160-240
Braces	115-160

In the Table 7 typical characteristics of the beams are presented.

Table 7. Sectional characteristics for beams

CHARACTERISTICS	PERMANENT LOADS	SEISMIC LOADS
A (m^2)	0.156	0.041
I (m^4)	0.0399	0.0168

The collapse mechanism of a braced steel frame with eccentrically placed braces is suggested by the diagrams of Fig. 6 in which capable bending moments on the frame beams are presented versus effective bending moments.

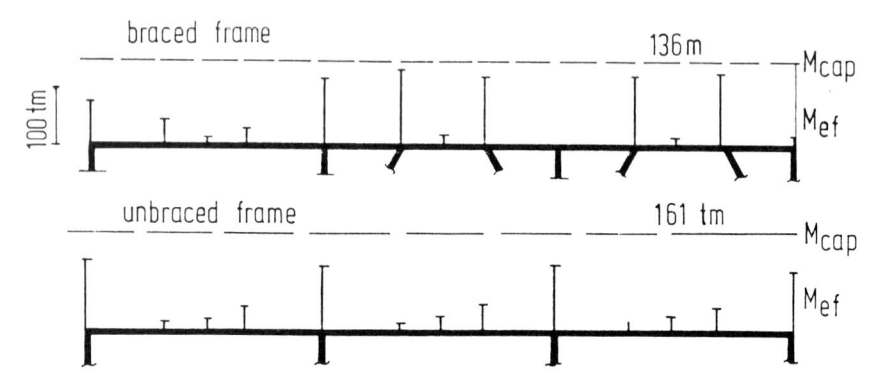

Fig. 6 Effective and capable bending moments

The squared shape of the floor slab has advantage in obtaining an low level inertial mass moment (Tab.8).

Table 8. Inertial characteristics of the structure

CHARACTERISTICS		LEVEL 1-4	LEVEL 5-9
M	(t)	1638	1344
J_m	(tm^4)	28600	34000
M_{teta}	(tm)	6270	6770

There are no important effects from second order analysis.

4 Effect of initial stresses

Erection technology generates initial stresses in the stiff reinforcement of the columns. Thus, initial compression stresses on the stiff reinforcement in the cross section is about 300 daN/cm², respectively an initial strain $\epsilon_i=0.15$ mm/m. This strain can be represented as a translation in the strain diagram (Fig.7).

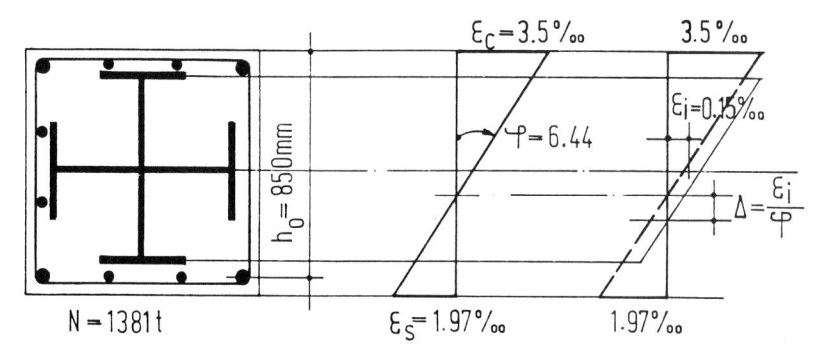

Fig. 7 Influence of the initial stresses

This translation can be quantified with two parameters: rotation φ and axial strain ϵ_i. Neutral axis will move down with Δ, increasing the height of the compression zone. The effect will be the increasing of the sectional ductility together with an reduction of the load bearing capacity of the column.

5 Nonlinear analysis

For nonlinear analysis a condensed model was used, in order to detect the nonlinear effects into the structural elements (Fig. 8).

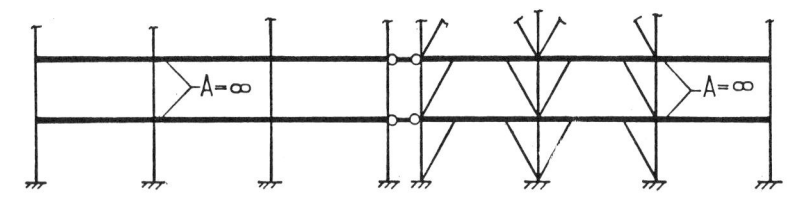

Fig. 8 Structural model for nonlinear analysis

The structural model was subjected to three different accelerograms, described in Tab. 9.

Table 9. Seismic accelerograms

ACCELEROGRAM	SEISMIC ZONE	CALIBRATION COEFFICIENT
1	Vrancea – 04.03.1977	0.81
2	Bucuresti – 31.08.1986	1.00
3	El Centro –	0.52

The first two accelerograms are represented in Fig. 9.

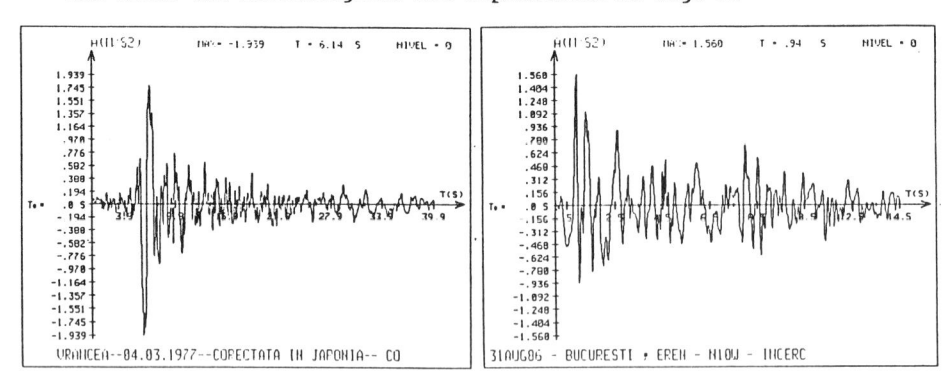

Fig. 9 Accelerograms 1 and 2

In the Table 10 a short synthesis of the nonlinear structural analysis is presented. In this table, A_{gr} is the ground acceleration, V_{gr} is the ground velocity, E_{ind} is the induced energy and E_{dis} is the dissipated energy, H is the total height of the structure,

Table 10. Structural response to seismic loads

VARIANT	LOAD (t)	LOAD POS. (H)	DISPL. (cm)	ROT. (1000*rad)	A/A_{gr}	V/V_{gr}	E_{ind} (tm)	E_{dis} (tm)
CODE	514.962	.56H	25.56	9.70				
ACC.1	1636.377	.39H	30.22	23.78	3.4	2.07	508.35	128.63
ACC.2	618.542	.48H	6.62	2.85	2.40	2.27	52.03	.00
ACC.3	600.555	.45H	7.79	2.75	2.36	1.77	44.73	.00

The level displacements and the required ductility in columns are represented in Fig. 10.

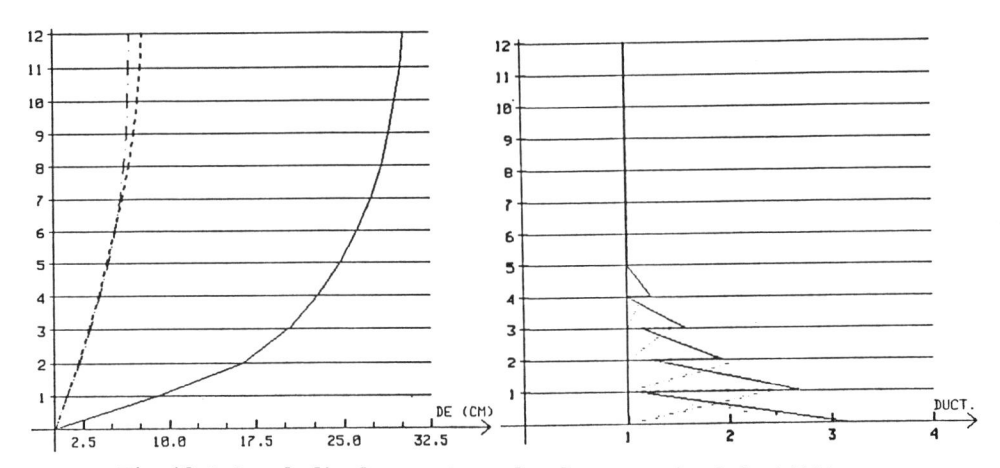

Fig.10 Lateral displacements and column required ductility

It is obviously that the structure has an typical frame structural behaviour, even in the presence of the braces. The structure is sensitive to Vrancea type accelerograms which has an reduce incidence in the region where the building is placed. Seismic loads and lateral displacements demonstrate that the reduction coefficient due to energy dissipation was well choose for the accelerograms 2 and 3 for which no plastic hinges were detected.

6 References

1 ANELISE – Time-history analysis of plane frames subjected to static and dynamic loads having elastic and plastic behaviour (1993), IPCT Bucuresti.
2 P100-92 – Romanian Seismic Code (1992), Bucuresti.

Author index

Abed, A. 211
Aiello, M. A. 191
Akiyama, H. 253
Amadio, C. 535
Angelescu, R. 727
Astaneh-Asl, A. 53, 310, 448, 495, 547

Ballio, G. 63
Balut, N. 717
Benussi, F. 535
Bernuzzi, C. 557, 568
Botici, A. 736
Bouwkamp, J. G. 221
Bugheanu, M. 478
Bursi, O. S. 371

Calado, L. 381
Calderoni, B. 111, 278, 333
Castiglioni, C. A. 63
Cosenza, E. 77, 390

Dalban, C. 458, 727
d'Amore, E. 557, 568
Dan, S. 149
de Stefano, M. 557, 568
de Luca, A. 557, 568, 631, 659
de Martino, A. 333, 390
Diaconu, A.-C. 357
Diaconu, D. 357
Dima, S. 727
Dimoiu, I. 89, 149, 736
Dinculescu, M. 472
Dubina, D. 580
Dunai, L. 159

Elnashai, A. S. 133

Faella, C. 590, 600

Faella, G. 659
Ferreira, J. 381
Fukumoto, Y. 159

Georgescu, D. 310, 472, 478, 495, 519
Ghersi, A. 278, 333
Gioncu, V. 3, 169, 182, 289, 485
Grecea, D. 580, 610

Imbimbo, M. 669
Ioan, P. 458
Iuhas, A. 169, 289
Ivan, M. 89

Kakaliagos, A. 401
Kasai, K. 97
Kato, B. 28

Landolfo, R. 333, 421
La Tegola, A. 411
Lenza, P. 691
Leonte, A. 727
Lu, L.-W. 45, 97

Magonette, G. 401
Mandara, A. 747
Manfredi, G. 77, 390
Mateescu, G. 169, 289, 610
Mazzolani, F. M. 111, 182, 300, 310,
    421, 485, 495, 747
Mele, E. 631, 659
Mera, W. 411
Moldovan, A. 717
Mustata, V. 727

Noe, S. 535

Ohi, K. 323

Ohtani, Y. 159
Olariu, I. 759
Ombres, L. 191

Pacurar, V. 201
Pagano, M. 691
Pasternak, H. 242
Piluso, V. 111, 300, 590, 600
Pinto, A. V. 401, 621
Plumier, A. 211, 441
Pradhan, A. M. 221
Psycharis, I. 231

Rauso, D. 278
Ricles, J. M. 97
Rizzano, G. 590, 600
Rossi, P. P. 691

Sandi, H. 507
Seifert, J. 431
Serino, G. 669, 703

Shibata, M. 121
Shimizu, K. 681
Shing, P. B. 371
Šimek, I. 431
Sophocleous, A. 344
Spanu, St. 458
Stoian, V. 759
Syrmakezis, C. 344

Takanashi, K. 323
Teramura, A. 681
Tiliouine, B. 211

Vayas, I. 231, 242, 344
Verzeletti, G. 401

Wald, F. 431

Zaharia, R. 580
Zandonini, R. 557, 568

Subject index

This index is compiled from the keywords assigned to the papers, edited and extended as appropriate. The page references are to the first page of the relevant paper.

Accelerogram 3, 289
Acceptance requirements 111
Active control 631
Algorithms, non-linear problems 371
Allowable stress design 401
Alternate static loads 610
Alternative methods, local ductility 182
Analysis 97
    cold-formed sections 333
Angle connections 568
Application rules 231

Base isolation 631, 691
    system 659, 669
Beam-column connections 149, 221, 448, 610
Beams, ductility 201
Behaviour factors 28, 278, 344, 717
Behaviour, semi-rigid connections 610
Bolted joints 221
Bolt slippage 557
Border columns 121
Braced frames 3, 310, 458
Braced moment-resisting frames 703
Braced wall 121
Bridges 53
Buckling failure, bearings 669
Buckling, local 63
Buckling resistance 211
Buildings
    dynamic measurement 411
    semi-rigid 547

Cantilever braced wall 121

Capacity design 133, 507
    criteria 344
Church, upgrading 747
Cladding panels 421
Classification 231
Codes of practice 3, 28, 111, 231, 448, 507, 441, 458, 485, 310, 621, 759
    Romania 478
Cold-formed steel sections 333
Collapse limit state 495
Collapse mechanisms 300, 485
Column-base connections 431
Column damper 681
Component method 431
Composite beam-columns 97
    joints 221
Composite columns 448
Composite connections 448
Composite construction 97
Composite elements 759
Composite frames 535
Composite structures 211, 441, 448
Computer model, ductility 289
Concentrically braced frames 310, 458
Concrete structures, strengthening 727
Connections 133, 458, 519, 600
    design 590
    semi-rigidity 519, 547
    strengthening 717
Constant amplitude tests 63
COST-C1 621
Criteria for classification 231
Cumulative inelastic deformation ratio 253

Subject index

Curvature 133
Cyclic actions 191
Cyclic behaviour 211, 231, 242
   experimental testing 390
Cyclic loading 159, 381
Cyclic response 221, 557

Damage accumulation models 63
Damage assessment 63
Damage classification 323
Damage concentration 253
Damage criteria 45, 77
Damageability limit state 495
Damaged structures 727
Damped braced moment-resisting frames 703
Database 323
Design 441
   algorithm 300
   charts, base isolation 669
   composite structures 448
   criteria 507
   industrial buildings 478
   frames 278
   method 77
   plate girders 472
   seismic resistant structures 3
Dissipative devices 747
Distributed inertial mass 89
D_s-value 253
Ductility 45, 169, 310, 600
   beams 201
   bridges 53
   local members 344
   unbraced frames 289
Dynamic analysis 45
Dynamic behaviour, light gauge frames 333
Dynamic deformability of materials 357
Dynamic response 89, 278

Earthquake behaviour, predictions 621
Earthquake engineering, experimental testing 390
Earthquake response, beam-column joints 221
Earthquakes, Romania 3

Eccentrically braced frames 97, 310, 458
Eigenperiod 580
Elastic modulus 149
Encased concrete 211
Energy concept 253
Energy dissipation 28, 631, 703
Equivalent ductility 77
Equivalent inertia mass matrix 89
Eurocode 3 231, 431, 568
Eurocode 8 231, 621
Experiment 97
Experimental error growth 371
Experimental investigation, beam-connection 610
Experimental methodologies 357
Experimental procedures 390
Experimental study 159
Experimental testing 357, 381

Failure criteria 45
Failure mode control 300
Finite element analysis 149, 421
Fixed end forces 89
Flange width-thickness ratio 169
Flexible connections 519
Frame selection 333
Full-scale testing 97, 401

General reports 45, 133, 253, 357, 441, 519, 631
General ductility 3
Global ductility 253, 289, 485, 600
Golden Gate Bridge 53

High-rise building, vibration control 681
Hinged joints 736
Hinge mechanism 580
History, seismic design 441
Human judgement 323
Hybrid mass damper 631
Hysteretic behaviour 133

Industrial buildings 478
Inelastic behaviour 111, 278
Inelastic cycles 77
Inelastic deformation ratio 253

Subject index 773

Inelasticity 133
Inertial mass 89
  matrix 89
Infilled plate panels 97
Irregularities 344
Isolation systems 631

Japan, codes of practice 28
Joints 736
  models 519

Knee joints 242

Laminated rubber bearings 669
Lattice structures 717
Lehigh University, research 97
Light gauge steel frames 333
Limit states 495
  design 28, 300
Liquid column damper 681
Local buckling 63, 133, 169, 182, 211
Local cyclic behaviour 231
Local ductility 3, 133, 182, 191, 253, 600
Local member ductility 344
Loma Prieta Earthquake 53
Longitudinal frames 478
Low-cycle fatigue concept 357

Masonry, strengthening 747
Maximum compression 121
Mechanical model 557
Mechanism equilibrium curves 300
Members, local ductility 191
Methodologies, experimental 357
Mixed connection 159
Mixed steel-concrete sections 201
Modal analysis 478
Modelling 441
  error study 323
Models, connections 568
Moment-curvature diagrams 191
Moment-resisting frames 3, 401, 458, 703
Moment-rotation behaviour, connections 547, 557
Monte Carlo simulation 111

Monumental building 747
Multistorey buildings 691
Multistorey steel structures 458

Natural vibration period 289
Non-linear dynamic analysis 221, 759
Non-structural elements 411
Northridge Earthquake 53
Numerical analysis, composite frames 535
Numerical applications 191
Numerical dissipation 371

Operator-splitting-based algorithms 371
Optimization 631
Overstrength 278

P-δ effect 253
P-Δ effects 300
Partially restrained frames 547
Passive control systems 631
Passive energy dissipation 703
Plastic buckling 169
Plastic rotation 169
  capacity 169
Plate girders 472
Prediction methods, connections 568
Predictor-multi-corrector algorithm 371
Principles and methods, repair 472
Probabilistic approaches 507
Probabilistic methods 111
Pseudo-dynamic testing 323, 401

Q-factor 253, 485, 580, 590
Quality control 45, 111
Quasi-static cyclic tests 390

Reinforced concrete cores 736
Repair, plate girders 472
Repairing, lattice structures 717
Resistance deterioration 323
Response analysis 121
Response spectra 28
Retrofit strategies, bridges 53
Retrofitting 747
Reversed pendulum 736
Rigid connections 519

Subject index

Rigid-plastic mechanism 169
Roll-out failure, bearings 669
Rolled joints 736
Romania, code of practice 478
Romania, seismic design 3
Roof bracing 478
Rotation and displacement ductility 133
Rotation capacity 28, 182
Rotation ductility 211
Rubber bearings 631

S-N curves 63
Safety levels, seismic design 495
San Francisco Bay Bridge 53
SCARLET Database 323
Secant 149
Second order effects 300, 600
Security 495
Seismic action, lattice structures 717
Seismic behaviour 45, 310, 535
   corrugated panels 421
   moment-resisting frames 401
   Romania 3
   semi-rigid joints 580
Seismic conditions 507
Seismic design 211, 344, 495
   code 458
   composite structures 448
   connections 590
   energy dissipation 703
   semi-rigid structures 547
Seismic isolation 669
Seismic performance, base isolation 659
Seismic research, steel bridges 53
Seismic-resistant design 45, 77, 441
Seismic response
   base isolation 691
   tall buildings 411
Seismic retrofitting 747
Seismic risk 507
Seismic simulators 357
Seismic working group 621
Semi-rigid behaviour 381
Semi-rigid composite frame 535
Semi-rigid connections 431, 557, 568, 610, 621
Semi-rigid connections and frames 547

Semi-rigidity 519
Serviceability limit state 495, 659
Shear behaviour, corrugated panels 421
Shear links 97
Shear panel thickness 221
Slip behaviour, connections 557
Stability coefficient, connections 600
Stability, semi-rigid joints 580
Statistical evaluation, behaviour factor 278
Steel bridges 53
Steel-concrete beams 201
Steel-concrete composite structure 759
Steel-concrete connections 159
Steel connections 381
Steel frames 310
   base isolation 659
   behaviour factor 344
   infilled plate panels 97
Steel members 231
   strengthening 727
Steel moment-resisting frames 401
Steel structures, strengthening 717
Stiffness deterioration 159
Stiffness distribution 344
Stiffness matrix 89
Strain hardening 568
Strength distribution 344
Strengthening
   concrete structures 727
   lattice structures 717
   plate girders 472
Stressed skin design 421
Structural behaviour factor 289
Structural design, steel concrete structure 759
Structural response control 631
Structural steel 97
Structural system 45
Supplemental damping 97
Suspended storeys 736
Territory seismicity 3
Testing procedures, overview 390
Tests, column-bases 431
Thin-walled knee joints 242
Transient algorithm 371
Transverse frames 478

Subject index **775**

Ultimate limit state 659
Ultimate plastic rotation 289
Umbrella-type building 736
Unbraced composite frames 535
Unbraced frames, ductility 289
USA, code of practice 448

Variable amplitude tests 63
Vibration, connections 600
Vibration control system 681
Vibrations, buildings 411

Visco-elastic dampers 97, 631

Water tank, vibration control 681
Web width-thickness ratio 169
Welded beam-column connections 149
Welded joints 221, 736
Welded knee joints 242
Width-to-thickness ratio 28

Yielding, semi-rigid joints 580